*TABULE RESOLUTE*

ALFONSINE ASTRONOMY

*STUDIES AND SOURCES*



ALFONSINE ASTRONOMY

This collection is created under the auspices of the European Research Council project ALFA: Shaping a European scientific scene, Alfonsine astronomy, CoG 723085.

# *Tabule resolute*

*The Central European Type of Alfonsine Tables*

*Edited by*
ALENA HADRAVOVÁ & PETR HADRAVA

BREPOLS

To the memory of our colleague and friend
José Chabás Bergón (1948–2024)

This work was carried out in the framework of the European Research Council (ERC) project *ALFA – Shaping a European Scientific Scene: Alfonsine Astronomy*, under the EU program Horizon 2020 (CoG 2016, agreement No 723085).

It has also been supported by Czech Academy of Sciences, Project RVO: 68378114 (Inst. Contemporary Hist.) and RVO: 67985815 (Astr. Inst.).

This book has been produced with the assistance of the database Czech Medieval Sources online, provided by the LINDAT/CLARIAH-CZ Research Infrastructure (https://lindat.cz), supported by the Ministry of Education, Youth, and Sports of the Czech Republic (Project No. LM2018101).



D/2026/0095/130
ISBN 978-2-503-61712-1
e-ISBN 978-2-503-61713-8
DOI 10.1484/M.ALFA-EB.5.144451

Printed in the E.U. on acid-free paper

# Table of Contents

## II
## Tables

## III
## Commentary

# Acknowledgements

During the years of preparation of this publication, we have been supported by many colleagues. They have helped us to photograph manuscripts in a number of libraries and have provided us with valuable information and advices. We are very grateful to all of them for their help. First and foremost among them are the members of the ALFA – *Alfonsine Astronomy* project team, with whom we shared many of the sub-phases of the book's preparation. These include in particular José Chabás (Universitat Pompeu Fabra, Barcelona), Richard L. Kremer (Dartmouth College, New Hampshire), Matthieu Husson (CNRS, Observatoire de Paris), Laure Miolo (University of Cambridge), Marie-Madeleine Saby (Université Grenoble Alpes), Samuel Gessner (University of Lisbon), Laura Fernández Fernández (Universidad Complutense, Madrid), Eleonora Andriani (CNRS, Paris), and Nick Jacobson (University of Wisconsin-Madison). We are also indebted to many other colleagues, especially Giancarlo Truffa (independent researcher, Milan) and Kristen Lippincott (art historian, London), as well as to staff of libraries and digitisation centres, whose help made it possible to work on this topic. These are mainly members of the Department of Manuscripts and Early Prints (National Library of the Czech Republic, Prague), Anna Sobańska (Jagiellonian Library, Cracow), Marta Czyżak (University Library, Toruń), Vilém Kaplan (Moravian Library, Brno), Miroslav Koudela (Regional museum, Mikulov) and Katrin Sturm (Universitätsbibliothek, Leipzig). We thank to Richard L. Kremer also for revision of the English version of our book.

Finally, we would like to thank both reviewers of the work, Charles Burnett (The Warburg Institute, London) and José Chabás (Universitat Pompeu Fabra, Barcelona), who have read the text very carefully and whose comments and suggestions have helped to improve the work considerably. However, the responsibility for any errors and inaccuracies lies, of course, solely with the authors.

1

# Introduction

## 1.1 Alfonsine Astronomy and its Journey through Europe

Among the most common and the most frequent works found in Latin astronomical manuscripts with a core in the fourteenth and fifteenth centuries and extending into the sixteenth century are large and usually multi-page numerical tables. Many codices, containing texts on a variety of topics and not always exclusively astronomical, include individual tables excerpted from a larger whole. These are usually the so-called Alfonsine Tables (*Tabule Alphonsine*).[1] While some manuscripts simply include a copy of the tables without any further explanation or commentary, others contain accompanying texts that explain the meaning of the tables and give instructions on how to use them. These accompanying texts are called 'canons' or 'rules' (*canones* or *proposiciones*) and they give more or less complete information on the principles behind the tables and instructions on how a medieval astronomer should use them in his work.

The Alfonsine Tables gradually replaced the previously used Toledan Tables (*Tabule Toletane*).[2] Just like them, they were created in Toledo, and they date to the second half of the thirteenth century. They were named after Alfonso X the Wise, the king of Castile and León (1221–1284), who gave the order to rework the older tables and to create new ones. He also ordered his astronomers to adapt the new tables to the year of his coronation, which was 1252. These new tables were written in Castilian and only the canons of the numerical tables are extant. They are now known as Castilian Alfonsine Tables (CAT). However, after the tables got to Paris in around 1320, they were translated into Latin. This new language version effectively saved the tables for future generations, making them immediately accessible to the intellectual community of the whole of Europe, for whom Latin was then the *lingua franca*. The most prominent Parisian astronomers who dealt with

1 José Chabás and Bernard R. Goldstein, *The Alfonsine Tables of Toledo* (Dordrecht: Kluwer, 2003).

2 The Toledan Tables (*Tabule Toletane*), compiled in the eleventh century (see Fritz S. Pedersen, *The Toledan Tables. A Review of the Manuscripts and the Textual Versions with an Edition.* [Copenhagen: C. A. Reitzels Forlag, 2002]), were based on Arabic sources. They replaced the previously used Arabic tables, *zijes* (Benno van Dalen, *Ancient and Mediaeval Astronomical Tables: Mathematical Structure and Parameter Values* [Utrecht: Universiteit Utrecht, 1993]), by al-Battānī and al-Khwārizmī, which in turn had been based on the Handy Tables by Ptolemy (*Πρόχειροι κανόνες*). See Anne Tihon, *Πτολεμαίου Πρόχειροι κανόνες. Les Tables faciles de Ptolémée, Volume Ia. Tables A1–A2* (Louvain-la-Neuve: Université Catholique de Louvain, Institut Orientaliste, 2011); Raymond Mercier, ed., *Πτολεμαίου Πρόχειροι κανόνες. Ptolemy's Handy Tables, Volume Ib. Tables A1–A2* (Louvain-la-Neuve: Université Catholique de Louvain, Institut Orientaliste, 2011); W. D. Stahlman, 'The Astronomical Tables of Codex Vaticanus Graecus 1291', unpublished PhD. thesis, Brown University, 1960; and of course Gerald J. Toomer, *Ptolemy's Almagest* (London: Duckworth, 1984).

the Latin Alfonsine Tables and created new, modified versions of the tables and canons were John Vimond (Johannes de Vimundi, Jean de Vimond, his tables originated *c.* 1320),[3] John of Murs (Johannes de Muris, Jean des Murs, Tables of 1321),[4] John of Lignères (Johannes de Lineriis, Jean de Lignères, Tables of 1322, *Tabule magne* 1322–25)[5] and John of Saxony (also Danko or Danck, Johannes de Saxonia). Among the tables, the Parisian Alfonsine Tables (hereafter PAT) resulting from the work of these and other astronomers are particularly notable. They are preserved in more than one hundred and eighty copies in European and overseas libraries and are thus the largest collection of Alfonsine tables. The comprehensive canons to the PAT, which start with words *Tempus est mensura motus*, were written by John of Saxony in 1327 and have been preserved in more than a hundred and twenty copies in various libraries around the world. The PAT with their canons *Tempus est mensura motus* were first printed in 1483 by Erhard Ratdolt in Venice.[6]

The Paris version of the Latin tables, PAT, spread quickly to other European centres of learning and to university towns in particular, so new local versions quickly emerged, some of them very far away from Paris.[7] The first Parisian tables were calculated for the Toledo meridian. However, as the tables spread through Europe, they were modified and adapted to different local meridians. Nevertheless, their parameters and general basic features remained unchanged.

One route of the Parisian Alfonsine Tables was to the British Isles. In 1336, a set of Alfonsine Tables was created in London.[8] A little later William Rede adapted the Parisian Alfonsine Tables to the Oxford meridian (with an epoch of 1340, i.e. to the year 1340), after

3 José Chabás and Bernard R. Goldstein, 'Early Alfonsine Astronomy in Paris: The tables of John Vimond (1320)', *Suhayl*, 4 (2004), 207–94.

4 Emmanuel Poulle, 'Jean de Murs et les tables alphonsines', *Archives d'Histoire doctrinale et litteraire du Moyen Age*, 47 (1980), 241–71; Beatriz Porres de Mateo and José Chabás, 'John of Murs's *Tabulae permanentes* for Finding True Syzygies', *Journal for the History of Astronomy*, 32 (2001), 63–72; Richard L. Kremer, 'John of Murs, Wenzel Faber and the Computation of True Syzygy in the Fourteenth and Fifteenth Centuries', *Acta Historica Leopoldina*, 54 (2008), 147–60.

5 Emmanuel Poulle, 'John of Lignères', in *The Dictionary of Scientific Biography* (New York: Charles Scribner's Sons, 1973), VII, pp. 122–28; Marie-Madeleine Saby, 'Les canons de Jean de Lignères sur les tables astronomiques de 1321. Édition critique, traduction et étude', unpublished thesis, Paris, École Nationale des Chartes 1987; a summary appeared as 'Les canons de Jean de Lignères sur les tables astronomiques de 1321', École Nationale des Chartes 1987: Positions des theses, pp. 183–90; Alena Hadravová and Petr Hadrava, 'John of Lignères, Quia ad inveniendum loca planetarum. An Edition and Translation', in *Alfonsine Astronomy. The Written Record*, ed. by Richard L. Kremer, Matthieu Husson, and José Chabás (Turnhout: Brepols, 2022), pp. 257–302; José Chabás and Marie-Madeleine Saby, *The Tables of 1322 by John of Lignères. An Edition with Commentary* (Turnhout: Brepols, 2022).

6 For an English translation of chosen parts see Victor E. Thoren and Edward Grant, 'Extracts from the Alfonsine Tables and Rules for their Use (John of Saxony)', in *A Source Book in Medieval Science*, ed. by Edward Grant (Cambridge: Harvard University Press, 1974), pp. 465–87. For a Latin edition and French translation see Emmanuel Poulle, *Les Tables Alphonsines avec les canons de Jean de Saxe* (Paris: Centre national de la recherche scientifique, 1984). See also Alena Hadravová and Petr Hadrava, 'John of Saxony', in *Medieval Science, Technology and Medicine: An Encyclopedia*, ed. by Thomas F. Glick, Steven J. Livesey, and Faith Wallis (New York, London: Routledge, 2005), p. 292.

7 For a brief summary of the history of the Alfonsine Tables, see e.g. José Chabás and Bernard R. Goldstein, *The Astronomical Tables of Giovanni Bianchini* (Leiden, Boston: Brill, 2009), pp. 1 and 9; Richard L. Kremer, 'Abbreviating the Alfonsine Tables in Cracow: The *Tabulae aureae* of Petrus Gaszowiec (1448)', *Journal for the History of Astronomy*, 38 (2007), 283–304, 283; José Chabás, *Computational Astronomy in the Middle Ages: Sets of Astronomical Tables in Latin* (Madrid: Consejo Superior de Investigaciones Científicas, 2019).

8 José Chabás, 'An Early Witness of Alfonsine Astronomy: The London tables for 1336', *Journal for the History of Astronomy*, 48 (2017), 324–28.

which William Batecombe recalculated them for the Oxford meridian in 1348.[9] Another version of the tables and rules for Oxford was produced by John Killingworth around 1445.

Another important branch of Alfonsine astronomy led to Italy, especially to its northern part. This branch of the PAT is represented by the tables calculated for Padua in 1428. The author of the modification in question was Prosdocimo de' Beldomandi, otherwise known as a music theorist. In 1442 Giovanni Bianchini (Johannes Blanchinus) presented tables for Ferrara in an innovative format; they were published in an early edition in the form of an incunabula in 1495[10] and have been studied in a recent monograph.[11]

The PAT also moved to Central Europe. John of Gmunden, an eminent astronomer at Vienna University, was working with the tables even before there was any news of the first north Italian versions. His adaptation of the tables to Vienna was completed around 1420 and was the subject of Beatriz Porres' extensive dissertation.[12] Within Central Europe, there are records of sets of tables in Leipzig, Magdeburg, Erfurt, Wrocław (Breslau), Prague, and Cracow, and these were all further developed and used in their respective locations. A compendium of all Central European tables, which is yet to be clearly defined, then became the base for the so called Resolved Tables (*Tabule resolute*), written in Cracow. While this large group of Central European tables is amply documented in manuscript copies and parts of it have been examined in various studies,[13] a full and comprehensive overview has not yet been published. The purpose of this monograph is to fill in this gap.

According to the information given in the manuscripts, the tables for Wrocław with the radix in 1424 were produced by Petrus Cruciferus. He later adapted them to Prague and to the year 1428. These tables were then developed by many Polish astronomers, e.g. Andrzej Grzymała of Poznań, Albertus of Brudzewo, John of Głogów, Petrus Gaszowiec – the author of the so-called Golden Tables (*Tabule auree*) in 1448 –, and adapted primarily to Cracow, where the local university had been founded in 1364 (some years after the Prague university, which had been founded in 1348 and a year before the Vienna University opened in 1365). The Cracow astronomical school was very active and highly regarded. And it would be the Polish astronomers who took the *Tabule resolute* to Spain and Italy and adapted them to the local meridians of Salamanca and Bologna.[14]

---

9 José Chabás and Bernard R. Goldstein, 'The Moon in the Oxford Tables of 1348', *Journal for the History of Astronomy*, 47 (2016), 159–67.

10 Giovanni Bianchini (Johannes Blanchinus), *Tabulae celestium motuum earumque canones*, ed. by Augustinus Moravus, additions by Johannes Basilius Augustonus (Venice: Simon Bevilaqua, 1495). – ISTC ib00697000, GW 4410.

11 Chabás and Goldstein, *The Astronomical Tables of Giovanni Bianchini.*

12 Beatriz Porres de Mateo, 'Les tables astronomiques de Jean de Gmunden: Édition et étude comparative', unpublished PhD thesis, Paris, École Pratique des Hautes Études, 2003. See also José Chabás, 'The Diffusion of the Alfonsine Tables: The Case of the Tabulae resolutae', *Perspectives on Science*, 10 (2002), 168–78.

13 E.g., Jerzy Dobrzycki, 'The Tabulae resolutae', in *De Astronomia Alphonsi Regis*, ed. by Mercè Comes, Roser Puig, and Julio Samsó (Barcelona: Universitat de Barcelona, 1987), pp. 71–77; reprinted in: *Jerzy Dobrzycki, Selected Papers on Medieval and Renaissance Astronomy*, ed. by Jarosław Włodarczyk, and Richard L. Kremer (Warsaw: Institute for the History of Science, 2010), pp. 129–35; Beatriz Porres de Mateo, 'Šíření středověkých astronomických tabulek ve střední Evropě v 15. století', in *Astronomie ve středověké vzdělanosti*, trans. and ed. by Alena Hadravová, and Petr Hadrava (Prague: ÚSD a AsÚ, 2003), pp. 39–51; Chabás, *Computational Astronomy in the Middle Ages*, pp. 311–20.

14 The Cracovian *Tabule resolute* were brought to Salamanca, Spain, around 1460 by the Pole Nicolaus of Tuchów, where Abraham Zacut used them around 1478 to create an Arabic version, see José Chabás, 'Astronomy in Salamanca in the Mid-Fifteenth Century: The Tabulae resolutae', *Journal for the History of Astronomy*, 29 (1998),

The main representatives of Alfonsine astronomy in the fifteenth century are undoubtedly Georg Peurbach (*Theorice nove planetarum* 1454, *Tabule eclipsium* 1460) and his disciple and follower Johannes Regiomontanus (*Tabula primi mobilis* 1463, *Tabule direccionum profeccionumque* 1467).[15] In the work of both astronomers, the development of Alfonsine astronomy in the pre-Copernican period is completed. By this brief overview we have attempted to outline how rich and varied European tradition of Alfonsine astronomy is.

It is generally accepted in the history of astronomy that Alfonsine astronomy was the most influential astronomical tradition of medieval Europe. It was taught in all universities and it served as a starting point for theoretical – and less frequently – also for practical investigations, studies and observations.

## 1.2 *Tabule resolute* and their Canons in the Context of Alfonsine Astronomy

Before *Tabule resolute* (hereafter TR) emerged and spread in Central Europe, the Parisian Alfonsine Tables, their canons by John of Saxony and works of John of Lignères (Tables of 1322, canons *Priores astrologi* and *Cuiuslibet arcus propositi* etc.) were mainly used in across Europe. Representative examples of manuscripts with PAT and John of Saxony's canons include, e.g., the so-called Wenceslavian codex (Vienna, ÖNB, Cod. 2352) which was prepared for the Czech King Wenceslas IV at the end of the fourteenth century. The choice of the PAT and the canons *Tempus est mensura motus* for the king shows that these texts were considered important and fundamental both for astronomers in that time and from a representative point of view. However, most of the manuscripts containing these and other works by Parisian astronomers come from the universities that were gradually founded from the middle of the fourteenth century to the beginning of the fifteenth century (Prague, Cracow, Vienna, Heidelberg, Erfurt, Leipzig ...).[16]

The basic task of the TR, like the PAT, was to enable the calculation of the positions of the planets for any moment in time, which was the starting point and a necessary condition for making astrological predictions. The authors of the canons to the TR usually emphasize in their introductions that King Alfonso's tables (by which they mean the most widely used Parisian version, the PAT) are the best tables that exist. At the same time, however, they complain that working with these tables is tedious and demanding, especially for less erudite users. Therefore, as they write, they present the TR as a user-friendly variant of tables that will provide comparably good results for the user more easily and with less effort. Thus, the TR were mostly compiled using the PAT and the values of the parameters

167–75; José Chabás and Bernard R. Goldstein, 'Some Astronomical Tables of Abraham Zacut Preserved in Segovia', *Physis*, 35 (1998), 1–10; Beatriz Porres de Mateo and José Chabás, 'Los cánones de las Tabulae resolutae para Salamanca: origen y transmisión', *Cronos*, 1 (1998), 51–83. Another Polish astronomer, Nicolaus Wodka of Kwidzyn, took them to Italy and recalcutated for Bologna meridian in around 1477.

15 Chabás, *Computational Astronomy in the Middle Ages*, pp. 365–89.

16 The University of Leipzig was founded in 1409 by scholars who left Prague University, see Alena Hadravová and Petr Hadrava, 'Spreading of treatises on astrolabe by Cristannus of Prachatice', in *Alfonsine Astronomy: Expanding the Scene*, ed. by José Chabás, Matthieu Husson, and Richard L. Kremer (Turnhout: Brepols, forthcoming).

associated with the models describing the movement of the celestial bodies used in them; however, they achieve better readability and reduce labor for the user by several layout changes, which are characteristic for the TR and allow them to be distinguished from other sets of tables.

The TR, in particular, abandon the consistent use of the sexagesimal system usually found in the PAT. The sexagesimal system made the PAT versatile and mathematically very advanced and consistent, but at the same time very abstract. All time data (dates in different calendar systems and hours) are first converted in PAT into sexagesimally expressed numbers of days and their fractions from the reference epoch, for which the so-called radices are given, i.e. the initial values of uniform (so-called mean) motions in the also sexagesimally expressed number of degrees. For any mean motion, then, one table of position increments for one to sixty days is sufficient to calculate the mean motion at the desired time by summing the radix and the corresponding increments shifted in their sexagesimal order. Instead, the TR return to an intuitively easier to understand tradition based on Ptolemy's *Almagest*, in which various tables give the increments of motions not only in days, but also in months, in individual years (so-called expanded years, *anni expansi*) and their multiples corresponding to the periodicity of the calendar (so-called collected years, *anni collecti*), and also in individual hours and their fractions. The TR give collected years in the form of so-called cyclical radices, that is sums of initial radices with increments of motions to the beginnings of twenty-year periods (i.e. five times the four-year period of the Julian calendar) in the range of several centuries of the user's expected interest (see Fig. 1). By shortening the period from the given radix to the target date, the TR reduced the demands on the accuracy and therefore the numerical complexity of the mean motion calculation. The TR also give the resulting positions of the planets usually in the zodiacal ('common') signs of thirty degrees instead of the sixty-degree 'physical' signs mostly used in the PAT for consistency of sexagesimal calculations. However, they are exceptions in this respect both in the PAT and the TR.

Another step to make it easier for astronomers to work with the TR is to adapt the radices to the meridian, i.e. to the local time of the particular city in which the tables are to be used. The PAT listed the radices generally for the Toledo meridian and the user had to recalculate them for his own meridian. Occasionally, additional corrections can also be found on the margins of some manuscripts with the TR which allow the tables to be used for other meridians.

In order to maintain a formally similar calculation procedure, TR replaced the calculation of the true motion of the apogees – which, according to the theories of some Arab astronomers, was supposed to be non-uniform but with a very long period – by a linear interpolation between the values calculated on the same dates for which the cyclical radices of the mean motions were given. Unlike the first TR compiled for Wrocław, subsequent versions of the TR computed for other cities facilitated linear interpolation in the equation tables (*tabule equacionum*, i.e. tables of corrections for non-uniform motions) by adding columns with differences between the subsequent values.

Various additional tables were added to the basic TR set to facilitate necessary calculations, e.g., multiplying sexagesimal numbers, preparing ephemerides, or performing other astronomical calculations that the TR user might need. Some of these were taken from other sets of tables.

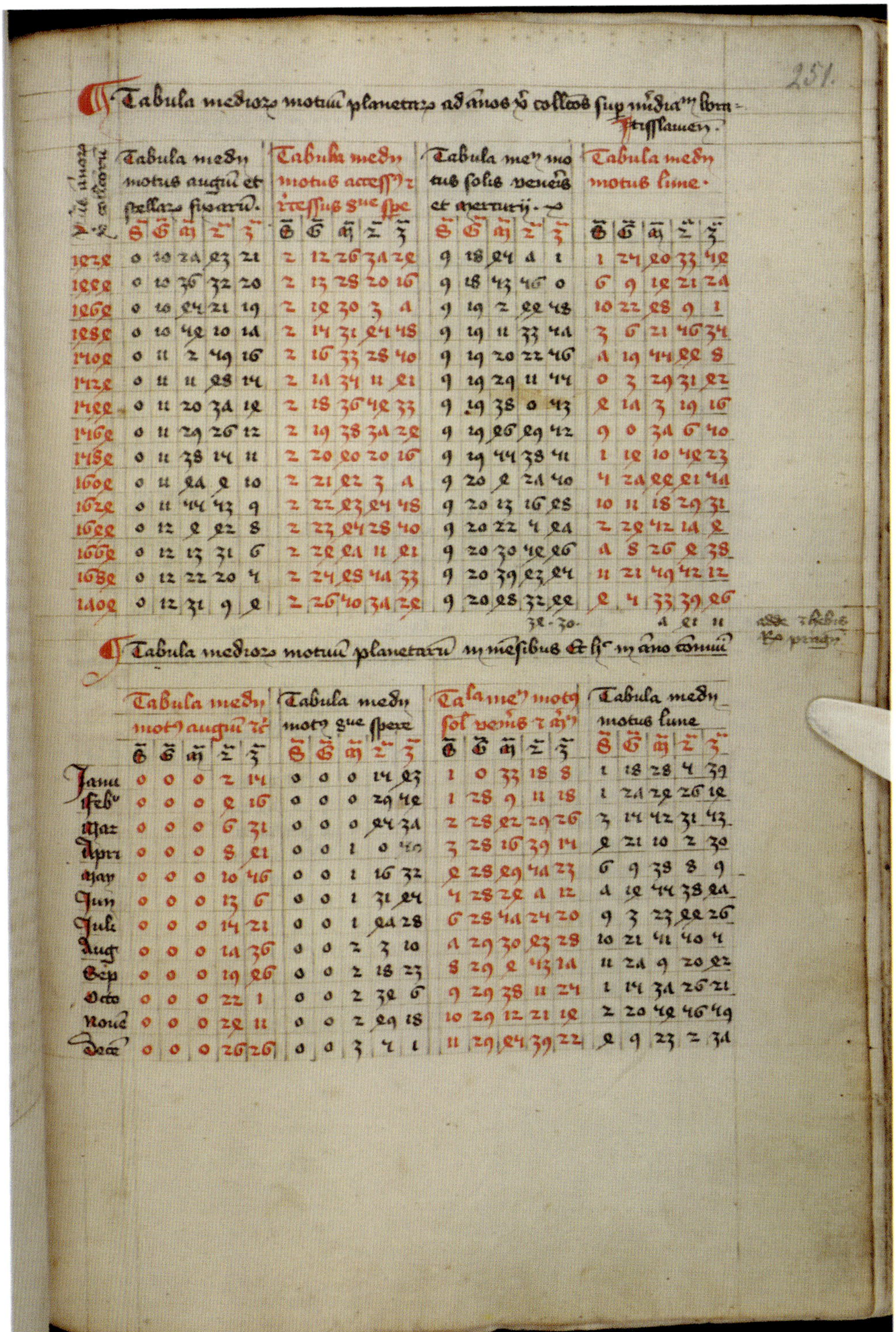

**Figure 1.** MS Toruń, BU, 74/III, f. 251r, the beginning of the *Tabule resolute*, which start with mean motion tables.

The title of the TR itself was assigned only afterwards by Polish astronomers. In the introduction of his canons *Pro (introduccione) generali*, John of Głogów explains that '*Tabule resolute* are named from resolving, decomposing' (*Tabule resolute dicte sunt a resolvendo*),[17] i.e. from the decomposition of the input time data into sub-intervals of years, months, days, hours and their fractions, from whose contributions the resulting value of the motion is then composed (see Chapters 8.2 and 8.3). In his canons to the TR *Quoniam plerique nostri temporis phisici*, Nicolaus Wodka of Kwidzyn also uses the term in the same sense as John of Głogów when he explains that when subtracting a larger sexagesimal digit from a smaller one, it is necessary to take a unit from the digit of the higher sexagesimal order and divide it into 60 units of the counted order: *De subtraccione. – Numerus subtrahendus maior si fuerit numero, a quo debuit fieri subtraccio, tunc unum quodque immediate est grossioris denominacionis in partes suas resolutum additur numero, a quo talis debeat fieri subtraccio, et a provenienti tunc numero numerus subtrahendus deponitur, residuo in suo loco manente.*[18] And in the same sense Andrzej Grzymała in his canons *Girum* also uses the verb *resolvere* to divide the whole circle into signs and their parts: *Si autem non posses subtrahere, tunc accomodares unam revolucionem, scilicet 12 signa, resolvendo ea in gradus et ceteras fracciones.*[19] The same meaning as *resolvo, -ere* in the canons of the TR is also sometimes given by the term *reduco, -ere*[20] which also records the canons *Girum* by Andrzej Grzymała, here synonymously titled *Tabule reducte*.

The basis of our overview of the TR and their canons is a list created by José Chabás[21] which we have revised and expanded. The chronology of the creation of the canons is as follows:[22]

1. *Mirabilis in altis Dominus*[23] (1424–1425);
2. *Girum recensendo* (1449);
3. *Nemo integre sapit* (*c.* 1450);
4. *Astrorum (observatores) sapientes* (after 1450);
5. *Quoniam Tabularum Alfonsi* (*c.* 1460);
6. *Quoniam plerique* (1477);
7. *Pro (introduccione) generali* (1486–1487).

The TR and their reworked canons were later published as early modern prints. A first edition was prepared by Ambrosius Lacher and published in Frankfurt an der Oder 1511. Several other editions followed: Johannes Schöner, Nuremberg 1536; Johannes Virdung, Nuremberg 1542; Johannes Schöner, Nuremberg 1551; Johannes Schöner, Wittenberg 1587 and 1588.

The earliest canons *Mirabilis in altis Dominus*, are for the TR calculated for Wrocław of 1424. The TR for Wrocław are preserved in only one copy, in the MS Toruń, BU, 74/III.

17 Vatican, BAV, Pal. lat. 1439, f. 165r.
18 London, BL, Add. 22317, f. 1v.
19 Cracow, BJ, 573, f. 230r–v.
20 Vienna, ÖNB, Cod. 5245, f. 34v.
21 Chabás, *Computational Astronomy in the Middle Ages*, pp. 317–20.
22 Updated on the basis of Dobrzycki, 'The Tabulae resolutae', pp. 71–77 (reprinted in 2010), pp. 129–35.
23 Canons are usually distinguished and cited by their incipits.

However, these TR for Wrocław were soon recalculated for the Prague meridian, but the canons for them were at first the same as for Wrocław: sometimes the mention of of "Wrocław" was only mechanically replaced by "Prague". Apparently the *Mirabilis* canons soon ceased to be suitable, certainly in the first half of the fifteenth century, so that especially in Poland many new versions of the canons began to appear, adapted to both the Prague and the Cracow meridians.[24] The new canons were written in Cracow shortly before 1450 by Andrzej Grzymała (canons *Girum recensendo*). Probably only a little younger are the anonymous canons with the incipit *Nemo integre sapit*. A much more elaborate version of the canons was written in the second half of the fifteenth century by Adalbertus (Wojciech) of Brudzewo (*Astrorum /observatores/ sapientes*).[25]

The Cracow TR were later – around 1460 – transferred to the University of Salamanca, Spain, by Nicolaus of Tuchów, therefore called Nicolaus Polonus (Polonius, Polonio) after his nationality, where in 1478 Abraham Zacut (1452–1515) created on their basis an Hebrew version called *ha-Ḥibbur ha-gadol* (The Great Compilation).[26] To the list of canons can therefore also be added Polonus's canons *Quoniam Tabularum Alfonsi*. Only a little later – around 1477 – the TR also reached Bologna, where they were recalculated for the Bologna meridian and new canons (*Quoniam plerique nostri temporis physici*) were written also by a Pole, Nicolaus Wodka of Kwidzyn.

The last manuscript canons for TR, written again in Cracow, come from the pen of John of Głogów (*Pro /introduccione/ generali*). They are the longest of all the extant canons, and this is because their author evidently felt the didactic necessity to explain to his students the principle of the use of tables in as much detail as possible, so that they would be able to understand everything. However, he did not modify or edit the tables in any way. His canons are characterised by a large number of examples for the years 1486 and 1487.

---

24 As far as the tables recalculated for the Prague meridian are concerned, we find even more of them stored in Cracow than in Prague – perhaps this is also a consequence of the turbulent times of the Hussite period in the first half of the fifteenth century and their effects. As a result of unsatisfactory political conditions, wars, confessional disputes and social upheavals in Hussite-era Bohemia the importance of the University of Prague gradually declined, while neighbouring universities (Cracow, Leipzig, Vienna) became more attractive and the scholars who went there naturally took their books with them.

25 The incipits of some canons (*Astrorum /observatores/ sapientes ingentem coeli tractantes magnitudinem* and *Pro /introduccione/ generali*) vary somewhat in the manuscripts and may be similar to those of other works, including non-astronomical, so that at least three or four words from the beginning should be quoted. See Lynn Thorndike and Pearl Kibre (hereafter ThK), *A Catalogue of Incipits of Medieval Scientific Writings in Latin* (Cambridge: The Mediaeval Academy of America, 1963), 158–59; 1132–33.

26 Chabás, 'Astronomy in Salamanca'; Chabás and Goldstein, 'Some Astronomical Tables of Abraham Zacut'; Porres de Mateo and Chabás, 'Los cánones de las Tabulae resolutae para Salamanca'.

# I

# Canons

# Canons *Mirabilis in altis Dominus* by Petrus Cruciferus

## 2.1 Manuscripts

There are currently seven known copies containing the canons *Mirabilis in altis Dominus* in six manuscripts, listed below in the tentative chronological order (the precise dating of the Prague manuscript is unknown):

Toruń, Biblioteka Uniwersytecka (BU), 74/III, ff. 268va–273ra. Prague, 1425
Cracow, Biblioteka Jagiellońska (BJ), 610, ff. 352ra–357vb. Cracow (?), 1440
Leipzig, Universitätsbibliothek (UB), 1476, ff. 34r–37v. Prague, XV/1
Oxford, Bodleian Library (Bodl.), Canon. misc. 499, ff. 122ra/112ra–125vb/115vb (upper/lower foliation). Prague, 1451
Prague, National Library (NK), X B 3 (Cim C 20), ff. 73ra–75va. Prague, 1435–1465
Mikulov, Regionální muzeum v Mikulově (RM), Mik 6361, ff. 20v–22v and ff. 91v–92v. Prague, 1470–1480.

In the opening to his canons, the author of the Toruń manuscript mentions Wrocław, for whose meridian (and the year 1424) both the tables and the canons were originally compiled: *Sed inter tabulas nunc repertas meliores sunt, quas composuit Alfoncius, rex Castelle, super civitatem Tholetum; quarum tamen tabularum operacio ita est prolixa et tediosa, quod vix aliqui deveniunt ad cognicionem motuum planetarum. Propterea quasdam tabulas compendiosas super annos Incarnacionis Christi et super radices et motus tabularum Alfoncii fundatas ad brevitatem et levitatem operis ordinavi et super meridianum civitatis Wratisslaviensis reduxi, que quidem civitas Wratisslavia est orientalior Tholeto in una hora et 28 minutis hore.* – 'From among the tables that are now available, the best ones are those compiled by King Alfonso of Castile in the town of Toledo. However, using those tables is such a lengthy and tedious process that some barely manage to arrive at the understanding of the motions of planets. To make the process shorter and easier, I compiled concise tables based on the radices and motions of the Alfonsine Tables for the years of the Incarnation of Christ and I transposed them to the meridian of the city of Wrocław, as this city of Wrocław lies one hour and 28 minutes east of Toledo.'

The tables were later recalculated for the meridian of Prague (and the year 1428), which can be seen in the different copies of the canons: while the first four manuscripts (Toruń, Cracow, Leipzig, Oxford) mention Wrocław, MS Prague, NK, X B 3 mechanically exchanged Wrocław for Prague, without changing the geographical location of Wrocław for the one of Prague in relation to Toledo. The first copy documented in manuscript Mikulov, RM, Mik 6361 (ff. 20v–22v) mentions both Wrocław and Prague, while the second copy in the same manuscript (91v–92v) only mentions Prague. This and other evidence shows beyond doubt that the tables were first calculated for Wrocław and later adapted to Prague.

### 2.1.1 *MS Toruń, Biblioteka Uniwersytecka (BU), 74/III (olim Königsberg, 1735)*

This parchment manuscript originated in Prague, 1424–25. It reached Toruń from Königsberg (now Kaliningrad, Královec in Czech), a town founded by King Otakar II of Bohemia from Přemyslid dynasty. In Königsberg this manuscript was registered under the shelfmark 1735 at the beginning of the nineteenth century. This is the oldest extant manuscript containing the canons *Mirabilis* to the tables for the meridian of Wrocław.[1]

- The canons *Mirabilis* (sometimes also referred to as *Canones Tabularum resolutarum ad meridianum Wratislaviensem*) are copied on ff. 268va–273ra: *Mirabilis in altis Dominus... × ... sic facere potero de ipsius introitu in alia signa* (see Fig. 2).
- The manuscript also contains the tables themselves (*Tabule resolute planetarum super meridianum Wratisslaviensem ad annos collectos 1424–1704*) with the radix in 1424, on ff. 251r–268r. They begin in 1424 and end in 1704 with a period of 20 years for mean motions. Marginalia in manuscript Toruń, BU, 74/III relate to the Prague meridian, added later, probably by the hand of the owner of the codex, Wenceslas Koranda the Younger of Pilsen.[2] On ff. 251r–252r, there is *Tabula mediorum motuum planetarum ad annos Christi collectos super meridianum Wratisslaviensem* – 'Table of mean motions of the planets to the collected years of Christ for the meridian of Wrocław'. At the end of f. 251r, the following note is added in the margin under the table: *34 (secunda), 30 (tercia) – 7 (minuta), 41 (secunda), 11 (tercia) adde et habebis radices Pragenses* (see Fig. 1). There is also an interlinear gloss on f. 251v: *radix Christi Prage 3, 1, 55, 41, 36, sed non servit huic tabule* and below in the margin: *adde et habebis radices Pragenses 7, 37, 17 – 7, 43, 2 – 1, 51*. Inscribed numerical values for the Prague meridian are also given on f. 252r. On f. 256v we can find *Tabula augium verarum omnium planetarum ad annos Christi collectos* (in the margin, there is written: *1 addatur, cum praxis sit ad meridianum Pragensem in omnibus augibus*).[3]
- On ff. 281v–285v[4] we find other tables related to Prague. On ff. 283v–284r are two tables from a second set of the TR, which in this case have cyclical radices starting in the year 1428 (and not in 1424 as the previous complete set of the TR for Wrocław). The contents of these tables is as follows: *Tabula augium planetarum ad annos Cristi <1428–1808>* (f. 283v); *Tabula secunda augium planetarum ad annos Christi expansos <1428–1728>* (f. 284r). These two tables belonging to the TR for 1428 are surrounded by three tables that depend on the Prague horizon, i.e. on the latitude of 50 degrees. These tables were not part of the TR sets: *Tabula ad sciendum numerum horarum et minutorum cuiuslibet noctis per circulum anni et progressus, quando Sol oritur quolibet die*

1 For more details on the manuscript see Dobrzycki, 'The Tabulae resolutae', p. 74 (reprinted in 2010, p. 132); Pavel Brodský, *Iluminované rukopisy českého původu v polských sbírkách. – The Illuminated Manuscripts of Czech Origin in the Polish Collections* (Prague: Archiv Akademie věd ČR, 2004), pp. 105–7; Marta Czyżak, Monika Jakubek-Raczkowska, and Arkadiusz Wagner, *Katalog rękopisów średniowiecznych Biblioteki Uniwersyteckiej w Toruniu* (Toruń: Wydawnictwo Naukowe Uniwersytetu Mikołaja Kopernika, 2016), pp. 307–16; Dieter Blume, Mechthild Haffner, and Wolfgang Metzger, *Sternbilder des Mittelalters und der Renaissance. Der gemalte Himmel zwischen Wissenschaft und Phantasie, Band II: 1200–1500*, 5 vols (Berlin: De Gruyter, 2016), II/1, pp. 447–52.

2 On his person see below.

3 For numerical analysis of these notes see Chapter 19.

4 Czyżak, Jakubek-Raczkowska, and Wagner, *Katalog rękopisów*, p. 314.

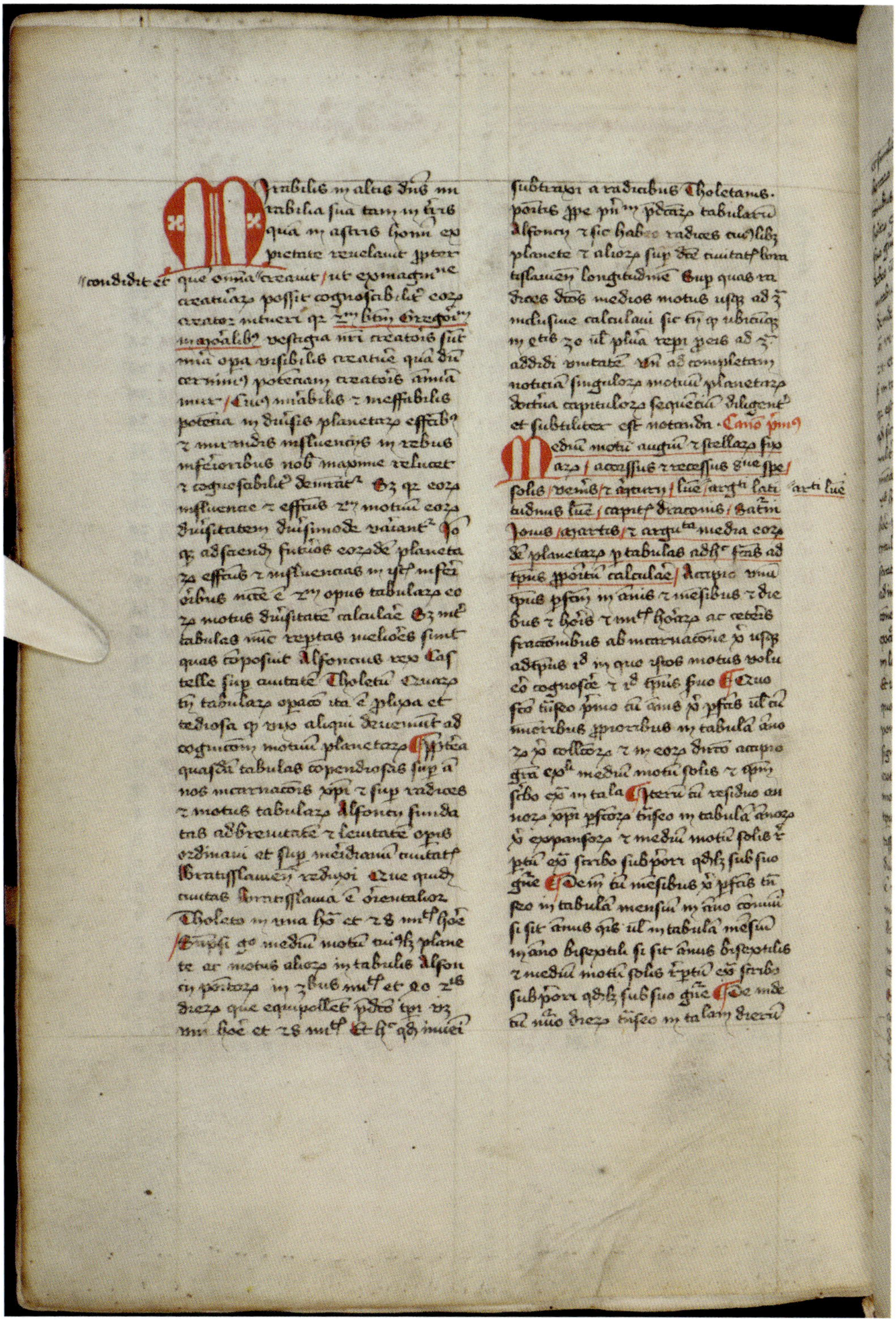

**Figure 2.** MS Toruń, BU, 74/III, f. 268v, the beginning of calligraphically written canons *Mirabilis*.

(f. 281v); *Tabula quantitatum noctuum, dierum et horarum meridiei et altitudinis Solis meridiane ad singulos gradus zodiaci in orisonte Pragensi* (f. 282v). The last table is *Tabula ad formandum quadrantem horarum super orisontem Pragensem* (f. 285v).

### *The author of the tables and the canons: Petrus Cruciferus or Petrus Rein Cruciferus?*

Jerzy Dobrzycki thought[5] that the author of the tables was Petrus Cruciferus,[6] who came from Prague and was a member of the Knights of the Cross with the Red Star (*Ordo militaris Cruciferorum cum rubea stella*) at the foot of the Charles Bridge, which was a Catholic order originating from Bohemia. It had been established in 1233 by St Agnes of Bohemia, a member of the Přemyslid dynasty and the daughter of the Czech King Otakar I. In 1252, St Agnes laid the cornerstone of the new hospital and church of the order at the Old Town side of the Judith Bridge, which Charles IV later replaced by the Charles Bridge; the seat of the order is now again at the foot of the bridge.[7] The order evolved from a lay brotherhood established to nurse the sick, poor and old. Membership in the order spread from Bohemia to Silesia and primarily to Wrocław,[8] where a commandery of the Knights of the Cross with the Red Star was established in the 1250s. Given the presence of the order in both Prague and Wrocław, it is possible that Petrus Cruciferus calculated the tables for the meridians of both cities as he moved between them in the uncertain times of the Hussite uprising.

However, two other manuscripts suggest another potential author of the tables, Petrus Rein (Reyn) de Zittavia. To quote MS Vienna, ÖNB, Cod. 5240 (dated 1440), f. 36r: *Expliciunt Tabule de mediis et veris motibus planetarum, compilate per fratrem Petrum Reyn,*[9] *commendatorem in Zittavia, ordinis sancti Iohannis Iherosolimitani. Et sunt scripte per Nicolaum Carnificis de Svalbenstat in Zittavia, Anno Domini M°CCCCXL° in estate.* – 'End of the tables of the mean and true motions of the planets, compiled by brother Petrus Reyn, commendator in Zittau, member of the order of Saint John of Jerusalem, and written by Nicolaus Carnificis of Swalbenstat in Zittau, in the summer of 1440'.[10] MS Cracow, BJ, 609, which was written in Prague and includes a commented copy of Sacrobosco's Sphere, contains a short treatise on phlebotomy on f. 234r–v. The heading of the work, written in rubric, reads *Sequitur Composicio minucionum, collecta per Petrum Reyn, cruciferum et comendatorem in Zittavia, ex diversis succinctim congregata* – 'Here follows the Composition of Bloodletting Calendars, succinctly compiled and collected from various sources by Petrus Reyn, cruciferus[11] and a commendator in Zittau' (see Fig. 3). Petrus Rein therefore

---

5 Dobrzycki, 'The Tabulae resolutae', pp. 74–75 (reprinted in 2010, pp. 132–33).

6 Petrus Cruciferus is also identified as the author of the canons in C. Philipp E. Nothaft, *Scandalous Error: Calendar Reform and Calendrical Astronomy in Medieval Europe* (Oxford: Oxford University Press, 2018), p. 197.

7 For more details on the history of the order, see *Kniha památní na sedmisetleté založení českých křižovníků s červenou hvězdou: 1233–1933* (Prague: Čs. akciová tiskárna, 1933); Milan M. Buben, *Rytířský řád křižovníků s červenou hvězdou* (Mělník: LINPA, 1996).

8 Josef Svátek, 'Organizace řeholních institucí v českých zemích a péče o jejich archivy', *Sborník archivních prací*, 20 (1970), 520.

9 In the secondary literature (see, e.g., *Tabulae codicum* IV, 1864–1893, p. 71, and the database maintained at www.manuscripta.at <http://18.235.151.129/detail.php?msid = 18597> [accessed: 12.03.2023]), one can encounter the transcription of the name *Reyn* as *Teyn*, but this is not correct reading.

10 See also Porres de Mateo, 'Šíření středověkých astronomických tabulek'.

11 It means a member of a religious order whose symbol is a cross, i.e. crusader.

belonged to a different religious order, the influential international military order *Ordo fratrum hospitalis sancti Ioannis Hierosolymitani*.[12] This order had been founded in the eleventh century at the territory of the Christian Kingdom of Jerusalem in the Holy Land. It is now usually called the Sovereign Military Order of Malta. Zittau lies on the border between Germany, Bohemia and Poland; it was given the town privileges in 1254 by King Otakar II of Bohemia. The commandery of the Order of Saint John was established in Zittau in 1300.

**Figure 3.** MS Cracow, BJ, 609, f. 234r.

As the discussion above suggests, the question of authorship has not yet been definitively answered. What is clear is that each of the two figures belonged to a different order and that the former Petrus had closer ties to Wrocław than the latter, but this argument alone is insufficient to attribute the authorship to him.

*Other selected works contained in this manuscript*

- On ff. 1v–9r: *Theorica planetarum* by Pseudo-Gerardus of Cremona.
- On ff. 30r–31v: Corrections of the Alfonsine tables (*Correcciones Tabularum Alfoncii*), which were according to the explicit written in the Charles' College, Prague, in the year 1460: *expliciunt correcciones Tabularum Alfoncii in principio novi anni 1460, post prandium, in collegio Caroli imperatoris gloriosi per manus W(enceslai) de P(ilzna)*. Without doubt, this refers to Wenceslas Koranda the Younger of Pilsen, the owner of the Toruń manuscript.
- On ff. 33r–55v: Parisian Alfonsine Tables with glossed radices for the meridian of Prague, Wrocław, Paris, Padua and Vienna.
- On ff. 56v–74r: tables from the Tables of 1322 by John of Lignères,[13] forming an eclectic miscellaneous collection of astronomical and astrological works.
- On ff. 105r–118v: John of Saxony, canons to the Parisian Alfonsine Tables with incipit: *Tempus est mensura motus*.
- On ff. 118v–129v: canons associated with the Tables of 1322 by John of Lignères with incipit: *Cuiuslibet arcus propositi*.
- On ff. 129v–148r: canons associated with the Tables of 1322 by John of Lignères with incipit: *Priores astrologi*.[14]
- On ff. 151v–181r: Tables of fixed stars. Incipit: *<T>abule stellarum fixarum verificatarum per dominum Alphoncium, quondam regem Hispanie, ex quibus sit sphera solida Ptholomei*.

12 Milan M. Buben, *Encyklopedie řádů, kongregací a řeholních společností katolické církve v českých zemích. I. díl: Řády rytířské a křižovníci* (Prague: Libri, 2002), pp. 21–54.

13 The Tables of 1322 were edited by Chabás and Saby, *The Tables of 1322 by John of Lignères*.

14 Both canons to the Tables of 1322 (*Cuiuslibet* and *Priores*) by John of Lignères were edited by Saby, 'Les canons de Jean de Lignères'.

> *Et additur super Ptholomeum 17 gradus et 8 minuta. Et ego, Hermannus Stilus, addo super Alphoncium anno 1360 completo 1 gradus, 32 minuta, 52 secunda, licet hic ponam earum loca secundum Alphoncium.* – 'The tables of fixed stars according to which Ptolemy's globe is to be made and that were corrected by Alfonso, once king of Hispania; he added 17 degrees and 8 minutes to Ptolemy. I myself, Hermann, add to Alfonso 1 degree, 32 minutes and 52 seconds for the end of the year 1360,[15] albeit that here I give the positions of the stars in accordance with Alfonso'. Hermann of Saxony, called Stilus, was probably born in Northeim, Lower Saxony, because the manuscripts mention his birthplace as 'Northem' or 'Nochem'. He was active in Paris around 1360.[16] The comparison of the numerical values of the star coordinates that we have conducted shows that the ecliptical longitude in the manuscript has genuinely been modified to take account of precession as of a date 1252, to the coronation of the Castilian king Alfonso X, which confirms Hermann's words. It seems that the Toruń codex, dated to 1425, offers a later copy of Hermann's work, probably produced as early as the 1360s.

Recently we have had the opportunity to study these Tables of fixed stars.[17] This is the translation of Ptolemy's Star Catalogue from Arabic into Latin, composed around 1175 by Gerard of Cremona; Gerard's translation is accompanied by Alfonsine coordinates. This catalogue is accompanied by an interesting set of drawings of personified constellations, as witnessed by the two unique examples described below:

Constellations of both the Northern Crown (*Corona Borealis*, CrB) and of Southern Crown (*Corona Australis*, CrA) take the form of a knotted (or twisted) veil on f. 155r and 180r, respectively (see Fig. 4). It is a well-known symbol of Wenceslas IV (1361–1419), King of Bohemia.[18] The inscription on f. 155r quite clearly says 'Crown of Ariadne' (*Corona Adriagnes*), thus referring to the Greek myth about a wedding crown or diadem given by Dionysos to Ariadne on the island of Naxos when Theseus abandoned her there on his return to Athens after killing the Minotaur. According to the myth, Hephaestus made the crown of gold and precious stones but in the MS Toruń, BU 74/III, it is depicted in a very different form which is rather a tribute to the King Wenceslas IV. The iconography

15 Statements of this kind in medieval manuscripts cannot be trusted a priori, but must always be verified. For example a similar formulation on Alfonso's verification of the stars found at the beginning of the *Catalogus stellarum fixarum* in MS Berlin, StM, KK, 78 D 12 (fourteenth century), is erroneous. In fact, what we find in this manuscript are al-Sūfī's coordinates adjusted for precession (al-Sūfī', *Catalogus stellarum fixarum*, ff. 1r–47r). The authors of the facsimile of the manuscript 78 D 12 (Alfonso X el Sabio, *Tablas de las Constelaciones – Alfonsine Tables – Alfonsinische Tafeln*, ed. by Ana Domínguez Rodríguez et al. [Valencia: Patrimonio, 2006]), however, did not subject the coordinates to investigation, and so mistakenly called their extensive edition, translated into several languages, *Alfonso X el Sabio, Tablas de las constelaciones*, though it is clearly al-Sūfī's Latin version of the *Catalogus stellarum fixarum*.

16 For details on Hermann of Saxony and his works, see José Chabás, 'Computational Practices in Alfonsine Astronomy: Hermann of Saxony's *practica motuum planetarum*', in *Alfonsine Astronomy: Computational Practices*, ed. by Matthieu Husson, José Chabás, and Richard L. Kremer (Turnhout: Brepols, forthcoming).

17 See description in Alena Hadravová and Petr Hadrava, *Sphaera octava – Mýty a věda o hvězdách IV. Katalogy hvězd a přemyslovský nebeský glóbus* (Prague: Artefactum – Academia, 2013), pp. 11–13.

18 This symbol of Wenceslas is also repeated several times in the rich illuminations in the copy of John of Saxony's canons to the Alfonsine Tables included in the Wenceslavian codex Vienna, ÖNB 2352, f. 34r (illustrations in the margins) and f. 36v (the initial of the letter 'M'). See the figure in Eric Ramírez-Weaver, 'Bohemian King Wenceslas IV's Copy of the Alfonsine Tables and their Place within his Astronomical and Astrological Corpus', in *Alfonsine Astronomy. The Written Record*, ed. by Richard L. Kremer, Matthieu Husson, and José Chabás (Turnhout: Brepols, 2022), p. 218.

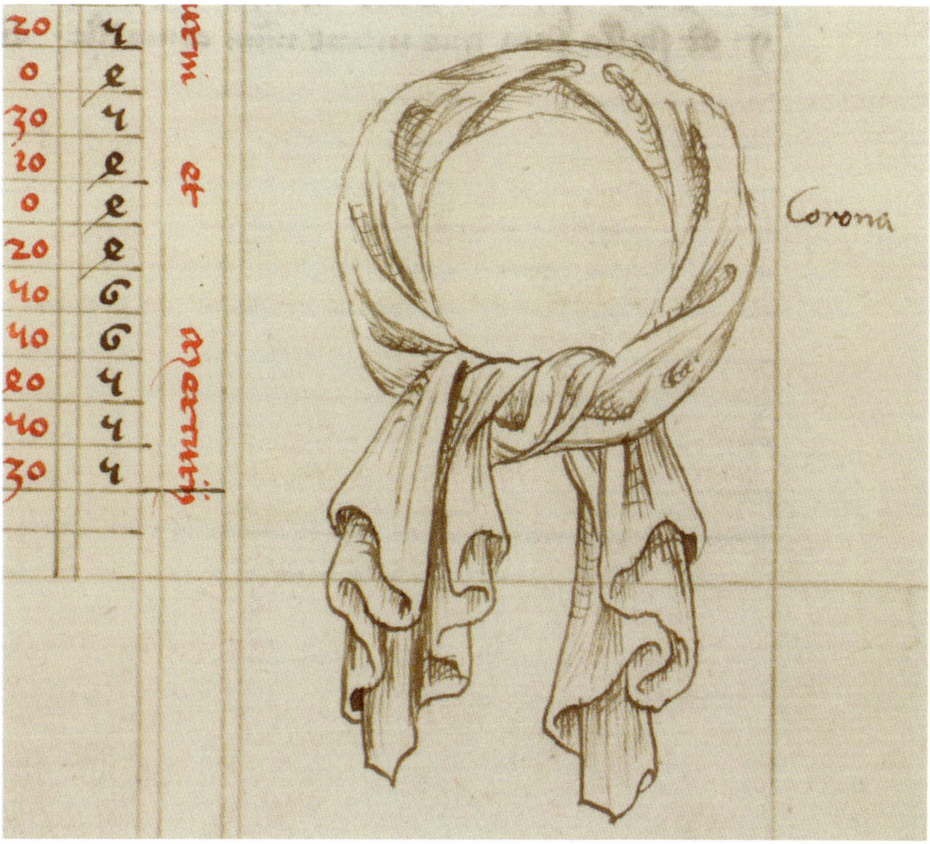


**Figure 4.** Constellation of the Southern Crown in the form of a knotted (or twisted) veil, the symbol of the King Wenceslas IV. MS Toruń, BU 74/III (1425), f. 180r. Gerard of Cremona's translation of Ptolemy's Catalogue from Arabic with Alfonsine coordinates.

of CrB and CrA is thus another convincing indication of this manuscript's Bohemian provenance.[19] In all the other drawings of the whole manuscript, stylistic features typical of Prague painting of 1380–90 are also recognised.[20]

The illumination of the Cassiopeia constellation in the Toruń manuscript (f. 157r) is also entirely unique in its interpretation of the text. The translation by Gerard of Cremona is strongly influenced by the Arabic source text. The name of Queen Cassiopeia, whose beauty was celebrated in Greek mythology, was paraphrased in the Arabic translation from Greek

19 Brodský, *Iluminované rukopisy*, p. 107; Hadravová and Hadrava, *Sphaera octava – Mýty a věda o hvězdách IV*, pp. 12–13; Alena Hadravová, *Sphaera octava – Mýty a věda o hvězdách III. Středověká pojednání o souhvězdích.* Chapter on iconography of planets by Lenka Panušková (Prague: Artefactum – Academia, 2013), p. 191. For more information on the other symbols of Wenceslas IV, see e.g., Milada Studničková, 'Drehknoten und Drachen: Die Orden Wenzels IV. und Sigismunds von Luxemburg und die Polysemantik ihrer Zeichen', in *Kunst als Herrschaftsinstrument: Böhmen und das Heilige Römische Reich unter den Luxemburgern im europäischen Kontext*, ed. by Jiří Fajt, and Andrea Langer (Berlin, München: Deutscher Kunstverlag, 2009), pp. 377–87.

20 See details in Czyżak, Jakubek-Raczkowska, and Wagner, *Katalog rękopisów*, p. 315.

**Figure 5.** Constellation of Cassiopeia. MS Toruń, BU 74/III (1425), f. 157r. Gerard of Cremona's translation of Ptolemy's Catalogue from Arabic with Alfonsine coordinates.

as *al-kaff al-hadib*[21] ('/a woman/ with a painted palm', i.e. whose palm was decorated with henna as a sign of beauty). This paraphrase then influenced Gerard's translation *stellatio Habentis palmam delibutam* ('the constellation of the One with a painted palm'), where it takes an appellative function. In the Christian medieval culture, the Arabic custom of painting the palm with henna as a sign of feminine beauty was unknown.[22] This is clearly demonstrated by a gloss next to a drawing of Cassiopeia in the MS Toruń, BU, 74/III with the following epithet: *Habens palmam delibutam, id est non ablutam*, i.e., 'The one with a grimy (dirty, painted) palm, i.e. unwashed' (see Fig. 5). One of the various interpretations of the constellation in the Christian culture (represented primarily by *Liber de signis* by Michael Scotus and its illuminated copies) took the opportunity of reinterpreting this for them unintelligible part of the text and making it into a stigma on Cassiopeia's palm. A number of illuminations, both in manuscript and early modern prints, therefore, show Cassiopeia bleeding heavily from her palm.[23]

21 Paul Kunitzsch, *Claudius Ptolemäus. Der Sternkatalog des Almagest. Die arabisch-mittelalterliche Tradition. II. Die lateinische Übersetzungen Gerhards von Cremona* (Wiesbaden: Otto Harrasowitz, 1990), p. 61, note 1.

22 Even the scribe of the Latin version of al-Sūfī's catalogue in Prague, Strahov, DA II 13 did not understand the expression *delibutam*; instead, he wrote the word *debelitam* (on f. 9r), which has no meaning in this context.

23 The case of Cassiopeia is described in more detail in Hadravová, *Sphaera octava – Mýty a věda o hvězdách III*, pp. 200–04.

- On ff. 182r–187v there follows composition and use of the astronomical instrument called *equatorium planetarum*.
- Another text on the composition and use of an astronomical instrument, in this case of *sphaera solida*, is contained on ff. 188r–199v. *Sphaera solida* (solid sphere) refers to a celestial globe, made from metal or wood, as it is said on f. 188v: *cum... volueris hoc instrumentum componere, facias primo pilam sphericam de metallo vel de ligno solido.*
- Messahalla's Composition of the astrolabe can be read on ff. 200r–204v.
- The next two texts contains Composition and Use of the astrolabe by Cristannus of Prachatice.[24] The first of them is the Composition on ff. 204v–215v and it has incipit and explicit: <Q>*uamvis de astrolabii composicione tam modernorum quam veterum dicta habentur pulcherrima... Cum ergo volueris facere astrolabium, recipe laminam eris vel stanni vel tabulam ligni solidi...* × ... *horum autem omnium pictura figure patule manifestat. Deo gracias. Et sic est finis composicionis astrolabii. Secuntur autem nunc ipsius astrolabii utilitates.*[25] The second text (ff. 216r–225r) is devoted to the Use of the astrolabe and its incipit and explicit are: <C>*um plurimi ob nimiam quandoque acurtacionem et magnam scriptorum sentenciam canones utilitates astrolabii declarantes intelligere...* × ... *eo, quod fidem facit de ibidem practicatis. Et sic est finis omnium instrumentorum in astrolabio positorum, et ut melius pateat legenti, figure possunt in margine depingi. – Si per astrolabium volueris scire, in quo gradu zodiaci sit Sol quolibet die anni secundum eius verum motum...* × ... *et si contingit regulam cadere in partem puncti, tunc operare, ut superius est ostensum. Et sic est finis huius operis.*[26]

Astronomer, mathematician, physician and theologian Cristannus of Prachatice (Křišťan z Prachatic) was born in southern Bohemia after 1360 and died in Prague1439. Cristannus studied at the Prague University (Bachelor of Arts in 1388, Master of Arts in 1390, then he studied medicine and theology). He spent his entire life at the university (dean of the Faculty of Arts in 1403–1404, rector in 1405, 1412–13, 1434 and 1437). He is known as a supporter of his younger countryman and friend, the reformer John Hus, during his studies. Later Cristannus became a conservative Calixtin. He is especially known for his works in natural sciences. His extant treatises on the Composition and Use of the astrolabe were written in 1407 on the basis of his university lectures and became influential worldwide. Their spread, popularity and importance can be seen from the high number of manuscript copies (the most recent of which date from the mid-sixteenth century) identified in libraries so far, about one hundred and twenty in total. They also became one of the first ever printed treatises on this astronomical instrument (Perugia: Petrus Petri de Colonia, Fridericus Ebert et Johannes Conradi 1477–79; Venice: Paganinus de Paganinis 1497–98 or already

24 Czyżak, Jakubek-Raczkowska, and Wagner, *Katalog rękopisów*, p. 312 erroneously attributes the authorship to Prosdocimo de' Beldomandi.

25 ThK 1164; Lynn Thorndike, 'Some Little Known Astronomical and Mathematical Manuscripts', *Osiris* 8 (1948), 41–72 (p. 56); Grażyna Rosińska, *Instrumenty astronomiczne na Uniwersytecie Krakowskim w XV wieku* (Wrocław etc.: Ossolineum, 1974), p. 105; Grażyna Rosińska, *Scientific Writings and Astronomical Tables in Cracow: A Census of Manuscript Sources (XIVth–XVIth Centuries)* (Wrocław etc.: Ossolineum, 1984), p. 336 (item 1715).

26 ThK 1456; Rosińska, *Instrumenty*, pp. 105–06; Rosińska, *Scientific Writings*, pp. 347–48 (item 1779).

1494?; Venice: Petrus Liechtenstein, 1512; Padua: Jacobus Fabrianus, 1549). The treatises were chosen for print due to their high didactical level and clarity of exposition. However, authorship of the work was attributed to Robertus Anglicus in the first incunabula from Perugia; later it was incorrectly attributed to Prosdocimo de' Beldomandi, and then even to Johannes Angelus (Engel) or Pietro Catena.[27] As a prominent figure in the Hussite revolution, Cristannus was *persona non grata* for Catholic Europe. This explains why the authorship of both his treatises was generally concealed and eventually forgotten, and how subsequently it was wrongly attributed to other authors, even though the texts' Prague origins are evident. The incorrect attributions persist in the secondary literature mostly up to the present, despite arguments published already in 2001 in favour of Cristannus's authorship and the text's Prague origin.[28] In the 1420s or 1430s, a version of Cristannus's treatises on the astrolabe was written by the Viennese master John of Gmunden, who borrowed it in its entirety and developed some of its passages. His treatises are known now in about sixteen copies.

- Treatises on astronomical instruments in MS Toruń, BU, 74/III continue with Profatius Iudeus's Composition of the new quadrant (incipit: *<C>ompositurus novum quadrantem accipe tabulam planam ad similitudinem quarte partis circuli factam*) on ff. 225v–228r and by Johannes Eligerus de Gunderslauen's Use of the new quadrant (incipit: *<U>tilitates quadrantis novi breviter colligere utile mihi visum*) on ff. 229v–232v.
- On ff. 239r–250r there appear Canons to the eclipse tables of the Sun and the Moon; the text begins with the words *Ad habendum eclipses sic procedendum est* and the author is anonymous. (This text is also known from the manuscript Cracow, BJ, 610, ff. 279r–289v, see below.)

### *The owner of the Toruń manuscript: Wenceslas Koranda the Younger of Pilsen*

Utraquist theologian Wenceslaw Koranda the Younger of Pilsen (Václav Koranda mladší z Plzně) was born after 1420 in Pilsen and died 1519 in Prague.[29] He was most likely related to the radical preacher Wenceslas Koranda the Elder. At the Utraquist University in Prague he became a bachelor in 1454 and master of liberal arts in 1458. He served at the University of Prague as a lecturer and in 1460 as a dean of the Faculty of Arts. He was a rector of the university in 1462–63, 1470–71 and 1513. He also worked as a diplomat in the service of the King George of Poděbrady. He was also active in literature. As a major figure in the third generation of the Czech Utraquism, Koranda authored a number of argumentative texts,

27 See Hadravová and Hadrava, 'Spreading of treatises on astrolabe by Cristannus of Prachatice'.

28 Hadravová and Hadrava, *Křišťan z Prachatic*: this is the critical edition of the Composition and Use of the astrolabe proving the Cristannus's authorship, supplemented by an annotated Czech translation, computer-simulated drawings, appendices and an English summary. An edition of a version by John of Gmunden is also added. See electronical version on the web-pages of the Centre for Medieval Studies in Prague (joint workplace of Charles University and the Academy of Sciences of the Czech republic): <https://sources.cms.flu.cas.cz/src/index.php?s = v&cat = 21&bookid = 1289>.

29 Jindřich Marek, *Václav Koranda mladší. Utrakvistický administrátor a literát* (Prague: NLN, 2017), pp. 114–16 and 211–12; Mlada Holá, Martin Holý et al., *Profesoři pražské utrakvistické univerzity v pozdním středověku a raném novověku (1457/1458–1622)* (Prague: Academia – Historický ústav, 2022), pp. 357–59 (Koranda's biography is written by Ivan Hlaváček).

mostly on religious and theological topics.[30] Some of his letters have also been preserved. Today 46 manuscripts and 25 incunabula are known from his apparently more extensive library, which he donated to Charles College. In addition to theological writings, his library included classical works of antiquity, but also works of natural science, especially astronomy, computistics and mathematics.[31] His interest in astronomy was probably influenced by already mentioned Master Cristannus of Prachatice; Cristannus was one of Koranda's teacher.

The fact that the manuscript Toruń, BU, 74/III, originating *c.* 1425, belonged to Koranda's library is evidenced by his handwritten notes in a number of places in the codex.

The back pastedown of MS Toruń, BU 74/III contains two of Koranda's monograms, which is evidence that he owned the book. Koranda's autograph is on ff. 30r–31v and he also wrote several notes in the margins. One of these notes (f. 31r) says: *Expliciunt correctiones Tabularum Alfonsii in principio novi anni 1460, post prandio in collegio Caroli imperatoris.* – 'Here end the corrections of the Alfonsine Tables at the beginning of the new year of 1460, after lunch at the College of Emperor Charles.' Another note on f. 118v says: *Nota, quod presentes canones super Tabulas magistri Iohannis de Lineriis compilavit Iohannes Danko, nacione Saxo, scolaris dicti magistri Iohannis de Lineriis, qui in calculacione dictarum tabularum fuit valde solers et expertus.* – 'Note that the canons to the Tables by Master John of Lignères contained herein were compiled by John Danko of the Saxon nation, a student of the said Master John of Lignères, who was very proficient and knowledgeable about calculating the said tables.'

### *2.1.2 MS Cracow, Biblioteka Jagiellońska (BJ), 610*

This manuscript originated partly in Prague, while another part was probably written in Cracow or another unknown location, 1440–77. The manuscript was owned by Stanislas de Sgrz (Zgierz) C(h)osus (Cosuch) and later by Johannes de Treubel.[32]

– The canons *Mirabilis in altis Dominus* are copied on ff. 352ra–357vb: *Mirabilis in altis Dominus ...* × *... sic facere potero de ipsius introitu in alia signa. – Expliciunt Tabule cum suis canonibus de mediis et veris motibus planetarum, Anno Domini MoCCCCoXLo, Sabbato ante Philippi.* At the end of folio 357vb, there is the following note in the hand of Józef Muczkowski, a nineteenth century librarian at the Jagellonian Library: *NB. Michaelis de Vratislavia De mediis et veris motibus planetarum.*[33] The manuscript does not contain any set of tables to which the canons could relate.

30 Jindřich Marek, 'Schriften des utraquistischen Konsistoriums aus den Jahren 1471–1489 in der Nationalbibliothek der Tschechischen Republik', in *Sacri canones servandi sunt*, ed. by Pavel Krafl (Prague: Historický ústav, 2008), pp. 545–56 (p. 552, note 32); Marek, *Václav Koranda mladší*.

31 For his library see Emma Urbánková, 'Zbytky knihovny mistra Václava Korandy mladšího v universitní knihovně v Praze', in *Ročenka universitní knihovny v Praze 1956* (Prague: Univerzitní knihovna, 1956), pp. 135–61. Four manuscripts from the Koranda library, now housed in the ÖNB in Vienna, which contain the writings of the ideological spokesmen of Hussitism, were identified by Pavel Spunar, 'Několik doplňků k rekonstrukci knihovnyVáclava Korandy ml.', *Listy filologické*, 91 (1968), 147–50. Additions to the library were also published by Marek, *Václav Koranda mladší*, pp. 108–46.

32 See a detailed description of the codex in the *Catalogus codicum manuscriptorum medii aevi Latinorum, qui in Bibliotheca Jagellonica (BJ) Cracoviae asservantur* (Wrocław: Ossolineum, 1988), IV, pp. 252–74.

33 This note by Muczkowski is mentioned in *Catalogus BJ*, IV, p. 273 and prior to that by Rosińska, *Scientific Writings*, p. 242 (item 1210), who erroneously attributed the note to Michael de Wratislavia rather than Muczkowski.

*Other selected works contained in this manuscript*

- Many texts belonging to Alfonsine astronomy are scattered throughout BJ, 610. Two of them concern the spread of the Alfonsine Tables to the British Isles. The oldest Alfonsine Tables written in London have been dated to 1336.[34] Several years later, William Rede adapted the Parisian tables to the meridian of Oxford (with an epoch date in 1340), and William Batecombe recalculated them for the meridian of Oxford in 1348.[35] Folios 1v–42r of the Cracovian manuscript contain a copy of the Tables of the true positions of planets (*Tabule verorum locorum planetarum*) with the incipit *Tabula proporcionum trium superiorum*, known as Oxford tables of 1348. On ff. 42v–43v there continue canons to the Oxford tables of 1348, also composed by Batecombe.
- A table for the lunar equations by Nicolaus Mülhus is contained on ff. 47v–77v. Only one copy of this table had been known (Vatican, BAV, Pal. lat. 1374, 87r–102r).[36] However, it now seems that this long table had a certain diffusion in the millieu of the Prague university and is frequently found together with Nicolaus de Heybech's Table for true syzygies (see below). There are currently known at least five copies of Mülhus's work: in addition to the copies in BJ, 610 and Vatican, BAV, Pal. lat. 1374, this work can also be found in Prague, NK, X B 3, ff. 42v–72v; Prague, NK, adlig. 14 B 10, ff. 1r–31v; and Mikulov, RM, Mik 6361, ff. 99r–129v.[37] In the Vatican copy, written after 1407 in Prague, the short text at the end of the table (f. 102r), which substitutes the canons, states: *Ad inveniendum breviter verum locum Lune per tabulas predictas, factas per Nicolaum Mülhus, lanificem et laicum morantem in Zittavia: Quere primo in aliis tabulis medium motum Lune × et nota, quod per parva 'a' et 'm' in istis tabulis congnoscitur, an secunda equacio est maior vel minor etc.* The otherwise unknown author, Nicolaus Mülhus, is identified here as a wool manufacturer and layman from Zittau. The calligraphic table in the manuscript Prague, NK, adlig. 14 B 10, from the fifteenth century, is in turn bound to the print of Peurbach's and Regiomontanus's works, published in Venice 1514.[38] As we shall see below,[39] Nicolaus Wodka of Kwidzyn constructed a similar double-argument Table of equations of the Moon for his version of the TR, which can be found in MS Milan, BA, C 278 b Inf., ff. 33v–39r. Only sketched and completely unfinished is also this table in MS London, BL, Add 22317, f. 20v.
- Folios 120r–125v contain a copy of the Table of equations of the twelve houses in the sky (*Tabula equacionum duodecim domorum celi*) with the beginning *Aries 1°*. The work is known under the title *Tabule primi mobilis* and its author is Giovanni Bianchini. Giovanni Bianchini (Johannes Blanchinus) recalculated the Alfonsine Tables for Ferrara

34 Chabás, 'An Early Witness of Alfonsine Astronomy'.

35 Chabás and Goldstein, 'The Moon in the Oxford Tables of 1348'.

36 See José Chabás and Bernard R. Goldstein, 'The Medieval Moon in a Matrix: Double Argument Tables for Lunar Motion', *Archive for History of Exact Sciences*, 73 (2019), 335–59.

37 See the descriptions of these manuscripts.

38 See Truhlář, *Catalogus codicum*, p. 371 (item 2760) <https://sources.cms.flu.cas.cz/src/index.php?s = v&cat = 36&bookid = 370&page = 375>. The scan of the manuscript is available at <https://www.manuscriptorium.com/apps/index.php?direct = record&pid = NKCR___-NKCR___14_B_000010_3PVPG66-cs> [accessed: 13.11.2023].

39 See Chapter 7, especially the Rule 16 of Nicolaus Wodka's canons, and Chapter 22.

in northern Italy around 1440. His tables were published as an incunable (in 1495)[40] and recently have been analysed.[41] One of Bianchini's predecessors in adjusting the Parisian Alfonsine Tables for Italy was Prosdocimo de' Beldomandi, also known as a music theorist. Prosdocimo recalculated the tables for Padua for an epoch of 1428. Bianchini's Table of equations was reduced to the meridian of Cracow on ff. 126r–131v of BJ, 610.

- Another well-known figure of early Alfonsine astronomy, John of Lignères, is also represented in the codex with several of his works. His Tables of planetary latitudes (*Tabule latitudinum planetarum*) with the incipit *Tabula continens veram latitudinem Saturni* are part of the Tables of 1322[42] and they are copied on ff. 141r–148r, 261r–268r, and 317r–338r. On ff. 247r–260v there are written canons to this Tables of 1322 by John of Lignères. Their incipit is *Priores astrologi*.
- The manuscript also contains a Table of the hours of night and sunrise above the meridian of Prague (*Tabula horarum noctis et ortus Solis super meridianum Pragensem*) on f. 159v with the names of the months in the table headings written in Old Czech: *leden* (January), *vnor* (February), *brzezen* (March), *duben* (April), *may* (May), *czrwen* (June), *czrwnecz* (July), *srpen* (August), *zarzi* (September), *rzieen* (October), *listopad* (November), *prosinecz* (December). That is why it has been assumed that at least some parts of the manuscript are of Bohemian origin, while others were probably written in Cracow.
- Tables of mean motions for the years 1469–1538 are contained on ff. 171r–246v.
- Nicolaus de Heybech's Table for true syzygies is copied on ff. 334r–335r. On f. 335v, there follows a version of Nicolaus's canon to his Table for true syzygies. Nicolaus de Heybech of Erfurt was active in the years 1384–1394 and based his work largely on the results of the earliest Paris Alphonsine astronomers.[43]

Seven works in BJ, 610 and Prague, NK, X B 3 are remarkably identical and indicate a possible dependence between both manuscripts:

1) *Eclipsis possibilitatem in Sole quantum ad aspectum terricolarum per tabulas invenire* = Cracow, BJ, 610, ff. 268v–269r; Prague, NK, X B 3, ff. 139va–142ra.
2) *Canones tabularum eclipsium Solis et Lune* (incipit: *Ad habendas eclipses sic procedendum est*) by anonymous author = Cracow, BJ, 610, ff. 279r–289v; Prague, NK, X B 3, ff. 16ra–28vb.
3) *De modo inveniendi coniunccionem et opposicionem* (incipit: *Pro vera coniunccione et opposicione invenienda per almanach*) = Cracow, BJ, 610, f. 311v; Prague, NK, X B 3, f. 15r.
4) Pseudo-Albertus Magnus, *Secretum de sigillo leonis* (incipit: *Ego Albertus, commentator in omni experiencia expertus*) = Cracow, BJ, 610, f. 312r; Prague, NK, X B 3, f. 15v.

40 Giovanni Bianchini, *Tabulae celestium motuum earumque canones*, GW 4410. – The editor and simultaneously the author of the preface to Bianchini book was Augustinus Moravus Olomucensis who was born in Moravian town Olomouc in 1467 and died there in 1513. He was a leading Moravian humanist (the author of the texts written both in prose and in verses), but he worked also as a royal clerk and provost. Augustinus dedicated his edition of Bianchini's tables to his mother's brother Andreas Stiborius (Ondřej Ctibor), canon in Olomouc and Kroměříž, who influenced and supported young Augustinus in his interests and studies of astronomy very much.

41 Chabás and Goldstein, *The Astronomical Tables of Giovanni Bianchini*.

42 See the edition by Chabás and Saby, *The Tables of 1322 by John of Lignères*.

43 See José Chabás and Bernard R. Goldstein, 'Nicholaus de Heybech and his table for finding true syzygy', *Historia Mathematica*, 19 (1992), 265–89; José Chabás and Bernard R. Goldstein, 'Nicholaus de Heybech of Erfurt, Table-maker and Copyist', *Aestimatio*, forthcoming.

5) *De virtute imaginum celestium vel excerpta de libris Picatricis*[44] (incipit: *Ego autem rogo altissimum Creatorem*) = Cracow, BJ, 610, ff. 312v–316r; Prague, NK, X B 3, ff. 13ra–14ra.
6) *Ad inveniendum gradum ascendentem* = Cracow, BJ, 610, f. 350v; Prague, NK, X B 3, f. 29ra.
7) *Mirabilis* = Cracow, BJ, 610, ff. 352ra–357vb; Prague, NK, X B 3, ff. 73ra–75va.

### 2.1.3 MS Leipzig, Universitätsbibliothek (UB), 1476

This manuscript originated in Prague, in the first half of the fifteenth century.

- On ff. 34r–37v are contained canons *Mirabilis*.[45] Prologue and incipit to them on f. 34r is as follows: *Incipiunt canones Tabularum Cruciferi excerptarum per eundem de Tabulis Alfoncii ad meridianum Wratislaviensem. Hic autem sunt redacte ad meridianum Pragensem et est idem modus operandi. <M>irabilis Deus in altis.* These introductory words are quite identical with the introduction of canons *Mirabilis* in the MS Oxford, Bodl., Canon. misc. 499, which was also written in Prague. The Prague origin of the Leipzig manuscript is attested by two other references to Prague:
- The table on f. 40r is labelled as follows: *Tabula diversitatis aspectus ad latitudinem civitatis Pragensis, que est 50 graduum, 6 minutorum.*
- On ff. 95v–96r we then read: *Tabula quantitatum noccium, dierum et horarum meridiei et altitudinis || Solis meridiane ad singulos gradus zodiaci in orisonte Pragensi.* Both manuscripts, preserved now in Leipzig and Oxford, are closely related; this is evident also from a few minor errors in the text, which are repeated in both manuscripts, and from minor additions known only from both of these copies (the Oxford manuscript records them in the marginalia). Thus it seems that both copies were copied from the same source. It is also interesting for the relationship between the manuscripts that the Leipzig manuscript, like both copies now in Mikulov, omitted half of the second canon.
- In the Leipzig copy of the canons *Mirabilis*, as in only one of our seven copies, a longer addition is inserted at the end of the third rule on conjunctions and oppositions; it is added here inappropriately because it refers to the tables of equations. The use of these tables for the Sun, Moon, and the planets is described in the canons *Mirabilis* in the Rules 4, 5, and 6, respectively. Similarly to the canons *Tempus est mensura motus* by John of Saxony,[46] this description assumes these tables in their simple form found in the PAT and basically preserved in the *Tabule resolute* for Wrocław.[47] However, the addition in the MS Leipzig deals with the usage of columns containing differences between the values in subsequent lines, which are added only in the later *Tabule resolute* for Prague, Cracow and other cities.[48] In the MS Leipzig this addition is contained in ff. 34v–36r:

44 See Jean-Patrice Boudet and Jean-Charles Coulon, 'La version arabe (Ghāyāt al-hakīm) et la version latine du Picatrix. Points communs et divergences', *Cahiers de recherches médiévales et humanistes / Journal of Medieval and Humanistic Studies*, 33 (2017), 67–101.
45 Dobrzycki, 'The Tabulae resolutae', p. 74 (reprinted in 2010, p. 132, note 11); <http://jordanus.badw.de/jordanus/ms/3729#133379> (Work 7) [accessed: 12.03.2023]. Scanned UB Leipzig.
46 Poulle, *Les tables Alphonsines*, pp. 64–80, Rules 17, 18, 20, and 21 (Ratdolt 1483, ff. A6v–A8v).
47 See Poulle, *Les tables Alphonsines*, pp. 145–83, Tables 26–32 (Ratdolt 1483, ff. E2v–G5v), and Chapter 12, the tables TRW-13 and TRW-15 – TRW-20.
48 An example of these enriched tables of equations is shown in Fig. 28.

*Notandum, quod in has tabulas equacionum ... × ... quarum superior in utraque margine respicit minuta proporcionalia, que sunt supra propior et alia litera respicit minuta, que sunt sub propior* (for the whole text see the critical apparatus of the edition in the Chapter 2.2). Although we do not know it from other copies of *Mirabilis*, it is included in the canons *Girum* by Andrzej Grzymała. Since the Leipzig copy was written after the creation of the *Girum* canons, it is possible that this addition came into the *Mirabilis* text from the *Girum* rules. We find it, for example, in the copy of the *Girum* text in MS Cracow, BJ, 2252, p. 44, where it forms part of the final rule, called *Canon de correccione tabularum*. It is also in the copies of the canons *Girum* in MS Prague, NK VIII G 24, ff. 61v–62r, and Cracow, BJ, 573, ff. 229v–230r. From Grzymała the wording was later adopted by Nicolaus of Tuchów (Polonus) to his version of the canons for the tables recalculated for Salamanca, which have incipit *Quoniam Tabularum Alfonsi* (see MS Oxford, Bodl., Canon. misc. 27, ff. 124vb–125ra; MS Cambridge, Trinity College, O.3.13, f. 76v; MS Lisbon, ANTT, ML, 2115, f. 12r). The text is also included in the canons *Astrorum* (see their Rule 18, e.g. in the MS Cracow, BJ, 600, f. 141v), in the canons *Pro (introduccione) generali* by John of Głogów (see his 16th Rule in the MS Vatican, 1439, ff. 175v–176r and in the MSS Cracow, BJ, 600, f. 153v, and Wrocław, BU, IV F 19, f. 253v) and also in the printed versions of the TR by Ambrosius Lacher and Johannes Schöner (the Rules 9 and 7, respectively). On the other hand, this paragraph is not contained in the canons *Nemo.*

### 2.1.4 MS Oxford, Bodleian Library (Bodl.), Canon. misc. 499

This manuscript originated in Prague in the middle of the fifteenth century.[49]

- Canons to the *Tabule resolute* with the incipit *Mirabilis in altis Dominus* by Petrus Cruciferus are included on ff. 122ra/112ra–125vb/115vb.[50]
- *Tabule resolute* for the meridian of Prague, also composed by Petrus Cruciferus, are contained on ff. 72r/62r–105r/95r. After that additional tables to the TR follow, e.g.: *Tabula minutorum proporcionalium pro diebus intermediis in almanach faciendum* on ff. 105v/95v–118r/108r. To the incipit of this work on f. 72r/62r (*Incipiunt Tabule de mediis et veris motibus planetarum super meridianum Pragensem et primo in annis collectis*) a younger hand added a misleading note *que appellantur Tabule aureе et resolute Alberti Bru<ze>*). We can also read a younger note in the upper margin: *Nota, quod iste tabule sunt facte ad meridianum Pragensem et Cracoviensem*. The note at the end of the whole work, written by the same younger hand on the margin of f. 118r/108r, is once again rather misleading: *Expliciunt Tabule aureе et resolute Alberti Bruze,* because there are TR by Petrus Cruciferus in the manuscript and not *Tabule aureе* by Albertus of Brudzewo.

Henry Coxe, author of the catalogue of manuscripts of the Bodleian Library,[51] gives the following incipit and explicit of the canons *Mirabilis*:

49 Short description of manuscript is provided by Henry O. Coxe, *Catalogi codicum manuscriptorum Bibliothecae Bodleianae pars tertia codices Graecos et Latinos Canonicianos complectens* (Oxonii: e typographo academico, 1854); see also Pedersen, *The Toledan Tables*, p. 141.

50 There and hereafter, we use both foliations of this manuscript, it means upper/lower foliation.

51 Coxe, *Catalogi codicum*, pp. 814–15.

> *Canones tabularum Cruciferi* ***Gaerparum*** *per eundem de tabulis Alfoncii ad meridianum Wratislaviensem, hic autem sunt* ***redactae*** *ad meridianum Pragensem. – Mirabilis in altis Dominus, mirabilia sua tam in terris ... × ... Expliciunt tabule Cruciferi* ***cum*** *suis canonibus de mediis et veris motibus planetarum reducte ad meridianum Pragensem,* ***scripte*** *Anno Domini* ***1471*** *per Iohannem* ***B. de Bazyn*** *in collegio beatissime* ***Marie*** *Virginis* ***nationis Boemorum*** *domus* ***Beczkonis****.*

However, both pieces of text – incipit and explicit (see Figs 6–9) – were misread, which led to the work's misconstrued perception in secondary literature. On top of that, several minor errors were added in recent years.[52] The erroneous readings are highlighted in bold. The following overview lists the erroneous readings and the corrected versions (again highlighted in bold):

> – *Gaerparum* → Coxe, *Catalogi codicum*, 814: 'Gaerparum'; ThK 876: 'Cruciferus Carparus, *Canones tabularum ad meridianum Wratislaviensem et ad meridianum Pragensem* (examples for 1424 and 1430 are given in the text)';[53] Rosińska, *Scientific Writings*, 242 (item 1210): 'Carparii'; Dobrzycki, 'The *Tabule resolute*', p. 74 (reprinted in 2010, p. 132): 'he (= the author) was identified as 'Cruciferus', 'Carparus', 'Petrus Cruciferus''; <www.manuscriptorium.com>: 'Caspar', 'Gaerparus'; *Catalogus BJ*, IV, 272, in the description of BJ, 610: 'Caspar Crucifer' etc. Correct reading: ***excerp<t>arum***
> – *redactae* → Coxe, *Catalogi codicum*, 814: 'redactae'; Rosińska, *Scientific Writings*, 242 (item 1210): 'reducte'; correct reading: ***redacte***
> – *cum* → Rosińska, *Scientific Writings*, 242 (item 1210): 'una cum'; correct version: ***cum***
> – *scripte* → Coxe, *Catalogi codicum*, 815: 'scripte'; Rosińska, *Scientific Writings*, 242 (item 1210): 'scripte'; correct reading: ***sc<r>ipte***
> – *1471* → Coxe, *Catalogi codicum*, 815: '1471'; correct version: ***1451***
> – *B. de Bazyn* → Coxe, *Catalogi codicum*, 815: 'B. de Bazyn'; Rosińska, *Scientific Writings*, 242 (item 1210): 'B. de Brzyn'; correct reading: ***b(accalaureum) de Krczyn***
> – *Marie* → Rosińska, *Scientific Writings*, 242 (item 1210): omitted; correct reading: ***Marie***
> – *nacionis Boemorum* → Rosińska, *Scientific Writings*, 242 (item 1210): omitted; correct reading: ***nacionis Boemorum***
> – *Beczkonis* → Coxe, *Catalogi codicum*, 815: 'Beczkonis'; correct reading: ***Reczkonis***.

Below is the complete correct reading:

> *Incipiunt canones Tabularum Cruciferi, excerp<t>arum per eundem de Tabulis Alfoncii ad meridianum Wratislawiensem. Hic autem sunt reducte*[54] *ad meridianum Pragensem. Et est idem modus operandi. – Mirabilis in altis Dominus... × ... sic facere potero de ipsius introitu in alia signa. – Expliciunt Tabule Cruciferi cum suis canonibus de mediis et veris motibus planetarum reducte ad meridianum Pragensem, sc<r>ipte Anno Domini 1451º per Iohannem, B(accalaureum) de Krczyn in collegio beatissime Marie Virginis nacionis*

52 Rosińska, *Scientific Writings*, p. 242 (item 1210).

53 In fact, the text only gives an example for the year of 1430, as 1424 is the first of the collected years (*anni collecti*) that are used as the base for the calculation for 1430 (interpolating between 1424 and 1444).

54 *ms. per err.:* redacte

*Boemorum domus Reczkonis*. – 'Here begin the canons to Cruciferus's Tables, which he excerpted from the Alfonsine Tables to the meridian of Wrocław. The way they are used is the same. – Marvellous is the Lord on high… × … this I can do with its entry in other signs. The end of the Cruciferus's Tables and their canons to the mean and true motions of the planets adapted to the meridian of Prague, copied in AD 1451 by Bachelor John (Jan, Johannes) Krčín in the College of the Blessed Virgin Mary of the Czech nation in the House of Reček.'

The corrected readings resolve several problems:

1) According to the text, the author of the Tables for the meridian of Wrocław (with an epoch date in 1424) is Cruciferus (a member of a religious order with the symbol of the cross). 'Gaerparus', 'Carpar(i)us' and 'Caspar' never existed.
2) Older literature was unclear on whether the tables were first calculated for Prague or for Wrocław. Ernst Zinner[55] assumed that they were first calculated for the meridian of Prague and then adapted to the meridian of Cracow. On the other hand, Jerzy Dobrzycki was of the opinion that the tables were first calculated for the Wrocław meridian to the year 1424 and later adapted to the Prague meridian in the year 1428.[56] The latter is now largely assumed to be true and is corroborated by the words of Krčín: the correct reading of the word *excerptarum* gives a clear answer as to which tables were prepared earlier. According to the text, the Tables for Wrocław came first and were excerpted from Alfonsine Tables, while the Tables for Prague were only compiled later from the Tables for Wrocław. (The internal textual logic of *Mirabilis* also corresponds with this order of events.)
3) The correct reading of the scribe's name – 'Krczyn', or 'Krčín' in modern Czech – instead of the non-existent 'Bazyn' or 'Brzyn' and the year '1451' of his copy are consistent with what is known about John Krčín, a Czech historical figure. He obtained the bachelor's title at the Prague university in 1450. His copy of the Alfonsine Tables and the canons *Mirabilis* corresponds with his interests in 1450, when he finished the study of seven liberal arts (*septem artes liberales*), including astronomy, at the Faculty of Arts.
4) The correct reading of the initial makes it possible to identify the location as the house of 'Reček', which was one of the colleges of the *studium generale* in Prague founded in 1438 by Jan Reček from Ledeč.[57]

As far as we know, John Krčín's signature as the scribe can be found in at least three other places in MS Oxford, Bodl., Canon. misc. 499, where the current reading of Coxe's catalogue and the secondary literature that relied on his reading needs to be revised. One of the texts in question is a copy of John of Lignères' Tables of 1322:

– Incipit: *Anno Domini millesimo quadringentesimo quinquagesimo secundo Tabule presentes magistri Iohannis de Lineriis scripte sunt per Iohannem de Krczyn* (Coxe, *Catalogi codicum*,

55 Ernst Zinner, *Leben und Wirken des Johannes Müller von Königsberg, genannt Regiomontanus*, 2nd rev. Ed. (Osnabrück: O. Zeller, 1968), p. 259.

56 Dobrzycki, 'The Tabulae resolutae', p. 74 (reprinted in 2010, p. 132).

57 For more details on the founder of the college see František Šmahel, 'Knihovní katalogy koleje Národa českého a koleje Rečkovy', *Acta Universitatis Carolinae – Historia Universitatis Carolinae Pragensis (AUC – HUCP)* 2/1 (1961), 59–85.

815, *per err.*: Bazyn) *in studio Pragensi, in collegio beatissime Marie Virginis domus Reczkonis* (Coxe, *Catalogi codicum*, 815, *per err.*: Beczkonis).[58]

- Explicit: *Expliciunt canones Tabularum magistri Iohannis de Lineriis finiti per Iohannem de Krzcyn* (Coxe, *Catalogi codicum*, 815, *per err.*: Bazyn) *in collegio beatissime Virginis Marie domo Reczkonis nacionis Bohemorum Anno Domini 1450* (Coxe, *Catalogi codicum*, 815, *per err.*: 1470).[59]
- Finally, Johannes de Krczyn is signed as the copyist of *Priores astrologi* on ff. 180ra/170ra–193vb/183vb, another set of canons by John of Lignères to his Tables of 1322. This copy was also made by John Krčín in Prague in 1450.

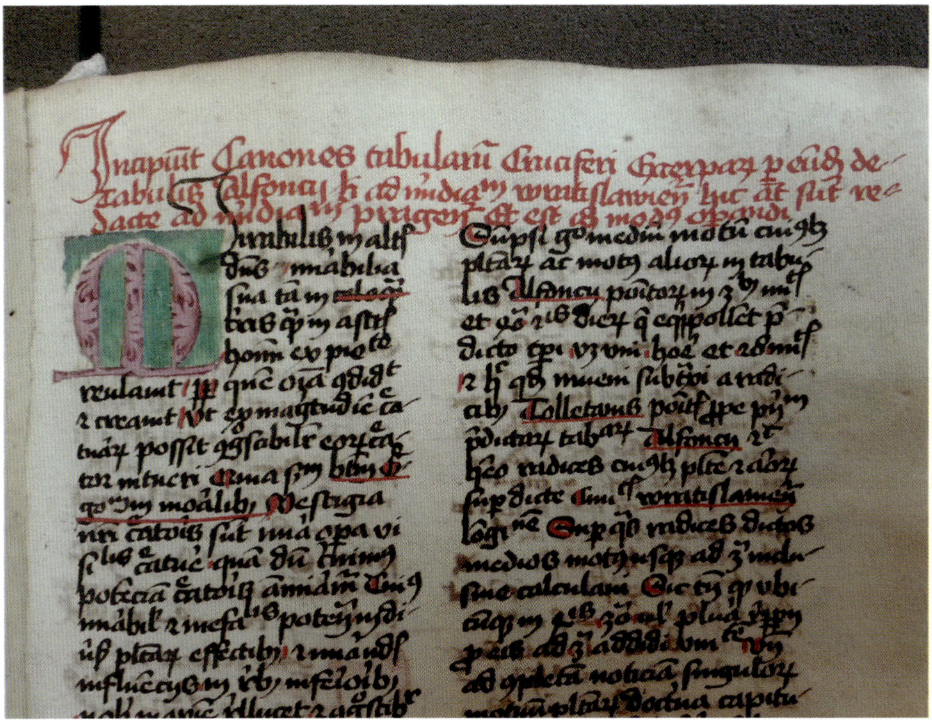

**Figure 6.** MS Oxford, Bodl., Canon. misc. 499, f. 122r/112r, incipit of the canons *Mirabilis*, scribe: John Krčín.

**Figure 7.** MS Oxford, Bodl., Canon. misc. 499, f. 122r/112r, a detail of the previous photo. Gaerparum (Carparus, Caspar) × excerp<t>arum.

58 F. 160r/150r.

59 F. 193vb/183vb. – We would like to thank Richard L. Kremer for providing us with the photographs of these folios.

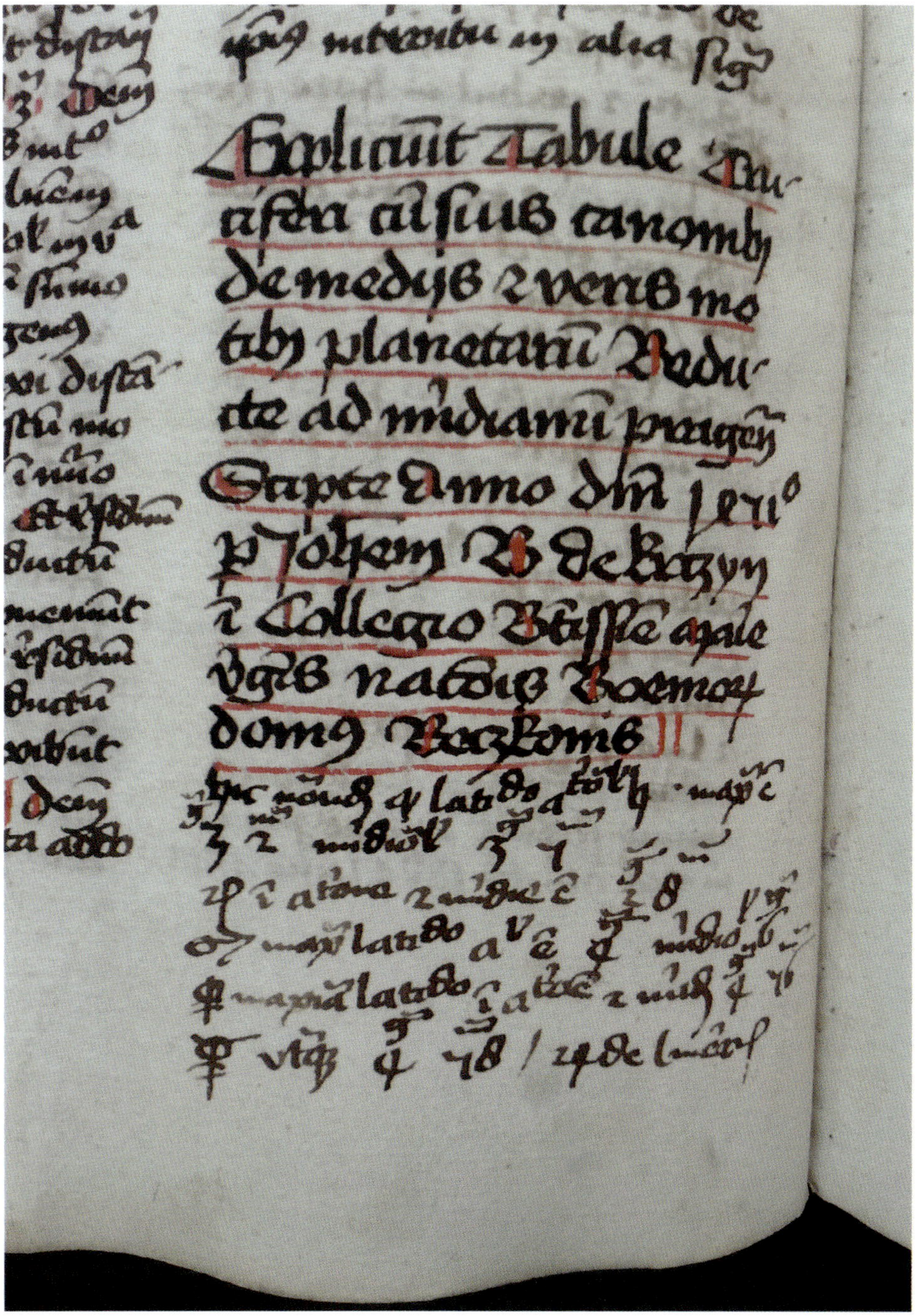

**Figure 8.** MS Oxford, Bodl., Canon. misc. 499, f. 125r/115r, explicit of the canons *Mirabilis* by the scribe John Krčín.

**Figure 9.** MS Oxford, Bodl., Canon. misc. 499, f. 125r/115r, explicit of the canons *Mirabilis* (detail of the previous photo): … *sc<r>ipte Anno Domini 1451°* *per Iohannem, B(accalaureum) de Krczyn.*

*The scribe of the canons: John Krčín*

John Krčín is a well-known and well-researched figure of the Czech medieval history. To the more recent studies that mention his person and life belongs František Šmahel's book devoted to Charles University in Prague and the biographical entry prepared by Blanka Zilynská for collective dictionary of professors of Prague Utraquist University in the late Middle Ages.[60]

John Krčín (Jan Krčín, Jan z Krčína) was born *c.* 1430 in the town Krčín above the river Metuje and died 1470 or 1471 in Prague. According to the Dean's Book (*Liber decanorum*), he earned his bachelor's degree at the University of Prague in 1450, which means that his copy of the tables and canons extant in the Oxford manuscript correlates with his studies at the time – the seven liberal arts (*septem artes liberales*). At that time he was staying in Reček's college. It is worth noting that the list of examiners contains the name of the respected astronomer John Borotín (Jan Borotín, Jan z Borotína):[61]

> *Item anno…* <1450>, *25. die Fe||bruarii… magistri examinatores baccalariandorum ad decanum prefatum facultatis arcium*[62] *iuncti hi fuerunt deputati: magister Ioannes Borotin, magister Ioannes* <de> *Jemnicz, magister Stanislaus de Gnezna, magister Martinus de Horzepnik, in quo quidem examine ad gradum baccalariatus examinati rigide et assumti hoc ordine: Ioannes Rodicz de Praga, Ioannes de Plana… Michael de Czaslavia, Tobias de Domazlicz, Andreas de Tyssnow, Ioannes de Basilea, Wenceslaus de Krzyzanow, Moravus, Mathias de Zluticz, Ioannes de Krczin, Ioannes de Mezyrziecz, Ioannes de Hassia, Petrus Nelinger, Nicolaus Sutoris de Praga, Wenceslas de Ohrazdiegewicz. … Item eodem anno, 13. die Aprilis sub reverendo ac venerabili pro tunc rectore, magistro Petro de Hradecz Regine, Ioannes de Plana, Ioannes de Krczyn et Nicolaus de Praga… gradumque baccalariatus legitime assumpserunt.*[63]

Another student who took his bachelor oath in the same year as John Krčín was the Moravian Wenceslas from Křižanov (Václav z Křižanova).[64] His name appears next to that of Krčín in several other places of the *Liber decanorum*. Clearly, they were close colleagues and probably friends, and their life stories developed in parallel and continued to have much in common. On 2 May 1454, during the deanship of John from Častolovice and under the supervision of Master Jiří (George) from Meziříčí, both John Krčín and Wenceslas from Křižanov earned their master's degree: *Wenceslaus de Krzizanow… et Ioannes de Krczin… sub mag(istro) Georgio de Mezirziecz… gradum sunt adepti magistralem.*[65]

60 František Šmahel, *Alma mater Pragensis. Studie k počátkům Univerzity Karlovy* (Prague: Karolinum, 2016), pp. 514, 532–33, 538; Holá, Holý et al., *Profesoři pražské utrakvistické univerzity*, pp. 362–63 (Krčín's biography is written by Blanka Zilynská); see also Ivana Čornejová et al., *Dějiny univerzity Karlovy I (1347/48–1622)* (Prague: Karolinum, 1995), p. 199.

61 See the partial edition of his notes on astronomy and astrology by Charles Burnett, 'Teaching the Science of the Stars in Prague University in the Early Fifteenth Century: Master Johannes Borotin', *Aither*, 2 (2014), 9–50.

62 His name was Petrus de Dwekaczowicz (Petr z Dvakačovic). In the manuscript Prague, NK VIII E 27, f. 52r (from around 1450) there is, for example, a note that Petr of Dvakačovice started to lecture Sacrobosco's Sphere at the Prague University. In the MS Prague, NK, XIII F 17, ff. 135r–170v, Sacrobosco's Sphere is written directly by his hand. See Hadravová and Hadrava, *Sféra Iohanna de Sacrobosco*, pp. 24, 25 and 358.

63 *Liber decanorum facultatis philosophicae Universitatis Pragensis ab anno Christi 1367 usque ad annum 1585*. Facsimile, ed. by Karel Beránek *et al.* (Pragae: Universitas Carolina, 1983), pp. 40–41.

64 Holá, Holý et al., *Profesoři pražské utrakvistické univerzity*, pp. 366–67 (Křižanovský's biography is written by Ivan Hlaváček).

65 *Liber decanorum*, p. 49.

However, the following years brought a striking change in the political climate at the university. The confessional disputes between the Catholic and the Utraquist masters about the future shape of the university became more heated and led to the eventual victory of the Utraquists. As a result, the university shut itself off from the world and became a local institution that did not support humanistic ideas and the exchange of knowledge with other universities. It was reduced to a single faculty, the Faculty of Arts, and only offered bachelor's education; anyone wishing to earn a master's degree had to go abroad. Under these circumstances, both John Krčín and Wenceslas from Křižanov, once colleagues from Reček's college, set out to continue their studies in Italy.

Before they left, however, their names appeared again in the records in the spring of 1456, when there was a dispute about accepting new members to the faculty council. Both Wenceslas from Křižanov and John Krčín were among the applicants. The masters split into two factions over the requests and the dispute eventually had to be resolved by Dean Stanislas from Velvary, who decided to accept all applicants who had finished their biennium (i.e. who earned their master's degree two or more years ago). This decision meant that both Wenceslas from Křižanov and John Krčín were accepted: on 1 May 1456, *facta est peticio magistrorum novellorum, magistri Wenceslai de Praga, Wenceslai de Krzizanow…, Ioannis de Krczin…, ut recipiantur ad concilium facultatis; pro quo, quod non erat intimacio facta, magistri in duas partes discordantes sunt divisi… decanus tamen,*[66] *ne gravamen petentibus ingerat magistrosque iterum fatiget, consensit uni parti, ut suscipiantur, maxime, qui biennium suum expleverunt, et hi, qui non expleverunt, ut eo diligencius explere conentur.*[67]

During the following three years, both studied at the prestigious University of Bologna: John Krčín focused on medicine and Wenceslas from Křižanov on theology (we found twelve theological works written by him in Prague and later in Bologna, stored now in the Library of the Metropolitan Chapter of Prague). Krčín also apparently owned the manuscript Prague, NK, VIII E 9, in which there are copies of medical works; on f. 65r Krčín is signed as scribe in the colophon of one of the copies: *explicit hoc totum per manus magistri Iohannis Krczin, rectoris Bononiensis artistarum et venerandorum medicorum, anno 1459* (see Fig. 10).[68] As can be seen from the colophon, he was elected rector of the artists and physicians in Bologna at that time.

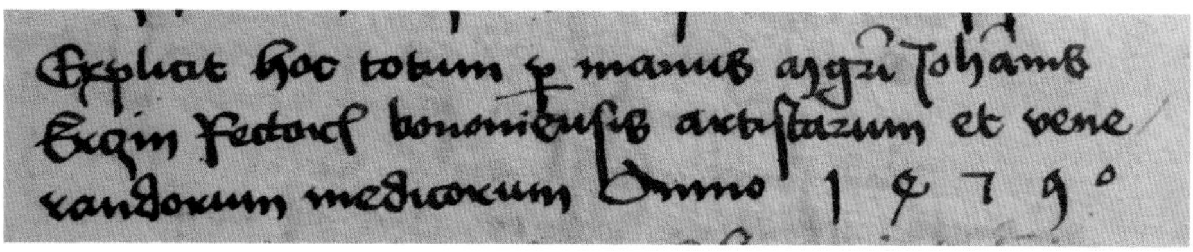

**Figure 10.** John Krčín's signature from the year 1459 in the MS Prague, NK, VIII E 9, f. 65r.

66 It means Stanislas from Velvary.

67 *Liber decanorum*, p. 53.

68 See Josef Truhlář, *Catalogus codicum manu scriptorum Latinorum, qui in c. r. bibliotheca publica atque Universitatis Pragensis asservantur I–II* (Pragae: Česká Akademie věd a umění, 1905–1906), pp. 564–65 (item 1534). <https://sources.cms.flu.cas.cz/src/index.php?s = v&cat = 36&bookid = 369&page = 584> [accessed: 12.03.2023]; <https://www.manuscriptorium.com/apps/index.php?direct = record&pid = AIPDIG-NKCR__VIII_E_9_____113Z2O2-cs> [accessed: 12.03.2023].

In the meantime, the political situation at the University of Prague escalated further. Both friends, doctor of medicine John Krčín and doctor of theology Wenceslas from Křižanov, returned to Prague in 1459, but it required a lot of time and effort to get their Italian education recognised at their home institution: *Item anno eodem <1460> ..., feria 6. quatuor temporum mensis Decembris, Wenceslaus de Krzizanow et Ioannes de Krczyn, studii Pragensis arcium liberalium magistri et post doctores studii Bononiensis, unus facultatis theologice, et alter medicine, postquam docuissent, se esse doctores studii universitatis Bononiensis, nondum tamen, ut sic, assumti sunt pro doctoribus nostrae universitatis Pragensis.*[69] The delay was caused by the negative attitude of the faculty towards the newly arrived doctors, who protested against the curtailment of the university and its disconnection from the rest of the world and refused to become proponents of Utraquism and communion under both kinds. For the same reason, John Krčín and other masters refused to recognise the newly elected Utraquist dean,[70] Wenceslas Koranda the Younger[71] (mentioned above as the owner of the MS Toruń, BU, 74/III). Moreover, the elections were influenced by the Hussite archbishop Jan Rokycana (Jan z Rokycan, John from Rokycany), who interfered in what should have been the internal decision-making process at the faculty.

One after another, the unwelcome masters and doctors were eventually forced to leave the faculty. Wenceslas from Křižanov left after a dispute over books borrowed from the university college, which was used as a pretext. John Krčín became the physician of King George of Poděbrady[72] who, although a Utraquist himself, was happy to hire a physician educated at a renowned Italian institution. King Jiří of Poděbrady gave Krčín a house in Prague Old Town in 1463 for his services. Another personal physician to the king was Pavel Žídek (Paulus from Prague, Paulerinus), who graduated from five universities (Vienna, Padua, Bologna, Prague and Cracow).[73] The University of Prague remained the strictly local and isolated Utraquist institution with much reduced numbers of both students and teachers until the Battle of White Mountain in 1620, after which the recatholization took place and Prague University was handed over to the Jesuits in 1622.

### *Other selected works contained in this manuscript*

It can be said that the whole MS Oxford, Bodl., Canon. misc. 499 contains tables and canons composed by prominent authors of the texts important for Alfonsine

69 *Liber decanorum*, p. 70.

70 *Liber decanorum*, p. 75.

71 Marek, *Václav Koranda mladší*, p. 23.

72 Gustav Gellner, *Jan Černý a jiní lékaři čeští do konce doby jagellovské* (Praha: Královská česká společnost nauk, 1934), p. 184.

73 For more information on his medical career see Ingrid Kästner, 'Paulus Paulirinus (Pavel Židek, 1413–1471) und die Medizin im Liber viginti arcium: ein Projektvorschlag', in *Padua als Europäisches Wissenschaftszentrum von der Renaissance bis zur Aufklärung*, ed. by Dietrich von Engelhardt and Gian Franco Frigo (Aachen: Shaker Verlag, 2017), pp. 153–66. For his astronomical tables, see Alena Hadravová and Petr Hadrava, 'Astronomy in Paulerinus's Fifteenth-Century Encyclopaedia Liber viginti arcium', *Journal for the History of Astronomy*, 38 (2007), 305–24, and Petr Hadrava and Alena Hadravová, 'Tabulka ekvací Měsíce z latinské encyklopedie Pavla Žídka a její srovnání s Almagestem, Toledskými a Alfonsinskými tabulkami', *Dějiny věd a techniky*, 41 (2008), 65–84. An edition of the natural science part of his encyclopaedia *Liber viginti arcium* (MS Cracow, BJ, 257) was published by Alena Hadravová, *Kniha dvacatera umění mistra Pavla Žídka. Část přírodovědná* (Praha: Academia, 2008); for his discussion of meteorology see Alena Hadravová, 'Meteorologie v Knize dvacatera umění mistra Pavla Žídka a ve vybraných pramenech zvláště české provenience', *Studie o rukopisech*, 44 (2014), 439–77.

astronomy. It is thus a very valuable codex. Several of these works were copied by John Krčín.

Works contained in this codex:

- F. 1v shows a concise table with geographical coordinates of cities, which is almost identical to the table in MS Mikulov, RM, Mik 6361, f. 23r (an edition of the table is published below in the description of the manuscript from Mikulov in Chapter 2.1.6). Furthermore, the notes to this table in both manuscripts (Oxford 499 and Mikulov 6361) similarly describe how to determine the geographical longitude of Prague in relation to Toledo.
- Parisian Alfonsine Tables are present on ff. 12r/2r–41r/31r and 48r/38r. The Table of sexagesimal multiplication (*Tabula proporcionum*) is inserted on ff. 41v/31v–47r/37r.[74]
- The first version of the Table for velocities by John of Gmunden (entitled *Tabula ad inveniendum motum Solis et Lune in una hora*) is written on ff. 51v/41v–52r/42r. This is one of several versions of tables which give the hourly motions (i.e. velocities) of luminaries (*luminaria*).[75] Another version of this table was included in the middle of the oldest version of the TR in MS Toruń, BU, 74/III,[76] but later versions of the TR no longer have it. In some cases, this table is included in codices in quite different places from the TR. Another copy is also present in the manuscript Oxford, Bodl., Canon. misc. 499, ff. 164v/154v–165r/155r which matches to the Table of hourly motion of the luminaries in an hour (*Tabula ad sciendum motum Solis et Lune in una hora*) contained in the manuscript Prague, NK, X B 3, f. 117r–v. Only slightly different are the other versions of this table included by John of Gmunden to his tables.[77]
- John of Saxony's canons to the Parisian Alfonsine Tables, with incipit *Tempus est mensura motus*, are on ff. 60ra/50ra–68va/58va.
- Incipit *Figuram celi ad quodcunque tempus coniunccionis* belongs to the work inscribed *Iste canon non spectat ad canones super Alphonsum*. It can be found on ff. 68vb/58vb–69vb/59vb.
- *Composicio almanach ingeniosissime* with incipit *In componendo almanach sic procede: ordina medios cursus omnium planetarum* is present on ff. 70r/60r–71r/61r.[78]
- Nicolaus de Heybech, Table for true syzygies (*Tabula equacionum temporum vere coniunccionis et opposicionis Solis et Lune*) is on ff. 120v/110v–121r/111r. The work originated *c.* 1392 and it is known in more then twenty other copies.[79]
- William Batecombe, Oxford Tables of 1348, ff. 126v/116v–133v/123v, a well known work from *c.* 1348, preserved in about thirty five manuscripts.
- Selection of works by John of Lignères from his Tables of 1322, ff. 134r/124r–151r/141r; ff. 152r/142r–154r/144r (see Fig. 11). In between, on f. 151v/141v, there is written the

74 See an edition of the beginning and the end of this table in Chapter 16, Table A-04.

75 See Chapter 22 for an explanation.

76 See the edition of this table in Chapter 12, the Table TRW-14.

77 See Tab. 2 in Chapter 22. For the history of this table see Porres de Mateo, 'Les tables astronomiques de Jean de Gmunden', pp. 565–66 (see also pp. 329–33, 536–37); Laure Miolo, 'Retracing the Tradition of John of Genoa's *Opus astronomicum* through Extant Manuscripts', in *Alfonsine Astronomy. The Written Record*, ed. by Richard L. Kremer, Matthieu Husson, and José Chabás (Turnhout: Brepols, 2022), pp. 343–80 (p. 352); Chabás and Saby, *The Tables of 1322 by John of Lignères*, pp. 108–13.

78 See also Rosińska, *Scientific Writings*, pp. 176–77 (item 853).

79 Estimates of the number of extant manuscripts here and below are based on the database of Alfonsine astronomy, prepared by the ALFA team.

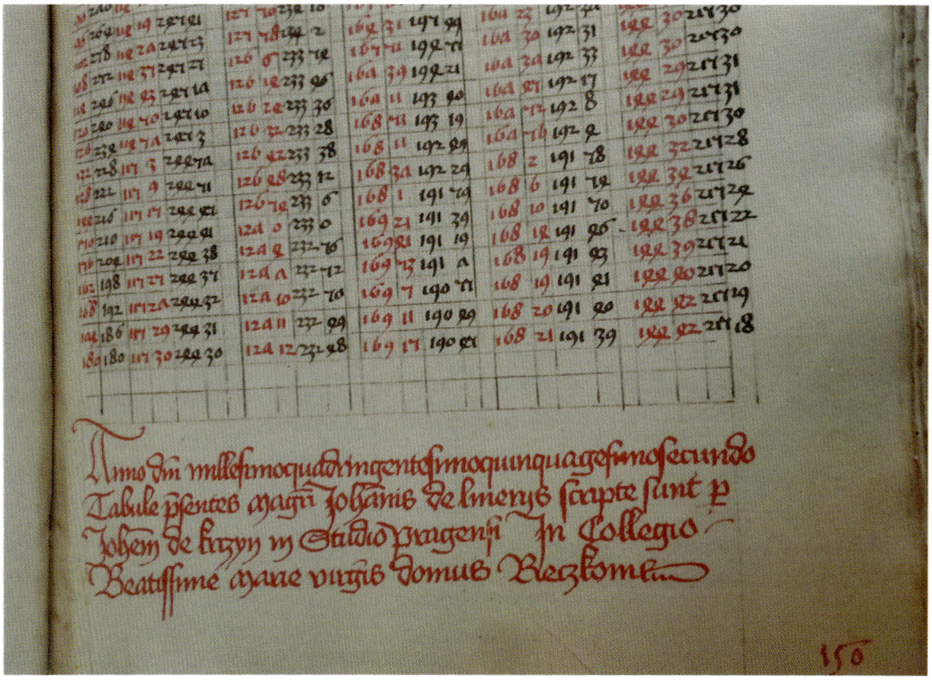

**Figure 11.** MS Oxford, Bodl., Canon. misc. 499, f. 150r/140r. Krčín's copy of John of Lignères' Tables of 1322, written in Prague, 1452 (incomplete).

Table for the hourly velocities and semidiameters of the luminaries by John of Genoa from *c.* 1330, which is known from at least thirty other manuscripts.

- John of Montfort, Table for solar and lunar velocities, can be seen on ff. 161v/151v–162v/152v.
- On f. 164v/154v there is written: *Hec tabula est magistri Iohannis de Lineriis et intratur in ipsam cum argumento Solis et cum argumento medio Lune cum duplici introitu, si oportet.* Tables are incomplete and were copied 1452 by Johannes de Krczyn in Prague, because he also signed another folio (f. 160r/150r), when he writes, that in the year 1452, he finished a copy of John of Lignères' tables: *Anno Domini millesimo quadringentesimo quinquagesimo secundo Tabule presentes magistri Iohannis de Lineriis scripte sunt per Iohannem de Krczyn*[80] *in studio Pragensi, in collegio beatissime Marie Virginis, domus Reczkonis.*[81]
- Canons on eclipses by John of Saxony (with incipit: *Eclipsis Solis quantitatem et duracionem per tabulas invenire*) are copied on ff. 170r/160r–171v/161v. They are known in more than fifty manuscripts.
- John of Lignères' canons associated with the Tables of 1322 with the incipit *Cuiuslibet arcus propositi*, are on ff. 172ra/162ra–179vb/169vb. On f. 172va/162va we read: *Instrumentum ad lineam meridianam… construere* (this is a description of Ptolemaic meridian instrument used to measure the solar altitude at noon). On f. 173va/163va there is: *Instrumentum ad capiendum altitudines Solis et stellarum observaciones faciendas construere. – Fac tres*

80 Coxe, *Catalogi codicum*, p. 815, *per err.:* Bazyn.
81 Coxe, *Catalogi codicum*, p. 815, *per err.:* Beczkonis.

*planas regulas de ligno vel cupro vel de ferro fortes et rigidas, ut de levi torqueri non possent* (it is a description of an astronomical instrument, so-called Ptolemaic ruler). The text *Cuiuslibet* was described by Emanuel Poulle, partly edited by M. Curtze and fully edited by M.-M. Saby.[82] This text contains 44 propositions, which correspond to the trigonometric part of the tables and deal with the circumstances connected with the daily motion of the Sun: trigonometric operations, determination of the ascendant and of the celestial houses, equal and unequal hours, and so on.

- John of Lignères' canons associated with the Tables of 1322 with incipit *Priores astrologi*, ff. 180ra/170ra–193vb/183vb. Both John of Lignères' tables and canons were also copied by John Krčín, in 1450 and then in 1452, as we know from two other places, in which John Krčín is signed as the scribe. At these other two places the reading of Coxe's catalogue and the related secondary literature must be revised. The explicit of John of Lignères' canons *Priores astrologi* copied by Krčín sounds (on fol. 193vb/183vb) as it is: *Expliciunt Canones tabularum magistri Iohannis de Lineriis, finiti per Iohannem de Krczyn*[83] *in collegio beatissime Virginis Marie, domo Reczkonis, nacionis Bohemorum, Anno Domini 1450*.[84]

It seems that Krčín's copies are not bound in the convolute in the chronological order in which its components were copied. After the canons *Mirabilis* from 1451 are first bounded the John of Lignères' tables, the copy of which was finished in 1452, and than his canons *Cuiuslibet arcus propositi*. The last part are canons *Priores astrologi*, copied by Krčín already in the year 1450, as already mentioned above.

- On ff. 202r–v (upper foliation), 205r–206r, and 213r–214v there are enlarged trigonometric tables by Giovanni Bianchini.
- Between the folia with Bianchini's work there is a copy of the treatise *De motu octave spere* with tables attributed to Thābit ben Qurra (ff. 206v–208v). Incipit: *Ymaginabor speram equacionis diei* (in other copies: *equinoccii*) *et tres circulos in ea signatos* ... × ... *quos habes additos super numerum minorem, cum quo intrasti in lineam numeri. Explicit Thebith De motu octave spere cum tabulis in parte secunda huius folii positis*.[85]
- The last work in the manuscript, *Summa super Tabulas astronomicas* by Martinus Król de Żurawica, can be found on ff. 222ra/212ra–258vb/248vb.[86] Król's commentary has the incipit *Ad honorem cunctipotentis, trine in personis*.[87]

82 Poulle, 'John of Lignères', pp. 122–28; Maximilian Curtze, 'Urkunden zur Geschichte der Trigonometrie im christlichen Mittelalter', in *Bibliotheca mathematica*, ser. 3, 1 (1900), pt. 7, pp. 321–416; on the pp. 390–413 there are 'Die Canones tabularum primi mobilis des Johannes de Lineriis', including the first nineteen canons (i.e. rules, pp. 391–403) followed by the titles of the succeeding canons, as well as by the Tables of sines and chords and of shadows and the *Tabula proporcionis* (pp. 411–13); canon 9 of the *Canones super Tabulas latitudinum planetarum et etiam eclipsium* (the second part of the canons of 1322) is also included (pp. 403–4); Saby, 'Les canons de Jean de Lignères', pp. 67–171 (esp. 89–92).

83 Coxe, *Catalogi codicum*, 815 has again the wrong reading Bazyn.

84 Coxe, *Catalogi codicum*, 815 has erroneously 1470.

85 See edition by Francis J. Carmody, *The Astronomical Works of Thābit ben Qurra* (Berkeley, Los Angeles: University of California Press, 1960).

86 Mieczysław Markowski, *Astronomica et astrologica Cracoviensia ante annum 1550* (Firenze: Leo S. Olschki, 1990), p. 120 (item 7); Rosińska, *Scientific Writings*, p. 426 (item 2228) and p. 24 (item 43).

87 ThK 43.

### 2.1.5 *MS Prague, Národní knihovna (NK), X B 3 (Cim C 20)*

This manuscript originated in Prague, 1435–1465.[88]

- The canons *Mirabilis in altis Dominus* are copied on ff. 73ra–75va. Incipit and explicit: *<M>irabilis in altis Dominus* ... × ... *sic facere potero de ipsius introitu in alia signa* (see Figs 12 and 13).
- The corresponding set of tables (*Tabule resolute*) for the Prague meridian can be found on ff. 79r–114r with incipit: *Incipiunt tabule de mediis et veris motibus planetarum super meridianum Pragensem. Et primo in annis collectis.* While the manuscript Toruń, BU, 74/III was written in 1425 and is only a year younger than the base year (epoch or *era*) of the tables calculated for the Wrocław meridian (1424), the Prague manuscript X B 3 dates to the middle of the fifteenth century and is penultimate in the chronological order of manuscript used in our edition of the canons *Mirabilis*, even though it could potentially be older than any of the two preceding manuscripts.[89]
- Several more tables follow on f. 114v and following folios. Some of them belong to frequent additions not only to the TR[90] but also to other sets of tables. After the Table of the motion of the luminaries in a minute of a day (*Tabula ad sciendum motum Solis et Lune in uno minuto diei*) on ff. 115v–116v, there follows Table of motion of the luminaries in an hour (*Tabula ad sciendum motum Solis et Lune in una hora*) on f. 117r–v.

For almost a century, manuscript Prague, NK, X B 3 was described in the Czech secondary literature as a codex written by Master Jan Ondřejův, called Šindel (Johannes Andreae, dictus Sindel), an important Czech astronomer of the early fifteenth century. John Šindel (1375, Hradec Králové in east Bohemia – 1455/58) became bachelor at Charles University in 1395 and master of arts in 1399. Later he studied medicine and obtained the title doctor of medicine and taught mathematics in Vienna. He was professor of astronomy and rector of Prague University. He also acted as a physician of King Wenceslas IV and from 1432 he was a physician of the Emperor Sigismund. Šindel's most important treatise is The rules for calculation of the Sun's and Moon's eclipses according to his own instrument (*Canones pro eclipsibus Solis et Lune per instrumentum ad hoc factum inveniendis Magistri Iohannis Šindel*) which are derived from Albion by Richard of Wallingford (fourteenth century) and from expanded version of it by John of Gmunden in Vienna (fifteenth century). Practically the same instrument as Šindel's can be found in Johannes Schöner's *Opera mathematica* from 1551 and 1561.[91]

---

88 See Truhlář, *Catalogus codicum*, pp. 42–43 (item 1832) <https://sources.cms.flu.cas.cz/src/index.php?s = v&cat = 36&bookid = 370&page = 44>. The scan of the manuscript is available at <https://www.manuscriptorium.com/apps/index.php?direct = record&pid = AIPDIG-NKCR__X_B_3________111XZSA-cs> [accessed: 12.03.2023].

89 This tentative chronological order is reflected in the designations (*sigla*) assigned to the manuscripts in this edition in alphabetical order.

90 We present editions of some of them in Chapter 16. In particular, there are the Table A-01 on f. 114v, A-02 on f. 115r, and A-05 on f. 144v of the manuscript Prague, NK, X B 3.

91 See Alena Hadravová and Petr Hadrava, 'The Eclipse Instrument by John Šindel', in *Studying the Arts in Late Medieval Bohemia: Production, Reception and Transmission of Knowledge*, ed. by Ota Pavlíček (Turnhout: Brepols, 2021), pp. 305–40.

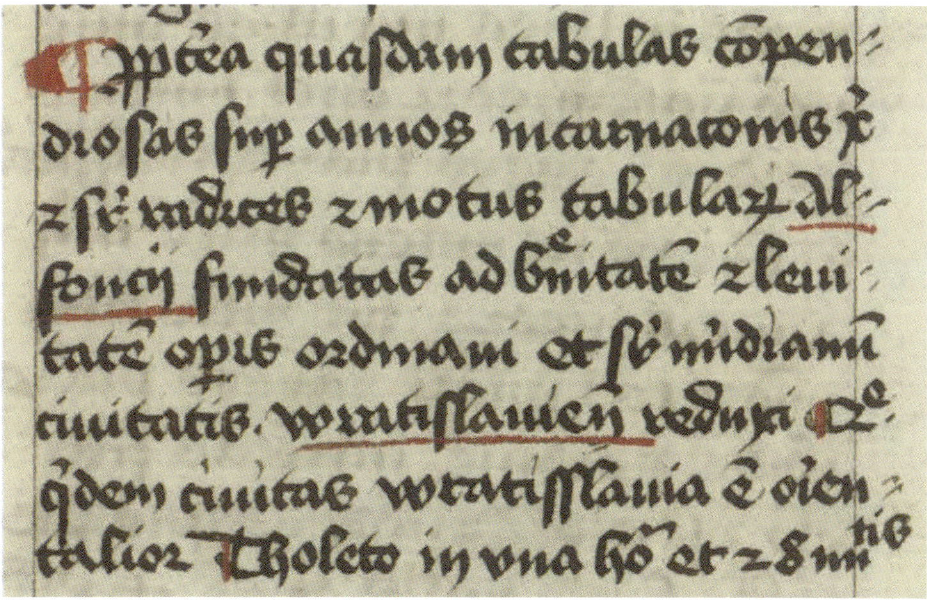

**Figure 12.** MS Cracow, BJ, 610, f. 352ra–rb, canons *Mirabilis* – detail with reference to the Wrocław meridian: *Propterea quasdam tabulas compendiosas super annos Incarnac<i>onis Christi et super radices et motus tabularum Alfoncii fundatas ad brevitatem et levitatem operis ordinavi et super meridianum civitatis* ***Wratislaviensis*** *reduxi, que quidem civitas* ***Wratisslavia*** *est orientalior Tholeto in una hora et 28 minutis || hore.*

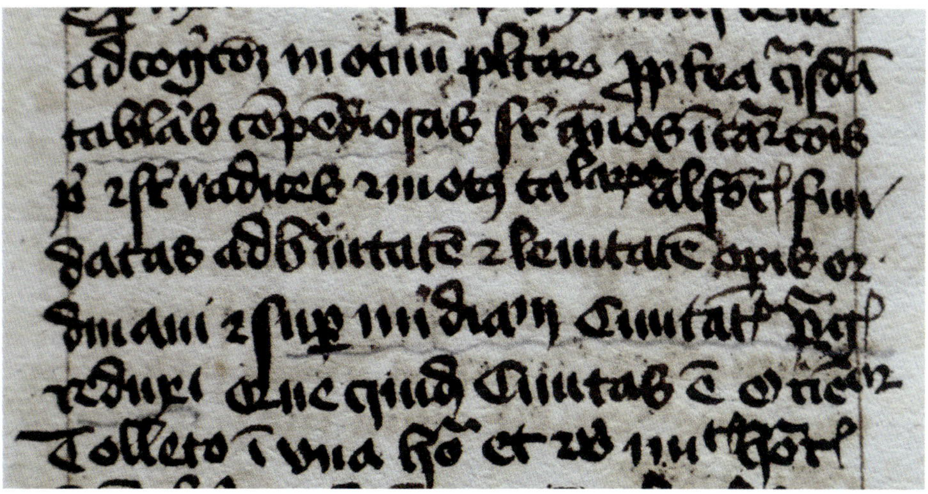

**Figure 13.** MS Prague, NK, X B 3, f. 73ra, canons *Mirabilis* – detail with reference to the Prague meridian: *Propterea quasdam tabulas compendiosas super annos Incarnacionis Christi et super radices et motus tabularum Alfoncii fundatas ad brevitatem et levitatem operis ordinavi et super meridianum civitatis* ***Pragensis*** *reduxi, que quidem civitas est orientalior Tolleto in una hora et 28 minutis hore.*

The authorship of MS Prague, NK, X B 3 and, most importantly, the included tables was attributed to Šindel in 1900 in an article published by the cataloguer Josef Truhlář.[92] The attribution was based solely on the received image of John Šindel as an eminent astronomer at the University of Prague in the first half of the fifteenth century. The tradition of knowing that John Šindel was an excellent astronomer is even much older. He was appreciated by Enea Silvio Piccolomini, later Pope Pius II. And Tycho Brahe himself even considered Šindel's astronomical observations, made available to him in Prague at 1599 by the Master of the Prague University Martin Bacháček of Neuměřice, to be more accurate than his own.[93]

In the 1960s, excerpts from Prague, NK, X B 3 together with their translation into Czech, were published in an academic review of older Czech literature, which was then considered a very prestigious publication, and were again attributed to John Šindel.[94] Occasionally, he is still noted as the author,[95] even though his authorship was challenged by Beatriz Porres de Mateo in the 1980s based on her knowledge of the wider European context of the work. As was later confirmed, she correctly identified manuscript Prague, NK, X B 3 as belonging to a group of Central European tables by Petrus Cruciferus or Petrus Rein Cruciferus, that is to the TR.[96]

Porres realised that the numerical tables in manuscript X B 3 are identical to those in manuscript Prague, NK, VIII G 24 (Cim K 120), ff. 2r–37r. The first base year, used in the tables in both manuscripts, is 1428. However, manuscript VIII G 24 only contains the tables without the corresponding canons by Petrus Cruciferus with the incipit *Mirabilis in altis Dominus*. A more detailed description of manuscript VIII G 24 is given in the explanation of the canons *Girum* in the Chapter 3.1.

- The canons *Mirabilis* are followed on ff. 75vb–78vb by a short astrological essay *De fortitudine… planetarum* ('On the effect… of planets'), which is not included in any of the other manuscripts that contain *Mirabilis*.

As regards some of the other works contained in the codex, it is worth noting that many of them are also included in manuscript Cracow, BJ, 610, as was noted above. This could mean that there were closer ties between the two manuscripts, which led to the similarities.

*Other selected works contained in this manuscript*

- F. 1ra–1vb: Prologue to John of Saxony's canons to the Parisian Alfonsine Tables (incipit: *In principio huius ardui operis Alphoncii quinque sunt diligenter consideranda*).[97]

---

92 Josef Truhlář, 'Paběrky z rukopisů Klementinských XLVII. Šindelovy astronomické tabulky', *Věstník České akademie*, 9 (1900), 473–74.

93 Alena Hadravová and Petr Hadrava, 'Tycho Brahe and Iohannes Šindel', in *Tycho Brahe and Prague: Crossroads of European Science*, ed. by Johan Robert Christianson, Alena Hadravová, Petr Hadrava, and Martin Šolc (Frankfurt am Main: Harri Deutsch Verlag, 2002), pp. 237–47.

94 See Zdeněk Horský and Dana Martínková, 'Jan Šindel: Úvod k tabulkám pohybu planet', in *Výbor z české literatury husitské doby II*, ed. by Bohuslav Havránek, Josef Hrabák, Jiří Daňhelka *et al.* (Prague: Nakladatelství ČSAV, 1964), pp. 591–97, 608–11.

95 See Pavel Spunar, *Repertorium auctorum Bohemorum provectum idearum post universitatem Pragensem conditam illustrans I*. Studia Copernicana 25 (Wrocław etc.: Ossolineum, 1985), p. 135 (item 365).

96 Porres de Mateo, 'Šíření středověkých astronomických tabulek'.

97 ThK 711; Thorndike, 'Some Little Known Astronomical and Mathematical Manuscripts'.

- Ff. 13ra–14ra: *De virtute imaginum celestium vel excerpta de libris Picatricis* (incipit and explicit: *Ego autem rogo altissimum Creatorem, quod iste noster liber non nisi ad manus perveniat sapientis ...* × ... *et scias, quod Aredafir /?/ est nomen domini istius mansionis*); ThK 486. See also Cracow, BJ, 610, ff. 312v–316r.
- F. 15r: *De modo inveniendi coniunccionem et opposicionem* (incipit and explicit: *Pro vera coniunccione et opposicione invenienda per almanach ...* × ... *accidit secundum quantitatem applicacionis Mercurii cum fortuna vel infortuna*). See also Cracow, BJ, 610, f. 311v.
- F. 15v: Pseudo-Albertus Magnus, *Secretum de sigillo leonis* (incipit and explicit: *Ego Albertus, commentator in omni experiencia expertus ...* × ... *secundum octavam speram secundum ymagines stellarum fixarum*). See also Cracow, BJ, 610, f. 312r.
- Ff. 16ra–28vb: *Canones tabularum eclipsium Solis et Lunae* (incipit and explicit: *Ad inveniendas eclipses sic procedendum ...* × ... *verus locus capitis a vero loco Lune*). See also Cracow, BJ, 610, ff. 279r–289v.
- F. 29ra: *Ad inveniendum gradum ascendentem ...* × ... *alias autem sex domos habebis per oppositum.* See also Cracow, BJ, 610, f. 350v.
- Ff. 30r–33v: canons to the Tables of 1322 by John of Lignères (incipit *Priores astrologi*, canons 35–39); incipit and explicit: *Tempus eclipsis Solis...* × ... *ostendit latitudinem eius ad finem.*
- Ff. 42v–72v: Table for the lunar equations (*Tabula de centro Lune*) by Nicolaus Mülhus.
- Ff. 139va–142ra: an anonymous canon on eclipses, not precisely identified. Incipit and explicit: *Eclipsis possibilitatem in Sole quantum ad aspectum terricolarum per tabulas invenire... Eclipsim Solis, quantitatem et duracionem per tabulas ad hoc factas sic invenire ...* × ... *a inicio ad medium et ad finem eclipsis.* The name of the city of Prague is mentioned at the beginning (see lines 11–12 on f. 139v). See also Cracow, BJ, 610, ff. 268v–269r.
- F. 142v: Nicolaus de Heybech's canons to the Table for true syzygies.
- Ff. 143v–144r: Nicolaus de Heybech's Table for true syzygies.

### 2.1.6 MS Mikulov, Regionální muzeum (RM), Mik 6361

This manuscript[98] originated in Prague between the years 1470–80. The dating of this manuscript of Bohemian provenance is based on an analysis of the writing and filigree in the manuscript. Several gatherings are missing from the beginning of the codex and some folios further on are out of order. This could have happened when the codex was newly bound in half-leather during the first half of the twentieth century, or even earlier.

- There are two copies of the canons *Mirabilis in altis Dominus* in this manuscript, which indicates that the volume consists of at least two parts that were originally separate and only later bound together. The first copy of the canons is on ff. 20v–22v: <M>*irabilis in altis Dominus ...* × ... *sic facere potero de ipsius introitu in alia signa* (see Fig. 14). The second copy, with the same incipit and explicit, is on ff. 91v–92v (see Fig. 15). The handwriting clearly shows that the texts were written by two different scribes. The

98 See the digitised manuscript at <https://www.manuscriptorium.com/apps/index.php?direct=record&pid=AIPDIG-RMM___MIK_6361_____3SE4HI4-cs> [accessed: 12.03.2023]. The database still includes the wrong name of 'Cruciferus Gaerparus' as one of the authors of the codex.

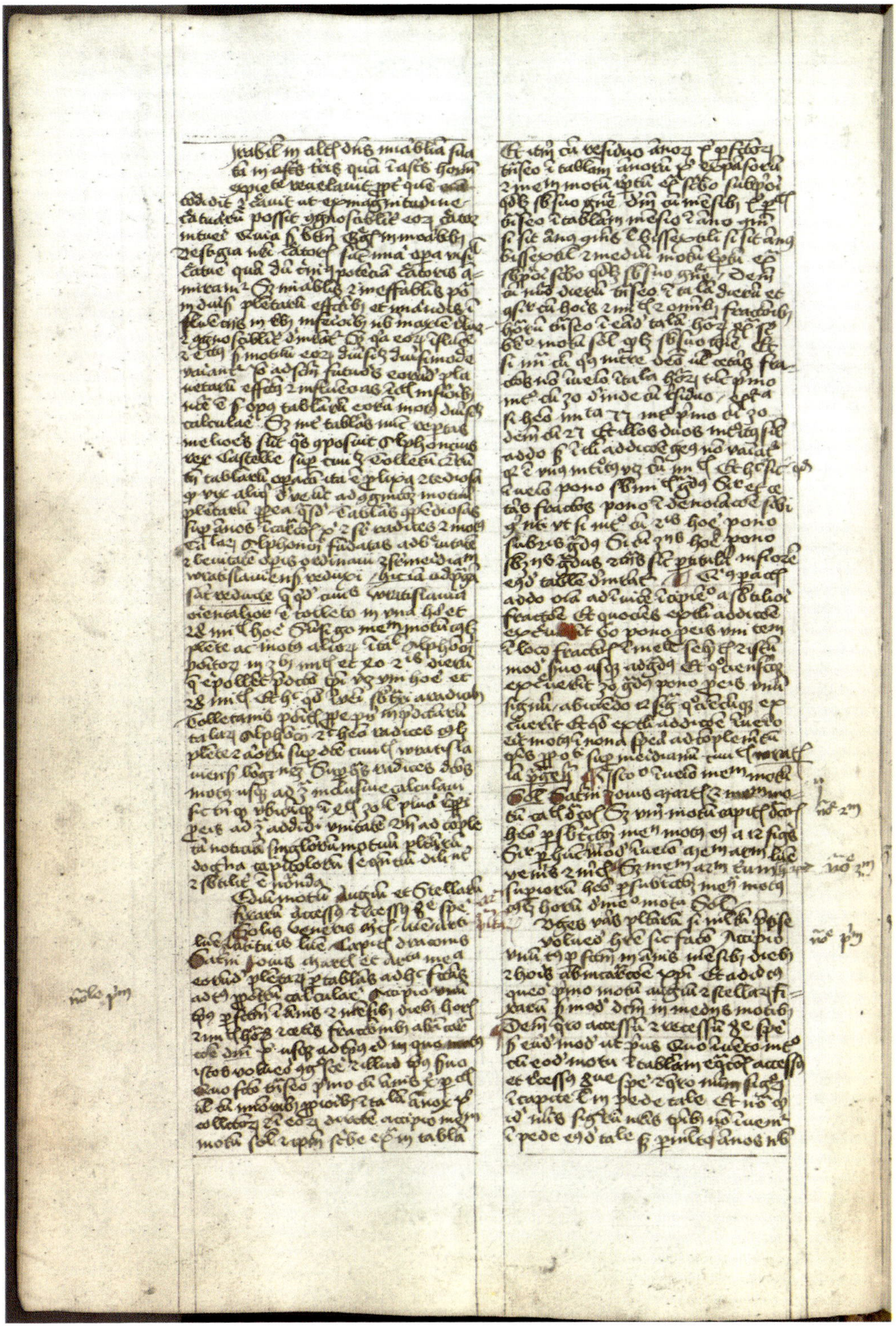

**Figure 14.** The beginning of the first copy of canons *Mirabilis* in the MS Mikulov, RM, Mik 6361, f. 20v.

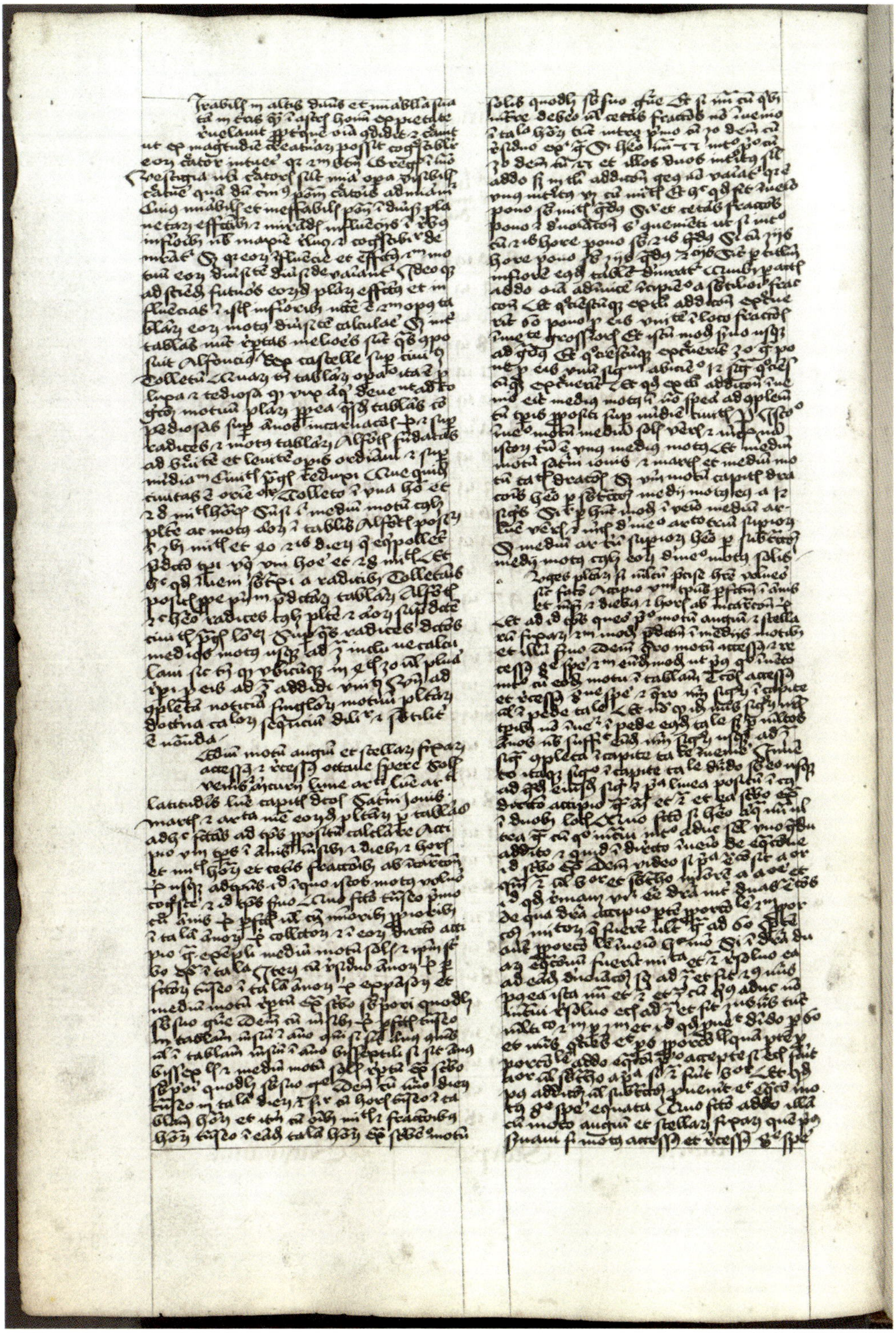

**Figure 15.** The beginning of the second copy of canons *Mirabilis* in the MS Mikulov, RM, Mik 6361, f. 91v.

first copy is inferior to the second; there are slight omissions in the text, as the scribe deliberately set out to simplify the work, while at the same time accidentally skipping some lines and duplicating others. The second copy is very good but lacks the first part of the second rule and the whole third rule.

- The manuscript also contains two sets of *Tabule resolute*: the first on ff. 47r–81r and the second on ff. 135r–171v. Both sets are calculated for the meridian of Prague with the radix in 1428.

*Other selected works contained in this manuscript*

- The manuscript was completed in the years immediately after the death of King George of Poděbrady, who died in 1471. The beginning of the manuscript contains the Parisian Alfonsine Tables to the canons of John of Saxony and on f. 1r, under a table called *Differencia diluvii et Alphoncii*, there is a note in the margins, which reads: *Differencia Nativitatis Christi et mortis Karoli, imperatoris Boemorum, anni 1378, mensis 10, dies 29 – 2, 19, 54, 7* (see Fig. 16). In sexagesimal notation, the numbers '2, 19, 54, 7' represent the 503,647 days from the beginning of the *Nativitatis Christi* era to the death of Charles IV. This number was used as the base for calculating the positions of the planets on the given day and so could have been used for an astrological reading. As we know – thanks in part to the work of Pavel Žídek, the only Czech author of a medieval Latin encyclopaedia, who died in the same year as King George of Poděbrady – there was an acute feeling of nostalgia in Bohemia for the enlightened rule of Charles IV about a hundred years after he died. This was caused by dissatisfaction with the social and political developments in the country; hence, this reference to Charles IV as a major historical figure right at the beginning of the Alfonsine Tables (ff. 1r–12v, 25r–36v) in MS Mikulov, RM, Mik 6361 is not out of place. The marginalia of the tables contain carefully noted radices for Prague: *radix reducta ad Pragam* (f. 8r); *radix Pragensis* (ff. 8v, 9r); *radix ad Pragam* (ff. 9v, 11v), and similarly on f. 25r.
- The Alfonsine Tables are meant to be used with the canons by John of Saxony *Tempus est mensura motus*, which is the following text copied in the manuscript (ff. 37r–44v).
- For a calculation of mean motions and times of syzygies knowledge of the geographical longitudes of cities was needed. Astronomical manuscripts often contain tables of geographical coordinates for various places. Such tables are also found in manuscripts with the TR. John of Głogów, for example, included a description of how to work with such tables in the Rule 36 of his canons *Pro (introduccione) generali*. On f. 23r, the first copy of *Mirabilis* in the Mikulov manuscript is followed by a short table of geographical locations. The description of the table ends with the following words: ... *quem motum subtrahe a radice Tolleti, quia civitas Pragensis est orientalior quam Tolletam, et quod post subtraccionem provenerit, erit radix Solis super meridianum Pragensem*. The description more or less matches the description and the content of the table on f. 1v in manuscript Oxford, Bodl., Canon. misc. 499.

*Tabula civitatum* according to the manuscript Oxford, Bodl., Canon. misc. 499, f. 1v, with variant readings from Mikulov, RM, Mik 6361, f. 23r:

| | longitudo | | latitudo | |
|---|---|---|---|---|
| | **gradus** | **minuta** | **gradus** | **minuta** |
| Egiptus | 55[99] | 0 | 30 | 0 |
| Allexandria | 60 | 27 | 30 | 18 |
| Ierusalem[100] | 66 | 30 | 31 | 50 |
| Damascus | 60 | 0 | 33 | 0 |
| Anthiochia | 63 | 20 | 35 | 30 |
| Armenia | 77 | 0 | 41 | 0 |
| Constantinopolis | 56 | 40 | 43 | 40 |
| Roma | 35 | 25 | 42 | 50 |
| Parisius | 23 | 51 | 44 | 50 |
| Toletum[101] | 11 | 0 | 49 | 54 |
| Cremania | 31 | 30 | 45 | 0 |
| Colonia | 27 | 20 | 45 | 55 |
| Maguncia | 28 | 24 | 50[102] | 9[103] |
| Herbipolis | 28 | 32[104] | 51 | 16 |
| Mardeburge[105] | 24 | 0 | 51 | 34[106] |
| Bruczwig[107] | 23 | 20 | 51 | 16[108] |
| Praga | 29 | 30 | 50 | 30 |
| Erfordia | 27 | 0 | 51 | 0 |
| Monspessulanus | 26 | 0 | 44 | 0 |
| Florencia | 23 | 25 | 42 | 30 |
| Pisa | 33 | 0 | 45 | 0 |
| Ethiopia | 63 | 30 | 35 | 30 |
| Ludania | 29 | 26 | 58 | 0 |
| Arym[109] | 9 | 0 | 0 | 0 |

Explanatory notes: *Egiptus* = Cairo; *Cremania* = Cremona; *Colonia*, i.e. *Colonia Agrippina* = Köln; *Maguncia* = Mainz; *Herbipolis* = Würzburg; *Bruczwig* = Braunschweig; *Erfordia* = Erfurt; *Monspessulanus* = Montpellier; *Ludania* = London; *Arym* = mythical city.[110]

99 *Mik 6361: per err.* 5.
100 *Mik 6361: Iherusalem.*
101 *Mik 6361: Tolletum.*
102 *Mik 6361:* 45.
103 *Mik 6361:* 55.
104 *Mik 6361:* 0.
105 *Mik 6361: Magdeburg.*
106 *Mik 6361 per err.:* 16.
107 *Mik 6361: Brunczink.*
108 *Mik 6361 per err.:* 34.
109 *Mik 6361: Arim.*
110 The mythical city of Arim (Arin, Arym) was often considered as the starting point of the calculation of positions and was located somewhere at the end of the inhabited world, sometimes identified with Ptolemy's Fortunate Islands (*Insule Fortunate*), i.e., the Canary Islands. According to Richard Lorch, 'The Treatise on the Astrolabe by Rudolf of Bruges', in *Between Demonstration and Imagination*, ed. by Lodi Nauta, and Arjo Venderjagt (Leiden, Boston, Köln: Brill, 1999), pp. 86 and 92, the name "Arim" originated from an erroneous transliteration of the Indian locality Ujjayan.

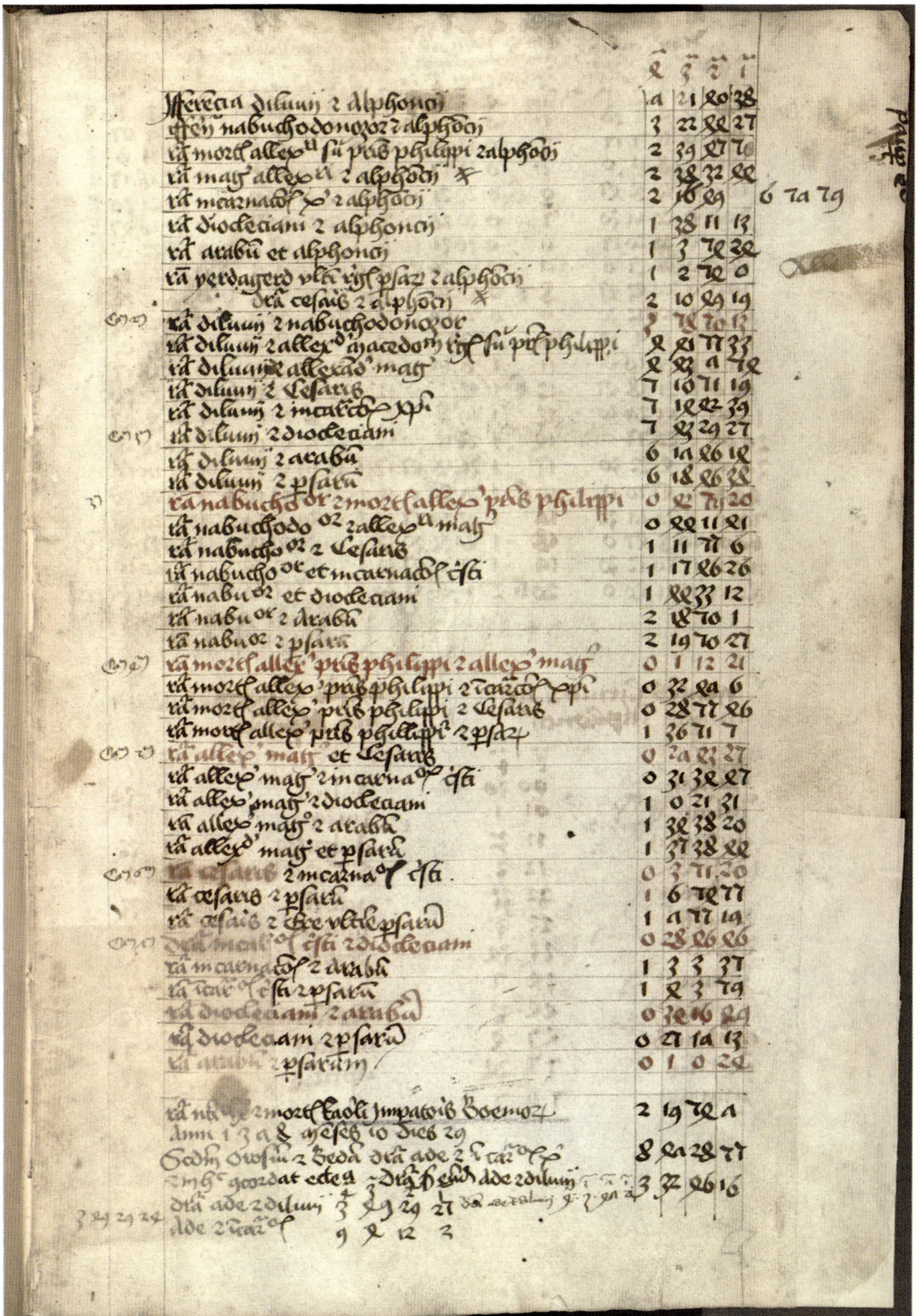

**Figure 16.** MS Mikulov, RM, Mik 6361, f. 1r: PAT with the marginal note to the death of the Emperor Charles IV (in sexagesimal notation).

Unlike the Oxford table, the Mikulov copy contains several errors in the numbers, so it was not copied from such a reliable source.

The same type of table (or only slightly different or extended) can be found in ten other astronomical manuscripts, not always containing *Tabule resolute*, but e.g. the Composition and Use of the astrolabe by Cristannus of Prachatice. Astronomers who worked with the astrolabe and other astronomical instruments also needed to know geographical positions of places, so tables of city coordinates were included in manuscripts. The same table is in the manuscript Cracow, BJ, 550, reverse of the front cover (*c.* 1440). In the table written by the scribe Kříž (Crux) from Telč[111] in the astronomical and medical manuscript Prague, NK, III C 2, f. 41r (fifteenth century), is missing Arim. In the MSS Cracow, BJ, 560, pp. 4–6 (*c.* 1448) and Cracow, BJ, 1852, p. 170 (*c.* 1464) there are added coordinates of Cracow. In the other manuscripts listed below, there are several other changes, but the core of the table is the same: Cracow, BJ, 1915, f. 4r (*c.* 1338 – this is the oldest copy of this type of geographical table in our list); Cracow, BJ, 551, f. 31v (*c.* 1388); Prague, NK, XIII F 25, f. 70r (1407–1408): as the last city is named Praga; Kalocsa, FK, 326, f. 13r (after 1434); Cracow, BJ, 1927, f. 154r (*c.* 1445); Vienna, ÖNB, Cod. 5145, f. 67vb (fifteenth century): two last cities are Praga and Wyenna.[112]

- Both copies of the canons *Mirabilis* are followed in the Mikulov codex by short explanations of the quadrant. The first of the two texts begins on f. 23v with incipit *Noticiam quadrantis assectans ex tribus premissis* and is supplemented with drawings on f. 24r. The second text starts on f. 94r–v with incipit *Quadratum geometricum. Accipe quatuor lamas ex ligno bene omnino equales.* According to Czech cataloguers, the second treatise was written by Havel ze Strahova (Gallus de Monte Sion), the first professor of astronomy and medicine at the University of Prague. He was the personal physician of King Charles IV and wrote a guide to a healthy lifestyle for the king.[113] Although his medical works are extant, his astronomical studies, if any, are probably lost.[114] He was the teacher of Master Cristannus of Prachatice.
- On ff. 99r–129v, is written the Table for the lunar equations by Nicolaus Mülhus.

---

111 Kříž (Crux) from Telč (1434–1504) was one of the most prolific scribes in medieval Bohemia. He is said to have copied or commented more than 25,000 pages of treatises of various contents. He wrote in Latin and Old Czech. His earliest copied texts are educational, from the time when he worked as a teacher in Žďár nad Sázavou (1454), Soběslav (1455–56, 1476–78), Roudnice nad Labem (1457) and Vyšehrad (1457–58). He studied at Charles University, namely at the King Wenceslas College (1459), but it is not known whether he completed his studies; the university was then calixtine and Crux was a catholic. In 1463 and 1478 he worked in Horšovský Týn, in 1463–66 he was a clergyman at Prague Castle, in 1468–72 he was a preacher in Plzeň, in 1472 there is a record of his stay in Planá near Plzeň (Pilsen) and in 1474 in Nepomuk. At the end of his life until his death (1475–1504) he was an Augustinian canon in Třeboň. He probably took care of the library there and brought all his copies with him. After the eighteenth-century dissolution of the monastery in Třeboň, most of Crux's writings were taken to the National Library in Prague, where they remain today. See Lucie Doležalová and Michal Dragoun, *Kříž z Telče (1434–1504). Písař, sběratel a autor* (Prague: Scriptorium, 2020), pp. 21–31.

112 See Hadravová and Hadrava, *Křišťan z Prachatic*, pp. 423–29.

113 See Milada Říhová, 'Na okraj latinské středověké lékařské literatury', in *Septuaginta Paulo Spunar oblata (70 + 2)*, ed. by Jiří K. Kroupa (Prague: KLP, 2000), p. 226.

114 See Spunar, *Repertorium*, p. 97.

*Baron Hoffman and Dietrichsteins: owners of the manuscript*

MS Mikulov, RM, Mik 6361 is held by the chateau library in the Moravian city of Mikulov, whose interesting and troubled history is worth a brief digression.[115] The current library is a fraction of what used to be the Dietrichstein family library. The library was established by Adam of Dietrichstein (1527–90), the emperor's envoy to the Spanish royal court, and was considerably expanded by his son František of Dietrichstein (1570–1636), bishop of Olomouc and cardinal. Some of his additions to the library included the libraries confiscated from non-Catholic noble families. He also obtained the library of the Jesuit college in Olomouc and the library of the monastery in Žďár nad Sázavou. In its heyday, the library in Mikulov contained about nine to ten thousand volumes, but in 1646 it was taken to Sweden as the spoils of war.[116] The modern library is comprised of what is left of the family libraries purchased or inherited by other Dietrichsteins, including the library of the humanist and president of the Hofkammer (court chamber) Baron Ferdinand Hoffman of Grünbüchel and Strechau (1540–1607) that Ferdinand of Dietrichstein had been gifted in 1678.[117] It is of note that Johannes Kepler stayed in Hoffman's palace for some time while he lived in Prague. Hoffman's library contained over four thousand volumes including 362 manuscripts. In 1769, the Mikulov library was further expanded with the libraries of another noble family, Pruskovští of Pruskov, from their chateaus in Nisa and Pruskov in Silesia,[118] and in the nineteenth century the library of the Piarist grammar school in Mikulov was incorporated into the chateau library.[119] However, misfortune struck again in the 1930s when a large part of the library's collection was sold at international auctions. The rest of the library was damaged at the end of Second World War, in May 1945, during a fire. The Mikulov chateau library currently houses 59 manuscripts, including 15 from the Middle Ages. The astronomical codex Mikulov, RM, Mik 6361 is one of the manuscripts of Bohemian provenance held by the library.

---

115 Vladislav Dokoupil, *Soupis rukopisů mikulovské dietrichsteinské knihovny. Catalogus codicum manu scriptorum bibliothecae dietrichsteinianae Nicolspurgensis* (Prague: SPN, 1958); Stanislav Petr, 'Rukopisné fondy zámeckých a hradních knihoven na Moravě a ve Slezsku', in *Problematika historických a vzácných knižních fondů Čech, Moravy a Slezska* (Brno: Sdružení knihoven ČR, 1995), p. 17; Stanislav Petr, 'Soupis rukopisů zámecké knihovny v Mikulově', in *Sborník Národního muzea v Praze,* řada C, 1994–95, sv. 39–40 (Prague: Národní muzeum, 1997), pp. 29–77.

116 See the portal <https://www.knizni-korist.cz/en/>, which has recently published the first results of the effort to catalogue the prints taken by the Swedish army from various libraries in Bohemia.

117 See Jaroslava Hausenblasová, 'Ferdinand Hoffman z Grünbühelu a Strechau – podnikatel, politik a mecenáš rudolfínské doby', in *Mezi kulturou a uměním*, ed. by Ivana Ebelová, Jiří Pešek, Tomáš Sekyrka, and Vít Vlnas (Prague: NLN, 2013), pp. 86–95.

118 One of the later members of the family was the well-connected chamberlain of Emperor Rudolph II, Oldřich Desiderius Pruskovský (died 1618). When Tycho Brahe moved to Prague, he gave Pruskovský a copy of his *Astronomiae instauratae mechanica* with a handwritten dedication. The book is now in the possession of the National Museum in Prague and is housed in the library of the Křivoklát Castle. See Alena Hadravová, Petr Hadrava, and Jole R. Shackelford (trans.), *Tycho Brahe, Instruments of the Renewed Astronomy* (Prague: KLP, 1996) pp. XII–XIII.

119 The major collection of astronomical and other scientific instruments from the property of the Piaristic grammar school was catalogued by Zdeněk Horský. His Czech and English catalogue with detailed photographs of the instruments was published twenty years after his death by the Regional Museum in Mikulov. See Zdeněk Horský, *Historické vědecké přístroje v mikulovských sbírkách. Katalog vědeckých přístrojů z 16. až 19. století ve sbírkách Regionálního muzea v Mikulově – Historical Scientific Instruments in the Mikulov Collections. Catalogue of Scientific Instruments from the 16$^{th}$ to 19$^{th}$ Centuries in the Collections of the Regional Museum in Mikulov*, ed. by Stanislava Vrbková *et al.* (Mikulov: Regionální muzeum, 2011).

## 2.2 Critical edition of canons *Mirabilis in altis Dominus* by Petrus Cruciferus

Sigla

**A** – Toruń, Biblioteka Uniwersytecka (BU), 74/III (*c.* 1425, Prague), ff. 268va–273ra[120]

**C** – Cracow, Biblioteca Jagellonica (BJ), 610 (1440–77, Poland or Bohemia), ff. 352ra–357vb

**L** – Leipzig, Universitätsbibliothek (UB), 1476 (XV/1, Prague?), ff. 34r–37v

**O** – Oxford, Bodl., Canon. misc. 499 (1451, Prague), ff. 122ra/112ra–125vb/115vb (upper/lower foliation)

**P** – Prague, Národní knihovna (NK) X B 3 (1435–65, Prague), ff. 73ra–75va

**R** – Mikulov, Regionální muzeum v Mikulově (RM), Mik 6361 (Prague, 1470–80), ff. 20v–22v

**S** – Mikulov, Regionální muzeum v Mikulově (RM), Mik 6361 (Prague, 1470–80), ff. 91v–92v

Mirabilis in altis Dominus, mirabilia sua tam in terris quam in astris homini ex pietate revelavit, propter quem omnia condidit et creavit, ut ex magnitudine creaturarum possit cognoscibiliter eorum Creator intueri, quia secundum beatum Gregorium in *Moralibus* 'vestigia nostri Creatoris sunt mira opera visibilis creature, quam dum cernimus, potenciam Creatoris ammiramur'.[121] Cuius mirabilis et ineffabilis potencia in diversis planetarum effectibus et mirandis influenciis in rebus inferioribus nobis maxime relucet et cognoscibiliter demonstratur. Sed quia eorum influencie et effectus secundum motuum eorum diversitatem diversimode variantur, ideoque ad sciendum futuros eorundem planetarum effectus et influencias in istis inferioribus necesse est secundum opus tabularum eorum motus diversitatem calculare. Sed inter tabulas nunc repertas meliores sunt, quas composuit Alfoncius, rex Castelle, super civitatem Tholetum; quarum tamen tabularum operacio ita est prolixa et tediosa, quod vix aliqui deveniunt ad cognicionem motuum planetarum. Propterea quasdam tabulas compendiosas super annos Incarnacionis Christi et super radices et motus tabularum Alfoncii fundatas ad brevitatem et levitatem operis ordinavi et super meridianum civitatis Wratisslaviensis reduxi, que quidem civitas Wratisslavia est orientalior Tholeto in una hora et 28 minutis hore. Sumpsi ergo medium motum cuiuslibet planete ac motus aliorum in tabulis Alfoncii positorum in tribus minutis et 40 secundis dierum, que equipollent predicto tempori, videlicet uni hore et 28 minutis, et hoc, quod inveni, subtraxi a radicibus Tholetanis

**1** Mirabilis *AC* : Incipiunt Canones tabularum Cruciferi, excerptarum (excerparum *O*) per eundem de Tabulis Alfoncii ad meridianum Wratislawiensem. Hic autem sunt redacte ad meridianum Pragensem. Et est idem modus operandi. Mirabilis *LO* : [ ]irabilis *PRS* in altis Dominus *ACOPRS* : Deus in altis *L* mirabilia *ALOR* : et mirabilia *CPS* **2** revelavit *ACLPRS* : revlavit *O* **5** cuius *ACLOPS* : sed *R* **8** ideoque *ACLOPS* : ideo *R* **11** Castelle *ACOPRS* : Castele *L* Tholetum *AC* : Toletum *L* : Tolletum *OPRS* **13** motuum planetarum *ALOPRS* : planetarum motuum (*corr. in*: motuum planetarum) *C* Incarnacionis *ALOPRS* : Incarnaconis *C* **15** civitatis *ACOPRS* : *om. L* Wratisslaviensis reduxi *ACLO* : Pragensis reduxi *PS* : Wratislaviensis reduxi, hic iam ad Pragam satis reducte *R* **16** civitas Wratisslavia *ACLOR* : civitas *PS* est orientalior *ACOPS* : orientalior est *LR* Tholeto *AC* : Toleto *L* : Tolleto *OPRS* et 28 minutis *ACLPRS* : [ ]tis *O* hore *ACLOR* : horariis *PS* **17** sumpsi *ACLOPR* : sumsi *S* ergo *ACLOR* : igitur *PS* planete *ACPRS* : planetarum *LO* **18** tribus minutis *ACOPRS* : quibus minutis 3 *L* **19** Tholetanis *AC* : Toletanis *L* : Tolletanis *OPRS*

120 The edition is based on the manuscript Toruń, BU, 74/III. This manuscript has been taken as the basic one and therefore marked as *A*.

121 *Moralia in Job*, 26, 17–18.

positis prope principium predictarum tabularum Alfoncii, et sic habeo radices cuiuslibet planete et aliorum super dicte civitatis Wratislaviensis longitudinem. Super quas radices dictos medios motus usque ad tercia inclusive calculavi, sic tamen, quod ubicumque in quartis 30 vel plura reperi, pro eis ad tercia addidi unitatem. Unde ad completam noticiam singulorum motuum planetarum doctrina capitulorum sequencium diligenter et subtiliter est notanda.

<1>

**Canon primus. Medium motum augium et stellarum fixarum, accessus et recessus octave spere, Solis, Veneris et Mercurii, Lune, argumenti Lune, argumenti latitudinis Lune, capitis draconis, Saturni, Iovis, Martis et argumenta media eorundem planetarum per tabulas ad hoc factas ad tempus propositum calculare**

Accipio unum tempus perfectum in annis et mensibus et diebus et horis et minutis horarum ac ceteris fraccionibus ab Incarnacione Christi usque ad tempus illud, in quo istos motus voluero cognoscere, et illud tempus servo. Quo facto transeo primo cum annis Christi perfectis vel cum minoribus propioribus in *Tabulam annorum Christi collectorum* et in eorum directo accipio gracia exempli medium motum Solis et ipsum scribo extra in tabula. Iterum cum residuo annorum Christi perfectorum transeo in *Tabulam annorum Christi expansorum* et medium motum Solis repertum extra scribo sub priori, quodlibet sub suo genere. Deinde cum mensibus Christi perfectis transeo in *Tabulam mensium in anno communi*, si sit annus communis, vel in *Tabulam mensium in anno bisextili*, si sit annus bisextilis, et medium motum Solis repertum extra scribo sub priori, quodlibet sub suo genere. Deinde cum numero dierum transeo in *Tabulam dierum* et similiter cum horis transeo in *Tabulam horarum* et iterum cum minutis et omnibus fraccionibus horarum transeo in eandem *Tabulam horarum* extra scribendo motum Solis, quodlibet sub suo genere. Et si minuta, cum quibus intrare debeo, vel ceteras fracciones non invenio in *Tabula horarum*, tunc intro primo cum 30, deinde cum residuo. Exempli gracia: si habeo minuta 55, intro primo cum 30, deinde cum 25, et illos duos introitus simul addo, sed in tali addicione genus non variatur, quia est unus introitus, videlicet cum minutis. Et hoc, quod sic invenio, pono sub minutis gradus. Similiter et ceteras fracciones pono in denominacione sibi convenienti, ut si intro cum secundis hore, pono sub secundis gradus. Si cum terciis hore, pono sub terciis gradus et sic de ceteris, sicut per tytulum inferiorem eiusdem tabule demonstratur. Quibus peractis addo omnia ad invicem incipiendo a subtiliori fraccione et quocienscumque ex tali addicione excreverint 60, pono pro eis unitatem in loco fraccionis inmediate grossioris

**21** super dicte *ACOR* : supradicte *LPS.* Wratislaviensis *ACLOR* : Pragensis *PS* **22** medios *ACLOPS* : *om. R* **25** et subtiliter *ACOPRS* : *om. L* **27** Canon primus. Medium *A* : Medium *CLO* : [ ]edium *PRS* **28** et *ACLOPS* : *om. R* argumenti[2] *ACLOPS* : *om. R* **31** perfectum *ACLOPR* : *om. S* et[1] *ACOR* : *om. LPS* et diebus et *ACOPS* : et diebus *L* : diebus *R* **32** ac *ACOP* : et *LRS* Christi usque *ACOPS* : Domini ad *L* : Domini Christi usque *R* **33** istos motus *ACLOPS* : motus istos *R* voluero cognoscere *ACOPRS* : cognoscere voluero *L* **35** directo *ACLOPS* : directe *R* gracia exempli *ACLOPS* : *om. R* **36** scribo *ACLOPS* : scribe *R* iterum *ACLOPS* : et iterum *R* **37** Solis *ACLO* : *om. PRS* **39** vel… mensium (mesium *C*)… bisextili *ACOPS* : vel in tabulam mensium in anno bisextili, bisextilium *L* : vel bisextili *R* **40** Solis *ACLOPS* : *om. R* scribo sub priori *ACLOPS* : sub priori scribo *R* **42** similiter *ACLOPS* : consimiliter *R* transeo… cum minutis (cum omnibus minutis *S*) *ACLOPS*: et minutis *R* **43** omnibus *ACLOPR* : *om. S* extra scribendo *ACOPRS* : et iterum extra scribe *L* **45** intro primo *ACLOPS* : primo intro *R* **46** residuo … cum[2] *ACOPRS* : *om. L* **49** et *ACOPRS* : sub gradibus et *L* **52** quocienscumque *ACLOPS* : quociens *R* **53** grossioris *ACLOPS* : sequentis *R*

et istum modum servo usque ad gradus; et quocienscumque excreverint 30 gradus, pono pro eis unum signum abiciendo 12 signa quocienscumque excreverint, et quod ex tali addicione invenio, erit medius motus in nona spera ad complementum temporis propositi super meridiem civitatis Wratisslavie. Isto modo invenio motum medium Solis, Veneris et Mercurii, nam istorum trium est unus medius motus, et medium motum Saturni, Iovis et Martis et medium motum capitis draconis, sed verum motum capitis draconis habeo per subtraccionem medii motus eius a 12 signis. Similiter per hunc modum invenio medium argumentum Lune, Veneris et Mercurii de medio argumento trium superiorum, sed medium argumentum trium superiorum habeo per subtraccionem medii motus cuiuslibet eorum de medio motu Solis.

<2>

**Canon secundus. Auges veras et centra media planetarum per tabulas ad hoc factas cognoscere**

Intro cum annis Christi perfectis in *Tabulam augium planetarum in annis Christi collectis*, et si annos meos in linea numeri illius invenero precise, in eorum directo invenio per ordinem auges planetarum quesitas. Si vero non invenero precise numerum annorum meorum, tunc intro cum minori tamen propiori et auges planetarum ibidem per ordinem inventas scribo extra et servo. Deinde accipio prope tabule finem in minutis et secundis et terciis in directo differenciam, que est motus augium in 20 annis, de illa differencia accipio partem proporcionalem, hoc modo accipio excessum annorum Christi perfectorum supra annos Christi collectos in tabula repertos, per quem excessum divido 20 et accipio numerum quocientem. Cum quo numero quociente divido differenciam resolutam in tercia et numerus quociens erit quesitum. Exempli gracia: volo auges planetarum invenire ad annum Christi 1430$^{m}$ completum, quos annos in *Tabula augium planetarum in annis Christi collectis* non invenio. Intro ergo cum numero minori propiori, videlicet cum annis 1424 completis, et auges ibidem repertas scribo extra. Deinde accipio excessum annorum ultra annos in tabula repertos usque ad annum propositum et sunt 6, per quem excessum divido 20 et in numero quociente habeo 3, sed in residuo remanent duo anni, quos ad partem reservo. Deinde accipio differenciam prope finem tabule ex directo annis Christi repertis et est ista 11 40 8 (*suprascr.:* minuta, secunda, tercia), que est motus augium in 20 annis, quam differenciam resolvo in tercia et sunt tot tercia 42008, que divido per predictum numerum quocientem, videlicet per tria, et erunt in numero quociente tot tercia 14002. Et hoc esset quesitum, si nichil in divisione annorum 20 fuisset residuum. Sed quia duo anni fuerunt residui, ideo pro eis sic facio: accipio differenciam differencie in fine tabule positam in directo differencie prius accepte et est ista 35, 0 (*suprascr.*: secunda, tercia), que est motus augium in uno anno, quam resolvo in tercia, et sunt tot tercia 2100, que multiplico per annos residuos,

**54** usque *ALOPRS* : usque usque *C* **56** medius *ACLOPS* : *om. R* **57** meridiem *ACLOPS* : meridianum *R* Wratisslavie (Wratislavie *L*) *ACL* : Wratislavie (*suprascr.*: Prage) *O* : Pragensis *PS* : Wratislavie, tam Pragensis *R* motum medium *ACLOPS* : medium motum *R* **58** Veneris… motum *ACLOPS* : *om. R* est unus *ACOPS* : unus est *L* **59** et[1] *ACLOPS* : *om. R* **60** draconis *ACOPRS* : *om. L* **61** de… superiorum *ACLOPS* : *om. R* **63** motu *ACLORS* : motus *P* **65/106** Canon … oportebit *om. LRS* **65** Canon secundus. Auges *A* : Secundus. Auges *C* : Auges *O* : [ ]uges *P* **71** tabule finem *ACO* : finem tabule *P* **72** et[1] *ACP* : *om. O* **75** quocientem *AOP* : quociente *C* **82** reservo *COP* : servo *A* **83** et est ista *ACP* : *om. O* **85** et sunt tot tercia *ACP* : *om. O* per tria *AC* : (*suprascr.*: per) tria *O* : 3 *P* **86** tercia *AOP* : tria *C* **89** 35, 0 (*suprascr.*: secunda, tercia) *ACP* : 3, 5, 0 (*suprascr.*: minuta, secunda, tercia) *O* **90** tercia[2] *AOP* : tria *C* videlicet per 2, et sunt tot 4200, que similiter divido per 3, et erunt in numero quociente tot tercia 1400, que subtraho a numero quociente prius invento et remanent tot tercia 12602, que dividendo per 60 reduco ad tercia, secunda et minuta, et proveniunt 3, 30, 2 (*suprascr.*: minuta, secunda, tercia) et hoc est quesitum. Nam cum hoc addo ad auges ex tabulis scriptas et servatas, habeo ad annum propositum auges omnium planetarum. Alio modo et facilius facio sic: accipio differenciam differencie in secundis et terciis, que est motus augium in uno mense, et illam differenciam differencie resolvo in tercia, quam sic resolutam multiplico per excessum predictum, et proveniet quesitum vel prope ipsum. Exempli gracia: in proposito differencia differencie est ista 35, 0 (*suprascr.*: secunda, tercia), quam resolvo in tercia, et sunt tot tercia 2100, prout superius demonstravi, que multiplico per excessum predictum, videlicet per 6, et erunt tot tercia 12600, que dividendo per 60 reduco ad tercia, secunda et minuta, et proveniunt 3 30 0 (*suprascr.*: minuta, secunda, tercia). Que addo super auges planetarum rescriptas ex tabula, et sic habeo auges planetarum equatas ad tempus propositum, videlicet ad inicium anni 1430[i] perfecti. Sed ad menses perfectos facio sic: accipio differenciam differencie, quam divido per 12, et in numero quociente exibit, quantum me pro quolibet mense addere oportebit.

Si autem multum precise auges planetarum habere voluero, sic facio:

Accipio unum tempus perfectum in annis et mensibus et diebus et horis ab Incarnacione Christi et ad illud tempus quero primo motum augium et stellarum fixarum secundum modum predictum in mediis motibus et illum servo. Deinde quero motum accessus et recessus octave spere secundum eundem modum, ut prius. Quo invento intro cum eodem motu in *Tabulam equacionis accessus et recessus octave spere* et quero numerum signorum in capite vel in pede tabule. Et nota, quod idem numerus signorum nostris temporibus non invenitur in pede eiusdem tabule, sed per multos annos nobis sufficit eundem numerum signorum usque ad tria signa completa in capite tabule invenire. Invento itaque signo in capite tabule descendo sub eo usque ad gradum eiusdem signi in prima linea positum, in cuius directo accipio gradus, minuta et secunda et ea scribo extra in duobus locis. Quo facto si habeo aliqua minuta ultra gradus, cum quo intravi, intro adhuc semel uno gradu addito, et quid in directo invenio de equacione, illud scribo extra. Deinde video, si prima sit equacio maior quam secunda vel minor, et subtraho minorem a maiore et illud, quod remanet, videtur esse differencia inter duas equaciones. De qua differencia accipio partem proporcionalem secundum proporcionem minutorum, que fuerunt ultra gradus ad 60. Partem autem proporcionalem invenio hoc modo: si in differencia duarum equacionum fuerint minuta et secunda et tercia, resolvo ea ad eandem denominacionem, scilicet ad tercia, et fit secundus numerus. Postea ista minuta et secunda et tercia, cum quibus adhuc non intravi, resolvo eciam ad

**98** resolutam *ACP* : resolvo *O* **102** reduco *ACP* : reducto *O* **108** Si… voluero *ACOP* : <A>uges (Auges *L*) veras planetarum si multum precise voluero habere (habere voluero *L*) *LR* : <A>uges planetarum si multum precise habere voluero, sic facio *S* **109** et mensibus et *ACOPS* : mensibus et *L* : mensibus *R* **110** primo motum *ACPRS* : primum motus *L* : primo motus *O* **111** predictum *ACLOPS* : dictum *R* et illum servo *ACLOPS* : *om. R* motum… recessus *ACLOPS* : accessum et recessum *R* **117** eiusdem signi *ACOPRS* : signi eiusdem *L* **120** intro *ACLOPS* : *om. R* **121** Deinde *ACLPRS* : Deinde deinde *O* prima sit equacio maior *ACLOP* : prima equacio maior sit *R* : prima equacio sit maior *S* **122** maiore *ACLOPS* : maiori *R* **125** et secunda et tercia *AC* : secunda, tercia *L* : secunda et tercia *OR* : et secunda *PS* **127** et[1] *ACOPS* : *om. LR*

tercia et fit tercius numerus. Tunc multiplico secundum per tercium et illud, quod provenerit, divido per 60 et numerus quociens erit pars proporcionalis, quam partem proporcionalem addo equacioni primo accepte, si secunda fuerit maior, vel subtraho a prima, si secunda fuerit minor. Et quod post addicionem vel subtraccionem provenerit, erit equacio motus octave spere equata. Quo facto addo illam cum motu augium et stellarum fixarum, quem prius servavi, si motus accessus et recessus octave spere, cum quo intravi in *Tabulam equacionis* eius, fuerit minor tribus signis, vel subtraho, si fuerit maior, et quod post addicionem vel subtraccionem invenio, erit aux in communi, quam addo radici Incarnacionis Christi posite in ultima linea cuiuslibet augis in *Tabula augium planetarum ad annos Christi collectos*, et sic habeo cuiuslibet planete rectissime verum locum augis.

Deinde augem cuiuscumque planete voluero: subtraho de medio motu eiusdem planete et exibit michi ipsius centrum medium. Isto modo invenio argumentum Solis et centrum medium Saturni, Iovis et Martis, Veneris et Mercurii, sed solum centrum Lune invenio per subtraccionem medii motus Solis a medio motu Lune et per duplacionem residui. Habitis ergo mediis motibus et mediis argumentis planetarum per canonem precedentem et mediis centris eorundem per canonem presentem eorum vera loca per canones sequentes potero invenire.

<3>

**Canon tercius. Mediam coniunccionem et mediam opposicionem Solis et Lune et possibilitatem eclipsis Solis et Lune in mense quolibet reperire**

Accipio omnes annos Christi connumerato illo anno, in quo volo coniunccionem vel opposicionem calculare et cum numero illorum annorum vel cum minori propiori intro *Tabulam collectam mediarum coniunccionum*, si quero coniunccionem, vel *Tabulam mediarum opposicionum*, si quero opposicionem. Nam idem est opus utrobique et illo de mediis coniunccionibus fiat sermo. Accipio ergo in directo istorum annorum, cum quibus intravi, tempus medie coniunccionis et medium motum Solis et Lune et medium argumentum Lune et medium argumentum latitudinis Lune, unde si numerum annorum Christi, cum quibus operor, precise inveni, tunc in eorum directo habeo tempus et motus predictos prime medie coniunccionis, videlicet in Ianuario. Sed si numerum istorum annorum precise non invenero, tunc cum annis residuis transeo in

**128** et[1] *ACOPS* : *om. LR* tercius *ACOPRS* : secundus *L* **129** provenerit *ACLOPS* : provenit *R* **130** secunda *ACLOPR* : eciam *S* a prima *ACLOPS* : *om. R* **132** erit *ACLOPS* : *om. R* motus *ACOPRS* : *om. L* illam *ACLOPS* : eam *R* **135** addicionem vel subtraccionem invenio (provenerit *R*) *APLRS* : addicionem invenio vel subtraccionem (*corr. in*: vel subtraccionem invenio) *C* : addicionem invenio vel subtraccionem *O* **136** ultima linea *ACOPRS* : linea ultima *L* **139** cuiuscumque planete *ACOPS* : cuiuslibet planete cuiuscumque *L* : cuiuslibet planete *R* voluero *ACLOPS* : *om. R* **140** planete *ACLOPS* : *om. et in mg. add.*: scilicet planete *R* exibit michi *ACPR* : exibit exinde *LO* : ex ibi mihi *S* ipsius… et *ACLOPS* : ipsum medium *R* **141** et[1] *ACOPS* : *om. LR* Mercurii *ACLORS* : Mercurii (*in mg. add.*: invenire centrum planete) *P* **143** planetarum… precedentem *ACLOPS* : *om. R* **144** eorundem *ACOPRS* : eorum *L* presentem *ACLOPS* : precedentem *R* **147** *cap. tertium ACLR* : Mediam coniunccionem et mediam opposicionem Solis et Lune et possibilitatem eclipsis Solis et Lune et cetera. Hic canon positus est post tabulas presentes inmediate post numerum proporcionalem, eciam ibi require predicta *O* : *totum cap. tertium om. PS* Canon tercius. Mediam *A* : Tercius. Mediam *C* : Mediam *L* : <M>ediam *R* mediam *ACR* : *om. L* **148** reperire *ACR* : poteris reperire *L* **150** propiori *ACR* : propinquiori *L* intro *ACR* : intro in *L* **151** mediarum *ACR* : tabularum *L* Tabulam mediarum opposicionum *ACL* : vel opposicionem *R* **153** mediis *ACR* : *om. L* **155** Lune et medium argumentum *ACR* : *om. L* **156** operor *ALR* : opereror *C*

*Tabulam annorum expansorum* et in eorum directo accipio singula, sicut prius. Postea tempus inventum in *Tabula annorum expansorum* addo tempori prius invento in *Tabulis annorum collectorum* et motus motibus, quodlibet ponendo sub suo genere. Sed de numero dierum proveniencium ex tali addicione subtraho, si potero, dies Decembris, id est 31 dies, et sic habeo tempus prime medie coniunccionis, videlicet in diebus Ianuarii cum suis motibus ad idem tempus. Si vero talis subtraccio a me fieri non potest, tunc isti tempori sic reperto addo tempus unius lunacionis medie, scilicet illud, quod ponitur in prima linea *Tabule mensium*, et motus ibidem repertos eciam addo motibus prius inventis. Et de diebus tunc ex tali addicione resultantibus subtraho dies 31 et dies remanentes cum horis et fraccionibus horarum erunt tempus medie coniunccionis in diebus Ianuarii cum motibus ad idem tempus requisitis, si operacio mea fuit de mediis coniunccionibus, aut proveniet prima media opposicio in diebus Ianuarii cum motibus suis ad idem tempus, si fuit operacio mea de mediis opposicionibus. Istam igitur mediam coniunccionem vel mediam opposicionem servo pro radice tocius anni. Si vero mediam coniunccionem vel mediam opposicionem alicuius alterius mensis voluero invenire, tunc dies, horas et ceteras fracciones prime coniunccionis sive prime opposicionis, quam servavi, pro radice anni scribo in tabula. Deinde intro *Tabulam medie coniunccionis et opposicionis Solis et Lune in mensibus*, accipiendo dies et horas cum ceteris fraccionibus. Similiter et motus predictos ex directo illius mensis, in cuius diebus coniunccionem vel opposicionem quero, et hec omnia addo ad primam coniunccionem sive opposicionem, videlicet tempus tempori et motus motibus, incipiendo a subtiliori fraccione. Et si ex tali addicione excreverint 60 vel plura, tunc pono pro eis unitatem in loco fraccionis inmediate grossioris. Et istum modum servo usque ad gradus et signa, de quibus facio secundum quod feci in mediis motibus. Similiter istum modum servo in tempore usque ad horas, quas simul addo, et si ex tali addicione excreverint plures quam 24, tunc pro 24 horis pono unum diem ad dies. Postea addo omnes dies lunares sive dies Lune in mensibus ad dies prime coniunccionis sive prime opposicionis, et si ex tali addicione plures dies provenerint, quam iste mensis habet, in cuius diebus coniunccionem vel opposicionem quero, tunc ab istis diebus lunaribus sic collectis subtraho, si potero, dies solares sive dies Solis positos in directo eiusdem mensis, de quo operor, et dies remanentes computo a principio mensis eiusdem. Et horas computo a meridie diei eiusdem. Si vero dies solares positos in directo illius mensis, de quo operor, non potero subtrahere a diebus lunaribus, quod contingit, quando aliquis mensis anni alicuius duas habet coniuncciones sive opposiciones, tunc accipio dies solares mensis precedentis, quos subtraho a diebus lunaribus, ut prius, et dies sic remanentes computo a principio eiusdem

---

**159** accipio *ACL* : accip<i>o *R* **160** Tabulis *ACL* : Tabula *R* **161** ponendo *ACL* : *om. R* **162** dies *ACL* : *om. et in mg. add. R* **163** dies *AC* : diem *LR* **164** cum… tempus *ACL* : *om. R* a me *ACL* : *om. R* **167** tunc *ACR* : *om. L* subtraho *ACL* : subtrahe *R* dies 31 *ACL* : 31 diem *R* **168** Ianuarii *ACL* : *om. et in mg. add. R* **169** mea *AC* : media *L* : *om. R* **170** cum… tempus *ACL* : *om. R* **171** fuit operacio mea *AC* : fuerit operacio media *L* : fuerit operacio *R* coniunccionem… opposicionem *ACR* : opposicionem vel mediam coniunccionem *L* **173** mediam *ACL* : *om. R* mensis voluero *AC* : volo *LR* **175** et *ACL* : vel *R* **177** predictos *ACL* : *om. R* mensis *ACR* : mensis, scilicet precedentis *L* **178** sive *ACL* : seu *R* **179** videlicet *ACL* : *om. R* et *ACL* : *om. R* si *ACL* : cum *R* **180** excreverint… plura *ACL* : provenerint 60 *R* pono pro eis *ACL* : pro illis pono *R* in… inmediate *AC* : inmediate fraccionis *L* : inmediate fraccione *R* **183** excreverint… 24 *AC* : plures dies quam 24 excreverint *L* : provenerint plures quam 24 *R* **184** dies[3] *ACL* : omnes dies *R* Lune *ACR* : *om. L* **185** sive prime *ACL* : seu *R* **186** provenerint… habet *ACR* : quam iste mensis habet, provenerint *L* **188** sive *ACL* : seu *R* **189** mensis eiusdem *AC* : eiusdem mensis *LR* **189/194** et horas… precedentis *ACL* : precedentis *R* 193 ut *ACR* : sicut *L* sic remanentes *ACR* : remanentes sic *L*

mensis precedentis. Similiter medius motus Solis et Lune et argumentum medium Lune et argumentum latitudinis Lune tunc inventi sunt in eodem tempore mensis precedentis. Sed de anno bisextili caute me considerare oportet, quia si annus fuerit bisextilis et cum hoc locus bisexti pertransivit, tunc ad dies solares addo unum diem et post talem addicionem subtraho dies solares a diebus lunaribus, ut dictum est, et sic habeo tempus medie coniunccionis, si operatus sum de coniunccione, vel tempus medie opposicionis, si operatus sum de opposicione. Habeo eciam ad idem tempus medium motum Solis et Lune, si operatus sum de coniunccione, vel medium motum Solis et nadir Lune, si operatus sum de opposicione. Habeo eciam ad idem tempus argumentum medium Lune. Similiter habeo ad idem tempus argumentum medium latitudinis Lune, et si idem argumentum fuerit 0 in signis et minus 12 in gradibus vel 5 in signis et plus quam 18 in gradibus vel 6 in signis et minus 12 in gradibus vel 11 in signis et plus quam 18 in gradibus, dico eclipsim fore possibilem. Si autem predictum argumentum extra illos terminos invenero, impossibile est eclipsim fieri naturalem.

<4>

**Canon quartus. Verum locum Solis reperire**

Accipio primo eius medium motum per primum canonem, deinde sumo eius medium argumentum per secundum canonem in fine, quibus habitis intro cum argumento Solis *Tabulam equacionis* eius, scilicet cum signo in capite vel in pede tabule. Et notandum est, quod quinque signa inveniuntur in capite, alia aut in pede tabule. Et quando signa inveniuntur in capite tabule, tunc gradus signi ponuntur in prima linea descendendo. Si vero signa inveniuntur in pede tabule, tunc gradus signi ponuntur in

---

**194** medius *ACR* : *om. L* **195** argumentum *ACL* : medium argumentum *R* inventi *ACL* : inventa *R* **196** considerare oportet *ACL* : oportet considerare *R* annus fuerit *ACR* : fuerit annus *L* **199/201** tempus medie opposicionis... vel *ACL* : *om. R* **199** tempus *ACR* : *om. L* **202** medium *AC* : *om. LR* **203** habeo ad idem tempus *ACL* : *om. R* idem[2] *ACL* : illud *R* **205** minus *ACL* : minus quam *R* **206** possibilem *ACR* : possibilem; eclipsis videlicet Solis, si fuerit coniunccio, vel eclipsis Lune, si fuerit opposicio *L* **207** naturalem *ACR* : naturalem. Notandum, quod in has tabulas equacionum intratur cum signis communibus, quorum sex sunt posita in superiori margine, videlicet 0, 1, 2, 3, 4, 5, quorum gradus ponuntur in prima linea descendendo versus sinistram et alia 6 sunt posita in inferiori margine, videlicet 6, 7, 8, 9, 10, 11 retrorsum, quorum gradus ponuntur in ultima linea ascendendo versus dextram. Si ergo signum, cum quo intratur, sit in superiori margine, tunc gradus signi accipiuntur versus sinistram descendendo et ex directo eiusdem gradus accipitur equacio sola, que scribitur extra. Deinde accipitur differencia, que in medietate sequitur in linea descendendo et consideratur litera scripta super differenciam in superiori margine, que litera scribitur extra super differenciam. Que litera, si fuerit 'a', significat, quod pars proporcionalis sit addenda equacioni accepte, si autem litera eadem fuerit 's', significat, quod pars proporcionali est subtrahenda ab equacione accepta. Si autem signum, cum quo intratur, si in inferiori margine, tunc gradus signi accipitur versus dextram ascendendo et ex directo eiusdem gradus versus sinistram accipitur equacio sive minuta proporcionalia sive diversitas diametri, que scribitur extra. Deinde accipitur differencia, que in medietate sequitur in linea ascendendo et consideratur litera sub differencia scripta in inferiori margine, que litera scribitur extra circa differenciam, que si fuerit 'a', significat, quod pars proporcionalis sit addenda equacioni accepte, ut dictum est. Et notandum, quod ubi in tabulis equacionum ponitur paragrafus rubeus vel niger, ibi in differencia tam superiori, quam inferiori margine ponuntur due linee *(sic)*, videlicet 'a', 's', vel 's', 'a' sic, quod una ponitur supra aliam, quarum superior litera in utraque margine respicit equaciones, que sunt super paragrafum, et inferior similiter in utraque margine respicit equaciones, que sunt sub paragrafum. Per hunc namque modum equantur Sol et Luna. De aliis autem planetis est notandum, quod ubi in tabulis ponitur propior, videlicet minutis proporcionalibus, ibi sunt in utraque margine due linee *(sic)*, una supra aliam, quarum superior in utraque margine respicit minuta proporcionalia, que sunt supra propior, et alia litera respicit minuta, que sunt sub propior *L* **209** Canon quartus. Verum *A* : Quartus. Verum *C* : Verum *LO* : [ ]erum *PRS* reperire *ACOPRS* : volens reperire *L* **210** eius medium[2] *ACLOPS* : ipsius *R* **211** in fine *ACLOPS* : *om. R* **212** in pede tabule *ACLOPS* : pede tabula *R* **213** aut *ACLOPS* : autem *R* **215** signa inveniuntur *ACLOPS* : inveniuntur signa *R*

ultima linea ascendendo. Transeo ergo eciam cum gradibus in uno latere correspondente usque sub signo in capite vel usque supra signum in pede tabule et accipio equacionem Solis ibidem inventam, et si cum signis et gradibus habuero aliqua minuta vel secunda, tunc intro iterum eandem tabulam uno gradu addito et equacionem Solis in suo directo loco accipio et differenciam inter istam equacionem et illam prius acceptam sumo et illam differenciam reductam in secunda multiplico per residuum argumenti ultra signa et gradus, cum quo non intravi, residuum dico eciam reductum in secunda, et considero denominacionem producti, quia multiplicando secunda per secunda proveniunt michi quarta in producto, multiplicando vero secunda per tercia proveniunt michi quinta in producto et sic de ceteris. Deinde illud productum divido per 60, quociens potero, et sic reducta ad ulteriora genera fraccionum exibit michi pars proporcionalis debita minutis et secundis argumenti propositi, quam partem proporcionalem addo equacioni Solis prius invente, si secunda fuerit maior, vel minuo eam ab eadem, si secunda fuerit minor. Tunc exibit equacio Solis examinata et hanc subtraho de medio motu Solis, si argumentum eius fuerit minus 6 signis, vel addo, si argumentum eius fuerit plus 6 signis, et exibit michi verus motus Solis in nona spera.

<5>

**Canon quintus. Verum locum Lune per tabulas ad hoc factas reperire**

Accipio primo eius medium motum per primum canonem, deinde accipio eius centrum per secundum canonem in fine, postremo accipio eius argumentum medium per tabulas ad hoc factas secundum doctrinam canonis primi. Quibus inventis cum centro medio intro *Tabulas equacionis* eius per similem modum, sicut intravi in *Tabulas equacionis Solis*, et quero equacionem centri cum duplici introitu, si opporteat. Similiter accipio ibidem in eius directo minuta proporcionalia eciam cum duplici introitu, si opporteat, id est, si in centro medio sint aliqua minuta vel secunda ultra gradus, cum quibus non intravi, et servo utrumque per se. Deinde considero centrum Lune, utrum fuerit maius 6 signis vel minus. Si minus, addo istam equacionem centri cum medio argumento Lune, si vero maius, tunc eandem minuo de eodem argumento et habeo argumentum Lune verum, cum quo argumento vero *Tabulas equacionis Lune* ingredior et equacionem argumenti et diversitatem dyametri in directo eius accipio, utrumque cum duplici introitu, si opporteat. Deinde de diversitate dyametri accipio partem proporcionalem secundum proporcionem minutorum proporcionalium servatorum prius ad 60 consimili modo, secundum quod dixi in canone de Sole, et illam partem proporcionalem addo equacioni prius invente et habeo equacionem examinatam, quam minuo de medio motu Lune, si

**217** usque[2] *ACLOPS* : *om. R* **218** aliqua *LOPRS* : aliquo *AC* vel *ACLOPS* : *om. R* **219** eandem *ACLOPS* : in eandem *R* in… accipio *ACLOPS* : in directo invenio *R* **220** prius acceptam *ACLOPS* : inventam *R* **223** quia *ACLOPS* : *om. R* **224** michi *ACLO* : *om. PRS* in producto *ACLOPS* : *om. R* vero *ACLOPS* : *om. R* **225** michi quinta *ACLO* : quinta *PRS* producto *ACLOPS* : producto, tercia per tercia proveniunt *R* illud *ACOPRS* : illum *L* **226** reducta *ACLPRS* : producta *(corr. in*: reducta*) O* **228** minuo *ACOPRS* : *om. L* **231** michi *ACLOPS* : *om. R* **233** Canon quintus. Verum *A* : Verum *CLO* : [ ]erum *PRS* reperire *ACOPRS* : cupiens reperire *L* **234** primum canonem *ACOPRS* : canonem primum *L* **236** inventis *ACLOPS* : habitis *R* centro medio *ACLOPS* : medio centro *R* **237** equacionis[2] *ACLOPS* : *om. R* **238** quero *ACLOPR* : quere *S* **239** ibidem… directo *ACLOPS* : ibi in directo *R* eciam *ACOPS* : *om. LR* **240** sint *ACLOPS* : fuerint *R* **244** tabulas *ACLOPS* : in tabulas *R* ingredior *ACLOPS* : *om. R* **245** cum *ACOPRS* : *om. L* **246** accipio partem proporcionalem *ACLOPS* : accipio partem proporcionalem accipio partem proporcionalem *R* secundum proporcionem *ACLOR* : *om. PS* **249** Lune *ACPRS* : *om. LO*

eius verum argumentum fuit minus 6 signis, vel addo, si fuit maius, et exibit michi verus locus Lune in nona spera.

<6>

**Canon sextus. Verum locum Saturni, Iovis, Martis, Veneris et Mercurii et eorum direccionem, stacionem et retrogradacionem reperire**

Accipio primo medium motum eius, cuius verum locum scire volo, per primum canonem. Deinde accipio eius centrum medium per secundum canonem in fine. Postea vero accipio eius argumentum medium per primum canonem in fine. Hiis inventis primo cum centro medio *Tabulam equacionis illius planete,* cuius verum locum scire volo, ingredior, et equacionem centri in eius directo cum duplici introitu accipio, supra quam scribo 'minuatur'. Et hoc, si centrum medium planete fuit minus 6 signis, et sic ipsam minuo de medio centro et addo medio argumento. Vel super ipsam scribo 'addatur' et hoc, si centrum medium planete fuit maius 6 signis, et sic illam equacionem centri addo centro medio et minuo de argumento medio et proveniet tam centrum, quam argumentum equatum sive verum. Cum isto igitur centro vero intro *Tabulam equacionis planete* et accipio minuta proporcionalia sibi correspondencia et supra ea scribo eorum tytulum, scilicet 'longior' vel 'propior'. Deinde cum argumento vero *Tabulas equacionis planete* intro et in eius directo accipio tam equacionem argumenti, quam diversitatem dyametri in longitudine longiori, si super minuta proporcionalia scripsi 'longior', vel in longitudine propiori, si scripsi 'propior', et operor videlicet cum duplici introitu utrumque, si oporteat, ut dictum est. Deinde de illa diversitate dyametri accipio partem proporcionalem secundum proporcionem minutorum proporcionalium prius servatorum ad 60 et illam partem proporcionalem minuo de equacione argumenti prius invente, si diversitas dyametri sit de longitudine longiori, vel eam addo equacioni argumenti prime, si diversitas dyametri sit de longitudine propiori, et exibit michi equacio argumenti proposito argumento correspondens, supra quam scribo 'addatur', si argumentum verum fuit minus 6 signis, vel scribo 'minuatur', si argumentum verum fuerit maius 6 signis. Considero igitur tam equacionem argumenti, quam equacionem centri, et si super utrumque sit scriptum 'addatur', illas simul iunctas addo medio motui planete, et si super utrumque sit scriptum 'minuatur', istas simul iunctas minuo de medio motu

**250** verum argumentum fuit minus *A* : argumentum verum fuit (fuerit *L*) minus *CLOP* : argumentum fuerit minus *R* : argumentum fuit *S* fuit[2] *ACOR* : fuerit *LPS* **253** Canon sextus. Verum *AC* : Verum *LO* : [ ]erum *PRS* **254** direccionem, stacionem *ACOPRS* : stacionem, direccionem *L* reperire *ACOPRS* : cupiens reperire *L* **255** primo *ACPRS* : prius *LO* locum *ACLOR* : loco *PS* **256** in fine *ACLOPS* : *om. R* **257** vero *ACLOPS* : *om. R* argumentum medium *ACOPRS* : medium argumentum *L* inventis *ACLOPS* : habitis *R* **258** tabulam equacionis *ACLOPS* : *om. R* verum locum *ACOPRS* : locum verum *L* volo *ACOPRS* : voluero *L* **259** supra *ACLOPS* : super *R* **260** minuatur *ACOPRS* : m *L* fuit *ACO* : fuerit *LPRS* **262** fuit *ACLOPS* : fuerit *R* sic *ACLOPR* : *om. S* equacionem *ACLOPS* : *om. R* **265** proporcionalia *ACLOPS* : sibi proporcionalia *R* scribo *ACLOPS* : scribor *R* **266** tabulas equacionis planete *ACLOPS* : tabulam equacionis *R* **267** diversitatem dyametri *ACLOR* : dyametri diversitatem *PS* **268** super *ACLOPS* : supra *R* in longitudine[2] *ACLOPS* : *om. R* **269** videlicet *ACOPS* : *om. LR* **270** est *ACLOPS* : *om. R* **272** proporcionalem *ACLOPS* : proporcionabilem *R* invente *ACOPRS* : inventi *L* **273** dyametri *ACLOPS* : *om. R* addo *ACLPRS*: adde *O* argumenti prime *ACLOPS* : *om.* (*add.*: prime) *R* **274** si diversitas dyametri (dyametri *om. PS*) … propiori *ACLOPS* : si sit de propiore *R* **276** verum[1] *ACLOPR* : *om. S* fuit *ACOPS* : fuerit *LR* argumentum… signis *ACLOPS* : si fuerit maius *R* **277** equacionem argumenti *ACLOPS* : equacionem tam equacionem (*in mg. add.*: argumenti) *R* et si super… planete *ACLOR* : *om. et in mg. add.*: et si super utramque sit scriptum 'addatur', tunc ambas equaciones simul iunctas addo medio motui eius *P* : *om. S* **279** istas *ACOPS* : illas *L* : tunc *R* motu *ACLOPR* : motui *S*

planete. Sed si super unam sit scriptum 'addatur' et super aliam 'minuatur', tunc minuo minorem de maiori et residuum addo vel minuo secundum quod super ipsum fuit scriptum, et proveniet verus locus planete in nona spera, de quo sum operatus.

De direccione et stacione et retrogradacione istorum quinque planetarum sic facio: Cum centro vero cuiuslibet eorum intro *Tabulam equacionis* eius et in directo illius centri accipio stacionem primam illius planete in signis, gradibus et minutis, quam exterius per se servo, et istam stacionem primam minuo de 12 signis et remanet michi stacio secunda, quam scribo sub prima. Deinde considero argumentum verum planete eiusdem, quod si fuerit minus stacione prima vel plus stacione secunda, erit idem planeta directus. Et si fuerit equale stacioni prime in signis, gradibus et minutis, erit planeta stacionarius in stacione prima, scilicet ut incipiat retrogradari. Si vero fuerit equale stacioni secunde, erit planeta stacionarius in stacione secunda, ut scilicet incipiat dirigi. Si autem idem argumentum fuerit plus stacione prima et minus stacione secunda, erit planeta retrogradus.

<7>

**Canon septimus. Tempus vere coniunccionis Solis et Lune et eorum opposiciones cognoscere**

Equare me oportet Solem et Lunam ad tempus medie coniunccionis vel medie opposicionis, quanto precisius potero. Et est tunc idem modus operandi in equacione Lune, sicut in equacione Solis eo, quod ad tempora mediarum coniunccionum et opposicionum nullum est centrum Lune. Si ergo vera loca earum fuerint eadem in signis, gradibus, minutis, secundis et terciis, non oportet me ultra calculare, quia in eodem tempore est vera coniunccio et media. Sed si verus locus Solis precessit verum locum Lune in ordine signorum, tunc scio, quod vera coniunccio adhuc est futura; si autem econverso verus locus Lune precessit verum locum Solis, scio, quod vera coniunccio est preterita. Ergo sive vera coniunccio fuerit futura sive preterita, subtraho minorem motum de maiori et remanebit michi distancia luminarium, quam reduco ad ultimum genus fraccionis, et productum servo et considero, cuius luminis est distancia. Quia si Sol precessit Lunam in ordine signorum, erit distancia Solis. Si econverso, erit distancia Lune. Deinde cum argumento Solis intro *Tabulam veri motus Solis et Lune in una hora* et in eius directo accipio verum motum Solis in una hora cum duplici introitu, sed super argumentum Lune addo medietatem distancie luminarium cum medietate duodecime partis eiusdem distancie. Sic ipsam distanciam divido per 12 et numerus quociens erit duode-

---

**280** aliam *ACLOPS* : alteram *R* minuo minorem *ACLOPS* : minorem minuo *R* **281** minuo *ACLOPS* : minuo de medio motu planete *R* super… scriptum *ACOPS* : fuerit super ipsum scriptum *L* : super ipsum scriptum erat addatur vel minuatur *R* **282** in nona spera *ACLOPS* : *om. R* **283** et[1] *ACLOPS* : *om. R* **284** eorum *ACLOPS* : illorum *R* **285** et *ACLOPS* : *om. R* **287** planete eiusdem *ACOPS* : planete *L* : eiusdem planete *R* **289** fuerit *ACLOPS* : erit *R* prime *ACLPRS* : prima *O* **290** scilicet *ACLOPS* : *om. R* **291** planeta *ACLOPS* : *om. R* **295** Canon septimus. Tempus *AC* : Tempus *LO* : [ ]empus *PR* : Posiciones *S* et eorum opposiciones *ACLOPS* : *om. R* **297** medie opposicionis *ACOPRS* : opposicionis medie *L* **299** in equacione *ACLOPS* : *om. R* **300** ergo *ACOPS* : *om. L* : vero *R* earum *ACLOPS* : eorum *R* eadem *ACLOPR* : *om. et in mg. add. S* **301** secundis et terciis *ACLOPS* : et secundis *R* **302** verus *ACLOPS* : *om. R* precessit *ACLOPS* : precesserit *R* **303** adhuc… coniunccio *ACOPRS* : *om. L* **305** futura sive preterita *ACLOPS* : preterita sive futura *R* **306** michi *ACLOPS* : *om. R* ultimum *ACLOPS* : unum *R* **307** luminis *ACLOPS* : luminaris *R* si *ACLOPS* : *om. R* **308** si *ACLOPS* : si vero *R* **309** tabulam *ACLOPS* : in tabulam *R* **310** motum *ACLOR* : motus *PS* **312** per *ACPRS* : in *LO* erit *ACOPRS* : est *L*

cima pars distancie, quam addo dicte distancie et illius tocius medietatem addo super argumentum Lune, si Sol precessit Lunam in ordine signorum, vel eandem medietatem minuo de argumento Lune, si Luna precessit Solem. Et cum argumento Lune sic proveniente tabulam predictam intro et in eius directo accipio verum motum Lune in una hora cum duplici introitu, si oporteat. Deinde subtraho verum motum Solis in una hora a vero motu Lune in una hora et remanenti addo tot secunda minus uno, quot sunt gradus in longitudine sive in distancia. Et hoc, si Luna fuerit in inferiori parte sui epicicli, quod scio per eius argumentum, vel minuo ab eodem tot secunda minus uno, si Luna fuerit in superiori parte sui epycicli, et proveniet michi superacio Lune equata in una hora. Quam reduco ad idem genus fraccionis, ad quas reduxi distanciam luminarium, et illam distanciam luminarium divido per productum et exibunt michi hore. Deinde, si aliquid fuerit residuum, ipsum multiplico per 60 et productum divido per idem, quo prius, et exibunt michi minuta hore. Adhuc, si aliquid sit residuum, ipsum multiplico per 60 et productum divido per idem, quo prius, et exibunt michi secunda hore. Adhuc, si voluero, multiplico residuum per 60 et divido productum per idem, quo prius, et exibunt michi tercia hore et sufficit. Deinde horas, minuta, secunda et tercia sic inventa coniungo tempori medie coniunccionis Solis et Lune, si Sol precessit Lunam in ordine signorum, vel illud tempus minuo, si fuerit econtra, et proveniet michi vera coniunccio Solis et Lune quesita.

Sol autem et Luna sic ducuntur ad coniunccionem sive opposicionem: Ipsam distanciam cum sua duodecima prius habita addo motui Lune in tempore coniunccionis vel opposicionis, solam autem duodecimam addo motui Solis. Et hoc, si fuit distancia Solis; si autem fuit distancia Lune, tunc ipsam distanciam cum duodecima sui minuo de motu Lune, solam autem duodecimam minuo de motu Solis et sic habeo vera loca ad tempus vere coniunccionis vel opposicionis et Sol et Luna tunc conveniunt, vel Luna est in nadir Solis. Vel si ad idem tempus vere coniunccionis equavero Solem et Lunam secundum *Tabulas et canones equacionum*, invenio eos eciam coniunctos vel per modicum distantes. Si vero in aliquo distant, video distanciam, secundum quod prius feci, et operor per omnia sicut prius. Sed non oportet me sic operari, nisi fuerit opus valde preci-

---

**313** dicte *ACPS* : *om. LOR* illius *ACOPRS* : illam *L* **316** motum *ACLOP* : locum *R* : motum ipsius *S* **317** oporteat *ACOPRS* : oportet *L* in una hora *ACLOPS* : *om. R* **318** hora *ACLOPR* : hora cum duplici introitu, si oporteat. Deinde subtraho verum motum Solis in una hora *S* **320** quod... argumentum *ACLOPS* : *om. et in mg. add. R* ab eodem *ACLOPS* : *om. et in mg. add. R* **321** epycicli *ACPS* : epicicli. Si argumentum medium Lune fuerit maior 95 gradibus usque ad 265 gradus, tunc Luna est in medietate sui epicicli inferiori *L* : *om. et in mg. add.*: epicicli. Si argumentum medium Lune fuerit maior 95 gradibus usque ad 265 gradibus, tunc Luna est in medietate inferiori epicicli sui. Si vero argumentum medium fuerit minus 95 gradibus vel plus 265 gradibus, tunc Luna est in superiori parte epicicli *O* : epicicli. Si argumentum medium Lune fuerit a 95 gradibus usque ad 265 gradus, tunc Luna est in medietate inferiori sui epicicli. Si vero argumentum medium Lune fuerit minus 95 gradibus vel plus 265, tunc Luna in superiori parte epicicli *R* **322** quas *ACLOPS* : quod *R* **323** productum *ALOPRS* : productum *(in mg. add.*: per longitudinem luminarium reductam ad ultimum genus fraccionis*) C* **324** aliquid *ACLOPS* : quid *R* **325** quo prius *ACOPS* : quod prius *L* : *om. R* **325/327** minuta ... hore *ACLO* : *om. et in mg. add.*: minuta hore et iterum residuum multiplico et exibunt *P* : minuta hore. Adhuc, si aliquid fuerit residuum, multiplico per 60 et divido per superacionem, ut prius, et exibunt michi *R* **325** Adhuc... secunda hore *om. S* sit *ACLOP* : fuerit *R* ipsum *ACLOP* : *om. R* **326** productum *ACLOP* : *om. R* idem, quo prius *ACLOP* : superacionem, ut *R* secunda hore *ACOPR* : secunde *L* **327** voluero *ACOPRS* : volo *L* divido... prius *ACLOPS* : per idem divido *R* **328** michi *ACLOPS* : *om. R* et² *ACLOPR* : quod *S* horas *ACLOPS* : hora *R* **329** coniungo *ACLOPS* : iungo *R* **330** econtra *ACLOPS* : econverso *R* **333** coniunccionis *ACLORS* : *(suprascr. add.*: vere*)* coniunccionis *P* **334** fuit *ACO* : fuerit *LPRS* **335** fuit *AOS* : fuerit *CLPR* duodecima sui *ACOPS* : sua 12 *L* : duodecima sui parte *R* **337** tunc *ACLOPS* : *om. R* **341** sic *ACLOPS* : *om. R* sum. Cuiusmodi sunt eclipses, coniuncciones, vel quando magna distancia fuerit inter loca luminarium equata ad tempus medie coniunccionis, quia tunc oportet verificari opus propter diversitatem motus Lune inter coniunccionem veram et mediam.

Si autem quesiero veram opposicionem, non diversificatur opus meum ab opere vere coniunccionis, nisi in illo, quando calculavi veram coniunccionem. Subtraxi verum locum Solis a vero loco Lune vel econtra. Sed pro opposicione habenda oportet me subtrahere verum motum Solis a nadir vero Lune vel econtra, secundum quod unus illorum fuerit maior altero et operor cum distancia, sicut prius feci in operacione vere coniunccionis. Et isto modo exibit michi vera coniunccio sive opposicio Solis et Lune diebus non equatis. Pro equacione autem dierum intro *Tabulam equacionis dierum* cum vero loco Solis in tempore vere coniunccionis sive opposicionis et in eius directo accipio equacionem dierum cum duplici introitu, si oportebit, in gradibus et minutis. Et pro quolibet gradu ibi reperto accipio quatuor minuta unius hore, multiplicando gradus per quatuor, et in producto habeo minuta hore. Et pro quolibet minuto ibi reperto accipio quatuor secunda hore, multiplicando minuta per quatuor, et in producto habeo secunda hore, que si fuerint plura quam 60, pono pro eis unum minutum ad minuta tociens, quociens potero. Que minuta et secunda addo super tempus vere coniunccionis sive opposicionis prius invente et exibit michi tempus vere coniunccionis sive opposicionis diebus equatis a meridie precedenti computando. Et ad illud tempus nunc inventum equanda est figura celi pro iudiciis faciendis.

<8>

**Canon octavus. Introitum Solis in Arietem vel in quodcumque signum voluero invenire**

Primo equare me oportet Solem ad tempus propinquum introitui eius in Arietem. Quo facto video distanciam veri loci Solis ab inicio Arietis, quam scio subtrahendo verum locum Solis a 12 signis, et residuum erit distancia, quam reduco ad secunda vel tercia. Deinde cum argumento Solis ad idem tempus intro *Tabulam veri motus Solis et Lune in una hora* et verum locum Solis in una hora in directo eius inventum sumo, quem reduco ad simile genus fraccionis, quo prius reduxi distanciam. Quam distanciam per istum motum sic reductum divido et in numero quociente exibunt michi hore. Et residuum, si sit, multiplico per 60 et productum divido per idem, quo prius, et proveniunt michi minuta hore. Adhuc residuum, si sit, multiplico per 60 et productum divido per

**342** eclipses *OR* : eclipsis *ACLPS* magna distancia *ACLOPS* : distancia magna *R* **343** verificari opus *ACLOPS* : verificacionem facere *R* **345** quesiero *ACLOPS* : quero *R* **347** pro *ACLOPS* : in *R* subtrahere *ACLOPS* : *om. R* **353** dierum *ACLOPS* : *om. R* oportebit *ACLOPS* : oporteat *R* **354** gradu ibi *ACLOPS* : ibi gradu *R* **354/356** minuta … quatuor[1] *ACLOPS* : *om. R* **357** tociens *ACLOPS* : *om. R* **359** opposicionis[1] *LPRS* : coniunccionis *ACO* prius invente *ACLOPS* : invente prius *R* michi *ACLOPS* : *om. R* **363** Canon octavus. Introitum *AC* : Introitum *LO* : [ ]ntroitum *PRS* vel *ACLOPR* : vel vel *S* voluero *ACOPS* : volens *L* : *om. R* **365** equare me oportet *ACOPS* : oportet me (me *om. L*) equare *LR* introitui *LOR* : introitu *ACPS* eius *ACLOPS* : *om. R* **367** Solis *ACLOPS* : Arietis *R* vel *APS* : vel (*del.*: ad) *C* : vel ad *LOR* **368** tabulam *ACOP* : in tabulam *LR* **369** et… directo *ACLOP* : et verum motum Solis in una hora accipio in directo *R* : in directo *S* **370** simile *ACLO* : idem *PRS* quo prius *ACLOPS* : ad quod *R* **371** motum *ACLOR* : modum *PS* sic reductum *ACOPRS* : *om. L* michi *ACLOPS* : *om. R* **372** per 60 *ACOPRS* : *om. L* quo prius *ACLOPS* : *om. R* proveniunt… hore *ACOPRS* : exibunt mihi minuta *L* **373** Adhuc *ACOPRS* : et adhuc *L* idem, quo prius, et exibunt michi secunda hore et sufficit. Deinde horas, minuta et secunda sic inventa addo tempori, ad quod prius equavi Solem, si Sol nondum intravit Arietem, vel minuo ab eodem, si Sol prius intraverit Arietem, et proveniet michi tempus introitus Solis in Arietem vel tempus sibi propinquum. Unde ad illud tempus nunc inventum equare me oportet Solem precisius, quam potero, et si invenero verum locum eius o in signis, o in gradibus, o in minutis et o in secundis, sufficit, quod tempus prius habitum est satis precisum. Si vero adhuc fuerit aliqua distancia sicut prius, per omnia operabor et facta operacione provenit tempus introitus Solis in Arietem diebus non equatis. Equabo ergo dies, sicut in precedenti capitulo dictum est, et exibit michi tempus introitus Solis in Arietem certissimum computando a meridie precedenti. Et ad illud tempus equandi sunt planete et facienda est equacio 12 domorum secundum Albumazar et Messehalle. Et simili modo, sicut nunc feci de introitu Solis in Arietem, sic facere potero de ipsius introitu in alia signa.

Expliciunt *Tabule cum suis canonibus de mediis et veris motibus planetarum.*

---

**374** quo *ACLOPS* : sicut *R* exibunt michi *ACLOPS* : in quociente exibunt *R* sufficit *ACLOPS* : sic consequenter si voluero usque ad tercia *R* horas *ACLOPS* : hora *R* **376** intraverit *ACS* : intravit in *L* : intravit *OP* : intraverit in *R* **377** Solis *ACLOPS* : *om. R* nunc inventum *ACLOPS* : *om. R* **378** equare me *ACLOPS* : me equare *R* verum locum *ACLOPS* : locum verum *R* **379** o in gradibus, o in minutis *ACLOPS* : o in minutis, o in gradibus *R* et *AC* : *om. LOPRS* **381** provenit *ACLOPS* : proveniet *R* **382** michi *ACLOPS* : *om. R* **384** Albumazar et Messehalle (Mesehale *R*) *ACOR* : Albumasar et Messehale *L* : Alb< > et Mesahal< > *PR* **386** ipsius introitu *ACOPRS* : introitu ipsius *L* **387** Expliciunt… planetarum *A* : Expliciunt Tabule cum suis canonibus de mediis et veris motibus planetarum, Anno Domini M°CCCC°XL°, Sabbato ante Philippi *C* : Hic notandum, quod latitudo septentrionalis Saturni maxima est 2, 2 (*suprascr.*: gradus, minuta), meridionalis 3, 5 (*suprascr.*: gradus, minuta), Iovis in meridie et septentrione est 2, 8 (*suprascr.*: gradus, minuta), Martis maxima latitudo septentrionalis 4 (*suprascr.*: gradus), meridionalis 6 (*suprascr.*: gradus), Mercurii maxima latitudo in septentrione et meridie est 4, 56 (*suprascr.*: gradus, minuta), Veneris utraque latitudo 4, 28 (*suprascr.*: gradus, minuta), 24 de Lineriis *L* : Expliciunt Tabule Cruciferi cum suis canonibus de mediis et veris motibus planetarum reducte ad meridianum Pragensem, sc<r>ipte Anno Domini 14510 per Iohannem, b(accalaureum) de Krczyn in collegio beatissime Marie Virginis nacionis Boemorum domus Reczkonis. *In mg. add.*: Hic notandum, quod latitudo septentrionalis Saturni maxima est 3, 2 (*suprascr.*: gradus, minuta), meridionalis 3, 5 (*suprascr.*: gradus, minuta), Iovis in septentrione et meridie est 2, 8 (*suprascr.*: gradus, minuta), Martis maxima latitudo septentrionalis est 4 (*suprascr.*: gradus), meridionalis 6 (*suprascr.*: gradus), Mercurii maxima latitudo in septentrione et meridie 4, 56 (*suprascr.*: gradus, minuta), Veneris utraque 4, 58 (*suprascr.*: gradus, minuta), 24 de Lineriis *O* : *om. PRS*

## 2.3 Commented Translation of Canons *Mirabilis*

Admirable is the Lord on high,[122] who out of love for man revealed his miracles both on the Earth as well as in the stars. For man he founded and created all things, to enable him to recognise and look at the Creator according to the magnificence of creation. Following St Gregory in the Books of the Morals 'the wonderful works of the visible creation are the footsteps of our Creator, when we see these things, we admire the power of the Creator'.[123] His admirable and ineffable power shines most and is recognisably demonstrated to us in diverse effects of planets and their influences on terrestrial matters. But because their influences and effects vary in different ways according to their different motions, to find the future effects and influences of these planets on these terestrial matters it is necessary to calculate their diverse motions according to tables. The best of the tables available to us are those composed by Alfonso, the king of Castille, for the city of Toledo. The work with these tables is, however, so lengthy and tedious that scarcely anyone comes to know the planetary motions. Therefore, to shorten and to facilitate the work, I have composed concise tables for the years of the Incarnation of Christ, based on radices and motions of the Alfonsine tables and reduced to the meridian of the city Wrocław. This city, Wrocław, is one hour and 28 minutes of an hour to the east from Toledo. I thus took the mean motion of each planet and other motions given in the Alfonsine tables in three minutes and 40 seconds of a day, which is equivalent to the above mentioned time, namely one hour and 28 minutes. And that what I had found, I subtracted from the Toledan radices given at the beginning of the abovementioned Alfonsine tables, and thus I have radices of any planet and of the others[124] for the longitude of the abovementioned city of Wrocław. Using these radices I calculated the mean motions up to tertias[125] inclusive, so that whenever I find

122 *Psalm*. 92,6.

123 *Moralia in Job* 26, 17–18. The same quotation derived from St Gregory (Gregorius Magnus) was also used by Wenceslas Faber of Budweis (Václav Faber z Budějovic) in the introduction to his commentary on Sacrobosco's *Sphaera mundi*, see Alena Hadravová and Petr Hadrava, *Sféra Iohanna de Sacrobosco. Středověká učebnice základů astronomie* (Prague: Akropolis, 2019), p. 125. It is possible that Wenceslas Faber found inspiration for his foreword in one of the Prague copies of the canons *Mirabilis*. At the beginning of his foreword to Sacrobosco, Wenceslas Faber listed six advantages and benefits that the knowledge of the treatise will bring to the reader, beginning with the theological: 'Prima cognita hac arte in cognicionem Dei altissimi manuducimur, quia iste tractatus de celo et eius magnitudine determinat. Celum autem enarrat, ut habet psal||mista gloriam Dei et alibi 'confitebuntur celi mirabilia tua' [*Psalm* 88,6]. Et Gregorius in *Moralibus*: 'vestigia nostri Creatoris sunt mira opera visibilis creature, que dum cernimus, potenciam Creatoris ammiramur' [*Moralia in Job* 26, 17–18]. Cameracensis autem *Contemplacio* conditoris in corporibus celestibus veluti in quodam speculo materiali manibus Dei scripto representata est. Non ergo sine racione Ptolomeus in principio *Almagesti* describendo astronomiam dicit: est scientia dirigens et ducens nos in cognicionem Dei et cetera.' See also Wenceslas Faber of Budweis, *Commentary on Iohannes Sacrobosco's Spera mundi* (Leipzig: Martin Landsberg, 1495), ff. A3v–A4r. Gábor Almási, 'Rethinking Sixteenth-Century Lutheran Astronomy', *Intellectual History Review*, 24 (2014), 5–20, used Faber's argument here to show that a perception of astronomy as a way to God was widespread over Europe in the sixteenth century, not only among Lutherans. In the canons *Mirabilis*, we can see a similar example already in the fifteenth century.

124 The 'others' obviously refer to mean motions of the eighth sphere, head of the dragon, and possibly also the anomalies of the Moon and planets.

125 It is common to translate the Latin terms *tercia*, *quarta*, etc. for the orders of sexagesimal notation into English as 'third', 'fourth', etc. It would be, however, confusing in the contexts of the canons to tables, because it could be misunderstood as fractions (in addition to the also frequently used sequence numbers; the 'third' could then mean either 1/3, or $60^{-3}$, or the $3^{rd}$). We thus use here the Latin terms *tertia(s)*, *quarta(s)*, etc., in designating the sexagesimal orders.

30 or more quartas, I add a unit for them to the tertias. For a complete understanding of individual motions of planets, the teaching contained in the subsequent chapters is thus to be carefully and finely comprehended.

<1>

**First rule. How to compute for a given time the mean motions[126] of apogees and fixed stars, access and recess of the eighth sphere, Sun, Venus and Mercury, Moon, anomaly of the Moon, argument of the latitude of the Moon, head of dragon, Saturn, Jupiter and Mars and mean anomalies of these planets using tables designed for this purpose[127]**

I take one completed time in years, months, days, hours, minutes of hours, and other parts which has elapsed from the Incarnation of Christ until the time in which I want to find the motions and I save this time. When it is done, I enter first with the completed years of Christ or with the previous smaller argument into the Table of collected years of Christ[128] and I receive there directly, for instance, the mean motion of the Sun and I write it aside. Then I enter with the remaining completed years of Christ into the Table of expanded years of Christ[129] and I write aside the mean motion of the Sun found there below the previous, each value below its type.[130] Then with the completed months of Christ[131] I enter into the Table of months in common year if the year is common,[132] or into the Table of months in leap year[133] if the year is leap, and I write again the found mean motion of the Sun below the previous, each value below its type. Then I enter with the number of days into the Table of days[134] and similarly I enter with the hours into the Table of hours[135] and I enter repeatedly with the minutes and all fractions of hours into the same Table of hours and I write down the motion of the Sun, everything below its type. And if I do not find minutes or the other fractions in the Table of hours with which I have to enter, I enter first with thirty and then with remaining. For example: if I have 55 minutes, I enter first with 30 and then with 25 minutes and I add together these two entries. The type is not changed in this addition because it is the same entry, namely with minutes. And I put below the minutes of degrees what I find in this way. Similarly I put other fractions to their corresponding names; hence if I enter with seconds of hours I put them below seconds of degree. If with tertias of hours, I put them below tertias of degree and so on, as it is shown in the

126 'Motion' denotes the angular position at the given time.

127 For the explanation of the Rule 1 see Chapter 18.

128 See Chapter 12, Table TRW-01. *Tabula annorum Christi collectorum* means a table compiled from the years after Christ's birth, the years of the Christian calendar (as opposed to the tables calculated according to the Muslim calendar); it is a table of periodic radices.

129 See Chapter 12, Table TRW-02.

130 The type (*genus*) means the signs or degrees and their sexagesimal parts of the found ecliptical longitude.

131 Year or month of Christ means year or month of the Julian calendar, contrary to the Arabic calendar used in the Toledan Tables.

132 See Chapter 12, Table TRW-03. Common year (*annus communis*) means a non-leap year.

133 See Chapter 12, Table TRW-04.

134 See Chapter 12, Table TRW-05.

135 See Chapter 12, Table TRW-06.

inscription below this table. When it is done, I add everything together, starting from the smallest fraction. And for each 60 resulting in this addition I put one in the position of the immediately greater fraction, and in this way I proceed up to the degree. And for each resulting 30 degrees I put one sign, discarding 12 signs whenever they are exceeded. And what results from this addition, it will be the mean motion in the ninth sphere at the completed time at the meridian of the city of Wrocław. In this way I find the mean motion of the Sun, Venus and Mercury, because these three bodies have the same mean motion, and the mean motion of Saturn, Jupiter and Mars, and the mean motion of the head of dragon, but I get the true motion of the head of dragon by subtracting its mean motion from 12 signs. Similarly I find in this way the mean anomaly of the Moon, Venus and Mercury. But for the mean anomaly of the three upper planets, I shall have the mean anomaly of the three upper planets by subtracting the mean motion of each of them from the mean motion of the Sun.

<2>

**Second rule. How to find true apogees and mean centres of planets using tables designed for this purpose**[136]

I enter with the completed years of Christ into the Table of planetary apogees in collected years of Christ.[137] If I find precisely my years in the column of numbers, I subsequently find there directly the desired planetary apogees. If, however, I do not find precisely the number of my year, then I enter with the lower but closest number, and I write aside and save the subsequently found planetary apogees. Then I find directly near the end of the table the differences in minutes, seconds, and tertias, which are the motions of apogees in 20 years.[138] I take the proportional part of this difference and in this way I find the excess of the completed years of Christ over the collected years of Christ found in the table. I divide 20 by this excess and I get the quotient. I divide the difference resolved in tertias by this quotient and the value of the quotient will be the sought amount.[139] For example: I want to find the planetary apogees for the completed year of Christ 1430, which year I do not find in the Table of planetary apogees in collected years of Christ. I thus enter with the closest smaller number, namely with the completed year 1424 and I write aside the apogees found there. Then I find the excess of the years over the years found in the table to the given year and there is 6 of them. I divide 20 by them and I will have the quotient 3, but there will remain 2 years, which I will save aside. Then I will take the difference found directly at the end of the table and it will be 11 minutes, 40 seconds, and 8 tertias, and this will be the motions of apogees in 20 years.[140] I convert this difference into tertias

136 For an explanation of the Rule 2 see Chapter 19.

137 See Chapter 12, Table TRW-07.

138 The differences are given in the column D of the Table TRW-07 (but not in the later versions of the TR).

139 Petrus Cruciferus refers here to the method of linear interpolation by division, which is then demonstrated in detail in the subsequent example. We present its explanation in Chapter 27.

140 This value is the difference between the years 1424 and 1444 and it is given in the column D (Chapter 12, Table TRW-07).

and there are 42008 of them. I divide it by the value of the abovementioned quotient, namely by three, and in the quotient it will be 14002. And that would be the desired result if there were no remainder in the division of twenty years. But, because there remained two years, I shall proceed with them as follows: I shall take the difference of differences at the end of the table in the direction of the previously taken difference and it is 35 seconds, 0 tertias,[141] what is the motion in one year. I shall convert them to tertias and there will be 2100 of these tertias. I shall multiply it by remaining years, namely by two, and it will be 4200, which I shall similarly divide by three and there will be in quotient 1400, which I shall subtract from the previously found quotient and there will remain 12602 tertias. Dividing them by sixty I shall convert them to tertias, seconds, and minutes, and there will result 3 minutes, 30 seconds, and 2 tertias which is the sought amount. If I add it to the apogees written down and saved from the tables, I shall get apogees of all planets to the given year. In another and simpler way I shall do it in this way: I shall take the difference of differences in seconds and tertias, which is the motion of apogees in one month,[142] and I convert this difference of differences to tertias, and I multiply this so converted by the above mentioned excess and there will result the required value or one close to it.[143] For example: the difference of differences in the given is 35 second and 0 tertias, which I will convert to tertias and there will be 2100 of these tertias as I showed above. And I shall multiply it by the excess, namely by 6, and there will result 12600 of these tertias. I shall divide them by 60 and convert to tertias, seconds, and minutes, and there will result 3 minutes, 30 seconds and 0 tertias. I shall add them to the planetary apogees written down from the table, and so I shall have the planetary apogees equated to the given time, namely to the beginning of the completed year 1430. Concerning the completed months, I shall do it in this way: I shall take the difference of differences, I shall divide it by 12 and in the quotient will result how much I have to add for each month.

If, however, I want to know the apogees very precisely, I have to make it in this way: I shall take one time in years, months, days, and hours which has passed since the Incarnation of Christ and for that time I find first the motion of apogees and fixed stars in the way described above in the mean motions and I shall save it. Then I search for motion of access and recess of the eighth sphere in the same way as the previous. When I find it, I enter with this motion into the Table of equation of the access and recess of the eighth sphere[144] and I search for the number of signs in the top or bottom heading of the table. And note that the same sign number is not found in our times at the bottom, but for our years it is sufficient to find the same sign number up to three complete signs at the top of the table.[145] Having found the sign in the top heading of the table, I go down below it up to the degree of that sign placed in the first column, and I shall take there directly the

141 This value is given in the rightmost column E (Chapter 12, Table TRW-07). This column is inscribed *differencia differencie*, but instead of second differences it gives always one twentieth of the difference in the previous column D – see the commentary on this table in Chapter 19.

142 Correctly should be 'in one year'.

143 This direct procedure has a larger error because of rounding of the motion in one year.

144 See Chapter 12, Table TRW-08.

145 The mean motion of the access and recess was about 2 signs and 12.5 degrees in the year 1424 (see column B of the Table TRW-01) and it was between 0 and 3 signs inscribed at the top of the table in the years AD 15–1765 (compare also the column B of Table TRP-01 in Chapter 13).

degrees, minutes, and seconds, and I shall write them aside at two places. Then, if I have some minutes in addition to the degrees with which I entered, I enter with one degree added and I shall write down separately what I have found from the equation. Then I have a look whether the first equation is greater than the second one or smaller,[146] and I subtract the smaller from the greater, and what remains seems to be the difference between the two equations. I shall take a proportional part from this difference according to proportion of the minutes which were over the degree to 60. I shall find the proportional part this way: If the difference of the two equations were minutes, seconds and tertias, I shall convert them to the same denominations, namely to the tertias, and it will become a second number. After that I shall convert to tertias the minutes, seconds and tertias with which I had not entered, and this will become a third number. Then I shall multiply the second number by the third, and what results I shall divide by 60, and the value of the quotient will be the proportional part. I shall add this proportional part to the first taken equation if the second one is greater, or I shall subtract it from the first one if the second one is smaller. And what results after the addition or subtraction will be the equated equation of the motion of the eighth sphere. When it is done, I shall add it to the motion of the apogees and fixed stars which I saved before if the motion of the access and recess of the eighth sphere with which I entered into the table of its equation was less than three signs, or I shall subtract it if it would be greater. And what I find after the addition or subtraction it will be a common apogee, which I shall add to the radices at the Incarnation of Christ placed in the last line of each apogee in the Table of planetary apogees in collected years of Christ and so I shall have the most correct true apogee of any planet.

Then I will want the apogee of any planet: I shall subtract it from the mean motion of the same planet and it will show me its mean centre. In the same way I shall find the anomaly of the Sun and mean centre of Saturn, Jupiter, Mars, Venus and Mercury. But only the centre of the Moon I shall find by subtracting the mean motion of the Sun from the mean motion of the Moon and by doubling the remainder. Thus, having the mean motions and mean anomalies of planets using the previous rule, and using this rule for their mean centres, I can find their true positions using the following rules.

<3>

**Third rule. How to find mean conjunction and mean opposition of the Sun and the Moon and the possibility of eclipse of the Sun and the Moon in any month**[147]

I shall take all the years of the Lord including the year in which I want to calculate the conjunction or opposition and with this number of years or with the closest smaller one enter the Table of collected mean conjunctions if I search for conjunctions, or Table of mean oppositions if I search for oppositions.[148] Since both have the same procedure, let us speak about mean conjunctions. I shall thus take directly, for the year with which I

146 In the above mentioned quadrant for 0 to 3 signs, the second equation is always greater than the first.

147 For an explanation of the Rule 3 see Chapter 20.

148 These tables are written in the MS Toruń, BU, 74/III, f. 257v; see Chapter 12, Tables TRW-09 and TRW-10.

entered, the time of mean conjunction and mean motion of the Sun and Moon, the mean anomaly of the Moon, and the mean argument of latitude of the Moon. If I find precisely the number of years of Christ with which I work, then I have directly the time and the aforementioned motions of the first mean conjunction, namely in January. If, however, I have not found precisely that number of years, then with the remaining years enter into the Table of expanded years[149] and there I take the particular data as before. Then I add the time found in the Table of expanded years to the time found before it the Table of collected years and motion to the motion, putting each one below its type. But, if possible, I shall subtract from the number of days resulting from this addition the days of December, i.e. 31 days, and then I shall have the time of the first mean conjunction, namely in the days of January with its motions for that time. If, however, it will not be possible for me to do this subtraction, then to the time which I found this way I shall add the time of one mean lunation, i.e., what is written in the first line of the Table of the months,[150] and the motions which I shall find there, I shall also add to the previously found motions. And from the days resulting from such addition I shall subtract 31 days, and the remaining days together with hours and fractions of days will be the time of mean conjunction in January, together with the motions found for this time if I searched for mean conjunctions, or there will result the first mean opposition in January with its motions at this time if I searched for mean oppositions. And I shall preserve this mean conjunction or mean opposition as a radix for the whole year. If I would like to find the mean conjunction or mean opposition of some other month, then I shall write the days, hours and other parts of the first conjunction or the first opposition which I have preserved as radices of the year into a table. I shall then enter into the Table of mean conjunction and opposition of the Sun and Moon in months and I shall take the days and hours with other fractions. Similarly I shall get directly the aforementioned motions of the month in the days of which I search for the conjunction or opposition, and I shall add all this to the first conjunction or opposition, namely the time to the time and motions to the motions, starting from the smallest fractions. And if there results 60 or more from such addition, I shall put for them a unit at the position of the immediately greater fraction. And I shall keep this way similarly up to degrees and signs with which I shall work in the same way as I did it with the mean motions. Similarly I proceed in this way in the time up to hours, which I add together, and if the result is more than 24, then for 24 hours I shall put one day to the days. Afterward I shall add all lunar days, i.e., the days of the Moon in months to the first conjunction or opposition and if there will result from such adding more days than are in the month in whose days I search for the conjunction or opposition, then I shall subtract from the so collected lunar days, if possible, solar days, i.e., the days of the Sun set directly in that months I work with and I shall count the remaining days from the beginning of that month. And I count the hours from the noon of that day. If I cannot subtract the solar days directly in the month I work with from the lunar days, as happens when a month of a year has two conjunctions or oppositions, then I shall take the solar days of the previous month, which I shall subtract from the lunar days as before, and I count the days found in this way from the beginning

149 See Table TRW-11.
150 See Table TRW-12.

of the previous month. Similarly, the mean motions of the Sun and Moon, mean anomaly of the Moon, and argument of latitude of the Moon found in this way are also in the same time of the previous month. However, I must consider carefully the leap year, because if there would be a leap year and the leap day has passed,[151] then I add one to the solar days and after this addition I subtract the solar days from the lunar days as it has been said, and so I shall have the time of mean conjuction, if I worked on conjunctions, or time of mean opposition, if I worked on oppositions. I shall also have the mean motion of the Sun and Moon at the same time, if I worked on conjunctions, or mean motion of the Sun and nadir of the Moon, if I worked on oppositions. I shall also have the mean anomaly of the Moon at the same time. Similarly, I have also the mean argument of latitude of the Moon at the same time, and if the argument would be 0 signs and less than 12 degrees, or 5 signs and more than 18 degrees, or 6 signs and less than 12 degrees, or 11 signs and more than 18 degrees, I say that an eclipse would be possible. However, if I find the aforesaid argument beyond these limits, at natural eclipse cannot occur.

<4>

**Fourth rule. How to find the true position of the Sun**[152]

I shall take first its mean motion according to the first rule and then I shall take its mean anomaly according to the end of the second rule. When I have it, I enter with the anomaly of the Sun into the Table of its equation, namely with the sign in the headings or footer of the table. And it should be noted that five signs[153] are in the headings and others in the footer of the table.[154] And if the signs are found in the headings, then the degrees of the sign are placed in the first column in descending order. However, if the signs are found in the footer, then the degrees of the sign are placed in the last column in ascending order. I thus go over also with the degrees on the corresponding side either below the sign in the headings or above the sign in the footer and I take the equation of the Sun found there. And if I have together with the signs and degrees also some minutes or seconds, then I enter again in the same table with one degree added and I take the equation of the Sun directly at its place. And I shall find the difference between this equation and the equation taken previously and I multiply this difference converted to seconds by the remainder over signs and degrees of the anomaly with which I did not enter. I say the remainder also converted to seconds, and I shall consider the denomination of the product, because a multiplication of seconds by seconds provides quartas in the product, multiplication of seconds by tertias provides quintas in the product, and so on. I shall then divide this product as many times I can by 60, and in this way the proportional part is converted to the last order of fractions debted for the minutes and seconds of the needed anomaly. I shall add this proportional part to the previously found equation of the Sun if the second one was greater, or I shall

151 The *locus bisextilis* is the beginning of the leap day, i.e. February 29 in the present context of counting the days.
152 For an explanation of the Rule 4 see Chapter 21.
153 The first sign, Ram (*Aries*), was counted as zero, so that Virgin (*Virgo*) was only the fifth sign.
154 There are six signs (from 0 to 5) in the headings and six (from 6 to 11) in the footer; see Table TRW-13 in Chapter 12.

subtract it from that if the second one is smaller. In this way the examined equation of the Sun will result, which I shall subtract from the mean motion of the Sun if its anomaly is less than 6 signs, or I shall add it, if the anomaly is greater than 6 signs. And this gives me the true motion of the Sun in the ninth sphere.

<5>

**Fifth rule. How to find the true position of the Moon using tables designed for this purpose**[155]

I shall take first its mean motion using the first rule, then take its centre using the end of the second rule and finally take its mean anomaly using the tables made for it according to the second teaching of the first rule. When I find it, I enter with the mean centre into the Table of its equation[156] in a similar way as I entered the Table of equation of the Sun and I search for the equation of centre with double entry, if needed. Similarly I take there directly the proportional minutes also with double entry, if needed. If there are in the mean centre some minutes or seconds over the degrees with which I have not entered, keep both of them. I shall then consider whether the centre of the Moon is greater than 6 signs or smaller. If it is smaller, add the equation of centre to the mean anomaly of the Moon; if it is greater, subtract it from the same anomaly to get the true anomaly of the Moon. With this true anomaly, I go into the Table of equation of the Moon and take there directly the equation of anomaly and the variation of diameter, both with double entry if needed. Then I take the proportional part from the variation of diameter according to the ratio of the previously saved proportional minutes to 60 as was specified in the rule about the Sun, and add this proportional part to the previously found equation to get the examined equation. Subtract it from the mean motion of the Moon if its true anomaly is less than 6 signs, or add it if it is greater, and this give the true position of the Moon in the ninth sphere.

<6>

**Sixth rule. How to find the true position of Saturn, Jupiter, Mars, Venus, and Mercury, and their direct motion, station, and retrograde motion**[157]

I shall take first the mean motion of the body whose true position I want to know according to the first rule. Then I take its mean centre according to the end of the second rule. Afterwards I take its mean anomaly according to the end of the first rule. Having found it, enter first with the mean centre into the Table of equations of that planet whose true position I want to know and take directly with double entry the equation of the centre, above which I write 'to be decreased'.[158] And if the mean centre of the planet is smaller

155 For an explanation of the Rule 5 see Chapter 22.
156 It means the Table of equations of the Moon, see Table TRW-15.
157 For an explanation of the Rule 6 see Chapters 23 and 24.
158 See Tables TRW-16 – TRW-20 in Chapter 12.

than 6 signs subtract it from the mean centre and add it to the mean anomaly. Or write above it 'to be added' and if the mean centre is greater than 6 signs then add this equation of centre to the mean centre and reduce it from the mean anomaly and there will result both the centre as well as the equated or the true anomaly. With this true centre enter into the Table of equation of the planet and take the proportional minutes corresponding to it and write its title above, namely 'farther' or 'nearer'. Then with the true anomaly enter into the Table of equation of the planet and take there directly both the equation of anomaly, and the variation of diameter in farther distance (apogee) if 'farther' is written above the proportional minutes, or in nearer distance if 'nearer' is written above.[159]

And of course I work for both with double entry if needed, as has been said. Then I take the proportional part from this variation of diameter according to the ratio of the previously saved proportional minutes to 60 and I subtract this proportional part from the previously found equation of anomaly if the variation of the diameter is from the farther distance or I add it to the first equation of anomaly if the variation of diameter is from the nearer distance. This shows me the equation of anomaly corresponding to the given anomaly, above which I shall write 'to be added' if the true anomaly is less than 6 signs or I shall write 'to be decreased' if the true anomaly is greater than 6 signs. I thus consider both the equation of anomaly and the equation of centre; if above both is written 'to be added' then I add both together to the mean motion of the planet. If there is written 'to be decreased' above both, I subtract those together from the mean motion of the planet. If, however, above one is written 'to be added' and above the other 'to be decreased', then I subtract the smaller from the greater and I add or subtract the remainder according to what was written above it, and it provides the true position in the ninth sphere of the planet with which I was working.

Concerning the direct motion, station, and retrograde motion of these five planets I work this way: I enter with the true centre of any of them into its Table of equations and I take directly at its centre the first station of this planet in signs, degrees, and minutes, which I save apart, and I subtract this first station from the 12 signs and there will remain to me the second station, which I write below the first one. Then I consider the true anomaly of the same planet and if it is less than the first station or greater than the second station, the planet will be direct. And if it equals the first station in signs, degrees, and minutes, the planet will be stationary in the first station, namely it will be starting to retrograde. If it equals the second station, the planet will be stationary in the second station, namely it will start to move directly. But if this anomaly would be greater than the first station and less than the second station, the planet will be retrograde.

159 The terms 'farther distance' (*longitudo longior*) and 'nearer distance' (*longitudo propior* or *propinquior*) generally denote the apogee and perigee of the path of a celestial body. In the context of equations of planets they refer to the upper and lower part of the path where the interpolation of the equation is performed between the mean distance and the apogee or perigee. See Chapter 23. – To the terminology see also Edward Grant, 'On Saving the Phenomena and the Reality or Unreality of Epicycles and Eccentrics', in *A Source Book in Medieval Science*, ed. by Edward Grant (Cambridge: Harvard University Press, 1974), pp. 516–29, esp. p. 523, first translated it as 'farther longitude' / 'nearer longitude', but later modified it to 'farther distance' / 'nearer distance', which we also use. See also Edward Grant, *Planets, Stars and Orbs* (Cambridge: Harvard University Press, 1994), p. 293. Pedersen, *The Toledan Tables*, pp. 694–95, uses the term 'far distance' / 'near distance', while he translates the terms *longitudo maior* / *longitudo minor* as 'apogee' / 'perigee'. José Chabás and Bernard R. Goldstein, *A Survey of European Astronomical Tables in the Late Middle Ages* (Leiden, Boston: Brill, 2012), pp. 166–73 use 'greatest distance' / 'least distance'.

<7>

## Seventh rule. How to find the time of true conjunction of the Sun and the Moon and their opposition[160]

I have to equate the Sun and Moon at the time of mean conjunction or mean opposition as precisely as possible. The procedure of equating the Moon is the same as for equating the Sun because at the time of mean conjunction or opposition there is no centre of the Moon. Therefore if their true positions will be the same in signs, degrees, minutes, seconds, and tertias, I need not calculate any more, because the mean and true conjunction occur at the same time. But if the true position of the Sun has gone before the true position of the Moon in the sequence of signs, then I know that the true conjunction is still in future. However, if just opposite the true position of the Moon has gone before the true position of the Sun, I know that the true conjunction is the past. Therefore whether the true conjunction is in future or the past, I subtract the smaller motion from the greater and there will remain the distance between the luminaries which I convert to the last type of fraction. I save the result, and I consider which one of them is the distance. Because if the Sun has preceded the Moon in the order of the signs, it is the distance of the Sun. If the opposite, it is the distance of the Moon. Then with the anomaly of the Sun enter into the Table of true motion of the Sun and Moon in one hour[161] in which I directly take the true motion of the Sun in one hour with double entry, but I add to the anomaly of the Moon half the distance of luminaries together with half of the twelfth part of this distance. So I divide the distance by twelve and the quotient is the twelfth part of the distance which I add to the found distance, and half of this total I add to the anomaly of the Moon if the Sun precedes the Moon in the order of the signs, or I subtract this half from the anomaly of the Moon if the Moon precedes the Sun. With the so obtained anomaly enter into the previously mentioned table and I take there directly the true motion of the Moon in one hour with double entry if needed. Then I subtract the true motion of the Sun in one hour from the true motion of the Moon in one hour and I add to the remainder so many seconds minus one as there are degrees in longitude or in distance if the Moon is in the lower part of its epicycle, as I recognise from its anomaly. Or I subtract from the same so many seconds minus one if the Moon is in the upper part of its epicycle.[162] And there will result the equated advance of the Moon in one hour, which I convert into the same type of fractions to which I have converted the distance of the luminaries. I divide this distance by the result to get the hours. Then, if there is some remainder, I multiply it by 60 and I divide the product by the same as before to find the minutes of an hour. If there is still some remainder, I multiply it by 60 and I divide the product by the same as before to find

160 For an explanation of the Rule 7 see second half of Chapter 22.
161 See Chapter 12, Table TRW-14.
162 The witness **O** adds in margin (and the witnesses **R** and **L** in an abbreviated form directly in the text) the following explanation: 'If the mean anomaly of the Moon is greater than 95 degrees up to 265 degrees, then the Moon is in the lower half of its epicycle. If, however, the mean anomaly is less than 95 degrees or more than 265 degrees, then the Moon is in the upper part of its epicycle.' – The equation of anomaly of the Moon reaches its maximum at the mean anomaly equal to 95°, where the line of sight is tangential to the epicycle (see Chapter 12, Table TRW-15).

the seconds of an hour. If I like, I multiply the remainder by 60 and divide the product by the same as before to find the tertias of an hour, which is sufficient. Then I join the hours, minutes, seconds, and tertias found in this way to the time of mean conjunction of the Sun and Moon if the Sun precedes the Moon in the order of the signs, or I subtract it from that time if it is on the contrary, and what results is the requested conjunction of the Sun and Moon.

But the Sun and the Moon are drawn at the conjunction or opposition in this way: I add the actual distance with its twelfth part which I already have to the motion of the Moon at the time of conjunction or opposition; however the twelfth part alone I add to the motion of the Sun if it was a distance of the Sun. However, if it was a distance of the Moon, then I subtract the actual distance with its twelfth part from the motion of the Moon; however, the twelfth part alone I subtract from the motion of the Sun, and I have thus the true positions at the time of true conjunction or opposition, and the Sun and Moon will then come together, or the Moon will be at nadir of the Sun. Or if I equate the Sun and Moon to that time of true conjunction according to the Tables and rules for equations, I shall find them either already conjuncted, or a short distance apart. If they are somewhat distant, I see the distance with which I worked previously and I make everything as before. But I need not to do this, unless there is a necessity of very precise work such as, for example, eclipses or conjunctions if there is a great distance between the positions of luminaries equated at the time of mean conjunction, because it is then necessary to verify the work due to the difference between the motion of the Moon at true and mean conjunction.

If, however, I search for true opposition, my work will not differ from that for the true conjunction, except that when I calculate the true conjunction, I subtract the true position of the Sun from the true position of the Moon or conversely. But for opposition, I must subtract the true position of the Sun from nadir of the true Moon or conversely according to which one of them is greater than the other, and I shall work with the distance as I did it before in searching for the true conjunction. And this will show me the true conjunction or opposition of the Sun and Moon in non-equated days. To equate the days I enter into the Table of equation of days[163] with the true position of the Sun at the time of true conjunction or opposition and I take there directly the equation of days with a double entry, if needed, in degrees and minutes. And for each degree found there I take four minutes of an hour, multiplying the degrees by four and in result I shall have minutes of an hour. And for each minute found there I take four seconds of an hour, multiplying the minutes by four and in result I shall have seconds of an hour. And if there are more than 60, I put for them one minute to the minutes as many times as I can. I add these minutes and seconds to the previously found time of the true conjunction or opposition and it will show to me the time of the true conjunction or opposition in equated days computed from the previous noon. And with this time found now, it is necessary to equate the figure of the sky for making the predictions.

---

163 The full name of the table is Table of equation of days with nights (*Tabula equacionis dierum cum noctibus*). It is referred in this context also in Rule 23 of canons *Quoniam plerique* by Nicolaus Wodka of Kwidzyn (see Chapter 7.2). This table is nowadays called Table of equation of time, see Table A-05 in Chapter 16; see also Chapter 25 for explanation.

<8>

**Eighth rule. Sun's entry into Aries or any other sign I would like to find**[164]

I must first equate the Sun at a time close to its entry into Aries. When this is done, I find the distance of the true position of the Sun from the beginning of the Aries, which I know by subtracting the true position of the Sun from 12 signs, and the difference will be the distance which I shall convert into seconds or tertias. I then enter with the anomaly of the Sun at that time into the Table of true motion of the Sun and Moon in one hour[165] and take there directly the true position of the Sun in one hour, which I shall convert to the same type of fractions to which I have converted previously the distance. I shall divide this distance by the motion converted in this way and the number of the quotient will show me the hours. And I multiply the residuum, if there is any, by 60 and divide the product by the same as before and it will provide me with the minutes of hours. The residuum, if there is any, I multiply again by 60 and I divide the product by the same as before and it will show to me the seconds of an hour, which is sufficient. I then add the hours, minutes and seconds to the time to which I previously equated the Sun if the Sun had not entered Aries yet, or I subtract it from that if the Sun had already entered Aries, and it will provide me the time of the Sun's entry into Aries or a time close to it. I then have to equate the Sun to that now found time more precisely as I can, and if I shall find its true position to be 0 signs, 0 degrees, 0 minutes, and 0 seconds, it is sufficient that the previously found time is precise enough. However, if there remains some distance, I shall repeat all steps as before, and making the operations will provide me the time of entry of the Sun into Aries in non-equated days. I shall then equate the days as described in the previous rule, and it will show me the most reliable time of entry of the Sun into Aries computed from the previous noon. And for this time the planets should be equated and the equation of 12 houses according to Albumasar and Messahalla should be found. I can repeat this procedure to find the Sun's entry into another sign.

The end of the Tables with their canons on mean and true motions of planets.

## 2.4 Notes on Canons *Mirabilis*

The canons *Mirabilis* by Petrus Cruciferus to the *Tabule resolute* are divided into eight chapters – rules. Only the witnesses **A** and **C** contain all eight rules; the copies **L**, **R** and **S** lack the first part of the second rule and copies **O**, **P** and **S** are missing the whole third rule of the work. These coincidences yield clues about relations between these copies and the manuscripts in which they are found.

Statistical treatment of the critical apparatus of the edition and computer-aided stematology can also indicate relationships between individual copies. In particular, it can be based on the following table of binary correlations between the pairs of witnesses, where

164 The procedure described in Rule 8 is a simpler analogy of the method of successive approximations used in the previous Rule 7 to search iteratively for true syzygies.

165 See Chapter 12, Table TRW-14.

the off-diagonal terms give the number of variant readings, in which the two witnesses are identical, diagonal terms give the number of variant readings in which the witness appears, and the last column **Σ** gives the sum of the off-diagonal terms in the line. Higher values of the off-diagonal terms thus indicate closer similarity of the witnesses. The highest agreement is between manuscripts **A** and **C**: they agree in 346 variant readings.[166]

**Table of binary correlations**

| | A | C | L | O | P | R | S | Σ |
|---|---|---|---|---|---|---|---|---|
| **A** | **366** | 346 | 223 | 260 | 256 | 108 | 230 | **1423** |
| **C** | 346 | **367** | 221 | 255 | 251 | 103 | 224 | **1400** |
| **L** | 223 | 221 | **357** | 199 | 179 | 57 | 164 | **1043** |
| **O** | 260 | 255 | 199 | **307** | 239 | 84 | 216 | **1253** |
| **P** | 256 | 251 | 179 | 239 | **307** | 92 | 262 | **1279** |
| **R** | 108 | 103 | 57 | 84 | 92 | **356** | 82 | **526** |
| **S** | 230 | 224 | 164 | 216 | 262 | 82 | **289** | **1178** |

The largest value **Σ** at the witness **A** confirms that the Toruń manuscript is the one closest to the expected original version of the canons. The largest correlation (346) is between **A** and **C**, which shows that **C** is the next most reliable copy. The second highest agreement (262) is for the pair **PS**, suggesting that the witness **S** was copied directly from manuscript **P**. The optimal-tree algorithm joins **S** to **P**, while all other witnesses **C**, **O**, **P**, **L**, and **R** are joined to the witness **A**, with which they agree better than with each other. The method of neighbour joining, which implicitly assumes existence of unpreserved templates from which the known witnesses had evolved, suggests a close relationship of the pair **AC**, then **PS** to which the witness **O** is next related, and finally the pair **LR**, which is the most different from the other copies.

Unlike the other canons to the TR, the canons *Mirabilis* describe work with the simplest version of these tables, which we have preserved only in the manuscript Toruń, BU, 74/III. This, together with their relative brevity and the age of some of the copies, suggests that the canons *Mirabilis* represent the earliest version, which was gradually developed by subsequent canons.

166 The method of binary correlations is explained by Petr Hadrava and Alena Hadravová, 'Cristannus de Prachaticz's Treatises on the Astrolabe', in *Certissima signa. A Venice Conference on Greek and Latin Astronomical Texts*, ed. by Filippomaria Pontani (Venezia: Edizioni Ca'Foscari, 2017), pp. 295–312.

3

# Canons *Girum recensendo* by Andrzej Grzymała of Poznań

Around the time when Petrus Cruciferus produced the first version of the *Tabule resolute* and their canons *Mirabilis*, Wenceslas Koranda the Younger was born in Pilsen, Bohemia, who later owned the first copy of it, and Andrzej Grzymała (1425–66) was born in Poznań, Poland, who continued to work on the TR. Andrzej Grzymała wrote his version of the canons, known according to the incipit as *Girum recensendo*, in Cracow 1449, when he was less than twenty-five years old. The first words of the text of the canons (***G**irrum **r**ecensendo **z**odiaci **i**nter **m**agnalia **a**stronomie **l**onge **a**lto*) are an acronym, i.e. a stylistic adornment, where the first letters of the words give the name, in this case the name of the author of the text: *Grzimala*. The work has been preserved in at least sixteen copies, now stored mostly in the Jagiellonian Library in Cracow. We have compared the text of the canons *Girum* in three manuscripts (Cracow, BJ, 573, ff. 227r–233r; Cracow, BJ, 2252, pp. 21–44; Prague, NK, VIII G 24, ff. 60v–64v) and we have found that the two Cracow manuscripts differ from the Prague one in many places and could not be placed in a single edition. Moreover, it is obvious that the scribe of the Prague copy of Grzymała's work tried to adapt the text to the Prague meridian, as can be seen, for example, in the Rule 4 titled *Radices ad quemcumque meridianum placuerit reducere*, where the Polish copies give the *meridianus Cracoviensis* and the Prague copy the *meridianus Prage* (or *Pragensis*).[1]

Andrzej Grzymała stands at the beginning of a roughly half-century period in which new versions of the canons to the TR were created. It then ends in the first half of the sixteenth century with the printing of the TR with canons. Andrzej Grzymała's studies and subsequent work are connected with the University of Cracow, founded 1364. The oldest building of the University of Cracow is the well-preserved *Collegium maius* (see Fig. 17), built 1400.

A teacher and colleague of Andrzej Grzymała was Marcin Król of Żurawica, founder of the chair in astronomy and astrology in Cracow; after receiving his master's degree at Cracow in 1445, he continued his studies in Prague, Leipzig, and Bologna. He died in 1459.[2]

Andrzej Grzymała became a master at the Cracow university in 1447. Then he lectured on rhetoric and poetry there. In 1453/54 and 1458 he was the dean of the Faculty of Arts. The years 1460–62 he spent in Italy, first in Perugia where in 1461 he obtained the doctorate of medicine. After his return to Cracow, he was again active at the University of Cracow, serving as dean at the Faculty of Medicine and then also as rector of the university. He died in 1466 during an epidemic in Cracow. The only extant work of his writings, mostly

1 Cracow, BJ, 573, f. 229r; Cracow, BJ, 2252, p. 40; Prague, NK, VIII G 24, f. 61v.

2 Paul W. Knoll, 'The Arts Faculty at the University of Cracow at the End of the Fifteenth Century', in *The Copernican Achievement*, ed. by Robert S. Westman (Berkeley: University of California Press, 1975), pp. 137–56 (146–47).

**Figure 17.** *Collegium maius* (courtyard) – one of the historical and architectural dominants of Cracow. Laima Gūtmane, Creative Commons.

devoted to poetry, rhetoric, and astronomy, are the canons to the *Tabule resolute* with incipit *Girum recensendo*. A few manuscripts from his library have also survived.[3]

## 3.1 Manuscripts

Berlin, SB-PK, 4° 175[4]

The scribe of this manuscript was Johannes Zyzno de Schydłów and he wrote it in Cracow 1474. – The canons *Girum* are contained on ff. 1r–9v.

Cracow, BJ, 570, pp. 29–30[5]

Individual works in this codex are dated to 1467, 1486, the end of the fifteenth and the beginning of the sixteenth century. It originated in Cracow and notes by Albertus of Brudzewo and Martinus Biem de Ołkusz were identified in it.

3 Aleksander Birkenmajer, 'Andrzej Grzymała z Poznania, astronom i lekarz z XV wieku', *Kwartalnik Historii Nauki i Techniki*, 3 (1958), 409–21; French trans.: Aleksander Birkenmajer, 'Andrzej Grzymała de Poznań, astronome et medecin du XVe siecle', in *Aleksander Birkenmajer, Études d'histoire des sciences et de la philosophie du Moyen Age*, Studia Copernicana IV (Wrocław: Ossolineum, 1972), pp. 515–26; see also <https://cac.historia.uj.edu.pl/osoba/1995846_Andrzej_Grzyma%c5%82a_z_Poznania> [accessed: 12.03.2023].

4 Renate Schipke, *Die lateinischen Handschriften in Quarto der Staatsbibliothek zu Berlin – Preußischer Kulturbesitz, Teil 1: Ms. lat. quart. 146–406* (Wiesbaden: Harrassowitz, 2007), pp. 157–59.

5 Rosińska, *Scientific Writings*, pp. 160–61 (item 769) and 57 (item 237); *Catalogus BJ*, IV, pp. 31–36.

- The manuscript contains only a fragment from the end of the canons *Girum* on the pp. 29–30, namely only the rule *Canon de aspectibus planetarum*, which corresponds, for example, to the copy in MS Cracow, BJ, 2252, but it is not present in the copy of Cracow, BJ, 573. The incipit of the rule on aspects is as follows: *I<esus> N<asarenus>, r<ex> I<udeorum>. – Aspectus planetarum in ordine ad Lunam*, and explicit: *quia planeta ante opposicionem respicit planetam ante se, post vero opposicionem post se etc.*

  Other selected works contained in this manuscript:
- John of Saxony's canons to the PAT with incipit *Tempus est mensura motus* can be found on pp. 3–19.[6]
- Tables for the planets by Giovanni Bianchini, recalculated for Cracow meridian, are on the pp. 20–23. Their incipit is *Tabule Iohannis Blanckini de gradu ascendente et equacione aliarum cardinum celi reducte et verificate ad latitudinem 50 graduum et ad meridianum inclite civitatis Cracoviensis*. The copy is dated to 1467. These tables are a simplified version of Bianchini's tables of domification for the seventh climate.[7] The equations of astrological houses are given here with a precision to integer degrees only, while the original Bianchini's tables give the values in degrees and minutes. Unlike the original, the cusps of the astrological houses here are referred to as *cardines* (that is cardinal points, which is appropriate only for the first, fourth, seventh, and tenth houses).
- Parisian Alfonsine Tables, with glossed radices for the meridians of Cracow, Paris, Prague, Prussia (*Prussicum*), and Vienna, are found on the pp. 31–74. Incipit of the work: *Incipiunt Tabule illustres (!) regis Alphoncii. Et primo Tabule differenciarum unius regni ad aliud.*

### Cracow, BJ, 573, ff. 227r–233r[8]

The scribe of the most works contained in this codex was Petrus de Chotków around 1457, when he obtained the master degree at the Cracow University. From Cracow he went to Bologna, where he received his doctorate in medicine.

- Canons *Girum* are on ff. 227r–233r written carefully but without the final rules. This version of canons *Girum* is close to Polonus's canons *Quoniam Tabularum Alfonsi* which accompany his TR calculated for Salamanca (see Chapter 6). TR are not included in the manuscript. From the other works in the codex we select:
- *Theorica planetarum* by Pseudo-Gerardus de Cremona introduces all treatises in the codex on ff. 1v–27r. The text is accompanied by colour astronomical diagrams and illustrations.
- It is followed by John of Lignères' *Algorismus de minutiis* on ff. 28r–43r.
- Sacrobosco's Sphere, the popular textbook of spherical astronomy, can be found on ff. 44v–77r.
- After several astrological texts, the volume includes on ff. 191r–208v extracts from Michael Scot's treatise on the constellations *De signis et imaginibus celi*, composed around 1220

6 ThK 1560; *Catalogus BJ*, IV, pp. 31–36; Rosińska, *Scientific Writings*, p. 427 (item 2231).
7 See Chabás and Goldstein, *The Alfonsine Tables of Toledo*, pp. 112–14, Table 86.
8 Rosińska, *Scientific Writings*, pp. 160–61 (item 769); Markowski, *Astronomica*, p. 25 (item 4). *Catalogus BJ*, IV, pp. 36–47.

at the court of the Emperor Frederick II of Hohenstaufen in southern Italy. This work is based, among other sources, on Hyginus's Astronomy, an unfinished work consisting of four books dealing with the fourty-eight Ptolemaic constellations and written at the beginning of the first century AD. Parts of book 3 of Hyginus's Astronomy are recorded in the manuscript at ff. 216r–221r. Both Scot's and Hyginus's treatises are accompanied by drawings of the constellations with the stars marked by a rubric.[9]

### Cracow, BJ, 1841, pp. 311–314[10]

The codex dates to the early sixteenth century. Like MS Cracow, BJ, 570, this manuscript contains only a rule on aspects (*Canon de aspectibus planetarum*) from the end of canons *Girum*.

- On the pp. 93–162 there are written canons by John of Głogów with incipit *Pro (itroduccione) generali* (see Chapter 8.1).[11]

### Cracow, BJ, 1852, pp. 361–379[12]

Manuscript originated *c.* 1464 and it was copied by Stanislas C(h)osus (Cosuch). The canons *Girum* are incomplete, lacking the introduction. Incipit and explicit: *Medium motum Solis et Lune et aliorum planetarum et eciam argumenta media Lune, argumentum latitudinis Lune…* × … *si sunt ambe occidentales sicut aput nos et Prage, et cuiuslibet civitatis longitudo fuerit maior, illa est orientalior, et cuiuscumque longitudo fuerit minor, est occidentalior. Et civitas Cracovia est orientalior quam Praga.*

- On the pp. 7–103 one can find the TR both for Prague and Cracow. Incipit on p. 7: *Tabule resolute Alphoncii regis. Incipiunt Tabule de mediis motibus super meridianum Pragensem verificate.* On p. 11 we can read: *Tabula de mediis motibus super meridianum Cracoviensem*; on p. 48: *Tabula medie opposicionis Solis et Lune cum suis motibus ad meridianum Cracoviensem in annis collectis*, for years 1429–1809.

### Cracow, BJ, 1864, pp. 47–70[13]

The manuscript is dated *c.* 1446–80. In the first 31 pages is copied *Theorica planetarum* by Pseudo-Gerardus of Cremona. The scribe signed himself as *Nicolaus de Ternaua, studens alme universitatis Cracoviensis*, in the year 1446.

9 The iconography of the constellations has a long and complicated tradition, but the style of the Cracow drawings resembles those of Bohemian provenance (e.g., illustrations of Michael Scot's *De signis* in the MS Prague, NK, III C 2, ff. 155ra–168vb). See Hadravová, *Sphaera octava – Mýty a věda o hvězdách III*, pp. 33–36 and passim. See also edition of Michael Scotus's *De signis* by Silke Ackermann, *Sternstunden am Kaiserhof. Michael Scotus und sein Buch von den Bildern und Ziechen des Himmels* (Frankfurt am Main: Peter Lang, 2009) and the edition of the treatise based on Michael Scot in the MS Prague, NK, XXVI A 3 from 1405 in Hadravová, *Sphaera octava – Mýty a věda o hvězdách III*, pp. 76–113.

10 Władysław Wisłocki, *Catalogus codicum manuscriptorum Bibliothecae Universitatis Jagellonicae Cracoviensis* (Kraków: Universitas Jagellonica, 1877–1881), p. 436; Markowski, *Astronomica*, p. 27 (item 15); p. 78 (item 122).

11 Wisłocki, *Catalogus*, p. 436; Markowski, *Astronomica*, p. 78 (item 122).

12 Wisłocki, *Catalogus*, 440; Rosińska, *Scientific Writings*, pp. 160–61 (item 769) and 238–39 (item 1196); Markowski, *Astronomica*, p. 275 (item 4).

13 Wisłocki, *Catalogus*, 445; Rosińska, *Scientific Writings*, pp. 160–61 (item 769); Markowski, *Astronomica*, p. 25 (item 3).

- The canons *Girum* on pp. 47–70 follow the canons *Astrorum* on pp. 32–38, 46 and 220. There are also the TR for Cracow meridian in this manuscript for the years 1428–1808, on pp. 71–141.[14] The explicit of the canons *Girum* is as follows: *inclite civitatis Cracoviensis… Finis canonum Tabularum resolutarum per magistrum Grzimala collectorum.*

### Cracow, BJ, 1865, ff. 58r–66v[15]

The codex was written *c.* 1460. Canons *Girum* are concluded by these words: *inclite civitatis Cracoviensis… habito ascendente statim et alie habebuntur.* The TR for Cracow follow on ff. 72r–111r.[16] There is also contained *Tabula de mediis et veris motibus planetarum super meridianum Pragensem* on f. 78v, written by the Cracovian bachelor Nicolaus de Nissa.

- On ff. 148r–158r there is a copy of the Composition of the astrolabe composed by Prague University master Cristannus of Prachatice in 1407–8. Its incipit: *Quamvis de astrolabii composicione tam modernorum, quam veterum dicta habentur pulcherrima.*[17]
- On ff. 160r–168r there is a copy of Cristannus's Use of the astrolabe with incipit *Quia plurimi of nimiam quandoque accurtacionem.*

### Cracow, BJ, 1916, pp. 47–48[18]

Like the above-mentioned MSS Cracow, BJ, 570 and Cracow, BJ, 1841, this manuscript contains only a rule on aspects (*Canon de aspectibus planetarum*) from the last quarter of the text of the canons *Girum.*

- However, the rule extracted from the *Girum* canons is preceded by the canons *Pro (introduccione) generali* by John of Głogów on pp. 1–45.
- The TR for Cracow meridian 1428–1808 can be found on pp. 57–128: *Incipiunt Tabule super meridianum Cracoviensem de mediis et veris motibus planetarum et primo in annis collectis.*

### Cracow, BJ, 1918, ff. 102r–109r[19]

The codex originated in Cracow in 1447–51 and, in addition to the canons *Girum*, the first approximately forty folios contain the three books of Euclid's Elements accompanied by commentaries, followed by Pseudo-Gerard Cremonensis's also commented *Teorica planetarum.*

### Cracow, BJ, 1920, pp. 183–201[20]

In addition to the canons *Girum* on pp. 183–201, this late fifteenth-century codex from Cracow contains the canons *Astrorum (observatores) sapientes* by Adalbertus (Wojciech) of

14 Markowski, *Astronomica*, p. 276 (item 12). He gives pp. 71–213, because he also counts additional tables to the TR.

15 Wisłocki, *Catalogus*, pp. 445–46; Rosińska, *Scientific Writings*, pp. 160–61 (item 769); Markowski, *Astronomica*, p. 25 (item 5).

16 Markowski, *Astronomica*, p. 276 (item 13).

17 Rosińska, *Instrumenty*, p. 105; Rosińska, *Scientific Writings*, pp. 336–37 (item 1715); Hadravová and Hadrava, *Křišťan z Prachatic*, p. 92.

18 Wisłocki, *Catalogus*, p. 458; Markowski, *Astronomica*, p. 26 (item 12).

19 Wisłocki, *Catalogus*, pp. 458–59; Rosińska, *Scientific Writings*, pp. 160–61 (item 769); Markowski, *Astronomica*, p. 25 (item 2).

20 Wisłocki, *Catalogus*, pp. 459–60; Rosińska, *Scientific Writings*, pp. 160–61 (item 769); Markowski, *Astronomica*, p. 27 (item 13).

Brudzewo on pp. 5–20, canons *Pro (introduccione) generali* by John of Głogów on pp. 207–235 (with a prologue on pp. 3a–3b), as well as the TR for Cracow meridian on pp. 25–98.

### Cracow, BJ, 2252, pp. 21–44[21]

Codex contained miscellaneous astronomical and astrological texts and was written partly in the fifteenth, partly in the sixteenth century. The version of the canons *Girum* on pp. 21–44 has incipit *Secuntur canones super Tabulas resolutas de veris et mediis motibus planetarum. Girum recensendo.* This version differs in a number of places from the version represented, e.g., by the manuscript Cracow, BJ, 573.

- On the pp. 47–120 can be found the TR for Cracow meridian.

### Prague, NK, VIII G 24, ff. 60v–64v[22]

On ff. 60v–64v we find canons *Girum recensendo*. This version is somewhat different from the others; it is also written with Prague in mind, not Cracow.

- The manuscript starts with the TR on ff. 2r–37r: *Tabule de mediis et veris motibus planetarum per Petrum Cruciferum compilate et ad meridianum Pragensem redacte* ('Tables of mean and true motions of the planets compiled and adapted to the meridian of Prague by Petrus Cruciferus'). Some of the information in the tables themselves makes it clear that they were calculated for the Prague meridian. In the very first table in the set, *Thabula augium et stellarum fixarum* (Table of apogees and fixed stars), there is a note in the margin of f. 2r which reads: *Iste sunt thabule facte et redacte ad meridianum Pragensem* ('Here are tables created and adapted to the meridian of Prague'). The explicit of this version of the tables also gives the name of the author: *Expliciunt thabule de mediis et veris motibus planetarum per P(etrum) Cruc(iferum) compilate* ('This is the end of the tables of mean and true motions of the planets by Petrus Cruciferus').
- On ff. 38r–60r there is *Tabula minutorum proporcionabilium pro diebus intermediis in almanach faciendis.*
- On ff. 70r–86r there is Johannes de Sacrobosco's *Computus ecclesiasticus* and on ff. 92r–123r there follows his *Computus philosophicus*, which contains a Czech phrase *chwala Bohu* meaning *laus Domini* in the explicit (*Compotus est sciencia* … × … *quod nos hinc fructificemus. Chwala Bohu. Explicit philosophicus Magistri Iohannis de Sacrobusto*).

### Vienna, ÖNB, Cod. 5227, ff. 26r–36v[23]

The manuscript originated in Vienna between 1482–1514. Among its former owners was the prominent Viennese humanist, astronomer and physician Georg Collimitius Tannstetter

21 Wisłocki, *Catalogus*, p. 540; Rosińska, *Scientific Writings*, pp. 160–61 (item 769); Markowski, *Astronomica*, p. 26 (item 8 – wrong pages).

22 Rosińska, *Scientific Writings*, pp. 160–61 (item 769); Markowski, *Astronomica*, p. 26 (item 7); Truhlář, *Catalogus codicum*, pp. 595–96 (item 1606).

23 *Tabulae codicum* IV, 1864–1893, p. 65; Franz Unterkircher, *Die datierten Handschriften der Österreichischen Nationalbibliothek von 1451 bis 1500. 1. Teil: Text. 2. Teil: Tafeln* (Wien, Verlag d. Österreichischen Akademie der Wissenschaften, 1974), p. 157; Rosińska, *Scientific Writings*, pp. 160–61 (item 769); Markowski, *Astronomica*, p. 320. See also <www.manuscripta.at>.

(1482–1535), see f. 116v: *Remuneraverunt me hiis tabulis devotissimi patres conventus ordinis predicatorum in Vienna, Anno etc. 1510. Liber Georgii Tannstetter Collimitii.*

- The first work to be encountered in this codex are canons *Astrorum* on ff. 1r–22r. The canons *Girum* are contained on ff. 26r–36v with explicit: *super meridianum inclite civitatis Cracoviensis… et tantum de canonibus Tabularum resolutarum.*
- The canons are followed by the TR for Prague and Cracow on ff. 37r–77v (*Tabule de mediis et veris motibus planetarum super meridianum Cracoviensem*).
- Among other works of Alfonsine astronomy in this manuscript we note the Oxford tables of 1348 by William Batecombe on ff. 117r–58r and 166r–69r, with their canons on ff. 158v–62v, as well as to the Table for true syzygies by Nicolaus de Heybech on ff. 163r–64r, accompanied by its canon on ff. 164v–65r. A copy of Regiomontanus's ephemerides for 1485–86 is on ff. 172r–83v.

### Vienna, ÖNB, Cod. 5240, ff. 67r–76r[24]

The codex dates to 1440 and originated in Zittau. Its scribe was Nicolaus Carnificis de Swalbenstat in Zittau, who mentioned his name in the explicit, as noted above (see the Chapter 2.1.1).

- TR calculated for Prague meridian from 1428 to 1888 are contained on ff. 1r–36r. The additional tables continue on next folios: the Tables of proportional minutes (*Incipit Tabula minutorum proporcionabilium pro diebus intermediis in almanach faciendis*) on ff. 37v–48r; *Tabula retrogradacionis, direccionis et stacionis planetarum et intratur cum argumento vero eorum* on f. 48v; *Tabula ad inveniendum motum Solis et Lune in una hora* on ff. 50v–51r; *Tabula equacionis dierum cum noctibus suis* on ff. 51v–52r; *Incipit Tabula ad inveniendum partem proporcionalem* on ff. 52v–59r; *Tabula ascensionum signorum in circulo recto* on ff. 59v–60r; *Tabula ascensionum signorum ultra medium septimi climatis, cuius latitudo est 50 graduum et eius dies longior 16 horarum, 20 minutorum* on ff. 60v–62r; *Tabula ascensionum signorum ultra finem septimi climatis, cuius latitudo est 51 graduum et eius dies longior 16 horarum, 25 minutorum* on ff. 62v–64r. Several astrological notes continue on ff. 64v–66v.
- Canons *Girum* follow on ff. 67r–76r (incipit and explicit: *Girum recensendo zodiaci* … × … *et tantum de canonibus tabularum resolutarum. Laus deo*).

### Vienna, ÖNB, Cod. 5245, ff. 27r–34v[25]

The scribe of some parts of this Cracovian manuscript from the middle of the fifteenth century was Michael Derslai de Krzanowice (Carnicze) in Silesia. His marks are on ff. 61v (*explicit per manus Derslai de Carnicze, in vigilia vigilie sancti Stanislai, 1450 anno*), 62r (*Derslai Michael de Carnicze*), and 129r (*Derslai Michael de Carnicze 1450*). Previous

---

24 *Tabulae codicum* IV, 1864–1893, p. 71; Rosińska, *Scientific Writings*, pp. 160–61 (item 769); Markowski, *Astronomica*, p. 27 (item 14). See also <www.manuscripta.at>.

25 *Tabulae codicum* IV, 1863–1893, p. 72; Franz Unterkircher, *Die datierten Handschriften der Österreichischen Nationalbibliothek von 1501 bis 1600. 1. Teil: Text. 2. Teil: Tafeln* (Wien: Verlag d. Österreichischen Akademie der Wissenschaften, 1976), p. 206; Rosińska, *Scientific Writings*, pp. 160–61 (item 769); Markowski, *Astronomica*, p. 25 (item 1). See Michela Malpangotto, *Theoricae novae planetarum Georgii Peurbachii dans l'histoire de l'astronomie* (Paris: CNRS Éditions, 2020), p. 346. See also <www.manuscripta.at>.

owners of the manuscript include the humanist Johannes Cuspinianus (1473–1529, active at Vienna University) and the humanist Johannes Fabri (Johann Heigerlein, 1478–1541, Viennese bishop from 1530).

- Folios 27r–34v support a copy of the canons *Girum* with explicit: ... *super meridianum inclite civitatis Cracoviensis. ... Et sic terminantur canones super Tabulas reductas ... per magistrum Andream Grzimala de Poznania in collegio artistarum alme universitatis studii Cracoviensis. ... Anno Domini millesimo quadringentesimo quinquagesimo, pro cuius consumacione laudetur Deus in personis et unus in essencia. Sic prius ostensum est. Sic est finis canonum astronomie.*
- On ff. 62r–98r are the TR for Cracow and partly for Prague meridian.

Other chosen works include:

- *Theorice nove planetarum* by Georg Peurbach on ff. 1r–26v.
- John of Saxony's canons *Tempus est mensura motus* continue on ff. 39r–51v (with explicit: *quem motum quere in tabula appropriata isti planete, cuius radicem reducis ad Cracoviam vel ad aliam civitatem*).
- Use of the astrolabe by Cristannus of Prachatice on ff. 52r–61v.[26] Cristannus's original text, containing a reference to Prague, was adapted by the scribe to its location and Prague was replaced by Cracow on f. 60r: *quas horas cum minutis computa ab occasu Solis secundum cursum horalogiorum in* Cracovia *currencium et habebis horas et minuta tempore introitus Solis in primum minutum Arietis.*
- The last work are the PAT (*Tabule astronomice regis Alphonsi*) found on ff. 129r–52v.

### Vienna, ÖNB, Cod. 5275, ff. 36r–42r[27]

This manuscript is believed to have been written in southeastern Germany. Once owned by the astronomer Johannes Schöner (1477–1547), the codex then entered the library of the Fugger family (Fugger-Bibliothek) in Augsburg.

- There are the canons *Girum* on ff. 36r–42r (incipit *<G>yrum recensendo*), followed by the geographical table on f. 42v.
- The TR are written on ff. 88r–123r.[28] They are calculated for Cracow meridian, but the Tables for mean conjunctions and mean oppositions of the Sun and Moon in the collected years are prepared for Prague meridian (ff. 151r–152v). Additional tables include: *Incipiunt Tabule minutorum proporcionalium pro diebus intermediis in almanach faciendis* on ff. 123v–135r, *Tabula ad inveniendum motum Solis et Lune in una hora* on ff. 135v–136r, *Tabula equacionum dierum cum noctibus* on ff. 136v–137r, *Tabula ad inveniendum partem proporcionalem* on ff. 137v–144r, *Tabula ascensionum signorum ultra medium septimi climatis, cuius latitudo est 51 gradus et eius dies longior 16 horarum, 25 minutorum* on ff. 144v–146r, *Tabula ascensionum signorum in circulo recto* on ff. 146v–147r, and *Tabula ascensionum ultra medium septimi climatis, cuius latitudo est 50 graduum et eius dies longior 16 horarum et 20 minutorum* on

26 Rosińska, *Scientific Writings*, pp. 336–37 (item 1715) incorrectly attributed to Prosdocimo de' Beldomandi; Hadravová and Hadrava, *Křišťan z Prachatic*, p. 96.

27 *Tabulae codicum* IV, 1863–1893, p. 82; Unterkircher, *Die datierten Handschriften der Österreichischen Nationalbibliothek von 1451 bis 1500*, pp. 210–11; Rosińska, *Scientific Writings*, pp. 160–61 (item 769); Markowski, *Astronomica*, p. 26 (item 10); Chabás, *Computational Astronomy in the Middle Ages*, p. 318.

28 Markowski, *Astronomica*, p. 278 (item 18).

ff. 147v–149r. Table on planetary anomalies and centres then continue on ff. 149v–150r, followed by the *Tabula retrogradacionis, direccionis et stacionis planetarum* on f. 150v, *Tabula medie coniunccionis Solis et Lune in annis Christi collectis cum motibus eorundem… quoad meridianum Cracoviensem* on f. 151r–v, *Tabula medie opposicionis* on f. 152r–v etc.

– Among other works contained in this codex we mention the Almanac for 1481, copied from Regiomontanus's printed ephemerides, on ff. 1r–14v and Sacrobosco's Sphere on ff. 43v–58v.

## 3.2 List of Rules

**Andrzej Grzymała, *Girum recensendo* – variant 1:**
(Cracow, BJ, 573 = **G1**)

1. Finding the mean motion of the Sun and Moon and other planets etc. (*Medium motum Solis et Lune ac aliorum planetarum et eciam argumenta media Lune, argumentum latitudinis Lune, similiter argumenta Veneris et Mercurii, motum augium et stellarum fixarum motumque accessus et recessus octave spere* /in mg. add.: *et motum capitis draconis*/ *per thabulas invenire*, **G1**, f. 227r)
2. Finding the time of the mean conjunction and opposition of the Sun and Moon and possibility of eclipsis of the Sun and Moon (*<T>empus medie coniunccionis vel opposicionis Solis et Lune possibilitatemque eclipsis Solis et Lune in mense quolibet per tabulas ad hoc factas reperire*, **G1**, f. 227v)
3. How to correct, how to create, or how to continue the Tables of the motions and time (*<M>odus corrigendi Tabulas motuum et temporum ipsasque de novo faciendi aut continuandi*, **G1**, f. 228v)
4. Converting radices to any meridian (*Radices ad quemcumque meridianum placuerit reducere*, **G1**, f. 229r)
5. Finding the true position of apogee of any planet using tables (*Verum locum augium cuiuslibet planete per thabulas invenire*, **G1**, f. 229r)
6. General way of entering Tables of equations (*Modum generalem in Tabulas equacionum <intrandi> proponere*, **G1**, f. 229v)
7. Finding the true position of the Sun using tables (*Verum locum Solis per tabulas invenire*, **G1**, f. 230r)
8. Finding the true position of the Moon using tables (*Verum locum Lune per tabulas invenire*, **G1**, f. 230r)
9. Finding the true position of the head of dragon (*Verum motum capitis draconis invenire*, **G1**, f. 230r)
10. Finding the true position of the three upper planets, i.e. Saturn, Jupiter, and Mars (*Verum locum trium superiorum, Saturni videlicet, Iovis et Martis, invenire*, **G1**, f. 230v)
11. Finding the true position of Venus and Mercury (*Verum locum Veneris et Mercurii invenire*, **G1**, f. 231r)
12. Finding whether a planet is stationary, retrograde, or direct (*Utrum planeta sit stacionarius, retrogradus aut directus, invenire*, **G1**, f. 231r)
13. Knowing whether a planet is ascending or descending (*Utrum planeta sit ascendens vel descendens cognoscere*, **G1**, f. 231v)

14. Finding the time of the true conjunction and opposition of the Sun and Moon (*Tempus vere coniunccionis et opposicionis Solis cum Luna invenire,* **G1**, f. 231v)
15. Finding the time of the true conjunction of Saturn and Jupiter and any other planet (*<T>empus vere coniunccionis Saturni et Iovis et quorumlibet aliorum planetarum invenire,* **G1**, f. 232r)
16. Finding the entry of the Sun into Aries or into any sign (*<I>ntroitum Solis in Arietem vel in quodcumque signum volueris invenire,* **G1**, f. 232r)
17. Creating an almanac for individual years (*<A>lmanach pro singulis annis invenire,* **G1**, f. 233r)

**Andrzej Grzymała, *Girum recensendo* – variant 2:**
(Cracow, BJ, 2252 = **G2**)

1. On the mean motion of the Sun and Moon and other planets (*Canon primus de medio motu Solis et Lune et aliorum planetarum,* **G2**, p. 21)
2. On finding the first conjunction and opposition (*Canon secundus ad inveniendum primam coniunccionem vel opposicionem,* **G2**, p. 23)
3. On finding of possible eclipse (*Canon de possibilitate eclipsis invenienda,* **G2**, p. 24)
4. On the conversion of radices to any meridian (*Canon de reduccionum radicum ad quemcumque meridianum,* **G2**, p. 24)
5. On the true positions of apogees as they move one degree in hundred years (*Canon de vero loco augium, ut moventur in centum annis uno gradu,* **G2**, p. 24)
6. On finding the true position of the Sun (*Canon ad inveniendum verum locum Solis,* **G2**, p. 25)
7. On finding the true position of the Moon (*Canon de vero loco Lune inveniendo,* **G2**, p. 26)
8. On the true position of the head and tail of dragon (*Canon de vero loco capitis et caude draconis,* **G2**, p. 27)
9. On the true position of the three upper planets, i.e. of Saturn, Jupiter, and Mars (*Canon de vero loco trium superiorum, vicelicet Saturni, Iovis et Martis,* **G2**, p. 28)
10. On the true position of Venus and Mercury (*Canon de vero loco Veneris et Mercurii,* **G2**, p. 29)
11. On the planets, whether they are direct, stationary, or retrograde (*Canon de planetis, cum sint directi, stacionarii aut retrogradi,* **G2**, p. 30)
12. On the true conjunction or opposition of the Sun and Moon (*Canon de vera coniunccione vel opposicione Solis et Lune,* **G2**, p. 31)
13. On the true conjunction of Saturn and Jupiter (*Canon de vera coniunccione Saturni et Iovis,* **G2**, p. 33)
14. On the entry of the Sun into Aries (*Canon de introitu Solis in Arietem,* **G2**, p. 34)
15. On the creation of an almanac (*Canon pro componendo almanach,* **G2**, p. 36)
16. On finding the degree of the ascendant (*Canon pro inveniendo gradu ascendente,* **G2**, p. 37)
17. On the aspects of planets (*Canon de aspectibus planetarum,* **G2**, p. 40)
18. On correction of tables (*Canon de correccione tabularum,* **G2**, p. 42)

**Andrzej Grzymała, *Girum recensendo* – variant 3:**
(Prague, NK, VIII G 24 = **G3**)

1. Finding the mean motion of the Sun and Moon and other planets etc. (*Medium motum Solis et Lune et aliorum planetarum et eciam argumenta media Lune, argumentum*

*latitudinis Lune, similiter argumenta Veneris et Mercurii, motum augium et stellarum fixarum motumque accessus et recessus octave spere per thabulas invenire,* **G3**, f. 60v)

2. Finding the mean conjunction and opposition of the Sun and Moon (*Tempus medie coniunccionis et opposicionis Solis et Lune in mense quolibet per tabulas ad hoc factas invenire,* **G3**, f. 60v)
3. How to correct, how to create, or how to continue the Tables of the motions and time (*Modus corigendi Thabulas motuum et temporum ipsasque de novo faciendi aut continuandi,* **G3**, f. 61r)
4. How to correct the conjunctions and oppositions of the Sun and Moon in expanded years (*Annorum expansorum coniuncciones et opposiciones Solis et Lune tali modo coriguntur,* **G3**, f. 61v)
5. Finding the true position of apogees of the planets using the tables (*Verum locum augium planetarum cuiuslibet planete per thabulas invenire,* **G3**, f. 61v)
6. Finding the true position of the Sun using the tables (*Verum locum Solis per thabulas invenire,* **G3**, f. 62r)
7. Finding the true position of the Moon using the tables (*Verum locum Lune per thabulas invenire,* **G3**, f. 62r)
8. Finding the true position of the head of dragon (*Verum locum capitis draconis invenire,* **G3**, f. 62r)
9. Finding the true motion of the tail of dragon (*Ad inveniendum verum motum caude draconis,* **G3**, f. 62r)
10. Finding the true position of the three upper planets, i.e. Saturn, Jupiet, and Mars (*Verum locum trium superiorum, Saturni videlicet, Iovis et Veneris* /in mg. corr. in: *Martis*/ *invenire,* **G3**, f. 62r)
11. Finding the true position of Venus and Mercury (*Verum locum Veneris et Mercurii invenire,* **G3**, f. 62v)
12. Finding whether the planet is stationary, retrograde or direct (*Utrum planeta sit stacionarius, retrogradus aut directus, invenire,* **G3**, f. 62v)
13. Finding whether the planet is ascending or descending (*Utrum planeta sit ascendens vel descendens cognoscere,* **G3**, f. 63r)
14. Finding the true conjunction and opposition of the Sun with the Moon (*Tempus vere coniunccionis et opposicionis Solis cum Luna invenire,* **G3**, f. 63r)
15. Finding the true conjunction of Saturn and Jupiter and any other planet (*Tempus vere coniunccionis Saturni et Iovis et quorumlibet aliorum planetarum invenire,* **G3**, f. 63r)
16. Finding the entry of the Sun into Aries or in any other sign (*Introitum Solis in Arietem vel in quodcumque signum volueris invenire,* **G3**, f. 63v)
17. The creation of an almanac for individual years (*Almanach singulis annis invenire,* **G3**, f. 63v)
18. Finding the degree of ascendant and other cardinal points of the sky[29] (*Gradum ascendentem aliosque celi cardines invenire,* **G3**, f. 64r)

---

29 Cardinal points of the sky (*cardines celi*) are the intersections of the ecliptic with the horizon or meridian, i.e. beginnings of the first, fourth, seventh, and tenth astrological house. As in the above described Bianchini tables of domification in MS Cracow, BJ, 570, pp. 20–23, the term *cardines* is incorrectly used for cusps (i.e.

**Andrzej Grzymała, *Girum recensendo* – variant 4:**
(Vienna, ÖNB, Cod. 5240 = **G4**; Vienna, ÖNB, Cod. 5275 = **G5**)

1. Finding the mean motion of the Sun and Moon and other planets etc. (*Medium motum Solis et Lune et aliorum planetarum, eciam argumenta media Lune, argumentum latitudinis Lune, similiter argumenta Veneris et Mercurii, motus augium et stellarum fixarum motumque recessus et accessus octave spere, insimul motum capitis et caude draconis per tabulas invenire*, **G4**, f. 67r; **G5**, f. 36r)
2. Finding the time of the mean conjunction or opposition of the Sun and Moon (*Tempus vero medie coniunccionis vel opposicionis Solis et Lune in mense quolibet per tabulas ad hoc factas reperire*, **G4**, f. 67v; **G5**, f. 36v)
3. How to correct, how to create, or how to continue the Tables of the motions and time (*Modus corrigendi Tabulas motuum et temporum ipsasque de novo faciendi aut continuandi*, **G4**, f. 68v; **G5**, f. 37r)
4. How to correct or how to continue the Tables of mean conjunctions of the Sun and Moon (*De mediis eciam coniunccionibus Solis et Lune tabule sic corriguntur aut continuari possunt*, **G4**, f. 68v–69r; **G5**, f. 37r)
5. How to convert radices to any noon or meridian (*Radices ad quemcumque meridiem vel meridianum*, **G4**, f. 69v; **G5**, f. 37v)
6. Finding the true position of apogee of any planet using tables (*Verum locum augium cuiuslibet planete per tabulas invenire*, **G4**, f. 69v; **G5**, f. 37v)
7. General way of entering the Tables of equations (*Modum generalem in Tabulas equacionum proponere*, **G4**, f. 70r; **G5**, f. 37v)
8. Finding the true position of the Sun using tables (*Verum locum Solis per tabulas invenire*, **G4**, f. 70v; **G5**, f. 38r)
9. Finding the true position of the Moon using tables (*Verum locum Lune per tabulas invenire*, **G4**, f. 70v; **G5**, f. 38r)
10. Finding the true position of the head of dragon (*Verum locum capitis draconis invenire*, **G4**, f. 71r; **G5**, f. 38v)
11. Finding the true position of the three upper planets, i.e. Saturn, Jupiter, and Mars (*Verum locum trium superiorum, Saturni, Iovis et Martis, invenire*, **G4**, f. 71r; **G5**, f. 38v)
12. Finding the true position of Venus and Mercury (*Verum locum Veneris et Mercurii invenire*, **G4**, f. 71v; **G5**, f. 39r)
13. Finding whether a planet is stationary, retrograde, or direct (*Utrum planeta sit stacionarius, retrogradus aut directus, invenire*, **G4**, f. 72r; **G5**, f. 39r)
14. Finding whether a planet is ascending or descending (*Utrum planeta sit ascendens vel descendens, cognoscere*, **G4**, f. 72v; **G5**, f. 39v)
15. Finding the time of the true conjunction and opposition of the Sun with the Moon (*Tempus vere coniunccionis et opposicionis Solis cum Luna invenire*, **G4**, f. 72v; **G5**, f. 39v)
16. Finding the time of the true conjunction and opposition of Saturn and Jupiter and any other planets (*Tempus vere coniunccionis Saturni et Iovis et quorumlibet aliorum planetarum invenire*, **G4**, f. 73r; **G5**, f. 40r)

---

beginnings) of all houses in the title of the Rule 18, as well as in the text of the Rule 16, version 1: *et proveniet ibi tempus diebus equatis tempore introitus Solis in Arietem, quod valet pro disposicione seu equacione 12 cardinum illo tempore* (Cracow, BJ, 573, f. 232v).

17. Finding the entry of the Sun into Aries or in any other sign (*Introitum Solis in Arietem vel in quodcumque signum volueris invenire,* **G4**, f. 73v; **G5**, f. 40v)
18. Creation an almanac for individual years (*Almanach pro singulis annis invenire,* **G4**, f. 74v; **G5**, f. 41r)
19. Finding the degree of the ascendant and other cardinal points of the sky[30] (*Buth / gradum/ ascendentem aliosque celi cardines invenire,* **G4**, f. 75r; **G5**, f. 41v)

## 3.3 Notes on Canons *Girum*

If we return to the earliest known manuscript of canons *Mirabilis* (that is MS Toruń, BU, 74/III) we can see that it is written in calligraphy and with great attention. Evidently there was an awareness of its significance and it was highly prized. A good formal level indicates a high appreciation of the importance of the content of the work. Other derived canons, starting with *Girum,* originated at universities and were mainly applications to other meridians. Sometimes they differ in several features, but they are derivatives. They are visibly more cursorily written, harder to read, contain more errors, etc. Their authors or scribes (students or lecturers at universities) copied them much more casually for practical and teaching purposes. This also implies that the preparation of critical editions from such canons is much more difficult. The texts diverge in many places, there are different versions that cannot be incorporated into one edition, and even if they have the same incipit, their text may be heterogeneous. They are also sometimes damaged (e.g. MS Leipzig, UB, 1482, 2r–9r, canons *Nemo,* has blurred writing on f. 7, probably the result of water damage).

In contrast to canons *Mirabilis,* canons *Girum* have rules on how to prepare an almanac for each year. Another addition of *Girum* with respect to *Mirabilis* is a rule on conjunctions of planets. However, this subject is dealt with in the canons *Tempus est mensura motus* by John of Saxony composed to the PAT so it seems that Andrzej Grzymała, author of the canons *Girum,* was inspired by both works.

Three of the inspected manuscripts with the canons *Girum* (Cracow, BJ, 570; Cracow, BJ, 1841; Cracow, BJ, 1916) are fragmentary and contain only the rule on planetary aspects. This rule gives instructions on how to determine from the true motions of the planets at the chosen time whether any of the prominent aspects (conjunction, opposition, quadrature, etc.) occur between a pair of planets. If the planets are not exactly in an aspect and are only close to it, the rule describes how to determine from the daily motions of the planets at what past or future time the aspect was or will be precise. Knowledge of aspects was important for astrological interpretation and prediction.

30 The term *cardines* is used here incorrectly similarly as in the title of the Rule 18 in version 3. Moreover, in MS Vienna, ÖNB, Cod. 5240 (= **G4**), f. 75r, there is incorrectly used term *buth* instead of *gradum* which is written in the MS Vienna, ÖNB, Cod. 5275 (= **G5**), f. 41v.

# 4

# Canons *Nemo integre sapit* by Anonymous Author

We do not know who authored the *Nemo* canons. According to various indications and relations to Prague, however, it seems likely that an archetype of these canons originated in Bohemia, probably in the environment of the Charles University founded in 1348 (see Fig. 18). With the exception of the copy Vatican, BAV, Pal. lat. 1409, all copies of the canons *Nemo* contain at their beginning a mention of the Prague meridian (Vatican, BAV, Pal. lat. 1381, f. 121r; Cracow, BJ, 614, f. 67va; Cracow, BJ, 1848, f. 89r; Leipzig, UB, 1482, f. 2v) so that they seem to have been first made for the meridian of the Bohemian capital and only later used and copied in Cracow, as evidenced by the explicits where Polish scribes are named.

The introductory words of canons *Nemo* according to MS Vatican, BAV, Pal. lat. 1381, f. 121r read:

> *Nemo integre sapit nec perfecte sapit testante Aristotile tercio Rethoricorum, quapropter nemo integram congnicionem de motibus corporum supercelestium habere potest, nisi tantum sola ymaginacio paralellorum in orbibus planetarum descriptorum sit sufficiens tam spectacionis, quam paucitas instruccionis magnum facit errorem in congnicione veritatis. Cum igitur animadvertendum quam plurimus studens vidi astronomiam pudiose reselire et ab ea declinare propter tedium et laborem querendi motus planetarum, qui tamen modicus est vel quasi nullus, ut plane patet illi, qui tale nectar degustavit succurrendum et consolendum eorum opinioni fore existimavi maxime, ut delectabilis sciencia astronomia, que inter quatuor correquisita ad divinum officium connumeratur. Teste beato Augustino in quadam auctoritate, qui ait in domo Dei quatuor sunt necessaria, scilicet grammatica, musica, ius canonicum, kalende, id est astronomia, ad aliqualem igitur, ut revocaretur ad memoriam hoc opus, bene feci et collegi ex Tabulis Alphoncii, serenissimi principis, felicis memorie, olim regem Castalie, ad* ***meridianum Pragensem,***[1] *in quo sine omni tedio et labore certitudinaliter loca planetarum secundum medium cursum eorum possunt haberi omnino in signis, gradibus et ceteris fraccionibus usque ad tercia inclusive secum concordando.*

1 The text of canons *Nemo* differ significantly in manuscripts, but the Prague meridian is mentioned in all four copies listed above.

**Figure 18.** Prague, *studium generale*. Karolinum, Gothic bay window.

## 4.1 Manuscripts

### Cracow, BJ, 614, ff. 67ra–73vb[2]

As indicated in the colophon on f. 73vb, this copy was written in Cracow by the scribe Paulus de Wywaras (Szachvar) from Hungary in 1461 when he was a student in Cracow. The initial owner was Johannes Cranghe de Stargardia, who in 1464 would earn his BA in Cracow.

- In addition to the canons *Nemo integre sapit* (ff. 67ra–73vb), the TR are copied at the beginning of the manuscript on ff. 1r–45v. They are standardly calculated for the Cracow meridian from the years 1428 to 1808.
- The TR are followed by the Table of proportional minutes (*Tabula minutorum proporcionalium pro diebus intermediis in almanach faciendis*) on ff. 46r–57v which was commonly added to different sets of astronomical tables, including the *Tabule resolute*.[3]
- Folios 58r–63v are devoted to the *Tabule magne* by John of Lignères.

### Cracow, BJ, 1848, ff. 88v–91r[4]

The manuscript was mostly written *c.* 1389 and contains descriptions of various types of algorithms and computus (e.g. *computus cyrometralis, computus orbicularis, computus Iudaicus, oycreper* (that is, 'repercio' written backwards).

- The table of radices of various motions of planets, added to the canons *Nemo* on ff. 88v–91r extend from 1388[5] to 1608, as does the table in the TR found with the canons *Nemo* in MS Vatican, BAV, Pal. lat. 1381. A somewhat different version of canons *Nemo* begins with the words: *Benedictus sit nomen Domini nostri Ihesu Christi et eius benedicte genitricis Virginis Marie in secula seculorum amen. – Nemo integre sapit nec qui perfecte sapit testante Aristotile tercio Rethoricorum suorum, quapropter nemo integram et perfectam cognicionem de motibus corporum supercelestium habere potest.* The text of the canons is not complete and soon turns into an astrological description of the twelve signs of the zodiac. Apparently there are several versions of the *Nemo* canons.

### Leipzig, UB, 1482, ff. 2r–9r[6]

The manuscript dates from the mid and late fifteenth century. Canons *Nemo* are found on ff. 2r–9r. We cite from the end of the canons (f. 9r): *et veniet longitudo inter meridianum Pragensem et Cracoviensem, scilicet 3 gradus, 5 minuta, 37 secunda, pro quibus 37 capiatur unitas ad minuta… addantur ergo isti 36 gradus cum 6 minutis ad longitudinem, que est inter Toletum et Pragam, et veniet longitudo inter Toletum et Cracoviam… et subtrahe istum motum inventum a radice Toletano, quia Cracovia est civitas orientalior quam Toletum, et habes*

2 ThK 910; Rosińska, *Scientific Writings*, p. 253 (item 1260); *Catalogus BJ*, IV, pp. 300–04; Markowski, *Astronomica*, p. 274 (item 1).

3 See Chapter 16, Table A-03.

4 Wisłocki, *Catalogus*, p. 439; Rosińska, *Scientific Writings*, p. 253 (item 1261).

5 The same beginning of the cyclical radices is also given in the early printed editions of the TR by Ambrosius Lacher and Johannes Schöner (from 1388 to 2008).

6 Markowski, *Astronomica*, pp. 274–75 (item 2); Rosińska, *Scientific Writings*, p. 253 (item 1260), she mentions ff. 2r–7v only; <http://jordanus.badw.de/jordanus/ms/3733> [accessed: 18.03.2023].

*radicem pro Cracovia secundum practicam etc.* The text thus gives instructions for finding the geographical longitude of Prague and Cracow in relation to Toledo.

- The tables on ff. 13r–25v deal with planetary mean motions. On ff. 13r–15v are tables which, in the tradition of *Tabule magne* by John of Lignères, give the collected years in increments of twenty and one-hundred years[7] and give the radices to the year 1400 for a longitude of approximately 17 degrees, 30 minutes east of Toledo.[8] In agreement with the TR, however, they also give mean motions for one to twenty-four hours (ff. 22r–24v). They are followed by an explanatory text ending with a mention of Leipzig (*Lypczk*) and on f. 30r by a colophon related to the Margraviate of Meissen, which also included Leipzig in that time: *In hiis itaque finite et complete sunt Tabule mediorum motuum omnium planetarum cum suis canonibus, extracte et calculate de Tabulis illustris regis Alphoncii, radicate super meridiano territorii inclitissimorum principum marchionum Misne, Anno Incarnacionis 1419.*

Vatican, BAV, Pal. lat. 1381, ff. 121r–123r

Ludwig Schuba[9] divides this manuscript into three parts, the first of which he locates in Franken/Oberpfalz and dates to 1451–1457, the second to the end of the fourteenth century and the third to the middle of the fourteenth century. He believes that the initial works of the volume (up to f. 136r) may have originated in Reichenbach (Saxony) and Kulmbach (Bavaria) further to the south, not far from the Bohemian border. Reichenbach was conquered by the Bohemians in 1336 and sold to the German Emperor and Bohemian King Charles IV together with the surrounding villages in 1367. In 1430, Reichenbach and its church were completely destroyed by Hussites. The Bavarian town of Kulmbach in Upper Franconia is also close to the Czech border. During the Hussite wars the town suffered the same fate; the Hussites completely destroyed it in 1430. Most of the works in the first part of the manuscript were written by Matthias of Kemnath (or Kemnatensis, *c.* 1429–76, graduated in Heidelberg),[10] but not the TR and canons *Nemo*. Of the more extensive works in this part of the manuscript we can mention the description of the astronomical instrument safea by al-Zarqālī, in Latinised form Arzachel (on ff. 30v–42v and 59r–63v), and the *Teorica planetarum* by Pseudo-Gerardus of Cremona (ff. 46r–58v). Some minor works, written in German, refer to the monastery in Reichenbach.

- In this Vatican manuscript, first the TR on ff. 113r–120v and then the canons *Nemo* on ff. 121r–123r are included. The TR are incomplete; only the mean motions have been copied. They are calculated for Prague (computed for a meridian 74 minutes east of

7 See Chabás, *Computational Astronomy in the Middle Ages*, p. 202; Matthieu Husson, 'Work Cohesion as a Test of Manuscipt Transmission: The Case of John of Lignères' Tabule magne', in *Alfonsine Astronomy. The Written Record*, ed. by Richard L. Kremer, Matthieu Husson, and José Chabás (Turnhout: Brepols, 2022), pp. 317–42 (p. 332).

8 This value agrees with the traditionally accepted geographical longitude of Leipzig (although it better fits to the modern longitude of Meissen, about one degree east of Leipzig). See Richard L. Kremer and Jerzy Dobrzycki, 'Alfonsine Meridians: Tradition versus Experience in Astronomical Practice c. 1500', *Journal for the History of Astronomy*, 29 (1998), 187–99.

9 Ludwig Schuba, *Die Quadriviums-Handschriften der Codices Palatini Latini in der Vatikanischen Bibliothek* (Wiesbaden: Dr Ludwig Reichert Verlag, 1992), pp. 115–25.

10 Klaus Graf, 'Matthias von Kemnat', in *Neue Deutsche Biographie*, 16 (1990), pp. 410–11.

Toledo, that is for Prague), starting unusually already from the year 1388 and finishing in 1668. (This does not mean, however, that this type of the TR is the oldest: it is just a mechanical shifting back of the tables, starting in 1428, for forty years.) The TR are indeed closely linked to the canons in this manuscript, as evidenced by the fact that the canons refer to the TR extending to 1668.

It should be noted that the values in the manuscript Vatican, BAV, Pal. lat. 1381 are given in physical signs, while the values in manuscript Cracow, BJ 614 are given in common (zodiacal) signs. Physical signs (*signum physicum*) have 60 degrees and are formed by two common (zodiacal) signs (*signum commune*) of 30 degrees each. On f. 113v we can see the radix of mean motion of the Moon for 1428: 3 (physical) signs, 46 degrees, 31 minutes, 0 seconds, 34 tertias – that is 226;31,0,34° which corresponds to Prague, while the previously described manuscript Cracow, BJ, 614, f. 1v has 7 (common) signs, 16 degrees, 24 minutes, 12 seconds, 7 tertias – that is 226;24,12,7° which corresponds to Cracow.

The scribe of the TR and the canons is probably the same person in this manuscript because in the titles of the rules he often refers to the tables in the codex (see the overview of rules below).

- Canons *Nemo* begin with the words: *Benedictum sit nomen Domini nostri Jesu Christi… Nemo integre sapit.* Explicit: *dic eclipsim fore possibilem.* At the beginning of canons *Nemo* the author/scribe laments that every student would prefer to get results without making any effort. He gives the analogy of a cat wanting to catch a fish but not willing to get wet. On f. 121r he tells: … *tamen sunt multi, qui scire volunt et discere nolunt: qui faciunt sicut cathus, qui amat piscem, sed non vult tangere flumen…* ('but still there are many people who want to know but do not want to learn: they do as a cat who loves fish but does not wish to dip his feet in the stream…').

Why was the year 1388 chosen the starting point of cyclical radices? In the MS Vatican, BAV, Pal. lat. 1381, f. 122v we read: *Propter habendas coniuncciones Solis et Lune posuit ciclum magnum, quod maior computi cyrometralis approbat perfectum, prout ibidem constantem ex 76 annis et hunc incepi anno Domini 1388 completo novo currente ita, quod tunc fuimus in primo anno huius cicli.* – 'For finding of conjunctions of the Sun and Moon one can work with the great cycle. The *computus cyrometralis* confirms that great cycle consists of 76 years, and this started when the year 1388 was over and when the year 1389 began – so then we were in the first year of this cycle.' The *cyclus magnus* ('great cycle') of 76 years is Callippic cycle, containing four Metonic cycles, each of 19 years. It was used from antiquity for equation of lunisolar cycles. The Metonic cycle was named after its discoverer, the astronomer Meton of Athens, active in the fifth century BC. The Metonic cycle was a nineteen-year cycle designed to align the solar cycle with the lunar cycle. However, since after 19 years there is still about a quarter of a day left for the two cycles to align exactly, the astronomer Callippus from Cyzicus in the fourth century BC established a 76-year cycle consisting of four Metonic cycles, which is already much more accurate. It is interesting to note that both cycles are inscribed, e.g., in spiral circles on the back of the so-called Antikythera mechanism, dating from 150–100 BCE.

The text of canons *Nemo* cited above refers to the *Magnus computus chirometralis* written by John of Erfurt *c.* 1330. A later version of this type of computus was written by John of

Głogów in Cracow (printed in 1507, revised edition in 1510, ed. Haller). John of Głogów uses for Callippic cycle the expression: *cyclus magnus* ***perfectus*** *Lune* = 'great **perfect** cycle of the Moon', for Metonic cycle the expression: *cyclus parvus* ***imperfectus*** *Lune* = 'small **imperfect** cycle of the Moon'. He claims that the first year of the Callippic cycle is the year 1330. If true, then the year 1388 in the canons *Nemo integre sapit* (that is in the manuscript Vatican, BAV, Pal. lat. 1381 only) would not be the beginning of the new Callippic cyclus. It would be the second year after the beginning of the third Metonic cycle.

- For the sake of completeness, let us add that the majority of the works in the second and third part of the manuscript belong to shorter descriptions of astronomical instruments; the codex contains, for example, Messahalla's treatise on Composition of the astrolabe and other anonymous minor works on the quadrant, sundial, cylinder and equatorium. Johannes Virdung's hand has been identified on folios 207v–208v.

Vatican, BAV, Pal. lat. 1409, ff. 52v–59v[11]

Canons *Nemo* are adapted to Cracow in this manuscript. The beginning of the manuscript (ff. 1r–52r) is filled by the ephemerides by John of Saxony (incipit: *Incipiunt Tabule almanach Johannis Danbois*, that is *Danconis*). Several other minor works are astrological. The manuscript originated in Germany, in the second half of the fourteenth century.

## 4.2 List of Rules

**Anonymous author, *Nemo integre sapit* – variant 1:**
(Leipzig, UB, 1482 = **N1**)

1. Finding the mean motion of the eighth sphere, the apogees, fixed stars etc. (*Medium motum octave spere, augium et stellarum fixarum, Solis, Veneris et Mercurii, Lune, capitis draconis, Saturni, Iovis et Martis et argumenta media eorum quolibet die et qualibet hora diei cupis invenire*, **N1**, f. 3v)
2. To find the true motion of the Sun (*Ad inveniendum verum motum Solis*, **N1**, f. 3v)
3. To find the true motion of the Moon (*Ad inveniendum verum motum Lune*, **N1**, f. 4r)
4. To find the true motion of the three upper planets, i.e. Saturn, Jupiter, and Mars (*Ad inveniendum verum motum trium superiorum, scilicet Saturni, Iovis, Martis*,[12] **N1**, f. 4v)
5. To find the true motion of Venus and Mercury (*Ad inveniendum verum motum Veneris et Mercurii*, **N1**, f. 5r)
6. On the direction, retrogradation, and station of planets (*De direccione, retrogradacione et stacione planetarum*, **N1**, f. 5r)
7. Finding the mean motion of the head of dragon (*Medium motum capitis draconis invenire*, **N1**, f. 5r)
8. Knowing the true motion of the tail of dragon (*Motum verum caude {caude} draconis scire*, **N1**, f. 5r)

11 Markowski, *Astronomica*, p. 275 (item 3) and 288; Schuba, *Die Quadriviums-Handschriften*, pp. 186–87; Rosińska, *Scientific Writings*, p. 253 (item 1260).
12 *Ms. per err.*: Mercurii.

9. You can find the time of the conjunction of Jupiter, Saturn, and any other planet in this way (*Tempus coniunccionis Iovis et Saturni et quorumlibet planetarum sic poteris invenire*, **N1**, f. 5v)
10. Finding the mean conjunction of the Sun and Moon (*Invenire mediam coniunccionem Solis et Lune*, **N1**, f. 5v)
11. Finding the mean conjunction and mean opposition of the Sun and Moon, and possibility of their eclipse in any month (*Mediam coniunccionem et mediam opposicionem Solis et Lune et possibilitatem eclipsis Solis et Lune in mense quolibet reperire volens*, **N1**, f. 6r)
12. Finding the mean conjunction or opposition in other month (*Si vero mediam coniunccionem vel opposicionem alterius mensis volueris invenire*, **N1**, f. 6r)
13. Finding the true conjunction of the Sun and Moon or the true opposition of the Sun and Moon (*Si vis invenire veram coniunccionem Solis et Lune vel veram opposicionem Solis et Lune*, **N1**, f. 6v)
14. Entry of the Sun into Aries or any sign (*Introitum Solis in Arietem vel in quodcumque signum*, **N1**, f. 7r)
15. The almanac is composed as follows (*Almanach componendo sic*, **N1**, f. 8r)
16. With the Moon, proceed as follows (*De Luna sic procede*, **N1**, f. 8r)
17. On dragon (*De dracone*, **N1**, f. 8r)
18. To convert radices of the Toledan meridian to radices of Cracovian meridian (*Ad reducendum radices meridiani Toletani in radices meridiani Cracoviensis*, **N1**, f. 8v)

**Anonymous author, *Nemo integre sapit* – variant 2:**
(Vatican, Pal. lat. 1409 = **N2**; the text is not divided into rules)

1. Finding the true motion of the Sun using tables present in this work (*Verum motum Solis per tabulas huius operis invenire*, **N2**, f. 53r)
2. Finding the true motion of the Moon using present work (*Verum motum Lune per hoc opus invenire*, **N2**, f. 53v)
3. Finding the true position of the Sun and Moon in any hour using the corresponding tables (*Verum locum Solis et Lune per huiusmodi tabulas qualibet hora… invenire*, **N2**, f. 53v)
4. The motion of any of them in one day (*Motus cuiuslibet illorum in uno die*, **N2**, f. 54r)
5. The motion of any of them in one minute (*Motum cuiuslibet illorum in uno minuto*, **N2**, f. 54r)
6. Finding the true conjunction of the Sun and Moon using the tables present in this work (*Coniunccionem veram Solis et Lune per tabulas huius operis invenire*, **N2**, f. 54r)
7. Finding the true position of the head of dragon using present work (*Verum locum capitis draconis per hoc opus invenire*, **N2**, f. 56v)
8. Finding eclipses of the Sun and Moon using present work (*Eclipses Solis et Lune per hoc opus invenire*, **N2**, f. 57v)
9. Knowing the true position of Saturn, Jupiter, Mars, Venus, and Mercury using tables (*Verum locum Saturni, Iovis, Martis, Veneris et Mercurii per tabulas scire*, **N2**, f. 58r)
10. Finding the true position of any of them in any hour (*Verum locum cuiuslibet istorum in quacumque hora volueris invenire*, **N2**, f. 59r)
11. Finding the conjunction of any of them with any other (*Coniunccionem cuiuslibet illorum cum quolibet invenire*, **N2**, f. 59r)

**Anonymous author, *Nemo integre sapit* – variant 3:**
(Vatican, BAV, Pal. lat. 1381 = **N3**; incomplete; the short text is not divided into rules)

After the beginning of the explanation, which we refer to as Rule 1 and in which the mean motions are briefly clarified, the text continues with remarks on the following topics, which cannot be considered as consistent rules and therefore are not numbered:

1. Finding the mean motion of the eighth sphere, the apogees, fixed stars etc. (*Medium motum octave spere, augium et stellarum fixarum, Solis, Veneris et Mercurii, Lune, capitis draconis, Saturni, Iovis et Martis et argumenta media eorum quolibet die et qualibet hora diei cupis invenire,* **N3**, f. 121v)
- If you want to know the motions for some months after the beginning of the year or for some days or hours (*Si vero volueris habere motus prescriptos ad aliquos menses post principium anni vel ad aliquos dies vel horas,* **N3**, f. 121v)
- A note on terminology. What they are the collected and expanded years (*Nota pro intellectu illorum terminorum. Anni collecti et anni expansi dicuntur,* **N3**, f. 122r)
- To continuous tables of radices of different planetary motions (*Ad perpetuas seu continuas tabulas radicum diversorum motuum planetarum hic positorum,* **N3**, f. 122r)
- To determine the conjunctions of the Sun and Moon etc. (*Propter habendas coniuncciones Solis et Lune posuit ciclum magnum, quam maior computi cyrometralis approbat perfectum,* **N3**, f. 122v)
- Finding the opposition (*Opposicionem vero sic invenies,* **N3**, f. 123r)
- Finding the possibility of an eclipse of the Sun and Moon (*Possibilitatem eclipsis Solis et Lune sic invenies,* **N3**, f. 123r)

**Anonymous author, *Nemo integre sapit* – fragment:**[13]
(Cracow, BJ, 1848 = **N4**)

1. Finding the mean motion of the eighth sphere, apogees, and fixed stars ets. (*Medium motum octave spere, augium et stellarum fixarum, Solis, Veneris et Mercurii, Lune, capitis draconis, Saturni, Iovis et Martis et argumenta media eorum quolibet die et qualibet hora diei cupis invenire,* **N4**, f. 90r)

## 4.3 Notes on Canons *Nemo*

Canons *Nemo* have no significant additions compared to *Girum*. They contain mean and true motions along with notes on conjunctions of the Sun, Moon and five planets. They also present explanations on direction, retrogradation, and station of planets and they discuss the preparation of almanacs. In MS Leipzig, UB, 1482, f. 8r, Rule 15 on the preparation of almanac is introduced by the words: *Hoc sequens pro secreto quidam tenebant, quare secrete est silendum* ('What follows has been kept secret by some authors because it is necessary to be secretly silent'). The author is probably referring to those canons which lack rules on how to compile an almanac.

13 The manuscript is incomplete, it has only one canon.

5

# Canons *Astrorum (observatores) sapientes ingentem celi tractantes magnitudinem* by Albertus (Adalbertus) of Brudzewo

Another important Polish astronomer and mathematician Wojciech z Brudzewa (Brudzewski, Albertus /Adalbertus/ of Brudzewo) was born *c.* 1445 in Brudzewo, east of Poznań, and died *c.* 1497 in Vilnius. In Cracow he studied with John of Głogów and obtained his master's degree in 1474; Nicolaus Copernicus probably studied with him.[1] He spent most of his active life at the university and taught there for two decades. Then he served as a secretary of Prince Alexander of Lithuania, later King of Poland. In 1495, Albertus of Brudzewo published his *Commentaria utilissima in Theoricis planetarum* in Milan,[2] a commentary on Georg von Peurbach's important treatise *Theorice nove planetarum*, known from many manuscript copies and early printed editions.

## 5.1 Manuscripts

Augsburg, UB, Cod. II 1 Quarto 74, ff. 239r–245v[3]

This codex dated to the fifteenth century contains several astronomical works.

- In addition to the canons *Astrorum* by Albertus of Brudzewo there are also canons *Pro (introduccione) generali* by John of Głogów (ff. 260r–280r) to the TR. All three parts of John's *Summa astrologie* are written on ff. 55r–158r.
- The Composition of the astrolabe by Cristannus de Prachatice can be found on ff. 208r–215v, his Use of the astrolabe follows on ff. 216r–227r.
- *Algorismus de minutiis* by John of Lignères is copied on ff. 191r–205r.

Cracow, BJ, 600, ff. 137r–147v[4]

This codex originated in 1486 at the University of Cracow and was completed in 1501. It seems that the manuscript first belonged to Albertus of Brudzewo. The scribe of parts of

1 Jürgen Hamel, 'Brudzewski, Albertus de', in *The Biographical Encyclopedia of Astronomers*, ed. by Thomas Hockey (New York: Springer, 2007), pp. 177–78; Knoll, 'The Arts Faculty', pp. 148–49.

2 See digital copy of this print <https://www.deutsche-digitale-bibliothek.de/item/7J7M3FIICRLLJG6K-GND5KT6UN7US2T4G> [accessed: 12.03.2023].

3 Hardo Hilg, *Lateinische mittelalterliche Handschriften in Octavo der Universitätsbibliothek Augsburg. Die Signaturengruppe Cod. I.2.8° und Cod. II.1.8°* (Wiesbaden: Harrassowitz, 2014), pp. 451–461.

4 Rosińska, *Scientific Writings*, pp. 63–64 (item 272); Markowski, *Astronomica*, p. 297. *Catalogus BJ*, IV, pp. 163–76. A digital copy see <https://polona.pl/item/opera-astronomica-et-mathematica,Nzg1MDUyNjM/37/#info:metadata> [accessed: 12.03.2023].

the text was Nicolaus of Wieliczka senior (he achieved the title of master in 1501, then he lectured mainly on astronomy and mathematics).

- The incipit of canons *Astrorum* by Albertus of Brudzewo on ff. 137r–147v sounds: *Astrorum sapientes ingentem celi tractantes magnitudinem nobilissima eius nec unquam sine admiracione quam maxima* … × … *iam itaque modum habet et exitum calculus exemplaris Tabularum resolutarum propter proposiciones has 38 etc.* The canons give examples for both Cracow and Prague. On f. 139r we read: *Eodem modo operari te convenit de radice cuiuslibet motus per suas tabulas ad meridianum Cracoviensem. – Radicem motus meridiani Cracoviensis cuiuscumque alteri meridiano adaptare… In exemplo: volo reducere radicem Cracoviensem medii motus Solis anni 1481 completi… quem addam radici Cracoviensi pro eo, quod longitudo Pragensis fuerit minor quam Cracoviensis, et prodibit radix hec pro Pragensi meridiano in Sole 9, 18, 56, 2, 1.* There are also several other examples for 1481 (e.g. on f. 139v, etc.); others refer to 1482 (on f. 139v; 141r; 142r; 144v). John of Głogów certainly drew inspiration for his rules from the canons of *Astrorum*: in *Astrorum*, for example, there are already mentioned *cautele* ('warnings' or 'cautions'), which John of Głogów then applied in many rules of his canons.
- John of Głogów's canons *Pro (introduccione) generali* are also copied in this manuscript (on ff. 149r–159v), see the Chapter 8.
- The TR on ff. 86r–100r are reduced for the Cracow meridian in its standard form for the years 1428 to 1808 (*Incipiunt Tabule resolute de mediis et veris motibus planetarum <super> meridianum Craco<viensem> verificate*).
- Many different, individual astronomical tables are scattered throughout the codex. Of the larger treatises, Regiomontanus's work deserves attention (*Tabule primi mobilis* on ff. 38r–82r, canons to them on ff. 1r–33v, *Sine table* on ff. 125r–134r and others) as well as Tables for the planets by Bianchini on ff. 115v–123r.

### Cracow, BJ, 605, ff. 97r–119v[5]

The scribe of this manuscript is unknown. The codex originated at the end of the fifteenth century, but the explicit of the canons *Astrorum* on ff. 97r–119v indicates a more recent writing: *iam itaque modum habet et exitum calculus exemplaris Tabularum resolutarum per proposiciones has a magistro Alberto de Brudowo* (sic) *in studio Cracoviensi… Valete. Anno Salvatoris nostri Iesu benedicti 1529, commutacione hiemali.* The text is hard to read in places, the writing is soggy.

- The standard TR for Cracow (starting in 1428 and finishing in 1808) are found on ff. 16r–53r.
- In addition to short texts and tables, the codex contains Tables for the planets by Bianchini on ff. 76r–85r and on ff. 87r–94v.

### Cracow, BJ, 1846, pp. 157–196[6]

The codex, dating *c.* 1500, contains on pp. 1–65 the TR calculated for Cracow from 1428 to 1808.

5 Rosińska, *Scientific Writings*, pp. 63–64 (item 272); Markowski, *Astronomica*, p. 16 (item 55), *Catalogus BJ*, IV, 1988, pp. 214–20.

6 Wisłocki, *Catalogus*, p. 438; Rosińska, *Scientific Writings*, pp. 63–64 (item 272); Markowski, *Astronomica*, p. 14 (item 42).

- Canons Astrorum in this copy (on pp. 157–196) end by the words: iam itaque modum habet et exitum calculus exemplaris Tabularum resolutarum per proposiciones has, a magistro Alberto de Brudzevo in studio Cracoviensi innovatus. Several examples in them are related to the year 1482.
- Canons Pro (introduccione) generali of John of Głogów are on pp. 203–227.

### Cracow, BJ, 1858, pp. 73–113[7]

- Canons Astrorum and the standard TR for Cracow meridan (from 1428 to 1808) are contained on pp. 73–113 and 121–191, respectively.

### Cracow, BJ, 1864, pp. 32–38, 46 and 220[8]

The canons Astrorum are continued by canons Girum on pp. 47–70.

- On pp. 71–141 there are contained the TR for Cracow meridian, written in 1446. They have the usual form for the years 1428 to 1808.

### Cracow, BJ, 1920, pp. 5–20[9]

This Cracow manuscript originated at the end of the fifteenth century. This manuscript is very rich in canons to the TR: in addition to the *Astrorum* on pp. 5–20 we find canons *Girum* on pp. 183–201 and *Pro (introduccione) generali* on pp. 3a–3b (prologue) and 207–235. The TR themselves are contained on pp. 25–98.

### Cracow, BJ, 1966, pp. 157–182[10]

The codex was written in the years 1515–20 by several hands. A standard TR for Cracow 1428–1808 are preserved on pp. 13–87. The canons *Astrorum* end: *Materie huius finis adest in die quinta post Conductum Paschatis 1515.*

### Cracow, BJ, 2478, ff. 73r–94v[11]

The codex dates from the turn of the fifteenth to the sixteenth centuries and is introduced by the TR on ff. 1r–38r. The canons *Astrorum* on ff. 73r–94v finish: *pro cuius complemento materie Tabularum resolutarum sit Deus benedictus in secula seculorum, amen. Per magistrum Martinum de Grodzijszko, arcium liberalium baccalaureum, protunc in schola Omnium sanctorum Cracoviensi officium locature tenentem, finita sunt feria secunda in crastino d. Andree apostoli, hora pene viginti vesperorum, anno partus Virginei 1522.* And another hand added to

---

7 Wisłocki, *Catalogus*, p. 541; Rosińska, *Scientific Writings*, pp. 63–64 (item 272); Markowski, *Astronomica*, p. 14 (item 46).

8 Wisłocki, *Catalogus*, p. 445; Rosińska, *Scientific Writings*, pp. 63–64 (item 272); Markowski, *Astronomica*, p. 14 (item 45).

9 Wisłocki, *Catalogus*, pp. 459–60; Rosińska, *Scientific Writings*, pp. 63–64 (item 272); Markowski, *Astronomica*, p. 15 (item 48).

10 Wisłocki, *Catalogus*, p. 481; Rosińska, *Scientific Writings*, pp. 63–64 (item 272); Markowski, p. *Astronomica*, 15–16 (item 53).

11 Wisłocki, *Catalogus*, p. 592; Rosińska, *Scientific Writings*, pp. 63–64 (item 272); Markowski, *Astronomica*, p. 16 (item 54).

it: *He tabule, quia pro meridiano Pragensi sunt facte, ad reducendum ergo motuum et tempus ad meridianum Cracoviensem 19 minuta, 36 sedunda in tempore sunt addenda, in motu vero subtrahenda, ex propria experiencia magistri Nicolai de Vijeliczka, huius artis indagatore optimo, sunt servanda et cum diligencia custodienda.*

Cracow, BJ, 2480, pp. 181–215[12]

The manuscript originated in 1506. Canons with incipit *Astrorum observatores sapientes ingentem celi tractantes magnitudinem* and with explicit *Finis Anno Domini 1506 in studio Cracoviensi* are contained on pp. 181–215.

- *Tabule resolute de mediis et veris motibus planetarum super meridianum Cracoviensem* to the years 1428–1808 are written on pp. 1–75.

Heidelberg, UB, cms 2503, ff. 1r–23r[13]

According to the explicit, the copy of canons *Astrorum* (ff. 1r–23r) originated in Vienna, 1498: *Valete, anno 1498 in vigilia Epiphaniarum Domini, M. S. T. Vienne*. On ff. 32r–97r there follow the TR for Cracow meridian to the years 1428–1808.

Jena, Thüringer Universitäts- und Landesbibliothek, Ms. El. f. 77, ff. 75r–89v[14]

This manuscript of Polish provenance originated in the sixteenth century. Canons Astrorum are present on ff. 75r–89v with incipit and explicit: In nomine Domini amen. Astrorum sapientes ingentem celi tractantes magnitudinem… × … iam itaque modum habet et exitum calculus exemplaris Tabularum resolutarum per proposiciones has a magistro Alberto de Brudzewo in studio Cracoviensi innovatus pro introducendis instituendisque utrumque huius artis sideralis studiosis auditoribus… studeant et deposcant valere in secula seculorum. Amen. Finis.

- Most of the manuscript is filled by the astrological work *Summa astrologie* by John of Głogów. It consists of three main parts: *Introductorium in artem astrologie* (ff. 12r–45v), *Introductorium astronomie in ephemerides* (ff. 46r–74r; printed in 1514 in Cracow), and *Introductorium in scienciam nativitatum* (ff. 6v–11v; 90r–112v).

Munich, BSB, Clm 51, ff. 49ra–58vb[15]

The manuscript was written in Cracow, in the second half of the fifteenth century (1485–87) and in 1503, mostly by the hand of the scribe Jacob of Würzburg.

12 Wisłocki, *Catalogus*, pp. 592–93; Rosińska, *Scientific Writings*, pp. 63–64 (item 272); Markowski, *Astronomica*, p. 15 (item 51).

13 Markowski, *Astronomica*, p. 14 (item 47) and p. 290.

14 Rosińska, *Scientific Writings*, pp. 63–64 (item 272); Markowski, *Astronomica*, p. 15 (item 49) and 290. See digital copy <https://collections.thulb.uni-jena.de/receive/HisBest_cbu_00015425>; <https://collections.thulb.uni-jena.de/receive/HisBest_cbu_00015409> [accessed: 12.03.2023].

15 Karl Felix Halm, *Catalogus codicum Latinorum Bibliothecae regiae Monacensis. Vol. 1.1: Clm 1–2329* (Munich: sumptibus Bibliothecae regiae, 1892), p. 11; Rosińska, *Scientific Writings*, pp. 63–64 (item 272); Markowski, *Astronomica*, p. 315. An overview of the astrological works written in this codex is given also by David Juste, *Catalogus codicum astrologorum Latinorum. I. Les manuscrits astrologiques latins conservés à la Bayerische Staatsbibliothek de Munich* (Paris: CNRS Éditions, 2011), pp. 51–52.

- Incipit of the canons *Astrorum* by Albertus of Brudzewo: *Astrorum sapientes ingentem celi tractantes magnitudinem sine admiracione quam maxima.*
- On ff. 124r–135r are the TR for Cracow, but in this case they start in 1408 and not in 1428 as is usually the case. They are incomplete, contain only mean motions and planetary equations.
- Regiomontanus's canons to the *Tabule direccionum profeccionumque* are contained on ff. 1ra–16vb.
- Peurbach's canons to the *Tabule eclipsium* are on ff. 37r–46r and his *Theorice nove planetarum* on ff. 72r–88v.
- Second part of *Summa astrologie* by John of Głogów is on ff. 99ra–109rb.

### Padua, BU, MS 643, ff. 53v–64r[16]

This copy was written in Buda in 1513, as stated in the explicit of canons *Astrorum*: *firmitudinem omnem preferencium procedere studeant et deposcant. Valete. Anno 1482 compilati, scripti anno 1513 in die undecim millium virginusm ac martyrum, hora vesperarum, Bude finem imposui.*
- The canons *Astrorum* precede the canons *Pro (introduccione) generali* by John of Głogów on ff. 41r–53v.
- *Tabule auree* by Petrus Gaszowiec are on ff. 67r–72v.

### Vienna, ÖNB, Cod. 5227, ff. 1r–22r[17]

The contents of the manuscript have already been described in the commentary on the canons *Girum* in the Chapter 3.1. It was written in Vienna 1482–1514.
- Canons *Astrorum* by Albertus of Brudzewo can be found on ff. 1r–22r. Their explicit is similar to that in Jena, Thüringer Universitäts- und Landesbibliothek, Ms. El. f. 77, f. 89v: *iam itaque modum et exitum habet calculus exemplaris Tabularum resolutarum per proposiciones has 38 a magistro Alberto de Brudzew in studio Cracoviensi innovatus pro introducendis instituendisque utrumque huius artis sideralis studiosis autoribus… studeant et deposcant. 1482.*
- The canons *Girum* follow on ff. 26r–36v as well as the TR for Prague and Cracow on ff. 37r–77v.
- Oxford tables of 1348 by William Batecombe on ff. 117r–158r and 166r–169r are supplemented by their canons on ff. 158v–162v.
- Table for true syzygies by Nicolaus de Heybech on ff. 163r–164r is accompanied by its canon on ff. 164v–165r.

### Vienna, ÖNB, Cod. 10460, ff. 1r–9r[18]

This manuscript from *c.* 1500 preserves an abbreviated version of the canons *Astrorum* on ff. 1r–9r. Its incipit and explicit: *Radicem anni Cristi uniuscuiusque motus medii primum*

16 Markowski, *Astronomica*, p. 15 (item 52). See also online catalogue: <https://cataloghistorici.bdi.sbn.it/file_viewer.php?IDCAT = 231&IDGRP = 2310009&LEVEL = 0&PADRE = &PROV> [accessed: 10.11.2022].

17 *Tabulae codicum* IV, 1864–1893, p. 65; Unterkircher, *Die datierten Handschriften der Österreichischen Nationalbibliothek von 1451 bis 1500*, p. 157; Rosińska, *Scientific Writings*, pp. 63–64 (item 272); Markowski, *Astronomica*, p. 13 (item 41).

18 *Tabulae codicum* VI, 1864–1899, p. 193; Markowski, *Astronomica*, p. 15 (item 50, redactio abbreviata).

*constituere etc. – Radix anni Christi nil aliud est, quam arcus alicuius orbis ... × ... per partem proporcionalem ad octo gradus. Hiis autem duobus inventis* [ ... ]. Other text is missing as well as the tables. The canons are introduced by a letter to the reader composed by humanist Johannes Aesticampianus (1457–1520). He studied in Cracow but later he was active at the University of Frankfurt/Oder where he lectured rhetoric and poetry. He knew the intellectual environments in Vienna, Cologne, Mainz, Rome (he obtained a doctorate in theology there), and Bologna.

## 5.2 List of Rules

**Albertus of Brudzewo: *Astrorum (observatores) sapientes***
(Cracow, BJ, 600 = **A1**; Munich, BSB, Clm 51 = **A2**):
On addition (*De addicione capitulum,* **A1**, f. 137v; **A2**, f. 49va)
On subtraction (*Capitulum de subtraccione,* **A1**, f. 137v; **A2**, f. 49va)
On multiplication by two (*De duplacione,* **A1**, f. 137v; **A2**, f. 49vb)
On division by two (*Capitulum de mediacione. Mediacio,* **A1**, f. 137v; **A2**, f. 49vb)
On multiplication (*De multiplicacione capitulum. Modum generalem multiplicacionis,* **A1**, f. 137v; **A2**, f. 49vb)
On division (*Capitulum de divisione,* **A1**, f. 138v; **A2**, f. 50rb)

1. Determining the radix of the year of Christ of each mean motion (*Radicem anni Christi uniuscuiusque motus medii primum constituere,* **A1**, f. 139r; **A2**, f. 51ra)
2. Determining the radix of the motion of the Cracow meridian (*Radicem motus meridiani Cracoviensis,* **A1**, f. 139r; **A2**, f. 51ra)
3. Finding the radix of apogee of any planet (*Radicem augis cuiuslibet planete reperire,* **A1**, f. 139r; **A2**, f. 51rb)
4. Finding the mean motion of any planet to a given time using tables (*Medium motum cuiuslibet planete per ipsius tabulas ad tempus propositum quodcumque invenire,* **A1**, f. 139v; **A2**, f. 51va)
5. To explore the apogee of any planet for any day of the year (*Augem cuiuslibet planete seorsum pro die intra annum quodcumque perscrutari,* **A1**, f. 140r; **A2**, f. 51vb)
6. Finding the anomaly of the Sun (*Argumentum Solis elicere,* **A1**, f. 140r; **A2**, f. 51vb)
7. Finding the centre of the Moon for any time (*Centrum Lune pro quocumque tempore inveniendum subiungere,* **A1**, f. 140r; **A2**, f. 51vb)
8. Finding the mean anomaly of each of the three upper planets for a given time (*Medium argumentum uniuscuiusque trium superiorum ad tempus datum educere,* **A1**, f. 140r; **A2**, f. 51vb)
9. Finding the mean centre of five planets for a given time (*Centrum medium in quinque planetis pro assignato tempore inveniendum explanare,* **A1**, f. 140v; **A2**, f. 52ra)
10. Method for verifying the correctness of mean motions in tables (*Modum, per quem medios motus in tabulis esse correctos experiantur, inserere,* **A1**, f. 140v; **A2**, f. 52ra)
11. Continuation the tables of the mean motion of the Sun, similarly of the Moon, and also of its anomaly for equal hours in the whole year (*Medium motum Solis, similiter Lune ac sui argumenti ad dies equales per totum annum continuare,* **A1**, f. 140v; **A2**, f. 52rb)
12. Extending the centre of the Moon in individual days of the year (*Centrum Lune per singulos dies anni extendere,* **A1**, f. 141r; **A2**, f. 52va)

13. Finding the mean motion of any of the three upper planets at steps of ten (*Medium motum cuiuslibet trium superiorum de 10 in de 10 perficere*, **A1**, f. 141r; **A2**, f. 52va)
14. Finding the equated anomaly of any of the three upper planets at steps of ten (*Argumentum equale cuiuslibet trium superiorum per denarios similiter ordinare*, **A1**, f. 141r; **A2**, f. 52vb)
15. Finding the mean centre of any of the three upper planets at steps of ten days (*Centrum medium eorundem trium alciorum per decem dies continue preparare*, **A1**, f. 141v; **A2**, f. 52vb)
16. The anomaly of the two lower planets (*Argumentum medium duorum inferiorum*, **A1**, f. 141v; **A2**, f. 53ra)
17. Forming the mean centre of the two lower planets (*Centrum medium eorundem inferiorum per ordinem formare*, **A1**, f. 141v; **A2**, f. 53ra)
18. Explanation of the Tables of equations (*Ordinem Tabularum in equacionibus exponere*, **A1**, f. 141v; **A2**, f. 53ra)
19. To reveal the common apogee (*Augem communem patefacere*, **A1**, f. 142r; **A2**, f. 53va)
20. Finding the true position of the Sun for a time (*Verum locum Solis pro tempore invenire*, **A1**, f. 142v; **A2**, f. 53vb)
21. To explore the true point of the zodiac, where the centre of the Moon is (*Punctum zodiaci verum, sub quo sit centraliter Luna, perscrutari*, **A1**, f. 142v; **A2**, f. 53vb)
22. Calculating the true position of the zodiac, where any of the five planets is located (*Verum locum zodiaci, sub quo sit quispiam quinque planetarum, calculare*, **A1**, f. 143r; **A2**, f. 54ra)
23. An explanation of the warnings to keep in mind when working (*Cautelas in operacione custodiendas explanare*, **A1**, f. 143v; **A2**, f. 54va)
24. To recognise if the planet is retrograde or direct (*Si planeta sit retrogradus aut directus cognoscere*, **A1**, f. 143v; **A2**, f. 54vb)
25. Finding of the true position of the head of dragon (*Verum locum bensahar /genzahar/ Lune reperire*, **A1**, f. 144r; **A2**, f. 54vb)
26. To learn how to quickly expand the table of the true position of the Sun (*Qualiter verus locus Solis per totum annum celeriter extendatur, edocere*, **A1**, f. 144r; **A2**, f. 55ra)
27. Continuation of the true motions in the almanac for the days between the five days (*Veros motus in quinque pro diebus intermediis in almanach continuare*, **A1**, f. 144r; **A2**, f. 55rb)
28. How to form the true motion of the head of dragon throughout whole year (*Qualiter formatur verus motus bensahar Lune per totum annum ostendere*, **A1**, f. 144v; **A2**, f. 55rb)
29. Finding the first and any other mean conjunction and opposition (*Primam coniunccionem primamque opposicionem mediam ac aliam quamcumque invenire*, **A1**, f. 144v; **A2**, f. 55rb)
30. To determine the mean conjunctions and oppositions which immediately follow in the course of the whole year (*Medias coniuncciones et opposiciones sese immediate sequentes per totum annum constituere*, **A1**, f. 145r; **A2**, f. 56rb)
31. To explore the time of the true conjunction or opposition of luminaries (*Tempus vere coniunccionis aut opposicionis luminarium explorare*, **A1**, f. 145v; **A2**, f. 56va)
32. Collecting the aspects of the Moon to the planets and the planets to each other in an almanac (*Aspectus Lune ad planetas planetarumque adinvicem in almanach colligere*, **A1**, f. 145v; **A2**, f. 56vb)

33. Finding the latitude of the Moon using its argument of latitude (*Latitudinem Lune per argumentum latitudinis eius investigare*, **A1**, f. 146r; **A2**, f. 57ra)[19]
34. Finding the latitude of the three upper planets (*Latitudinem trium superiorum reperire*, **A1**, f. 146r; **A2**, f. 57rb)
35. Finding the latitude of Venus (*Latitudinem Veneris exquirere*, **A1**, f. 146v; **A2**, f. 57va)
36. Finding the latitude of Mercury (*Latitudinem Mercurii perscrutari*, **A1**, f. 147r; **A2**, f. 58ra)
37. Finding the time of the entry of the Sun into the first minute of Aries or any other sign (*Tempus, quo Sol intrat primum minutum Arietis aut alterius cuiuscumque signi, invenire*, **A1**, f. 147r; **A2**, f. 58va)
38. Finding the twelve horizontal houses on the sky for any hour (*Duodecim celi domos orisontales pro quotacumque hora scire*, **A1**, f. 147v; **A2**, f. 58vb)

## 5.3 Notes on Canons *Astrorum*

Compared to previous canons, the canons *Astrorum* are innovative in some respects. For example, the individual rules are preceded by chapters on addition, subtraction, multiplication and division; presumably this addition was necessitated by teaching practice. This was then followed by Nicolaus Wodka's *Quoniam plerique nostri temporis physici* and later also by Ambrosius Lacher and Johannes Schöner in their editions of the canons to the *Tabule resolute*. They too begin with similar explanations. It is therefore not impossible that canons similar to the text of Albertus of Brudzewo's *Astrorum* could have been the source or model for the early printed editions.

Judging by their similarity, it also seems that the canons *Astrorum* served as a basis for canons *Pro (introduccione) generali* by John of Głogów. Albertus of Brudzewo elaborates in detail individual rules for the steps necessary to compile almanacs. He also adds rules for the calculation of planetary latitudes and of the ascendant and astrological houses.

19 Canons 33–36 deal with tables which are already beyond the *Tabule resolute*.

# Canons *Quoniam Tabularum Alfonsi laboriosa difficultas* by Nicolaus of Tuchów (Polonus)

Just as the Wrocław tables, represented by the manuscript from Toruń, BU, 74/III, were recalculated for the Prague and then for the Cracow meridian and new versions of the canons were created for them, we can trace at least two other lines of *Tabule resolute* diffusion in later years, namely to Salamanca in Spain and Bologna in Italy. The scope of the *Tabule resolute* thus expanded again. The Spanish branch is represented by the tables and canons by Nicolaus of Tuchów (Polonus) around 1460, and the Italian branch by the tables and canons by Nicolaus Wodka of Kwidzyn around 1477. Let us notice them in more detail.

Firstly to the author of canons for Salamanca, to Mikołaj z Tuchówa (Nicolaus of Tuchów), known in abroad as *Polonus* ('Polish').[1] Nicolaus of Tuchów was a fifteenth century astronomer and physician who studied in Cracow and Paris. From the time he worked in Paris comes the manuscript Paris, BnF, lat. 7427, in which his Ephemerides for 1455 for Cracow are recorded on ff. 43v–55r. In a note on f. 43r we can read these words: *he tabule sunt almanach motuum planetarum anni ere Incarnacionis Christi 1455 ad latitudinem civitatis inclite Cracovie, calculate per Nicolaum de Thuchovia, in benedicto ac mihi multum complacito monasterio Sancte Trinitatis.* Examples of this ephemerides are given in Figs 20 and 21. We have used them to illustrate the detailed instructions by Nicolaus Wodka on how to make a *tacuinum* (i.e. almanac). Nicolaus of Tuchów then became the first head of the astronomy department in Salamanca University, Spain. The reason why Nicolaus of Tuchów left Salamanca before 1464 is unknown, but we know that after a short stay in Nantes he worked as an astronomer at the court of the Dukes of Burgundy between 1464 and 1472 and from 1476 he was a physician at the court of the Castilian Kings Isabella I and Ferdinand II of Aragon. He died before 1484.[2]

It was to the meridian of Salamanca, that Nicolaus of Tuchów adapted the *Tabule resolute* and also adapted the canons *Girum recensendo* by Andrzej Grzymała to canons *Quoniam Tabularum Alfonsi* around 1460. Andrzej Grzymała was, in fact, Nicolaus's teacher during his studies in Cracow. If, for example, we compare the copy of the canons *Quoniam Tabularum Alfonsi* by Nicolaus of Tuchów in MS Oxford, Bodl., Canon. misc. 27, ff. 122v–129r, with the copy of the canons *Girum recensendo* by Andrzej Grzymała in MS Cracow, BJ, 573,

1 An edition of Polonus's canons based on the MS Oxford, Bodl., Canon. misc. 27 was prepared by Porres de Mateo and Chabás, 'Los cánones de las Tabulae resolutae para Salamanca'. See also Chabás, 'Astronomy in Salamanca'. A description of these canons is missing in Rosińska, *Scientific Writings*, and Markowski, *Astronomica*.

2 Stanisław A. Sroka, 'Mikołaj z Tuchowa. Zapomniany astronom i lekarz z XV wieku oraz jego europejskie peregrynacje', *Roczniki historyczne*, 86 (2020), 177–85.

ff. 227r–233r, we find that their text is at least ninety percent identical.[3] Nicolaus of Tuchów has a different incipit and introductory text, then both canons are identical, the differences are only within the standard variant readings. In the rule dealing with radices, Grzymała indicates Cracow, of course, and Nicolaus has Salamanca. In the copy of Grzymała's text in MS Cracow, BJ, 573, the penultimate rule (*Gradum ascendentem aliosque celi cardines invenire*) is missing, but the determination of the ascendant and the domification is briefly mentioned at the end of the Rule 16 on the entry of the Sun into Aries, again with an incorrect reference to cusps as cardinal points.[4] Both authors share a rule on almanacs at the end of the work, but Grzymała's text ends a few sentences earlier. Like the canons *Girum*, canons *Quoniam* depend to some extent on John of Saxony's canons *Tempus est mensura motus*: their Rules 1–3 correspond more or less to Rules 12–15 of the *Tempus*, Rules 5–11 match more or less to Rules 16–21, and Rules 14–16 more or less to Rules 22–23.

## 6.1 Manuscripts

### Cambridge, Trinity College, O.3.13, ff. 74v–80v[5]

It is a well-read humanistic manuscript from the sixteenth century, containing various astronomical but mainly astrological texts.

Incipit and explicit of the canons *Quoniam Tabularum Alfonsi*: *Tabularum resolutarum canones feliciter expliciuntur. Quoniam Tabularum Alfonsi laboriosa difficultas... × ... sed in capite draconis differencia semper subtrahi debet, ex quo non retrogradatur neque dirigitur. Expliciuntur Tabul{e}arum resolutarum canones. Deo gracias.*[6] The TR are not present in this manuscript.

- Johannes de Sacrobosco's Sphere precedes the canons on ff. 58v–74r which are followed by Alcabitius's Introduction to astrology on ff. 80v–97v.[7]
- On ff. 10v–21r are the canons of John of Saxony, *Tempus est mensura motus*.
- Folios 26r–48r contain Parisian Alfonsine Tables.

### Lisbon, ANTT, ML, 2115, ff. 8r–18v[8]

The text of the canons *Quoniam Tabularum Alphonsi* begins at f. 8r as Polonus's adaptation of Grzymała's *Girum* to Salamanca (incipit and explicit: *Quoniam Tabularum Alfonsi*

3 The texts of the canons *Girum* in MS Cracow, BJ, 2252, pp. 21–44, and Prague, NK, VIII G 24, ff. 60v–64v, differ significantly in many places from manuscript Cracow, BJ, 573 and cannot therefore be compared here.
4 See the note on cardinal points in Chapter 3.2, Andrzej Grzymała, *Girum recensendo* – variant 3, Rule 18.
5 Montague Rhodes James, *The Western Manuscripts in the Library of Trinity College, Cambridge. A Descriptive Catalogue. I–IV* (Cambridge: Cambridge University Press, 1900–1904), p. 1185.
6 See <https://mss-cat.trin.cam.ac.uk/Manuscript/O.3.13/UV> and description <https://ptolemaeus.badw.de/ms/414> [accessed: 12.10.2022].
7 See edition by Charles Burnett, Keiji Yamamoto, and Michio Yano, *Al-Qabīṣī (Alcabitius): The Introduction to Astrology*, Editions of the Arabic and Latin texts and an English translation (London, Turin: The Warburg Institute – Nino Aragno Editore, 2004), p. 161.
8 See a general description and images at: <https://digitarq.arquivos.pt/details?id = 4248799> [digitarq.arquivos.pt].

*laboriosa difficultas* ... × ... *nec dirigitur*). Folio 8v is vacant but no text is missing, and it is followed directly by the rule 1 (incipit: *Medium motum Solis et Lune*). However, the text ends earlier (see f. 17r: *videlicet diebus non equatis et proveniet tibi tempus diebus equatis tempore introytum Solis in Arietem*). The rule on the composition of the almanac is omitted.

- On ff. 19r–56v there are contained the TR for Salamanca to the years already from 1348 (!) to 1788 with incipit: *Tabule de mediis et veris motibus planetarum super meridianum Salamantinum ordinate*; on ff. 59v–103r there continue different additional tables to the TR.

Oxford, Bodl., Canon. misc. 27, ff. 122v–129r

The edition of Nicolaus Polonus's canons was published in modern times, as mentioned above, on the basis of this manuscript.[9]

- On ff. 33r–62r there are the TR calculated for the meridian of Salamanca with starting point in the year 1348.
- Among other astronomical works in this manuscript, the canons *Priores astrologi* and *Cuiuslibet arcus propositi* by John of Lignères are found on ff. 130r–138r and 138r–145v, respectively.
- A collection of drawings with zodiacal constellations is contained on ff. 152r–157v. Their stars are marked in red colour and the surrounding description is written in Castilian.[10]

## 6.2. List of Rules

**Nicolaus of Tuchów (Polonus), *Quoniam Tabularum Alfonsi laboriosa difficultas*:** (Cambridge, Trinity College, O.3.13 = **Q1**; Lisbon, ANTT, ML, 2115 = **Q2**; Oxford, Bodl., Canon. misc. 27[11] = **Q3**)

1. On the mean motion of the Sun etc. (*De medio motu Solis. Medium motum Solis et Lune et aliorum planetarum et argumenta media Lune, argumentum latitudinis Lune, similiter argumenta Veneris et Mercurii, capitis drachonis et medium motum augium et stellarum fixarum, accessus et recessus octave spere per tabulas invenire,* **Q1**, ff. 74v–75r; **Q2**, f. 9r–v; **Q3**, ff. 122vb–123ra /pp. 61–62/)
2. On the mean conjunction and opposition of the Sun and Moon and on the possibility of eclipse (*De media coniunccione et opposicione Solis et Lune et possibilitate eclipsis. Tempus medie coniunccionis vel opposicionis Solis et Lune possibilitatemque eclipsis in mense quolibet per tabulas ad hoc factas reperire,* **Q1**, f. 75r–v; **Q2**, f. 9v–10v; **Q3**, f. 123rb–vb / pp. 62–64/)
3. On the correction of tables (*De correccione tabularum. Modum corrigendi Tabulas motuum et temporum ipsasque de novo faciendi aut continuandi subiungere,* **Q1**, ff. 75v–76r; **Q2**, f. 10v–11r; **Q3**, ff. 123vb–124rb /pp. 63–66/)

9 See Porres de Mateo and Chabás, 'Los cánones de las Tabulae resolutae para Salamanca', 60–83.

10 For details of the manuscript see Laura Fernández Fernández, 'Alfonsine Astronomy, Salamanca and the Oxford Canon. misc. 27', in preparation.

11 Prepared according to the edition by Porres de Mateo and Chabás, 'Los cánones de las Tabulae resolutae para Salamanca', 60–83.

4. On converting the radices (*De reduccione radicum. Radices Salamantinas ad meridianum tuum reducere… × … secundum doctrinam canonum astrolabii, scilicet per eclipses,* **Q1**, ff. 76r; **Q2**, f. 11r–v; **Q3**, ff. 124rb–124va /pp. 66–67/)[12]
5. On the true position of apogees (*De vero loco augium. Verum locum augis cuiuscumque planete per tabulam invenire,* **Q1**, f. 76r–v; **Q2**, ff. 11v–12r; **Q3**, ff. 124va–124vb /pp. 67–68/)
6. General way of entering Tables of equations (*Modum generalem in Tabulas equacionum intrandi proponere,* **Q1**, ff. 76v–77r; **Q2**, f. 12r–v; **Q3**, ff. 124vb–125rb /pp. 68–69/)[13]
7. On the true position of the Sun (*De vero loco Solis. Verum locum Solis per tabulas invenire,* **Q1**, f. 77r; **Q2**, f. 12v; **Q3**, f. 125rb /pp. 69–70/)
8. On the true position of the Moon (*De vero loco Lune. Verum locum Lune per tabulas invenire,* **Q1**, f. 77r; **Q2**, f. 13r; **Q3**, f. 125rb /p. 70/)
9. On the true position of the head of dragon (*De vero loco capitis draconis. Verum motum capitis draconis per tabulas invenire,* **Q1**, f. 77r; **Q2**, f. 13r; **Q3**, ff. 125rb–125va /p. 70/)
10. On the true position of the three upper planets, i.e. Saturn, Jupiter, and Mars (*De vero loco trium superiorum. Verum locum trium superiorum, Saturni scilicet, Iovis et Martis, invenire,* **Q1**, f. 77r–v; **Q2**, ff. 13v–14r; **Q3**, ff. 125va–125vb /pp. 71–72/)
11. On the true position of Venus and Mercury (*De vero loco Veneris et Mercurii. Verum locum Veneris et Mercurii invenire,* **Q1**, f. 77v; **Q2**, f. 14r–v; **Q3**, ff. 125vb–126ra /p. 72/)
12. On direction, retrogradation, and station of the planets (*De direccione, retrogradacione et stacione planetarum. Utrum planeta sit stacionarius, retrogradus aut directus, invenire,* **Q1**, ff. 77v–78r; **Q2**, f. 14v; **Q3**, ff. 126ra–126rb /p. 73/)
13. If the planet is ascending or descending (*Utrum planeta sit ascendens vel descendens,* **Q1**, f. 78r; **Q2**, f. 15r; **Q3**, f. 126rb /p. 74/)
14. On the true conjunction and opposition (*De coniunccione et opposicione veris. Tempus vere coniunccionis et opposicionis Solis cum Luna invenire,* **Q1**, f. 78r–v; **Q2**, f. 15r–16r; **Q3**, ff. 126rb–126vb /pp. 74–75/)
15. On the conjunction of any two planets (*De coniunccione duorum planetarum quorumcumque. Tempus coniunccionis Saturni, Iovis et quorumlibet planetarum invenire,* **Q1**, f. 78v–79r; **Q2**, f. 16r–v; **Q3**, ff. 126vb–127ra /pp. 75–76/)
16. On the entry of the Sun into any sign (*De introitu Solis in quemcumque signum. Introitum Solis in Arietem vel in quodcumque signum volueris invenire,* **Q1**, f. 79r–v; **Q2**, f. 16v–17r; **Q3**, ff. 127ra–127va /pp. 76–78/)
17. On finding of the ascendant and other cardinal points (*De invencione ascendentis et aliorum. Gradum ascendentem aliosque celi cardines invenire,* **Q1**, ff. 79v–80r; **Q2**, f. 17r–18r; **Q3**, ff. 127va–128rb /pp. 78–80/)
18. On the composition of an almanac for individual years (*De composicione almanach. Almanach pro singulis annis invenire,* **Q1**, f. 80r–v; **Q2**, f. 18r–v; **Q3**, ff. 128rb–128vb / pp. 80–81/)

12 In **Q2** and **Q3**, the text of the Rule 4 is significantly expanded compared to **Q1**. The text of **Q2**: *Nota, quod Gades Herculis… × … longitudinem earum ab oriente*; the text of **Q3**: *Que sequuntur, non sunt de canone… × … longitudinem earum a vero oriente.*

13 This rule is inserted in addition to the canons *Girum* by Andrzej Grzymała.

## 6.3 Notes on Canons *Quoniam Tabularum Alfonsi*

It is known that Nicolaus of Tuchów (Polonus) wrote his canons *Quoniam Tabularum Alfonsi* in Salamanca on the basis of Grzymała's canons *Girum*, which he knew from Cracow. By a detailed collation of the texts, we have found that Polonus's canons are closest to the version of the *Girum* canons represented by MS Craców, BJ, 573 (i.e. variant 1). They differ from them only in the opening paragraph and the conclusion; the core of the two canons shows only very slight differences at the level of variant readings.

In the chapter on radices, Polonus of course lists Salamanca, Grzymała Cracow. Polonus adds a rule on the ascendant and cardinal points. This chapter is missing in Grzymała, but both texts have a common chapter on almanacs at the end, only in Grzymała it ends a little earlier.

# Canons *Quoniam plerique nostri temporis phisici* by Nicolaus Wodka of Kwidzyn

The author of the canons *Quoniam plerique* was the Polish astronomer Mikołaj Wodka z Kwidzyna[1] (Nicolaus Wodka of Kwidzyn). Sometimes he was also called *Abstemius*, that is 'Abstinent', because his name 'Wodka' is a diminutive of the word 'water' in Polish. Nicolaus Wodka is identical with Nicolaus de Prusia, named in MS London, BL, Add 22317. Nicolaus Wodka was the first astronomer who brought the Cracow TR to the Apennine Peninsula and who recalculated them for the local meridian at the University of Bologna before 1480 (probably 1477).[2] His work was therefore of similar importance to that of Nicolaus of Tuchów (Polonus) in Salamanca. As far as we know, Wodka's *Tabule resolute* and canons to them are attested only in the two manuscripts mentioned below.

Nicolaus Wodka was born *c.* 1442. In 1462 he enrolled at the University of Cracow and studied there until 1467. He then continued his studies at the Universitas Istropolitana in Pressburg (today's capital of Slovakia, Bratislava) and in the second half of the 1470s his stay in Bologna is documented. He worked at the university as a professor and he also received his doctorate in medicine here. He met and collaborated with a number of important contemporary astronomers, but also studied older works, such as those of Bianchini. Like Nicolaus of Tuchów, he was sometimes called Nicolaus Polonus because of his origin. After 1480 he returned to Poland, first to Poznań and later to Włocławek in central Poland where he worked as an astrologer and physician. In Włocławek he prepared the young Nicolaus Copernicus for his studies at the University of Cracow and together they made a sundial for the south side of the local cathedral in 1490 (see Fig. 19).[3] He died 1494/95.

Wodka's canons *Quoniam plerique* are somewhat different from other sets of rules for using the *Tabule resolute*. For example, he included detailed procedures and practices on how to make an almanac of the planets. In the TR canons, authors often refer to the creation of almanacs as a practical application of theoretical tables. Almanacs were analogous to astronomical calendars with computed positions of the planets for each day or only for selected days. Advice on how to compute almanacs were mentioned already by Andrzej

1 See Ludwik A. Birkenmajer, *Mikołaj Wodka z Kwidzyna zwany Abstemius. Lekarz i astronom Polski XV-go stulecia* (Toruń: Towarzystwo naukowe, 1926), p. 56; see also *Corpus Academicum Cracoviense (CAC)*, <http://cac.historia.uj.edu.pl/osoba/819945_Miko%C5%82aj_Wodka_z_Kwidzyna_syn_Grzegorza> [accessed: 12.03.2023].

2 Markowski, *Astronomica*, p. 160a; Rosińska, *Scientific Writings*, pp. 354–55 (item 1818); L. A. Birkenmajer, *Mikołaj Wodka z Kwidzyna*. – These tables will be discussed below in Chapter 15.3.

3 L. A. Birkenmajer, *Mikołaj Wodka z Kwidzyna*, pp. 104–5.

**Figure 19.** Sundial on the wall of the Basilica Cathedral in Włocławek, created by Nicolaus Wodka of Kwidzyn and Nicolaus Copernicus in 1490. Reconstructed at the end of the nineteenth century. Jolanta Dyr, Creative Commons.

Grzymała and later by John of Głogów[4] and others, but Nicolaus Wodka describes in detail the process of how to prepare an almanac in step-by-step stages. While Grzymała and John of Głogów deal with the computation of the almanacs, Wodka also describes their layout, arrangement, the form and the process of drawing the grid into which the individual values were then entered (see his Rule 14). In Rule 16 he describes the work with a kind of table which does not belong to the TR set but is helpful for calculations of the daily motion of the Moon in the almanac. We have found that this double-argument table is similar to the table of Nicolaus Mülhus, which is known to us from five copies. Overall Wodka's text provides a detailed guide to creating an almanac.

However, neither the term 'almanac' nor term 'ephemerides' appears in Wodka's canons: he uses the synonymous expression 'tac(h)uinum'.[5] The term *tac(h)uinum* (from *taqwīm* in Arabic) means table, resp. arrangement into a table, into an almanac. The word *tacuinum* was used by an eleventh century Baghdad scholar Abū l-Hasan (Ibn Buṭlān) and this term became widely adopted. Wodka may have encountered the term *tacuinum* in northern Italy. At the University of Bologna (founded as early as 1088), almanacs in the

4 See his canons below, Chapter 8.

5 Lynn Thorndike and George Sarton, 'Question No. 10 – Tacuinum', *Isis* 10 (1928), 489–90; 'Answer No. 10 – Tacuinum, taqwīm', *Isis*, 10 (1928), 490–93.

form of *tacuinum* arrived as early as the thirteenth century.[6] An overview and typology of the pre-alfonsine tacuins is given by C. Philipp E. Nothaft.[7] Most relevant to *tacuinum* are Wodka's Rules 14–20, but also 25 and others.

To the partial translation and comments of our edition of the canons by Nicolaus Wodka based on both surviving manuscripts, we attach for illustration two examples (see Fig. 20 and 21 below) from the almanac whose author was Nicolaus of Tuchów (Polonus), known by his version of the *Tabule resolute* calculated for Salamanca. The almanac is in the MS Paris, BnF, lat. 7427. For comparison we also used MS Vatican, BAV, Vat. lat. 4572, ff. 1r–102v which contains an *almanach planetarum* for the years 1293–1303. Concerning texts, another short treatise titled *De sciencia faciendi almanac, id est tacuinum* from the fifteenth century is found in MS Vatican, BAV, Vat. lat. 3099, ff. 9v–10v. The manuscript Brno, MZK, A 64, f. 192r–193v, from the second half of the fifteenth century, provides further very brief notes on this subject.

## 7.1 Manuscripts

Milan, BA, C 278 b Inf., ff. 65r–76v[8]

Incipit and explicit of the canons: Quoniam plerique nostri temporis phisici ab ea superiorum sciencia celorum laborum onustate tumque tabularum prolixitate… × … volumus tamen quatenus more nostro probacionem facias. The canons comes from the year 1477, as it is also indicated on f. 76r: Introitus Solis in Arietem et revolucio annorum mundi pro meridiano Bononiensi Anno Domini 1477 currente. In the past, the authorship of the accompanied TR was erroneously attributed to Bianchini. At the bottom of the first folio is noted: Liber Domini Grimani, Cardinalis S. Marci;[9] the book was owned by the cardinal Domenico Grimani (1461–1523). Other fifteen manuscripts from his possession are preserved in the Milan, Biblioteca Ambrosiana.

– On ff. 1r–31v, there are the TR for Bologna meridian, designed as usual for years 1428 to 1808.[10] On f. 12v, there is a *Tabula civitatum*.[11] From ff. 32r–64r additional tables of the TR continue and after the blank folio 64v the canons to the TR are present. Starting from f. 32r, there are firstly tables that are needed for the creation of *tacuinum*[12] that are

6 The term *tacuinum* also penetrated other literary genres: in northern Italy, for example, a text called *Tacuinum sanitatis* (Tables of health) was written, containing instructions for a healthy lifestyle, often mixed with astrological predictions. A Latin version spread from this region to the rest of Europe and was popular for several centuries.

7 C. Philipp E. Nothaft, 'Ephemerides in high medieval Europe: The textual evidence', *Journal for the History of Astronomy*, 51 (2021), 33–52.

8 ThK 1293.

9 *Inventario ceruti dei manoscritti della Biblioteca Ambrosiana*, Vol. 1 (Trezzano: Editrice Etimar, 1973), p. 355. We are indebted to Giancarlo Truffa for the detailed description of the manuscript he prepared for us at a time when we did not yet have a scan of the manuscript at our disposal.

10 Markowski, *Astronomica*, p. 160a.

11 See also Lynn Thorndike, 'Notes on Some Astronomical, Astrological and Mathematical Manuscripts of the Bibliothèque nationale, Paris', *Journal of the Warburg and Courtauld Institutes*, 20 (1957), 138.

12 F. 32r: *Tabule moderantes prolixitates laboris tachuini: Tabula Solis*; f. 32v: *Tabula Lune in diebus post coniuncciones Solis et Lune medias*; f. 33r: *Tabula horarum distancie coniunccionis a meridie*; f. 33v–39r: *Tabula equacionis Lune in diebus lunacionum*; f. 39v (left): *Tabula argumenti Lune in horis*; (right): *Tabula argumenti Lune in minutis*

not found in the TR of other authors; other additional tables continue. However, the almanac itself, called *tacuinum*, the composition of which Nicolaus Wodka describes in his canons, is not attached to the TR in the Milan manuscript.

### London, BL, Add 22317, ff. 1r–18r

The copy was probably written shortly after 1477. In several places in his description, for example, the scribe skipped lines and omitted text. It is interesting to note that in some places where the London copy uses a different word from the Milan manuscript (in fol. 5r, e.g., the original words *super addito* are substituted for the word *residuo*) scratched-out words *super addito* that are identical to the wording of the Milan manuscript seem to shine through under the altered word. The London copy was therefore probably corrected in several places after it was made.

Incipit and explicit of the canons: *Incipiunt canones Tabularum resolut{iv}arum ad meridianum Bononie, Italie, et ad quemcumque alium locum compositi per egregium doctorem, magistrum Nicolaum de Prusia. – Quoniam plerique nostri temporis phisici ab ea superiorum sciencia celorum laborum onustate tum tabularum prolixitate ...* × ... *volumus tamen quatenus more nostro probacionem facias.*

- On ff. 25r–89v are the TR calculated for Bologna meridian.[13] A geographical table inscribed *Tabula locorum insignium Europe*, beginning with the Pillars of Hercules (*Gades Herculis*), is incorporated on f. 36v.
- The manuscript contains also Regiomontanus's Calendarium with tables that had been printed in 1474.[14]

---

*horarum*; f. 39bis r: *Tabula Saturni*; f. 39bis v: *Tabula Iovis*; f. 40r: *Tabula Martis*; f. 40v: *Tabula Veneris*; f. 41r: *Tabula Mercurii*.

13 Markowski, *Astronomica*, p. 160a.

14 *Catalogue of Additions to the Manuscripts in the British Museum in the year 1854–1860. Additional Mss. 19,720–24,026* (London: The Trustees, 1875), p. 632; L. A. Birkenmajer, *Mikołaj Wodka z Kwidzyna*, p. 57 and 132–33.

## 7.2 Edition of Canons *Quoniam plerique* by Nicolaus Wodka of Kwidzyn

Sigla

**A** – Milan, Biblioteca Ambrosiana (BA), C 278 b Inf., ff. 65r–76r

**B** – London, British Library (BL), Add 22317, ff. 1r–18r

Incipiunt canones *Tabularum resolut{iv}arum* ad meridianum Bononie, Italie, et ad quemcumque alium locum, compositi per egregium doctorem, magistrum Nicolaum de Prusia.

**Prohemium**

Quoniam plerique nostri temporis phisici ab ea superiorum sciencia celorum laborum onustate tumque tabularum prolixitate sopiti resiliunt, curavimus nonnullos eorum opitulante nobis Domino Deo astrorum discipline zelotipos efficere facilem his modum, quo tachuinum conficere possunt, ex diversis tabulis diffusisque amphoristice consertum hoccine propinantes.

**Divisio orbis**

Orbis signorum quemadmodum spere circulus omnis in senas apud veteres scinditur partes, quas signa phisica nominamus. Horum quodlibet in sexaginta gradus secarunt. Signorum quopiam unumquodque phisicorum in binas modernas partes libuit dividere, quas duodecim signa communia in zodiaco nominibus animalium in circulis aliis animali denominacione nuncupamus, que signa communia pro cursibus superiorum dinoscendis ad presens quam magis ponderabimus.

Ex tempore quoque diem anni porcionem in 24 partes dividimus, quas horas nominamus. Hunc diem unam revolucionem celi continentem diem equalem et eius horas horas equales dicimus, eum quoque diem cum suis horis in tabulis nostris audiemus. Rursus dies equalis in duo rescinditur, in partem videlicet, qua Sol supra orizontem, et in partem, qua Sol sub orizonte moratur. Pars prima dies inequalis, altera nox inequalis in omni regione citra equatorem appellatur. Dies itidem inequalis, nox quoque inequalis duodecim suas habent horas inequales, que hore inequales ab horis equalibus alie planetarum hore dicuntur. Rerum ceterum umbras et res ipsas luminariumque corpora duodenis punctis porcionibus maioribus mensuramus. Porro gradus et horas quaslibet rerum puncta in 60 minuta, minutum unumquodque in 60 secunda, secundorum quodlibet in 60 tercia, terciorum singulum in 60 quarta et sic de reliquis minutatim subdividimus. Est et alia apud Alphonsium equalis diei sive horis in 60 particio, que minuta diei nominantur. Hec eciam minuta in alias porciunculas minimas, ut reliquarum porcionum minuta profluunt. Cordam eciam maioris arcus et dyametrum in 300 minuta secamus, que apud quosdam 120 gradus ferme continent. Harum parcium ut antefatorum ulterior in minima fit refraccio, cuiuslibet eciam arcus corde medietatem synum rectum appellamus. Pars dyametri, que est intra arcum et cordam et synum rectum cordam et arcum dimidians, sinum versum intitulamus. Sinus sine adieccione cum dicimus sinum rectum denotamus. Porciones, que ante minuta sunt integra, vocari consuevimus minu-

**1/4** Incipiunt… Prohemium *B* : *om. A* **6** tumque *A* : tum *B* **7** his *A* : hiis *B* **9** hoccine *A* : huccine *B* **10** Divisio orbis *B* : *om. A* **11** senas *A* : scenas *B* **19** horas *A* : *om. B* **35** denotamus *A* : appellamus *B*

ta et subdivisiones 'ceteras fracciones' appellamus. Integra a 0, minuta ab 1, secunda a 2, tercia a 3, quarta a 4 et sic de ceteris denominamus, quos numeros una cum 0 numeros denominacionum vocari statuimus.

Hiis generalibus perspectis de arismetrica, qua usus tabularum facilius habeatur, aliquid seminemus.

**De addicione**

In omni addicione, ubi porciones valentes, porcionem in medietate grossiorem aut plures apparuerint, eis porcionibus ab eo loco depositis additur pro eis unitas porcionibus in medietate grossioribus et residuum, si quod fuerit, in loco proprio conservatur, si non, in eo loco cifra reponatur.

**De subtraccione**

Numerus subtrahendus maior si fuerit numero, a quo debuit fieri subtraccio, tunc unum quodque immediate est grossioris denominacionis in partes suas resolutum additur numero, a quo talis debeat fieri subtraccio, et a provenienti tunc numero numerus subtrahendus deponitur, residuo in suo loco manente.

**De multiplicacione**

Ex multiplicacione numeri provenientis eiusdem quandoque plerisque diverse sunt denominacionis, quorum denominaciones facile sic deprehendes. Amborum numerorum numeri, videlicet multiplicantis et numeri multiplicandi, denominacionis numeros considera, qui numeri insimul iuncti producent numerum, a quo numerus ex multiplicacione proveniens denominatur, ut minuta in minuta multiplicata producunt secunda, minuta in secunda faciunt tercia et sic de aliis. Integra in integra ducta conficiunt integra et integra in fracciones et econverso multiplicata constituunt fracciones. Est eciam una regula communis, que pro alleviacione practice multiplicacionis usu tenetur. Cum volueris numeros per articulum multiplicare, illis numerus figuras nichili istius articuli antepone et numerum productum per digitum articuli multiplica et multiplicacionis summam leviter accipies. Numeros diverse denominacionis per multiplicacionem ad eandem denominacionem reducere: multiplica numerum grossioris denominacionis per 30, si habueris signa et gradus, et provenientibus tunc gradibus adde gradus, quos habes, et sic tocius numeri summa graduum resultabit. Si habueris dies et horas, multiplica dies per 24 et resultantibus horis adde horas, quas habes, et tocius numeri horarum summam procreabis. Si autem habueris integra cum minutis, integra in 60 multiplica et provenientibus tibi minutis minuta, que habes complicata, tocius numeri summa minutorum emanabit. Si autem habueris minuta et secunda minutis in 60 multiplicatis habebis secunda, quibus adde secunda, que habes sicque unius denominacionis numeros ubique conficies. Quando aliquis numerus absque denominacione numeri multiplicantis multiplicari demandatur, numerus resultans cum numero multiplicando eandem servabunt denominacionem, ut si multiplico 10 minuta per 5, proveniunt minuta 50, sic eciam si multiplico 3 gradus per 11, facio gradus 33.

---

**39** Hiis *A* : hic *B* **40** seminemus *A* : tractabimus *B* **57** faciunt *A* : producunt *B*

## De divisione

Numeri quocientis in divisione eciam diversus est respectus eiusdem etenim et quandoque diverse a divisore fit denominacionis. Numeri igitur quocientis denominacionem siccine percipias: subtrahe numerum denominacionis divisoris a numero denominacionis numeri dividendi, videlicet minorem a maiori. Numerus tunc remanens erit numerus, a quo numerus quociens denominabitur. Ut si diviseris secunda per minuta, tunc numerus quociens erit minuta. Si tercia per minuta, tunc quociens erit secunda et sic de aliis. Eodem modo integra per integra partita fit quociens integra, fracciones per integra si diviseris, tunc in quociente fracciones invenies. Residuum numeri dividendi a denominacione numeri dividendi non erit denominacionis alterius. Quando aliquis numerus, qui dividi debeat, diverse denominacionis fuerit, opus erit eum prius ad eandem denominacionem reducere. Quando numerus dividendus erit minor numero divisore, eum ad porciones subtiliores reducere oportebit et sic dividere. Quandocunque aliquis numerus sine denominacione divisoris dividitur, numerus quociens et numerus, qui divisus est, eandem habebit denominacionem, ut si divido minuta per 5 vel per 10, in quociente erunt minuta.

Summam fraccionum in porciones grossiores per divisionem convertere: divide fracciones per 60, si plures fuerint, et numerus quociens erit fraccio immediate grossior. Residuum quoque eiusdem erit denominacionis, cuius fuit ante divisionem, ut si divido secunda per 60, in quociente erunt minuta. Si divido minuta, in quociente erunt integra. Et si integra fuerint gradus, eos divido per 30, si plures extiterint, et in quociente erunt signa et in residuo gradus. Si vero integra fuerint plures quam 24 hore, eis per 24 divisis quociens erunt dies et residuum hore et sic de reliquis.

<1>

## <De medio cursu>

Cum cuiuslibet planete et capitis draconis medium cursum, argumenta quoque media Lune, Veneris et Mercurii libuerit invenire, annorum Christi Domini perfectorum numerum in *Tabula annorum collectorum* cuiusvis eorum perquire vel eo minorem tamen propinquiorem, et quod in eius directo ex signis et gradibus, minutis atque secundis inveneris, sigillatim eo ordine, quo sunt in tabula, extra scribe. Deinde si qui remanserint anni in *Tabula annorum collectorum* non inventi, eis in *Tabula annorum expansorum* planete illius inventis ea, que in directo eorum ex signis, gradibus, minutis atque secundis inveneris, omnia ordinatim sub prius extra scriptis pone, signa videlicet sub signis, gradus sub gradibus, minuta sub minutis, secunda sub secundis. Deinde omnia eiusdem generis observatis regulis addicionum in unum collige, quodque ex colleccione tali provenerit, erit motus medius planete, quem volebas, ad meridiem ultime diei Decembris e dictorum annorum pro meridiano alme civitatis Bononiensis, cuius longitudo et latitudo in *Tabula* clarent *regionum*. Quos medios motus, scilicet medios cursus, et argumenta media planetarum pro radicibus tocius anni sequentis observabis. Ceterum si aliquot istorum ad aliquem diem anni sequentis, scilicet anni

**79** videlicet *B* : vel econverso *A* **80** tunc *B* : vel econverso *A* **81** tunc *B* : vel econverso *A* **83** tunc *B* : vel econverso *A* **89** habebit denominacionem *A* : denominacionem habebit *B* **91** divisionem *A* : divisorem *B* **107** inveneris *A* : inventis *B* **111** e *A* : uni e *B*

Christi incomplete, habere placuerit, intrabis primo cum mensibus, si qui perfecti fuerunt, in *Tabulis mensium communium aut bisextilium*, secundum quod annus evenerit. Postea intrabis cum diebus in *Tabulam dierum*, si autem nullus mensis fuerit completus, operare dumtaxat cum diebus, et que eis debentur ex signis et gradibus et fraccionibus e directo positis, radici sue coniunge et in aggregato medius motus planete, quem queris, apparebit ad meridiem diei propositi. Porro si hec ad aliquas horas post meridiem perquirere volueris, cum eis horis intra *Tabulam horarum*, et que ex gradibus et fraccionibus inveneris e directo, adde mediis motibus diei et proveniet tibi medius cursus planete ad horas propositas aut argumentum medium planete. Rursus si cum horis fuerint minuta horarum, cum eis eandem *Tabulam horarum* ingredere, et quod ex minutis et secundis eis debentibus reperieris, mediis motibus horarum adiungito, et hoc modo medii motus ad minuta horarum mox apparebunt. Hic quibusdam cautelis opus est uti: *Tabula horarum* numerum 30 non excedit. Si ergo minuta eo plura fuerint, prius intra eandem tabulam cum 30 minutis, postea cum residuis, sic quoque fac cum secundis, si habueris qua. Et nota intrando *Tabulam horarum cum minutis horarum* ex directo minutorum horarum in columpna graduum erunt minuta, deinde in columpna minutorum erunt secunda et in columpna secundorum erunt tercia, et postea quarta quemadmodum supposita tabularum earumdem loricacio admonet. Annum bisextilem futurum ex introitu tabularum hoc modo considerabis: si anni expansi perfecti fuerint 3, 7, 11, 15 aut 19, noveris annum futurum fore bisextilem, quem et nota, licet in tabulis posita prodit, et tunc cum mensibus perfectis anni incompleti *Tabulam mensium bisextilium* intrare iuberis.

<2>

### De augibus planetarum

Auges planetarum si placuerit agnoscere: intra cum annis Christi perfectis vel eis minoribus propinquioribus in *Tabula augium planetarum annorum collectorum* et auges planetarum e directo repertas extra scribe. Deinde si qui fuerint anni residui in tabula non inventi, cum eis intra *Tabulam augium annorum expansorum* et minuta et secunda, que inveneris e directo sub annis, sub quibus auges acceperas, unicuique augi coniunge et habebis auges propositas. Aux Veneris eadem est cum auge Solis. Item habita auge Solis ceterorum planetarum auges cum addicione illius, quod in parva tabula in margine apparet, per alium modum poteris habere. Inventaque auge uniuscuiusque planetarum ad principium anni hec pro tocius anni calculo valens erit eo, quod notabiliter ob tarditatem motus octave spere modo non variatur.

<3>

### De mediis coniunccionibus planetarum, scilicet Solis et Lune

Tempus medie coniunccionis Solis et Lune si volueris indagare, intra cum annis Christi completis *Tabulam annorum collectorum medie coniunccionis Solis et Lune*, quos si inveneris precise in directo eorum, habebis coniunccionem, que erit in Ianuario anni sequentis. Habebis eciam medium motum Solis et Lune et aggregatum Lune ad idem tempus. Si autem annos Christi precise non inveneris, tunc illud, quod in directo anno-

---

**132** futurum *A* : *om. B* **133** perfecti fuerint *A* : fuerunt perfecti *B* **138** De… planetarum *B* : *om. A* **145** parva tabula *A* : tabula parva *B* **150** De… Lune *B* : *om. A* **154** Lune et aggregatum Lune *A* : Lune *B*

rum Christi proxime minorum in eadem tabula reperieris ex tempore et motibus, extra scribe. Demum cum residuis annis Christi in tabula non inventis expansorum tabulam mox sequentem ingredere et illud, quod de tempore et motibus istic offertur tempori et motibus annorum collectorum adunabis, et quod tunc collectum fuerit ex tempore de diebus, horis, minutis atque secundis, erit tempus medie coniunccionis Solis et Lune. Similiter quod colligitur ex mediis motibus Solis et Lune et argumento Lune, erunt medii motus ad tempus coniunccionis pretactum. Si hec coniunccio cum annis Christi inventa habuerit dies minus 31, ipsa erit ante inicium anni sequentis in diebus videlicet Decembris anni perfecti. Si autem fuerint dies precise 31, erit coniunccio talis ultima die Decembris anni completi. Si vero fuerint dies plures 31, ab eis depone 31 dies et remanebunt dies Ianuarii anni sequentis, scilicet anni incompleti, in quibus talis coniunccio perficitur. Item ex dictis coniunccionibus ea, que fit in Ianuario, erit prima coniunccio anni et radix omnium aliarum coniunccionum sequencium illius anni in anno non embolismali et in anno embolismali ante embolismi coniunccionem. Si autem evenit coniunccio ante inicium anni in diebus Decembris, unam lunacionem, quam in margine invenias, ei coniunccioni coniunge, tempus videlicet tempori et motus motibus, et ex tali collecto sine alio subtrahe dies 31 et proveniet anni coniunccio prima cum motibus suis in diebus Ianuarii anni incompleti, que erit radix, ut predictum est omnium aliarum coniunccionum pro meridiano Bononiensi. Coniunccionem Solis et Lune in diebus cuiuscunque mensis usualis sive solaris invenire: intra *Tabulam mensium coniunccionis Solis et Lune cum mense perfecto sive completo*, qui est inmediate ante mensem, in cuius diebus coniunccionem queris, et tempus cum suis motibus illic compertum adde radici coniunccionum et a collectis taliter diebus Lune subtrahe dies solares anni communis ex directo apparentes, si fuerit annus communis, vel dies solares anni bisextilis, si fuerit annus bisextilis, et remanens cum motibus suis erit coniunccio mensis illius quesita.

Habita coniunccione Solis et Lune media, si opposicionem et Lune mediam quesieris, adde tempus cum motibus suis a coniunccione ad opposicionem, quod in margine notatur ad tempus coniunccionis, cuius vis opposicionem, et proveniet tempus opposicionis Solis et Lune cum motibus suis quesitum. Quod si proveniens, hoc tempus opposicionis habuerit dies solares sive usuales minus, quam habet mensis, in quo fit coniunccio, tunc coniunccio et opposicio fiunt in diebus unius mensis usualis. Si vero tempus opposicionis excreverit dies mensis usualis, in quo fit coniunccio talis, ab eius diebus opposicionis sic excrescentibus remove tot dies usuales, quot habet mensis usualis, in quo talis coniunccio fuit, et proveniet tempus opposicionis cum motibus suis in diebus usualibus mensis sequentis. Eritque notandum, si in practica coniunccionum dies solares lunaribus diebus maiores evenerint ita, quod dies solares a diebus lunaribus, prout mandatum fuit, minime subtrahi poterunt, tunc in eodem mense completo due celebrantur coniuncciones, quod in annis embolismalibus solet evenire, quarum prima fit in primis diebus illius mensis completi. Secunda, que est mensis sequentis, in ultimis eiusdem mensis diebus instat, in quibus autem diebus ultimis hoc fit, sic scire poteris. Quere coniunccionem precedentem, que fit in primis diebus illius mensis completi, cui adde

**157** Christi *A* : *om. B* **158** offertur *A* : offertum *B* **168** non *A* : *om. B* **174** Bononiensi *A* : Bononie *B* **175** invenire *A* : *om. B* **178** solares *A* : usuales *B* **179** solares *A* : usuales *B* **187** diebus opposicionis *A* : opposicionis diebus *B* **188** remove *A* : move *B* **192** celebrantur *A* : celebrabuntur *B* **193** prima *A* : prima *(in mg. add.*: que est embolismalis*) B* **194** sequentis *B* : embolismalis *A* **195** instat *A* : *om. B*

unam lunacionem cum motibus suis, et apparebit coniunccio sequens videlicet cum diebus et horis et motibus suis quesita, que coniunccio embolismalis debet concordare cum horis et motibus, quos habuisti, quando dies solares subtrahere non valebas. Adde eciam unam lunacionem cum motibus suis ad radicem coniunccionum, et quod provenerit nullo die sublato, erit radix omnium illarum aliarum coniunccionum post coniunccionem embolismalem. Cum qua coniuncciones relliquas modis nominatis investigabis. Cum habueris omnes coniuncciones, que sunt in anno, et dubitaveris, cuius mensis sint, scilicet an illius, in quo fit coniunccio, an sequentis, de hac hesitacione modo ruralium sic certificaberis. Vide coniunccionem Marcii, que semper est coniunccio illa, que est proxima ante Carnisprivium, vel illa, que precesserit Quadragesimam inmediate. Ab ea itaque coniunccione Marcii baptiza ceteras coniuncciones, scilicet Aprilis, Maii, Iunii et sic de aliis, et ubi duas coniuncciones in uno mense usuali reperieris, secunda coniunccio, que nullius mensis erit, dicetur embolismalis. Qua transacta baptizacionem aliarum coniunccionum usque ad finem anni continuabis. Item habita prima coniunccione anni coniuncciones relliquorum mensium eciam sine *Tabula mensium* hoc modo poteris habere: adde ad primam coniunccionem unam lunacionem cum motibus suis, et si addicione facta excreverint dies usuales Ianuarii, eis sublatis proveniet coniunccio sequens in Februario, cui iterum adde lunacionem. Et si excreverint dies Februarii, ipsis sublatis a diebus excrescentibus remanebit coniunccio, que fit in Marcio, et sic de aliis coniunccionibus usque ad finem anni, semper dies mensis, in quo facta est coniunccio, si post addicionem unius lunacionis excreverit a diebus excrescentibus deponendo. Is modus poterit priorem et econverso certificare.

<4>

## De medio argumento latitudinis Lune ad tempus medie coniunccionis

Medium argumentum latitudinis Lune ad tempus medie coniunccionis conficere. Medium cursum capitis draconis ad idem tempus constitue, quem adde medio motui Lune, proveniens erit argumentum latitudinis Lune medium. Habito quoque capite draconis ad unam coniunccionem, ipsam ad reliquas coniuncciones facile poteris ducere per addicionem, videlicet motus capitis draconis a coniunccione ad coniunccionem, qui in margine apparet. Sic eciam ad opposicionem per eius motum et ex hoc argumentum latitudinis Lune medium ad omnem coniunccionem vel opposicionem componitur.

---

**197** sequens videlicet *B* : embolismalis *A* **198** suis… motibus *A* : *om. et in mg. add.*: suis quesita, que coniunccio sic inventa debet concordare cum horis et motibus *B* **208** secunda *A* : prima *B* **209** erit *A* : *om. B* **210** continuabis *A* : continuabis *(in mg. add.*: vel aliter: in quo mense solari mensis lunaris suos terminat dies illius mensis solaris, est econverso mensis lunaris, qui preteriit, quoniam omnia a fine denominare iubemur, ut si duo menses lunares in uno mense solari finiuntur, secundus dicetur embolismalis, cuius coniunccio fuit in primo vel secundo vel tercio die mensis eiusdem. Item vel sic illius mensis est nova Luna, in quo eadem inveterata finem recipiet*)* *B* **220** De… coniunccionis *B* : *om. A* **221** medie coniunccionis *B* : coniunccionis medium *A* **222** Medium *A* : *om. B* **223** latitudinis Lune medium *A* : medium latitudinis *B* **225** motus : *mss. per err.*: motis

<5>

**De anno embolismali**

Item annum embolismalem futurum ex notis embolismis in *Tabula annorum expansorum coniunccionum* intelligere poteris.

Medios cursus planetarum et tempus coniunccionum Solis et Lune aut revolucionum ad alia loca regionum sive civitatum extra Bononiam Italie reducere. In hac re hec primo sunt attendenda. Civitas, cuius longitudo in *Tabula regionum* inventa erit minor longitudine Bononie, hec ipsa civitas respectu Bononie erit occidentalis. Civitas autem, cuius longitudo erit maior longitudine Bononie, illa erit orientalis. Si igitur volueris medios cursus et tempora coniunccionis et opposicionis Solis et Lune aut alia tempora, que ad Bononiam sunt, ad meridianum alicuius alterius civitatis reducere, vide differenciam inter gradus longitudinis illius civitatis et Bononie, quam differenciam graduum in 4 minuta multiplica, et proveniet tempus longitudinis, quod adde ad tempora Bononie pro civitate orientali, et subtrahe a tempore Bononie pro civitate occidentali, et habebis tempora prefata ad civitatem regionis illius. Rursus quere, quantum movetur Sol in tanto tempore longitudinis. Similiter quantum movetur Luna in eodem tempore longitudinis, et sic de reliquis planetarum motibus. Quos motus temporis planetarum subtrahe a radicibus vel mediis motibus planetarum supra Bononiam pro civitate orientali, cuiuslibet planete longitudinem a motu vel radice sua, vel adde pro civitate occidentali et habebis motus planetarum ad civitatem, quam volueris. Item mutato tempore coniunccionis vel opposicionis Solis et Lune a Bononia ad aliam regionem cave illius temporis medios motus Solis et Lune et argumentum Lune mutare, qui in universa terra – licet sub diversis temporibus – non variantur.

<6>

**Cognitis motibus et temporibus mediis pro meridiano Bononie vel reductis ad meridianum civitatis regionis alterius, prout est demonstratum eorum, nunc veros seu certos motus et tempora vera restat per Tabulas equacionum calculare, pro quibus duximus notandum**

Si in eis *Tabulis equacionum* signa in superiori parte tabule invenerimus, numerum graduum in parte sinistra tabule descendendo querimus. Cumque signa in parte inferiori tabule invenerimus, gradus in dextra parte tabule ascendendo requirimus. Illud quoque, quod unicuique gradui in tabula debetur sub signo illius gradus, acceperimus, quod acceptum sit, addendum, si in parte signi scriptum fuit 'adde' vel 'a', aut erit subtrahendum, si scriptum fuit 'subtrahe', aut 'minue' vel 's'. Differencie, que fuerit ex subtraccione duorum sibi proximorum numerorum, inter duos numeros in lineis ponuntur et tam descendendo, quam ascendendo deserviant. Cum et enim equaciones vel aliquod tale per gradus descendendo recepimus, differencia, que est in linea inmediate sub equacione recipienda, per oppositum fit ascendendo, reperiuntur quandoque in numeris differenciarum quidam paragraphi, et tunc in superiori parte differenciarum ponuntur due littere et in inferiori parte tabule similiter due, 'subtrahe' aut 'adde' significantes. Cum in

**230** De… embolismali *B* : *om. A* **233** aut revolucionum *A* : ac revolucionis *B* **243** in *A* : in Tauro tempore longitudinis, similiter quantum movetur Sol in *B* **247** cuiuslibet *B* : quemlibet *A* **268** similiter due *A* : 2 similiter *B*

superiori parte tabule ponitur 's/a', significat, quod fit subtraccio usque ad paragraphum et sub paragrapho fit addicio. Si autem poneretur 'a/s', quod fit addicio descendendo usque ad paragraphum et sub paragrapho subtraccio. Quod eciam ex augmentacione vel diminucione equacionum perpenditur, quam si augmentaciones vel equaciones augmentantur usque ad paragraphum, tunc fit addicio, et si minuuntur, fit subtraccio. In inferiori autem parte tabule 'a/s' signat, quod fit subtraccio ascendendo usque ad paragraphum et addicio supra paragraphum et sic de reliquo. Id eciam usui nostro volumus inservire numerum, cum quo tabulam intrare iubemur. Numerum intrandum vocabimus sive precise, sive non in tabula inveniatur. Et numerum, qui e directo numeri intrandi in tabula accipi precipitur, numerum accipiendum nominabis. Illud autem, quod de numero intrando tabulam non invenitur, residuum libuit nominare. Partem proporcionalem vocabimus numerum correspondentem numero super addito. Differenciam vocabimus id, quod relinquitur post subtraccionem duorum sibi proximorum numerorum in tabula positorum. Ubicumque differencie in tabulis circa numeros adnotate fuerint, eas in propria forma accepimus. Ubi vero invente non fuerint, eas per subtraccionem invenimus. Porro quandocunque numerus intrandus in tabula precise non invenitur, opus erit intrare cum numero, intrando in minori tabula invento et de residuo partem proporcionalem accipere. Pars proporcionalis de residuo numero hac regula conficitur: multiplica residuum numeri intrandi per differenciam numeri accipiendi et inde proveniens divide per differenciam tabularum numeri intrandi. Quodque post divisionem apparuerit, erit pars proporcionalis addenda vel subtrahenda, veluti patebit. Ubi autem numerus intrandus non habebitur, alius erit operandi modus, ut in parte proporcionali minutorum proporcionalium cum diversitate dyametri. Ibi etenim minuta proporcionalia ducuntur in diversitatem et proveniens erit pars proporcionalis ad 60 minuta. Prima regula partis proporcionalis erit communis ad omnes partes proporcionales, sive fuerint ad 60 minuta, sive ad 30 minuta, sive ad 15 minuta, sive ad 30 gradus, sive ad 6 gradus et sic de aliis quemadmodum in *Tabulis ecclipsium* sepe evenit. Unde notandum quandoque numeri intrandi in columpnis intermediis tabularum querimus et numerum accipiendum, tunc in latere accipimus et fit ferme per modum regule econversa, sicuti fit in arcuacione numerorum et in *Tabulis ascensionum*. Unde si habuerimus aliquam equacionem per partem proporcionalem equatam et voluerimus scire, cui numero correspondet hec equacio, accipimus eam equacionem, que prius erat numerus accipiendus, pro numero intrando et querimus eam in intermediis lineis tabularum inter numeros equacionum vel ea minorem. Et accipimus numerum accipiendum, qui prius erat numerus intrandus eidem equacioni correspondentem in latere tabule et facimus secundum regulam et producemus numerum, cui tali equacioni correspondet et reddibit nobis idem numerus, cum quo equacionem prius querebamus. Is est modus, per quem habilitantur practicantes in introitibus tabularum et parcium proporcionalium et multum necessarius in tabulis elevacionum signorum et in direccionibus ylegiorum in nativitatibus.

Partem proporcionalem ad 60 minuta per tabulam ad hoc factam. Quere numerum multiplicantem in superiori parte tabule et numerum multiplicandum in latere sinistro

**271** subtraccio *A* : fit addicio *B* Quod *A* : si autem poneretur 'a/s', quod *B* **280** super addito *A* : residuo *B* **285** in minori tabula *B* : minore in tabula *A* **289** veluti… ut *A* : velut *B* **293** communis *A* : communior *B* omnes *A* : omnes alias *B*

tabule descendendo et in eorum angulo communi erit pars proporcionalis, cuius partis proporcionalis fraccio minor illius erit denominacionis quam numerus multiplicans, et numerus multiplicandus per suos numeros denominacionum conficiunt. Item proporcionalis ad tot erit accipienda, ad quot tabula in lineis numeri intrandi augmentabitur, ut si augmentatur per 60 minuta pars proporcionalis accipitur ad 60 minuta. Si augmentatur per duos gradus, pars proporcionalis accipitur ad duos gradus. Similiter si ad 6 vel ad 30 gradus et sic de aliis, que omnia in prima regula partis proporcionalis complectuntur. Item in omni loco tam mediorum motuum, quam verorum et equacionum et sic de aliis, ubi ad 30 vel plura secunda evenerint pro eis addimus unitatem in minutis. Si vero secunda minus fuerint 30, de eis non curabimus. Hunc eciam modum in posicione *Tabularum mediorum motuum* in terciis observabimus ideoque eis tabulis examinatis quandoque in secundis habundare videbuntur.

<7>

**De vero loco Solis**

Verum sive certum locum efficere. Augem Solis a medio eiusdem Solis motu depone, remanens erit argumentem Solis. Cum eo intra *Tabulam equacionis Solis* et equacionem e directo eius sub signo argumenti inventam accipe, signando eam cum nota ‘s’, si habuerit ‘subtrahe’, vel cum nota ‘a,’ si habuerit <‘adde’>. Et si cum argumento fuerint minuta, accipe differenciam equacionis ibidem compertam et constitue partem proporcionalem et adde equacioni Solis, si differencia equacionis habuerit ‘a’, vel subtrahe, si habuerit ‘s’. Et quod post addicionem vel subtraccionem evenerit, erit equacio Solis equata, quam adde medio motui Solis, si equacio habuit ‘adde’, vel ‘subtrahe’ ab eo, si habuerit subtrahe, et quod resultabit, erit quesitus verus locus Solis.

<8>

**De vero loco Lune**

Verum locum Lune invenire: subtrahe Solis medium motum a medio motu Lune et residuum dupla. Duplatum erit centrum Lune, cum quo intra *Tabulam equacionis Lune* et accipe equacionem centri et minuta proporcionalia. Equacionem centri adde vel subtrahe, ut titulus tabule admonet, ab argumento medio Lune et consurget argumentum Lune equatum, cum quo argumento Lune equato accipe in eadem tabula equacionem argumenti notando cum suo adde vel subtrahe et cum eodem eciam argumento equato accipe diversitatem dyametri. De diversitate dyametri et minutis proporcionalibus fac partem proporcionalem ad sexaginta minuta, quam partem proporcionalem adde equacioni argumenti prius accepte et proveniet equacio argumenti equata. Hanc equacionem argumenti equatam iunge medio motui Lune, si fuerit addenda, vel subtrahe ab eodem, si fuerit subtrahenda, et egredietur locus Lune verus. Porro que in calculacione verorum motuum habenda et queve retinenda, queve abicienda, ipsa practica docet, et ea, que in practica non explicantur, tituli tabularum manifestant.

**311** decendendo *B* : descendens *A* **324** De… Solis *B* : *om. A* **326** et equacionem *A* : equacionemque *B* **335** De… Lune *B* : *om. A* **336** et… duplatum *B* : residuum duplatum *A* **340** in *A* : et in *B*

<9>

## Trium superiorum planetarum, Saturni videlicet, Iovis et Martis, vera loca cognoscere

Subtrahe medium motum planete a medio motu Solis, remanens erit argumentum medium planete. Subtrahe eciam augem planete a medio motu planete, residuum erit centrum medium planete. Cum medio centro *Tabulas equacionum* illius planete ingredere et accipe equacionem centri. Equacionem centri si fuerit addenda, adde centro medio et proveniet centrum equatum planete. Eandem eciam equacionem centri, si fuerit subtrahenda, subtrahe ab argumento medio planete et residuum erit argumentum planete equatum. Si autem equacio centri fuerit minuenda, minue eam a centro medio planete et adde medio argumento et habebis omnia equata, que in suo loco una cum equacione centri conservabis. Cum centro equato in eadem tabula accipe minuta proporcionalia cum suo titulo 'propior' vel 'longior', quem supra minuta inveneris. Cum eodem eciam centro accipe stacionem primam, quam subtrahe a 12 signis, remanens erit stacio secunda. Hec omnia cum prioribus erunt retinenda. Ceterum cum argumento equato accipe equacionem argumenti et diversitatem dyametri. Diversitatem dyametri accipe in longitudine propiori, si minuta proporcionalia prius servata habuerint 'propior', vel in longitudine longiori, si habuerint 'longior'. De diversitate dyametri et minutis proporcionalibus fac partem proporcionalem ad 60 minuta, quam adde equacioni argumenti, si minuta proporcionalia sunt 'propior', vel subtrahe, si sunt 'longior', et erit equacio argumenti equata. Rursus si equacio argumenti equata et equacio centri prius servata ambe fuerint addende, iunge eas simul et totum iunge cum medio motu planete. Vel si ambe fuerint minuende, iunge eas simul et totum collectum minue ex medio motu planete. Si autem una fuerit addenda et altera minuenda, subtrahe minorem de maiori et residuum adde cum medio motu planete, si equacio maior fuerit addenda, vel subtrahe ab eodem, si fuerit minuenda, et resultabit verus locus planete, quem quesivisti.

In Venere et Mercurio idem erit operandi modus, nisi quod eorum media argumenta ex tabulis haberi oportebit.

<10>

## Cum autem quesieris, utrum planeta sit stacionarius, vel retrogradus, vel directus

Stacionem eius primam et secundam considera, ponendo ordine primo stacionem primam et sub ea stacionem secundam. Demum si argumentum eiusdem planete equatum fuerit equale stacioni prime in signis, gradibus atque minutis, planeta stacionarius erit stacione prima, scilicet ut retrogradari incipiat. Et si idem argumentum equatum fuerit equale stacioni secunde, erit planeta stacionarius stacione secunda, ut se dirigere incipiat. Si autem dictum argumentum fuerit inter stacionem primam et secundam, planeta erit retrogradus. Extra hos terminos erit directus. Item quando argumentum equatum equabitur argumento inicii retrogradacionis, quod in margine notatur, planeta

**359** adde *A* : adde *(supra scr.*: eam*)* *B* **361** propior *A* : proprior *B* **365** propiori *A* : propriori *B* propior *A* : proprior *B* **368** propior *A* : proprior *B* **371** iunge *A* : iunge eas simul et totum iunge cum medio motu planete, vel si ambe fuerint minuende, iunge *B* **373** ab eodem *A* : *om. B* **375** In… oportebit *A* : *om. B* **384** prime *(corr. in*: secunde*)* *A* : prime secunde *B*

incipit retrogradari, et quando erit simile fini retrogradacionis, planeta incipit dirigi. Et quando argumentum equatum inter hec duo argumenta compertum fuerit, planetam retrogradum nunciat. Extra ea erit directus. Quod si planeta fuerit retrogradus et volueris scire, quot dies fluxerunt ab inicio retrogradacionis, stacionem eius primam de eius argumento equato minue, et quod inde remanserit, per motum argumenti planete unius diei divide; numerus quociens erit numerus dierum ab inicio retrogradacionis. Residuum quoque in 24 multiplica et per eundem motum argumenti unius diei partire, numerus quociens erunt hore, que cum diebus suis erunt tempus ab inicio retrogradacionis. Si vero planeta fuerit directus et volueris scire tempus inicii direccionis, stacionem eius secundam ab argumento equato depone et residuum, ut prius, per motum argumenti unius diei partire et exibit dierum et horarum numerus ab inicio direccionis. Argumenta autem unius diei sub argumentis planetarum in tabulis moderantibus reperies.

<11>

**Latitudinem trium superiorum, Veneris eciam et Mercurii, adnotare**

Subtrahe verum locum genzahar planete, quod in margine invenias, a vero loco planete et habebis argumentum latitudinis planete. Argumentum vero Veneris et Mercurii sic invenies: iunge argumentum eorum equatum cum medio cursu et a collecto verum locum zenzar eorum subtrahe et surget argumentum latitudinis eorum quesitum. Quod si argumentum latitudinis cuiuslibet eorum fuerit ab uno gradu in 3 signa, erit planeta septentrionalis ascendens, et a tribus signis in 6 signa, erit septentrionalis descendens. Et a sex signis in novem, erit meridionalis descendens, et a novem in duodecim, erit meridionalis ascendens. Rursus quia gradus latitudinis pluris sunt laboris quam utilitatis, eos in tachuinum ponere non consuevimus nec eciam de eis curavimus hiccine doctrinare.

<12>

**Verum locum capitis draconis sive zenzar Lune conficere**

Subtrahe medium cursum capitis draconis de 12 signis, remanens erit verus eius locus.

<13>

**Verum argumentum latitudinis Lune constituere**

Locum zenzahar Lune equatum de vero loco Lune minue et habebis argumentum latitudinis Lune equatum.

<14>

**Post hec ad tacuinum constitucionem progredere**

Pro quo papira sic adaptabis: accipe 13 folia papiri et in secunda plana primi folii duc lineam longitudinalem, scilicet longitudinem papiri, prope extremitatem longitudinis

---

**396** stacionem… prius *A* : *om. B* **407** latitudinis cuiuslibet *A* : cuiuslibet quesitum latitudinis *B* **411** consuevimus *A* : conservimus *B* **412** doctrinare *A* : determinare *B* **422** tacuinum *B* : tachium *A* constitucionem *A* : constituendum *B*

exteriorem competenti margine dimisso, duc et aliam equedistantem prope extremitatem longitudinis interiorem et spacium, quod hiis duabus lineis clauditur, continebit vera loca planetarum. Consimiles lineas fac e directo in prima plana secundi folii et spacium hiis lineis interceptum habebit aspectus planetarum. Ceterum prope extremitatem latitudinis superiorem duorum digitorum spacio dimisso duc lineam latitudinalem euntem per ambas planas usque ad extremitates planarum, duc aliam sub ea lineas longitudinales plane verorum motuum non excedentem et spacium confectum erit pro karacteribus planetarum. Iterum fac consimilem spacium pro karacteribus signorum, fac eciam tercium pro titulis retrogradacionum et direccionum, fac quartum pro latitudinibus planetarum, fac quintum, cuius linea transibit per ambas planas usque ad extremitates, sicut linea prima, et spacium interveniens continebit verorum motuum denominaciones, scilicet gradus, minuta. Spacium, quod ex tali duccione in secunda plana, scilicet in prima secundi folii confectum est, continebit titulos aspectuum Lune ad planetas et planetarum inter se. Postea fac spacia dierum pro Ianuario 31 sub spaciis prioribus, quorum linee transibunt per ambas planas ita, quod preter ultimam non excedant ultimas etremales longitudinales. Eo modo lineabis in ceteris binis et binis planis pro mensibus duodecim, faciendo tot spacia dierum, quot quilibet mensis habet dies. Lineis latitudinalibus sic perfectis duc lineam longitudinalem in plana verorum motuum equedistantem prime linee longitudinali, secantem omnes lineas latitudinales, euntem usque ad duas lineas latitudinales extremales, constituentem sufficiens spacium pro numero dierum. Fac consimile spacium pro litteris dierum et pro sillabis dierum. Residuum spacium magnum secundum longitudinem divide in octo partes distinctas lineis rubeis longitudinalibus, transeuntibus ut prius usque ad extremitates linearum extremalium latitudinis. Spacia emergencia continebunt vera loca planetarum. Prima pars erit pro Sole, altera pro Luna, tercia pro Saturno, quarta pro Iove, quinta pro Marte, sexta pro Venere, septima pro Mercurio, octava pro capite draconis. Iterum spacium cuiuslibet planete secundum longitudinem in duo partire cum linea transeunte ab inferiori linea extremali latitudinis per omnes lineas latitudinales dierum, spacium denominacionum non excedentem, et erit columpna una pro gradibus, altera pro minutis. Imperfecta est una plana. Simili modo in secunda plana duc lineam longitudinalem equedistantem prime linee longitudinali, secantem eciam omnes lineas latitudinales euntem eciam usque ad duas lineas extremales latitudinales, constituentem sufficiens spacium pro numero dierum. Fac consimilem spacium, sed modice maius, pro aspectibus Lune ad Solem, fac tercium consimilem pro aspectibus Lune ad Saturnum, item quartum Lune ad Iovem, quintum Lune ad Martem, sextum Lune ad Venerem, septimum Lune ad Mercurium, octavum pro coniunccione et opposicione Lune cum capite draconis. Ultimo relinquetur spacium grossum pro aspectibus planetarum inter se. Hoc modo pro duodecim mensibus lineabis. Quomodo autem spaciis tituli inscribendi erunt, iam ipsa lineacio docuit. Inscriptis itaque numero dierum et litteris dierum et sillabis dierum cum titulis sibi suprapositis et posito nomine mensis in margine latitudinali superiori, similiter in positis karacteribus planetarum, quemlibet in suo spacio, et denominacionibus graduum et minutorum in spaciis propriis et in altera plana posito numero dierum cum titulo sibi supposito et in sequenti spacio supra lineas latitudinales dierum posito

**426** interiorem *A* : exteriorem competenti margine dimisso, duc et aliam equedistantem prope extremitatem longitudinis interiorem *B* hiis *A* : de hiis *B* **449** Iove *A* : Sole alias Iove *B* **467** titulo sibi supposito *A* : titulis sibi suppositis *B*

titulo aspectus Lune ad Solem et in altero aspectus Lune ad Saturnum et sic de aliis. Et ultimo aspectus planetarum inter se in singulis mensibus, sic faciendo, karacteribus autem signorum et titulis direccionum et retrogradacionum et latitudinibus inscribendis in postea relictis iam habes papirum pro thachuino recte adaptatum.

<15>

**Vera loca Solis in tachuinum pro die anni quolibet ordinare**

Medium motum Solis et argumentum Solis per tabulas predictas ad meridiem ultime diei Decembris anni transacti pro Bononia vel reductum ad regionem, ad quam vis tachuinum calculare, constitue. In quibus motibus si apparuerint 30 vel plura secunda, pro eis adde unitatem minutis eosque pro anni radicibus serva. Si vero secunda fuerint minus 30, non curabis de eis. Hunc eciam modum in planetis ceteris insequentibus observabis. Quibus radicibus habitis cum eis confice verum locum Solis modo superius posito, et tum verum locum Solis extra in aliqua carta serva. Rursus in *Tabulam Solis* auctam de quinque in quinque ingredere, et quod e directo 5. dierum Ianuarii inveneris, radicibus, medio videlicet motui Solis et argumento Solis, adiunge et cum eis, que provenerint, ut prius, verum locum Solis exquire. Et tum verum locum Solis nunc inventum papiro adaptato more prefato e directo quinque dierum ad columpnam Solis conscribe et habebis verum locum Solis ad 5. diem Ianuarii. Iterum intra predictam tabulam, et quod e directo 10. dierum Ianuarii inveneris, radicibus prenominatis adde et cum predictis iterum verum locum Solis confice et habebis verum locum Solis ad 10. diem Ianuarii, quem eciam in directo 10. dierum Ianuarii ad tachuinum collocabis, et sic fac de aliis diebus mensium usque ad finem tabule et habebis verum locum Solis ad omnes dies tabule adauctos. Porro annus bisextilis cum advenerit, non id, qui in directo 28 dierum Februarii apparuerit, sed id, qui in margine inferioris partis tabule circa 29 dies ponitur, radicibus iungendum erit, et cum aggregato verus locus Solis ad finem Februarii conficiendus. Et tunc in eodem anno bisextili, antequam in dies Marcii processeris, radicibus pretactis motum unius diei, qui in margine notatur, adiunge et collecta pro radicibus aliquorum verorum motuum tocius anni habeto. Veri quoque motus taliter inventi distabunt per dies quinque et in fine mensium habencium dies 31 distabunt per dies sex. In fine autem Februarii in anno communi distabunt per dies tres, in bisextili per dies quatuor.

Habitis veris motibus ad dies adauctos inter mediorum dierum vera loca conficere. Subtrahe verum locum Solis, quem extra in carta servasti, a vero loco Solis e directo 5. dierum Ianuarii posito et remanet illorum verorum motuum differencia, quam divide per quinque; in quociente erit verus locus Solis unius diei. Hunc verum motum Solis unius diei adde vero loco Solis in carta servato et proveniet verus locus Solis ad meridiem primi diei Ianuarii. Hunc adauge cum vero motu unius diei usque quinque ad quintum diem Ianuarii et habebis veros motus ad quartum diem Ianuarii, quos inscribe tachuino et in quinto die Ianuarii ex tali auccione redibit tibi numerus verus motus quinti diei, quem prius habes. Et sic facias de aliis diebus intermediis per omnes menses usque ad finem anni. Hec dumtaxat animadvertere oportet: quando verorum motuum

---

**470** autem signorum *A* : signorum autem *B* et[3] *A* : in *B* **475** anni transacti *A* : ultimi transacti anni *B* ad[1] *A* : ad bo *B* **480** carta *A* : parte carte *B* serva *B* : observa *A* **484** directo *A* : directio *B* **487** Solis confice *A* : confice Solis *B* **488** diem *A* : *om.* *B* **503** loco *A* : motui *B* **504** adauge *A* : adiunge *B* **505** inscribe *A* : inscribe in *B*

sex dierum distancia erit, tunc differencia per sex erit dividenda, si trium, per tres, si 4, per 4, quod in fine mensium sepe evenit. Quibus perfectis habebis vera loca Solis per annum in tachuino. Quorum verorum locorum karacteres signorum in principio mensis in spacio signorum supra vera loca reponantur. Cum vero Sol in medio mensis mutaverit karacter, signum illius, quod intrat in apice cellule, ibidem conscribatur.

<16>

**Lunam per annum de die in dies in tachuinum collocare**

Quere primam coniunccionem Solis et Lune mediam cum motibus suis, ut ante didicisti, eam videlicet coniunccionem primam, que in Ianuario evenerit, a quo Ianuario cum inicio annorum Christi velut Christiani tachuinum incipere consuevimus. Pariformiter et omnes coniuncciones medias omnium mensium tocius anni una cum motibus Solis et Lune et argumento Lune perquire, quas omnes cum prima coniunccione in carta studiose reserva. Illis coniunccionibus mediis reservatis accipe primam coniunccionem mediam, et considera, si fuerit ante meridiem vel post. Si etenim eius hore fuerint 12 precise vel minus, erit coniunccio post meridiem. Si vero fuerint plures quam 12, ipsa erit ante meridiem, scilicet sequentis diei. Postea considera distanciam coniunccionis per horas a meridie. Si enim hore fuerint post meridiem, ipse erunt distancie coniunccionis. Si vero fuerint ante meridiem, ipsius a 24 horis sublatis, residuum erit distancia coniunccionis a meridie. Hac distancia notata reduc medium motum ad meridiem hoc modo: intra cum distancia horarum coniunccionis a meridie in tabulam, que eis debetur, et gradus et minuta, que e directo earum inveneris, extra scribe. Deinde si qua cum horis fuerint minuta, cum eis *Tabulam minutorum* pro eis factam intrabis et minuta, que ex directo reperieris, gradibus et minutis prioribus adiunge. Et si fuerint in distancia minuta plura quam 30, intrabis prius cum 30 et ceterum cum residuis. Gradus quoque et minuta sic producta adde medio motui Lune, si coniunccio fuerit ante meridiem, vel subtrahe, si fuerit post meridiem, et habebis medium motum Lune reductum ad meridiem. Quem pro radice serva usque ad dies 30, et argumentum Lune tempore medie coniunccionis erit radix argumenti Lune similiter ad 30 dies. Habitis itaque tribus hiis distancia, videlicet horarum, et medio motu Lune pro meridie et argumento Lune tempore coniunccionis, intra cum argumento Lune *Tabulam equacionis Lune in diebus lunacionum* et accipe equacionem argumenti, que argumento et suis minutis in ea tabula debetur. Accipe eciam augmentum horarum, quod ibi inveneris, equacionem subtrahe a medio motu Lune, si fuerit equacio subtrahenda, vel adde, si fuerit addenda, et quod sic ex vero loco provenerit, serva. Demum cum horis distancie a meridie intra *Tabulam augmenti Lune in horis*, et quod ex gradibus et minutis sub augmento tunc invento reperieris, nota. Quod si cum horis distancie fuerint minuta, intrabis cum eis more solito *Tabulam augmenti Lune in minutis horarum* et minuta sub augmento inventa gradibus et minutis noviter acceptis coniunge, aggregatum quoque adde vero loco Lune proxime invento, si coniunccio fuerit ante meridiem, vel subtrahe, si fuerit post, et proveniet verus et equatus locus Lune ad meridiem diei, in quo facta erat coniunccio. Hunc verum locum loca in tachuinum ad columpnam Lune e directo diei coniunccionis. Ceterum intra *Tabulam argumenti Lune equati* in diebus post coniunccionem et argumentum

**512** signorum supra *A* : supra signorum *B* **518** annorum Christi *A* : Christi annorum *B* **522** mediam *A* : scilicet mediam *B* **529** gradus : *mss. per err.*: gradibus

equatum, quod in directo unius diei apparet, iunge cum argumento radicis et proveniet argumentum equatum in uno die post coniunccionem, cuius equacionem et augmentum in *Tabula equacionis Lune* prius inquire. Huic equacioni adde partem proporcionalem, quam sub prima die Lune e directo gradus argumenti inveneris, et erit equacio argumenti equata, quam adde medio motui Lune in uno die post coniunccionem, quem in tabula propria inveneris. Si equacio fuerit addenda, adde eam, vel subtrahe, si fuerit subtrahenda, et verum cursum Lune inde provenientem nota. Postea cum augmento nunc invento et cum horis distancie fac, ut prius, et proveniet verus locus Lune, correspondens uni diei post mediam coniunccionem, quem adde medio motui Lune in radice et habebis verum locum Lune ad meridiem in primo die post coniunccionem, que sub priori ad tachuinum repone. Iterum intra *Tabulam argumenti equati* et argumentum, quod duobus diebus debetur, adde argumento radicis et habebis argumentum Lune equatum pro secunda die post coniunccionem mediam. Cum quo argumento equato iterum accipe coniunccionem et augmentum equacioni argumenti adde iterum partem proporcionalem, quam ex directo argumenti sub secunda die Lune inveneris, et erit equacio argumenti equata, quam adde, si fuerit addendum, vel subtrahe, si fuerit subtrahenda, a medio motu Lune invento in *Tabula dierum medii motus Lune* post coniunccionem ex directo duorum dierum et verum locum Lune, ut prius, considera. Postea fac cum augmento et horis distancie et proveniet verus locus Lune correspondens duobus diebus post coniunccionem, quem adde medio motu radicis, et habebis verum locum Lune ad meridiem in secundo die post coniunccionem, quem eciam sub prioribus ad tachuinum convenit reponere. Et sic de aliis diebus post coniunccionem usque ad 30 dies, quod est usque ad illam coniunccionem mediam singula singulis compensando. Cumque expleveris dies unius coniunccionis, eodem modo cum secunda coniunccione media operare pro aliis 30 diebus, et sic de tercia coniunccione et postea de quarta usque quo Lunam per totum annum ad tachuinum collocabis. Item si prima coniunccio ceciderit inter medios dies Ianuarii, tunc per eam verum locum Lune ad primum diem Ianuarii non habebis, sed per precedentem coniunccionem, que fit in Decembri anni precedentis. Hoc modo subtrahe unam lunacionem a prima coniunccione et remanebit coniunccio in diebus Decembris. Si dies subtrahere non volueris, adde prime coniunccioni dies 31 et demum subtrahe. Deinde considera, quota dies Lune huius coniunccionis cadit in primum diem Ianuarii et ad illum diem modo predicto quere verum locum Lune ad meridiem, et tunc reliquam, scilicet doctrinam faciendo omnes veros motus usque ad primam coniunccionem invenies, quos in tachuinum ordina ex directo eius correspondentium dierum. Preterea non te perturbet practica gloriosi Iohannis de Klarissimis, que quandoque tecum in quibusdam minutis veri motus Lune discrepabit, eo quod in ea parte omnimodam precisionem in Luna ut ipsemet perhibet de se non observavit, ea tamen discrepante quando fit sine notabili variacione in opere viciosa non erit. Nostre denique tabule praxis precisior erit quam descensum epicicli Lune in ecentrico per argumentum equatum pensilius adnotavimus. Item signa, in quibus Luna erit per karacteres supra vera loca ponenda erunt, scilicet in principio mensis in spacio signorum deputato consignantur per augmentum quoque, de quo presolenter mencionem fecimus illud, quod Luna in una hora utra 30 minutis percurrit, intelligimus.

---

**558** distancie *A* : distancies *B* **582** quere verum *A* : querum *B* **588** observavit *A* : observabit *B*

<17>

## Vera loca aliorum planetarum in tachuinum conserere

Medium motum et centrum eius medium et argumentum medium ad meridiem ultimi Decembris pro regione, ad quam fabricas tachuinum, per tabulas premissas constitue, que pro radicibus anni retineto. Cum illis eciam radicibus verum locum conficias et ipsum in aliqua carta serva. Ceterum intra cum 10 diebus Ianuarii in *Tabulam planete* auctam de decem in decem dies et medios motus, quos ibi repereris, adde ad radices anni pretactas, medium motum, scilicet planete, ad medium motum in radice, et eundem eciam medium motum adde ad centrum in radice et argumentum adde ad argumentum in radice et consurgent medii motus, quilibet in genere suo. Cum quibus mediis motibus inquire verum locum planete nunc inventum in tachuinum e directo decem dierum Ianuarii in columpnam planete repone. Iterum intra cum 20 diebus Ianuarii in *Tabulam planete* priorem et medios motus ibi compertos adde iterum ad radices et cum mediis motibus evenientibus quere verum locum planete pro 20. die Ianuarii, quem in tachuinum ex directo 20 dierum Ianuarii consigna. Et sic operare de omnibus aliis diebus mensium in tabula notatis usque ad finem tabule et habebis verum locum planete ad omnes dies tabule adauctos per totum annum. Porro sic operando considerandum erit. In anno bisextili non accipiantur medii motus ex directo 28 dierum Februarii, sed illi, qui in margine circa dies 29 inveniuntur, qui addantur radicibus, et postea verus locus planete more solito perquiratur. Et tunc antequam dies Marcii aggredieris, iunge prius radicibus motus planete unius diei in margine signatos, quemlibet motum unius diei motui suo in radicibus, et producta usque ad anni complecionem pro radicibus teneto. Rursus si planeta erit directus, donec eius verus motus in tachuino augeatur. Retrogradus fit, quando eius verus motus minuitur. Postquam planeta fuerit directus et eius verus locus minui reperitur, tunc planeta incipit retrogradari. Quot autem dies ab inicio retrogradacionis transiverunt, modis antedictis considera et tot dies completos numera ascendendo a vero loco planete, diem in questione a tali vero loco planete secludendo, et ubi talis computacio desierit, scribe super cellulam diei 'retrogradus'. Si autem planeta fuerit retrogradus et eius verus motus augeri inceperit, iam planeta incepit dirigi, quot autem dies fluxerunt ab inicio direccionis, ut prius considera, quibus a vero loco planete pariformiter sursum numeratis, in eorum termino scribe 'directus'.

<18>

## Animadverte eciam latitudinem planete,

scilicet si fuerit meridionalis ascendens vel descendens, aut septentrionalis ascendens vel descendens, quam latitudinem in tachuinum per titulos iam dictos in spaciis eis debentibus supra veros motus consigna et hoc in principio mensis et intermediis diebus invariabitur. Signa autem planetarum, ubi reponi ex prioribus practicis, poteris perpendere.

---

**599** serva *A* : observa *B* **601** eundem *A* : eum et eundem *B* **602** motum *A* : modum *B* **603** quilibet *A* : quibus *B* **608** tachuinum *A* : tachuino *B* consigna *A* : signa *B* **611** motus *A* : *om. B* **618** motus *(corr. in:* locus*) A* : motus alias locus *B* **620** a tali vero loco *B* : talis verus locus *A* **630** planetarum *A* : *om. B*

Omnibus peractis habes veros motus planetarum per annum auctos per decem dies et in fine mensium habencium dies 31 erunt aucti per dies 11, in fine autem Februarii in anno communi sunt aucti per dies 8 et in anno bisextili per dies 9. Vera loca planetarum ad dies intermedios tachuini siccine perficias: verum locum planete in carta servatum, quem per radices invenisti, a vero loco planete decimi diei Ianuarii vel minorem a maiori subtrahe, residuum erit horum verorum motuum differencia, quam divide per 10, et proveniet verus motus unius diei, per quem adauge verum locum planete in carta servatum usque ad dies decem, si planeta fuerit directus, vel minue, si fuerit retrogradus, et sic operando redibit tibi verus locus planete decime diei in decima addicione vel diminucione, quam prius habuisti. Alios autem veros motus novem pone in tachuinum, primum scilicet ad primum diem Ianuarii, non eum, qui est in carta, sed eum, quem ex prima addicione vel diminucione produxisti. Similiter et reliquos veros motus scribe secundum ordinem ex directo dierum eis debencium. Eodem modo fac de omnibus aliis differenciis verorum motuum tachuini sibi proximorum. Sed in fine mensium habencium dies 31 differencia per 11 dividenda erit, in fine eciam Februarii in anno communi per octo, in anno autem bisextili per novem dividitur, ut motus unius diei verus habeatur.

Quod si planeta fuerit directus et inter duos veros motus retrogradacio supervenerit, tunc postquam per motum unius diei expleveris vera loca planete dierum intermediorum precedencium, adde eciam eundem verum motum unius diei continuo pro aliis diebus intermediis, qui sunt inter sequentes veros duos motus usque ad diem retrogradacionis exclusive tunc inter verum locum, qui erit ante diem retrogradacionis, et verum locum proxime sequentem. Nota differenciam, quam divide per distanciam dierum illorum verorum motuum abinvicem, et proveniet verus motus unius diei, quem tunc minue continue veluti nosti, donec repleas veros motus usque ad diem alterius veri motus compleas. Quid autem agendum sit, si planeta fuerit retrogradus et inter dies intermedios sequentes occurrit directio, ex predictis percipiatur. Quibus omnibus completis veri motus omnium planetarum in tachuino pro anno currente perficiuntur.

Sciendum Venus et Mercurius de decem in decem equati in intermediis diebus non secundum omnimodam precisionem poterunt haberi, sed quia hii duo planete semper non remote a Sole discurrunt, ideo discrepancia talis in intermediis diebus in iudiciis astrologicis non vituperatur. Sufficit ergo, quod sit precise de decem in decem dies secundum Iudeum Profacium. Si autem alicui is modus visus fuerit non sufficere, propter direcciones cum aliquis ipsorum duorum fuerit ylech in nativitate equet eos de 5 in 5 dies. Hoc modo adde radicibus eorum et mediis motibus tabule eorum aucte per 10 motum medium 5 dierum, et sic de quinque in quinque veros motus inveniendo, quemadmodum actum fuit in Sole de diebus intermediis similiter sicut in Sole ingeniando, nisi quod retrogradacio vel direccio, cum in diebus intermediis apparuerit, ut in aliis planetis de 10 in 10 equatis erit observanda.

**633** dies 11 *A* : 11 dies *B* **634** dies 8 *A* : 8 dies *B* **662** ideo *B* : ideoque *A* **667** motum medium *A* : medium motum *B*

<19>

## Verum locum capitis draconis in tachuinum proicere

Quere eius medium motum ad meridiem ultime diei Decembris, quem subtrahe a 12 signis, et consurget verus locus capitis draconis ad meridiem eiusdem diei, quem pro radice tocius anni teneto. Demum intra *Tabulam medii motus capitis draconis* et id, quod in directo cuiuslibet mensis inveneris, subtrahe a radice capitis draconis et habebis vera loca capitis draconis ad finem cuiuslibet mensis, que in tachuinum repone. Si fuerit annus communis, age cum mensibus anni communis. Si bisextilis, fac cum mensibus bisextilibus. Post hoc accipe medium motum capitis draconis 10 dierum, quem subtrahe a radice et proveniet 10. verus locus eius pro 10. die Ianuarii. A quo iterum subtrahe motum 10 dierum, et apparebit verus locus capitis pro 20. die Ianuarii. Eodem modo fac pro aliis mensibus, subtrahendo motum decem dierum a motu finis precedentis mensis, et proveniet motus eius verus ad 10. diem mensis sequentis, a quo iterum subtrahe motum 10 dierum, et proveniet motus ad 20. diem, et ita habebis verum locum capitis draconis ad omnes dies adauctos. Postea a vero loco in verum locum diminue 3 minuta, que sunt locus verus capitis draconis unius diei; diminuendo procede et eo modo vera eius loca in diebus intermediis habitis in tachuinum e directo dierum eis debencium signa. Et apparebit tunc tachuini verorum planetarum motuum finis.

<20>

## Aspectus planetarum ex veris motibus, qui in tachuino sunt, elicere

Pro huiusmodi negocii explecione opere precium est aliqua considerare. Aspectus sunt equales gradus, hoc est scilicet gradus zodiaci in tachuinum ponere consuevimus. Mediam cellulam dierum in tachuino meridiem representare volumus et linee supra et infra medium cellule noctis medium insinuant. Si igitur karacteris medium alicuius aspectus in medio cellule compertum fuit, aspectum talem in meridie fore dicimus. Si autem fuerit in aliquo linearum, fit talis aspectus in medio noctis. Si autem medium karacteris fuerit ante medium cellule prope lineam, tunc dicimus talis aspectus est post medium noctis mane. Sed si fuerit prope medium, dicimus is aspectus est ante meridiem. Similiter si tale medium karacteris fuerit post medium cellule, plus declinans ad medium cellule quam ad lineam, dicimus aspectum esse post meridiem. Si vero ad lineam plus flecteretur ante medium noctis, talem aspectum pronunciamus. Notamus eciam aspectus per cancellos digitis indicis sinistri in secundum equidem eius cancellum aspectum sextilem. In tercium aspectum quartum et in quartum cancellum trinum aspectum reponimus. Coniunccioni autem et opposicioni, quia note sunt eis, cancellos assignare pretermissimus. Cum itaque aspectum duorum planetarum cupiveris indagare, considera veros motus eorum, quo die erunt similes in gradibus et minutis vel fere similes, sive tales planete fuerint in eodem signo, sive in diversis signis. Si tales planete fuerint in eodem signo similes in gradibus et minutis vel fere similes, tunc eodem die erit eorum coniunccio. Si autem taliter dispositi fuerint in oppositis signis, tunc eodem die erit eorum opposicio. Si autem nec in eodem, nec in oppositis signis fuerint, considera signum planete eius, qui est minor in signis, vel eius, qui precedit alium motu diur-

**680** subtrahe *A* : *om. B* **685** diminue : *mss. per err.*: divide per **693** Mediam cellulam : *mss. per err.*: medium cellulum **707** signis *A* : *om. B* **711** in *A* : *om. B*

no, eo signo secluso numera cetera signa per cancellos digiti predicti usque ad signum planete alterius. Si tunc tale signum alterius planete ceciderit super cancellos sextilis aspectus, eodem die erit talis aspectus sextilis. Si autem ceciderit ad cancellum quarti aspectus, eodem die erit quartus aspectus et sic de reliquis. Extra predicta loca non aspiciunt se planete, eciam si in gradibus et minutis concordes fuerint. Item taliter operantes in motibus Lune pocius gradus quam minuta consideramus ipsam etenim Lunam, eciam si a motu alterius planete in 6 gradibus diversa fuerit, eam cum tali planeta in gradibus fere similem dicimus. Cum autem aspectum duorum planetarum et eius aspectus diem congnoveris, in qua parte diei talis aspectus futurus sit, scilicet an in meridie, an ante vel post meridiem, an in medio noctis, aut ante vel post, optaveris scire, eorum veros motus modo predicto considera. Si eorum veri motus in gradibus et minutis precise concordes fuerint, dic talem aspectum precise in meridie fore. Si vero precise non concordaverint, per intellectum tuum discrepanciam eorum verorum motuum considera. Considera eciam, quos istorum planetarum velocioris sit motus, tunc pensa per ingenium tuum, in quot horis secundum propinquam estimacionem talis planeta velocior illam discrepanciam percurrere possit. Si tunc idem planeta, qui est velocior, erit minor in gradibus et minutis, aut in minutis tantum, aspectus talis in tot horis post meridiem est futurus. Si autem planeta velocior fuerit maior in gradibus et minutis aut minutis tantum, aspectus talis preterit et fuit ante meridiem in tot horis. Quibus cognitis loca medium caracteris, aspectus talis preteriit et fuit ante meridiem in tot horis. Quibus cognitis loca medium karacteris aspectus talis in tachuinum ad cellulam diei debentem sub planeta, cuius est aspectus ad reliquum, scilicet propinquum horarum quantitatem ante meridiem vel post meridiem. Si tales hore estimacionis propinque fuerint plures quam sex, karacteris talis aspectus locabitur plus versus lineam medie noctis. Si vero tales hore fuerint minus sex horis, medium karacteris versus meridiem se tenebit. Si autem volueris talem locacionem karacteris observare, tunc fac punctum circa karacterem in talem partem diei, qui punctus indicabit tibi, in qua parte diei aspectus talis sit venturus. Et faciendo aspectus talem ordinem observabis. Primo fac aspectus Lune ad Solem eundo per omnes menses tocius, postea Lune ad Saturnum, similiter eundo per omnes menses tocius anni, postea Lune ad Iovem, postea Lune ad Martem et sic de reliquis planetis. In capite quoque, quando iungitur Lune aut opponitur ei, advertenciam habemus, alios aspectus eius non curavimus. Hos omnes aspectus per karacteres proprios signa in cellulis eis deputatis. Ceterum fac Solis ad Saturnum, iterum Solis ad Iovem, iterum Solis ad Martem, et sic ad Venerem et Mercurium. Iterum quando Sol capiti coniungitur vel opponitur, postea fac Saturni et Iovis, postea Saturni et Martis, Saturni et Veneris, Saturni et Mercurii. Postea quando Saturnus est cum capite vel in opposito, rursus fac Iovis et Martis, Iovis et Veneris, Iovis et Mercurii et coniunccionem et opposicionem capitis cum Iove. Item Martis et Veneris, Martis et Mercurii, Martis cum capite vel in eius opposito. Item coniuncciones Veneris et Mercurii et eorum aspectus sextiles. Ceteri aspectus in hiis duobus planetis non fiunt, eciam sextilis in eis rarissime occurrit. Horum omnium aspectuum karacteres loca ad cellulas eis debentes sub titulo aspectus planetarum inter se. Et ecce: tachuini finis advenit. Porro precisionem horarum aspectuum non curamus, preterquam in coniunccionibus et opposicionibus

---

**720** sit *A* : *om. B* **729** fuerit maior *A* : fuerit maior fuerit *B* **736** hore *A* : karacteres *B* **742** iungitur *A* : coniungitur *B* **744** deputatis *A* : deputamus *B*

luminarium et in coniunccionibus ponderosorum planetarum, de quibus per patulo sermonem faciemus.

<21>

### Declinacionem Solis conficere et cuiuslibet gradus zodiaci

Intra cum gradu Solis aut cum gradu nominato in *Tabulam declinacionis Solis* et declinacio sub signo illius gradus inventa erit declinacio quesita, que erit meridionalis ascendens vel descendens, aut septentrionalis ascendens vel descendens, veluti tituli docent.

<22>

### Latitudinem Lune congnoscere

Cum argumento latitudinis Lune intra *Tabulam latitudinis Lune* et eius latitudinem mox sub signo argumenti videbis, que cuius partis sit, ibidem circa signa conspicies.

<23>

### Tempus coniunccionis vere vel implecionis sive opposicionis luminarium

et coniunccionis vel opposicionis alias implecionis certum locum siquidem cupiveris investigare, fac verum locum Solis, fac eciam verum locum Lune ad tempus coniunccionis medium, verum locum Solis cum argumento Lune, verum locum Lune cum argumento Lune tempore medie coniunccionis intrando cum eo ad *Tabulam equacionis Lune in diebus lunacionum* aut ad *Tabulam equacionis Lune* propriam, quod argumentum Lune nulla eget equacione. Eorum verorum motuum subtrahe minorem a maiori, residuum erit longitudo, que longitudo erit Solis, si verus locus Solis fuerit maior vero loco Lune, aut erit longitudo Lune, si verus locus Lune fuerit maior vero loco Solis. Eius longitudinis duodecimam partem sic quere: multiplica longitudinem in quinque minuta et proveniens erit longitudinis duodecima pars, vel divide longitudinem in quinque minuta et proveniens erit longitudinis 12 partes, vel divide longitudinem per 12 et in quociente eadem duodecima pars, que prius, apparebit. Aggrega longitudini eius duodecimam partem et collectum longitudo cum sua duodecima vocabitur. Dimidia longitudine cum sua duodecima et unam medietatem iunge cum argumento Lune, si fuerit longitudo Solis, vel minue ab eo, si fuerit longitudo Lune, et producetur argumentum horarum Lune. Cum argumento horarum Lune intra *Tabulam motus Solis et Lune* equalis in una hora sub signo eiusdem argumenti comperto et eum verum motum Lune in una hora nota cum prioribus. Divide longitudinem cum sua duodecima per verum motum Lune in una hora, et que inde provenerint hore, serva. Residuum quoque in 60 multiplica et numerum inde surgentem ut prius per eundem motum divide et videbis minuta horarum, que cum horis pretactis erunt tempus inter mediam coniunccionem et veram, quod adde ad tempus medie coniunccionis, si fuerit longitudo Solis, vel minue ab eo, si fuerit longitudo Lune. Quodque addendo vel minuendo provenerit, erit tempus vere coniunccionis, pro quo laborasti. Ceterum certum locum luminarium, hoc est verum locum vere coniunccionis, taliter habebis: adde longitudinem cum sua duodecima ad verum locum Lune et duodecimam tantum ad verum locum Solis tempore medie

---

**759** gradu[2] *A* : gradu cumque *B* **768** opposicionis alias implecionis *B* : implecionis *A* **770** verum[1] … Lune A : *om. B* **773** Eorum *A* : *om. B* **777** in… longitudinem *A* : *om. B* **782** horarum *A* : horarium *B* **783** horarum *A* : horario *B*

coniuccionis, si fuerit longitudo Solis, vel ea subtrahe ab eis, si fuerit longitudo Lune, et habebis Solem et Lunam precise in eisdem gradibus et minutis, que erunt luminarium et coniunccionis verus locus. Quod si tempore medie coniunccionis verus locus Solis et verus locus Lune in eisdem gradibus et minutis evenerint, quod paucissime contingit, tunc in eodem tempore fit coniunccio media et vera. Item de implecione seu opposicione vera idem erit operandi modus. Cum enim habueris mediam opposicionem, nota medium Lune motum, cuius nadir, id est gradus oppositus, semper erit in opposito, scilicet in coniunccione Solis, et sic tempus opposicionis Solis et Lune medium erit tempus coniunccionis medie Solis cum nadir Lune. Quere ergo per modum superiorem tempus vere coniunccionis Solis et Lune accipiendo tempus medie coniunccionis et medios motus Lune et argumentum Lune, que tempore medie opposicionis inveniuntur. Cumque habueris tempus vere coniunccionis Solis cum nadir Lune, ad ipsum erit tempus vere opposicionis Solis et Lune. Et quando verus locus Solis et verus locus Lune erunt in eisdem gradibus et minutis, Luna erit in opposito eorum in eisdem gradibus et minutis et hoc erit certus locus vere opposicionis, si Luna fuerit super Terram, aut verus locus Solis erit opposicionis verus locus, si Sol fuerit super Terram.

Verum argumentum latitudinis Lune, per quod possibilitas ecclipsium dinoscitur ad tempus vere coniunccionis vel opposicionis invenire. Verum locum capitis draconis tempore vere coniunccionis vel opposicionis de vero loco Lune eiusdem temporis minue. Quod tunc remanet, erit verum argumentum latitudinis Lune, quod queritur. Idque sciendum: occurrit tempus vere coniunccionis vel opposicionis ex premissa praxi, productum erit tempus vere coniunccionis vel opposicionis inequatum sive tempus vere coniunccionis vel opposicionis diebus non equatis. Pariformiter fit eciam et de omni tempore vero, quod per horas equinoccialis circuli ex tabulis trahitur, ut est tempus verum revolucionum annorum et nativitatum et sic de aliis. Quapropter huiuscemodi tempora te equare oportet. Intra cum gradu Solis eiusdem temporis veri inequati ad *Tabulam equacionis dierum cum suis noctibus*, et que minuta et secunda equacionis dierum sub signo illius gradus inveneris, adde tempori vero equato et proveniet tempus verum diebus equatis. Rursus loca planetarum vera ad tempora vera inequata querimus, scilicet gradum ascendentis, et domorum equaciones pro equatis solum modo temporibus erigimus. Tempora, que per instrumenta accepimus pro equatis temporibus, retinemus eo, quod in certo puncto emisperii locum Solis observant.

<24>

**Verum locum Solis in una hora et verum locum Lune in una hora quovis tempore percipere**

Cum gradu argumenti Solis intra *Tabulam veri motus Solis et Lune equalis in una hora* et motum Solis, quem sub signo argumenti inveneris, erit motus verus Solis in una hora. Similiter cum gradu argumenti Lune equati eandem tabulam ingredere et motus Lune sub signo argumenti Lune compertus erit verus motus Lune in una hora. Sic inventus non omnino precisus erit, non tamen notabilem in opere faciet variacionem. Si autem eundem precisiorem habere volueris, quere eum cum argumento horarum, sicuti supra doceris.

---

**799** enim *B* : etenim *A* **806** Solis[1] *A* : *om. B* **808** locus vere opposicionis *B* : vere opposicionis locus *A* verus locus *B* : locus verus *A*

<25>

**<T>empus vere coniunccionis Iovis et Saturni vel Martis cum aliquo istorum ex veris locis tachuini comprehendere**

Considera, quo die sibi sunt in zodiaco viciniores. Demum subtrahe verum locum eiusdem diei minorem unius istorum a maiori et remanebit longitudo illius, cuius motus est maior. Quod si pro eodem tempore motus planete tardioris cursus fuerit maior, tunc coniunccio erit futura post meridiem. Si autem planeta velocioris cursus fuerit maior, tunc coniunccio preteriit ante meridiem. Ceterum inter verum locum cuiuslibet eorum presentis diei et verum locum sequentis diei differenciam divide per 24 et verum motum in una hora uniuscuiusque eorum inde provenientem considera. Subtrahe verum motum in una hora minorem a maiore, residuum erit superacio. Per superacionem divide longitudinem et horas in quociente orientes nota. Residuum longitudinis in 60 multiplicatum et iterum per superacionem divisum in quociente horarum minuta producit. Has horas et minuta longitudinis adde ad meridiem diei coniunccionis, si coniunccio fuerit futura, vel subtrahe a meridie, hoc est a 24 horis, si coniunccio preteriit ante meridiem, et habebis tempus vere coniunccionis planetarum diebus non equatis, quod voluisti. Quod tempus equa, si gradum ascendentis et figuram celi volueris erigere. Verum motum unius istorum in una hora in horas longitudinis multiplica et proveniens iunge cum vero loco illius planete, si coniunccio fuerit futura, vel subtrahe, si preteriit, et emerget tibi coniunccionis certus locus. Quod si die nominato planete in eisdem gradibus et minutis fuerint, tunc coniunccio eorum erit precise in meridie eiusdem diei.

<26>

**Ascensiones gradus Solis aut cuiuslibet gradus zodiaci in circulo recto congnoscere**

Intra cum numero istius gradus in *Tabulam ascensionum* in circulo directo et ascensiones sub signo gradus illius reperte erunt ascensiones illius gradus, qui sunt gradus equinoccialis a principio Carpicorni usque ad gradum Solis.

<27>

**Sinum rectum cuiuslibet arcus sive porcionis circuli investigare**

Intra cum gradibus illius arcus seu porcionis in *Tabulam sinus* et sinum eius rectum sub signo illius gradus invenies.

<28>

**<S>i cuiuslibet arcus sinum versum volueris** et fuerint gradus arcus pauciores 90, eos de 90 minue et eius, quod remanserit, sinum rectum invenias, quem de 150, qui est tocius sinus rectus, minue. Quod tunc remanserit, erit sinus versus, quem queris. Si autem gradus arcus fuerint plures 90, ab eis gradibus 90 depositis residui sinum rectum confice, quem adde 150 et habebis sini versi summam, quam voluisti. Si vero arcus fuerit 90 gradus precise, tunc sinus rectus et sinus versus eiusdem erunt quantitatis.

---

**842** erit *A* : est *B* **845** hora *B* : hora minorem *A* **847** horas *A* : hore *B* **865** arcus *A* : gradus *B* sinum : *mss. per err.*: sinus **871** arcus *A* : arcus, quem queris *B*

<29>

**Si per altitudinem Solis cuiuslibet rei umbre recte quantitatem placuerit indagare**, cum gradibus altitudinis Solis intra *Tabulam umbre* et puncta cum minutis, que ex directo gradus altitudinis apparuerint, umbre recte quantitatem indicabunt, per que si 144 diviseris, umbre verse quantitatis eveniet.

<30>

**Altitudinem rei per umbre quantitatem apprehendere**

Divide umbram in tot partes, quod umbre fuerint puncta. Que si precise 12 fuerint, tunc umbra et altitudo rei eiusdem sunt quantitatis. Si autem umbra fuerit maior 12 partibus, de eis partibus accipe 12, que elevate rei altitudinem significabunt. Si vero umbra fuerit minor 12 partibus, tunc unam partem extende in 12 vel accipe eam duodecies et consurget altitudo rei quesita.

<31>

**<C>um autem altitudinem rei sine umbra per aldidode elevacionem scire volueris,**

*Tabulam umbre* cum altitudine aldidode ingredere et puncta e directo comperta considera. Postea spacium ab ipsa re elevata usque ad medium plante pedis tui loci, in quo steteras, aldidodam elevando in tot partes divide, quot habueris puncta, et cum eis partibus altitudinem rei ut prius pronunciabis. Hoc adiecto cum iam quantitatem altitudinis rei designaveris, adde ad eam quantitatem, que est ab occulo tuo usque ad terram, et dic tanta est rei istius altitudo.

<32>

**Si putei sic et vallis profunditatem libuerit percipere,**

sta in una parte putei et vide per defossionem aldidade *Tabulam umbre* ingressus. Accipe puncta, que ei debentur. Postea in tot partes, quot fuerint puncta, dyametrum putei divide, eam videlicet dyametrum, que est ab oculo tuo usque ad putei partem oppositam, que eciam dyameter erit eiusdem quantitatis cum dyametro superficiei aque putei, et cum eis partibus profunditatem putei, que est ab oculo usque ad aquam, quemadmodum altitudinem rei per umbram pronunciabis.

<33>

**Planiciei quantitatem mensurare**

Vide per depressionem aldidode extremitatem planiciei, cuius depressionis puncta in *Tabula umbre* considera. Postea quantitatem, que est ab oculo in terram, in 12 partes divide. Ceterum si puncta plura quam 12 fuerint, eas partes in tot partes, quot erunt puncta, extende et habebis planiciei quantitatem. Si vero minora 12 fuerint, tunc de 12 partibus accipe tot partes, quot fuerint puncta, et proveniet quantitatis planiciei certitudo.

---

**878** umbre *A* : *om.* **B** **881** puncta *A* : scilicet puncta *B* 12 fuerint *A* : fuerint 12 *B* **882** eiusdem sunt *A* : sunt eiusdem *B* **891** cum *A* : *om.* *B* **894** rei istius *A* : istius rei *B* **896** profunditatem libuerit *A* : libuerit profunditatem libuerit *B* **901** profunditatem putei *A* : putei profunditatem putei *B* ad *A* : *om.* *B*

<34>

**Elevacionem capitis Arietis et Libre meridianam in omni regione invenire**

Subtrahe elevacionem poli eiusdem regionis a 90 gradibus, remanens erit elevacio capitis Arietis et Libre in tali regione, que eciam erit elevacio Solis meridiana Sole existente in pincipio Arietis aut Libre. Cuius elevacionis umbram in *Tabula umbrarum* investigabis et eam pro usu sequencium retineto. Cum ad ceteros gradus elevacionem Solis meridianam desideraveris, adde declinacionem Solis ad elevacionem Arietis pro signis septentrionalibus, vel subtrahe pro meridionalibus et habebis elevacionem Solis in meridie pro omni gradu Solis et die anni, qua volueris. Habita quoque Solis cuiuscumque elevacione meridiane umbre quantitas meridiana ex prioribus haud difficilimum exquiritur.

<35>

**Altitudinem poli et latitudinem regionis per altitudinem Solis meridianam ex instrumentis congnoscere**

Accipe per aliquod instrumentum altitudinem Solis meridianam, quam subtrahe a 90 gradibus. Remanens erit poli altitudo et regionis latitudo, si Sol in principio Arietis vel Libre fuerit. Si vero fuerit alibi, subtrahe declinacionem Solis ab altitudine meridiana, si Sol fuerit in signis septentrionalibus, vel adde, si fuerit in signis meridionalibus, et cum altitudine meridiana proveniente fac ut prius.

<36>

**<C>um ascensionis uniuscuiusque signi tabulam in qualibet regione libuerit componere,**

umbram altitudinis Solis meridiane Sole existente in principio Arietis vel Libre in tua regione in id, quod in directo unius gradus inveneris in *Tabula differencie ascensionum*, multiplica et provenientis tibi sinus in *Tabula sinus* porcionem circuli quere, que erit differencia primi gradus Arietis, quam deme ab elevacione primi gradus Arietis in circulo recto et remanens tibi numerus est, quod cum primo gradu Arietis in tua regione elevatur. Eam eciam differenciam adde ellevacioni primi gradus Libre in circulo recto et habebis elevaciones gradus eiusdem in tua regione. Similiter eandem umbram multiplica in id, quod in directo duorum graduum in *Tabula differencie ascensionum* inveneris, provenietque sinus, quem ut prius arcuabis. Qui arcus erit differencia duorum graduum Arietis, quam ex elevacione duorum graduum circuli recti minuas, et remanens est, quod elevatur cum duobus gradibus Arietis in regione tua. Et hanc eciam differenciam adde elevaciones duorum graduum Libre circuli recti et proveniens erit, quod in tua regione cum duobus gradibus Libre oritur. Et sic fac usque ad gradus 90 et habebis *Tabulam medii circuli*, scilicet sex signorum a principio videlicet Arietis usque ad finem Geminorum et a principio Libre usque ad finem Sagittarii, si *Tabula circuli directi* ab Ariete principium habuerit, que siquidem a Capricorno fluxerint. Tunc a quibuslibet graduum ascensionibus eiusdem circuli recti a Capricorno fluenti 90 gradus depone et habebis in remanente computum ascensionum *Tabule circuli recti ab Ariete*, cum quibus,

**916** elevacionem *A* : elevacionum *B* **936** quam… Arietis *A* : *om. B* **946** circuli *A* : arcus *B*

que dicta sunt, perficies. Reliquam quoque partem aliorum sex signorum hoc modo cudas: pone pro fine Piscium 360 gradus, 0 minuta, et pro fine Virginis pone 180, 0 minuta. Ceterum subtrahe singulas ascensiones Arietis noviter factas a 360 gradibus et provenient omnes ascensiones Piscium. Sic quoque fac cum ascensionibus Libre et habebis ascensiones Virginis. Itaque ex ascensionibus primi gradus Arietis provenient ascensiones 29 graduum Piscium et ex ascensionibus primi gradus Libre produces ascensiones 29 gradus Virginis. Et ex ascensionibus secundi gradus Arietis fluunt ascensiones 28 gradus Piscium, et ex ascensionibus secundi gradus Libre fluunt ascensiones 28 gradus Virginis, et sic retrograde usque ex ascensionibus 29 gradus Arietis ascensiones primi gradus Piscium producantur et ex ascensionibus 29 gradus Libre ascensiones primi gradus Virginis apparebunt. Et ex ascensionibus 30 gradus Arietis ascensiones 30 Aquarii conficiuntur et ex ascensionibus 30 gradus Libre 30 gradus Leonis emanant, et sic de reliquis signis et gradibus, quamadmodum *Tabula differencie ascensionum* per suprascripta graduum columpnarum admonet. Postremo invento arcu diurno eum in 12 partire et proveniens numerus quociens partes horarum inequalium indicabit, que e directo graduum ascensionum locari consueverunt. Habitis partibus horarum ad medium circuli subtrahe singulas a 30 gradibus et habebis partes horarum oppositarum ad aliam partem circuli vel partis horarum alterius medietatis circuli sic perficitur. Partes horarum primi gradus Arietis erunt partes horarum 29 Virginis et partes primi Tauri erunt partes 29 Leonis et partes primi Geminorum erunt 29 Cancri et partes primi gradus Libre erunt 29 Piscium et partes primi Scorpionis erunt partes 29 Aquarii et partes primi Aquarii erunt partes 29 Capricorni et sic de aliis gradibus intermediis retrogradando sicuti in *Tabula differencie ascensionum* iuxta se comperiuntur eo quoque ordine proveniet, quod partes horarum 30 gradus Arietis erunt partes 30 gradus Leonis et partes 30 gradus Libre erunt partes 30 Aquarii. Et partes 30 Tauri erunt partes 30 Cancri, et partes 30 Scorpionis erunt partes 30 Capricorni eruntque in fine Virginis et in fine Piscium partes horarum in omni regione 15 gradibus, 0 minutis, et in hoc erit *Tabula ascensionum regionis tue*, que in *Tabula circuli obliqui* in tua regione completa.

<37>

**Ascensiones gradus Solis aut cuiuslibet gradus zodiaci in circulo obliquo tue regionis per tabulam ad hoc factam invenire**

Cum eo gradu intra *Tabulam ascensionum seu elevacionum signorum in circulo obliquo* tue regionis et ascensiones, quas sub signo illius gradus inveneris, pro quesito teneto.

<38>

**<P>artes horarum gradus Solis aut alterius gradus cuiuscumque**

sunt gradus ascensionum, qui elevantur in una hora inequali, consimili introitu accipiuntur.

---

**951** que… perficies *A* : *om. B* **952** cudas *A* : *om. B* o minuta[1] *A* : *om. B* o minuta[2] *A* : *om. B* **956** ascensionibus *B* : ascensiones *A* **965** indicabit *A* : *om. B* **977** o minutis *A* : minutis o *B*

<39>

**<A>rcum diurnum cognoscere**

Minuas ascensiones gradus Solis regionis tue de ascensionibus nadir eius, remanens erit arcus diei quesitus, vel multiplica partes horarum gradus Solis in 12 et idem invenies. Ipsum quoque arcum diurnum a 360 gradibus deme, in residuo habebis arcum noctis gradus Solis.

<40>

**<D>iei quantitatem percipere**

Arcum diurnum per 15 divide, in quociente habebis horas et residuum in 60 multiplicatum per idem rescinde et habebis minuta horarum, que cum horis suis indicant diei quantitatem, quam a 24 horis si deposueris, quantitas noctis remanebit.

<41>

**Horas equales in ascensionem horarum convertere**

Eas in 15 multiplica. Horas inequales in ascensiones ducere: eas per partes horarum multiplica. Horas equales in horas inequales convertere: ascensiones horarum per partes horarum divide. Horas inequales ad horas equales permutare: ascensiones horarum inequalium per 15 divide.

<42>

**Gradum ascendentis et principia 12 domorum figure celi per Tabulas ascensionum invenire**

Ascensiones gradus Solis circuli directi imprimis invenias, quibus adde ascensiones horarum, ad quas ascensiones desideras, et provenient ascensiones in *Tabula elevacionis signorum gradus ascendentis in circulo obliquo*, quas quere inter ascensiones in *Tabula elevacionum signorum regionis tue*, et gradus zodiaci equalis, qui e directo apparuerit, erit gradus ascendens in illa hora, qui illius erit signi, quod supra ascensiones situatur. Ad principia aliarum domorum sic procede: ascensiones gradus ascendentis in circulo obliquo tene pro ascensionibus medii celi seu decime domus et reliquarum domorum ascensiones ceterum sic invenias. Accipe partes horarum gradus ascendentis, quas duplatas adde ascensionibus gradus medii celi et productum erit ascensiones undecime domus. Item ascensionibus undecime domus adde easdem partes duplatas et producentur ascensiones 12 domus. Quibus iterum adde easdem partes duplatas et provenient ascensiones gradus ascendentis in circulo recto, que erunt alie ab ascensionibus gradus ascendentis circuli obliqui prius habitis, et nunc pro ascensionibus medii celi in circulo recto positis. Ceterum subtrahe easdem partes duplatas a 60 gradibus, residuum erit partes horarum noctis duplate, quas adde ad ascensiones ascendentis circuli recti, et habebis ascensiones secunde domus et iterum ascensionibus secunde domus adde easdem partes horarum noctis duplatas et surgent ascensiones tercie domus et iam habes ascensiones circuli recti ad sex domos, quas quere inter ascensiones circuli recti et gradus equales, que eis debentur, erunt inicia domorum earundem et horum graduum

---

**1001** ducere *A* : *om. B* **1002** ascensiones *A* : in ascensiones *B* **1009** desideras *A* : *om. et in mg. add.* queris *B*

signa ponuntur supra ascensiones in principio tabule. Item proveniet eciam idem gradus ascendentis cum ascensionibus gradus ascendentis circuli recti, qui inventus erit cum ascensionibus circuli obliqui, et ita habes principia sex domorum, quorum nadir sunt reliquarum sex oppositarum domorum inicia, que solito modo figuram celi conscribe. Porro traximus quasdam tabulas ex labore ingeniosi Blankini, per quas ascendens et ceteros undecim cardines perfacile velut sequitur, cognoscimus.

<43>

**Horas gradus Solis invenire**

Intra cum gradu Solis *Tabulam ad inveniendum ascendens climatis tui* et hore a meridie, que e directo obiciuntur, sunt hore gradus Solis a meridie et hore horologii, quas ibidem reperieris, sunt hore gradus Solis horologii.

<44>

**Gradum ascendentis qualibet hora invenire**

Si hore tue, ad quas ascendens cupis invenire, fuerint hore a meridie, adde cum eis horas gradus Solis a meridie. Si vero hore tue fuerint hore horalogii, que ab occasu Solis suputantur, iunge cum eis horas gradus Solis horalogii et utrobique provenient tibi hore gradus ascendentis. Perquire in eadem tabula sub horis ascendentis et gradum, quem e directo earum sub titulo ascendentis inveneris, est gradus ascendentis, qui cuius signi fuerit, ibidem conspicies. Ceterum ex directo gradus ascendentis, qui est inicium prime domus, in *Tabula equacionum domorum tui climatis* invenies reliquarum quinque domorum inicia. Quibus iniciis sex domorum habitis nadir eorum oppositorum domorum erunt principia. Rursus conneximus minuta gradibus quarte domus, ex eis quoque minuta decime domus oriuntur propter direcciones nativitatum et revolucionum annorum, in quibus minuta quatuor cardinum celi, ad quos direcciones celi fiunt, quam plurimum attenduntur. In aliis autem domibus minuta non curamus.

<45>

**Habito gradu ascendentis horam invenire**

Gradum ascendentis in *Tabula ad inveniendum ascendens tui climatis* sub titulo 'ascendens' considera, ab eo versus sinistrum hore ascendentis apparebunt, a quibus horis ascendentis subtrahe horas gradus Solis meridiei et provenient hore a meridie, vel subtrahe horas gradus Solis horalogii et habebis horas horalogii gradui ascendentis correspondentes.

<46>

**Horas a meridie in horas horalogii convertere**

Adde cum eis horas gradus Solis a meridie et a productis horis ascendentis subtrahe horas gradus Solis horalogii. Remanens erit hore horalogii.

---

**1029** nadir sunt *A* : sunt nadir *B* **1031** ingeniosi *A* : laboriosi *B* Blankini *A* : Blanchini *B* **1032** perfacile *A* : facile *B* **1044** ascendentis[1] *B* : ascendens *A*

<47>

**Horas horalogii in horas a meridie transfigurare**

Iunge cum eis horas gradus Solis horalogii et provenientibus horis ascendentis deme horas gradus Solis a meridie. Residuum erit hore a meridie.

<48>

**Qua hora horalogii fit meridies, cognoscere**

Subtrahe horas horalogii gradus Solis a meridie et residuum manifestabit, qua hora horalogii fit meridies.

<49>

**Quantitatem alio modo quam supra invenire**

Subtrahe horas, in quibus fit meridies, a 24 horis, residuum erit quantitas medii diei, quam vero eam eciam medietatem diei dupla, et totam diei quantitatem habebis. Que 24 horis deposita noctis quantitatem indicabit.

<50>

**Qua hora horologii oritur Sol,**

hore quantitatis noctis indicabunt, vel subtrahe medietatem quantitatis diei ab horis horalogii, in quibus fit meridies, remanens erit hore horalogii ortus Solis et noctis quantitas.

<51>

**Qua hora a meridie occidit Sol,**

medietas quantitatis diei docet. Item subtrahe medietatem quantitatis diei a 24 horis et in residuo habebis, qua hora meridiei Sol oriatur. Que omnia rectoribus horalogiorum et temporum observatoribus sunt pensiliter attendenda.

<52>

**Vera loca quarumdam famosarum stellarum fixarum agnoscere**

Intra cum nominibus earum in *Tabulam stellarum fixarum*, signa et gradus et minuta e directo proxime apparencia vera earum loca significabunt, que vera loca sunt eciam earum longitudo. Alia ceterumque, que eadem tabula per titulos admonet post vera loca, seriatim considerabis.

<53>

**Introitum Solis in primum minutum sive punctum Arietis**

est tempus revolucionis annorum mundi invenire. Quere verum locum Solis ad tempus propinquum, quo estimes Solem in talem punctum intrare, quam verus locus Solis,

---

**1065** horalogii *A* : horologii, remanens erit hore horologii *B* **1066** hore *A* : hore horologii *B* **1068** Qua… meridies *A* : Horas horologii in meridie *B* **1074** quam… diei *A* : *om.* *B* **1077** hora horologii *B* : autem hora hora horalogii *A* **1085** temporum *A* : ipsorum *B* **1088** stellarum *A* : *om.* *B* **1093** minutum sive punctum *A* : punctum sive minutum *B*

si fuerit o in signis precise, scilicet quod eciam o erit in gradibus et minutis, tunc eo tempore Sol intrat pretactum punctum. Si vero non fuerit o in signis, considera distanciam seu longitudinem Solis a predicto puncto, hoc est considera verum locum Solis, qui est ultra o in signis, ut hic motus, qui est ultra punctum quesitum, dicitur longitudo seu distancia subtrahenda, aut considera, quantum Sol ad hoc currere debeat ad hoc, quod eius verus motus o in signis fiat precise, quod scire potes subtrahendo eius verum motum a 12 signis, et residuum, quia est ante punctum pretactum, dicitur longitudo seu distancia addenda. Demum longitudinem divide per verum motum Solis in una hora et horas et minuta horarum longitudinis in quociente provenientes subtrahe a tempore estimato, quo putabas Solem Arietem intrare, si longitudo fuerit subtrahenda. Vel adde easdem horas tempori estimato, si longitudo fuerit addenda, et proveniet tempus revolucionis annorum mundi, in quo Sol primum minutum Arietis aggreditur diebus non equatis. Quod si probare volueris, equa Solem ad idem tempus revolucionis et proveniet eius verus locus o in signis precise, si non errasti. Et sufficit nobis in hoc negocio precisio in minutis. Si autem fuerit diversitas in secundis, non curamus. Istud tempus diebus non equatis equabis cum fine Piscium, qui est principium Arietis, per modum superius dictum intrando *Tabulam equacionis dierum* et proveniet tempus revolucionis annorum mundi diebus equatis, ad quod tempus diebus equatis solent gradus ascendentis et cetera domorum gradus pro iudiciis anni inveniri. Pariformiter invenitur introitus Solis in Cancrum, quando Sol est precise 3 in signis, et similiter, quando intrat Libram, scilicet quando est precise sex in signis. Eodem modo, quando intrat Capricornum, scilicet quando est precise 9 in signis, et sic de aliis signis et gradibus in quibuslibet principiis et nativitatibus, quarum revoluciones practicantur.

Introitus Solis in Arietem et revolucio annorum mundi pro meridiano Bononiensi anno Domini 1477 curente:

| **Dies** | **Hora** | **Minuta** | **Secunda** | |
|---|---|---|---|---|
| 10 | 16 | 44 | 11 | Marcii diebus non equatis – meridiei |
| 10 | 16 | 52 | 51 | Marcii diebus equatis – meridiei |
| 10 | 10 | 44 | 11 | Marcii diebus non equatis – horalogii |
| 10 | 10 | 52 | 51 | Marcii diebus equatis – horalogii |

| **Gradus** | **Minuta** | |
|---|---|---|
| 28 | 37 | Aquarii ascendens tempore revolucionis |

<54>

**<T>empus revolucionum annorum mundi seu nativitatis vel alterius principii cuiuscumque per tabulas invenire**

Nativitatis vel alterius principii radicem et locum Solis hora ipsa diebus non equatis rectifica. Postmodum considera, si annus ipsius radicis fuerit bisextilis aut primus, secundus vel tercius post bisextum et quere in *Tabula revolucionum annorum* tabulam ipsi anno suppositam. Demum considera distanciam anni, cuius revolucionem queris, ad annum radicis, cum qua tabulam supradictam intra et in linea annorum quere nume-

**1111** equabis *A* : equatis *B* **1115** intrat *A* : Sol intrat *B* **1131** cuiuscumque… principii *A* : *om. B* **1133** primus *A* : primus vel *B*

rum similem illi anni distancie, et quod in eius directo inveneris de horis, minutis et secundis radicis, si scriptum fuerit 'minue', ipsas minue de horis radicis, et quod per addicionem vel diminucionem provenerit, erit tempus revolucionis radicis nec equatur anni dati, sed quia tabula ipsa non excedit numerum 40 annorum. Si autem distancia anni revolucionis ad annum radicis excederet annos 40, tunc quod in directo annorum 40 inveneris, accipe primo. Postmodum in annis, qui supersunt, iterum tabulam ipsam intra et quere horas, minuta et secunda eius in directo, quas adde vel minue cum aliis horis, prout scriptum est in eius directo, et habebis tempus predictum non equatum. Iterum vide gradum Solis, in quo signo est tempore predicto, cum quo gradu intra *Tabulam equacionis revolucionum* in linea supposita signo, in quo Sol reperitur, et secunda, que ibi e directo gradus illius signi iveneris, multiplica per annos distancie revolucionis ad annum radicis et predictum adde seu minue de horis et minutis et secundis temporis revolucioni non equati superius inventi et habebis tempus revolucionis equatum diebus non equatis, id est tempus precise, quo Sol ingreditur punctum, in quo erat hora radicis. Et hec secundum tradicionem domini Iohannis de Blankinis, que licet satis precisa sit, volumus tamen quatenus more nostro probacionem facias.

---

**1138** per addicionem vel *B* : addicionem ut *A* **1143** quas… directo *A* : *om. B* **1151** domini Iohannis de Blankinis *A* : Iohannis Blanchini *B*

## 7.3 Partial Translation and Commentaries on Canons *Quoniam plerique* by Nicolaus Wodka of Kwidzyn[15]

Begin the canons to the *Tabule resolute* for the meridian at Bologna, Italy, and for any other place, compiled by the eminent doctor, Master Nicolaus de Prusia.

### Introduction

As most physicists in our time are turning away from the science of the celestial heights, being dulled by the difficulty of the work and the tediousness of the tables, we have taken care, with God's help, to provide an easy way for some of the ardent men interested in the science of the stars to make a tacuinum and to present it in a concise and coherent form, compiled from various and scattered tables.[16]

### Division of the sphere

*– Nicolas Wodka first explains the difference between the division of the ecliptic into 6 physical signs (signa phisica) of 60 degrees and 12 common zodiacal signs of 30 degrees. He attributes the former method to old astronomers and denotes the latter style as modern. Wodka then explains the division of the day and night into 24 equal hours used in the tables, as well as the division of the day from the sunrise to sunset and the night from the sunset to sunrise each into 12 unequal hours. He also mentions the division of day into 60 minutes, used in the PAT, and the subsequent division of any minute into 60 seconds and the next sexagesimal division into tertias, quartas etc. For the trigonometric functions of chord, sine, and versine, Wodka divides the diameter of circle into 300 parts. In the following four paragraphs, Wodka describes the basic arithmetic operations required for working with the tables.*

### On addition

*– In the case of addition (e.g. in the sexagesimal system, see Chapter 1.2), the carrying of a digit of the sum to the higher-order digits is specified.*

### On subtraction

*– In the case of subtracting a larger digit from a smaller one, the borrowing of a unit of higher order is described.*

15 We present a translation of main rules on the creation and use of tacuinum as well as several other chosen rules. In the remaining cases, the translations of the rules are replaced by commentaries with a brief description and paraphrase of their content. These commentaries are printed in italics. – The seemingly high number of rules is compensated by their brevity; especially the second half of the text is rather just a syllabus for a longer explanation.

16 Already in the introduction Nicolaus Wodka's interest in enriching his canons with a detailed description of the tacuinum is expressed.

### On multiplication

*– To multiply two sexagesimal numbers (or those combined with signs or hours), it is reccommended to convert them first to the order of the smallest fraction. The denomination of the product is given by the sum of sequence numbers of the factors, e.g., a multiplication of minutes with number –1 and seconds with number –2 results in tertias with the sequence number –3.*

### On division

*– Conversely to multiplication, the denomination of the quotient in division is given by the difference of the sequence numbers of the dividend and the divisor. Converting a high integer number of some fractions into the fractions of higher order is also explained.*

### <1> <**On the mean motion**>

*– In the first Rule Nicolaus Wodka summarizes the calculation of mean motions in signs, degrees, minutes, and seconds, using the tables in collected years, expanded years, months, days and hours. He explicitly mentions here the meridian of Bologna, to which he calculated his tables. Further references to Bologna are in Rules 3, 5, 6, 15 and 53. In Rule 1 the term 'cautela' (warning) is used for the first time, which was later consistently used by John of Głogów in his canons.*

### <2> On the apogees of the planets

*– Nicolaus Wodka advises how to find the apogees of the planets for the beginning of the year using the corresponding tables in the collected and expanded years. He considers the motion of the apogees in the course of the year to be negligible.*

### <3> On the mean conjunctions of the planets, that is of the Sun and Moon

*– The rule describes computation of the mean conjuncions of the luminaries as well as their oppositions. Wodka's deviations from procedures common in other Tabule resolute are described in Chapter 15.3.*

### <4> On the mean argument of the latitude of the Moon to the time of the mean conjunction

*– Nicolaus Wodka gives instructions on how to calculate the argument of the latitude of the Moon, because he does not give this value in his tables. In the TR, most other authors included the argument of the latitude of the Moon in the Table of conjunctions of the Sun and Moon. Another copy, where this value is not indicated in the TR, is MS Leipzig, BU, 1473, f. 126r, computed for Leipzig (see Chapter 15.1, Table TRL-09).*

### <5> On the intercalary year

*– The intercalary years (anni embolismales) were already mentioned in Wodka's Rule 3. Here only a reference to the Table of conjunctions in expanded years is added, where these years are*

*denoted similarly as in the Table TRW-11. The rest of Rule 5 deals with a correction of the time of conjunctions to meridians different from that of Bologna.*

<6> **Having known the motions and mean times for the meridian of Bologna, or having reduced them to the meridian of another city, as has been shown, it now remains to calculate the true or certain motions and true times by means of the Tables of equations**

– *Rule 6 describes general procedures for searching and interpolating in the Tables of equations.*

<7> **On the true position of the Sun**

– *A brief description of using the Table of equation of the Sun to find true position of the Sun is presented.*

<8> **On the true position of the Moon**

– *A brief description of using the Table of equations of the Moon to find true position of the Moon is given.*

<9> **Find the true positions of the three upper planets, i.e., Saturn, Jupiter, and Mars**

– *Rule 9 describes using the Tables of equations of the Saturn, Jupiter, and Mars to find true positions of the three upper planets.*

<10> **To find out if a planet is stationary, retrograde or direct**

– *In Rule 10, Nicolaus Wodka describes the calculation of the stations and retrogradations of the planets in a verbally different but essentially identical way, as did his teacher Andrzej Grzymała in Rule 12 of his canons Girum recensendo (see Chapter 3.2), and after him Nicolaus of Tuchów.*[17] *He refers, among other things, to a Table of the daily motions of the anomalies of planets, which is missing in the copy with Wodka's canons in the manuscript London, BL, Add 22317, but can be found in Grzymała's canons Girum, for example, in MS Cracow, BJ, 573, f. 231r:*

| signa | gradus | minuta | secunda | tercia | argumenta <planetarum> in uno die |
|---|---|---|---|---|---|
| 0 | 0 | 57 | 7 | 44 | Saturni |
| 0 | 0 | 54 | 9 | 4 | Iovis |
| 0 | 0 | 27 | 41 | 41 | Martis |
| 0 | 0 | 36 | 59 | 27 | Veneris |
| 0 | 3 | 6 | 24 | 8 | Mercurii |

17 See Porres de Mateo and Chabás, 'Los cánones de las Tabulae resolutae para Salamanca', 73–74.

*The values for Venus and Mercury are contained also in the Table TRW-05. Values for the upper planets are calculated by subtracting the daily motions of the planets from the daily motion of the Sun.*

<11> **Find the latitude of the three upper planets, Venus and Mercury**[18]

Subtract the true position of the genzahar[19] of the planet, which you find on the margin, from the true position of the planet and you will have an argument of the latitude of the planet. You find the argument of Venus and Mercury as follows: add their equated anomaly to the mean motion, subtract the true position of the genzahar from the result, and you have the argument of the latitude you are seeking. If the argument of any latitude was from one degree to three signs, the planet would be northern ascending; from three signs to six signs it would be northern descending. And from six signs to nine it would be southern descending, and from nine to twelve it would be southern ascending. And since the degrees of latitude are more laborious than they are useful, we will not include them in the *tacuinum*, nor will we be more interested in them.

<12> **Know the true position of the head of dragon, i.e. genzahar**

Subtract the mean motion of the head of dragon from 12 signs and the true position of the head of dragon remains.

<13> **Determine true argument of the latitude of the Moon**

Subtract the equated genzahar of the Moon[20] from the true position of the Moon and you have the true argument of the latitude of the Moon.

<14> **We now proceed to the creation of the *tacuinum***

To do this, prepare the paper as follows: take 13 sheets of paper and draw a longitudinal line[21] on the second page of the first folio,[22] namely along the length of the paper, at the outer edge of the long side, and leave some margin. Then draw another longitudinal line equally distant from the inner edge. The space delimited by these two lines will contain the true positions of the planets. The same lines run on the first side of the second folio[23] and the space that lies between these lines will contain the aspects of the planets. Finally, at the top edge, draw a transverse line,[24] leaving a space of two digits, and the line will run along the width of both pages to the edge of the page. Make another line below it, not exceeding the longitudinal

18 Tables of planetary latitudes were not a standard part of the *Tabule resolute*.
19 'Genzahar' is an ascending node.
20 I.e., the true position of the head of dragon.
21 I.e., a vertical line.
22 I.e., on the verso of the folio.
23 I.e., on the recto of the following folio.
24 I.e., a horizontal line.

lines of the space for true motions, and the created space will be for the symbols of planets. Again, make a similar space for the symbols of signs, make a third space for the names of retrogradations and directnesses, make a fourth space for the latitudes of the planets, make a fifth space, the line of which will pass through both pages to the edges as the first line did, and the space between them will contain the names of the true motions, namely degrees and minutes. The space bounded by these lines on the second page, that is, on the first side of the second folio, will contain the names of the aspects of the Moon to the planets and of the planets to each other. Then make under the first spaces the spaces for the days, for the 31 days of January; their lines will pass through both pages so that they do not cross the last marginal longitudinal lines except the last. In this way you will draw lines on the other pairs of pages for the twelve months, making as many spaces for days as there are days in each month. Once you have the transversal lines set up like this, draw a longitudinal line on the page of true motions that is equidistant from the first longitudinal line, intersects all the transversal lines and goes to the two outermost longitudinal lines, and creates a space sufficient for the numbers of days.[25] Similarly, create space for the day letters[26] and for the syllables of the days.[27] Divide the remaining large space longitudinally into eight parts, separated by red longitudinal lines that go – as before – to the edges of the transverse lines. Divide the remaining large space lengthwise into eight distinct parts with red longitudinal lines, passing as before to the ends of the outermost transverse lines. The emerged spaces will contain the true locations of the planets. The first part will be for the Sun, next one for the Moon, the third for Saturn, the fourth for Jupiter, the fifth for Mars, the sixth for Venus, the seventh for Mercury, the eigth for the head of dragon. Divide again the space of each planet longitudinally into two parts by a line passing from the bottom outermost latitudinal line across all latitudinal lines of days, not exceeding the space of names, and it will be a column for degrees and the other for minutes.[28] One page will remain incompleted.[29] In a similar way draw on the second page a longitudinal line, equidistant from the first longitudinal line, which will also cross all latitudinal lines up to the two outermost latitudinal lines, and it will create a space sufficient for the numbers of days. Make a similar space, but slightly larger,

---

25 I.e., numbers 1–30 (or 31, 28) according to the number of days in a month.

26 The letters of day (*littere dierum, littere calendarum*) denote the days cyclically by letters 'a, b, c, d, e, f, g', starting each year from the 1st of January and then again from January 8 till the end of the year. The dominical letters of the year, which correspond to Sunday (*feria prima*), are denoted in almanacs by rubric (see Fig. 20 and other almanacs in the same MS Paris, BnF, lat. 7427, or in Vatican, BAV, Vat. lat. 4572, ff. 1r–102v, containing almanacs for the years 1293–1303).

27 Wodka was referring to the abbreviations of days according to some version of so-called cisiojans. The cisiojans usually took the form of syllabic couplets for each month of the year, composed from the initial syllables of the names of the main fixed feasts of the Christian year. They originated as a mnemonic tool for remembering the names of the days of the year. They were used both in Latin and in some vernacular languages from the thirteenth to the seventeenth centuries. The cisiojan is contained in the almanac by Nicolaus of Tuchów (see Fig. 20), but it is missing in most other fifteenth-century almanacs.

28 Nicolaus of Tuchów gives the positions of planets with the precision to seconds, which is exceptional in almanacs. He also writes the physical signs in sexagesimal notation in addition to the supralinear symbols of the zodiacal signs. To have a sufficient space for the seconds, he thus shifted the last two columns for the Mercury and the head of dragon to the recto page (see Fig. 21).

29 If Nicolaus Wodka takes for the preparation of the *tacuinum* 13 sheets of the paper and begins on the verso of the first folio, after inscribing twelve months, spaced always on the verso of one folio and on the recto of the following folio, at the end of the work he will be left with a free verso of the 13th folio.

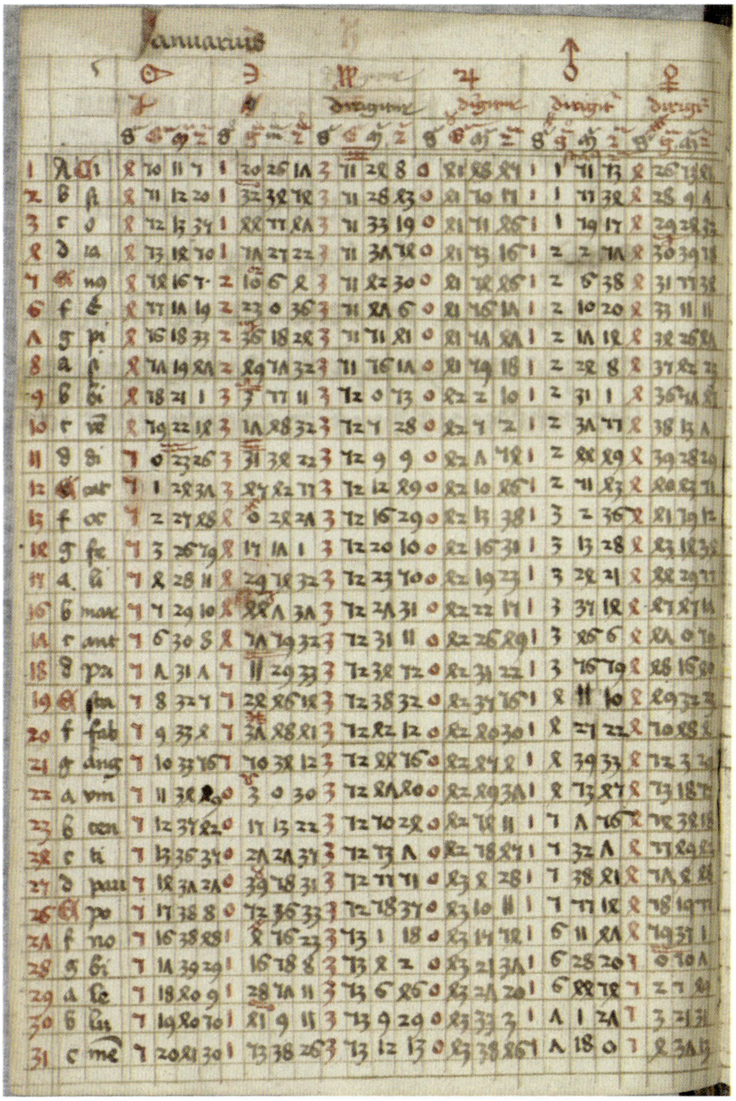


**Figure 20.** Almanac for the year 1455 (January). Nicolaus of Tuchów, Paris, BnF, lat. 7427, f. 43v.

for the aspects of the Moon to the Sun, make a third similar for the aspects of the Moon to Saturn, then a fourth for the Moon's aspects to Jupiter, a fifth for the Moon's aspects to Mars, a sixth for the Moon's aspects to Venus, a seventh for the Moon's aspects to Mercury, and an eighth for the Moon's conjunction and opposition to the head of the dragon. Finally, a large space is left for the aspects of the planets to each other. This way you will line up for the twelve months. And how the headings are to be written into the spaces, it has already been taught by doing the linking itself. So therefore the numbers of days are inscribed, and the letters of the days, and the syllables of the days with their titles superimposed upon them, and the name of the month in the upper latitudinal margin, similarly the symbols of the planets are placed each in its own space, and the denominations of the degrees and minutes

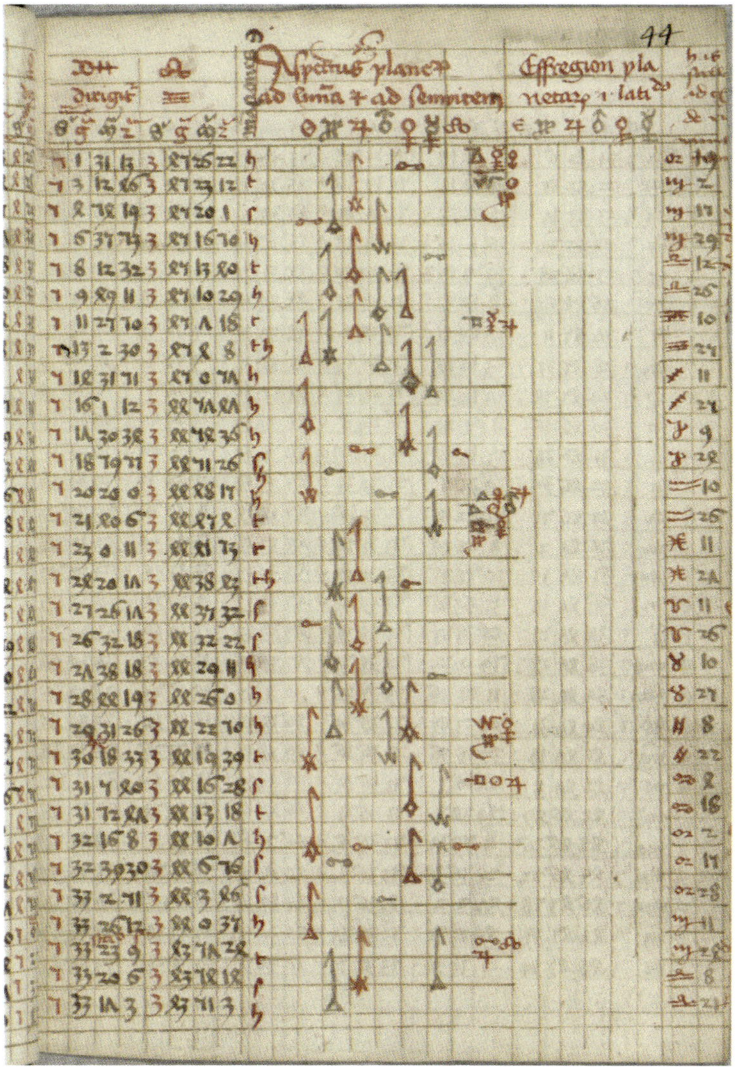

**Figure 21.** Planetary aspects in Almanac for January of the year 1455. Nicolaus of Tuchów, Paris, BnF, lat. 7427, f. 44r.

in their proper spaces, and on the second page the numbers of days with their titles, and in the next space the title of the aspect of the Moon to the Sun is placed in the space above the transverse lines, in the second space the aspect of the Moon to Saturn, and so on with the others. And finally, the aspect of the planets to each other in each month is written. In this way you also inscribe the symbols of signs and the names for directnesses, retrogradations, and latitudes, and then you have the paper for the *tacuinum* correctly adapted.

## <15> Arrange the true positions of the Sun in *tacuinum* for each day of a year

Determine the mean motion of the Sun and the anomaly of the Sun using the tables mentioned earlier as of noon on the last day of December of the past year for Bologna or for the Bologna region for which you want to calculate the *tacuinum*. If there are 30 or more seconds in these motions, add one minute for them and store them as the radices of the year. If there are fewer than 30 seconds, do not worry about them. Keep this method in mind for the other planets also. When you have obtained the radices in this way, finish finding the true position of the Sun with them in the manner described above, and then keep the Sun's true position aside on a paper. Go again into the Table of the Sun at steps of five days and add what you find right at the fifth day of January to the radices, namely to the mean motion of the Sun and the anomaly of the Sun, and with the result find the true position of the Sun as before. Then you write the true position of the Sun now found on a paper, adjusted in the mentioned manner, at steps of five days in the column of the Sun, and thus obtain the true position of the Sun on the fifth day of January. Again, go into said table and add what you find right at the tenth day of January to the previously named radices and with that sum again look for the Sun's true position and you will have the Sun's true position at the tenth of January. Place this also directly to the tenth day of January in the *tacuinum*, and do so with the other days of the months to the end of the table, and you will have the true position of the Sun to all the augmented days[30] of the table. Further, when the leap year arrives, not that which appears directly at the 28$^{th}$ day of February, but the one that appears at the 29$^{th}$ day at the bottom of the table, there must be added to the radices, and with that the true position of the Sun at the end of February is completed. And then, in the same leap year, before you proceed to the day of March, add to the previously determined radices the motion of one day which is noted in the margin. And having been added to the radices of the true motions for the whole year, the true motions found in this way will differ by six days. They will differ by three days at the end of February in the common year, and by four days in the leap year.

By means of the true motions to the augmented days, the true positions at the intermediate days are made. Subtract the true position of the Sun, which you have kept aside on the paper, from the true position of the Sun placed directly at the five days of January, and there remains the difference of those true motions, which you divide by five; in which quotient will be the true position of the Sun of one day. Add this true motion of the Sun for first day to the true position of the Sun as recorded on the paper, and the result will be the true position of the Sun at noon on the first day of January. Add this to the true motion of one day to the five up to the fifth day of January and you will have the true motion for the four days of January, which you enter in the *tacuinum*, and on the fifth of January the number of the true motion at the fifth day that you had before will be returned to you from such an increment. And work in this way on the other intermediate days throughout all the months until the end of the year. At the same time, it is necessary to notice that when the true motions are six days apart, then the difference will be divided by six, if three by three, if four by four, which happens easily at the end of the months. When completed, you will have the true positions of the Sun for all the year in *tacuinum*. The symbols of signs of these true positions are placed at the beginning of the month in the space above the signs of true motions.[31] When the Sun changes the sign in the middle of a month, the sign in which it is entering is written at the top of the cell.

30 I.e., for time intervals of five days.

31 I.e., above the letters denoting the degress and minutes.

### <16> To place the Moon in *tacuinum* day by day for a year

*– Due to the rapid motion of the Moon, it was necessary to calculate its position for an almanac or tacuinum exactly for each day. In Rule 16, Nicolaus Wodka describes how this calculation was carried out on a lunation-by-lunation basis. In doing so, he refers to the Table of equations of the Moon in the days of the lunations, which can be found in MS Milan, BA, C 278 b Inf., ff. 33v–39r. This table is also inscribed but unfinished in MS London, BL, Add 22317, f. 20v and is similar to the above mentioned table by Nicolaus Mülhus*[32] *in Prague, NK, X B 3, ff. 42v–72v; Prague, NK, adlig. 14 B 10, ff. 1r–31v; Mikulov, RM, Mik 6361, ff. 99r–129v; Cracow, BJ, 610, ff. 47v–77v, and Vatican, BAV, Pal. lat. 1374, ff. 87r–102r. This is a double-argument table, more complex than the TR tradition. Such tables have been described by Chabás and Goldstein.*[33] *The table by Nicolaus Wodka is another example of the double-argument format: on one line for each degree of the true anomaly it gives the equation of the anomaly with its difference; the products of variation of diameter with the proportional minutes are presented in thirty columns for the elongation of the Moon from the Sun in each day of the lunation.*[34] *It thus combines the tables of Nicolaus Mülhus and Giovanni Bianchini.*

### <17> Placing the true positions of other planets in *tacuinum*

Using the above tables, determine the mean motion, its mean centre and mean anomaly at noon of the final day of December for the locality for which you are making the *tacuinum*, and keep this as the radices of the year. With these radices you will find the true position and note it on some paper. Then enter with the tenth day of January into the Table of planets at steps of ten days, and to the said radices of the year add the mean motions which you find there, that is to say, add the mean motion, namely, of the planet, to the mean motion in radix, and then add this and that mean motion to the centre in radix and add the anomaly to the anomaly in radix, and the mean motions appear, which is in its kind. With these mean motions now found, search for the true positions of the planet, and place it in the *tacuinum* directly at the tenth day of January in the column of the planet. Enter again with the twentieth day of January in the previous Table of planets and add the mean motions found there to the radices again, and with the resulting mean motions search for the true position of the planet for the 20th of January, which note in the *tacuinum* directly at the 20th day of January. And so you must work for all the other days of the month noted in the table until the end of the table, and you will have the true positions of the planet added to all the days throughout the whole table. Furthermore, working in this way, it must be taken into account that in a leap year the mean motions are not taken directly for February 28, but those found in the marginalia for February 29 are taken. They are added to the radices, and then the true position of the planet is sought in the usual way. And then before you go to the days of March, add to the radices first the motion of the planet of one day, each motion of one day to its own radix, and keep the result for the radices until the year is completed. Again, the planet will be direct, until its true position

32 See Chapter 2.1.2 (other selected works in the MS Cracow, BJ, 610).

33 Chabás and Goldstein, 'The Medieval Moon in a Matrix', pp. 338–46.

34 See Eq. (22.5) in Chapter 22.

increases in the *tacuinum*. It is retrograde when its true motion is decreasing. After the planet has been direct and its true motion or position is found to decrease, then the planet begins to retrograde. Now, to find how many days have passed since the beginning of the retrogradation, consider in the aforementioned ways and count so many complete days by going up from the true position of the planet, excluding the day in question from such a true position of the planet, and where such calculation stops, write on the cell of the day 'retrograde'.[35] If, however, the planet has been retrograde and its true motion begins to increase, then the planet already has started to be direct, and to know how many days have passed since the beginning of the direction, count as before, numbering upwards from the true position of the planet, and write 'direct' in its cell.

<18> **Pay attention also to the latitude of the planet,**

that is, whether it is southern, ascending or descending, or northern, ascending or descending. Note this latitude in the previously described titles in the *tacuinum*, in its corresponding space above the true motions at the beginning of the month and it will be invariable in the intermediate days. And you may judge from previous practice where to place the signs of planets.

When you have done all this, you will have the true motions of the planets for the year at steps of ten days. At the end of months having 31 days there will be step of eleven days, but at the end of February there will be the step of 8 days in a common year and of 9 days in a leap year. Complete the true positions of the planets for the intermediate days in *tacuinum* in this way: subtract the true position of the planet saved in the paper, which you found as the radix, from the true position of the planet at the tenth day of January, or the smaller from the greater. The remainder will be the difference of these true motions, which you divide by 10, and there will result the true motion of one day, which you add to the true position of the planet kept in the paper up to ten days, if the planet was direct, or subtract, if it was retrograde. And by working in this way, the true position of the planet for the tenth day that you had before will emerge for you in the tenth addition or subtraction. Now put into the *tacuinum* the nine other true motions, starting with the first day of January, not the one that is on the paper, but the one which you obtained from the first addition or subtraction. Similarly write the remaining true motions, according to the order of the days that belong to them. Treat similarly all the other differences of the true motions in the *tacuinum* which are closest to each other. But at the end of the month having 31 days, the difference will be divided by eleven, at the end of February in the common year divide it by eight, and at the end of a leap year by nine, to obtain the true motion of one day.

And if a planet was direct and a retrogradation occurred between the two true motions, then after you have found the true positions of the planet in the preceding intermediate days using the motion in one day, add the same true motion of one day successively to the motions of other intermediate days which are between the following two true motions up to the day of retrogradation, excluding that between the true

35 By this formulation, Nicolaus Wodka was trying to explain how to determine the first station of the planet, where the retrogradation began. The corresponding date may precede the day at which the precise true motion is sampled.

position which will be before the day of retrogradation and the nearest following true position.[36] Note: divide the difference by the distance of the days of those two true motions and the result will be the true motion in one day, which you then subtract continuously as you know, until you completely fill the true motions up to the other day of true motion. Now to know what is to be done if the planet is retrograde and during the following intermediate days it becomes direct, it may be understood from the aforesaid. When all this is completed, the true motions of all the planets in the *tacuinum* for the current year are completed.

Remember that Venus and Mercury may not be considered to be equidistant with any degree of precision in the intermediate days at steps of ten days, namely because these two planets do not always run far from the Sun; therefore, such a discrepancy between the intermediate days is not to be attributed to astrological judgements. It is sufficient, therefore, that it be precise at steps of ten days according to Profatius Judeus. But if this method were seen to be insufficient to anyone because of the directions when any of the two planets would be *ylech* in the nativity,[37] he would equate them at steps of 5 days. In this way, add to their radices and their mean motions in the table, which is at steps of ten, the mean motion of 5 days, and thus find true motions at steps of 5 days, just as was done for the Sun with the intermediate days, as was done in correcting the Sun, except that the retrogradation or directness appears in the intermediate days as will be observed for other planets corrected at steps of 10 days.[38]

### <19> **Adding the true position of the head of dragon to the *tacuinum***

Look for its mean motion at noon on the last day of January, subtract it from the twelve signs, and the true position of the dragon's head at noon of the same day will be found, which you shall keep as the radix of the whole year. Then go to the Table of mean motion of the head of dragon and subtract, what you find directly for that month from the radix of the head of dragon and you will have the true positions of the head of dragon at the end of any month; these are to be stored in the *tacuinum*. If it is a non-leap year, work with the months of the non-leap year. If it is a leap year, work with the leap months. Then take the mean motion of the head of dragon on the tenth day and subtract that from the radix to get his tenth true position for the tenth day of January. Subtract the motion of the tenth day from it again and find the true position of the head for the twentieth day of January. Work in this way in the other months as well, and thus subtract the motion of the tenth day from the motion of the end of the previous month, and its

36 Nicolaus Wodka suggests here to extrapolate the true motion from the previous ten days until the station of a planet, if it appears between the dates for which the true motion is calculated precisely. In the almanac by Nicolaus of Tuchów in Fig. 21 we can see the first station of Mercury denoted at the date January 23, despite the maximum longitude having been achieved one day earlier. The increments of true longitude are constant here always for four consecutive days.

37 *Ylech* or *hylech* is a term of Persian and Arabic origin important in the nativity horoscopes.

38 Steps of 10 days for the outer planets and 5 days for the inner planets are also reccommended by John of Głogów. However, in his Rule 26, he prefers to use 10 days for Saturn, 8 for Jupiter, 6 for Mars, and 4 for Venus and Mercury to compose a precise almanac. Just these steps are used by Nicolaus of Tuchów in his almanac for 1455 shown in Figs 20 and 21.

true motion on the tenth day of the following month is found. Subtract from it again the motion of the tenth day and the motion of the twentieth day will come out, and thus you will have the true position of the head of dragon for all augmented days. Then subtract from the true position in its place 3 minutes, which are the true motion of the dragon's head in one day. Then proceed by subtraction in the same manner, and write down the true positions in the intermediate days contained in the *tacuinum* directly against the corresponding days. And then the end of the true motions of the planets in the *tacuinum* will appear.

### <20> **Finding the aspects of the planets using the true motions that are in *tacuinum***

*– In Rule 20, Nicolaus Wodka explains how to fill the table of aspects on the recto pages of the tacuinum. The greater portion of these pages is divided for aspects of the Moon with other planets and the head of the dragon, as these occur most often due to the fast motion of the Moon. The remaining portion holds the mutual aspects of the other planets. According to Wodka, the vertical position of the centre of the aspect symbol within the appropriate box (between the upper and lower lines marking midnight at the beginning and end of the day) indicates when, during the day, the aspect occurs. Other authors give the hour of the aspect numerically. In determining this time, Wodka refers to the reader's judgement based on the discrepancy of positions and the difference in velocity of the two bodies. Rule 25 provides a more precise algorithm for calculating the time of conjunction of two planets. The day on which an aspect between two objects occurs is recognised by the (at least approximate) agreement of the numerical values of the degrees and minutes of their true longitudes written in the table on the verso pages of the tacuinum. In the case of the Moon, a correspondence of at least six degrees is sufficient. Which aspect takes place is deducted by the difference in the order of the zodiacal signs in which the bodies are located – 0 indicates conjunction, 2 sextile, 3 square, 4 trine, and 6 opposition.*

### <21> **Determine the declination of the Sun and of any degree of the zodiac**

Enter with the degree of the Sun or with the mentioned degree of the zodiac to the Table of the declination of the Sun and the declination found under the marker of that degree will be the declination you are looking for. And it will be either southern ascending or descending, or northern ascending or descending, as the headings indicate.

### <22> **Know the latitude of the Moon**

Enter the Table of the latitude of the Moon[39] with the argument of the latitude of the Moon and you will see its latitude under the heading of argument. To which part it belongs,[40] you can find there at the sign.

---

39 The Table of the lunar latitude is traditionally part of various sets of tables, see Chabás and Saby, *The Tables of 1322 by John of Lignères*, p. 80–83. For details see also Rule 37 of the canons by John of Głogów (Chapter 8).

40 For the sequence numbers of signs 0–5 the table indicates northern (*septemtrionalis*) part, for signs 6–11 southern (*meridionalis*) part. For signs 0–2 and 9–11 the Moon is ascending (*ascendens*), and for signs 3–8 is descending (*descendens*).

### <23> Time of the true conjunction or opposition of luminaries

*– Nicolaus Wodka describes the calculation of the time and position of syzygies basically in the same way as does Petrus Cruciferus in his Rule 7 of the canons Mirabilis or John of Głogów in his Rule 31 of the canons Pro (introduccione) generali (see also Chapter 22 for explanation). However, Wodka refers again, like in his Rule 16, to the Table of equations of the Moon in days of lunation. This table, which is not present in the TR for other meridians, is used here as an alternative to the standard Table of equations of the Moon. In Rule 23, Wodka also describes calculation of the argument of latitude of the Moon, from which the possibility of an eclipse can be inferred.*

### <24> Know the true motion of the Sun in one hour and the true motion of the Moon in one hour at any time

Enter the Table of the equated true motion of the Sun and Moon in one hour[41] with the degree of the anomaly of the Sun and the motion of the Sun that you find under the heading of the anomaly will be the true motion of the Sun in one hour. Similarly, enter the same table with the equated degree of the anomaly of the Moon and the motion of the Moon found under the Moon anomaly heading will be the true motion of the Moon in one hour. The value found in this way will not be accurate but it will not cause a noticeable change in the work. If you want it more accurate, look for it with the anomaly of hours, as we taught above.

### <25> To know the time of the true mutual conjunctions of Jupiter, Saturn or Mars from their true positions in the tacuinum

Consider on which day the planets are nearest to each other in the zodiac. Then subtract their true position at the same day, the lesser from the greater, and there will remain the longitude of the one whose motion is greater. If, for the same time, the motion of that planet is slower, then the conjunction will be in the afternoon. If, however, the velocity of the planet has been greater, then it will pass the conjunction before noon. Then divide the difference between the true position of each of them at the present day and the true position at the following day by 24, and consider the true motion in one hour of each of them resulting from it. Subtract the true motion in one hour of the lesser from the greater, and the remainder will be the excess of velocity. Divide the difference of longitude by this excess and note the hours in the quotient. The remainder of the longitude multiplied by 60 and again divided by the excess produces minutes of hours in the quotient. Add these hours and minutes to noon of the day of conjunction, if the conjunction is to come, or subtract from noon, that is, from 24 hours, if the conjunction has passed before noon, and you will have the time of the true conjunction of the planets on the non-equated days, which you wanted. Equate that time, if you want to find the degree of the ascendant and

41 See Table TRW-14. Such a table is also included in MS Milan, BA, C 278 b Inf., f. 42r between the auxiliary tables for preparation of *tacuiunum*.

the figure of the sky. Multiply the true motion of one of these planets in one hour by the hours of longitude, and add the result to the true position of that planet, if the conjunction is to come, or subtract it, if it has passed, and the true position of the conjunction emerges for you. But if the planets at the named day were in the same degrees and minutes, then their conjunction will be precisely at noon of the same day.

<26> **To know the right ascension of the degree of the Sun or of any degree of the zodiac**

Enter with the number of that degree into the Table of right ascensions, and the ascensions you find under the heading of that degree will be the ascensions of that degree, which are the degrees of equator from the beginning of Capricorn to the degree of the Sun.[42]

<27> **Find the right sine of any angle or part of a circle**

Enter with the degrees of that angle or portion in the Table of sine and you will find its right sine directly under the sign of that degree.

– *To his set of the Tabule resolute, Nicolaus Wodka added also tables of trigonometric functions sine and shadow (cotangent)*[43] *which were not common in earlier TR, but they were often a part of other sets of even pre-Alfonsine tables. Correspondingly, in his canon he included rules for these tables and descriptions of their use for practical, geodetic applications (see Rules 27–33). Similar rules can be found in canons on the use of the astronomical instruments like the astrolabe or quadrant. For instance, an analogy of Wodka's Rule 29 can be found in the rule 51 of the treatise On the use of the astrolabe by Cristannus of Prachatice, Wodka's Rule 30 in Cristannus's Rule 53, Wodka's Rule 31 in Cristannus's Rule 55, Wodka's Rule 33 in Cristannus's Rule 57, and Wodka's Rule 32 in the Rule 51 of the version On the use of the astrolabe based on Cristannus's text and enriched by John of Gmunden.*[44]

<28> **If you want to know the versed sine of some angle**

If you want to know the versed sine of any arc and there are fewer degrees of the arc than 90, subtract them from 90 and find the right sine of what is left. Subtract it from 150, which is the right sine of the whole right angle, and what remains is the versed sine which you are seeking. If, however, the degrees of the arc which you are seeking are more than 90, after having subtracted 90 degrees from them, make the right sine of the remainder and add 150, and you will have in the sum the versed sine which you wanted.[45] If, however, the arc is exactly 90 degrees, then the right sine and the versed sine will be the same quantity.

---

42 See Table A-06 and Chapter 25.

43 In the manuscript Milan, BA, C 278 b Inf., see f. 43v.

44 See Hadravová and Hadrava, *Křišťan z Prachatic*, pp. 271–73; 274–75; 276–78; 280–81; 373 resp.

45 It is obvious from this sentence that Nicolaus Wodka is using the al-Khwārizmī's Table of sine normalized to the unit equal to 150, which was included, e.g., in the Toledan Tables; see Pedersen, *The Toledan Tables*, pp. 950–52, or Daniel P. McCarthy, and John G. Byrne, 'Al-Khwārizmī's Tables and a Western Table with the

<29> **If it is necessary to know the size of the direct shadow (cotangent) of some object by the altitude of the Sun,**

enter the Table of shadow with the Sun's degrees of altitude and the points along with the minutes that appear next to the degrees of altitude will indicate the size of direct shadow. When you divide 144 by that, the reversed shadow (tangent) found.[46]

<30> **Know the height of the object by the size of the shadow**

Divide the shadow[47] by as many parts, namely points, as the shadow[48] would have. If there are exactly 12, then the shadow and the height of the object will be of the same size. If the shadow is greater than twelve parts, take a twelfth of those parts, and that will indicate the height of the object. If the shadow is smaller than twelve parts, then stretch one part to 12, or take it twelve times, and the height of the object will be found.

<31> **If you want to know the height of some object without a shadow using the elevation of the alidade,**

enter the Table of shadow with the altitude of the alidade and look directly at the points there. Then, having rised the alidade,[49] divide the space from the elevated object up to the middle of the foot at your place where you are standing into as many parts as you would have points, and using its parts determine as before the height of the object *(in mg. add.:* that is, take the 12 parts that are the height of the object if the points are more than 12, or take one-twelfth if they are less than 12*)*. Now, in addition to the height of the object you have found, add the distance from your eye to the ground, and this is the height of the object.

<32> **If you want to know the depth of a well and a valley,**

stay on one side of the well and look through the slit of the alidade. Enter the Table of shadow and find the points that correspond to it. Then divide the diameter of the well by as many parts as there would be points, namely, the diameter that is from your eye to the opposite side of the well. This diameter will be the same size as the diameter of the surface of the water in the well, and with these parts you will report the depth of the well from your eye to the water, in terms of the height of the object using the shadow.

---

Hindu Norm of R = 150', *Archive for History of Exact Sciences*, 57 (2003), 243–66 for other references. While the sine tables which we can find in most of manuscripts with the TR are calculated for the unit equal to 60 parts, the tables for unit of 150 parts are written in the manuscript Milan, BA, C 278 b Inf., f. 43v, with Wodka's canons and the TR calculated for Bologna.

46 Both direct shadow and reversed shadow are normalized to 12 parts (points, digits) at the angle of 45 degrees.

47 The shadow cast by the object on a horizontal plane.

48 The cotangent of the altitude of the Sun found in the tables.

49 To point toward the top of the object.

### <33> **Measure the size of a plane**

Look over the alidade slit to the edge of the plane and find the slit points in the Table of shadow. Then divide by twelve the distance from the eye to the ground. If there are more than 12 points, expand these parts into as many parts as there are points, and you will have the size of the plane. If there are less than 12, then take the 12 parts to be as many parts as there would be points, and you will find with certainty the size of the plane.

### <34> **Finding the meridian elevation of the head of Aries and Libra in any region**

Subtract the altitude of the pole of that region from ninety degrees and the remainder will be the elevation of the head of Aries and Libra in that region, which will also be the meridional elevation of the Sun, when the Sun is at the beginning of Aries or Libra. In the Table of shadows you will find a shadow of this elevation and keep that for future use. If you want a meridional elevation of the Sun for other degrees, add the declination of the Sun to the elevation of Aries for the northern signs, or subtract for the southern signs, and you would have the elevation of the Sun at noon for every degree of the Sun and day of the year you want. When you have the meridional elevation of the Sun, the meridional size of the shadow is easy to find.

### <35> **Know the pole altitude and latitude of the region using the meridional altitude of the Sun, detected by an instrument**

Find the meridional altitude of the Sun with some instrument and subtract it from ninety degrees. The remainder will be the altitude of the pole and the latitude of the region if the Sun is at the beginning of Aries or Libra. If it is not there, subtract the declination of the Sun from the meridional altitude if the Sun is in the northern signs, or add it if it is in the southern signs, and treat the meridional altitude as before.

### <36> **If you want to make a table of the ascensions of signs in any region**

*– In Rule 36, Nicolas Wodka describes a compilation of the Table of oblique ascensions, an example of which is presented in the Table A-07. He uses an algorithm given by Eq. (26.5), derived in Chapter 26. For the calculation, he refers to the Table of differences of ascensions ('in universa terra') which he added to his set of the TR.*[50] *This table gives values of the tangent of declination $\delta$ for a point of the ecliptic with a longitude $\lambda$. The declination can be calculated according to (Eq. 26.2). Unlike the table with the same title by John of Lignères,*[51] *Wodka's Table of the ascensional differences is normalized to 12.5* $\tan(\delta(\lambda))$*. This is consistent with the fact that Wodka uses a sine table normalized to 150 parts at 90° instead of the usual 60 and the shadow* (cotan) *normalized to the usual 12 parts. Wodka uses the Ptolemaic inclination of ecliptic 23;51°. He also describes the calculation of the parts of hours according to Eq. (26.6).*

50 See MS Milan, BA, C 278 b Inf., f. 44r.
51 See Chabás and Saby, *The Tables of 1322 by John of Lignères*, pp. 58–61.

**<37> Find the ascension of the degrees of the Sun or of a degree of the zodiac on the oblique circle of your region using a table created for this purpose**

With that degree, go into the Table of ascensions or elevations of signs on the oblique circle of your region and keep the ascensions you find under the sign of that degree as those you are seeking.[52]

**<38> The parts of an hour of the degree of the Sun or another degree**

are degrees of ascension which rise in one unequal hour; they are obtained by similar entry.

**<39> Know the diurnal arc**

Subtract the ascensions of the degree of the Sun in your region from the ascensions of its nadir and remainder will be the daily arc you are looking for. Or multiply the parts of hours of the Sun's degree by twelve and you find the same. Subtract this daily arc from 360 degrees and you have the nocturnal arc of the Sun's degree.[53]

**<40> Know the length of the day**

Divide the diurnal arc by 15 and you will have hours in the quotient. Multiply the remainder by sixty and divide by the same and you will have the minutes of the hours, which together with their hours show the length of the day. When you subtract it from 24 hours, you will have the length of the night.

**<41> Convert equal hours to ascension of hours**

Multiply them by 15. To convert the unequal hours to ascensions, multiply them by fractions of hours. To convert equal hours to unequal hours, divide the ascension of hours by fractions of hours. To convert unequal hours to equal hours, divide the ascensions of hours by 15.[54]

**<42> Find the degree of the ascendant and the beginnings of the twelve houses in the horoscope using the Tables of ascendants**

– *In this rule, Nicolaus Wodka refers to Bianchini's tables of domification (a part of his Tables for the planets), and like Andrzej Grzymała before him and also Nicolaus of Tuchów, he names the cusps of astrological houses incorrectly as 'cardines' (although in Rule 44, this term is correctly used for the four cardinal points). This shows that Bianchini's tables were known*

52 See Table A-07 and Chapter 26.

53 See Table A-07 and Chapter 26.

54 Fifteen degrees of the equator rise in one equal hour. The unequal hours can be converted to the ascension if an equivalent of one unequal hour in equal hours and their minutes are known. However, the Table A-07 reffered to in the Rule 38 gives directly the ascension in one unequal hour.

*in Cracow very soon after their origin in 1442*[55] *and that this mistake apparently originated in Cracow; Andrzej Grzymała seems to have been the first to make it. Nicolaus of Tuchów and Nicolaus Wodka took the error from him, and it later appears in their canons, written both in Salamanca, and Bologna.*

*For the reception of Bianchini's Tables for the planets in Cracow two manuscripts are crucial: Cracow, BJ, 555 and Cracow, BJ, 557. Both are written in the hand of a Polish scholar Johannes Zmora de Leśnica.*[56] *BJ, 555 was written in 1453 in Perugia, and the manuscript contains Bianchini's canons to the Tables for the planets (on pp. 1–20) followed by the Tables for the planets themselves (on pp. 25–254).*[57] *The tables of domification are written on pp. 231–236. BJ, 557 was written in Cracow in 1456, after Johannes's return from Italy. Its contents is identical to the manuscript that Johannes had previously produced in Perugia; on ff. 1r–10v we find Bianchini's canons to the Tables for the planets and on ff. 13r–131v Bianchini's Tables for the planets. His tables of domification can be found on ff. 120r–122v.*[58]

*Bianchini's tables of domification can also be found in Cracow, BJ, 570, pp. 20–23 (incipit: 'Tabule Iohannis Blanckini de gradu ascendente et equacione aliarum cardinum celi, reducte et verificate ad latitudinem 50 graduum et ad meridianum inclite civitatis Cracoviensis').*[59] *It is dated just to the year 1467 when Wodka was leaving Cracow. This manuscript already contains the above mentioned error.*

*Nicolaus Wodka mentions Bianchini once more in his Rule 54.*

### <43> **Find the hours of the degree of the Sun**

Enter with the degree of the Sun into the Table for finding the ascendant of your climate. The meridional hours which are directly opposite are the meridional hours of the degree of the Sun, and the horological hours which you will find there are the horological hours of the degree of the Sun.[60]

### <44> **Find the degree of the ascendant at any hour**

If your hours, at which you wish to find the ascendant, are the hours from noon, add to them the meridional hours of the degree of the Sun. If your hours are the horological hours, which are counted from the sunset, join with them the horological hours of the degree of the Sun, and in both cases you will get the hours of the degree of the ascendant. Search in the same table under the hours of the ascendant, and the degree which you find directly at it under the heading ascendant is the degree of the ascendant and you will also see there its sign. Moreover, from the degree of ascendant, which is the beginning of the first house, you will find in the Table of equations of the houses of your climate

55 Chabás and Goldstein, *The Astronomical Tables of Giovanni Bianchini*, p. 2.
56 Leśnica is a Silesian town.
57 *Catalogus BJ*, III, 1984, pp. 374–76.
58 *Catalogus BJ*, III, 1984, pp. 382–83.
59 ThK 1560; *Catalogus BJ*, IV, 1988, pp. 31–36; Rosińska, *Scientific Writings*, p. 427 (item 2231).
60 Nicolaus Wodka refers to the table entitled *Tabula ad inveniendum ascendens in quinto climate.* It can be found in MS Milan, BA, C 278 b Inf., ff. 47v–48r and it gives both meridional and horological hours.

the beginnings of the remaining five houses. Having these beginnings of the six houses, their nadirs will be the beginnings of the opposite houses. Then we have to connect the minutes of the degree of the fourth house, from which also the minutes of the tenth house arise. The minutes of the four cardinal points are of a particular importance for casting horoscopes of births and of the revolutions of the years.[61] But in other houses we do not care about minutes.

<45> **Find the hour of the ascendant when I have the degree of the ascendant**

Consider the degree of the ascendant in the Table for finding the ascendant of your climate under the heading 'ascendant'. The hours of the ascendant will appear to the left. Then you subtract the meridional hours of the degree of the Sun from the hours of the ascendant to find the meridional hours, or subtract the horological hours of the degree of the Sun and you will have the horological hours which correspond to the degree of the ascendant.

<46> **Convert the meridional hours to the horological hours**

To the meridional hours add the meridional hours of the degree of the Sun and from the resulting hours of the ascendant subtract the horological hours of the degree of the Sun. The remainder will be the horological hours.

<47> **Convert horological hours to the meridional hours**

To the horological hours add the horological hours of the degree of the Sun. From the resulting hours of the ascendant subtract the meridional hours of the degree of the Sun and the remainder will be the hours counted from noon.[62]

<48> **Recognise which horological hour will be noon**

Subtract the horological hours of the degree of the Sun from the meridional ones, and the remainder will show at which horological hour noon occurs.

<49> **Find the quantity in a different way than above**[63]

Subtract the hours in which noon occurs from 24 hours and the remainder will be half of the length of the day. Double this half of the day to get the total length of the day. Subtract it from 24 hours and you get the length of the night.

---

61 The importance of the cardinal points for astrological predictions is also mentioned, e.g., by Cristannus of Prachatice in his Use of the astrolabe. See Hadravová and Hadrava, *Křišťan z Prachatic*, pp. 226–29.

62 I.e., meridional hours.

63 Contrary to its title, this rule does not provide an alternative method to the previous rule, but continues it by calculating other quantities.

<50> **At what horological hour the Sun rises,**

this is indicated by the hours of the length of the night. Or subtract half of the length of the day from the horological hours in which the noon occurs, and there remain horological hours of the sunrise and the length of the night.

<51> **At what meridional hour the Sun sets,**

this is taught by half of the length of the day. Similarly, subtract half of the length of the day from 24 hours and you will have in the remainder at which meridional hour the Sun rises. All this must be carefully watched by custodians of the clocks and by observers of time.

<52> **Know the true positions of some well-known fixed stars**

Enter with their names in the Table of the fixed stars and the signs, degrees and minutes that are listed for them will mark their true positions. These true positions are also their longitudes. The other values in the table are considered successively according to their headings after the true positions.

<53> **To find the entry of the Sun into the first minute or point of Aries,**

this is the time of the revolution of the years of the world.[64] Search for the true position of the Sun at the time close to which you estimate the Sun will reach such a point. Enter with this true position of the Sun; if it is precisely 0 signs, that is to say, it will also be 0 in degrees and minutes, then at that time the Sun enters the predetermined point. But if it is not 0 in the signs, consider the distance or longitude of the Sun from the aforesaid point. That is, consider the true position of the Sun, which is beyond 0 in the signs, as the motion which is beyond the point in question, i.e., the longitude or distance to be subtracted. Or consider how much the Sun must run to this, so that its true motion be 0 in signs precisely, which you can know by subtracting its true motion from 12 signs and the remainder, because it is before the above mentioned point, must be added to the longitude or distance. Finally, divide the longitude by the true motion of the Sun in one hour, and subtract the resulting hours and minutes of the longitude in the quotient from the estimated time at which you thought the Sun would enter Aries, if the longitude was to be subtracted. Or add the same hours to the estimated time, if the longitude is to be added, and the result will be the time of the revolution of the years of the world, at which the Sun approaches the first minute of Aries in unequated days. If you want to prove it, equate the Sun at the same time of the revolution and, if you are not mistaken, its true position will be precisely 0 signs. And a precision of minutes is sufficient for us in this business. If there is a difference in seconds, we do not care. To equate this time in unequated days to the end of Pisces, which is the beginning of Aries, enter in the above stated method into the Table of equation of days and the result will be the time of the revolution of the years of the world in equated days,

64 See Chabás and Goldstein, *A Survey of European Astronomical Tables in the Late Middle Ages*, pp. 218–20.

at which time in equated days the degrees of the ascendant and the other degrees of the houses are found for the judgments of the year. The entry of the Sun into Cancer is found in the same way, when the Sun is exactly 3 in the signs, and similarly, when the Sun enters Libra, that is, when it is exactly 6 in the signs. In the same way, when it enters Capricorn, that is, when it is precisely 9 in the signs, and in the same way with other signs and degrees in any beginnings and birth horoscopes, whose revolutions are practiced.

Entry of the Sun into Aries and revolution of the years of the world at the meridian of Bologna in the year of the Lord 1477 occurs:

| Day | Hour | Minute | Second | |
|---|---|---|---|---|
| 10 | 16 | 44 | 11 | March non-equated day – meridional |
| 10 | 16 | 52 | 51 | March equated day – meridional |
| 10 | 10 | 44 | 11 | March non-equated day – horological |
| 10 | 10 | 52 | 51 | March equated day – horological |
| **Degree** | **Minute** | | | |
| 28 | 37 | ascendant of Aquarius at the time of revolution | | |

### <54> To find the time of the revolutions of the years of the world or the nativity, or of an other epoch using the tables

Rectify the radix of the nativity or of the other epoch and the place of the Sun at the hour in unequated days. Then consider, if the year of the radix is a leap year, or the first, or the second, or the third after the leap year, and search in the Table of revolutions of the years below the heading of the same year.[65] Finally, consider the distance of the year, whose revolution you are seeking, from the year of radix, with which you enter to the above-mentioned table, and in the line of years find a number similar to the distance of that year. And what you find in its direct line in the hours, minutes and seconds of the radix, subtract it from the hours of the radix, if there is written 'subtract', and what is obtained by addition or subtraction will be the time of the revolution of the radix. But it is not equal to the given year, because the table itself does not exceed the number of 40 years. If the distance between the year of the revolution and the year of the radix should exceed 40 years, then take first what you find in the direct line of 40 years. Then enter again the table itself with the years that remain and find directly their hours, minutes and seconds. Add them with other hours or subtract them as it is inscribed, and you will have the aforementioned unequated time. Then look at the degree of the Sun, in which sign it is at the aforesaid time. With that degree enter into the Table of the the equation of revolutions[66] to the column below the sign in which the Sun is found, and multiply the seconds that you find there for the degree of that sign by the distance in years of the

65 Nicolaus Wodka refers here to the table in collected years, which can be found in MS Milan, BA, C 278 b Inf., f. 57v. See also Chabás and Goldstein, *The Astronomical Tables of Giovanni Bianchini*, pp. 119–20.

66 This table is included in Wodka's set of tables in MS Milan, BA, C 278 b Inf., f. 58r. See also Chabás and Goldstein, *The Astronomical Tables of Giovanni Bianchini*, pp. 120–22.

revolution to the year of the radix and add or subtract the aforesaid from the hours and minutes and the seconds of unequated time of revolution found above, and you will have the equated time of the revolution in the unequated days, that is, the precise time at which the Sun enters the point at which the hour of the radix was found. And this is according to the tradition of Master Giovanni Bianchini, for which, although it is quite precise, we still want you to verify according to our method.

## 7.4 Notes on Canons *Quoniam plerique*

The canons usually give advice on how to use the tables, but not how to compose them. An exception can be found in Rule 36 by Nicolaus Wodka of Kwidzyn where he gives an algorithm for composing a Table of oblique ascensions. In Rule 14 and following, he also describes in detail how to design and create an almanac (*tacuinum*).

Nicolaus Wodka significantly extends the standard set of the TR as well as the canons for these tables. He includes in the canons a brief summary of arithmetics needed for the work with tables. He also describes (in Rule 22) how to calculate the latitude of the Moon using the table included in the Tables of 1322 by John of Lignères, while for the other planets he dealt only with the orientation of their latitude and the direction of its change. Another substantial extension, in which Wodka's tables resemble those of Lignères, is the inclusion of tables of trigonometric functions and their applications in geodesy and in various problems of spherical astronomy required for astrological purposes. The works of Giovanni Bianchini, to whom Wodka explicitly refers, also had a significant influence on Wodka's tables and canons. Another interesting feature in Wodka's tables and canons is the presence of a table resembling those by Bianchini and Nicolaus Mülhus for finding lunar positions. Nicolaus Wodka thus appears to be a good example of a scholar who contributed to the dissemination and exchange of received astronomical knowledge among European universities and centres of learning.

# 8

# Canons *Pro (introduccione) generali* by John of Głogów

The last and the most extensive handwritten version of canons to the TR was produced around 1486–87 by yet another Polish author, John of Głogów. He stands at the end of the line of the manuscript tradition of the TR and his work deserves an edition for many reasons. We can thus compare Petrus Cruciferus's first canons to the TR, *Mirabilis*, with the last version from the pen of John of Głogów.

Jan z Głogówa (Johannes de Glogovia, John of Głogów), astronomer and philosopher,[1] was born *c.* 1445 in the town of Głogów about 80 km northwest of Wrocław in Lower Silesia. It is one of the oldest towns in Poland, founded in the tenth century by the Piast dynasty. In 1329 it came under the administration of the Bohemian King John of Luxembourg and in 1506 was incorporated into the Kingdom of Bohemia. It was situated on several trade routes, from which the city's wealth and showpiece architecture flowed. Unfortunately, during the Second World War many monuments were destroyed, including the Gothic collegiate church of the Assumption of the Virgin Mary, where John of Głogów began attending school. However, it is currently being rebuilt after the destruction (see Fig. 22). John of Głogów began his university studies in Cracow where he received his bachelor's and then his master's degree (in 1468). With the exception of the years 1497–98, when he stayed at the University of Vienna, he spent the next forty years until his death at the University of Cracow. His surviving extensive work covers many disciplines of the seven liberal arts. Also known are his commentaries on Aristotle, Johannes de Sacrobosco, Alexander de Villa Dei, etc. John of Głogów's lectures at the University in 1486 were attended by the eminent astronomer and astrologer Johannes Virdung, who authored a version of the TR published posthumously in 1542. In 1491, Nicolaus Copernicus enrolled at the University of Cracow and stayed there until to 1495. John of Głogów was one of his teachers. Copernicus is known to have attended his lectures on geography. At that time, John of Głogów also collaborated with Albertus of Brudzewo, the author of the canons *Astrorum (observatores) sapientes* to the TR.

John of Głogów was one of the outstanding teachers in Cracow. His didactic skills and his desire to make material accessible to his students clearly shine through in the style and presentation of his canons to the TR. He also helped students in practical life. For those who came to study from his native Silesia, he built a dormitory called *Bursa nova* in 1484 near the *Collegium maius* in Cracow. Students coming originally from Głogów were the largest group at the university, along with those from Wrocław. John of Głogów died 1507 in Cracow and is buried in the local Church of St Florian.

1 See e.g. André Goddu, 'Johannes de Glogovia', Routledge Resources Online, Medieval Studies, 31 Oct. 2022. DOI: 10.4324/9780415791182-RMEO16-1; Knoll, 'The Arts Faculty', pp. 147–49.

**Figure 22.** Głogów, Gothic Collegiate church of the Assumption of Virgin Mary, where John of Głogów started his studies. In his canons *Pro (introduccione) generali* the examples for calculation were chosen from the years 1486 and 1487, mostly for the day of Assumption of Virgin Mary (e.g. *ad annum Domini 1486 currentem, ad diem Assumpcionis Marie... volo scire medium motum Solis*). Andrzej Otrębski, Creative Commons.

## 8.1 Manuscripts

Augsburg, UB, Cod. II 1 Quarto 74, ff. 260r–280r[2]

This codex is already mentioned in the Chapter 5.1 because it also contains the canons *Astrorum* (on ff. 239r–245v).

Cracow, BJ, 600, ff. 149r–159v[3]

This manuscript was already described in the Chapter 5.1 in connection with the canons *Astrorum* which are contained on ff. 137r–147r. On ff. 86r–100r there are TR for Cracow.

- The canons *Pro (introduccione) generali* has incipit: *Pro generali Tabularum motuum planetarum... resolutarum dicimus introductorium* × *et manebit ultimo latitudo Mercurii ista: 1 gradus, 2 minuta, 39 secunda, 45 tercia, septentrionalis. Finis... canonum exemplarium super Tabulas resolutas ad meridianum Cracoviensem anno quo supra.* The text is very close to the copy in the MS Vatican, BAV, Pal. lat. 1439, which we have taken as the basis for our edition; even the examples are related to the same years 1486 and 1487.

2 Hilg, *Lateinische mittelalterliche Handschriften in Octavo*, pp. 451–461.

3 Rosińska, *Scientific Writings*, p. 323 (item 1651); Markowski, *Astronomica*, p. 77 (item 113); *Catalogus BJ*, IV, pp. 163–76.

### Cracow, BJ, 1841, pp. 93–162[4]

This manuscript was already mentioned in the Chapter 3.1 where copies of canons *Girum* are described. It dates to the beginning of the sixteenth century and probably originated in Cracow, since other minor astronomical-astrological works contained in this codex refer to that city.

- John of Głogów's canons have incipit and explicit: *Pro generali in Tabulas motuum planetarum, quas resolutas vocant, aliqua sunt premittenda*... × ... *si diversorum colorum, erit septemtrionalis et tantum de isto.*

### Cracow, BJ, 1846, pp. 203–227[5]

The manuscript of Cracovian provenance was produced at the end of the fifteenth century, several parts at the beginning of the sixteenth century. On pp. 1–65 can be found the TR for Cracow; on pp. 157–196 are written Albertus of Brudzewo's canons *Astrorum* and on pp. 203–227 the canons *Pro (introduccione) generali.*

### Cracow, BJ, 1916, pp. 1–45[6]

The fifteenth-century manuscript preserves the canons *Pro (introduccione) generali* on pp. 1–45. On p. 37 we can read: *Pro radice autem habenda Cracoviensi*... *est alter eciam modus querendi radices Cracovienses per distanciam a Praga*...

- On pp. 47–48 we find a rule on aspects (*Canon de aspectibus planetarum*) which is considered to be an extract from the last quarter of the text of the canons *Girum.*
- The TR for Cracow 1428–1808 are on pp. 57–128.

### Cracow, BJ, 1917, ff. 84r–97r[7]

The manuscript was produced *c.* 1492 and includes the TR for Cracow 1428–1808 on ff. 1r–34v, followed by additional and other tables (up to f. 68v). The *Tabule auree* by Petrus Gaszowiec can be found on ff. 70r–77v. The John of Głogów's canons *Pro (introduccione) generali* follow on ff. 84r–97r, incipit and explicit: *Pro generali in Tabulis motuum planetarum, quas resolutas dicimus, introduccione, aliqua premittenda sunt*... × ... *Finis harum Tabularum resolutarum, 1492.*

### Cracow, BJ, 1920, pp. 3a-3b;[8] 207–235[9]

At the beginning of this manuscript, which originated in Cracow at the end of the fifteenth century, is written a prologue to the canons *Pro (introduccione) generali* on the pp. 3a–3b

4 Wisłocki, *Catalogus*, p. 436; Rosińska, *Scientific Writings*, p. 323 (item 1651); Markowski, *Astronomica*, p. 78 (item 122).

5 Wisłocki, *Catalogus*, p. 438; Rosińska, *Scientific Writings*, p. 323 (item 1651); Markowski, *Astronomica*, p. 78 (item 120).

6 Wisłocki, *Catalogus*, p. 458; Rosińska, *Scientific Writings*, p. 323 (item 1651); Markowski, *Astronomica*, pp. 77–78 (item 117).

7 Wisłocki, *Catalogus*, p. 458; Rosińska, *Scientific Writings*, p. 323 (item 1651); Markowski, *Astronomica*, p. 77 (item 116).

8 Wisłocki, *Catalogus*, pp. 459–60; Markowski, *Astronomica*, p. 78 (item 119) – prologue only.

9 Wisłocki, *Catalogus*, pp. 459–60; Rosińska, *Scientific Writings*, p. 323 (item 1651); Markowski, *Astronomica*, p. 78 (item 118).

and on pp. 5–20 the canons *Astrorum* by Albertus of Brudzewo. Pp. 25–98 are devoted to the TR calculated for Cracow meridian and the years 1428–1808. Pp. 183–201 give the canons *Girum* by Andrzej Grzymała. In this codex, Alfonsine Central European astronomy includes a full copy of John of Głogów's canons *Pro (introduccione) generali* on pp. 207–235.

Cracow, BJ, 2650, pp. 3–64[10]

This Cracovian manuscript written *c.* 1486–87 brings canons *Pro (introduccione) generali* on pp. 3–64 and the usual TR for Cracow (from 1428 to 1808) on pp. 75–154.

Padua, BU, MS 643, ff. 41r–53v[11]

The manuscript originated *c.* 1513. In addition to the canons of Albertus of Brudzewo *Astrorum* on ff. 53v–64v, it preserves the canons *Pro (introduccione) generali* by John of Głogów with prologue and incipit: *Pro generali in Tabulas motuum planetarum, quas resolutas dicimus, introduccione aliqua premittenda sunt… Canon primus. Quomodo radix cuiuslibet motus medii debeat invenire ad principium anni… Pro huius canonis intelligentia sciendum primo, quod ex Tabulis illustris Alphonsi, Romanorum et Castelle regis, ad meridianum nostrum Cracoviensem dicuntur reducte.* Explicit: *et illas, que fuerunt septentrionales, adde, similiter meridionales etc. Et sic est finis canonum super Tabulas resolutas.*

- The *Tabule auree* by Petrus Gaszowiec are found on ff. 67r–72v.

Vatican, BAV, Pal. lat. 1439, ff. 165r–194r[12]

It is an astronomical-astrological codex, written partly in Cracow, e.g., the canons *Pro (introduccione) generali*[13] from *c.* 1486–87, and partly in Leipzig between 1487 and 1493. According to a note on f. 1r, the owner of the manuscript was Johannes Virdung.

- On ff. 13r–39r, are canons to the *Tabule direccionum profeccionumque* by Johannes Regiomontanus.
- On ff. 79r–102v, we find a commentary on Sacrobosco's Sphere by John of Głogów. On ff. 122r–152r and 211r–239v, is John of Głogów's *Summa astrologie* (first part only).
- Copies of the two Cristannus of Prachatice treatises on the Composition and Use of the astrolabe are also contained in this manuscript and are of Cracow origin. The Composition (incipit: *Quamvis de astrolabii composicione*) is written on ff. 279r–285r, the Use (incipit: *Quia plurimi ob nimiam*) on ff. 293r–304v.[14] The scribe of both texts

10 Wisłocki, *Catalogus*, pp. 627–28; Rosińska, *Scientific Writings*, p. 323 (item 1651); Markowski, *Astronomica*, p. 77 (item 114).

11 Markowski, *Astronomica*, p. 78 (item 121). See also online catalogue: <https://cataloghistorici.bdi.sbn.it/file_viewer.php?IDCAT = 231&IDGRP = 2310009&LEVEL = 0&PADRE = &PROV> [accessed: 10.11.2022].

12 Markowski, *Astronomica*, p. 77 (item 115) and 288; Schuba, *Die Quadriviums-Handschriften*, pp. 236–44. See digital copy <https://digi.ub.uni-heidelberg.de/diglit/bav_pal_lat_1439/0728> [accessed: 12.03.2023].

13 The text of the canons begins in MS Vatican, BAV, Pal. lat. 1439, f. 165r with the words *Pro introduccione generali*. The incipit of the work, however, is given in accordance with other copies as *Pro generali*; the reason is, among other things, that the longer version of the incipit (*Pro introduccione generali*) is identical with the incipit of several other works, according to ThK 1133.

14 Rosińska, *Scientific Writings*, pp. 336–37 (item 1715) and pp. 347–48 (item 1779); Markowski, *Astronomica*, p. 188 (item 1) and 288; Schuba, *Die Quadriviums-Handschriften*, p. 242 (all these three researchers attributed incorrectly the authorship of Cristannus's works to Prosdocimo de' Beldomandi); Hadravová and Hadrava, *Křišťan z Prachatic*, p. 96.

was Johannes Virdung from Hassfurt during his studies in Cracow; the copies date to 1488. Notes on f. 302v, written by the same hand, mention the year 1492, on f. 303r the year 1487.

- *Relacio Johannis Virdung de studiis suis*, Virdung's handwritten report on his studies, is preserved on ff. 331v–335r. Some other parts of the codex are also believed to be written in Virdung's hand (e.g. ff. 67r–75r, 148r–152r, 211r–239r): he was, after all, the owner of the codex during his studies in Cracow (1484–88). On f. 39r, he writes, among other things: *Item a magistro Johanne de Glogovia audivi Tabulas resolutas tempore estatis et iudicia… tempore hyemali anno Cristi 1486, cui pro laboribus exposui florenum Ungaricalem. Item anno salutis 1487 audivi Utilitates astrolabii a magistro Stanislao de Cracovia, qui gratis resumpsit in collegio. … Item a… magistro Alberto de Cracovia audivi Tabulas Alphonsii anno 1488, tempore estatis.* After 1492, Virdung moved to Heidelberg. In 1542, there was posthumously published his version of the TR in Nuremberg.

### Wrocław, BU, IV F 19, ff. 247v–264v

This fifteenth-century manuscript contains all forty rules of canons *Pro* (*introduccione*) *generali* by John of Głogów.[15] The incipit is unusual because the copy of the canons in this manuscript begins with the second sentence of the text, not with the words *Pro* (*introduccione*) *generali* (this is probably why the text is not found in the ThK): *Canones super Tabulas resolutas antecedentes. In nomine Domini Ihesi Cristi amen. Aliqua premittenda sunt, que ad postea intellligendorum*. The explicit is dated to 1494: … *septentrionalis. Finis canonum feria 2a post festum Penthecostes, anno 1494*. It is noteworthy that all the examples in this copy have been recalculated to the years 1493, 1494, or 1495. The text of the rules is occasionally abbreviated, some parts are omitted, and in some places there are even spaces left on the folios. The copy is not very suitable for a critical edition.

- According to the handwritten manuscript catalogue of the Wrocław University Library,[16] there are several other works in the manuscript: the TR for Prague for the year 1428 are on ff. 47r–75v.
- John of Saxony's canons *Tempus est mensura motus* are contained on ff. 193r–204v; the PAT themselves are on ff. 169r–190v.
- On ff. 138r–142v canons with incipit *Cuiuslibet arcus propositi* to the Tables of 1322 by John of Lignères are found.
- Georg Peurbach's *Theorice nove planetarum* are on ff. 223r–239r.
- Canon to the Table for true syzygies by Nicolaus de Heybech with incipit *Tempus vere coniunccionis et opposicionis Solis et Lune invenire* is copied on f. 32v while his Table for true syzygies can be found on ff. 31v–32r.

15 See the poor quality scan of the microfilm at these addresses: <https://www.bibliotekacyfrowa.pl/dlibra/publication/39216/edition/54076/content>; <https://pinax.bu.uni.wroc.pl/prod/?id=1118>; a copy of it is also at the database Manuscriptorium.com: <https://www.manuscriptorium.com/apps/index.php?direct=record&pid=ULW___-BUW___IV_F_19______3BQF05D-pl>.

16 See <https://glam.uni.wroc.pl/index.php?s=RKP_AKC_1967_2_8_1198&p=60&x=0.6920663> [accessed: 11.01.2023]. We thank the librarian Adam Poznański for the information.

## 8.2 Edition of Canons *Pro (introduccione) generali* by John of Głogów

Sigla

**A** - Vatican, BAV, Pal. lat. 1439, ff. 165r–194r (*c.* 1486–87)

**B** - Cracow, BJ, 600, ff. 149r–159v (*c.* 1486–87)

**C** - Wrocław, BU, IV F 19, ff. 247v–264v (*c.* 1493–95)[17]

|**f. 165r:**| Pro introduccione generali in *Tabulas planetarum, quas resolutas* dicimus, in introduccione aliqua premit{aliqua premit}enda sunt, que ad postea dicendorum intelligencia necessaria sunt et oportuna. Sciendum itaque hee *Tabule resolute* dicte sunt a 'resolvendo', quia ex *Tabulis illustri Alphonsi*, Romanorum et Castelle regis, ad meridianum nostrum Cracoviensem sunt reducte. Illud enim, quod in *Tabulis Alphonsi* magno labore et tedio queritur, in istis *Tabulis* facili via et brevi invenitur. Unde recte *resolute* sunt nominate. Quis autem sit modus resolucionis et reduccionis nedum ad meridianum Cracoviensem, sed ad quemcumque alium, inferius patebit sufficienter.

Notandum ergo, quod in braxi astronomie et tabularum duplici via procedimus: via resolucionis et divisionis primo, via composicionis et colleccionis secundo. Primo: in praxi astronomie procedimus via resolucionis. Zodyacus enim totus dividitur in 12 signa, quorum quodlibet a nomine alicuius animalis denominacionem accipit et sunt Aries, Taurus, Gemini, Cancer, Leo, Virgo, Libra, Scorpio, Sagittarius, Capricornus, Aquarius, Pisces. Unum signum dividitur in 30 gradus, unus gradus dividitur in 60 minuta, unum minutum in 60 secunda, unum secundum in 60 tercia consequenter usque ad sexta ascendendo. Secundo: in braxi astronomie et tabularum procedimus via colleccionis; pro 60 enim tercia capimus unum in secundis et pro 60 secunda capimus unum minutum. Et pro 60 minuta excrescentibus capimus unum gradum et pro 30 gradus capimus unum signum. Quando vero in praxi excreverunt 12 signa, subtrahimus 12 et residuum in loco suo tenemus, ut dicetur inferius. Racio autem istius quare *(sic)* astronomi praxes suos numero sexagenario perficiunt. Astronomus enim considerat motum stellarum et planetarum et tempora coniunccionum et aspectuum. Motus autem et tempus sunt de numero continuorum. Continuum autem licet potest suscipere divisionem quantamcumque est enim divisibile insuper divisibilia, tamen quia nullus numerus sub cento est ita aptus divisioni, nisi 60. Dividitur enim in duas partes, ut in 30 et 30, in tres partes, ut in ter 20. Dividitur in quatuor partes, ut in quater |**f. 165v:**| 15, dividitur in 5 partes, ut in quinquies 12 et dividuntur in sex partes, ut in sexies decem. Et sic consequenter merito practicandas tabulas suas praxas *(sic)* modo sexagenario perficiunt.

Notandum eciam, quod in his tabulis duplices anni inscribuntur, collecti et expansi. Collecti hii dicuntur, qui descripti in tabulis per 20 annos a se inficem distant, ut in principio tabularum facile haberi potest. Addendo enim ad primos annos collectos 20, statim excrescent sequentes. Et quid ex opposito illorum inscriptum est, tituli ad sensum ostendunt.

17 The edition was prepared according to manuscript **A**. Manuscripts **B** and **C** were taken into account unsystematically, especially in ambiguous parts of the text. However, manuscript **C** could not be used for comparison in a number of places which are missing in it, where wording in **C** differs significantly from both **A** and **B**, or for the examples given in **A** and **B** for the years 1486 and 1487 which in **C** has been made for the years 1493–95.

Anni vero expansi hii dicuntur, qui per 20 extensi se continue sine intermissione sequuntur, ut clare in *Tabula annorum expansorum* videtur. In qua *Tabula annorum expansorum* semper ex opposito quarti anni ponitur littera 'b' designans annum bisextilem. Annus vero sequens, primus post bisextilem, iungitur secundus post bisextilem, tercius tercio et sic consequenter, ut manifeste apparet in *Tabula annorum expansorum* ad sensum.

Sciendum eciam, quod annus in praxi harum tabularum et generaliter in astronomie incipitur ab ultima die Decembris in meridie ab illa sillaba 'Sil', ad quam diem et eius meridiem queruntur omnes motus planetarum, qui radices anni tunc incipientis dicuntur, ut postea dicitur. Astrologi autem suum annum incipiunt, quoad iudicia, ab ingressu vel introytu Solis in Arietem.

Menses vero incipiunt a Ianuario. Dies autem a meridie diei precedentis incipiunt usque ad meridiem diei sequentis durans. Meridies enim est pars nobilior diei, inquit Cameracensis propter vim magnam et fortitudinem Solis in illa parte diei. In meridie enim Sol est in domo decima, hoc est in medio celi, que pars celi est fortissima eo, quod tunc radius Solis forcius his inferioribus infigitur. Merito astronomi diem a meridie inchoant. Et motus planetarum ad meridiem precipue inquirunt, inquit Cameracensis. Cum eciam hic nihil in signis invenitur, 12 signa completa sunt, sed primum imperfecte. Cum autem unum signum invenitur completum, est primum signum perfecte. Sic similiter de horis et aliis fraccionibus est intelligendum, ut ergo |**f. 166r:**| harum tabularum praxis facilis sit et pro iunioribus plana. Canones cum exarari praxi levissime stilo plano scribere institui. Nunc que in Cristo Iesu Salvatoris nomine eiusque genitricis gloriose propositum incipio.

<1>

**<C>anon primus, quomodo radix {quomodo radix} cuiuslibet motus medii debeat inveniri ad principium anni, ad meridiem ultime diei Decembris**

Pro huius canonis intelligencia sciendum, quod radix motus alicuius planete est motus medius alicuius planete, inventus ad ultimam diem Decembris ad meridiem in illa sillaba 'Sil', ut radix Solis est motus Solis inventus ad meridiem ultime diei Decembris. Similiter radix Saturni est medius motus Saturni ad meridiem predicte diei inventus. Et sic similiter intelligitur de aliis motibus planetarum. Queruntur autem hee radices ad meridiem civitatis Cracoviensis. Quomodo autem ad alias civitates radices reducende sunt, dicetur in fine, in canone speciali. Dicuntur autem motus planetarum inventi in ultima die Decembris radices secundum quandam similitudinem. Sicut enim radix in arbore est fundamentum crescencie in arbore et fructuum et presupponitur radix in arbore, sic motus planetarum inventi ad ultimam diem Decembris in meridie sunt fundamentum invencionis motuum in aliis diebus et mensibus per totum annum. Invenire ergo radicem alicuius planete est invenire medium motum planete ad diem ultimam Decembris.

**Modus autem et canon, quomodo radices cuiuslibet motus inveniende sunt, est iste.** Primo: scias annum Cristi completum, cum quo anno Cristi completo intra *Tabulam annorum collectorum*, quem annum Cristi completum in eadem tabula quere. Si eum sic inveneris, statim ex opposito illius anni completi radicem cuiuslibet planete ad ultimam diem Decembris sine ulteriori operacione invenies: in *Tabula augium et stella-*

*rum fixarum* radicem motus earundem, in *Tabula Solis* radicem motus Solis invenies et sic in aliis est intelligendum. Secundo: si annum Cristi completum in *Tabula annorum collectorum* non invenies, tunc intrabis cum |**f. 166v:**| anno Cristi minori propinquiori tuis annis currentibus et ea, que ex opposito illius anni invenies in gradus et minutis, extra scribe in tabula distincte. Tercio: annum hunc propinquiorem minorem, quem in *Tabula annorum collectorum* invenisti, subtrahe ab annis tuis Cristi completis, et quot post illam subtraccionem manebit, erit numerus annorum expansorum, qui aduc superest usque ad annum tuum Cristi completum. Quarto: cum illo residuo ex subtraccione tali remanente intra *Tabulam annorum expansorum*, et quod ex opposito illius anni invenies, accipe et adde ad illa, que prius accepisti ex opposito annorum collectorum. Signa scribenda sub signis, gradus sub gradibus, minuta sub minutis, secunda sub secundis et sic de aliis. In quolibet planete motu radicem queris. Quinto et ultimo: adde ea, que accepisti circa annos expansos ad ea, que invenisti circa annos collectos incipiendo a subtiliori fraccione, ut tercia ad tercia, secunda ad secunda, minuta ad minuta, gradus ad gradus, signa ad signa, et operare fac, ut predictum est. Et quid remanserit in quolibet motu, est radix motus illius Decembris. **Exempli gracia**: Volo invenire radicem motuum augium et stellarum fixarum ad principium anni Cristi currentis 1486. Accipio primo annum Cristi completum, scilicet 1485, hic enim completus est. Principium sequentis quero ita: hunc annum Cristi completum in *Tabula annorum collectorum*, sed cum {cum} non invenio, secundo considero annum Cristi propinquiorem elapsum et est 1468, ex cuius opposito capio 0 in signis, 10 in gradibus, 47 in minutis, 7 in secundis et 6 in terciis. Tercio: postquam subtraho annum Cristi propinquiorem in *Tabula annorum collectorum* inventum, scilicet 1468, ab annis Cristi completis, manent 17 in residuo, qui sunt anni medii inter annum 1468 et 1485 completum. Cum his ergo decem septem annis intrabo *Tabulam annorum expansorum* eciam sub titulo augium et stellarum fixarum et invenio ex opposito 17: 0, 0, 7, 59, 37. Quarto: addo hec omnia simul secundum modum superius dictum, incipiendo a fraccione subtiliori, et inveniet radix motus augium et stellarum fixarum ad |**f. 167r:**| annum Cristi completum 1485, quod est ad principium anni 1486 currentis, ad ultimam diem Decembris, ad meridiem.

| **S** | **G** | **M** | **S** | **T** | | {a} |
|---|---|---|---|---|---|---|
| 0 | 10 | 54 | 36 | 43 | | augium et stellarum fixarum |
| 2 | 15 | 34 | 50 | 59 | | motus accessus et recessus octave spere |
| 9 | 18 | 57 | 17 | 16 | | motus Solis, Veneris et Mercurii |
| 7 | 15 | 45 | 51 | 54 | | motus Lune |
| 11 | 25 | 41 | 49 | 34 | radix | medii argumenti Lune |
| 7 | 29 | 52 | 22 | 29 | | medii argumenti latitudinis Lune |
| 0 | 14 | 6 | 30 | 35 | | capitis draconis |
| 8 | 2 | 34 | 53 | 16 | | motus Saturni |
| 8 | 28 | 25 | 12 | 19 | | motus Iovis |
| 8 | 13 | 43 | 49 | 11 | | motus Martis |
| 2 | 25 | 0 | 6 | 22 | | medii argumenti Veneris |
| 10 | 8 | 9 | 1 | 3 | | medii argumenti Mercurii |

<2>

**<C>anon secundus de medio motu cuiuslibet planete ad aliquod <tempus> propositum per proprias tabulas invenire**

Motum cuiuscumque planete medium si scire desideras, primo tempus aliquot propositum habeas, ad quod medium motum illius planete scire optas. Et considera, si illud tempus propositum menses completos, dies, horas, minuta, secunda, tercia in se contineat. Hoc itaque tempus in partem scribe. Secundo: radicem anni eiusdem motus in tabula scribe. Tercio: si aliquis mensis elapsus est, ad *Tabulam mensium* cum mense completo intrabis sub titulo cuiusdem motus, quem queris. Si annus communis fuerit, in tabulam superiorem mensium intrabis. Si bisextilis, in tabulam inferiorem. Et quod ibi ex opposito mensis invenies, ad radicem adde, quodlibet ad suum genus, ut dictum est in primo canone. Quarto: cum diebus, si eos habueris, ad *Tabulam eiusdem planete in diebus* intrabis, et quod ex opposito illius diei invenies, eciam ad prius inventum adde. Quinto: cum horis, si aliquas habueris, ad *Tabulam motus illius planete in horis* intra|**f. 167v:**|{tra}bis, et quod ex opposito tue hore presignate invenies, ad prius inventum adde. In eadem eciam tabula, in qua horas accepistis, accipe minuta, secunda et tercia denominacione tum variata, ut inferius in cautelis dicetur. Et quidquid ex opposito minutorum, secundorum, terciorum invenies, totum ad prius inventum adde. Sicque cum toto tempore proposito intrabis, quod evenerit, medium motum planete, quem quesivisti, designat. **Cautela prima:** Cum tempore proposito minuta, secunda, tercia habueris, per eandem tabulam minuta, secunda, tercia invenies, per quam motum in horis accepisti. Cum qua ergo fraccione tabulam intrabis, semper primus numerus eandem denominacionem optinebit sive ciphra, sive numerus servativus fuerit respectu illius numeri, cum quo intrasti. Ut si intras cum minutis, primus numerus minuta erunt, postea secunda et ultimo tercia. Si cum secundis intrabis, primus secunda, deinde tercia. Si cum terciis intrabis, primus numerus tercia, deinde quarta. Quod si plura quarta quam 30 evenient, propter precissionem unum adde ad terciam. Et hec *Tabule loricate* dicuntur, in quibus numerus horarum ultra 24 extenditur. Nota ideo, quod numerus horarum ad triginta protenderetur vel quod plures hore in die essent quam 24. Sed ideo ad 30 tabula eadem extenditur, quia non solum per eam hora, sed et minuta, secunda et tercia inveniuntur, que ultra 24 extenduntur. Non sint autem necessarie specialem tabulam pro minuta, secunda, tercia scribere, quia tabularum inutilis esset multiplicacio. Sed per eandem tabulam, per quam horas accipimus, per eandem minuta, secunda et tercia inveniemus denominacione tamen, ut dictum est, variata. **Cautela secunda:** Si in tempore proposito plura minuta, secunda vel tercia quam 30 habueris, primo cum 30 minutis, secundis vel terciis intrab{r}is, deinde cum residuo, et quod post completum introytum eveniet, ad prius inventa, ut dictum est, adiunge. **Cautela tercia:** Si in tempore tuo proposito nullus mensis completus esset, ad *Tabulam mensium* te intrare non oportet, sed demissa *Tabula mensium* cum tempore proposito dierum ad *Tabulam dierum* intrabis. Sic si horas in tempore tuo proposito non habueras, ultra dies nec minuta nec secunda nec tercia tabulam harum fraccionum ingredi non oportebit. **Exempli gracia:** Ad annum Domini 1486 currentem, ad diem Assumpcionis Marie, que est dies Augusti 15, ad horam quartam post meridiem, minu|**f. 168r:**|tum quadragesimum octavum, secundum 56, tercium 20, volo scire medium motum Solis. Primo: radicem anni eiusdem presuppono et in tabula scribo, scilicet 9 in signis, 18 gradus, 57 minuta, 17 secunda, 16 tercia. Secundo: considera, si aliqui menses completi sunt a principio anni usque ad Augustum, et patet, quod septem. Intrabo itaque ad *Tabulam mensium* superiorem in Sole, quia annus communis est, et ex opposito numeri accipio motum Solis 6, 28, 57, 25, 20, quem motum addo ad radicem Solis et veniet post addicionem 4, 17, 54, 42, 36. Et quia in tempore proposito ultra menses completos habeo 15 in diebus, intrabo *Tabulam Solis in diebus*, ex opposito quindecim dierum accipio motum correspondentem et est 0, 14, 47, 4, 54, quam motum addo ad prius inventum et veniet 5, 2, 41, 47, 30. Et quia in tempore proposito ultra dies habeo 4 horas, intrabo cum 4 horis ad *Tabulam medii motus Solis in horis* et invenio 0 gradus, 9 minuta, 51 secunda, 23 tercia, quod addo ad prius inventum et venient 5, 2, 51, 38, 53. Et quia in tempore proposito ultra horas habeo 48 minuta, debemus intrare ad eandem tabulam cum 48 minutis. Sed quia ea non invenio, intrabo primo cum 30 minuta iuxta cautelam secundam et remanebunt aduc 18 minuta residua. Cum ergo intrabo cum 30 minutis, invenio in opposito ex parte motus 1 minutum, 13 secunda, 55 tercia. Deinde intrabo cum 18 minuta et invenio motum ex opposito 0 minuta, 44 secunda, 21 tercia. Quos motus correspondentibus minutis cum addo ad prius inventum, veniunt 5, 2, 53, 37, 9, et quia ultra minuta aduc in tempore proposito habeo 56 secunda, intrare debemus cum 56 secundis. Sed ea simul non invenio, intrabo primo cum 30 secundis et accipio 1 minutum, 13 tercia. Deinde intrabo cum residuis secundis 26 et accipio 1 minutum, 4 tercia, que omnia si addo ad prius inventum, venient 5, 2, 53, 39, 27. Et quia ultra secunda in tempore proposito habeo 20 tercia, intrabo ad *Tabulam medii motus Solis* ex opposito 20 terciorum et invenio 0 tercia, 49 quarta, pro quibus quartis propter precisionem addo 1, id est ad tercia priora, et erunt 28. Inveni itaque iam complete medium motum Solis ad tempus propositum, scilicet 15 die Augusti, hora 4, minuta 48, 56 secunda, 20 tercia post meridiem, scilicet 5, 2, 53, 39, 28. Et quemadmodum in Sole calculatum est, ita pariformiter de motibus mediis aliorum planetarum agendum. |**f. 168v:**|

<3>

**<C>anon tercius de radice augis cuiuslibet planete ad principium anni**

Similiter de auge cuiuslibet planete quodlibet die anni inveniendo. Augem alicuius planete ad certum tempus cum scire desideras, primo radicem augis planete invenies iuxta doctrinam datam in canone secunda de mediis motibus planetarum. Secundo considera, si aliquis mensis completus est respectu tui diei proposite. Tunc cum mense completo intrabis, et quod ex opposito illius mensis invenies, adde ad radicem augis. Sic similiter de diebus operandum erit iuxta canonem secundum de mediis motibus planetarum. Si autem nullus mensis completus est, similiter cum numero dierum propositorum ad *Tabulam dierum* intrabis, et quod ex opposito dierum de motibus invenies, ad radicem adde, et quod post addicionem proveniet, motus erit augis illius planete, de quo intendis ad tempus propositum. **Cautela:** Si precisiore praxi intendere volueris in anno bisextili, si post Februarium tempus propositum fuerit, non ad diem propositam in diebus augium intrabis, sed ad sequentem propter viam excrescentem superfluum, nullus tamen error sensibilis eveniet, si directe cum die proposita intrabis. Cautela tamen si tenebitur, precisior operacio erit. Cum horis in *Tabulis augium* operare non est

---

**165** 57 *B* : 59 *A per err.*

necessarie, quia motus exiguus et parvus est et nihil nisi in quartis aliquid eveniret, quibus non utimur querendo motus in resolutis. **Exempli gracia:** Volo scire auges planetarum ad diem Assumpcionis Marie, anno Domini 1486 currente. Primo radices augium habere est necessarie. Radices itaque augium sunt hee:

| <S> | <G> | <M> | <S> | <T> | | |
|---|---|---|---|---|---|---|
| 3 | 1 | 2 | 50 | 16 | | Solis |
| 8 | 13 | 1 | 9 | 16 | | Saturni |
| 5 | 23 | 14 | 27 | 16 | radix augis | Iovis |
| 4 | 14 | 49 | 40 | 16 | | Martis |
| 7 | 0 | 17 | 0 | 16 | | Mercurii |

Habita itaque radice intrabo ad *Tabulam mensium* complete cum Iullio complete et demum cum 15 diebus, et cum motus ex opposito temporis propositi addidero ad radicem propriam planete, habebo statim motus augium in quolibet planeta ad diem Assumpcionis Marie et sunt isti 3, 1, 3, 11, 50.

**|f. 169r:|**

| S | G | M | S | T | | |
|---|---|---|---|---|---|---|
| 3 | 1 | 3 | 11 | 50 | | Solis |
| 8 | 13 | 1 | 30 | 50 | | Saturni |
| 5 | 23 | 14 | 48 | 50 | aux | Iovis |
| 4 | 14 | 50 | 1 | 50 | | Martis |
| 7 | 0 | 17 | 21 | 50 | | Mercurii |

<4>

**<C>anon quartus de argumento Solis ad aliquod tempus propositum inveniendum**

Cum argumentum Solis ad tempus tuum propositum scire volueris, primo medium motum Solis ad tempus propositum inquire. Secundo: augem Solis ad idem tempus propositum habe. Tercio: subtrahe augem Solis temporis tui propositi a motu Solis eciam in tempore proposito. Et quod post subtraccionem manet, argumentum erit Solis ad tempus propositum. **Cautela:** Si subtraccionem facere non posses eo, quod maior a minori subtrahi non potest, acomodare te oportet, ut si in tercia subtraccio facere non posset, acomodabis, id est <in> secundis. Et si in secundis subtraccio facere non potest, acomodabis, id est in minutis. Si in minutis non potest facere, acomodabis, id est in gradus. Si in gradus subtraccio facere non posset, acomodabis, id est in signis. Si autem subtraccionem in signis facere non poteris, unam totam revolucionem, hoc est 12 signa, acomodabis. Et hec cautela in omni braxi astronomie est memoranda. **Exempli gracia:** Ad annum nostrum currentem, ad diem Assumpcionis Marie, volo scire argumentum Solis. Primo debeo scire motum Solis et est 5, 2, 41, 47, 30. Aux vero Solis ad eundem diem Assumpcionis Marie est 3, 1, 3, 11, 50. Post subtraccionem manet 2 in signis, 1 gradus, 38 minuta, 35 secunda, 40 tercia.

<5>

**<C>anon quintus de centro Lune ad aliquod tempus propositum invenire**

Si centrum Lune ad aliquod tempus habere desideras, primo medium motum Lune ad illud tempus propositum inquire. Secundo motum Solis ad idem tempus scias. Tercio eundem motum Solis medium subtrahe a medio motu Lune iuxta cautelam presuppositam. Centrum Lune est distancia {a} Lune a Sole. In coniunccione sunt simul Sol et Luna. Et illud, quod post subtraccionem provenerit, dupla, et quod post duplacionem provenerit, centrum erit. **Exempli gracia:** Ad annum nostrum currentem, ad diem Assumpcionis Marie, volo scire centrum Lune. Primo ad id tempus motum Lune habe et est iste: 11, 6, 48, 21, 39. Secundo: debes scire motum Solis ad eundem diem et est iste: 5, 2, 41, 47, 30. |**f. 169v:**| Tercio: motum Solis predictum debeo subtrahere a motu Lune et manebit 6, 4, 6, 34, 9. Quarto: illud debeo duplare et veniet centrum Lune ad diem Assumpcionis Marie 0, 8, 13, 8, 18.

<6>

**<C>anon sextus de argumento planetarum trium superiorum, ut Saturni, Iovis, Martis, ad tempus propositum inveniendo**

Si argumentum planetarum trium superiorum scire desideras, primo motum Solis ad id tempus propositum habe. Secundo motum illius planete, cuius argumentum scire volueris. Tercio motum planete a medio motu Solis subtrahe, et quod post residuum fuerit, agumentum erit planete ad tempus tuum propositum. **Exempli gracia:** Ad annum nostrum currentem, ad diem Assumpcionis Marie, volo scire argumentum Saturni. Primo motum Solis inquiro et est iste: 5, 2, 41, 47, 30. Secundo motum Saturni ad diem Assumpcionis Marie invenio et est iste: 8, 10, 11, 6, 47. Tercio motum planete subtrahe a medio motu Solis et residuum erit argumentum Saturni hoc: 8, 22, 30, 40, 43. Medius motus Iovis ad diem Assumpcionis Marie 9, 17, 17, 23, 47. Argumentum Iovis ad eundem diem Assumpcionis Marie 7, 15, 24, 23, 43. Argumentum Martis ad diem Assumpcionis Marie 4, 30, 0, 10, 1. Argumenta vero Veneris et Mercurii per modum prius dictum per proprias tabulas invenies.

<7>

**<C>anon septimus de centro medio quinque planetarum ad tempus propositum invenire**

Primo: medium motum planete ad tempus propositum inquire. Secundo: augem planete ad idem tempus a medio eius motu subtrahe, et quod post subtraccionem remanserit, erit medium centrum planete. **Verbi gracia:** Ad diem Assumpcionis Marie volo scire medium centrum Saturni. Primo: eius medium motum debeo scire et est 8, 10, 11, 6, 47. Secundo: debeo eius augem scire et est 8, 13, 1, 30, 50, quam augem debeo subtrahere a motu Saturni, et quod manserit, est centrum medium Saturni et est 11, 27, 9, 35, 57. Sic similiter operare in Iove, Marte, Mercurio et Venere.

**284** 9 *B* : *om. A*

<8>

**<C>anon octavus de motibus rectificandis et corrigendis et de novo inveniendis in tabulis iam premissis**

Ut autem huius canonis processus levior sit et planior de modo continuandi tabulas et corrigendi, placuit eum in quatuor particulas |**170r:**| distinguere. **Particula prima de motibus cuiuscumque continuandis, corrigendis et de novo formandis in annis collectis.** Cum motum planetarum in *Tabulis collectis* descriptum continuare volueris, prima riga ex *Tabulis Alphonsi* presupposita correcta, cui primo si addideris motum ex oppositum 20 annorum expansorum, eveniet tibi secunda. Cui secunde si addideris id, quod ex opposito 20 annorum in motibus, exibit tercia. Cui tercie si addideris illud idem, exibit quarta et cetera consequenter. Iste est modus continuandi motus in annis collectis. **Exempli gracia:** In annis collectis in motu Solis medio ex opposito annorum Cristi 1428 est 9, 18, 46, 56, 45. Cui si addideris motum Solis, qui est ex opposito 20 annorum in *Tabulis expansis*, scilicet 0, 0, 8, 48, 59, mox veniet motus in secunda riga in annis collectis ex opposito annorum Cristi 1448, scilicet 9, 18, 55, 45, 44 et cetera consequenter. **Modus corrigendi tabulas motuum in annis collectis est iste:** Captis duabus rigis sibi inficem *(sic)* proximis si subtraxeris minorem a maiore, debet manere in residuo motus, qui ponitur ex opposito 20 annorum expansorum. Quod si sic fuerint tabule, correcte erunt. Si sic non fueri<n>t, incorrecte et male sunt debesque eas per continuacionem rectificare, ut dictum est. **Exempli gracia:** Volo scire, si secunda riga in medio motu Solis in annis collectis sit correcta. Capio primam rigam, scilicet 9, 18, 46, 56, 45, capioque secundam et proximam rigam in Sole, que est 9, 18, 55, 45, 44, subtrahoque primam a secunda, quia prima minor est quam secunda, et remanet id, quod ex opposito 20 annorum de medio motu Solis in annis expansis, scilicet 0, 0, 8, 48, 59 et cetera. Operaberis in correccione aliarum rigarum et tabularum. **Particula secunda de correccione et continuacione tabularum motuum in annis expansis.** Si *Tabulas motuum planetarum in annis expansis* continuare volueris et de novo facere, presuppone primam rigam correctam ex Alphonso. Cui prime si addideris se ipsam vel prima duplata, quod idem est, statim exibit secunda. Cui secunde si addideris primam, exibit tercia. Cui tercie si addatur prima cum motu planete, |**f. 170v:**| quantum movetur una die, exibit quarta. Et sit ista addicio propter annum bisextilem, in quo additur una dies. **Exempli gracia:** In *Tabula annorum expansorum* ex opposito unius et primi anni in Sole reperitur 11, 29, 45, 39, 22. Cui si addideris se ipsam vel eam duplicaveris, exibit secunda 11, 29, 31, 18, 44. Cui tercie si addideris primam cum motu Solis in una die, scilicet 0, 0, 59, 8, 20, exibit quarta 0, 0, 1, 45, 48 et sic consequenter. Continuabis in Sole usque in finem in annis expansis. Eodem modo operare in motibus annorum expansorum aliorum planetarum. **Modus corrigendi tabulas motuum in annis expansis est iste:** Quecumque riga tibi suspecta fuerit, minue precedentem ex sequenti, et si manet motus prime rige, recte sunt tabule. Si autem non manet, sunt incorrecte. Et corrige per continuacionem. Cautela tamen ista adiuncta, quod in anno bisextili [...][18] precedentem rigam ex sequenti oppositi te eciam cum hoc minuere de residuo motu unius diei, et si tunc manet motus prime rige, tabule sunt correcte. **Exempli gracia:** Postquam accipio in *Tabula annorum expansorum* in medio motu Solis ex opposito secundi anni, scilicet 11,

18 The Latin text is corrupted in the copies available to us, but its meaning is clear (see the translation).

29, 31, 18, 44, et subtraho ex motu, qui est ex opposito tercii anni, scilicet 11, 29, 16, 58, 6, manet motus rige prime ex opposito primi anni. Et sic eciam alios motus corrige in Sole et aliis planetis ad annum tamen bisextilem, ut dictum est habendo respectum. **Particula tercia de continuacione et correccione motuum in *Tabulis mensium*.** Cum *Tabulas mensium* in motibus cuiuscumque planete continuare et de novo facere volueris, hoc scias, quod prima riga uniuscumque tabule in Ianuario est ultima riga de *Tabula dierum*. **Exempli gracia:** Prima riga in Ianuario in anno communi in medio motu Solis est 1, 0, 33, 18, 8. Et idem motus invenitur in *Tabula dierum Solis* ex opposito 31 vel in fine eiusdem *Tabule Solis*. Sic similiter in aliis planetis et eorum motibus, quoad Ianuarium intellige. Si autem motus aliorum mensium et rigas invenire volueris, primam rigam Ianuarii presuppone, cui adde motum tot dierum, quot dies usuales mensis iste continet, |**f. 171r:**| cuius rigam queris. Ut si velles habere motum Februarii in Sole, adderes motui in Ianuario reperto vel in prima riga mensium motum Solis medium ex opposito 28 dierum, scilicet 0, 27, 35, 53, 9, et veniet riga Februarii in motu Solis in anno communi ista: 1, 28, 9, 11, 18. Et sic in aliis rigis aliorum mensium in quocumque planeta volueris procedere et rigas motus cuiuslibet mensis invenies. **Cautela:** In anno bisextili pro motu Solis in Februario habendo motum 29 dierum super primam rigam Ianuarii adde, quia una dies superflua ibi imponitiva et veniet in Februario 1, 29, 8, 19, 37. Unde et rigis aliorum mensium in anno communi repertis in qualibet riga cuiuscumque planete motum unius diei addideris motus mensium omnium in tabula inferiori bisextilo anno subserviente exibunt, ut ex opposito Marcii in anno communi ponitur 2, 28, 42, 29, 46. Cui si addideris 0, 0, 59, 8, 20, exibit riga Marcii in anno bisextili in tabula inferiori 2, 29, 41, 37, 46. Et sic in aliis est intelligendum. **Particula ultima de continuacione et correccione motuum in Tabulis dierum, horarum et minutorum.** Cum *Tabulas dierum* in motibus planetarum continuare et de novo facere volueris in diebus per motum unius diei planete illius, cuius motum continuare volueris, perficies similiter in horis planete per motum in una hora. Unde prima riga presupposita correcta ex Alphonsi. Si eandem super se ipsam addideris, exibit secunda. Si hunc secunde primam addes, evenit tercia. Si tercie primam addideris, exibit quarta et sic consequenter tam in diebus, quam in horis. **Exempli gracia:** In *Tabula Solis in diebus* ex opposito prime diei ponitur 0, 0, 59, 8, 20. Hoc idem si addideris super se ipsam, veniet riga secunda, videlicet 0, 1, 58, 16, 39. Et sic consequenter in motibus aliarum rigarum, sic eciam in horis operaberis. **Modus corrigendi Tabulas dierum et horarum.** Si *Tabulas motuum dierum* in planetis corrigere volueris, subtrahe precedentem a sequenti et manebit tibi prima riga. Si sic fuerit, tabule sunt correcte. Si non manebit prima, sunt incorrecte debesque eas emendare per continuacionem. **Verbi gracia:** Capio in Sole in diebus quintam et est 0, 4, 55, 41, 38. Capio et sextam rigam et est 0, 5, 54, 49, 58. Subtrahere precedentem a sequenti, hoc est minorem a maiore, et post subtraccionem manebit prima riga |**f. 171v:**| 0, 0, 59, 8, 20. Et similiter operandum est in aliis. Ex dictis itaque apparet *Tabularum de mediis motibus planetarum et augibus* praxis 'continuacio' et 'correccio'. Unde nunc ad veros motus planetarum calculandos est accedendum, antequam tamen hoc fiat, medii motus planetarum, quomodo continuandi sunt pro almanach faciendo, ne postea tedium et labor superfluus oriatur, hic ostendere dignum videlicet et rectum.

---

**329** tercii *Had* : secundi *AB per err.* **364** 41 *B* : 45 *A per err.*

<9>

### <C>anon nonus, quomodo medius motus Solis, Lune et argumenti Lune singulis diebus per totum annum debeat continuari

Cum motus Solis et Lune, similiter et argumentum Lune continuare volueris per singulos dies tocius anni, primo radicem anni illius planete habe, cuius dies continuare intendis. Secundo: motum eiusdem planete accipe unius diei de tabula dierum et radici adde et tunc habebis motum eiusdem planete ad primam diem Ianuarii. Hunc itaque motum prime diei Ianuarii in sexterno tuo scribe ad hoc disposito. Tercio: motui prime diei adde motum unius diei et veniet motus secunde diei Ianuarii, cuiusdem motum eciam inventum in sexterno scribe. Et si motui secunde diei addideris motum unius diei, habebis ad terciam diem et sic consequenter continuando usque ad 31 dies, in quibus est finis Ianuarii. **Exempli gracia:** Ad annum Domini 1487 currentem si volo continuare motum Solis medium incipiendo a prima die Ianuarii usque in finem anni pro almanach anni 1487 currentis, faccione primo quere radicem eiusdem anni currentis iuxta prius dicta et est: 9, 18, 42, 56, 38. Secundo: eidem radici adde motum Solis in uno die, ut 0, 0, 59, 8, 20. Et post addicionem venit tibi motus Solis ad primam Ianuarii iste: 9, 19, 42, 4, 59. Cui motui prime diei si addideris eundem motum unius diei, veniet motus Solis ad secundam diem Ianuarii 9, 20, 41, 13, 19 et sic consequenter usque in finem mensis Ianuarii operaberis. Similiter in Luna et argumento Lune quolibet die. **Modus corrigendi continuacionem tuam.** Cum iam Ianuarii dies in motu Solis et Lune et argumenti Lune perfeceris et, si bene fecisti, desideras scire, hoc modo operare: Accipe motum Solis vel Lune vel argumenti Lune ex opposito Ianuarii positum secundum exigenciam anni communis vel bisextilis et eundem motum adde radici Solis, si operaberis de Sole, aut radici Lune, si operaris de Luna, aut radici argumenti Lune, si de eodem operaris. Si tunc eveniet tibi motus ad ultimam Ianuarii, per continuacionem inventus, et maxime in signis, gradus, minutis, de terciis et quartis non sit cura, tunc |**f. 172r:**| recte fecisti et bene continuasti. Si autem motus primus inventus per continuacionem et per correccionem repertus discordarent in minutis, male continuasti. Oportet ergo te querere illum diem, in quo errasti, et eum cum sequentibus emenda. **Exempli gracia:** Cum ad annum Domini 1487 currentem omnes motus Solis medios usque in finem Ianuarii perfeceris, si radici eiusdem anni, scilicet 9, 18, 42, 56, 38, addes illum motum, qui est in *Tabula mensium* sub titulo Solis in anno communi ex opposito Ianuarii, scilicet 1, 0, 33, 18, 8, veniet tibi finis Ianuarii eciam eodem anno, quam eciam per continuacionem tuam similiter debuisti invenire: 10, 19, 16, 14, 47. **Consideracio habenda.** Quicumque motus ex addicione motus alicuius mensis ad radicem eveniet, eum scribe pro ultima die tue mensis et cum ipso motus dierum sequentis mensis continuabis. Precisius enim est sic practicare quam per continuacionem et in fine cuiuslibet mensis diligenter in Sole et Luna et argumento Lune memorie est commendandum.

**389** 20 *B* : 28 *A per err.*

<10>

**<C>anon decimus, quomodo centrum Lune per singulos dies tocius anni pro almanach faciendo est continuandum**

Primo radicem centri Lune habe, quam sic invenies: quere radicem motus Lune ad principium anni. Similiter habe radicem motus Solis et radicem Solis subtrahe a radici Lune, et quod post subtraccionem unius radicis ab alia venerit, dupla et habebis post duplacionem centrum Lune. Et eidem adde motum centri Lune in una die et habebis motum centri ad primam diem Ianuarii, cui motui prime diei invento si addideris hoc idem, veniet motus centri secunde diei et sic consequenter invenies autem motum centri Lune singulis diebus et mensibus in tabula intitulata motus centri Lune in diebus. In correccione autem si bene operatus es, eodem modo age, sicut dictum est in canone precedenti, ultimam diem mensis cuiuslibet examinando. **Exempli gracia:** Ad annum Domini 1487 currentem volo scire radicem centri Lune. Primo: quero radicem motus Lune et est 11, 25, 8, 54, 32. Secundo: scio radicem motus Solis et est 9, 18, 42, 56, 39. Tercio: radicem motus Solis subtrahe a radice Lune, et quod post subtracionem veniet, dupla et duplum erit radix centri Lune et erit 4, 12, 51, 55, 48. Cui radici centri si addo id, quod ponitur in *Tabula dierum centri Lune*, e parte prime diei veniet centrum Lune ad primam diem Ianuarii. Illud, quod ponitur ex parte unius diei, est 0, 24, 22, 53, 23. Et veniet motus centri Lune prima die Ianuarii: 5, 7, 14, 49, 9. Et sic continuabis usque in finem cuiuslibet mensis et in correccione praxis modum tene superius dictum. |**f. 172v:**|

<11>

**<C>anon undecimus. Medium motum trium superiorum, Saturni, Iovis et Martis, pro almanach componendo de decem in decem invenire**

Medium motum trium superiorum, Saturni, Iovis et Martis, pro almanach faciendo sufficit de decem in decem continuare. Eis enim inventis dies intermedii pro mediis et veris motibus inveniendis facillime inveniuntur, ut docebit inferius. Cum ergo medium motum trium superiorum continuare de decem in decem volueris, primo radicem, cuius motum continuare volueris, prescribe. Secundo: in *Tabulam dierum illius planete*, cuius motum continuare intendis, intrabis et motum ex opposito decem dierum accipe. Tercio: hunc motum ex parte 10 dierum inventum adde ad radicem, et quod post addicionem provenerit, erit motus planete ad decimum diem Ianuarii et in sexterno scribe. Quarto: motui invento medio 10. die Ianuarii iterum adde motum ex opposito decem dierum acceptum et ad motum decime diei adde et veniet motus vigesime diei, quam iterum in sexterno scribe. Quinto: motum ex opposito 10 dierum inventum iterum adde ad motum, quam vigesima die invenisti, et post addicionem habebis motum diei trigesime. Cui si addideris motum unius diei, si necessarie fuerit, cum scilicet mensis 31 dies habuerit, veniet motus ad 31. diem et iterum motum hunc in sexterno signa. Et sic continuabis per menses omnes, motus planetarum trium superiorum. **Exempli gracia:** Ad annum Domini 1487 currentem volo extendere motum Saturni medium de decem in decem pro almanach faciendo. Primo radicem Saturni pro principio eiusdem anni quero et est ista: 8, 14, 48, 27, 58. Secundo: eidem radici adde motum Saturni, quem invenies ex opposito 10 dierum, scilicet 0, 0, 20, 5, 53. Et post addicionem venit tibi motus medius Saturni ad 10. diem Ianuarii iste: 8, 15, 8, 33, 51. Tercio: si motui decime diei invento addideris ex opposito 10 dierum acceptum, veniet medius motus Saturni ad vigesimam diem

Ianuarii iste: 8, 15, 28, 39, 44. Cui si addideris idem, veniet motus medius ad 30. diem iste: 8, 15, 48, 45, 37. Et sic similiter operaberis in Iove et Marte. **Cautela prima:** Diligenter adverte, quot dies quilibet mensis usualis habeat. Si 30 habeat dies, res perfecta est. Si autem 31 dies habuerit mensis, tum motum trigesime diei invenies aduc te addere, videlicet motum unius diei, ut compleantur 30 et unus dies perfecte, ut contingit |**f. 173r:**| in Ianuario, Marcio et aliis. **Cautela secunda:** In Februario cum braxim feceris et diei vigesime medium motum planete habueris pro motu aliorum dierum habendo non it, quod ex opposito accepisti, decem addes, quoniam Februarius scilicet 28 in anno communi habet. Intrabis itaque cum 8 diebus et motum ex opposito octo dierum inventum ad motum vigesime diei respectu adde sicque Februarium perficies. In anno vero bisextili propter diem superfluam cum 9 intrabis et motum ex opposito 9 dierum inventum adde ad motum vigesime diei sicque in anno bisextili Februarium continuabis.

<12>

**<C>anon duodecimus. Argumentum equale trium planetarum superiorum, ut Saturni, Iovis et Martis, de decem in decem pro almanach faciendo continuare**

Cum argumentum planetarum trium superiorum de decem in decem pro almanach faciendo continuare volueris, primo radicem argumenti habe, quam sic invenies: accipe radicem motus Solis et radicem motus planete, de quo intendis. Secundo: subtrahe radicem motus planete, de quo intendis, a radice motus Solis, et quod post subtraccionem manebit, erit radix argumenti planete, de quo intendis. Et ab hac radice sic inventa continuabis argumenta trium superiorum de decem in decem tali modo et forma {tibi argumentum planete decem dierum hoc modo}: recipe medium motum Solis decem dierum et eciam motum planete 10 dierum, deinde subtrahe motum planete decem dierum a medio motu Solis 10 dierum et id, quod manebit, erit argumentum medium planete decem dierum. Quod argumentum planete decem dierum adde ad radicem argumenti planete et habebis argumentum planete, de quo intendis, ad decimam diem Ianuarii, quod argumentum scribe in sexterno tuo, quemadmodum scripsisti medium motum planete. Tunc iterum adde argumentum planete, de quo intendis, et ultimus age, quemadmodum dictum est in canone precedenti, et sic queres motum agumenti in planetis tribus superioribus per totum annum in singulis mensibus. **Exempli gracia:** Ad annum Domini 1487 currentem volo extendere et continuare argumentum medium Saturni de decem in decem pro almanach faciendo. Primo oportet me habere radicem motus Solis et est 9, 18, 42, 56, 39. |**f. 173v:**| Debeo eciam scire radicem motus Saturni et est 8, 14, 48, 27, 58, quam radicem Saturni cum subtraxero a radice Solis, manebit radix argumenti Saturni 1, 3, 54, 28, 41. Secundo: quero mihi argumentum Saturni 10 dierum et capio motum Solis 10 dierum medium et est 0, 9, 51, 23, 16. Deinde capio motum Saturni 10 dierum et est 0, 0, 20, 5, 53. Et subtraho motum Saturni 10 dierum a medio motu Solis eciam 10 dierum et remanet argumentum Saturni 10 dierum 0, 9, 31, 17, 23. Quod addo radici argumenti Saturni et veniet argumentum Saturni 10. die Ianuarii 1, 13, 25, 46, 4. Quod scribam in sexternum ex opposito decime diei Ianuarii penes medium motum Saturni et continuabo, ut prius dictum est, addendo argumentum 10 dierum ad id, quod provenit, et habebo ad diem vigesimam Ianuarii, et est 1, 22, 57, 3, 27. Cautelas tamen duas predictas in precedenti canone mente habe de fine mensium et Februario. Sic de Iove et Marte operare.

<13>

**<C>anon 13. Centrum medium planetarum trium superiorum de decem in decem pro almanach faciendo continuare**

Cum centrum medium planetarum trium superiorum de decem in decem continuare volueris pro almanach faciendo, primo radicem motus centri illius planete inquire, de quo intendis, quam sic invenies: inquire inprimis radicem motus planete, deinde quere radicem augis illius planete, de quo intendis, et subtrahe radicem augis planete a radice medii motus planete, et quod post subtraccionem manet, erit radix medii centri planete, de quo intendis. Secundo: radici augis planete adde motum augium decem dierum et illo erit aux planete pro 10. die Ianuarii. Quam augem pro 10. die Ianuarii inventam debes subtrahere eciam de medio motu planete. Quod decima die Ianuarii invenies et quod post subtraccionem manebit, erit centrum planete ad 10. diem Ianuarii. Et scribe illud in sexterno iuxta motum et argumentum planete, de quo intendis. Tercio: centri radicem planete subtrahe de centro invento ad decimam diem Ianuarii, et quod remanet, erit centrum medium planete decem dierum, per quod centrum decem dierum omnia centra in eodem planeta per circuitum anni continuabis eodem in sexterno tuo collocando. **Exempli |f. 174r:| gracia:** Ad annum Domini 1487 currentem volo continuare et extendere medium motum centri Saturni per illum annum de 10 in decem pro almanach faciendo. Primo: oportet me habere radicem medii centri Saturni, quam sciam subtrahendo radicem augis illius planete, de quo intendo, a radice motus Saturni. Radix motus Saturni est 8, 14, 48, 27, 58. Radix augis Saturni est 8, 13, 1, 43, 59. Et post subtraccionem manet radix centri Saturni 0, 1, 46, 44, 59. Secundo: radici augis Saturni adde motum augium decem dierum, scilicet 0 in secundis, 57 terciis, et habebis augem Saturni pro 10. die Ianuarii istam: 8, 13, 1, 43, 56. Quam augem subtrahe a medio motu Saturni decime diei Ianuarii, qui est 8, 15, 8, 33, 51, et veniet centrum Saturni ad decimam diem Ianuarii hoc: 0, 2, 6, 49, 55. Et hoc scribe in sexterno tuo penes motum et argumentum Saturni. Tercio: radicem centri Saturni, sciliet 0, 1, 46, 44, 59, subtrahe a centro Saturni decima die inventum, scilicet 1, 2, 6, 49, 55, et quod post subtraccionem manebit, erit centrum Saturni 10 dierum et veniet 0, 0, 20, 4, 56. Per quod centrum Saturni decem dierum omnia centra in Saturno continuabis per totum annum, eodem in sexterno tuo collocando iuxta motus et argumenta Saturni.

<14>

**<C>anon quartus decimus de argumento medio duorum inferiorum, Veneris et Mercurii, pro almanach faciendo**

Cum argumentum Veneris per totum anni circulum continuare volueris de quinque in quinque, operacio tua erit hoc modo: Accipe inprimis radicem argumenti Veneris, cui adde argumentum Veneris ex opposito 5 dierum in *Tabula Veneris* repertum et veniet argumentum Veneris pro 5. die Ianuarii. Cui argumento Veneris quinta diei, si hoc idem addideris, erit argumentum Veneris pro 10. die Ianuarii, et sic continuabis usque in finem Ianuarii et per omnes menses tocius anni. **Exempli gracia:** Ad annum Domini 1487 currentem volo continuare argumentum Veneris per totum annum per singulos menses. Primo radicem argumenti Veneris inquire et est hoc: 10, 10, 1, 48, 3. Cui radici argumenti Veneris si addo motum argumenti Veneris, qui est ex opposito 5 dierum in *Tabula argumenti Veneris*, ut 0, 3, 4, 57, 17, et veniet 10, 13, 6, 45, 20. Cui iterum si addo motum argumenti Veneris 5 dierum in tabula repertum, veniet motus argumenti Veneris pro 10 die, scilicet 10, 16, 11, 42, 37, et sic continualiter argumentum Veneris per singulos menses tocius anni. |**f. 174v:**| **Cum vero argumentum Mercurii ostendere volueris per totum annum de tercia in terciam diem, operacio tua erit hoc modo:** Accipe imprimis radicem argumenti Mercurii, cui radici motum argumenti ex opposito trium dierum in *Tabula argumenti Mercurii* repertum adde et habebis argumentum Mercurii pro tercia die Ianuarii. Hoc idem si addideris argumento dierum trium invento, habebis pro sexta die Ianuarii, et sic continuabis usque in finem Ianuarii et per alios menses. **Exempli gracia:** Ad annum Domini 1487 currentem volo continuare et extendere argumenta Mercurii pro almanach faciendo. Primo debeo scire radicem argumenti Mercurii et est 0, 2, 5, 47, 57. Cui radici si addo motum argumenti Mercurii ex opposito trium dierum in *Tabula argumenti Mercurii* repertum, scilicet 0, 9, 19, 12, 23, et post addicionem veniet argumentum Mercurii pro tercia die Ianuarii hoc scilicet: 0, 11, 25, 0, 20. Cui secundo si iterum addes motum argumenti Mercurii trium dierum, habebis motum argumenti Mercurii pro sexta die et est 0, 20, 44, 12, 43. Et sic continuabis argumentum Mercurii per omnes menses tocius anni.

<15>

**<C>ano<n> quindecimus de centro medio Veneris et Mercurii pro almanach faciendo per totum annum continuando**

Cum centrum Veneris extendere volueris per singulos menses tocius anni pro almanach componendo, primo habe radicem centri Veneris, quam hoc modo invenies: subtrahe radicem augis Solis a radice motus Solis, et quod post subtraccionem manserit, erit radix argumenti Veneris. Secundo: accipe motum Solis 5. die Ianuarii. Similiter augem Solis 5. die Ianuarii, quem augem quinte diei Ianuarii si subtraxeris a quinta die motus Solis Ianuarii, in residuo manebit centrum Veneris pro 5. die Ianuarii. Tercio: a centro Veneris 5. die minue radicem centri Veneris et in residuo manebit centrum Veneris quinque dierum. Per quod centrum quinque dierum continuabis centra Veneris per singulos menses tocius anni iuxta regulas et cautelas prius positas superius. **Exempli gracia:** Ad annum Domini 1487 currentem volo continuare per singulos menses anni centrum Veneris pro almanach componendo. Primo radicem centri Veneris oportet me habere |**f. 175r:**| et hoc sic inveniam: primo medium motum Solis habe, qui est radix eiusdem anni et est 9, 18, 42, 56, 39. Deinde sciam radicem augis Solis et est 3, 1, 3, 23, 59. Quam radicem augis Solis cum subtraxero a radice motus Solis, manet radix centri Veneris et est 6, 17, 39, 32, 40. Secundo: debeo scire motum medium Solis pro quinta die Ianuarii et est 9, 23, 38, 38, 17. Debeo eciam scire augem Solis quinta die Ianuarii et est 3, 1, 3, 24, 28. Quem augem quinta die Ianuarii subtrahe a medio motu Solis quinte diei Ianuarii et in residuo manebit centrum Veneris ad quintam diem Ianuarii et est 6, 22, 35, 13, 49. Tercio: a centro Veneris quinte diei minue radicem centri Veneris et in residuo manebit centrum Veneris quinque dierum et est 0, 4, 55, 41, 9, per quod centrum quinque dierum continuabis centra Veneris per singulos menses tocius anni. **In Mercurio similiter operandum est per augem suam et motum Solis in respectu ad motus trium dierum.** Cum ergo centrum Mercurii extendere volueris per singulos menses pro almanach faciendo, primo habe radicem centri Mercurii, quam hoc modo invenies: subtrahe radicem augis Mercurii a radice motus Solis, quia motus Mercurii est idem cum motu Solis, et quod post subtraccionem provenerit, erit radix centri Mercurii. Secundo:

accipe motum Solis vel Mercurii, quod idem est, de tercia in terciam diem. Similiter augem Mercurii de tercia in terciam, quam augem Mercurii trium dierum si subtraxeris a motu Solis vel Mercurii trium dierum Ianuarii, in residuo manebit centrum medium Mercurii pro tercia die Ianuarii. Tercio: a centro Mercurii tercie diei Ianuarii minue radicem centri Mercurii et in residuo manebit centrum Mercurii trium dierum, per quod centrum trium dierum continuabis centra Mercurii per singulos menses tocius anni iuxta regulas et cautelas superius positas. **Exempli gracia:** Ad annum Domini 1487 currentem volo continuare per singulos menses anni centrum medium Mercurii pro almanach componendo. Primo radicem centri Mercurii habe, quam radicem sic invenies: primo debes habere radicem augis Mercurii et est iste: 7, 0, 17, 33, 59. Secundo habe radicem motus Solis et est 9, 18, 42, 56, 39. Subtrahe itaque radicem augis Mercurii a radice motus Solis, et quod post subtraccionem venerit, erit |**f. 175v:**| {erit} radix centri Mercurii et est hec: 2, 18, 25, 22, 40. Secundo: accipe motum Solis vel Mercurii, quod idem est, de tercia die in terciam et erit motus Solis tercia die Ianuarii iste: 9, 21, 40, 21, 38. Similiter quere augem Mercurii de tercia in terciam diem et est 7, 0, 17, 34, 16. Et subtrahe hanc augem a motu Solis vel Mercurii trium dierum et veniet centrum Mercurii ad terciam diem Ianuarii, hoc est 2, 21, 22, 47, 22. Tercio: a centro Mercurii tercia die Ianuarii subtrahe radicem centri Mercurii et in residuo manebit centrum Mercurii trium dierum hoc: 0, 2, 57, 24, 42. Per quod centrum trium dierum continuabis centra Mercurii per singulos menses tocius anni iuxta regulas et cautelas superius positas. Et hec dicta sufficiunt pro continuacionem motuum planetarum pro almanach.

<16>

**<C>anon sedecimus de generali in Tabulas equacionum planetarum de veris motibus calculandis introytum, ut intellectus pariter et introytus facilis in Tabulas equacionum de veris motibus planetarum pateat**

Primo sciendum et animadvertendo, quod in *Tabulis equacionum* ex opposito graduum lateralium equaciones scribuntur. Differencie vero equacionum non aliud sunt nisi excessus maioris equacionis ad minorem. Supra lineas inter equaciones inscribuntur circa has equaciones et differencias. Tam supra, quam infra est numerus signorum ponunturque ibi littere 'a', 's' vel 'm'. 'A' addicionem equacionis scilicet designat, 's' vero vel 'm', quod equacio subtrahi debet, designat. Tamen in aliquibus tabulis "adde" expresse scribitur, vel 'subtrahe', vel 'minue'. Advertendum est eciam, quod quando in tabulis equacionum ponitur paragraphus rubeus vel niger, tunc ibi in differencia tam in superiori, quam inferiori margine ponuntur due littere, scilicet 'a', 's' vel 's', 'a' sic, quod una ponitur supra aliam, quarum superior in utroque margine respicit equaciones, que sunt supra paragraphum. Et inferior similiter in utroque margine respicit equaciones, que sunt sub paragrapho. Et per hunc modum equantur Sol et Luna. Advertendum est eciam, quod in *Tabulis equacionum aliorum planetarum* ponitur propior in minutis proporcionalibus, ibi eciam in utroque margine ponuntur due littere, una superior et alia inferior. Quarum propior in utroque margine respicit minuta proporcionalia, que sunt supra propior, et alia littera minuta |**f. 176r:**| respicit, que sunt sub propior.

De modo intrandi ad *Tabulas equacionum*. In *Tabulas equacionum* intratur cum signis communibus, quorum sex ponuntur in superiori margine, scilicet 0, 1, 2, 3, 4, 5, eorumque gradus ponuntur in prima linea descendendo versus sinistram. Et alia sex ponuntur in margine inferiori, scilicet 6, 7, 8, 9, 10, 11 retrorsum. Et eorum gradus ponuntur in ultima <linea> ascendendo versus dextram. Si ergo signum, cum quo intrabis, sit in margine vel spacio superiori, tunc gradus signi accipiuntur versus sinistram descendendo et ex directo eiusdem gradus invenitur equacio, quam eciam scribe. Deinde differenciam invenies, que inmediate scribitur in linea descendendo, et consimiliter littera scripta supra differenciam in superiori margine, que scribitur extra circa differenciam et eam litteram. Si fuerit 'a', significat, quod pars proporcionalis sit addenda equacionis accepte. Si autem fuerit 'm', significat, quod pars proporcionalis sit subtrahenda ab equacione. Si autem signum, cum quo intratur, sit in margine inferiori, tunc gradus signi accipitur versus dextram partem ascendendo et ex directo eiusdem gradus versus sinistram accipitur equacio sive minuta proporcionalia sive dyametri diversitas, que scribitur extra. Deinde incipitur differencia, que inmediate sequitur in linea ascendendo et consimiliter littera scripta sub differencia in margine inferiori, que littera scribitur extra circa differenciam. Et eadem littera, si fuerit 'a', significat, quod pars proporcionalis sit addenda equacioni accepte. Si 's', significat, quod sit subtrahenda. **Sciendum eciam**, quod per minuta proporcionalia vel per differenciam equacionis queritur pars proporcionalis per tabulam specialem, cuius titulus est *Tabula ad inveniendum minuta proporcionalia*, que tabula extenditur numero sexagenario tam in margine superiori, quam a latere. Immo de eius introytu est sciendum, quod postquam numerus unus ducitur in alium, debet videri denominator, quem ostendunt quidem tituli superius scripti. Tum sub intelleccione aliarum fraccionum more *Tabularum de horis in mediis motibus* intratur ergo in eandem tabulam cum duobus numeris, cum uno in linea numeri a parte scripta habendo respectum ad alium in superiori margine, et sic tunc secundum denominacionem resolve ad propositum. **Consideracio prima.** In *Tabula equacionum octave sphere* solum tres tabelle sunt descripte, quia ille equaciones ad quartam partem circuli *(in mg. add.:* se extendunt*)*, hoc est |**f. 176v:**| tria signa. Tum ille equaciones acomodantur et tabelle omnibus duodecim signis videlicet de numero signorum. In superiori enim parte supra tabellas ponuntur a nihil usque ad tria exclusive cum eorum oppositis, infra vero continuantur a tribus usque ad sex exclusive similiter cum eorum oppositis, ut ad sensum apparet in tabellis. **Consideracio secunda:** In tabulis et equacionibus Solis ponuntur sex tabelle parciales, que sunt extensa pro medietate una zodyaci et in superiori parte vel margine ponuntur signa a nihil usque ad sex exclusive. In inferiori vero parte alia signa descripta inveniuntur a sex usque in undecim complete. Ad quas tabellas cum introytum feceris, equaciones Solis cum differencia equacionis accipies iuxta modum superius dictum. **Consideracio tercia.** In Luna et aliis planetis sub quolibet signo plures descripte equacionum sunt tabelle propter hoc, quia Luna et alii planete diversos habent motus et plures, quod aliqui sit racione eccentrici, aliqui racione epicicli, aliqui racione longitudinum. In tabulis eciam aliorum planetarum in fine *Tabulam stacionum primam* continentem invenies planete. Et quomodo eciam stacionem secundam exinde extrahere possumus, dicetur inferius in canone, quando docebitur, quomodo congnosci debeat, si planeta est stacionarius vel directus.

---

**632** ascendendo *C* : descendendo *AB* **640** accipitur *BC* : accipio *A* **641** que *C* : qui *AB* **655** descripte *BC* : descripti *A* **660** equacionibus *C* : adequacionibus *AB*

<17>

**<C>anon decimus septimus de auge communi invenienda**

Cum augem communem invenire volueris ad aliquod annum, primo motum augium et stellarum fixarum ad eiusdem annum inquire per modum dictum in canone primo. Secundo: motum accessus et recessus octave circuli per eundem modum inquire. Tercio: cum signis et gradibus, que in motu octavi circuli invenisti, in *Tabula<m> equacionum* eiusdem circuli octavi intrabis et equacionem pro signis et gradibus accipe et scribe circa eam 'a' vel 's', ut in tabulis eisdem videbis. Quarto: differenciam equacionis accipe et circa eam scribe 'a' vel 's', ut in tabulis circa equacionem videbis. Quinto: partem proporcionalem iuxta modum in canone precedenti dictum inquire, ducendo differenciam in residuam numeri, quemadmodum dictum est in canone precedenti. Sexto: partem proporcionalem inventam adde vel subtrahe ab equacione prius inventa et exibit equacio equata et perfecta. Ultimo: eandem equacionem equatam vel perfectam adde vel subtrahe a motu augium et stellarum fixarum, prout te titulus docebit, et post addicionem vel subtraccionem augem communem habebis. Si autem augem specialem cuiuslibet planete eodem modo |**f. 177r:**| habere volueris: augem communem cuilibet augi speciali {specialis} cuiuslibet planete seorssim tempore Nativitatis Christi adde, et quod post addicionem provenerit, aux erit specialis cuiuslibet planete ad tempus tuum propositum. Radices autem augium pro tempore Nativitatis Christi stabuntur in tabella sequenti.

**Radices augium ad tempus Nativitatis Cristi:**

| S | G | M | S | T | | |
|---|---|---|---|---|---|---|
| 2 | 11 | 25 | 23 | 4 | | Solis et Veneris |
| 7 | 23 | 23 | 42 | 4 | | Saturni |
| 5 | 3 | 37 | 0 | 4 | aux | Iovis |
| 3 | 25 | 12 | 13 | 4 | | Martis |
| 6 | 10 | 39 | 33 | 4 | | Mercurii |

**Exempli gracia:** Ad annum Domini 1487 currentem volo scire augem communem. Primo: debeo querere motum augium et stellarum fixarum iuxta modum in primo canone dictum ad annum propositum et est 0, 10, 55, 3, 9. Secundo: debeo querere motum accessus et recessus octavi circuli ad eundem annum et est 2, 15, 37, 56, 0. Tercio: debeo intrare cum motu octavi circuli, scilicet cum 2 signis et 15 gradus, et capere equacionem in tabula ad hoc disposita. Et postquam intrabo cum numero predicto, in equacione invenio 8 gradus, 41 minuta, 28 secunda, cuius titulus est 'a'. Differencia autem circa equacionem scripta est 2 minuta, 22 secunda, cuius titulus est 'a'. Quarto: eandem differenciam debeo ducere in residuum, cum quo non est intratum, et debeo querere partem proporcionalem et venient 1 minutum, 29 secunda, 47 tercia, que pars proporcionalis debet addi ad equacionem primo inventam, scilicet ad 8 gradus, 41 minuta, 28 secunda, et post addicionem veniet equacio equata et perfecta ista: 8 gradus, 42 minuta, 57 secunda, 47 tercia. Quam equacionem ita perfectam et adequatam ultimo debeo addere ad motum augium et stellarum fixarum prius inventum et veniet aux communis ad principium anni ista: 0, 19, 38, 0, 56. Cui augi communi si addidero radicem augis cuiuslibet planete ad tempus Christi, habebo augem cuiuslibet planete ad annum propositum meum, ut ad principium anni 1487 aux Solis veniet ista: 3, 1, 3, 24, 0, et sic invenies augem omnium planetarum.

<18>

### <C>anon 18 de vero motu Solis inveniendo

Cum verum motum Solis habere volueris, primo medium motum Solis, ut supra patuit, inquire. Secundo: augem eciam eius habe ad eundem diem, ut eciam supra dictum est. Tercio: argumentum Solis ad eundem diem inquire, quod – ut supra dictum est – invenies subtrahendo augem Solis a motu Solis. Quarto: cum argumento ad *Tabulas equacionum* intrabis et equacionem centri Solis cum differencia accipies; circa ea scribendo titulos, ut tabula docebit. Quinto: partem proporcionalem inquire iuxta modum in canone sedecimo dictum, quam partem proporcionalem adde vel subtrahe ab equacione argumenti iuxta titulorum designacionem et veniet equacio perfecta |**f. 177v:**| et equata. Ultimo eandem equacionem adde vel subtrahe a motu iuxta litteram circa equacionem sitam vel locatam et habebis verum motum Solis ad tempus tibi propositum. **Exempli gracia:** Volo scire verum motum Solis ad diem Assumpcionis Marie, anno Domini 1486 currente. Primo: debeo habere motum Solis ad eundem diem et est 5, 2, 41, 47, 30. Secundo: debeo habere augem Solis ad eandem diem et est 3, 1, 3, 11, 50. Tercio: debeo habere argumentum Solis ad eundem diem, quod sciam, cum subtraxero augem Solis a motu Solis, et est hoc: 2, 1, 38, 35, 40. Quarto: cum eodem argumento ad *Tabulas equacionum* intrabis et veniet equacio ista: 1 gradus, 51 minuta, 51 secunda, circa quam ponitur littera 's'. Differencia autem est 1 minuta, 5 secunda, cuius titulus est 'a'. Quinto: partem proporcionalem inquire et venient 0 minuta, 41 secunda, 49 tercia, quam addo ad equacionem primo inventam et veniet equacio perfecta ista: 1 gradus, 52 minuta, 32 secunda, 49 tercia. Que equacio perfecta debet subtrahi a <medio> motu et veniet postea <verus> motus Solis iste: 15, 0, 49, 14, 41 ad diem Assumpcionis Marie.

<19>

### <C>anon 19 de vero motu Lune inveniendo

Cum verum motum Lune ad aliquod tempus habere desideras, primo eius motum inquire, quemadmodum dictum est in canone secundo. Secundo: argumentum Lune inquire per tabulam propriam ad hoc ordinatam, ut dictum est circa motus planetarum. Tercio: centrum Lune habe, et quomodo invenire debeat, dictum est in canone quinto, scilicet subtrahendo motum Solis a motu Lune, et residuum dupla. Quarto: cum centro Lune medio ad diem propositam *Tabulam Lune* ingredere et equacionem centri accipe et circa eam litteram 'a' vel 's' scriptam considerabis. Et differenciam similiter accipies et eius equacionem, ut dictum est superius. Similiter partem proporcionalem iuxta modum dictum invenies. Quinto: equacionem centri perfectam adde vel subtrahe a medio argumento Lune et verum argumentum Lune proveniet. Sexto: cum medio eciam centro minuta proporcionalia invenies iuxta modum superius dictum, que cum inventa fuerint, serventur ad partem. Septimo: cum vero argumento Lune *Tabulam* ingredere *equacionum Lune* et equacionem argumenti Lune inquire cum eius differencia, circa que scribe litteras illas, ut te tabella docebit. Et quod post operacionem tuam provenerit, erit equacio prima examinata. Octavo: cum eo vero argumento queratur diversitas dyametri cum sua differencia et partem proporcionalem iuxta modum superius dictum inquire. Nono: per diversitatem dyametri partem proporcionalem inquire per minuta proporcionalia ad partem servata modo generali superius dicto ponendo minuta proporcionalia superius et diversitatem dyametri inferius. Decimo: partem proporcionalem inde provenientem adde |**f. 178r:**| equacioni prima examinata et veniet equacio secunda examinata. Undecimo: equacionem nunc ultimo examinatam adde vel subtrahe a motu Lune, secundum quod titulus equacionis requirit, et habebis verum motum Lune ad tempus tuum propositum. **Cautela prima:** Considerandum est et diligenti animo advertendum, quod cum in Luna et ceteris planetis si in minutis proporcionalibus nihil est in minutis, nihil in secundis, tunc non est opus querere partem proporcionalem, nam tunc equacio prima examinata erit equacio secundo examinata. **Cautela secunda:** Cum 60 proveniunt in minutis proporcionalibus et nihil in differencia, tunc diversitas dyametri est pars proporcionalis. **Verbi gracia:** Volo scire motum verum Lune ad annum Domini 1486 currentem, ad diem sancte Marie Magdalene. Primo debeo scire motum Lune et est iste: 0, 20, 34, 21, 9. Secundo: debeo scire argumentum Lune et est hoc: 4, 7, 53, 23, 8. Tercio: debeo scire centrum Lune ad eundem diem et est hoc: 4, 23, 3, 46, 58. Quarto: cum centro Lune *Tabulam equacionum Lune* intrabo et equacionem centri Lune invenio 10 gradus, 43 minuta. Cuius titulus est 'a'. Differencia 10, cuius titulus est 's', pars proporcionalis est 0 minuta, 37 secunda, 50 tercia, que debet subtrahi ab equacione centri et subtraccione facta manebit equacio centri perfecta ista: 10 gradus, 42 minuta, 22 secunda, 10 tercia, cuius titulus est 'a'. Quinto: equacionem centri hanc perfectam debeo addere ad medium argumenti Lune et veniet verum agumentum Lune hoc: 4, 18, 35, 45, 18. Sexto: cum eodem centro Lune minuta querantur proporcionalia et eius differencia. Minuta proporcionalia venient 53 et in differencia 0, que minuta proporcionalia debent servari ad partem. Septimo: cum vero argumento Lune *Tabulam equacionum Lune* ingredere et equacionem argumenti Lune inquire et differenciam similiter. Equacio argumenti erit 3 gradus, 31 minuta, 3 secunda, cuius titulus est 's'. Differencia vero 3 minuta, 53 secunda. Pars autem proporcionalis est hec: 2 minuta, 18 secunda, 51 tercia, que debet subtrahi ab equacione argumenti et veniet equacio prima examinata: 3 gradus, 28 minuta, 44 secunda, 9 tercia, cuius titulus est 's'. Octavo: cum eodem vero argumento diversitas dyametri queratur cum sua differencia et parte proporcionali. Diversitas dyametri est 2 gradus, 5 minuta. Differencia 2 minuta, cuius titulus est 's', pars proporcionalis est 1 minutum, 11 secunda, 31 tercia, que debet subtrahi a diversitate dyametri et veniet diversitas dyametri hec: 2 gradus, 3 minuta, 48 secunda, 29 tercia. Nono: per diversitatem dyametri et minuta proporcionalia prius ad partem servata partem proporcionalem inquire modo generali, minuta proporcionalia ponere superius et diversitatem dyametri inferius et veniet pars proporcionalis ista: 1 gradum, 49 minuta, 21 secunda, 50 tercia. Decimo: partem proporcionalem |**f. 178v:**| provenientem adde equacioni primo examinata et veniet equacio secunda examinata ista: 5 gradus, 18 minuta, 5 secunda, 59 tercia, cuius titulus est 's'. Ultimo: equacionem hanc ultimo examinatam adde vel subtrahe a motu Lune, secundum quod titulus equacionis requirit, et habebis verum motum Lune ad tempus tuum propositum et est iste: 0, 15, 16, 15, 10.

<20>

### <C>anon vigesimus de vero loco capitis draconis inveniendo

Cum verum motum capitis draconis habere volueris, primo eius medium motum inquire. Secundo: medium motum capitis draconis habes, ipsum a 12 signis subtrahe et

---

**796** 59 *B* : 19 *A per err.*

residuum erit verus motus eius. Tercio: cum verum motum capitis draconis habes, in signo opposito in simili gradu, minuta, secunda et tercia motum caude computabis. **Verbi gracia:** Si volo scire verum motum capitis draconis ad annum Domini 1486 currentem, ad diem Assumpcionis Marie, primo debeo scire motum capitis draconis et est iste: 0, 26, 7, 44, 49. Secundo: debeo motum capitis draconis subtrahere a 12 signis et remanebit eius verus motus iste: 11, 3, 52, 15, 11. Tercio: in opposito signo in simili gradu, minuta, secunda et tercia motum caude draconis computabis.

<21>

**<C>anon 21 de argumento latitudinis Lune inveniendo**

Cum verum argumentum latitudinis Lune scire desideras, primo verum locum Lune habe. Secundo: verum locum capitis draconis inquire. Tercio: verum locum capitis draconis a vero loco Lune subtrahe, et quod remanet, erit argumentum latitudinis verum Lune. **Exempli gracia:** Volo scire argumentum latitudinis Lune ad diem Assumpcionis Marie, ad annum Domini 1486 currentem. Primo: debeo habere verum motum Lune et est 11, 1, 59, 4, 43. Secundo: verum motum capitis draconis scire debeo et est 11, 3, 52, 15, 11. Tercio: verum motum capitis draconis debeo subtrahere a vero loco Lune et remanet argumentum latitudinis hoc: 11, 28, 6, 49, 32.

<22>

**<C>anon 22 de vero loco Iovis, Saturni, Martis inveniendo**

Cum verum motum Saturni, Iovis aut Martis scire volueris, primo motum illius ad diem propositam inquire, cuius verum motum scire desideras. Secundo: augem eiusdem planete ad eundem diem propositam habe. Tercio: augem eiusdem planete a motu medio planete ad diem propositam subtrahe et residuum erit centrum medium planete, quod iuxta medium motum serva. Quarto: ad eundem diem quere motum Solis, a quo motu Solis subtrahe motum planete et residuum erit argumentum medium planete, quod argumentum eciam serva iuxta medium centrum planete et motum planete. Quinto: cum medio centro planete tabulam equacionum eiusdem planete ingredere cum signis et gradibus et cape equacionem centri et differenciam, cum qua partem proporcionalem inquire et titulus 'a' vel 's' tam circa equacionem, |**f. 179r:**| quam differenciam considera. Sexto: partem proporcionalem inventam adde vel subtrahe ab equacione centri iuxta titulorum 'a' vel 's' stacionem et erit equacio centri perfecta. Septimo: eandem equacionem centri perfectam adde centro medio, si circa eam scriptum fuit 'a', vel subtrahe, si circa eam scriptum fuit 's', et habebis verum centrum eiusdem planete ad diem propositum. Octavo: si equacio centri est addita medio centro, tunc eadem equacio centri debet subtrahi a medio argumento et manebit verum argumentum. Si vero e contrario equacio centri est subtracta a medio centro, tunc eadem equacio centri debet addi medio argumento et manebit verum argumentum planete ad diem propositam. Nono: cum vero centro quere minuta proporcionalia cum sua differencia et parte proporcionali et scribe ea ad partem cum titulo 'longior' vel 'propior', ut docebit te titulus in tabula repertus. Decimo: cum vero argumento intra *Tabulam equacionum* cum signis et gradibus et quere equacionem argumenti cum sua differencia et accipe partem proporcionalem et exibit equacio prima examinata, circa quam signa 'a' vel 's', ut te

---

**837** addita *BC* : addida *A*

tabula docebit. Undecimo: iterum intra cum vero argumento et accipe diversitatem dyametri in altera longitudinum. Si enim supra minuta proporcionalia scriptum fuerit 'propior', tunc accipe diversitatem dyametri in longitudine propinquiori. Si autem circa eam scriptum fuerit 'longior', tunc accipe diversitatem dyametri in longitudine longiori. Duodecimo: per diversitatem dyametri et minuta proporcionalia prius servata quere partem proporcionalem. Tredecimo: partem proporcionalem inde provenientem adde equacioni primo examinata, si diversitas dyametri capta est in longitudine propinquiori, vel subtrahe, si diversitas dyametri capta est in longitudine longiori, et veniet equacio secundo examinata. Quartodecimo: vide ambas equaciones, scilicet centri et argumenti secundo examinati, si circa ambas scriptum est 'a', tunc ambe debent addi simul et tunc residuum inde proveniens debet addi medio motui planete et veniet verus motus ad tempus tuum propositum planete. Si autem circa ambas equaciones sit scriptum 's' vel 'm', tunc ambe debent addi simul et totum residuum debet subtrahi a motu planete et habebis verum motum planete ad tempus propositum. Si autem circa unam equacionem scriptum fuerit 'a' et circa aliam 's' vel 'm', tunc subtrahe minorem a maiori, residuum addendo vel subtrahendo a motu secundum titulum 'a' vel 's', scriptum circa maiorem equacionem, a qua subtraxisti minorem, et habebis verum motum planete ad tempus tuum propositum. **Exempli gracia:** Volo scire |**f. 179v:**| verum motum Saturni ad annum Domini 1486 currentem, ad diem Assumpcionis Marie. Primo: debeo scire medium motum Saturni et est 8, 10, 11, 6, 47. Secundo: debeo habere augem Saturni vel planete, de quo intendis, ad eundem diem habere et est ista: 8, 13, 1, 30, 50. Tercio: augem eiusdem planete a motu planete eadem die subtrahe et residuum erit medium centrum hoc: 11, 27, 9, 35, 57. Quarto: debeo scire medium motum Solis eadem die et est iste: 5, 2, 41, 47, 30. A quo medio motu Solis debeo subtrahere motum planete et residuum erit medium argumentum Saturni hoc: 8, 22, 30, 40, 43. Quinto: cum medio centro planete *Tabulam equacionum Saturni* ingredere querendo equacionem centri et differenciam et partem proporcionalem. Equacio centri erit: 0 gradus, 20 minuta, cuius titulus est 'a', differencia 6, cuius titulus est 's'. Pars proporcionalis erit hec: 0 minuta, 57 secunda, 36 tercia. Sexto: partem proporcionalem subtrahe ab equacione centri et erit equacio centri perfecta; est 0 gradus, 19 minuta, 2 secunda, 24 tercia, cuius titulus est 'a'. Septimo: eandem equacionem centri perfectam debeo addere medio centro et veniet verum centrum 11, 27, 28, 38, 21. Octavo: eandem equacionem centri debeo subtrahere a medio argumenti et veniet verum argumentum Saturni hoc: 8, 22, 11, 38, 19. Nono: cum vero centro quere minuta proporcionalia cum sua differencia et parte proporcionali. Minuta proporcionalia veniunt 60, cuius titulus est 'longior longitudo'. Differencia 0. Decimo: intro cum vero argumento et accipe equacionem argumenti cum differencia et parte proporcionali. Equacio argumenti veniet ista: 6 gradus, 13 minuta, cuius titulus est 'm', differencia 0. Undecimo: iterum intrabo cum vero argumento et accipio diversitatem dyametri sub longitudine longiori et veniet ista: 0 gradus, 20 minuta, differencia nihil. Duodecimo: quererem partem {partem} proporcionalem per diversitatem dyametri et minuta proporcionalia, sed hoc hic non oportet iuxta cautelam secundam in 19. canone dicta<m>. Tredecimo: subtraho diversitatem dyametri ab equacione argumenti prima examinata et manebit equacio argumenti secunda examinata ista: 5 gradus, 53 minuta, cuius titulus est 'm'. Ultimo considero ambas equaciones, scilicet centri et argu-

---

**867** planete[1] *B* : planeta *AC* **870** medio *BC* : vero *A* **888** 53 *B* : 51 *A*

menti, et subtrahe minorem a maiori et manet residuum hoc: 5 gradus, 33 minuta, 57 secunda, 36 tercia. Quod residuum debeo subtrahere a motu Saturni et veniet verus motus Saturni ad diem Assumpcionis Marie ad meridiem iste: 8 signa, 4 gradus, 37 minuta, 9 secunda, 11 tercia, 9 quarta. |**f. 180r:**|

<23>

**<C>anon 23 de vero motu Veneris et Mercurii inveniendo**

Cum verum locum Veneris aut Mercurii habere desideras, eodem modo est operandum, sicut dictum est in canone precedente, hoc solum attendendo, quod cum medium centrum Veneris aut Mercurii habere desideras, primo motum Solis habe ad tempus tuum propositum, qui motus Solis idem est cum motu Veneris et Mercurii. Secundo: subtrahe augem Veneris, que est eadem cum auge Solis, a motu Veneris et Solis, et habebis centrum Veneris. In Mercurio autem augem propiorem Mercurii subtrahe a motu Mercurii vel Solis, quod idem est, et habebis verum centrum Mercurii. Argumenta autem media tam Mercurii, quam Veneris per proprias tabulas inquire. In aliis autem agendum est eodem modo, quoadmodum dictum est in canone precedenti. **Cautela prima:** In tabula sexta Veneris sive Martis circa differenciam equacionis argumenti gradus inveniuntur et minuta gradus incomplete. Minuta vero complete, que omnia, cum partem proporcionalem inquires, more solido collocantur, gradus sub gradibus et minuta sub minutis. **Cautela secunda:** Cum in Marte aut in Venere in equacione argumenti fuerint plures gradus quam 30 et non poterit scilicet subtraccio a motu planete, tunc acomodandum est unum signum et resolvendum est in 30 gradus et a toto collecto in gradibus subtrahenda est equacio argumenti. Et quod postea provenerit, erit verus motus planete. In addicione autem quociens plures gradus quam 30 excreverint, sunt removendi 30 et pro subtraccione 30 graduum unum signum ad signa est apponendum. **Cautela tercia:** Cum paragraphum in equacione centri aut argumenti evenerit, tunc duo tituli in differencia sequenti in parte superiori et inferiori tabule occurrunt. Si enim intratur cum signis in parte superiori acceptis, primus titulus usque ad paragraphum ostendit. Secundus titulus a paragrapho usque ad finem tabule, quid agendum sit, designat. Si autem signa inferius accipiuntur, primus titulus, occurrens usque ad paragraphum, ostendit 29 in finem tabule, quid agendum sit, demonstrat. **Quarta cautela:** Cum in argumento aut centro aliquid fuerit in signis et nihil minuta, gradus et signa in parte superiori essent tabule, resolvendum est unum signum in 30 gradus et intrandum est cum signis residuis et 30 gradibus in *Tabulas equacionum*. |**f. 180v:**| **Cautela quinta:** Cum in centro vel argumento habetur 0 in signis, 0 in gradibus, tunc acomodanda est una revolucio 12 signorum, a quibus 12 signis subtrahendum est unum signum, quod signum in 30 gradus reducatur, et sic cum 11 et 29 in *Tabulas equacionum* intrabis.

<24>

**<C>anon 24. Si planeta directus sit, retrogradus aut stacionarius**

Cum leviter scire volueris, an planeta directus sit, stacionarius aut retrogradus, primo cum centro equato eiusdem planete *Tabulas equacionum* ingredere et ex opposito gradus centri accipe stacionem primam. Secundo: eandem stacionem primam inventam subtra-

---

**914** in *Cautela tercia*, the manuscript *A* four times repeats the paragraph mark which the scribe of the manuscript *B* substituted directly by the word *paragraphum* on f. 155r; we adopt his solution : *C om.*

he a 12 signis, unum signum resolvendo in 30 gradus, unum gradum in 60 minuta et sic consequenter, ut sepe superius est ostenssum, et quid post subtraccionem manebit, stacio secunda est planete. Tercio: habita stacione prima et secunda argumentum planete vide, si illud est equale stacione prima, planeta erit stacionarius stacione prima. Si argumentum sit equale stacione secunda, planeta erit stacionarius stacione secunda. Si autem argumentum equatum sit minus stacione prima et eciam secunda, planeta erit directus. Si eciam argumentum equatum fuerit plus stacione prima et secunda, eciam erit directus. Si autem argumentum equatum est maius stacione prima et minus stacione secunda, planeta erit retrogradus. Possunt eciam scire staciones directos et retrogradaciones planetarum per tabulam, que intitulatur *Tabula direccionis, retrogradacionis et stacionis planetarum*. Intrando in eam cum vero argumento planete et per inicia et fines stacionis et retrogradacionis considerando argumentum equatum planete. **Verbi gracia:** Ad annum Domini 1486 currentem, ad diem Assumpcionis Marie, volo scire, si Saturnus est directus, retrogradus aut stacionarius. Debeo cum centro equato, quod est 11, 27, 28, 38, 21, intrare *Tabulam equacionum Saturni* et ex opposito gradus centri capere stacionem primam, que est 3 signa, 22 gradus, 44 minuta. Secundo: eandem stacionem primam subtraho a 12 signis et manebit stacio secunda ista: 8 signa, 7 gradus, 10 minuta. Tercio: debeo videre argumentum Saturni equatum, id est verum, quod est 8 signa, 22 gradus, 11 minuta, |**181r:**| tunc si illud esset equale stacioni prime aut secunde Saturni, esset stacionarius. Si eciam argumentum equatum sit minus stacione prima et eciam secunda, ipsa esset directus. Similiter si argumentum est plus prima et secunda, ipse est directus. Quia ergo argumentum equatum Saturni est maius stacione prima et eciam secunda, patet, quod ipse est directus.

<25>

### <C>anon 25. Utrum planeta sit ascendens vel descendens

Si hoc cognoscere volueris centrum planete equatum, considera, quod si fuerit ab uno gradu in sex signa, planeta erit descendens in suo ecentrico, a 6 autem in 12 signa erit ascendens. Si autem argumentum equatum planete fuerit ab uno gradu in sex signa, erit planeta descendens in epiciclo. Et si a 6 in 12, erit ascendens in epiciclo suo.

<26>

### <C>anon 26. De continuacione verorum motuum in almanach reponendis

Cum veros motus planetarum continuare volueris, per singulos dies anni eos in almanach reponendo, quere primo veros motus planetarum cum capite draconis ad principium anni, ad primam diem Ianuarii, in signis, gradibus, minutis, secundis, terciis. Secundo: quere veros motus eorum post aliquot dies pretemittendo, ut in Sole de quinque in quinque. In Luna vero quolibet die eius verum motum inquire. Et in Saturno et capite draconis, Iove et Marte de decem in decem veros motus inquire. Si tamen distinccius agere volueris et cercius, in Iove de octo in octo, in Marte de sex in sex veros motus inquire, in Venere autem et Mercurio de quarto in quartam diem. Si vero cercius operare volueris, de tercia in terciam diem veros motus Veneris et Mercurii inquire. Tercio: cum veros motus dierum a se distancium habueris, minorem a maiori subtrahe et post subtraccionem in gradibus, minutis, secundis et terciis manebit tibi differencia unius veri motus ab alio per dies distans. Quarto: cum hac differencia *Tabulam minutorum proporcionalium pro diebus intermediis* in almanach intitulatus ingredere et quere numerum consimilem graduum et minutorum, sicut in differencia habes. Quinto: considera, quot dierum est differencia, et tunc sub tali numero dierum |**f. 181v:**| gradus, minuta, secunda et tercia inquire, que in tabula serva. Sexto: si in differencia duorum verorum motuum ultra gradus et minuta sunt secunda vel tercia, tunc variando denominaciones intra iterum tabulam sub simili numero dierum, quanta est talis distancia in ea parte tabule, ubi ciphras invenies. Tunc ciphra habet denominacionem minuti, et quod sequitur, erunt secunda et tercia, que adde prioribus. Pari modo cum terciis intrabis. Septimo: hec omnia adde adinvicem et totum agregatum ostendit, quantum movetur talis planeta uno die, et sic correspondebit uni diei intermedie. Octavo: hunc motum uni diei correspondentem adde ad motum priorem et habebis verum motum planete secunde diei. Et post adde hoc idem motui secunde diei et habebis verum motum ad terciam diem et sic consequenter usque ad diem, ubi verum motum habuisti quesitum, et sic almanach verorum motuum in omnibus planetis perficies. **Cautela prima:** Diligenter fines mensium, quot quilibet dies finales habeat, attende, quia secundum hoc veros motus distans inquire et secundum hoc eciam differencia erit querenda et post in *Tabulam minutorum proporcionalium* sub tot diebus erit intrandum. **Cautela secunda:** Diligenter advertendum est, si planeta directus aut retrogradus fuerit, quoniam si directus est, ad motus veros habendos id, quod uni diei totum respondet, addendum est. Si vero est retrogradus, hoc idem erit subtrahendum. **Cautela tercia:** In tribus superioribus dum retrogradacioni aut direccioni apropinquant et in tota retrogradacione de quinque in quinque agere caucius esset. Alias autem cum directi sunt, de decem in decem operare cercius existit. In Luna eciam si quis precisionem magnam non attenderet, de tercia in terciam operacio fieri posset atque almanach verorum motuum facile perficies. **Verbi gracia:** Volumus continuare almanach verorum motuum ad annum Domini 1487 currentem. Primo habeamus verum motum Solis ad principium anni huius, quod motus vocatur radix et est 9, 19, 23, 47, 19. Secundo: debeo habere verum motum Solis ad quintam diem sequentem et est iste 9, 24, 30, 2, 29. Tercio: debeo subtrahere motum precedentem verum a motu vero ad quintam diem et manebit differencia ista: 5 gradus, 6 minuta, 15 secunda, 10 tercia. Quarto: cum ista differencia debeo intrare in *Tabulam minutorum proporcionalium* sub quinque et capere proporcionem et introytu completo veniet 1 gradus, 1 minuta, |**f. 182r:**| 15 secunda, 2 tercia. Quod si addemus ad verum motum Solis ultima die Decembris, qui radix est, veniet ad primam Ianuarii verus motus iste: 9, 20, 25, 2, 21. Quinto: si addideris id, quod in tabula invenisti, ad motum prime diei Ianuarii, veniet verus motus Solis ad secundam diem iste: 9, 21, 26, 17, 23. Et sic continuabis veros motus Solis per totum annum.

<27>

**Canon 27 de vero motu capitis draconis levissime in almanach reponendo**

Cum faciliter verum motum capitis draconis continuare volueris, tali ordine procede: a fine anni usque ad eius principium primo ad finem anni tui invenies medium motum capitis draconis. Secundo: ipsum de 12 signis subtrahe, et quod post subtraccionem venerit, verus locus capitis draconis erit ad finem anni tui currentis, hoc est ad ultimam diem Decembris. Tercio: illi vero motui capitis draconis, reperti ad ultimam diem Decembris, si addideris motum capitis uni diei, exibit ad penultimam Decembris et sic continuabis continue per dies et menses verso ordine. **Exempli gracia:** Volo continuare verum motum capitis draconis ad annum Domini 1487 currentem. Primo debeo scire motum capitis draconis ad finem eiusdem anni, scilicet ad ultimam diem Decembris, et est iste: 1, 22, 45, 54, 23. Secundo: eundem motum debeo subtrahere a 12 signis, et post subtraccionem manebit verus motus capitis draconis ad finem anni ad ultimam diem Decembris et veniet iste: 10, 7, 14, 5, 37. Tercio: eidem motui capitis draconis ad ultimam diem Decembris reperto debeo addere motum capitis draconis unius diei, scilicet 3, 10, 38, et veniet verus motus capitis draconis ad penultimam Decembris 10, 7, 17, 16, 15. Cui motui si addo motum capitis draconis unius diei, veniet verus motus capitis ad diem precedentem iste: 10, 7, 20, 26, 53. Et sic continuabis motum capitis draconis per dies et menses verso ordine per totum annum.

<28>

**Canon 28 de coniunccionibus et opposicionibus mediis per totum annum calculando**

Cum primam coniunccionem anni invenire volueris, primam dicam, que in diebus usualibus Ianuarii contingit, et hec 'radix coniunccionum' dicitur, primo cum annis Cristi currentibus in *Tabulis coniunccionum annorum collectorum* ingredere, quos si precise inveneris, statim ex opposito illorum annorum collectorum coniunccionem primam illius anni in diebus Ianuarii, que Februarii est, invenies |**f. 182v:**| in diebus, horis, minutis, secundis et terciis. Et ex opposito illius coniunccionis invenies motum Solis et Lune, motum argumenti Lune et motum argumenti latitudinis Lune. Secundo: si annum tuum currentem precise non invenisti, tunc cum anno propinquiori, quem in *Tabulis coniunccionum* invenies, intrabis et ex opposito illius tempus coniunccionis accipe. Similiter medium motum Solis et Lune, medium argumentum Lune et medium argumentum latitudinis Lune. Tercio: cum annis residuis a tuis annis currentibus distantibus intrabis in *Tabulas annorum expansorum*, et quod ex opposito annorum invenies in diebus, horis et minutis, similiter in motu Solis et Lune, argumenti Lune et argumenti latitudinis Lune quodlibet ad suum genus adde, quemadmodum dictum fuit in capitulo primo. Quarto: de diebus provenientibus ex tali addicione 31 dies subtrahe et habebis primam coniunccionem anni in diebus Ianuarii. Que radix dicitur coniunccionum cum suis motibus ad idem tempus. Quinto: si a diebus provenientibus ex tali addicione non poteris subtrahere 31 dies, iterum adde ipsius unius lunacionis, scilicet 29 dies, 12 horas, 44 minuta, et motus ex opposito unius lunacionis repertus eciam adde motibus prius inventis, quodlibet ad suum genus. Sexto: a toto collecto subtrahe 31 dies, et quod manebit in diebus, horis et minutis, erit tempus prime coniunccionis in diebus Ianuarii cum suis motibus. Ista itaque coniunccio erit radix omnium coniunccionum eiusdem anni. **Verbi gracia:** Volumus invenire primam coniunccionem ad annum Domini 1487 currentem. Intrabimus primo cum eodem anno currente in *Tabulas coniunccionum ad annos collectos*, et si annum currentem precise non invenimus, intrabis cum propinquiori, scilicet 1469, et opposito eius invenimus duodecim in diebus, 18 in horis, 13 in minutis, 31 secundis, 21 terciis, 38 quartis, in medio autem motu Solis et Lune invenio 10 signa, 1 gradus, 39, 9, 11. In argumento Lune 1 signum, 11 gradus, 53, 6, 19. In argumento latitudinis Lune 11, 17, 38, 37, 19. Tercio: cum annis residuis, scilicet 18, intra *Tabulas annorum expansorum* et ex opposito eiusdem anni 18 accipio 12 in diebus, 18 horas, 59 minuta, 17 secunda, 1 tercia, 46 quarta. In motu Solis invenio 11, 11, 41, 31, 2. In argumento Lune accipio 11, 1, 20, 9, 20. In argumento latitudinis invenio 10, 28, 50, 56, 33. Que omnia debeo addere simul, quodlibet ad suum genus, et addicione facta in tempore eveniet 25 diebus, 13 in horis, 12 minuta, 48 secunda, 23 tercia, 25 quarta. In motu Solis 9, 13, 20, 40, 15. In argumento Lune 0 signa, 13 gradus, 13, 15, 39. In argumento latitudinis Lune 10, 16, 29, 33, 52. Quarto: quia a diebus provenientibus ex tali addicione non possum subtrahe 31 dies, ergo debet me addere tempus unius lunacionis cum motibus ex opposito positis in tempore 29 diebus, 12 horas, |**f. 183r:**| 44 minuta, 3 secunda, 2 tercia, 59 quarta. In motu Solis 0, 29, 6, 24, 12. In argumento Lune 0, 25, 49, 0, 35. In argumento latitudinis Lune 1, 0, 40, 13, 46. Postquam addo quodlibet ad suum genus, in tempore venient 24 dies, 1 hora, 56 minuta, 51 secunda, 26 tercia, 24 quarta. In motu Solis 10, 12, 27, 4, 27. In argumento Lune 1, 9, 2, 16, 14. In argumento latitudinis Lune 11, 17, 9, 47, 38. Hec itaque coniunccio cum suis motibus erit radix coniunccionum omnium ad annum Domini 1487 currentem. **Cautela prima:** Cum in praxi coniunccionum aut opposicionum plures hore quam 24 excereverint, tunc 24 horas subtrahe et residuum erit tempus coniunccionis aut opposicionis in horis et pro his 24 horis subtractis unum adde ad dies. **Cautela secunda:** Coniunccio in aliquo mense inventa non est illius mensis, in quo invenitur, sed sequentis, ut coniunccio Februarii invenitur in diebus Ianuarii.

<29>

**<C>anon 29. Quomodo alie coniuncciones Solis et Lune medie per anni circulum sunt inveniende**

Cum coniuncciones alias medias Solis et Lune in aliquo anno invenire volueris, primo dies coniunccionis prime, horas, minuta, secunda et tercia, quas pro radice habes, in tabulam scribe. Secundo: intra *Tabulam medie coniunccionis Solis et Lune* et accipe dies, horas, minuta, secunda et tercia. Similiter motus ex directo illius mensis, in cuius dies coniunccionum queris. Et hec omnia adde ad primam coniunccionem, videlicet tempus tempori, hoc est dies ad dies, horas ad horas etc. Tercio: si ex tali coniunccione plures dies venerunt quam mensis iste continet, in cuius dies coniunccionem queris, tunc ab ipsis diebus lunaribus sic provenientibus subtrahe dies solares, si potes, scilicet dies Solis positos in directo mensis, in quo coniunccionem queris, et dies remanentes computa ad principium cuiusdem mensis et horas a meridie. Quarto: si dies solares indirecti illius mensis, de quo operaris, subtrahere non possunt, quod contingit, quando aliquid mensis duas habet coniuncciones, tunc subtrahe dies mensis precedentis et dies remanentes eiusdem mensis computa. Similiter motum Solis et Lune, motum argumenti Lune et argumentum latitudinis Lune et ceteras coniuncciones per totum annum continuabis. **Cautela prima:** In anno bisextili te cautum esse oportet, quia si annus bisextilis fuerit et bisextus pertransivit, tunc ad dies solares positos super unum diem adde et dies in tabula mensium repertos et post talem addicionem subtrahe dies solares a diebus lunaribus et tempus coniunccionis habebis. Lunares dies sunt, que veniunt per praxim, si isti, que sunt ex opposito mensium, sunt solares. |**f. 183v:**| **Cautela secunda:** Si in diebus usualibus Ianuarii due coniuncciones evenirent, prima embolismalis erit et secunda Februarii dicenda. Embolismus tamen verus per coniuncciones veras et non medias indicandus erit. **Cautela tercia:** Cum in diebus usualibus cuiuscumque mensis due coniuncciones eveniunt, una circa principium et alia circa finem, prima embolismalis erit, secunda erit mensis sequentis. **Exempli gracia:** Ad annum Domini 1487 volo scire coniunccionem Marcii, que erit in diebus Februarii. Primo debeo habere coniunc-

---

**1070** $0^{1}$ *Had* : 1 *AB*

cionem primam eiusdem anni, que radix est cum suis motibus. Radix autem coniunccionis est 24 in diebus, 1 gradus, 56, 51, 26, 24. Secundo: intrabo *Tabulam medie coniunccionis Solis et Lune* ex directo illius mensis, in cuius diebus coniunccionis dies quero, et accipio ex opposito Februarii 29 in diebus, 12 in horis, 44 minuta, 3 secunda, 2 tercia, 59 quarta. In motu Solis et Lune accipio 0 signa, 29 gradus, 6, 24, 12. In motu argumenti Lune 0, 25, 49, 0, 35. In argumento latitudinis Lune 1, 0, 40, 13, 46. Et postquam addo quodlibet ad suum genus, veniet coniunccio Marcii hec cum suis motibus. Prius tamen debeo subtrahere ab illis diebus lunaribus dies solares in directo eiusdem mensis positos, in quo coniunccionem quero, hoc est: 31, 22 dies, 14 horas, 40 minuta, 54, 29, 23. Medius motus Solis et Lune ad idem tempus 11, 11, 33, 28, 39, medium argumentum Lune 2, 4, 51, 16, 49, medium argumentum latitudinis Lune 0, 17, 50, 1, 24.

<30>

**<C>anon 30 de mediis coniunccionibus et opposicionibus Solis et Lune continuandis**

Cum coniunccionem Solis et Lune primam in anno habueris, que radix est, et eius opposicionem statim habere volueris ad tempus coniunccionis inventum, adde tempus, quod est a coniunccione ad opposicionem, scilicet 14 dies, 18 horas, 22 minuta. Similiter in motibus et habebis opposicionem mensis, cuius coniunccionem habes cum suis motibus. Secundo: habita opposicione si vis habere coniunccionem mensis inmediate sequentis, hoc idem tempus, scilicet 14 dies, 18 horas, adde ad tempus opposicionis inventum et similiter motus ad motus et habebis coniunccionem inmediate sequentem cum motu Solis et Lune, medio argumento Lune, et argumento latitudinis Lune. **Cautela consideranda:** In tali continuacione cum continuaciones has opposicionum et coniunccionum examinare volueris ad tempus coniunccionum, quod ex canone precedenti invenisti, motum Solis inquire, quod si concordaverit cum motu Solis per continuacionem invento, certus sis et continuacionem esse regularem |**f. 184r:**| et te recte iuxta canonem egisse intelliges.

<31>

**<C>anon 31 de veris coniunccionibus et opposicionibus Solis et Lune inveniendis**

Cum tempus vere coniunccionis aut opposicionis scire volueris, primo tempus medie coniunccionis aut opposicionis habe iuxta canonem predictum. Similiter motum Solis et Lune, argumentum Lune et argumentum latitudinis Lune. Secundo: ad hoc tempus coniunccionis medie vel opposicionis verum motum Solis inquire. Tercio: quere verum motum Lune cum vero argumento Lune, cum quo queratur equacio argumenti, que addatur vel subtrahatur a motu Lune. Non oportet te hic quere partem proporcionalem, sed dumtaxat per argumentum Lune est operandum, quod ad partem tene. Quarto: cum verum motum Solis et Lune habueris, subtrahe minorem a maiori quoad gradus et alias fracciones et non quoad signa. Et considera, cuius est differencia sive quis illorum sit maior in motu, et si fuerit longitudo Solis, scribatur circa longitudinem signum Solis. Si autem longitudo est Lune, scribatur circa eam signum Lune. Quinto: si longitudo est Solis, coniunccio vera aduc est ventura post mediam coniunccionem. Si vero longitudo

---

**1111** 59 *B* : 54 *A* **1113** 40 *B* : 4 *A*

sit Lune, iam coniunccio vera precessit mediam. In opposicione autem, si Luna aduc non venit ad nadir Solis, aduc opposicio est futura. Si autem preterit, iam opposicio est preterita. Sexto: longitudinem hanc sive Solis, sive Lune reduc ad secunda multiplicando per 60. Septimo: postquam longitudo reducta est ad secunda, divide longitudinem sic reductam ad secunda per 12, et si aliquid manserit in residuo ultra medietatem de 12, tunc addatur numero quocienti unitas, qui numerus quociens est 12 longitudinis. Octavo: idem numerus quociens addatur longitudini reducte ad secunda et capiatur eius medietas per mediacionem. Nono: eadem medietas reducatur ad minuta dividendo ea per 60. Decimo: eadem medietas longitudinis cum sua 12 reducta ad minuta addatur argumento Lune tempore medie coniunccionis vel opposicionis, si fuerit longitudo Solis, vel subtrahatur ab eodem argumento, si fuerit longitudo Lune, et habebitur argumentum horarium. Undecimo: cum eodem argumento horario quere motum Lune in una hora in suo duplici introytu, videlicet uno gradu habito ad habendum differenciam. Duodecimo: cum argumento Solis tempore medie opposicionis vel coniunccionis quere eciam motum Solis in una hora per eandem tabulam eciam duplici introytu, |**f. 184v:**| si ultra gradus minuta habueris. Tredecimo: subtrahe motum Solis in una hora a motu Lune in una hora et residuum, quod manebit, vocatur 'superlacio' vel 'superacio' secundum, quantum Luna plus movetur Sole in una hora. Quartodecimo: hanc superlacionem reduc ad eandem denominacionem, ad quam fuit reducta longitudo simplex, scilicet ad secunda. Quindecimo: divide longitudinem simplicem per superacionem et in numero quociente veniunt hore. Sedecimo: residuum multiplica per 60 et divide per superacionem et in numero quociente veniunt minuta. Decimo septimo: residuum multiplica per 60 et divide per superacionem et venient secunda. Decimo octavo: si residuum extendit se ultra medietatem superlacionis, tunc addatur unum secundum secundum inventis et habebis tempus extractum. Decimo nono et ultimo: illud tempus extractum adde tempori medie coniunccionis vel opposicionis, si fuerit longitudo Lune, et habebis tempus vere coniunccionis Solis cum Luna vel vere opposicionis Solis cum Luna. **Exempli gracia:** Volo scire ad annum Domini 1487 currentem primam coniunccionem anni, que est coniunccio Februarii in diebus Ianuarii. Primo debeo scire tempus medie coniunccionis et est hoc: 24 in diebus, 1 in horis, 56 minuta, 51, 26, 24. Medius motus ad idem tempus est 10, 12, 27, 4, 27. Medium argumentum Lune est 1, 9, 2, 16, 14. Medium argumentum latitudinis Lune est 11, 17, 9, 47, 38. Secundo: ad hoc tempus medie coniunccionis debeo scire verum motum Solis. Aux Solis est 3, 1, 3, 26, 16. Argumentum Solis est 7, 11, 23, 38, 11. Verus motus Solis est 10, 13, 55, 35, 1. Tercio: debeo scire verum motum Lune et est iste: 10, 9, 32, 41, 38. Quarto: cum subtraxero minorem a maiore, manebit longitudo Solis ista: 4 gradus, 22 minuta, 53 secunda, 23 tercia. Quinto: coniunccionem veram scio esse aduc venturam, quia longitudo est Solis. Sexto: longitudinem Solis debeo reducere ad secundam et veniet 15773. Septimo: postquam divido longitudinem reductam ad secunda per 12, veniet duodecima longitudinis ista: 1314. Octavo: cum 12 longitudinis addo longitudini in secundis, veniet longitudo cum sua 12 in secundis ista: 17087. Nono: accipio medietatem longitudinis cum sua duodecima et veniet 8543. Decimo: eandem medietatem reduco ad minuta et venient 16 minuta, 23 secunda. Undecimo: eandem medietatem reductam ad minuta adde agumento Lune tempore medie coniunccionis et veniet argumentum horarium hoc: 1 signum, 9 gradus, 18 minuta, 39, 14. |**f. 185r:**| Duodecimo: cum eodem argumento horario quere motum Lune in una hora in propria tabula duplici introytu. Primus introytus: 30 minuta, 14 secunda. Secundus introytus: 30 minuta, 16 secunda. Differencia: 2 secunda. Motus autem Lune perfectus in una hora est iste: 30 minuta, 14 secunda, 37 tercia. Tredecimo: cum argumento Solis tempore medie coniunccionis quere motum Solis in una hora et venient 2 minuta, 32 secunda. Quarto decimo: cum subtraho motum Solis in una hora a motu Lune in una hora, manebit superacio Lune iste: 27 minuta, 42 secunda, 37 tercia. Quinto decimo: reduco hanc superacionem ad secunda et veniet 1662. Sedecimo: cum divido longitudinem simplicem reductam ad secunda per superacionem eciam reducta<m> ad secundam, in quociente veniunt 9 hore et residuum manebit 815. Decimo septimo: illud residuum multiplico per 60 et iterum divido per superacionem in secundis et in quociente venient 29 minuta, residuum manebit 702. Decimo octavo: si illud residuum multiplicabo per 60 et divido per superacionem in secundis, in quociente veniunt 25 secunda et resiuum manet 570. Erit itaque tempus extractum, per quod media coniunccio distabat a vera 9 hora, 29 minuta, 25 secunda. Ultimo: hoc tempus extractum addo ad coniunccionem mediam et venient vera ista: 24 dies, 11 hora, 26 minuta, 17 secunda.

<32>

**<C>anon 32 ad gradum ascendentem et alias domus celi inveniendas**

Cum gradum ascendentem ad tempus coniunccionis vel opposicionis vel aliquod ipsius aliud invenire volueris, sive hoc sint meridiei, hoc est a meridie computate, vel hore horalogii, primo tempus tuum propositum, ad quod ascendens hore volueris, equa tali modo: intra *Tabulam equacionum dierum cum noctibus* cum vero gradu Solis illo tempore et accipe ex opposito gradus Solis et sub illo signo, in quo est Sol, equacionem dierum cum noctibus, gradus et minuta. Secundo: pro quolibet gradu accipe quatuor minuta hore, hoc est multiplica gradus, quos invenies, per 4, et habebis minuta horarum. Tercio: pro quolibet minuto accipe quatuor secunda hore, multiplica minuta per 4. Quarto: si excreverint plura quam 60, divide ea per 60 et in quociente veniunt minuta, que adde prioribus minutis et residuum erunt secunda hore. Quinto: minuta proveniencia ex equacione et secunda adde tempori tuo proposito. Sexto: quere gradum ascendentem hoc modo: intra tabulas illas cum vero motu Solis sub illud signum, in quo est Sol. Si in tempore proposito habes horas meridiei, easdem accipe. Si autem in tempore tuo proposito habes horas horalogii, eciam easdem accipe. Septimo: quod ex opposito |**f. 185v:**| gradu<s> Solis invenies, sive in horis meridiei, sive horalogii, ad tempus tuum propositum adde. Octavo: id, quod post addicionem provenerit sub horis horalogii, quere, et ubi talem numerum horarum et minutorum invenies aut propinquiorem, ibi ex cuius opposito gradum ascendentem cum aliis domibus quinque invenies, scilicet secunda, tercia, quarta, quinta, sexta. Nono: cum primos sex domos habueris ex opposito earum in simili gradu, alias sex domos computabis. **Exempli gracia:** Volo scire gradum ascendentem ad tempus coniunccionis Ianuarii per canonem precedentem inventam. Primo: tempus propositum, scilicet 11 horas, 26 minuta, 17 secunda, debeo {primo} equare {tempus} intrando ad *Tabulam equacionum dierum cum noctibus* cum vero motu Solis tempore coniunccionis, scilicet 13. Aquarii, et inveniam scilicet 0 gradus, 6 minuta. Secundo: cum multiplicaveris 6 minuta per 4, veniunt 24 secunda. Tercio: illa 24 secunda addam ad secunda in tempore proposito et veniet tempus equatum hoc: 11 hore, 26 minuta, 41 secunda. Quarto: cum vero motu Solis intravero in

**1239** secunda *B* : minuta *A*

*Tabulas ascendencium* {cum vero motu Solis}, scilicet 13. Aquarii, sub horis meridiei inveniam 3 horas, 2 minuta. Quinto: hoc addam ad tempus meum propositum, minuta ad minuta et horas ad horas, et venie<n>t. 14 horas, 28 minuta, 41 secunda. Sexto: hoc totum quero sub horis horalogii, et ubi hoc invenero vel numerum propinquiorem, ex cuius opposito inveniam sex domos celi, scilicet ascendens, secundam, terciam, quartam et quintam, sextam. Septimo: cum intrabo ad *Tabulas ascensionum* cum 14 horis, 28 minutis vel cum propinquioribus, inveniam ascendens. Et si ambo numeri fuerint equales, capiatur sequens 27. Libre cum aliis quinque domibus. Alii autem sex domus ex opposito illorum sex servati. Erit itaque figura celi in 12 domibus ista:[19] **Cautela:** Circa canonem istum habenda quecumque in horis excreverint plures hore quam 24, subtrahende sunt 24 hore.

<33>

**<C>anon 33 de reducendis coniunccionibus et opposicionibus in horis |f. 186r:| meridiei ad horas horalogii**

Cum aliquam coniunccionem aut opposicionem, hoc est tempus coniunccionis vel opposicionis, reducere volueris ad horas horalogii et maxime integri, primo ad tempus tuum propositum tempus meridiei scias. Quod tempus meridiei scias ex tabula infra scripta, que *Tabula meridiei* tempus per singulos dies anni ostendet. Secundo: tempus hora meridiei in tuo die proposita ad tempus tuum inventum ex praxi adde, et quod post addicionem provenit, tempus coniunccionis aut opposicionis vere in horis horalogii integri docebit. Tercio: si hoc probare volueris, ad has horas horalogii integri provenientes, quere ascendentem cum ceteris domibus iuxta canonem predictum. Quarto: si idem ascendentis gradus cum ceteris domibus eveniet, scias te ipsum recte reduxisse. **Cautela:** Cercius est querere ascendens cum horis meridiei non reductis quam cum horis reductis. **In horologio autem medio** facile est hoc congnoscere. Si autem in tempore tuo calculato excessus ultra 12 non fuerit, tunc est eadem hora post meridiem, quia horalogium medium una hora post meridiem unum tangit. Si autem plures hore in tempore tuo proposito quam 12 excreverint, subtrahe 12 et residuum facile computabis. **Verbi gracia:** Volo reducere coniunccionem Februarii, que erit anno Domini 1487 currente 24. die Ianuarii 11 horas, 26 minuta, 17 secunda. Primo debeo scire tempus meridiei ad eandem diem 24. et invenio ex *Tabula horarum* 19, 23 minuta. Secundo: easdem horas meridiei cum addo ad tempus propositum, veniet 6 in horis, 49 minuta, 17 secunda, quod est tempus coniunccionis ad horas horalogii integri. Ad quod tempus cum quero ascendens, veniet 27 Libre cum aliis domibus, quemadmodum dictum est in canone precedenti.

---

**1261** ascendentem *Had* : ascendens *ABC per err.*

---

19 Attached is a simple and unfilled drawing of the horoscope in the manuscript *A*, a filled drawing in *B*, and *C om.* – The drawing in the manuscript *B* brings this text (written mostly in abbreviations and symbols): Oriens, Septentrio, Occidens, Meridies; Coniunccio Februarii in diebus equatis vera 24 d(ies), 11 h(ora), 26 m(inuta), 17 s(ecunda); 27 Libre, 2 Sagittarii, 3 Capricorni, 5 Aquarii, 1 Piscium, 29 Piscium, 27 Arietis, 2 Geminorum, 3 Cancri, 5 Leonis, 1 Virginis, 29 Virginis. In the tenth house there is added: the Sun and Moon in 13 Aquarii.

<34>

### <C>anon 34 de introitu Solis in Arietem et in alia signa celi cardinalia

Cum introytum Solis in Arietem vel in aliud signum cardinale aut quodcumque aliud signum celi desideras, primo verum motum Solis ad tempus propinquum tue estimacioni, in qua Sol non remote distat ab Ariete vel alio signo cardinali, inquire. Secundo: habito vero motu Solis subtrahe ipsum a 12 signis et sic, si vis habere introytum Solis in Arietem. Si vero in Cancrum, tunc |**f. 186v:**| {tunc} subtrahe verum motum Solis a tribus signis. Si in Libra, tunc a 6 signis. Si in Capricornum, tunc a novem signis. Et remanet distancia Solis ab Ariete. Tercio: distanciam manentem reduc ad eandem denominacionem, videlicet secunda vel tercia, precisius tamen est reducere a<d> tercia. Quarto: scias motum verum Solis post unam horam querendo medium Solis et verum post unam horam. Quem sic invenies: quando habes medium vel verum motum Solis, tunc intra ad *Tabulas loricatas* et capias illud, quod stat ex opposito unius hore, et adde vero vel medio motui et sic habebis ad unam horam. Quinto: subtrahe verum motum Solis primo inventum a vero motu Solis secundo inventum et in residuo manebit verus motus Solis in una hora. Sexto: motum Solis in una hora reduc ad eandem denominacionem, ad quam reduxisti distanciam Solis ab Ariete, ut ad secunda vel tercia, precisius tamen est ad tercia. Septimo: divide distanciam Solis ab Ariete in terciis per motum Solis in una hora eciam in terciis et in residuo veniunt hore. Octavo: illud residuum multiplica per 60 et divide per motum Solis in una hora in terciis et in residuo habebis minuta. Nono: illud residuum multiplica per 60 et in numero quociente veniunt secunda. Decimo: illud residuum multiplica per 60 et divide per idem, quod prius, et in quociente veniunt tercia. Undecimo: tempus hoc extractum in horis, minutis, secundis et terciis adde ad tempus tuum estimatum et veniet introytus Solis in Arietem. Duodecimo: ad illud tempus tuum, scilicet introitum Solis in Arietem, quere verum motum Solis et veniet tibi 0 signum, 0 gradus, 0 minuta, 0 secunda, 0 tercia. Introitus Solis recte est calculandus. Tredecimo: ad hoc tempus figuram celi erige et duodecim domos celi, ut dictum est in canone 32., et planetas ad figuram talem, similiter loca et ordina et habebis figuram celi et constellacionem ad principium et revolucionem anni, que est, cum Sol intrat primum punctum Arietis. **Exempli gracia:** Ad annum Domini 1487 currentem volo scire introytum Solis in Arietem. Primo debeo scire tempus estimatum, ergo Sol non remote distat ab Ariete, et accipio hoc: 11 in diebus, 0 horas, 59 minuta, 31 secunda, 25 tercia. Secundo: debeo scire verum motum Solis ad illud tempus estimatum. Medius motus Solis est 11, 27, 45, 6, 13. Aux Solis 3, 1, 3, 30, 38. Argumentum Solis 8, 26, 41, 31, 35. Equacio Solis 2 gradus, 10 minuta, 0 secunda. Verus motus Solis ad illius tempus estimatum 11, 29, 55, 6, 13. Tercio: debeo scire distanciam Solis ab Ariete et est hec: 4 minuta, 53 secunda, 47 tercia. Quarto: eandem distanciam cum reduco ad tercia, venient 17627. Quinto: cum quero motum Solis post unam horam verum, medius motus |**f. 187r:**| Solis erit 11, 27, 47, 34, 4. Verus motus Solis post unam horam: 11, 29, 57, 34, 4. Sexto: cum subtraxero verum motum Solis, minorem a maiore, veniet motus Solis in una hora 2 minuta, 27 secunda, 51 tercia. Septimo: cum reduco motum Solis in una hora ad tercia, veniet 8871. Octavo: divide distanciam Solis ab Ariete in tercia per motum Solis in una hora eciam in tercia, in quociente veniet 1 hora. Et residuum manebit 8756. Nono: illud residuum cum multiplico per 60 et divido per motum Solis in una hora in terciis, in quociente veniunt 59 minuta. Residuum 1971. Decimo: cum illud residuum multiplico per 60 et divido per idem, quod prius, in quociente venient 13 secunda, residuum 2937. Undecimo: cum illud residuum multiplico per 60 et divido per idem, quod prius, in quociente veniunt 19 tercia et residuum manebit 7671. Pro quo residuo adde unum ad tercia et veniunt 20 tercia. Duodecimo: illud tempus extractum, scilicet 1 hora, 59 minuta, 13 secunda, 20 tercia, addo ad tempus estimatum et veniet introitus Solis in Arietem 11. die Marcii, hora secunda, minuta 58, 44 secunda, 45 tercia. Tredecimo: quero verum motum Solis ad illud tempus. Motus Solis erit 11, 27, 50, 0, 0. Cum medio motui debeo adde<re> equacionem prius inventam, scilicet 2 gradus, 10 minuta, et veniet verus motus Solis ad introitum Solis in Arietem 0, 0, 0, 0, 0 tercia. Quarto decimo: quero ascendens cum aliis 12 domibus celi et exibit configuracio celi ista.[20] **Cautela:** Cum tempus estimatum ad introitum Solis in Arietem vel in aliud signum cardinale scire volueris, primo radicem introitus Solis in Arietem vel aliud signum habe, quam *(sic)* radices habes ad annum Domini 1468, 10 in diebus, 12 in horis, 22 minuta, 41 secunda, 49 tercia. Secundo: intra tabulam infrascriptam ab uno anno incipientem et 800 terminantem et intra cum distancia a radice anni usque in annum tuum currentem, ut pro anno Domini 1487 currente intrabis cum 19 annis et ex opposito accipe horas, minuta, secunda et tercia et adde ad radicem. Et si plures hore excreverint |**f. 187v:**| quam 24, subtrahe 24 et adde unum ad dies. In anno tamen bisextili in simili hora diei precedentis introitum computabis.

<35>

**<C>anon 35 de coniunccione aliorum quinque planetarum**

Cum coniunccionem aliquorum planetarum scire desideras, primo amborum vera loca ad tempus propositum coniunccioni secundum tuam existimacionem inquire. Secundo: subtrahe minorem a maiori et manet longitudo, quam longitudinem signabis. Tercio: eandem longitudinem reduc ad eandem denominacionem, ut ad secunda. Quarto: quere motum utriusque post unum diem. Quinto: subtrahe motum cuiuslibet primo inventum a motu secundo invento, si ambo fuerint directi. Vel econverso, motum secundo inventum a motu primo invento <subtrahe>, si fuerint ambo retrogradi, et quod remanserit, erit motus diei cuiuslibet. Sexto: subtrahe motum diei tardioris a motu diei velocioris, ut si motus in Iove (?) et Marte, subtrahe motum Iovis in una die a motu Martis in una die, et quod remanet, est superlacio. Septimo: eandem superlacionem reduc ad eandem denominacionem, ad quam longitudo fuit reducta. Octavo: divide longitudinem per superlacionem, si potes, et si non potes, tunc scribe 0 dies et in quociente veniunt dies. Nono: si aliquid manserit post divisionem, servetur in propria forma non reducendo. Superlacio vero diei dividatur per 24 et habebis superlacionem hore. Decimo: per superlacionem hore divide residuum hore, quod prius servasti in propria forma, et in numero quociente veniunt hore. Si autem dividi non possunt, tunc loco horarum ponatur ciphra et designat nullam horam. Undecimo: residuum multiplica per 60 et dividatur per superlacionem hore et in quociente veniunt minuta hore. Duodecimo: residuum multiplica per 60 et divide per superlacionem hore et in quociente veniunt secunda. Tredecimo: hoc tempus extractum adde tempori estimato et habebis coniunccionem planetarum veram, si ambo planete fuerint directi vel retrogra-

20 Attached is a simple, incomplete drawing of the horoscope in manuscript *A*, a completed drawing in *B*, and *C om.* – The drawing in the manuscript *B* brings this text (written mostly in abbreviations and symbols): 1487 Sol in Arietem 11 d(ies), 21 h(ora), 58 m(inuta), 44 s(ecunda), 45 t(ercia); 29 Leonis, 26 Virginis, 23 Libre, 19 Scorpionis, 23 Sagittarii, 25 Capricorni, 29 Aquarii, 26 Piscium, 23 Arietis, 19 Tauri (ms. per err.: 29), 23 Geminorum, 25 Cancri.

di. Quarto decimo: motum amborum quere ad idem tempus et concordabunt in signis, gradibus, minutis, secundis et terciis. **Cautela:** Si autem unus planeta, cuius coniunccio queris, fuerit directus et alter retrogradus, tunc quere motum directi in una die, scilicet post primam diem, quemadmodum dictum est. Similiter quere motum retrogradi in una die, scilicet post primam diem. Tunc secundus motus erit minor priori, id est motus retrogradi secunda die est minor quam eius motus |**f. 188r:**| prima die. Cum enim planeta est retrogradus, semper eius motus minuitur. Subtrahe igitur motum secundum ipsius planete retrogradi a motu planete diei precedentis et manebit, quantum retrocedit una die. Deinde iunge illos motus simul, scilicet motum directi et retrogradi, et quod provenerit, per illud divide longitudinem. Et prius reduc ad secunda, quemadmodum longitudinem reduxisti, et tempus cum suis fraccionibus extrahe, sicut dictum est ante, quod tempus extractum adde ad tempus estimatum, si levior aduc non precessit tardiorem, vel subtrahatur, si levior precessit tardiorem. **Exempli gracia:** Volo scire ad annum Domini 1486 currentem coniunccionem Saturni et Mercurii, que erit 13. die Decembris. Primo debeo scire veros motus amborum ad hoc tempus estimatum. Motus Saturni est 8, 15, 38, 29, 38. Motus autem Mercurii verus est iste: 8, 15, 22, 55, 33. Longitudo autem est 15 minuta, 34 secunda, 5 tercia. Que longitudo cum reducta fuerit ad secunda, erit 934. Motus autem Saturni post unum diem est iste: 8, 15, 45, 8, 35. Motus autem Mercurii post unum diem est iste: 8, 16, 33, 24, 1. Cum itaque subtraho motum Saturni prime diei a motu Saturni secunde die, manebit motus Saturni in una die 6 minuta, 38 secunda, 57 tercia. Cum eciam subtraho motum Mercurii prima die inventum a motu Mercurii secunda die inventum, veniet motus Mercurii in uno die 1 gradus, 10 minuta, 28 secunda, 28 tercia. Deinde cum subtraho motum Saturni in uno die a motu Mercurii in uno die, manet superlacio ista, scilicet quantum Mercurii motus plus Saturni uno die: 1 gradus, 3 minuta, 49 secunda, 31 tercia. Quam superlacionem reduco ad secunda, multiplicando per 60 et veniet 3829. Deinde longitudinem debeo dividere per superlacionem, sed non possum, est ergo signum, quod nihil erit in diebus addendum. Deinde superlacionem diei divido per 24 et in numero quociente veniet superlacio hore 159. Et residuum 13. Deinde per superlacionem hore divide illud, quod antea in propria forma servavi, et in quociente veniunt 5 hore. Residuum 139. Deinde hoc residuum multiplico per 60 et divido per superlacionem hore et in numero quociente veniunt 52 minuta. Residuum 72. Deinde illud residuum multiplico per 60 et divido per superlacionem hore et in numero quociente venient 27 secunda. Residuum 27. Erit itaque extractum tempus ad tempus estimatum coniunccionis addendum 0 in diebus, 5 hore, 52 minuta, 27 secunda. Erit itaque coniunccio vera |**f. 188v:**| Saturni et Mercurii 13. die Decembris, hora quinta post meridiem, 52 minuta, 27 secunda etc.

<36>

### <C>anon 36 de reduccione radicum ad quemcumque alium meridianum

Experiencia docet civitates notabili distancia a se inficem *(sic)* esse constructas. Quedam enim sunt propinquiores orienti, quedam vero propinquires occidenti. Orientales itaque prius habent ortum Solis, meridiem et occasum quam occidentales. Necesse itaque est modum ostendere reducendo tabulas pro certo meridiano factas ad alium meridianum. Cum itaque hoc facere volueris, scilicet radices unius civitatis reducere ad radices alterius civitatis, primo scias longitudines illarum civitatum, quas poteris scire ex *Tabula longitudinum civitatum*, que communis est. Est autem longitudo civitatis distancia zenith civitatis ab oriente vel occidente. In civitatibus enim orientalibus longitudo computatur vero ab oriente, vel a Gadibus Alleyri, in civitatibus vero occidentalibus, sicut apud nos, ab occidente, sive a Gadibus Herculis, longitudo civitatis computatur. Secundo: habitis longitudinibus duarum subtrahe minorem a maiori et manebit differencia longitudinum, ita scilicet una civitas plus distat ab oriente vel occidente quam alia. Tercio: eandem distanciam vel differenciam reduc ad fracciones tali modo. Si fuerint in differencia plures gradus quam 15, tunc semper pro 15 gradibus accipe unam horam eo, quod in una hora 15 gradus equinocciales oriuntur. Et pro quolibet gradu residuo infra 15 accipe minuta 4 hore. Hoc est gradus infra 15 multiplica per 4 et habebis minuta horarum. Quarto: si fuerint minuta graduum, in tali distancia pro quolibet minuto graduum accipe 4 secunda hore, multiplicando illa per 4. Quinto: si in minutis vel secundis plura excreverint quam 60, per 60 divide et in quociente excrescit fraccio grossior precedens. Si autem non est excessus in minutis vel secundis ultra 60, in forma propria dimitte. Sexto: cum fraccionibus tali distancie correspondentibus queratur medius motus planete, huius, cuius radicem verificare volueris. Si habes horas, intra cum horis, si non, similiter intra |**f. 189r:**| cum minutis et secundis. Septimo: motum inventum tali distancie correspondentem subtrahe a radice illius planete, in cuius tabula eum invenisti, si civitas tua fu<er>it orientalior, vel adde, si fuerit occidentalior. **Verbi gracia:** Si vis reducere Pragenses radices ad meridianum Cracoviensem, scias primo longitudines ambarum civitatum. Longitudo civitatis Cracoviensis est 32 gradum, 36 minuta, longitudo civitatis Pragensis est 29 gradus, 30 minuta. Secundo: subtrahe minorem a maiori et manebit differencia unius civitatis ab alia, hoc est, quod gradus equinocciales intercipiuntur inter Cracoviam et Pragam, et manebit in differencia 3 gradus, 6 minuta. Tercio: quia isti gradus non excedunt 15, tunc nulla hora erit distancia inter meridianum Cracoviensem et Pragensem. Quarto: gradus 3 multiplicabo per 4 et veniunt 12 minuta hore. Multiplicabo eciam minuta graduum per 4 et veniunt 24 secunda hore. Pertinguitur distancia inter Cracoviam et Pragam in tempore 0 hora, 12 minuta, 24 secunda. Licet alii resolucius loquentes dicunt, quod sit nulla hora, 12 minuta, 23 secunda, 59 tercia, 35 quarta. Quod si addas ad radicem Pragensem in coniunccionibus, exibit radix Cracoviensis in eisdem. Unde radix Pragensis est iste: 5 in diebus, 2 hore, 35 minuta, 57 secunda, 46 tercia, 30 quarta. Addita itaque predicta distancia, scilicet 12 minuta, 23 secunda, cum aliis fraccionibus, exibit radix Cracoviensis in mediis coniunccionibus ista: 5 dies, 2 hore, 48 minuta, 21 secunda, 46 tercia, 5 quarta. Pro radice autem habenda Cracovie in motibus annorum collectorum intra cum 12 minuta et 24 secunda et non cum aliis fraccionibus in tabulam planete, cuius motum vis reducere, ut in Solis tabula invenies: 0 minuta, 30 secunda, 33 tercia. Et illa subtrahe a radice Pragensi, que est 9 signa, 18 gradus, 47 minuta, 27 secunda, 18 tercia, et veniet radix Cracoviensis ista: 9 signa, 18 gradus, 46 minuta, 56 secunda, 45 tercia. **Est eciam alter modus querendi radices Cracovienses per distanciam a Praga.** Considera, quot sunt miliaria inter Cracoviam et Pragam modo precisiori, quo potes. Et quia communiter dicitur, quod sunt 60

---

**1422** *A, f. 189r, in mg. add.*: Longitudo Cracovie 32 gradus, 36 minuta, longitudo Heydelberge 27 gradus, 40 minuta. Distancia 4 gradus, 56 minuta, que cum multiplicatur per 4, veniet 0 hora, 19 minuta, 44 secunda. Distancia inter Cracoviam et Heydelbergam in horis et minutis subtrahenda a radicibus Cracoviensibus venit ad Heidelbergam, et hoc in gradus et minutis. In motibus autem planetarum adde ad Cracovienses radices ipsius, quod correspondet in motibus, 0 horas, 19 minuta, 44 secunda. Alii ponunt longitudinem Heidelberge 25 gradus, 38 minuta. Differencia inter Cracoviam et Heidelbergam sunt 6 gradus, 58 minuta, que faciunt 0 horas, 27 minuta, 52 secunda.

miliaria, tum quasi vie sepius obliquantur propter valles et colles, recipio ergo 50 miliaria |**f. 189v:**| communia. Si ergo esset transitus per directum a Cracovia ad Pragam, tot essent miliaria. Pro quibuslibet ergo miliaribus 16 capias 1 gradum et habebis 3 gradus in quociente dividendo per 16 et in residuo manebunt 2 miliaria. Et quia cuilibet miliari correspondent 3 minuta graduum et 45 secunda, tunc capio pro quolibet miliari 3 minuta et habebis 6 minuta. De secundis non est curandum. Vel si vis operacionem precisiorem facere, accipe 7 in minutis, sed non est necessarie, quia ergo in gradibus non est excessus ultra 15, nec in minuta ultra 16. Illud utrumque multiplica per 4 iuxta predicta et habebis 12 minuta hore et 24 secunda hore, quemadmodum autem invenisti.

<37>

### Canon 37 de latitudine Lune invenienda

Cum latitudinem Lune invenire volueris, primo argumentum latitudinis Lune inquire. Hoc autem argumentum latitudinis Lune duplici scire poteris. Si enim subtraxeris verum motum capitis draconis a vero motu Lune, residuum erit argumentum latitudinis Lune equatum, aut si addideris verum motum Lune cum motu capitis draconis, eciam verum argumentum latitudinis Lune habebis. Secundo: cum argumento latitudinis Lune *Tabulam latitudinis Lune* duplici introytu ingredere et latitudinem Lune in directo illius signi positam accipe. Tercio: subtrahe minorem a maiori propter differenciam habendam. Quarto: differenciam duc in minuta, secunda et tercia in argumento latitudinis Lune et quere partem proporcionalem. Quinto: eandem partem proporcionalem adde vel subtrahe a primo introytu et latitudinem Lune habebis. Hec itaque latitudo septemtrionalis erit et ascendens, si argumentum fuerit minus tribus signis. Si autem fuerit a tribus signis usque ad 6 signa, erit septemtrionalis descendens. Si autem fuerit a 6 signis usque ad 9, erit meridionalis descendens. Si autem fuerit a 9 signis usque ad 12 signa, erit meridionalis ascendens. **Verbi gracia:** Volo scire latitudinem Lune ad diem Assumpcionis Marie, ad annum Domini 1486 currentem. Primo scias argumentum latitudinis Lune et est hoc: 11, 28, 6, 49, 32. Secundo: cum isto argumento *Tabulam latitudinis Lune* ingredere duplici introytu propter precisionem. Primus introitus: 0 gradus, 10 minuta, 27 secunda. Secundus introytus est 0 gradus, 5 minuta, 33 secunda. Tercio: subtrahe unum ab alio et manebit |**f. 190r:**| differencia 4 minuta, 54 secunda. Quarto: illam differenciam duc in residuum argumenti latitudinis Lune, querendo partem proporcionalem, et venient 33 secunda, 26 tercia. Quam partem proporcionalem cum subtraho a primo introytu, manebit latitudo Lune 0 gradus, 9 minuta, 53 secunda, 34 tercia. Et patet, quod latitudo Lune est meridionalis ascendens, quia est a 9 signis usque ad 12 etc.

<38>

### <C>anon 38 de latitudine trium superiorum ut Saturni, Iovis et Martis invenienda

Cum latitudinem trium superiorum invenire volueris, cuiuslibet centrum verum primo inquire. Secundo: centrum verum, cuius latitudinem queris, eius signa in gradus reduc et gradus tuos superfluos, quos habes, eciam adde. Tercio: facta reduccione, si latitud<inem> Saturni habere volueris, centro eius reducto ad gradus, 50 gradus adde. Si Iovis latitudinem queris, 20 gradus de centro reducto subtrahe. Si vero Martis latitudinem queris, centrum reductum in propria forma serva. Quarto: cum centro reducto in *Tabulam latitudinis quinque planetarum* ingredere, cuius si centrum reductum in propria forma inveneris ex opposito eius in ultima columna sub titulo minutorum proporcionalium, minuta proporcionalia accipe duplici introytu propter differenciam habendam. Quinto: si gradus vel centrum reductum in propria forma non invenies, cum gradibus minoribus propinquioribus intra et considera, quod gradus manent, cum quibus non est intratum, et similiter in ultima columna sub titulo 'minuta proporcionalia' duplici introytu ingredere utrumque tamen introytum considera, qui maior vel minor fuerit. Sexto: minuta proporcionalia minora subtrahe a maioribus et differenciam habebis. Septimo: differenciam reduc in residuum graduum, minutorum, secundorum et terciorum, si habueris querendo partem proporcionalem. Octavo: partem proporcionalem divide per sex et habebo partem proporcionalem ad sex gradus. Nono: partem proporcionalem adde ad minuta proporcionalia prius inventa vel subtrahe iuxta communem modum. Si enim primus introytus fuerit maior secundo, partem proporcionalem subtrahe. Si primus minor, adde et habebis minuta proporcionalia equata. Signa circa eas 'superior' vel 'inferior' iuxta titulum in tabula inventum. Decimo: cum vero argumento reducto in gradus et aliis |**f. 190v:**| {aliis} gradibus additis intra in *Tabulam latitudinum* vel cum minori propinquiori. Et si circa minuta signatum est 'superior', accipe latitudinem planete sub titulo 'effregion septentrionalis'. Si vero circa minuta est signatum 'inferior', accipe latitudinem sub titulo meridionali et intra duplici introytu et considera, quis introytus sit maior. Undecimo: cum intrasti duplici introytu, accipe differenciam et illam differenciam duc in residuum graduum, si manserit aliquid in minuta, secunda et tercia, si ea habes, et quere partem proporcionalem. Duodecimo: partem proporcionalem inventam divide per 6 gradus et habebis partem proporcionalem ad 6 gradus. Tredecimo: hanc partem proporcionalem adde vel subtrahe a primo introytu, secundum quod titulus designat, et exibunt gradus et minuta equata. Quarto decimo: id, quod exibit ultimo, multiplica per minuta prius inventa, que quesita fuerint cum gradibus veri centri, et exibit latitudo planete per te inquirenda. **Cautela prima:** Cum minutis proporcionalibus 50 gradus addideris ad gradus centri reductos, si plures excreverint quam 360 gradus, quos in tabula non habes, subtrahe 360 gradus et cum residuo vel propinquiori, ut dictum est, intrabis. **Cautela secunda:** Cum latitudinem ad alium diem sequentem queris et latitudinem maiorem inveneris quam precedentem, illa ascendens latitudo dicitur. Si vero minorem invenies, descendens dicenda erit. Et tunc ipsum utraque, scilicet tam maior quam minor, est septentrionalis. Si autem utraque fuerit meridionalis et sequens maior, planeta est descendens et econverso ascendens dici debet, cum latitudo meridionalis sequens decrescit aut minuitur. **Verbi gracia:** Volo scire latitudinem Saturni ad annum Domini 1486 currentem, ad diem Assumpcionis Marie. Primo debeo scire centrum verum et equatum Saturni et est hoc: 11, 27, 28, 38, 21. Secundo: cum reduco signa in gradus aliis gradibus additis similiter adiunctis aduc 50 gradus veniunt 407. Et quia excedunt 360, ideo subtraho 360 et manebunt 47 gradus. Tercio: cum gradibus residuis vel cum propinquioribus intrabo ad *Tabulam latitudinum* sub minutis in ultima columna et |**f. 191r:**| primo introytu invenio 44 minuta, 24 secunda. Secundo introitu invenio 40 minuta, 0 secunda. Primus maior. Deinde cum subtraxero minorem a maiori, manent in differencia 4 minuta, 24 secunda. Quarto: minuta proporcionalia cum duco in residuum graduum, minutorum, ut in 5 gradus, 28 minuta, 38 secunda, 21 tercia, veniet pars proporcionalis ista: 24 minuta, 6 secunda, 0 tercia. Deinde cum hanc partem

---

**1505** additis C : addidis *A* : addidis *(corr. in:* additis*) B*

proporcionalem divido per sex, habebo partem proporcionalem ad 6 gradus istam: 4 minuta, 1 secunda. Que pars proporcionalis debet subtrahi a primo introytu et manebunt minuta proporcionalia equata ista: 40 minuta, 23 secunda. Que debeo servare ad partem et circa eas signare 'superior', quia minuta proporcionalia sunt capta in superiori parte tabule. Deinde debeo scire verum argumentum Saturni ad diem Assumpcionis Marie et est hoc: 8, 22, 11, 38, 19. Quod cum reduco ad gradus, venient 262 gradus, cum quibus aut minoribus propinquioribus, scilicet 252, intrabo ad *Tabulam latitudinis*, et qui circa minuta est signatus titulus 'superior', accipio latitudinem Saturni sub titulo septentrionali et primo introytu invenio 2 gradus, 39 minuta. Secundo introytu 2 gradus, 42 minuta. Primus minor. Deinde cum subtraho minorem a maiori, veniet differencia 3 minuta. Deinde differenciam debeo reducere in residuum graduum et alias fracciones, scilicet 10 gradus, 11minuta, 38 secunda, 19 tercia, et veniet pars proporcionalis ista: 30 minut, 34 secunda, 55 tercia. Post hoc partem proporcionalem divide per sex et veniet pars proporcionalis ad sex gradus, 5 minuta, 5 secunda, quam addo ad primum introytum et exibunt gradus et minuta equata. Ultimo idem, quod ultimo exibit et in gradibus et minutis equatis cum multiplicavero per minuta prius inventa, que quesita fuerunt, unde gradus veri centri Saturni exibit latitudo Saturni in die Assumpcionis Marie ista: 1 gradus, 50 minuta, 26 secunda, 14 tercia, latitudo septentrionalis.

<39>

**<C>anon 39 de latitudine Veneris invenienda**

Cum latitudinem Veneris scire volueris, tres latitudines parciales ipsius invenies, ex quibus finaliter unam completam habebis. Prima enim dicitur 'declinacio', secunda 'refleccio', tercia 'deviacio'. Latitudinem Veneris, que declinacio dicitur, sic invenies: primo signa in argumento Veneris vero in gradus reduc gradibus |**f. 191v:**| residuis additis et cum his autem minoribus propinquioribus *Tabulam Veneris* ingredere et accipe declinacionem duplici introytu propter differenciam et considera, quis introitus sit maior vel minor. Secundo: differenciam duc in residuum graduum, minutorum, secundorum et terciorum, ut dictum est superius, querendo partem proporcionalem. Tercio: illam partem proporcionalem inventam adde vel subtrahe a primo introytu iuxta modum communem et habebis declinacionem Veneris. Quarto: centro vero Veneris adde 60 gradus, et si plures excreverint quam 360, subtrahe 360 et cum residuo aut propinquiori intra ad tabulam et ex opposito eorum recipe minuta proporcionalia duplici introytu ut ante. Quinto: subtrahe minora minuta a maioribus et cape differenciam. Sexto: eandem differenciam duc in residuum graduum, minutorum, secundorum et terciorum et accipe partem proporcionalem. Septimo: partem proporcionalem divide per 6 et habebis partem proporcionalem ad 6 gradus. Octavo: ipsam partem proporcionalem adde vel subtrahe a primo introytu iuxta modum communem. Nono: his duobus inventis, scilicet declinacione Veneris et minutis proporcionalibus, multiplica unum per aliud, et quod post multiplicacionem veniet, est latitudo Veneris, que dicitur declinacio, quam serva cum suo scilicet titulo ad partem. **Cautela prima:** Si tam in centro, quam in argumento in eandem partem tabule intrasti, declinacio erit meridiana. Si vero cum argumento in unam partem tabule et cum centro in aliam partem intrasti, latitudo erit septentrionalis. **Cautela secunda:** Si in centro vero 270 gradus habueris, cum quibus

**1540** 22 *Had* : 29 *AB* **1552** 50 *B* : 5 *A* **1558** additis *C* : addidis *AB*

intrandum esset vel propius ad 360 gradus, in declinacione nihil invenies, nam erit latitudo declinacionis, unde nunc cum centro 60 gradus addendo intrare oportet. **Verbi gracia:** Volo invenire latitudinem Veneris, que declinacio dicitur, ad annum Domini 1486 currentem, ad diem Nativitatis Marie. Primo debeo scire argumentum Veneris et est hoc: 8, 1, 53, 49, 59. Et postquam reduco signa in gradus, venient 241 gradus. Secundo: cum his gradibus vel prope maioribus minoribus cum intravero tabulam Veneris secundi introytu declinacionis, primus introytus 0 gradus, 59 minuta. Secundus introytus 1 gradus, 13 minuta. Primus minor. Differencia 14 minuta. Deinde cum reduco differenciam in residuum graduum, veniet pars proporcionalis ista: 26 minuta, 33 secunda, 39 tercia. Quam partem proporcionalem cum divido per 6, veniet pars proporcionalis ad 6 gradus ista: 4 minuta, 25 secunda, |**f. 192r:**| que debent addi ad primum introytum et veniet declinacio Veneris ista: 1 gradus, 3 minuta, 25 secunda. Deinde debeo scire centrum verum Veneris ad diem Nativitatis Marie et est hec: 2, 23, 8, 53, 14, quod eciam reduco ad gradus et ei addo 60 gradus et veniunt 143, cum quibus vel propinquioribus intro ad *Tabulam minutorum proporcionalium* duplici introytu. Primus introytus 44 minuta, 24 secunda. Secundus 48 minuta, 24 secunda. Primus minor. Differencia 4 minuta, 0 secunda. Cum duco differenciam in residuum graduum, scilicet 5 gradus, 8 minuta, 53 secunda, 14 tercia, veniet pars proporcionalis ista: 20 minut, 35 secunda, 33 tercia. Que si divido per 6, veniet pars proporcionalis ad 6 gradus ista: 3 minuta, 26 secunda, que debent addi ad primum introytum, scilicet 44 minuta, 24 secunda, et venient minuta proporcionalia equata ista: 47 minuta, 50 secunda. Ultimo declinacionem Veneris debeo multiplicare per minuta proporcionalia et exibit latitudo Veneris, que dicitur declinacio ista, scilicet 0 gradus, 50 minuta, 33 secunda, 26 tercia, septentrionalis.

**Latitudinem Veneris, que 'refleccio' dicitur, sic invenies.** Intra primo cum gradibus veri argumenti signis reductis in gradus residuos, qui si fuerint, addendo, vel cum minoribus propinquioribus et refleccionem in tabula Veneris accipe duplici introytu. Secundo: introytum minorem subtrahe a maiori et erit differencia. Tercio: duc differenciam in residuum graduum, minutorum, secundorum et terciorum et habebis partem proporcionalem. Quarto: hanc partem proporcionalem divide per sex et habebis partem proporcionalem ad sex gradus. Quinto: hanc partem proporcionalem adde vel subtrahe a primo introytu iuxta modum communem et habebis refleccionem Veneris. Sexto: a gradibus centri Veneris deme 60, quos si subtrahere non posses, adde 360 et postea 60 subtrahe. Septimo: cum illis gradibus veri centri intra duplici introytu, querendo minuta proporcionalia et subtrahe minorem a maiori querendo differenciam, quam duc in residuum graduum, minutorum, secundorum et terciorum et veniet pars proporcionalis. Quam partem proporcionalem divide per sex et habebis partem proporcionalem ad 6 gradus. Quam partem proporcionalem adde vel subtrahe a primo introytu iuxta modum communem et habebis minuta proporcionalia equata, que servabis ad partem. Ultimo: per minuta proporcionalia multiplica refleccionem inventam et veniet secunda latitudo Veneris, que dicitur latitudo refleccionis equata. **Cautela prima:** Si centrum, cum quo nunc recepisti minuta proporcionalia, ceciderit in superiorem partem tabule et verum argumentum fuerit minus 180 gradus, refleccio inventa |**f. 192v:**| est septentrionalis. Si vero argumentum est plus 180 gradus, meridionalis esse debet. **Cautela secunda:** Si centrum verum ceciderit in partem inferiorem, argumentum verum fuerit minus 180 gradus, meridiana est refleccio. Si vero argumentum est plus, septentrionalis dicenda est.

Hanc itaque latitudinem sic inventam serva ad partem. **Exempli gracia:** Si volo invenire latitudinem refleccionis Veneris ad annum Domini 1486 currentem, ad diem Nativitatis Marie, primo intra cum gradibus veri argumenti Veneris reducto in gradus vel cum minoribus propinquioribus accipiendo refleccionem Veneris duplici introytu. Argumentum autem verum Veneris est hoc: 8, 1, 53, 49, 59, quod cum reduco ad gradus, veniunt 241. Primus introitus 2 gradus, 25 minuta, secundus introitus 2 gradus, 28 minuta. Primus minor. Differencia 3. Pars proporcionalis veniet ista: 5 minuta, 41 secunda, 30 tercia. Minuta proporcionalia ad sex gradus, 0 minuta, 52 secunda, que debent addi ad primum introytum, et veniet latitudo refleccionis Veneris ista: 2 gradus, 25 minuta, 52 secunda. Deinde a gradibus centri cum subtraho 60 gradus, manent 23 gradus, cum quibus vel minoribus propinquioribus quero minuta proporcionalia duplici introytu. Primus introytus 57 minuta, 0 secunda, secundus introytus 54 minuta, 36 secunda. Primus maior. Differencia 2 minuta, 24 secunda. Que cum duco in residuum graduum et minutorum, veniet pars proporcionalis ista: 12 minuta, 21 secunda, 19 tercia. Que cum divido per sex, veniet pars proporcionalis ad sex gradus ista: 2 minuta, 3 secunda, 33 tercia, que debent subtrahi a primo introytu et venient minuta equata ista: 54 minuta, 56 secunda, 27 tercia. Ultimo cum multiplicavero refleccionem per minuta proporcionalia, veniet secunda latitudo Veneris, que dicitur latitudo refleccionis equata ista: 2 gradus, 13 minuta, 34 secunda, 1 tercia, meridionalis.

**Latitudinem Veneris, que 'deviacio' dicitur, invenire.** Cum latitudinem Veneris terciam scire volueris, que deviacio dicitur, minuta proporcionalia servata ad partem in latitudine precedenti refleccionis divide per 6 et veniet tibi latitudo tercia, scilicet latitudo deviacionis, et illam invenere semper est septentrionalis. Secundo: considera, si iste tres latitudines sunt in eodem titulo, iunge eas simul et inventa est altitudo Veneris completa. Si vero sunt de diversis titulis, tunc adde illas latitudines adinficem, que sunt eiusdem tituli, et tercio minorem subtrahe a maiori residuumque maioris erit latitudo Veneris cum suo **|f. 193r:|** titulo quesita. **Verbi gracia:** Minuta proporcionalia ad partem servata in latitudine refleccionis sunt 54 minuta, 56 secunda, 27 tercia. Que cum divido per sex, veniet latitudo deviacionis Veneris ista: 0 gradus, 9 minuta, 9 secunda, 24 tercia, septentrionalis. Et quia latitudines sunt de diversis titulis duas septentrionales, addo simul et postea, cum subtraxero minorem a maiori, veniet ista latitudo Veneris completa: 1 gradus, 13 minuta, 51 secunda, 11 tercia, meridionalis.

<40>

### <C>anon 40 et ultimus de latitudine Mercurii invenienda

In latitudine Mercurii similiter tres latitudines sint inveniende, quarum prima dicitur 'declinacio', secunda 'refleccio', tercia 'deviacio'. Et latitudinem Mercurii, que declinacio dicitur, invenire volueris, primo signa et gradus in argumento Mercurii reduc. Similiter et cum gradibus istis minoribus vel propinquioribus intra tabulam et recipe declinacionem Mercurii duplici introytu. Differenciam similiter reduc in residuum graduum, minutorum, secundorum et terciorum, capiendo partem proporcionalem, quam divide per 6, et habebis partem proporcionalem ad 6 gradus, quam adde vel subtrahe, secundum quod primus introytus maior vel <minor> fuerit, et habebis declinacionem Mercu-

**1638** 33 *B* : 31 *A*

rii. Secundo: cum vero centro Mercurii reducto in gradus illis gradibus adde 270, et si plures gradus provenerint quam 360, hoc subtrahe et cum residuis intra *Tabulam minutorum* duplici introytu et differencia habita reduc eam in residuum graduum, minutorum, secundorum et terciorum querendo partem proporcionalem, quam divide per 6 et habebis minuta proporcionalia ad 6 gradus, que adde vel subtrahe a primo introytu iuxta modum communem, et habebis minuta equata. Tercio: declinacionem multiplica per minuta proporcionalia equata et habebis primam latitudinem Mercurii, que dicitur latitudo Mercurii declinacionis. **Cautela:** Si tam argumentum, quam centrum Mercurii, in quo invenisti minuta proporcionalia, in eandem partem tabule ceciderint, declinacio inventa est meridiana. Si in diversis erit, septentrionalis. Tene itaque istam latitudinem ad partem cum suo titulo. **Verbi gracia:** Volo scire latitudinem declinacionis Mercurii ad annum Domini 1486 currentem, sancte Lucie. Prius debeo scire verum argumentum Mercurii et est hoc: 7, 2, 44, 53, 23. Quod cum reduco ad gradus, veniunt 212 gradus, cum quibus intro vel minoribus propinquioribus, scilicet 210, et erit primus introytus 3 gradus, 7 minuta. Secundus 3 gradus, 26 minuta. Primus minor. Differencia 19 minuta. Pars proporcionalis veniet ista: 0 gradus, 52 minuta, 12 secunda, 54 tercia, quam cum divido per 6, veniet hec |**f. 193v:**| pars proporcionalis addenda ad primum introytum: 8 minuta, 42 secunda, et veniet declinacio Mercurii 3 gradus, 15 minuta, 42 secunda septentrionalis. Deinde debeo scire verum centrum Mercurii et est hoc: 1 signum, 28 gradus, 42 minuta, 41 secunda, 42 tercia. Quod cum reduco in gradus, additis 270 gradibus, venient 328 gradus, cum quibus vel minoribus quero minuta duplici introytu. Primus introytus 48 minuta, 24 secunda. Secundus 44 minuta, 24 secunda, {24}. Primus maior. Differencia 4. Pars proporcionalis veniet 18 minuta, 50 secunda, 46 tercia, que cum divido per sex, veniunt minuta proporcionalia 3 minuta, 8 secunda, que debent subtrahi a primo introytu et veniunt minuta equata ista: 45 minuta, 16 secunda. Ultimo cum multiplico declinacionem per minuta equata proporcionalia, veniet prima latitudo Mercurii et declinacionis ista: 2 gradus, 27 minuta, 38 secunda, 41 tercia, septentrionalis.

**Latitudinem Mercurii, que refleccio dicitur, invenire.** Primo signa et gradus argumenti Mercurii reduc in unum et cum ipsis vel minoribus intra *Tabulam Mercurii* capiendo eius refleccionem duplici introytu. Differenciam reduc in residuum querendo partem proporcionalem, quam divide per sex, et quod inde provenitur, adde vel subtrahe a primo introytu iuxta modum communem et habebis refleccionem Mercurii. Secundo: cum centro Mercurii intra *Tabulam minutorum*, prius 180 gradus addendo, et si plus excreverit quam 360 gradus, hoc subtrahe et cum residuo intra vel minori propinquiori duplici introytu. Differenciam duc in residuum querendo minuta proporcionalia, que divide per 6 et partem proporcionalem adde vel subtrahe a primo introytu iuxta modum communem, que minuta serva ad partem, quoniam per ipsa tercia latitudo est invenienda. Ultimo multiplica refleccionem per minuta equata et veniet latitudo secunda Mercurii, que dicitur latitudo refleccionis. **Cautela prima:** Si centrum, cum quo minuta invenissem, cadit in superiorem partem tabule, argumentum verum fuerit minus 180 gradibus, refleccio inventa erit septentrionalis. Si vero argumentum erit plus, erit meridiana. **Cautela secunda:** Cum centrum, cum quo minuta invenisti, cadit in inferiorem partem tabule et argumentum minus 180 gradibus fuerit, refleccio est meridiana. Si plus, septentrionalis est. **Exempli gracia:** Volo scire latitudinem refleccionis Mercurii ad diem

---

**1685** additis *Had* : adidis *A* : addidis *B* **1692** 38 *Had* : 8 *A* : 18 *B*

supradictam. Primo reduco signa et gradus argumenti in unum et veniunt 212, cum quibus vel propinquioribus cum intro ad *Tabulam refleccionis Mercurii* duplici introitu, veniet primus introytus iste: 1 gradus, 45 minuta. Secundus |**f. 194r:**| 1 gradus, 49 minuta. Primus minor. Differencia 4. Pars proporcionalis est ista: 10 minuta, 59 secunda, 34 tercia. Quam cum divido per 6, habebo partem proporcionalem ad 6 gradus istam: 1 minutum, 50 secunda. Addendam et veniet refleccio ista equata: 1 gradus, 46 minuta, 50 secunda. Deinde debeo reducere centrum Mercurii in gradibus ei addendo 180 gradus et venient 238 gradus, cum quibus vel minoribus propinquioribus intra ad *Tabulam minutorum* et veniet primus introytus iste: 35 minuta, 12 secunda. Secundus introytus 40 minuta, 0 secunda. Differencia 4 minuta, 48 secunda. Pars proporcionalis veniet ista: 22 minuta, 36 secunda, 55 tercia, quam cum divido per 6, veniet pars proporcionalis ista: 3 minuta, 46 secunda. Addendo ad primum introytum veniunt ergo minuta equata ista: 38 minuta, 58 secunda. Tercio: cum multiplico refleccionem per minuta equata, veniet latitudo secunda Mercurii, que refleccionis dicitur, ista: 1 gradus, 9 minuta, 22 secunda, 56 tercia, meridionalis.

**Latitudinem Mercurii, que deviacio dicitur, invenire.** Cum latitudinem terciam Mercurii habere volueris, tunc de minutis proporcionalibus servatis accipe quartam partem et dimidiam, hoc est divide illa minuta per quatuor, et in quociente veniet quarta pars isti, ergo quarti partis. Ipsius medietatem adde et erit tercia latitudo inventa, que in Mercurio semper est meridiana. Secundo considera, si omnes latitudines sunt eiusdem tituli, iunge eas simul et inventa est latitudo Mercurii completa. Si vero fuerint de diversis titulis, minorem a maiori subtrahe residuumque maior erit latitudo Mercurii cum suo titulo quesita. **Verbi gracia:** Si volo scire latitudinem Mercurii, que deviacio dicitur, debeo dividere minuta primo inventa per 4 et in quociente veniet quarta pars ista: 9 minuta, 44 secunda et in residuo manebit 2 in superfluo. Cuius quarte partis cum accipimus medium, erit hoc: 5 minuta, 22 secunda. Quam medietatem cum addo ad quartam partem, veniet latitudo deviacionis Mercurii ista: 0 gradus, 15 minuta, 6 secunda meridionalis. Ultimo latitudines duas Mercurii meridionales, scilicet deviacionis et refleccionis, addo simul et postea minorem subtrahe a maiori et manebit ultimo latitudo Mercurii ista: 1 gradus, 2 minuta, 39 secunda, 45 tercia, septentrionalis. Finis Cracovie.

## 8.3 Commented Translation of the Canons *Pro (introduccione) generali* by John of Głogów

For a general introduction to the planetary tables which we call *resolute*, it is necessary to preface an introduction of what will be necessary and convenient for understanding what will be said later. It is necessary to know that these *Tabule resolute* are named from *resolvendo* because they are derived from the Tables of the illustrious Alfonso, the Roman and Castilian King, and reduced to our Cracovian meridian. Everything that is sought in the Alfonsine tables with a great and tedious effort, can be found easily and quickly in these tables. Hence they are correctly named *resolute*. However, the method to resolve and reduce them not only to the meridian of Cracow but to any other place, will be sufficiently clarified below.

It should be noted that we proceed in two ways in astronomical and tabular practice: firstly by the way of decomposition and division, secondly by the way of composing and collecting. First, we proceed in astronomical practice by decomposition. The whole zodiac is divided into twelve signs,[21] each of which is named for some animal: Aries, Taurus, Gemini, Cancer, Leo, Virgo, Libra, Scorpio, Sagittarius, Capricornus, Aquarius, Pisces.[22] One sign is divided into thirty degrees, one degree is divided into sixty minutes, one minute into sixty seconds, one second into sixty tertias, and so on up to the sextas.[23] Second, in astronomical and tabular practice we follow the way of collection (*colleccionis*). For sixty tertias, therefore, we take one second, and for sixty seconds we take one minute. And for increasing sixty minutes we take one degree, and for thirty degrees we take one sign. When twelve signs are exceeded in the work, twelve are subtracted and the rest is left in place, as will be noted below. This is because astronomers carry out their practice in sexagesimal numbers. The astronomer considers the motion of the fixed stars and planets, and the times of conjunctions and aspects. However, motion and time are calculated continuously. And although the continuum is divisible, yet no number below 100 is so suitable for division as the number 60, because it is divisible better than all divisible numbers. For it is divisible into two parts, namely, 30 and 30; further into three parts, namely, three times 20. It is divisible into four parts, namely four times 15, it is divisible into five parts, namely five times 12, and it is also divisible into six parts, namely six times ten. And consequently this is also how one can usefully work in the sexagesimal way when using tables.[24]

Also note that the years are written twice in these tables, as collected and expanded years. The collected years are those that are written at steps of twenty years, as you can

---

21 Here and in the following text, the 'sign' means the so called 'natural sign', which has thirty degrees. In the contrary, in the PAT there are mostly used 'physical signs', which have sixty degrees to keep the strictly sexagesimal counting of angles. The whole circle of the zodiac thus contains six physical signs only, each one consisting of two thirty-degree 'natural signs'.

22 The Bull, Twins, Crab, Lion, Virgin, Balance, Scorpion, Archer, Goat, Water Bearer, and the Fishes.

23 The authors of the original PAT composed them and wrote the canons on their use for learned astronomers; it was accepted that whoever used the tables must understand them. Later – and this is a remarkable feature of the *Tabule resolute* – astronomers wanted to explain the tables to wider audience, especially to students at universities. They wanted them to be brought into practice as much as possible (en masse). Hence the unprecedented solicitude of John of Głogów's description.

24 The established use of the sexagesimal system was common in general, not only in astronomy. Commodities sold in medieval marketplaces and much later were counted by the dozen, by the threescore, by gross etc.

easily see at the beginning of the tables. When you add twenty to the first collected years, the subsequent years are immediately increased by that amount. And what is written beside them is clearly explained by the headings.

The expanded years are those that follow up to twenty, one after the other without any interruption, as can be clearly seen in the Table of expanded years. In this Table of expanded years, the letter 'b' is always entered beside the fourth year,[25] which indicates a leap year. The following year is the first after the leap year; then follow the second and third after the leap year, and so on, as can be clearly seen in the Table of expanded years.

It should also be known that the year in the practice of these tables – and in astronomy in general – begins with the last day of December at noon, from the syllable 'Sil';[26] to this day and its noon all the motions of the planets are sought, which are called the radices of the year then beginning, as will be indicated later. For their predictions, however, astrologers begin their year with the entry of the Sun into Aries.

The months start from January. The days start from noon of the previous day and last until noon of the following day. For noon is the most noble part of the day, as Cameracensis[27] says, because of the great strength and power of the Sun at this part of the day. At noon the Sun is in the tenth house, that is, in the midheaven. This part of the heavens is the strongest because the Sun's ray then shines more strongly than do more inclined rays. Astronomers, therefore, properly start the day at noon. And they look for planetary motion especially towards noon, as Cameracensis says. When no signs are found, the twelve signs are completed, but the first sign is incomplete. When one sign is found complete, it is the first completed sign. The hours and other fractions should be thought of in a similar way, so that the use of these tables may be easy and intelligible to the young. I have therefore taken the trouble to write the rules in as simple and clear a style as possible. I now begin what I have set out to do in the name of Jesus Christ the Saviour and His glorious Mother.

<1>

### The first rule on how to find a radix of any mean motion to the beginning of a year, the noon of the last day of December

To understand this rule it is necessary to know that the radix of the motion of a planet is the mean motion of a planet found at noon on the last day of December, namely, at the syllable 'Sil', that is, the radix of the Sun is the motion of the Sun found at noon on the last day of December. Similarly, the radix of Saturn is the mean motion of Saturn found at noon of the previously mentioned day. And so similarly it is to be understood for the motions of the other planets. These radices are found at noon for the city of Cracow. In what way the radices are transformed to other cities will be treated in a special rule. The motions of the planets found on the last day of December are also called radices: just as

25 The letter 'b' means *bisextilis*, that is the leap year.
26 Cisiojanus, Silvester.
27 French philosopher, theologian, and astrologer Petrus de Alliaco (1350–1420) served as cardinal of Cambrai (*cardinalis Cameracensis*).

the root[28] of a tree is the basis for the growth of a tree and its fruit and the root is assumed in a tree, so to the motions of the planets found at noon on the last day of December are the basis for finding motions in other days and months throughout the year. So to find the radix of a planet is to find the mean motion of that planet on the last day of December.

**The way and the rule how the radices of a motion are to be found is this.** First, you must know the completed year of Christ. With this completed year of Christ, enter the Table of collected years and look for the completed year of Christ in this table. When you have found it in this way, you will immediately find beside this completed year without any further operations the radix of any planet on the last day of December. In the Table of apogees and fixed stars you will find the radix of their motions; in the Table of the Sun you will find the radix of the motion of the Sun; and the same with the other planets. Second: if you do not find the completed year of Christ in the Table of collected years, then enter with the smaller year of Christ nearest to your current years, and what you find beside that year in degrees and minutes, write separately outside the table. Third: subtract the nearest lesser year which you found in the Table of collected years from your completed years of Christ, and what remains after that subtraction will be the number of expanded years still remaining to your completed year of Christ. Fourth: with this remainder left after subtraction, enter into the Table of expanded years, and take what you find beside that year and add it to what you have previously obtained beside the collected years – writing signs under signs, degrees under degrees, minutes under minutes, seconds under seconds, and so on – and thus seek the radix of the motion of any planet. Fifth and last: add what you have found in the expanded years to what you have found in the collected years, and start from the smallest fraction, that is, tertias to tertias, seconds to seconds, minutes to minutes, degrees to degrees, and work as before. And what results in each motion is the radix of the motion for the end of December. **For example:** I want to find the radix of the motion of the apogees and fixed stars for the beginning of the current year of Christ 1486. First I will take the completed year of Christ, namely 1485, for it is completed. I search for the beginning of the following as follows: I look in the Table of collected years for the completed year of Christ, but if I do not find it, then – secondly – I take into account the nearest completed year of Christ, and it is 1468, and beside it I obtain 0 signs, 10 degrees, 47 minutes, 7 seconds and 6 tertias.[29] Third, after subtracting the nearest year of Christ found in the Table of collected years, namely 1468, from the completed years of Christ, I am left with 17, which are the years between 1468 and the completed year 1485. With these seventeen years I enter the Table of expanded years, which also has the inscription Table of apogees and fixed stars, and beside the seventeen I find $0^{s}\ 0;7,59,37^{\circ}$.[30] Fourth: I add it all together in the manner described above, starting with the smallest fractions. I get the

28 The root is *radix* in Latin.

29 See the third line of column A in Table TRC-01 of mean motions in collected years for the meridian of Cracow. The values for this slow motion are the same also in the Table TRP-01 for the meridian of Prague (see also the Fig. 25).

30 See column A of the Table TRW-02 of mean motions in expanded years. The correct value is $0^{s}\ 0;7,29,37^{\circ}$ which is in agreement with the resulting value given in the table below. This same error is also in MS Cracow, BJ, 600, f. 149v. In the variant of John of Głogów's canons contained in MS Wrocław, BU, IV F 19, ff. 247v–264r, the same example (f. 278r) is updated to the year 1493 with the radix $0^{s}10;55,56,6^{\circ}$ for the year 1488, an increment of $0^{s}\ 0;2,12,14^{\circ}$ for five years, resulting in $0^{s}10;58,8,20^{\circ}$.

radix of the motion of the apogees and fixed stars to the completed year of Christ 1485, which is the beginning of the current year 1486, and to the last day of December, noon.

| S | G | M | S | T | | |
|---|---|---|---|---|---|---|
| 0 | 10 | 54 | 36 | 43 | | apogees and fixed stars |
| 2 | 15 | 34 | 50 | 59 | | motion of access and recess of the 8th sphere |
| 9 | 18 | 57 | 17 | 16 | | motion of the Sun, Venus, and Mercury |
| 7 | 15 | 45 | 51 | 54 | | motion of the Moon |
| 11 | 25 | 41 | 49 | 34 | radix of mean | anomaly of the Moon |
| 7 | 29 | 52 | 22 | 29 | | argument of latitude of the Moon |
| 0 | 14 | 6 | 30 | 35 | | head of dragon |
| 8 | 2 | 34 | 53 | 16 | | motion of Saturn |
| 8 | 28 | 25 | 12 | 19 | | motion of Jupiter |
| 8 | 13 | 43 | 49 | 11 | | motion of Mars |
| 2 | 25 | 0 | 6 | 22 | | anomaly of Venus |
| 10 | 8 | 9 | 1 | 3 | | anomaly of Mercury |

<2>

## The second rule on finding the mean motion of any planet at any proposed time using the appropriate tables

If you wish to know the mean motion of a planet, you must first have some predicted time for which you wish to know the mean motion of that planet. And consider whether that predicted time includes completed months, days, hours, minutes, seconds, and tertias. Write this time aside. Second: write down in the table the radix of the motion of the same year. Third: if a month has passed, enter the Table of months with the month ended below the title of motion you are looking for. If it is a non-leap year, enter the upper table of months. If it is a leap year, you will go to the bottom table. And what you find beside the month, add to the radix, all to its kind, as was said in the first rule. Fourth: with the days, if you have them, enter into the Table of the same planet in days, and what you find beside that day, add that also to what you have found before. Fifth: with the hours, if you have any, enter into the Table of the motion of that planet in hours, and what you find beside that supposed hour, add that also to what you have found before. And in that table in which you have obtained the hours, find the minutes, seconds, and tertias according to the change of name, as mentioned in the cautions below. And what you find beside the minutes, seconds, and tertias, add it all to what you found before. And if you go in with the full estimated time, then what results indicates the mean motion of the planet you were seeking. **First caution:** when you have minutes, seconds, and tertias in your predicted time, use the same table you used to find the motion in hours to find the minutes, seconds, and tertias. With that fraction, you enter the table and the first number will always get the same name, whether it be a zero or a significant number, according to the number with which you entered. If you enter with minutes, the first number will be minutes, then seconds and finally tertias. If you will enter with seconds, the first number will be seconds, then tertias. If you will enter with tertias, the first number will be tertias, then quartas. And if you come out with more than 30 quartas, add one to the tertias for accuracy.[31] And these tables in which the number of hours is greater than 24 are called Loricated tables (see Fig. 23).[32] Note therefore, that the number of hours extends to thirty, or that there are more hours in the day than 24. But for that reason this table is extended to thirty, because with its help not only the hours, but also minutes, seconds, and tertias are found, which exceed beyond 24. It is thus not necessary to write separate tables for minutes, seconds, and tertias, since a greater number of tables would be useless. But using this table with which we obtained the hours, we can also find the minutes, seconds, and tertias, as mentioned, just with the names shifted. **Second caution:** if we would have more minutes, seconds, or tertias than 30 at the estimated time, enter first with 30 minutes, seconds, or tertias, and then with the remainder, and add what results after the complete entry to what was found earlier, as has been said. **Third caution:** if there is no completed month in your estimated time, there is no need to enter the Table of months, but skip the Table of months and enter with the days of the estimated time into the Table of days. And likewise, if there are no hours, nor minutes, nor seconds, nor tertias above the days in your estimated time, there will be no need to enter into the Table of these fractions. **For example:** for the current year of the Lord 1486, for the day of the Assumption of Mary,[33] which is the 15$^{th}$ of August, for the fourth hour after noon, 48 minutes, 56 seconds and 20 tertias, I want to know the mean motion of the Sun. First: I assume the radix of the same year and write it in the Table, namely, 9 signs, 18 degrees, 57 minutes, 17 seconds and 16 tertias.[34] Second: consider if there are some months completed from the beginning of the year until August, and it is clear that there are seven. And since it is a non-leap year, I will therefore enter the upper Table of months for the Sun, and beside the number I find the motion of the Sun $6^s$ 28;57,25,20°.[35] This motion I will add to the radix of the Sun, and when added up it will come out $4^s$ 17;54,42,36°. And since in the expected time I still have 15 days over the completed months, I enter the Table of the Sun in days and beside the fifteen days I get the corresponding motion, and it is $0^s$ 14;47,4,54°.[36] I add this motion to what was found earlier, and it comes to $5^s$ 2;41,47,30°. And since I have four hours over the days in the expected time, I enter the four hours in the Table of the mean motion of the Sun in hours, and find 0 degrees, 9 minutes, 51 seconds, and 23 tertias,[37] which I add to what I found before, and find $5^s$ 2;51,38,53°. And since I have 48 minutes extra over the day in the predicted time, we must enter the same

31 The value is to be rounded, not truncated, to the tertias.

32 *Tabula loricata* is the table TRW-06 of mean motions in hours which is extended beyond 24 to be usable also for finding the values in minutes and smaller sexagesimal fractions of an hour. It is called 'loricated' in analogy with the additional lines of loricated armour, consisting of separate rows of metal strips fastened one below the other.

33 John of Głogów began his education at the local school of the Collegiate Church of the Assumption, so perhaps that is why he chose the example for this date.

34 This value can be calculated in the same way as in the example in Rule 1, but using the columns C of the Tables TRC-01 and TRW-02. It is given in the table at the end of the first rule. In the MS Vatican, BAV, Pal. lat. 1439, f. 168r there is written 59 minutes by an error, while it is correct in the MS Cracow, BJ, 610, f. 150r.

35 See the Table TRW-03 of mean motions in months in non-leap year, the line for July, column C.

36 See the Table TRW-05 of mean motions in days, column C.

37 See the Table TRW-06 of mean motions in hours and their fractions, column C.

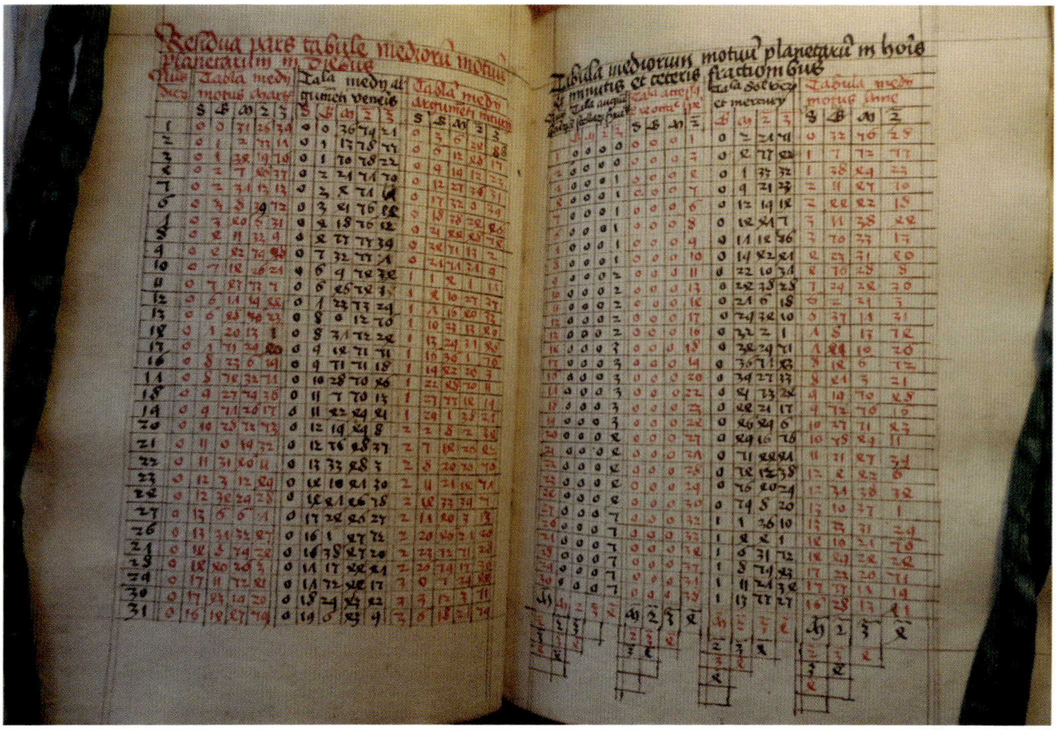

**Figure 23.** *Tabula loricata* (Loricated table) in the MS Cracow, BJ, 2252, p. 63 (canons *Girum*).

table with 48 minutes. But since I can't find them, I'll enter directly with 30 minutes according to the second caution, leaving me with 18 minutes left. So when I enter with 30 minutes, I find beside the motion 1 minute, 13 seconds and 55 tertias. Then I enter with 18 minutes and find beside the motion 0 minutes, 44 seconds and 21 tertias. When I add these motions to the corresponding minutes found earlier, I find $5^s$ 2;53,37,9°,[38] and since I have 56 seconds extra over the minute in the expected time, we must enter with 56 seconds. However, I don't find these together; I first enter with 30 seconds and get 1 minute and 13 tertias. Then I enter with the remaining 26 seconds and get 1 minute and 4 tertias, which, if I add everything to what was found before, results to $5^s$ 2;53,39,27°.[39] And since I have 20 tertias over the seconds in the expected time, I enter the Table of the mean motion of the Sun with 20 tertias and beside I find 0 tertias and 49 quartas. For the sake of precision I will add one for these quartas, that is, to the preceding tertias, and it will be 28. And thus I have already found the complete mean motion of the Sun at the supposed time, namely, at the 15th of August, the 4th hour, the 48th minute, the 56th second, and the 20th tertia after noon, namely, $5^s$ 2;53,39,28°. And as it was calculated for the Sun, so too are the mean motions of the other planets dealt with.

38 In this addition 40 quartas are truncated.

39 If the quartas are ignored, there should be in the sum 26 tertias. However, the motion for 56 seconds should yield additional 59 quartas.

<3>

**The third rule for the radix of apogee of any planet at the beginning of a year**

A similar procedure should be followed when looking for the apogee of a planet. When you wish to know the apogee of a planet at a certain time, first find the radix of the apogee of the planet according to the teaching given in the second rule of mean planetary motions. Second, consider whether there is a completed month relative to your intended day. Then enter with the completed month and add what you find beside that month to the radix of apogee. You will need to work similarly with the days, according to the second rule on the mean motions of the planets. If, however, no month is completed, then with the number of intended days enter the Table of days and add that motion which you find beside the days to the radix, and what results from the addition will be the motion of the apogee of that planet which you seek to the expected time. **Caution:** if you wish to work more precisely in a leap year, and if the intended time is after February, do not enter the Table of apogee in days for the predicted day, but for the following day because of the increasing excess, owing to the gradually growing excess. Still, no perceptible error will result if you enter directly with the intended day. However, if this caution is held, the work will be more precise. It is not necessary to work with hours in the Tables of apogee because the motion is a minute and small, and nothing results except in the quartas, which we do not use when we are looking for a motion in *Tabule resolute*. **For example:** I wish to know the apogees of the planets on the day of the Assumption of Mary, in the current year of the Lord 1486. First it is necessary to know the radices of the apogees.[40] And so the radices of the apogees are:

| S | G | M | S | T | | |
|---|---|---|---|---|---|---|
| 3 | 1 | 2 | 50 | 16 | | Sun[41] |
| 8 | 13 | 1 | 9 | 16 | | Saturn |
| 5 | 23 | 14 | 27 | 16 | radix of apogee of the | Jupiter |
| 4 | 14 | 49 | 40 | 16 | | Mars |
| 7 | 0 | 17 | 0 | 16 | | Mercury |

So if I have a radix, I enter the Table of completed months with completed July and finally with 15 days, and when I add the motion beside the intended time to the appropriate radix of the planet, I will immediately have the motion of the apogee of any planet for the day of Assumption of Mary,[42] and they are $3^s$ 1;3,11,50°.

40 The (annual) radices of apogees can be computed using the Table TRP-07 of apogees of planets in collected years and the Table TRP-07a of apogees of planets in expanded years. For instance for the Sun we find in the collected year 1468 in the Table TRP-07, column B the value $3^s$ 0;53,13,34°, and in 17 expanded years in the Table TRP-07a, column 1468 the value $0^s$ 0;9,36,42°. Their sum is the first line of the following table.

41 *The radix of apogee of the Sun and of Venus are identical.*

42 The longitudes of apogees at this date given in the following table are increased for $0^s$ 0;0,21,34° with respect to the radices of that year given in the previous table. This would correspond to an annual increase of about $0^s$ 0;0,34,40°. A table of apogees in months and days with such a value can be found, for instance, in MS Cracow, BJ, 600, f. 90r (see Chapter 14, Tables TRC-07b and TRC-07c). In these tables we find the value $0^s$ 0;0,20,8,28°

| S | G | M | S | T | | |
|---|---|---|---|---|---|---|
| 3 | 1 | 3 | 11 | 50 | | Sun |
| 8 | 13 | 1 | 30 | 50 | | Saturn |
| 5 | 23 | 14 | 48 | 50 | apogee of the | Jupiter |
| 4 | 14 | 50 | 1 | 50 | | Mars |
| 7 | 0 | 17 | 21 | 50 | | Mercury |

<4>

### The fourth rule on finding the anomaly of the Sun at any proposed time

If you wish to know the anomaly of the Sun for your proposed time, first look for the mean motion of the Sun to the proposed time. Second: get the apogee of the Sun for the same proposed time. Third: subtract the apogee of the Sun at your proposed time from the motion of the Sun at the same proposed time. And what remains after the subtraction will be the anomaly of the Sun to the proposed time. **Caution:** if you can't subtract because the greater can't be subtracted from the smaller, you need to make an adjustment. If it is not possible to subtract in tertias, you adjust, that is, in seconds.[43] And if it is not possible to subtract in seconds, you adjust, that is, in minutes. If it is not possible in minutes, you shall adjust, that is, in degrees. If it is not possible to subtract in degrees, you shall adjust, that is, in signs. But if you would not be able to subtract in signs, you shall adjust one whole revolution, that is, 12 signs. And this caution is to be remembered in all astronomical practice. **For example:** for our current year for the day of Assumption of Mary, I want to know the anomaly of the Sun. First I must know the motion of the Sun,[44] and it is $5^{s}$ 2;41,47,30°. The apogee of the Sun on the same day of Assumption of Mary is $3^{s}$ 1;3,11,50°. After subtraction, there remain 2 signs, 1 degree, 38 minutes, 35 seconds and 40 tertias.

<5>

### The fifth rule on finding the centre of the Moon at any proposed time

If you wish to have the centre of the Moon for some time, first look for the Moon's mean motion to that proposed time. Second: find the motion of the Sun to the same time. Third: subtract this mean motion of the Sun from the mean motion of the Moon according to the previous caution. The centre of the Moon is the distance of the Moon from the Sun. In conjunction the Sun and Moon are together. And double what remains after the subtraction, and what results after the doubling by two is the centre. **For example:** I want

for July and $0^{s}$ 0;0,1,25,30° for 15 days. The sum is thus $0^{s}$ 0;0,21,33,58°, which rounded to the tertias gives the value used by John of Głogów. It should be noticed, however, that according to the Table TRP-07a, the increase in the course of the year 1486 was only $0^{s}$ 0;0,33,43°. See Chapter 19 for detailes.

43 The adjustment means to add one unit from the fraction of higher order.

44 See the example in Rule 2.

to know the centre of the Moon for our current year, for the day of Assumption of Mary. First, find the motion of the Moon at that time, and it is: $11^{s}$ 6;48,21,39°. Second: you must find the motion of the Sun[45] on the same day, and it is: $5^{s}$ 2;41,47,30°. Third: I must subtract the motion of the Sun from the motion of the Moon, and the remainder is: $6^{s}$ 4;6,34,9°. Fourth: I must multiply this by two, and the centre of the Moon on the day of the Assumption of Mary is $0^{s}$ 8;13,8,18°.

<6>

**The sixth rule on finding the anomaly of the three upper planets, namely the Saturn, Jupiter, and Mars, at the proposed time**

If you wish to know the anomaly of the three upper planets, first find the motion of the Sun at the proposed time. Second: determine the motion of the planet whose anomaly you want to know. Third: subtract the motion of the planet from the mean motion of the Sun, and what is then left will be the planet's anomaly for your proposed time. **For example:** for our current year on the day of Assumption of Mary, I want to know the anomaly of Saturn. First, I find the motion of the Sun[46] and it is: $5^{s}$ 2;41,47,30°. Second: I find the motion of Saturn[47] on the day of Assumption of Mary and it is: $8^{s}$ 10;11,6,47°. Third: subtract the motion of the planet from the mean motion of the Sun and the remainder will be the anomaly of Saturn and it is: $8^{s}$ 22;30,40,43°. The mean motion of Jupiter on the day of Assumption of Mary is: $9^{s}$ 17;23,47°. The anomaly of Jupiter to the same day of the Assumption of Mary is: $7^{s}$ 15;24,23,43°. The anomaly of Mars on the day of the Assumption of Mary is: $4^{s}$ 30;0,10,1°. Find the anomalies of Venus and Mercury using the method mentioned earlier through the respective tables.

<7>

**The seventh rule on finding the mean centre of the five planets at the proposed time**

First, look for the mean motion of the planet to the proposed time. Second: subtract the apogee of the planet at the same time from its mean motion, and what remains after subtraction will be the mean centre of the planet. **For instance:** on the day of the Assumption of Mary, I want to know the mean centre of Saturn. First: I need to know the mean motion of Saturn,[48] which is: $8^{s}$ 10;11,6,47°. Second: I need to know its apogee,[49] which is $8^{s}$ 13;1,30,50°. I must subtract this apogee from the motion of Saturn, and what remains is the centre of Saturn, and this is $11^{s}$ 27;9,35,57°.Work similarly with Jupiter, Mars, Mercury and Venus.

45 See the example in Rule 2.
46 This example is again to the year 1486 – see the example in Rule 2.
47 Adding the corresponding values in columns H in the Tables TRC-01, TRW-02, TRW-03 and TRW-04 yields $8^{s}$ 10;11,6,46°.
48 See the example in Rule 6.
49 See the second table in Rule 3.

<8>

## The eighth rule on rectification and correction of motions in previously made tables and on making new tables

In order to make the work with this rule on the method of continuation and correction of the tables easier and smoother, it is advantageous to separate it into four parts. **The first part is on continuing, correcting, and newly creating any motion in collected years.** If you wish to continue the motion of the planets described in the Tables of collected years, it is necessary to take the first line corrected from the Alfonsine tables. And if you add to this first line the motion beside the 20 expanded years, you will get the second line. If you add to this second line that which is in motion beside the 20 years, the third line will result. If you add the same to the third line, the fourth one will result, and so on. This is the method of continuing the motion in the collected years. **For example:** in the Table of collected years, the mean motion of the Sun beside the year of Christ 1428 is $9^s$ 18;46,56,45°.[50] If you add to this the motion of the Sun, which is beside the 20 years in the Table of expanded years, namely $0^s$ 0;8,48,59°,[51] the motion in the second line beside the year of Christ 1448 in the collected years will immediately result, namely $9^s$ 18;55,45,44°,[52] and so on consecutively. **The method of correcting the tables of motions in the collected years is as follows:** take the two lines closest to each other and if you subtract the smaller from the larger, there must in the residual remain the motion which is given beside the 20 expanded years. If it is so, the tables are correct. If it is not be, they are incorrect and bad, and you must rectify them by the continuation, as has been said. **For example:** I want to know if the second line of the mean motion of the Sun in collected years is correct. I take the first line, namely $9^s$ 18;46,56,45°, and I take the second and nearest line in the Sun, which is $9^s$ 18;55,45,44°, and I subtract the first from the second, because the first is smaller than the second, and there remains that which is beside the 20 years of the mean motion of the Sun in expanded years, namely $0^s$ 0;8,48,59°, and so on.[53] And you will work in this way to correct other lines and tables. **The second part on the correction and continuation of the Tables of motion in expanded years.** If you wish to continue the Tables of planetary motions in expanded years and make them anew, presuppose the first line corrected from Alfonso. If you add the first line to itself or you double it, which is the same thing, the second line will immediately result. If you add the first to the second line, a third line will result. If to this third line is added the first together with the motion of the planet in one day, the fourth line will result. And let this addition be due the leap year in which one day is added. **For example:** in the Table of expanded years is found beside the first year for the Sun $11^s$ 29;45,39,22°.[54] If you add it to itself or you double it, the second line will be $11^s$ 29;31,18,44°.[55] If you add to the

50 See the first line of the column C in the Table TRC-01 of mean motions in collected years for the meridian of Cracow.

51 See the last line of the column C in the Table TRW-02 of mean motions in expanded years.

52 See the second line of the column C in the Table TRC-01.

53 See again the same numbers of the Tables TRC-01 and TRW-02 as above.

54 See the first line, column C of the Table TRW-02 of mean motion in expanded years.

55 See the second line, column C of the Table TRW-02. One circle, i.e. 12 signa, must be subtracted from the sum.

third line the first with the motion of the Sun in one day, namely 0ˢ 0;59,8,20°, the fourth line will be 0ˢ 0;1,45,48°, and so on.[56] Continue in this way for the Sun until the end in the expanded years. This same method should be used for motions in the expanded years of the other planets. **The method of correcting the tables of motions in expanded years is as follows:** if you suspect any line, subtract the previous line from the next line, and if there remains the motion in the first line, the tables are correct. But if it does not remain, they are incorrect. And correct them by the continuation. However, a caution is to be added: if in a non-leap year the preceding line for a leap year is subtracted from the following one and with this you subtract from the residual the motion beside the one day and if there remains the motion in the first line, the tables are correct. **For example:** if I take in the Table of expanded years the mean motion of the Sun beside the second year, that is 11ˢ 29;31,18,44°, and subtract from it the motion beside the third year, that is 11ˢ 29;16,58,6°, there remains the motion from the first line beside the first year.[57] And so I will correct the other motions of the Sun and other planets, respectively, to the leap year, as has been noted. **The third part on the continuation and correction of the motions in the Tables of the months.** If you want to continue and redo the Table of the motions of any planet in months, you should know that the first line of any table in January is the last line of the Table of days.[58] **For example:** the first line for January in the common year in the mean motion of the Sun is 1ˢ 0;33,18,8°. And the same motion is found in the Table of the days of the Sun beside 31, that is at the end of the same table of the Sun.[59] And it is similar for the other planets and their motions for January. But if you want to find the motion in the lines of other months, take the first line of January, add to it the motion of as many days as are contained in the month whose line you are seeking. If you would like to have the motion of the Sun in February, you should add to the motion found in January, that is in the first line of months, the mean motion of the Sun beside 28 days, namely 0ˢ 27;35,53,9°, and this line will give the motion of the Sun in February in the non-leap year: 1ˢ 28;9,11,18°.[60] And so continue in the other lines of other months with whatever planet you want and you will find the motion of any month. **Caution:** if you wish to find the motion of the Sun in February of a leap year, take the motion of 29 days and add it to the first line of January, because you have to put in one extra day, and it results 1ˢ 29;8,19,37° in February.[61] When the lines of other months in the non-leap year are found in this way, you add the motion of one day in any line of any planet and the motion of all the months in the lower auxiliary table of the leap year results.[62] So, beside the March in the non-leap year, one places 2ˢ 28;42,29,46°.

---

56 The first line, column C of the Table TRW-05 of mean motions in days (which is a rounded value of the corresponding basic parameter of Alfonsine Tables) is to be added to the third line of the Table TRW-02 and two circles subtracted to get the fourth line.

57 See the Table TRW-02.

58 It means that the first line of the Table TRW-03 and TRW-04 is the last line of the Table TRW-05.

59 This value is at the top of the column C in the Table TRW-03 and at the bottom of column C in the Table TRW-05.

60 See columns C in the Tables TRW-05 and TRW-03.

61 That is because of the intercalary (*embolismalis*) day, the February has 29 days in a leap (*bisextilis*) year – see column C in the Table TRW-04.

62 The Tables of mean motions in non-leap and leap years are written on the top and bottom of the same page in many manuscripts, however in some other manuscripts they are dispersed to other pages (e.g. below the tables for collected and expanded years in MS Toruń, BU, 74/III) to save a space.

Adding to this $0^{s}$ 0;59,8,20°, the line for March in the leap year in the lower table becomes $2^{s}$ 29;41,37,46°.[63] And so it is to be understood in other cases. **The last part on continuing and correcting motions in the Tables of days, hours and minutes**. If you want to continue and to make anew the Tables of days for planetary motions in days, through the one-day motion of the planet whose motion you want to continue, do it similarly in the planet's hours using the one-hour motion. From there the first supposedly correct line from Alfonso is taken. If you add the line to itself, the second line results. If you add the first to the second, the third line results. If you add the first to the third, the fourth results, and so on, both in days and in hours. **For example:** in the Table of the Sun in days, beside the first day, there is $0^{s}$ 0;59,8,20°. And so you will work gradually in the motion of the other lines, as well as in the hours. **A way of correcting the Tables of days and hours.** If you want to correct the Tables of daily motions of the planets, subtract the previous from the next line and you will be left with the first line. If it agrees, the tables are correct. If the first one does not remain, they are wrong and you must correct them one by one. **For instance:** I take for the Sun in days the fifth day and it is $0^{s}$ 4;55,41,38°. I take the sixth line, and it is $0^{s}$ 5;54,49,58°. Subtract the former from the latter, that is, the lesser from the greater, and the first line remains $0^{s}$ 0;59,8,20°.[64] And the others are to be treated similarly. From what has been said, the practice of continuing and correcting the Tables of the mean motions of the planets and apogees follows. It is now necessary to proceed to the computation of the true motions of the planets. Before this can be done, the mean motions of the planets must be determined, which is the way in which the almanac is compiled, in order to avoid tedious and unnecessary work later on; it is therefore proper and right to call attention to this here.

<9>

**The ninth rule, how the mean motion of the Sun, Moon, and the lunar anomaly should be continued for each day throughout the year**

If you wish to continue to find the motion of the Sun and Moon, and similarly the anomaly of the Moon for each day of the year, first take the radix of the year of the desired planet. Second: take the motion of the same planet for one day from the Table of days and add it to the radix, and you will have the motion of the same planet for the first day of January. Write this motion for the first day of January in your notebook, which is prepared for this purpose. Third: add the motion in one day to this motion of the first day, and the motion of the second day of January will result. Write this motion also in your notebook. And if you add to the motion of the second day the motion in one day, you will have the motion for the third day, and so gradually you can go on for 31 days, after which January ends. **For example:** if I want to continue to the year of the Lord 1487 with the mean motion of the Sun starting from the first day of January until the end of the year for the almanac of the

63 See columns C in the Tables TRW-03, TRW-04, and TRW-05.

64 John of Głogów chose an example of the fifth and sixth line in the column C of the Table TRW-05, the difference of which is really equal to the first line, but he avoided the difference, e.g., between the second and first line, or the fourth and the third, which are for one tertia smaller, because in reality the tables are not created by a simple multiplication of the first rounded value but by a rounding of values computed with a higher precision.

current year 1487, look first for the radix of the current year according to what was said before, and it is: $9^s$ 18;42,56,38°. Secondly, to this radix add the motion of the Sun in one day,[65] namely, $0^s$ 0;59,8,20°. And after adding it, you will find the motion of the Sun for the first of January: $9^s$ 19;42,4,58°. If you add to this motion of the first day the motion of the Sun for one day, the motion of the Sun for the second day of January will come out: $9^s$ 20;41,13,18°. Similar is the case with the Moon and the anomaly of the Moon for any day. **A method of correcting your procedure.** When you have already completed the January days for the motion of the Sun, Moon and anomaly of the Moon and you want to know if you have done it well, proceed in this way: take the motion of the Sun or of the Moon or the anomaly of the Moon placed beside January, according to whether it is a common or a leap year,[66] and add this motion to the radix of the Sun if you have worked with the Sun, or to the radix of the Moon if you have worked with the Moon, or to the radix of the anomaly of the Moon if you have worked with it. If the motion to the end of January results, which you have found by the successive addition, especially in signs, degrees and minutes – no need to worry about tertias and quartas[67] – then you have worked properly and have proceeded well. If, however, the motion you found first by successive addition disagrees in minutes with the one you found when correcting, you continued badly. It is therefore necessary to seek that day in which you made a mistake and that you correct it, together with those that follow. **For example:** if to the current year of the Lord 1487 you complete all the mean motions of the Sun up to the end of January, then if you add to the radix of the same year, namely $9^s$ 18;42,56,38°, the motion which is in the Table of months under the inscription of the Sun in the common year beside January, namely $1^s$ 0;33,18,8°, you will find the end of January of the same year, which you similarly had to find by your continuation, that is this: $10^s$ 19;16,14,47°. **A necessary consideration.** Every motion that results after adding the motion of any month to the radix is written down as the last day of your mont. Continue with it while looking for the motion in the days of the following month. It is in fact more accurate to proceed in this way than by successive addition, and at the end of each month it is recommended to carefully check the Sun, Moon and anomaly of the Moon.

<10>

### The tenth rule, how the centre of the Moon is to be continued for each day of the year for the almanac

First, find the radix of the centre of the Moon. Find it this way: look for the radix of the motion of the Moon at the beginning of the year. Similarly, find the radix of the motion

65 The radix to the year 1487 is the sum of cyclical radix for Cracow 1468 in the Table TRC-01 and the motion in 18 expanded years in the Table TRW-02. The motion in one day agrees with the value in the Table TRW-02, but the subsequent sums are incorrectly ending by 59 and 19 tertias in both MSS Vatican, BAV, Pal. lat. 1439, f. 171v and Cracow, BJ, 600, f. 151v.

66 That is, in the corresponding Table TRW-03 or TRW-04.

67 The quartas are not taken into account at all in the described procedure of continuation. The tertias are not likely to agree because the tables in months are derived by rounding of values computed with a higher precision. A more precise procedure would be to add to the radix the values in corresponding number of days from the Table TRW-05 instead of the successive addition of the rounded value for one day.

of the Sun, and subtract the radix of the Sun from the radix of the Moon, and double what results when you subtract one radix from the other, and after the doubling you will have the centre of the Moon. And to this add the motion of the centre of the Moon in one day, and you will have the motion of the centre on the first day of January. If you add the same to this found motion of the first day, the motion of the centre of the second day will result and so on. Find further the motion of the centre of the Moon in each day and month in the table called Motion of the centre of the Moon in days. And in correcting, whether you have worked well, proceed in the same way as was given in the previous rule, examining the last day of each month. **For example:** for the current year of the Lord 1487, I want to know the radix of the centre of the Moon. First I search for the radix of the motion of the Moon and it is: $11^{s}$ 25;8,54,32°.[68] Second: I know the radix of the motion of the Sun and it is $9^{s}$ 18;42,56,39°.[69] Third: subtract the radix of the motion of the Sun from the radix of the Moon,[70] and double what will result after the subtraction, and the double will be the radix of the centre of the Moon[71] and it will be $4^{s}$ 12;51,55,48°. If I add to that radix of the centre that given in the Table of the centre of the Moon in days[72] for the first day, the centre of the Moon will be found for the first day of January. That which is given for one day is $0^{s}$ 24;22,53,23°. And the motion of the centre of the Moon on the first day of January will be:[73] $5^{s}$ 7;14,49,9°. And so you will continue until the end of each month and to practice the correction follow the above-described method.

<11>

## The eleventh rule. Finding the mean motion of the three upper planets, Saturn, Jupiter, and Mars, for composing an almanac at steps of ten days

To find the mean motion of the three upper planets, Saturn, Jupiter and Mars, to make an almanac, it is sufficient to proceed in ten by ten.[74] Once found, the mean and true motions for intermediate days are very easy to determine, as will be taught below. So if you want to continue the mean motion of the three upper planets in ten by ten, first prescribe the radix of the one whose motion you want to continue. Second: enter the Table of days of the planet whose motion you want to continue and find the motion beside the ten days.

68 It is got adding the entries for the year 1468 in the Table TRC-01, column D, and for 18 years in the Table TRW-02.

69 See the example in the Rule 9. In MS Cracow, BJ, 600, f. 152r, the final number of 38 tertias obtained there is rewritten to 39, so that both readings are possible.

70 The difference is $2^{s}$ 6;25,57,54° for final 38 tertias in the radix of the Moon.

71 In the MS Cracow, BJ, 600, f. 152r, the final 48 tertias are rewritten to 46 tertias, which would correspond to 39 in the radix of the Moon.

72 Such a table is not a part of standard set of the *Tabule resolute* but it is a part of the Table A-02 which is added to many copies of the TR. This means to give the motions at steps of ten days. This Table A-02 gives the mean motions with a precision up to quartas and it is computed with even higher precision, or it could also be extracted from the Tables of mean motions in Parisian Alfonsine Tables. In particular the centre of the Moon is doubled elongation of the Moon from the Sun (i.e. the difference of mean motions of the Moon and Sun), so that its values can be found there for the doubled number of days; see Poulle, *Les tables Alphonsines*, p. 138, Table 19 (Ratdolt 1483, p. D7r).

73 The final number of 9 tertias corresponds to 39 tertias in the radix of the Moon.

74 That means at steps of ten days.

Third: add this motion found beside the ten days to the radix, and what results after the addition will be the motion of the planet on the tenth day of January, and write it in the notebook. Fourth: to the mean motion found on the tenth day of January, again add the motion found beside the ten days and add to the motion of the tenth day, and the motion of the twentieth day will come out, which again write in the notebook. Fifth: add again the motion found beside the ten days to the motion found on the twentieth day, and when you have added it up you will have the motion of the thirtieth day. When you add to it the motion of one day, if necessary, that is, when the month has 31 days, the motion of the thirty-first day will be found, and you shall again note this motion in the book. And so you shall continue through all the months to find the motion of the three upper planets. **For example:** for the current year of the Lord 1487 I want to extend the mean motion of Saturn in ten by ten to create an almanac. First I will look for the radix of Saturn for the beginning of that year[75] and it is: $8^s$ 14;48,27,58°. Second: to this radix add the motion of Saturn found beside the ten days,[76] namely $0^s$ 0;20,5,53°. And after adding them together, the mean motion of Saturn on the tenth day of January is: $8^s$ 15;8,33,51°. Third, if you add to the motion found on the tenth day what will be beside the ten days, the mean motion of Saturn on the twentieth day of January will be: $8^s$ 15;28,39,44°. If you add the same to it, this mean motion results for the thirtieth day: $8^s$ 15;48,45,37°. And so similarly you will work with Jupiter and Mars. **First caution:** notice carefully how many days each common month has. If it has 30 days, the matter is perfected. But if the month has 31 days, then find the motion of the 30th day and add to this motion of one day to complete thirty-one days; this is the case in January, March, and other months. **Second caution:** if you work with February and had the mean motion of the planet on the twentieth day, you would not add to it what you would have beside, that is, the ten days, because February has only 28 days in the non-leap year. And so you enter with eight days, and add the motion found beside the eight days to the motion of the twentieth day, and so you complete February. In a leap year, because of the extra day, you must enter with nine and add the motion found beside the nine days to the motion of the twentieth day. And so you will continue after February of the leap year.

<12>

### The twelfth rule. Continuing in the equated anomaly of the three upper planets, namely Saturn, Jupiter and Mars, at steps of ten days, for making the almanac

When you want to continue with the anomaly of the three upper planets in ten by ten to make an almanac, first find the radix of the anomaly, which you find as follows: take the radix of the motion of the Sun and the radix of the motion of the given planet. Second: subtract the radix of the motion of the given planet from the radix of the motion of the Sun, and what remains after subtraction will be the radix of the anomaly of the given planet. And from this radix thus found you will proceed with the anomalies of the three upper

75 This value can be obtained by adding the values from columns H from the Tables TRC-01 and TRW-02.
76 See the Table TRW-05, column H.

planets in ten by ten in the following manner and procedure: take the mean motion of the Sun in ten days and also the motion of the planet in ten days, then subtract the motion of the planet of ten days from the mean motion of the Sun of ten days, and what remains will be the mean anomaly of the planet of ten days. Add this planet anomaly of ten days to the radix of the planet anomaly and you will have the given planet's anomaly for the tenth day of January, and write this anomaly in your notebook, just as you wrote down the mean motion of the planet. Then add again the anomaly of the given planet, and finally work as in the previous rule, and so find the motion of the anomaly of the three upper planets in each month of the whole year. **For example:** for the current year of the Lord 1487 I want to extend and continue the mean anomaly of Saturn in ten by ten to create an almanac. First I must have the radix of the Sun's motion[77] and it is: $9^s$ 18;42,56,39°. I must also know the radix of Saturn's motion[78] and it is: $8^s$ 14;48,27,58°. If I subtract this radix of Saturn from the radix of the Sun, I am left with the radix of the Saturn anomaly: $1^s$ 3;54,28,41°. Second: I search for the anomaly of Saturn in ten days and take the mean motion of the Sun in ten days and that is: $0^s$ 9;51,23,16°. Then I find the motion of Saturn in ten days[79] and that is: $0^s$ 0;20,5,53°. And I subtract the motion of Saturn in ten days from the mean motion of the Sun also in ten days, and the anomaly of Saturn in ten days remains: $0^s$ 9;31,17,23°. I add this to the radix of the anomaly of Saturn and the anomaly of Saturn in the tenth day of January results: $1^s$ 13;25,46,4°. I shall write this in my notebook beside the tenth day of January at the mean motion of Saturn, and proceed as before said, adding the anomaly of the ten days to what results, and I will have it for the twentieth day of January, and it is: $1^s$ 22;57,3,27°. However, keep in mind the two cautions noted in the previous rule about the end of months and about February. And so work with Jupiter and Mars as well.

<13>

### Rule 13. Continuing the mean centre of the three upper planets at steps of ten days for making an almanac

If you wish to continue the mean centre of the three upper planets in ten by ten to make an almanac, first search for the radix of the motion of the centre of the given planet, and you will find it as follows: look first for the radix of the planet's motion, then look for the radix of the apogee of the given planet, and subtract the radix of the apogee of the planet from the radix of the mean motion of the planet, and what remains after the subtraction will be the radix of the mean centre of the given planet. Second: to the radix of the apogee of the planet, add the motion of the apogee in the ten days, and that will be the apogee of the planet for the tenth day of January. This apogee, found for January 10, must be subtracted from the mean motion of the planet. What you find on January 10 and what remains after subtraction will be the centre of the planet on January 10. And write it in your notebook next to the motion and anomaly of the given planet. Third: subtract the radix of the planet's centre from the centre found for January 10, and what remains will be the

---

77 See the Rule 9, where, however, the value $9^s$ 18;42,56,38° is found.

78 See the example in the previous Rule 11.

79 See the values in columns C and H of the Table TRW-05 for the Sun and Saturn, respectively.

centre of the planet for ten days. Using this ten day centre, continue with all the centres of the same planet throughout the year and write it in your notebook. **For example:** for the current year of the Lord 1487, I want to continue and extend the mean motion of the centre of Saturn in this year in ten by ten, for making an almanac. First: I must have the radix of the mean motion of the centre of Saturn, which I shall get know by subtracting the radix of the apogee of the given planet from the radix of the motion of Saturn. The radix of Saturn's motion[80] is $8^s$ 14;48,27,58°. The radix of Saturn's apogee[81] is $8^s$ 13;1,43,59°. And after the subtraction, the radix of the centre of Saturn remains $0^s$ 1;46,44,59°.[82] Second: to the radix of the apogee of Saturn add the motion of the apogee of the ten days, namely 0 seconds, 57 tertias, and you have this apogee of Saturn for the $10^{th}$ day of January: $8^s$ 13;1,43,56°. Subtract the apogee from the mean motion of Saturn on the $10^{th}$ of January,[83] which is $8^s$ 15;8,33,51°, and the centre of Saturn for the $10^{th}$ of January will result: $0^s$ 2;6,49,55°. And write this in your notebook next to the motion and anomaly of Saturn. Third: subtract the radix of the centre of Saturn, namely $0^s$ 1;46,44,59°, from the centre of Saturn found on the $10^{th}$ day, namely $1^s$ 2;6,49,55°, and what remains after the subtraction will be the centre of Saturn in the ten days and will come out $0^s$ 0;20,4,56°. Using Saturn's ten-day centre, continue with all of Saturn's centres throughout the year, and write it in your notebook next to Saturn's motion and anomalies.

<14>

**Rule fourteen on mean anomaly of the two lower planets, Venus and Mercury, for making an almanac**

If you want to continue with the anomaly of Venus throughout the whole circle of the year in five by five,[84] you have to work this way: first get the radix of the anomaly of Venus, add to it the Venus anomaly beside the five days found in the Table of Venus, and the anomaly of Venus for the fifth day of January results. If you add the same to the anomaly of Venus on the fifth day, the Venus anomaly for the $10^{th}$ day of January results, and so you continue to the end of January and all the months of the year. **For example:** for the current year of the Lord 1487 I want to continue with the Venus anomaly throughout the whole year month by month. First find the radix of the Venus anomaly[85] and it is: $10^s$ 10;1,48,3°. If I add to this radix of the Venus anomaly the motion of the Venus anomaly that is beside the five days in the Venus anomaly table,[86] namely $0^s$ 3;4,57,17°, it results $10^s$ 13;6,45,20°. If I again add to this the motion of the Venus anomaly for the 5 days found in the table,

80 See the example in Rule 11.
81 The radix of Saturn's apogee for 1468 is $8^s$ 12;51,32,34° according to the column C of the Table TRP-07 and its motion in 18 years is $0^s$ 0;10,10,25° according to the column C of the Table TRP-07a, so that the radix of the apogee for 1487 should be $8^s$ 13;1,42,59° and not 43 seconds.
82 The difference is calculated with the correct radix of apogee.
83 See the example in Rule 11.
84 I.e., at steps of five days.
85 The radix for the year 1487 is the sum of the cyclical radix for Cracow 1468, which is $7^s$ 7;3,19,57° in column K of the Table TRC-01, and the increase in 18 years in the Table TRW-02, which is $3^s$ 2;58,28,7°, so the sum is $10^s$ 10;1,48,4°.
86 See column K of Table TRW-05.

the motion of the Venus anomaly for the $10^{th}$ day results, namely $10^s$ 16;11,42,37°, and so I continue the anomaly of Venus for each month of the whole year. **If you want to show the anomaly of Mercury throughout the year in three by three days, you have to work in this way:** first, find the radix of Mercury anomaly. Add to this radix the motion of the anomaly found beside the three days in the Table of Mercury anomaly and you will have the Mercury anomaly for the $3^{rd}$ day of January. If you add the same to the anomaly found for the $3^{rd}$ day of January, you get the anomaly for the $6^{th}$ day of January, and so you will continue until the end of January and throughout the other months. **For example:** I want to continue and extend the Mercury anomaly for the current year of the Lord 1487 for the creation of the almanac. First I need to know the radix of Mercury anomaly[87] and it is $0^s$ 2;5,47,57°. If I add to this radix the motion of the anomaly of Mercury found beside the three days in the Table of Mercury anomaly,[88] namely $0^s$ 9;19,12,23°, when added together, the anomaly of Mercury for the $3^{rd}$ day of January results, namely $0^s$ 11;25,0,20°. If I again add to this radix the motion of the anomaly of Mercury for the three days, you will have the motion of the anomaly of Mercury for the $6^{th}$ day, and it is $0^s$ 20;44,12,43°. And so continue with the Mercury anomaly for all the months of the year.

<15>

### Rule fifteen on the mean centre of Venus and Mercury for making an almanac by continuing through the whole year

If you want to extend the Venus centre over the months of the year to make an almanac, first find the radix of the Venus's centre, and you find it in this way: subtract the radix of the apogee of the Sun from the radix of the motion of the Sun, and what remains after subtraction will be the radix of the Venus anomaly. Second: take the motion of the Sun on the $5^{th}$ of January. Similarly, take the apogee of the Sun on the $5^{th}$ of January. And if you subtract this apogee of the $5^{th}$ day of January from the motion of the Sun on the $5^{th}$ day of January, the result will be the centre of Venus for the $5^{th}$ day of January. Third, subtract the radix of the centre of Venus from the centre of Venus on the $5^{th}$ day, and in the result there remains the centre of Venus in five days. Using this five-day centre, you will continue with the centre of Venus through the months of the entire year according to the rules and cautions previously established above. **For example:** for the current year of the Lord 1487, I want to continue through the individual months of the year of the centre of Venus to compose an almanac. First: I must have the radix of the centre of Venus, and I find it as follows: first find the mean motion of the Sun, which is the radix of that year,[89] and it is $9^s$ 18;42,56,39°. Then I find the radix of the apogee of the Sun[90] and it is $3^s$ 1;3,23,59°. This radix of the apogee of

87 I.e., the sum of the radix for Cracow 1468, which is $3^s$ 8;38,7,3° in Table TRC-01, column L, and the increment in 18 years, which is $8^s$ 23;27,40,54° in Table TRW-02, column L.

88 See Table TRW-05, column L.

89 See Rule 9.

90 Similarly to the example in Rule 3 for the beginning of the year 1486, the radix of apogee of the Sun and Venus for the year 1487 is now $3^s$ 0;53,13,34° plus its increment $0^s$ 0;10,10,25° in 18 years (see Table TRP-07a, column C inscribed 1468).

the Sun when I subtract it from the radix of the motion of the Sun, I am left with the radix of the centre of Venus and it is 6ˢ 17;39,32,40°. Second: I must know the mean motion of the Sun for the 5th day of January,[91] and this is 9ˢ 23;38,38,17°. I must also know the apogee of the Sun on the 5th day of January,[92] and this is 3ˢ 1;3,24,28°. Subtract this apogee of the 5th day of January from the mean motion of the Sun on the 5th day of January, and the result is the centre of Venus on the 5th day of January, and this is 6ˢ 22;35,13,49°. Third, subtract the radix of the Venus centre on the 5th day from the Venus centre on the 5th day and in the result you are left with the increase of Venus's centre in the five days and it is 0ˢ 4;55,41,9°. Using this centre of Venus in the five days you will continue with the Venus centres for each month of the year. **Similarly should Mercury be treated, using its apogee and the motion of the Sun in relation to the motion of the three days.** If, therefore, you want to extend the centre of Mercury through each month for making an almanac, first take the radix of the centre of Mercury, which you will find in this way: subtract the radix of the apogee of Mercury from the radix of the motion of the Sun, because the motion of Mercury is identical with the motion of the Sun, and what comes out after subtraction will be the radix of the centre of Mercury. Second: take the motion of the Sun or Mercury, which is the same, in three by three days. Similarly the apogee of Mercury in three by three. If you subtract this apogee of Mercury in three days from the motion of the Sun or Mercury of three days in January, you are left in the result with the mean centre of Mercury for the third day of January. Third: subtract the radix of Mercury's centre from Mercury's centre on the third day of January, and the result leaves the increment of Mercury's centre in three days. Using this centre in the three days, you will continue with the centre of Mercury for each month of the year, following the rules and cautions noted above. **For example:** for the current year of the Lord 1487, I want to continue through the months of the year with mean centre of Mercury for compiling an almanac. First find the radix of Mercury's centre, and you find this radix as follows: first you must have the radix of Mercury's apogee,[93] and it is: 7ˢ 0;17,33,59°. Second, you must have the radix of the motion of the Sun and it is: 9ˢ 18;42,56,39°. Then subtract the radix of the apogee of Mercury from the radix of the motion of the Sun and what results after the subtraction is the radix of the centre of Mercury and it is: 2ˢ 18;25,22,40°. Second, take the motion of the Sun or Mercury, which is the same, in three by three days,[94] and the motion of the Sun on the third day of January will be this: 9ˢ 21;40,21,38°. Similarly seek the apogee of Mercury in three by three days,[95] and this is: 7ˢ 0;17,34,16°. And subtract this apogee from the motion of the Sun or Mercury of three days, and the centre of Mercury

91 This can be calculated adding to the radix the motion in five days 0ˢ 4;55,41,38° from column C of the Table TRW-05.

92 This value is obtained here by adding to the radix of apogee the motion of 28 tertias and 30 quartas in five days from the Table TRC-07c, column A, rounded to 29 tertias. However, the proportional part of the annual increase in the year 1487 according to Table TRP-07a should be 27 tertias and 44 quartas.

93 The radix of Mercury's apogee in 1487 is its cyclical radix 7ˢ 0;7,23,34° in 1468 (according to Table TRP-07, column H) plus its increment in 18 years, which is 0ˢ 0;10,10,25° as above.

94 That is 0ˢ 2;57,24,59° from the column C of Table TRW-05 which is to be added to the radix 9ˢ 18;42,56,39° of the mean motion of the Sun.

95 That is 17 tertias according the Table TRW-07c to be added to the above given annual radix of Mercury's apogee.

on the third day of January will come out, that is $2^{s}$ 21;22,47,22°. Third, from the centre of Mercury on the third day of January, subtract the radix of the centre of Mercury, and in the result the centre of Mercury in three days will remain as follows: $0^{s}$ 2;57,24,42°. Using this centre in the three days you shall continue with the centre of Mercury for each month of the entire year according to the above rules and cautions. And these words are sufficient to continue the motions of the planets for the almanac.

<16>

**Rule 16 on a general entry into the Tables of planetary equations for calculations of true motions as shown both by reason and by easy entry in the Tables of equation for true planetary motions**

First of all, it is important to know and pay attention to the fact that in the Tables of equations, the equations are written beside the degrees on the side. The differences in equations are nothing else than the excess of the larger equation over the smaller one.[96] The differences between the equations are written above the lines with the equations. Both above and below there is a number of signs and there are placed the letters 'a', 's', or 'm'. 'A' means that the equation is to be added, 's' or 'm' means that the equation is to be subtracted.[97] However, in some tables is explicitly written 'add' or 'subtract' or 'decrease'. Care should also be taken that when a red or black paragraph is in the tables of equations, then two letters, namely 'a' and 's', or 's' and 'a', are put there in the margin, both at the top and bottom, so that one is written above the other, the upper one on both margins referring to the equations which are above the paragraph. Similarly, the lower letter on both margins refers to those equations which are below the paragraph. And in this way the Sun and Moon are equated. It is also to be noticed that in the Tables of the equations of the other planets the nearer distance[98] is given in proportional minutes; there, too, two letters are put on both margins, one upper and the other lower. Of these, the nearer on both edges refers to the proportional minutes that are above the nearer, and the other letter refers to the minutes that are below the nearer.

About how to enter Tables of equations. The Tables of equations are entered with the common signs, six of which are placed in the top margin, namely 0, 1, 2, 3, 4, 5, and their degrees are placed in the first column on the left, going down.[99] And the other six are placed in the bottom margin, namely, 6, 7, 8, 9, 10, 11, and in reverse order. And their degrees are placed in the last column on the right, going up. If, therefore, the sign with which you enter is on the top margin or in the upper space, then the degrees of the sign are found descending to the left and directly in relation to the degree the equation

96 These differences are the increments between the tabulated values. Their sign is indicated by inscriptions above and below their columns. They are written shifted for half of the line height in later copies of the *Tabule resolute* but not in MS Toruń, BU, 74/III.

97 The letter 'a' is an abbreviation for *adde*, 's' for *subtrahe*, 'm' for *minue*.

98 See Fig. 26 for the indication of the nearer or farther distance (*longitudo propior* or *longior*) in the Tables of equations.

99 The degrees for the first six signs in the first leftmost column are increasing downwards from the top of the folio, see Fig. 26.

is found which you also write down. Then you find the difference, which is directly written in the column descending, and similarly the letter written above the difference in the upper margin, which is written separately at that difference and that letter. If it is 'a', this indicates that the proportional part is to be added to the obtained equation. If it is an 'm', it means that the proportional part is to be subtracted from the equation. If, however, the sign to be entered is on the bottom margin, then the degree of the sign is taken at right, ascending, and directly beside the degree to the left the equation or the proportional minutes or the variation of diameter[100] is obtained, which are noted separately. Then begins the difference, which follows immediately in the column ascending, and similarly the letter written below the difference at the lower margin; this letter is written aside at the difference. And that letter, if it be 'a', indicates that the proportional part is to be added to the obtained equation. If there is an 's', it indicates that it is to be subtracted. **It is also necessary to know** that by using the proportional minutes or by using the difference of the equation, the proportional part is found in a special table, which is called the Table for finding proportional minutes, which table is extended by the sexagesimal number both on the upper edge and on the side. What must be known about its entry is that after one number is led into another, the denominator must be seen, which is indicated by the inscriptions written at the top. After knowing the other fractions according to the Tables of mean motions in hours, one enters the same table with two numbers; with one written in the column of numbers on the side related to the other one in the upper margin, and is then resolved according to the denomination the assignment. **First consideration.** In the Table of equations of the eighth sphere only three tables are described,[101] because these equations extend over a quarter of a circle, that is, over three signs. Then these equations are also adapted to the table of all twelve signs, namely to the number of signs. In the upper part above the tables they are placed from only zero to three together with their opposite signs. At the bottom they continue from only three to six similarly with the opposites, as can be seen in the tables.[102] **Second consideration.** In the tables for the equations of the Sun, six sub-tables are given, extending for one half of the zodiac, and at the upper part or margin, signs from zero up to six are given.[103] At the bottom part, the other signs are found written from six to eleven. If you have to enter to these tables, you will get the equations of the Sun with the difference of the equation according to the method described above. **Third consideration.** In the Moon and other planets, many tables of equations are written under each sign because the Moon and other planets have several different motions, because some are eccentric, some epicyclic, some longitudinal. In the tables of the other planets you will find at the end The table of first station of the planet. And how to derive from it the second station, it will be shown below in the rule concerning how to recognise whether a planet is stationary or direct.

---

100 Variation of diameter is *diversitas diametri*.

101 This means that only three subtables are written, each one for one sign.

102 See Table TRW-08. The numbers of signs inscribed at the top run from 0 to 2 only (and their opposites from 6 to 8) and at bottom from 3 to 5 (and 9 to 11).

103 Latin expression *ad sex exclusive* means 'up to six without six', i.e., 'up to five' because the value for six signs is the last in the column for degrees of the fifth sign.

<17>

### Rule 17 on finding the common apogees

If you want to find the common apogee of a year, first find the motion of the apogees and the fixed stars of the same year, using the method explained in the first rule. Second: find the motion of the access and recess of the eighth circle[104] in the same way. Third: with the signs and degrees which you have found in the motion of the eighth circle, you will enter the Table of equations of the same eighth circle and take the equation for the signs and degrees and write at it 'a' or 's', as you will find in the same tables. Fourth: take the difference of the equation and write 'a' or 's' next to it, as you will find in the Table of the equation. Fifth: find the proportional part in the manner described in the previous rule, converting the difference to the remainder of the number, as stated in the previous rule. Sixth: add or subtract the proportional part found from the equation previously found and you will have an equated and finished equation. Finally: add or subtract this equated and finished equation from the motion of the apogees and fixed stars, as the inscription will teach you, and after this addition or subtraction you will obtain the common apogee. If, however, you want to have the special apogee of any planet, add in this way the common apogee to the special apogee of each planet separately at the time of the Nativity of Christ, and what results after addition will be the special apogee of each planet at the time you intend. The radices of the apogees for the time of the Nativity of Christ are in the following table:[105]

**Radices of apogees at the time of Nativity of Christ:**

| S | G | M | S | T | | |
|---|---|---|---|---|---|---|
| 2 | 11 | 25 | 23 | 4 | | Sun and Venus |
| 7 | 23 | 23 | 42 | 4 | | Saturn |
| 5 | 3 | 37 | 0 | 4 | apogee of | Jupiter |
| 3 | 25 | 12 | 13 | 4 | | Mars |
| 6 | 10 | 39 | 33 | 4 | | Mercury |

**For example:** I want to know the common apogee for the current year of the Lord 1487. First, I must look for the motion of the apogees and the fixed stars for the year by the method given in the first rule,[106] and this is $0^{s}$ 10;55,3,9°. Second: I must look for the motion of the access and recess of the eighth circle for the same year, and this is $2^{s}$ 15;37,56,0°. Third: I must enter with the motion of the eighth circle, namely, with two signs and 15 degrees,

104 The term 'eighth circle' (*circulus octavus*) means 'eighth sphere', i.e. the sphere of stars in the Ptolemaic model.

105 These radices are extracted from the PAT. We can find them among other radices in Poulle, *Les tables Alphonsines*, pp. 124–26, Table 9 (Ratdolt 1483, ff. C8r–D1r), or in many manuscripts (e.g. Prague, NK, X A 23, f. 4r–v, Cracow, BJ, 550, ff. 5v–6r, Vienna, ÖNB, Cod. 2352, ff. 56v–57v, or Vatican, BAV, Pal. lat. 1374, ff. 6r–7r). These radices are also written below Table TRW-07 of true planetary apogees in the MS Toruń, BU, 74/III. While the radix of apogee of the Sun is truncated to seconds in the PAT, in the TR it is given to 4 tertias like the radices of other planets.

106 The example in Rule 1 was done to the previous year 1486, so the difference here is to take in Table TRW-02 the motions in 18 expanded years.

and look at the equation in the table which is prepared for it. And after entering with the aforementioned number, I find in the equation 8 degrees, 41 minutes, 28 seconds, whose inscription is 'a'. The difference written for the equation is 2 minutes and 22 seconds; its inscription is 'a'.[107] Fourth: I must convert this difference to the remainder with which it is not entered, and I must look for the proportional part, and 1 minute, 29 seconds and 47 tertias results.[108] This proportional part must be added to the equation first found, namely 8 degrees, 41 minutes and 28 seconds, and after adding it, this equated and finished equation becomes: 8 degrees, 42 minutes, 57 seconds and 47 tertias. I have to add this finished and equated equation to the motion of the apogee and fixed stars previously found, and this common apogee at the beginning of the year becomes: $0^{s}$ 19;38,0,56°. If I add to this common apogee the radix of the apogee of any planet at the time of Christ, I will have the apogee of any planet for the year I have assumed. So that at the beginning of the year 1487, this apogee of the Sun is: $3^{s}$ 1;3,24,0° and in this way you will find the apogees of all the planets.

<18>

**Rule 18 on finding the true motion of the Sun**

If you want to have the true motion of the Sun, first find the mean motion of the Sun as above. Second: take also its apogee on the same day, as also stated above. Third: search for the anomaly of the Sun for the same day, which, as stated above, you will find by subtracting the apogee of the Sun from the motion of the Sun. Fourth: with the anomaly you will enter the Table of equations and get the equation of the Sun's centre with the difference; at this you write the inscriptions, as the table will teach you. Fifth: find the proportional part by the method told in the sixteenth rule. Add or subtract this proportional part from the equation of the anomaly according to the directions of the inscriptions, and the equated and finished equation will come out.[109] Finally, add or subtract this equation from the motion according to the letter written or placed by the equation and you will have the true motion of the Sun for the time you expect. **For example:** I want to know the true motion of the Sun to the day of the Assumption of Mary in the current year of the Lord 1486. First: I must have the motion of the Sun on that day, and it is $5^{s}$ 2;41,47,30°. Second: I must have the apogee of the Sun for that day, and it is $3^{s}$ 1;3,11,50°. Third: I must have the anomaly of the Sun for that day,[110] and this I get by subtracting the apogee of the Sun from the motion of the Sun and it is: $2^{s}$ 1;38,35,40°. Fourth: with this anomaly, you will enter the Tables of equations and this equation results: 1 degree, 51 minutes and 51 seconds, at which the letter 's' is written. The difference is 1 minute and 5 seconds and its caption is 'a'. Fifth: search for the proportional

107 See Table TRW-08. The Tables of equations in MS Toruń, BU, 74/III do not give the differences, but the value $0^{s}$ 0;2,22° can be calculated as the difference between the equations for $2^{s}$ 16° and $2^{s}$ 15°.

108 It can be found in Table A-04 that 0;37 × 0;2° = 0;1,14°, 0;37 × 0;0,22° = 0;0,13,34°, 0;0,56 × 0;2° = 0;0,1,52°, and 0;0,56 × 0;0,22° = 0;0,0,20,32°, which added together gives 0;1,29,46,32° and after rounding to the tertias the presented value.

109 I.e., Table TRW-13.

110 See the example in Rule 4.

part and you get 0 minutes, 41 seconds and 49 tertias,[111] which you add to the equation you found first and you will get this finished equation: 1 degree, 52 minutes, 32 seconds and 49 tertias. This resulting equation must be subtracted from the mean motion and then the true motion of the Sun for the day of the Assumption of Mary is $15^{s}$ 0;49,14,41°.

<19>

**Rule 19 on finding the true motion of the Moon**

If you want to have the true motion of the Moon for a given time, first look for its motion as stated in the second rule. Second: find the anomaly of the Moon using the table that is provided for it, as described for the motion of the planets. Third: find the centre of the Moon, following the fifth rule, namely, by subtracting the motion of the Sun from that of the Moon and doubling the remainder. Fourth: with the mean centre of the Moon for the expected day, enter the table of the Moon and find the equation of the centre, noting whether it is marked with an 'a' or a 's'. And similarly you will take the difference and its equation as mentioned above. Similarly you will find the proportional part according to the mentioned method. Fifth: add or subtract the finished equation of the centre from the mean anomaly of the Moon and the true anomaly of the Moon will result. Sixth: with the mean centre, you will also find the proportional minutes according to the method stated above, which when they are found are kept aside. Seventh: with the true anomaly of the Moon, enter into the Table of the equations of the Moon, and find the equation of the anomaly of the Moon with its difference and write those letters as given in the table. And what results after your operation will be the first examined equation. Eighth: with that anomaly, the variation of diameter is investigated with its difference, and the proportional part is sought according to the method stated above. Ninth: using the variation of diameter and the proportional minutes previously set aside, find the proportional part according to the general method, placing the proportional minutes above and the variation of diameter below. Tenth: add the proportional part resulting from it to the first examined equation and the second examined equation will result. Eleventh: add or subtract the now finally examined equation for the motion of the Moon, as the inscription of the equation requires, and you will have the true motion of the Moon for your intended time. **First caution:** it should be considered and carefully taken into account that if there are no minutes or seconds in the proportional minutes for the Moon and other planets, then there is no need to look for the proportional part because then the first examined equation will be the second examined equation. **Second caution:** if 60 results in proportional minutes and nothing in the difference, then the variation of diameter is a proportional part. **For instance:** I want to know the true motion of the Moon in the year of the Lord 1486 for the Day of Saint Mary Magdalene.[112] First, I must know the motion of the Moon,[113] and it is:

---

111 Using Table A-04 of sexagesimal multiplication we find that 0;1,5 × 0;38,35,40 = 0;0,38,35,40 + 0;0,3,12,58,20 = 0;0,41,48,38,20.

112 The feast day of St Mary Magdalene is July 22.

113 The mean motions for the date 1486, July 22 can be found adding the values for the collected year 1468 in the Table TRC-01, for 17 expanded years in Table TRW-02, for June in Table TRW-03, and for 22 days in Table TRW-05.

$0^s$ 20;34,21,9°. Second: I must know the anomaly of the Moon, and it is: $4^s$ 7;53,23,8°. Third: I must know the centre of the Moon on the same day,[114] and it is: $4^s$ 23;3,46,58°. Fourth: with the centre of the Moon, I enter the Table of equations of the Moon and find the equation of the Moon's centre 10 degrees, 43 minutes.[115] Its inscription is 'a'. The difference is 10 and its inscription is 's'; the proportional part is 0 minutes, 37 seconds, 50 tertias.[116] These must be subtracted from the equation of the centre, and after subtraction the finished equation of the centre remains: 10 degrees, 42 minutes, 22 seconds, 10 tertias, whose inscription is 'a'. Fifth: I have to add this finished equation of the centre to the Moon's mean anomaly and this true anomaly of the Moon is: $4^s$ 18;35,45,18°. Sixth: with the same mean centre of the Moon find the proportional minutes and its difference. Proportional minutes will be 53 and in the difference 0, which proportional minutes must be kept aside. Seventh: with the true anomaly of the Moon, enter the Table of the equations of the Moon and find the equation of the anomaly of the Moon and the difference in the same way. The equation of the anomaly will be 3 degrees, 31 minutes, 3 seconds and its inscription is 's'.[117] The difference is 3 minutes, 53 seconds. The proportion is this: 2 minutes, 18 seconds, 51 tertias.[118] This is to be subtracted from the equation of anomaly and the first examined equation results: 3 degrees, 28 minutes, 44 seconds, 9 tertias, whose inscription is 's'. Eighth: with the same true anomaly, the variation of diameter is investigated with its difference and the proportional part. The variation of diameter is 2 degrees, 5 minutes. The difference is 2 minutes and its inscription is 's'. The proportional part is 1 minute, 11 seconds, 31 tertias,[119] and this is to be subtracted from the variation of diameter and this variation of diameter results: 2 degrees, 3 minutes, 48 seconds, 29 tertias. Ninth: using the variation of diameter and the proportional minutes previously set aside, find the proportional part according to the general method. Put the proportional minutes above and the variation of diameter below[120] and that proportional part will be: 1 degree, 49 minutes, 21 seconds, 50 tertias.[121] Tenth: add the resulting proportional part to the first examined equation first and the second examined equation will be: 5 degrees, 18 minutes, 5 seconds, 59 tertias, whose inscription is 's'. Finally: add or subtract this last examined equation of the motion of the

---

114 The mean motion of the Sun to that date is $4^s$ 9;2,27,40° and its difference from the mean motion of the Moon is $8^s$ 11;31,53,29°, which, multiplied by two, gives the mean centre.

115 This is the value for $4^s$ 23° in Table TRW-15.

116 This value is 0;3,47 × 10 minutes for which the value 10;43° should be decreased in the interpolation with respect to the subsequent value for $4^s$ 24°. Regarding the fact that the equation is rounded to minutes, the precision to tertias is unnecessary. Interpolating between more precise values, the finished equation of centre would be $0^s$ 10;43,12,50°.

117 See Table TRW-15.

118 I.e., 0;35,45,18 × 0;3,53°.

119 I.e., 0;35,45,18 × 0;2°.

120 Andrzej Grzymała in Rule 8 of his canons *Girum recensendo* similarly writes *ponendo minuta proporcionalia superius, diversitatem diametri inferius* (Cracow, BJ, 600, f. 230r). The two quantities are to be multiplied – see Eq. (22.5). This can be done using Table A-04 of sexagesimal multiplication, as it is described, e.g., in the Rule on multiplication by Johannes Schöner (1536, f. A3r–v). Because the proportional minutes usually have fewer sexagesimal digits (here just one) than the variation of diameter (four digits here), it is convenient to find the column with the digit of the proportional minutes in the header at the upper edge, and below it to find the multiples in the lines corresponding to the digits of the variation of diameter (written on the left edge).

121 I.e., 0;53 × 2;3,48,29°.

Moon according to what the inscrition of the equation requires, and you will have the true motion of the Moon at your intended time, and it is: $0^{s}$ 15;16,15,10°.

<20>

### The twentieth rule on finding the true position of head of dragon

If you want to have the true motion of the head of dragon, search first for its mean motion. Second: if you have the mean motion of the head of dragon, subtract it from twelve signs and the reminder will be its true motion. Third: with the true motion of the head of dragon you have the tail of dragon in the opposite sign in the similar degree, minutes, seconds, and tertias.

**For instance:** if I wish to know the true motion of the head of dragon for the day of Assumption of Mary in the current year of Lord 1486, I have first to know the motion of the head of dragon[122] and it is: $0^{s}$ 26;7,44,49°. Second: I must subtract the motion of the head of dragon from 12 signs and there will remain its true motion $11^{s}$ 3;52,15,11°. Third: compute the motion of the tail of dragon in the opposite sign, similar degree, minute, second, and tertias.

<21>

### Rule 21 on finding the argument of latitude of the Moon

If you want to know the true argument of latitude of the Moon, first get the true position of the Moon. Second: find the true position of the head of dragon. Third: subtract the true position of the head of dragon from the true position of the Moon, and what remains will be the true argument of latitude of the Moon. **For example:** I want to know the argument of latitude of the Moon on the day of Assumption of Mary in the current year of the Lord 1486.

First: I need to have the true motion of the Moon and it is: $11^{s}$ 1;59,4,43°. Second: I must know the true motion of the head of dragon[123] and it is: $11^{s}$ 3;52,15,11°. Third: I must subtract the true motion of the head of dragon from the true position of the Moon and that leaves the argument of the latitude of the Moon and it is: $11^{s}$ 28;6,49,32°.

<22>

### Rule 22 on finding the true positions of Jupiter, Saturn, and Mars

If you want to know the true motion of Saturn, Jupiter, and Mars, first find the mean motion on the intended day of the one whose true motion you want to know. Second: get the apogee of the same planet on the same intended day. Third: subtract the apogee of

---

122 This value results from adding the corresponding entries from columns G in Tables TRC-01, TRW-02, TRW-03, and TRW-05.

123 See the previous Rule 20.

that planet from the mean motion of the planet on the intended day, and the remainder will be the mean centre of the planet, which you keep next to the mean motion. Fourth: look for the motion of the Sun on the same day, and from this motion of the Sun subtract the motion of the planet, and the remainder will be the mean anomaly of the planet. Keep this anomaly also at the mean centre of the planet and the motion of the planet. Fifth: with the mean centre of the planet enter the Table of equations of the same planet with signs and degrees and take the equation of the centre and the difference, with which find the proportional part and consider the inscription 'a' or 's' both at the equation and the difference. Sixth: add or subtract the found proportional part from the equation of centre according to the inscriptions 'a' or 's' and the equation of centre will be finished. Seventh: add this finished equation of centre to the mean centre if 'a' was written at it, or subtract if 's' was written at it, and you will have the true centre of the same planet at the intended date. Eighth: if the equation of the centre is added to the mean centre, then this equation of centre must be subtracted from the mean anomaly and the true anomaly remains. If, on the other hand, the equation of centre is subtracted from the mean centre, then this equation of centre must be added to the mean anomaly and the true anomaly of the planet at the intended day remains. Ninth: with the true centre, look for the proportional minutes with their difference and proportional part and set them aside with the inscription 'farther' or 'nearer', as found in the table. Tenth: with the true anomaly, enter the Table of equations with signs and degrees and look for the equation of the anomaly with its difference and find the proportional part and the first examined equation will come out, at which note the 'a' or 's' as given in the table. Eleventh: enter again with the true anomaly and get the variation of diameter in one or the other distance. And if 'nearer' is written above the proportional minutes, then take the variation of diameter in the nearer distance. But if 'farther' is written at it, then take the variation of diameter in the farther distance. Twelfth: using the variation of diameter and the proportional minutes previously saved find the proportional part. Thirteenth: add the proportional part resulting from there to the first examined equation, if the variation of diameter is taken at a nearer distance, or subtract it, if the variation of diameter is taken at the farther distance, and there will come out the second examined equation. Fourteenth: look at the two equations, namely, the equation of the centre and of the second examined anomaly; if 'a' is written at both, then the two must be added together and the result must be added to the mean motion of the planet and the true motion of the planet for your intended time will result. However, if at both equations are written 's' or 'm', then both must be added together and the result must be subtracted from the planetary motion and you will have the true motion of the planet at the intended time. If, however, 'a' is written at one equation and an 's' or 'm' at the other, then subtract the smaller from the larger one, and adding or subtracting the result from the motion according to the 'a' or 's' written at the larger equation from which you have subtracted the smaller, you will have the true motion of the planet at your intended time. **For example:** I want to know the true motion of Saturn in the year of the Lord 1486 for the day of the Assumption of Mary. First: I must know the mean motion of Saturn[124] and it is $8^{s}$ 10;11,6,47°. Second: I must have the apogee of Saturn or the planet you are concerned

124 See the example in Rule 6.

with,[125] for the same day and it is: $8^{s}$ 13;1,30,50°. Third: subtract the apogee of the same planet from the motion of the planet on the same day and the remainder will be the mean centre: $11^{s}$ 27;9,35,57°. Fourth: I must know the mean motion of the Sun[126] on the same day and it is this: $5^{s}$ 2;41,47,30°. From which mean motion of the Sun I must subtract the motion of the planet and the remainder will be the mean anomaly of Saturn: $8^{s}$ 22;30,40,43°. Fifth: with the mean centre of the planet enter the Table of equations of Saturn searching for the equation of the centre, and the difference, and the proportional part. The equation of the centre[127] will be: 0 degrees, 20 minutes, whose inscription is 'a'; the difference is 6, whose label is 's'. The proportional part will be[128] as follows: 0 minutes, 57 seconds, 36 tertias. Sixth: subtract the proportional part from the equation of the centre and the equation of centre will be finished; it is 0 degrees, 19 minutes, 2 seconds, 24 tertias, whose inscription is 'a'. Seventh: I must add this finished equation of centre to the mean centre and the true centre will be: $11^{s}$ 27;28,38,21°. Eighth: I must subtract the same equation of centre from the mean anomaly and this true anomaly of Saturn will be: $8^{s}$ 22;11,38,19°. Ninth: with the true centre look for the proportional minutes with their difference and proportional part. Proportional minutes are 60, whose inscription is 'farther distance'. The difference is 0. Tenth: enter with the true anomaly and take the equation of the anomaly with the difference and the proportional part. The equation of the anomaly will be: 6 degrees, 13 minutes, whose inscription is 'm', difference 0. Eleventh: I will enter again with the true anomaly and take the variation of diameter under the farther distance and it will be: 0 degrees, 20 minutes, no difference. Twelfth: I look for the proportional part using the variation of diameter and proportional minutes, but this is not necessary here according to the second caution mentioned in the $19^{th}$ rule. Thirteenth: I subtract the variation of diameter from the first examined equation of anomaly and there will remain this second examined equation of anomaly: 5 degrees, 53 minutes, whose inscription is 'm'. Finally, I consider both equations, namely of centre and of anomaly, and subtract the smaller from the larger and the remainder will be: 5 degrees, 33 minutes, 57 seconds, 36 tertias. I must subtract this remainder from the motion of Saturn and this true motion of Saturn at noon of the day of the Assumption of Mary is: 8 signs, 4 degees, 37 minutes, 9 seconds, 11 tertias, 9 quartas.

<23>

### Rule 23 on finding the true motion of Venus and Mercury

If you wish to have the true position of Venus or Mercury, you must work in the same way as described in the preceding rule, only paying attention to this: when you wish to have the mean centre of Venus or Mercury, first find the motion of the Sun at your intended time, which motion of the Sun is the same as the motion of Venus and Mercury. Second:

---

125 See the second table in Rule 3.

126 See the example in Rule 2.

127 See Table TRW-16, column A for the input value $11^{s}$ 22°.

128 With the above obtained numbers would be 0;6 × 0;9,35,57 = 0;0,57,35,42. The current result can be obtained with the rounded value of the mean centre, 0;6 × 0;9,36 = 0;0,57,36. However, the precision of the equation of the centre and its increments rounded to minutes of a degree is insufficient for interpolate to the tertias.

subtract the apogee of Venus, which is the same as the apogee of the Sun, from the motion of Venus and the Sun, and you will have the centre of Venus. But for Mercury, subtract the nearest apogee of Mercury from the motion of Mercury or the Sun, which are the same, and you will have the true centre of Mercury. Find the mean anomalies of both Mercury and Venus in their own tables. But finding the others is to be done in the same way as described in the preceding rule. **First caution:** in the sixth table of Venus or Mars, at the difference of the equation of anomaly, the degrees and incomplete minutes of degrees are found at the difference of the equation of anomaly.[129] But when you search for the proportion part, complete the minutes, everything in its place, degrees under degrees, minutes under minutes. **Second caution:** if in Mars or Venus there are more than 30 degrees in the equation of anomaly and it cannot be subtracted from the motion of the planet, then one sign must be adjusted and decomposed into 30 degrees and the equation of anomaly must be subtracted from the whole in degrees. And what then comes out will be the true motion of the planet. In addition, whenever the degrees have increased to more than 30, 30 must be removed and for the subtraction of 30 degrees, one sign must be added to the signs. **Third caution:** when in the equation of centre or of anomaly a paragraph mark[130] appears, then the two inscriptions occur in the following difference in the upper and lower part of the table. Indeed, if it is entered with the signs received in the upper part, the first inscription displays down to the paragraph mark. The second inscription specifies what is to be done from the paragraph mark to the end of the table. But if the signs at bottom are accepted, the inscription occurring first shows what is to be done up to the paragraph mark, the second one up to 29 at the end of the table.[131] **Fourth caution:** if there is something in the signs in the anomaly or centre and there are no minutes, degrees and signs in the upper part of the table, one sign must be resolved into 30 degrees and it must be entered with the remaining signs and 30 degrees into the Tables of equations.[132] **Fifth caution:** if in the centre or anomaly there is 0 in signs, 0 in degrees, then one revolution of 12 signs must be accommodated, from 12 signs of which must be subtracted one sign, which sign is resolved to 30 degrees, and so you will enter in the Tables of equations with 11 signs and 29 degrees.[133]

---

129 Because of a steep decrease of the equation of anomaly at 6 signs, the increments of the equation of anomaly of these two planets exceed one degree and the column with the differences is split into degrees and minutes, despite there being only minutes written in its heading (see Fig. 26).

130 In the tables, the paragraph mark (usually of reversed-P shape) denotes the maximum of the equation. The differences above this mark are thus positive increments while below they are negative and the proportional parts are to be subtracted in the interpolation. Such paragraph marks can be also seen in Fig. 26 in the column of *longitudo longior* between lines 4 and 5, and in the column of *longitudo propior* between lines 8 and 9. (The maxima of the equations are in the tables of signs 4 and 5.)

131 The letter 'a' for *adde*, and below it 's' for *subtrahe*, is inscribed in the top heading of the column and inversly 's' above 'a' in the bottom margin of the column. For input values from 0 to 5 signs, which are indicated by the upper heading and degrees in the left column, addition is to be applied down to the paragraph mark and subtraction below it. For entries from 6 to 11 signs given in the bottom and degrees in the right column, addition should be done for degrees from the bottom up to the paragraph mark, and subtraction from it up to the 29 degrees at the top of table.

132 This means that, for instance, the value $2^s\ 0°$ is entered as $1^s\ 30°$, i.e., at the bottom line of the first table.

133 The input value $0^s\ 0°$ is missing in the tables of equation, while the value 6 signs can be entered in two ways, either as $5^s\ 30°$ or $6^s\ 0°$. But the recommendation in this fifth caution is misleading, because the entry $11^s\ 29°$ is equivalent to $0^s\ 1°$ and not to zero. However, it is obvious that for the value zero both equations and both

<24>

### Rule 24. If the planet is direct, retrograde, or stationary

If you want to easily know whether a planet is direct, stationary or retrograde, first enter the Tables of equations with the equated centre of that planet and beside the degree of the centre find the first station. Second: subtract this first station found from 12 signs, resolve one sign into 30 degrees, one degree into 60 minutes, and so on, as often shown above, and what remains after the subtraction will be the second station of the planet. Third: having the first and second stations, look at the anomaly of the planet; if it is equal to the first station, the planet will be stationary at the first station. If the anomaly is equal to the second station, the planet will be stationary at the second station. However, if the equated anomaly is less than both the first and the second station, the planet will be direct. If the equated anomaly is greater than both the first and second stations, it will also be direct. If the equated anomaly is greater than the first station and less than the second station, the planet will be retrograde.

The stations, direct motion, and retrogradation of planets can also be known from a table called the Table of directions, retrogradation and stations of planets.[134] Enter it with the true anomaly of the planet, and using the beginnings and ends of stations and retrogradations, consider the equated anomaly of the planet. **For instance:** for the current year of the Lord 1486, for the day of the Assumption of Mary, I want to know if Saturn is direct, retrograde or stationary. I must enter the Table of equations of Saturn with the equated centre, which is[135] $11^{s}$ 27;28,38,21°, and beside the degree of the centre to take the first station, which is 3 signs, 22 degrees, 44 minutes.[136] Second: I subtract this first station from 12 signs and the second station will be: 8 signs, 7 degrees, 10 minutes.[137] Third: I must see the equated anomaly of Saturn, which is 8 signs, 22 degrees, 11 minutes;[138] then if it were equal to the first or second station of Saturn, it would be stationary. If the equated anomaly were less than the first station and also the second, it would be direct. Similarly, if the anomaly is greater than the first and the second, it is direct. Therefore, since the equated anomaly of Saturn is greater than the first station and also the second, it is clear that it is direct.

<25>

### Rule 25. Whether the planet is ascending or descending

If you want to know the equated centre of the planet, consider that from one degree to six signs the planet will be descending in its eccentricity, but from six to 12 signs it will

---

distances (the farther and the nearer one) are also zero, while the proportional minutes are 60 and the first station has the limiting value which is in the presented precision numerically equal to the same value as in the neighbouring degrees.

134 It is the additional Table A-01.

135 See the sevenths step of example in Rule 22.

136 See Table TRW-16.

137 Should be 16 minutes.

138 This is a truncated value found in the eighth step of example in Rule 22.

be ascending. If, however, the equated anomaly of the planet is from one degree to six signs, the planet will be descending in the epicycle. And if from six to twelve, it will be ascending in its epicycle.

<26>

**Rule 26. For placing the continuation of true motions into an almanac**

If you want to place the true motions of the planets for each day of the year into an almanac, look first for the true motions of the planets with the head of the dragon at the beginning of the year, on the first day of January, in signs, degrees, minutes, seconds, and tertias. Second: look for their true motions by always omitting a few days, that is, for the Sun in five by five days. But for the Moon, inquire into its true motion every day. And for Saturn and the head of the dragon, for Jupiter and Mars, search for their motions in ten by ten.

If, however, you wish to act more distinctly and more certainly, look for the true motions of Jupiter in eight by eight, of Mars in six by six, and of Venus and Mercury in four by four days. But if you wish to work more certainly, look for the true motions of Venus and Mercury in three by three days. Third: when you have the true motions in days at a distance from each other, subtract the smaller from the greater, and after the subtraction you will have the difference in degrees, minutes, seconds, and tertias, of one true motion from another separated by several days. Fourth: with this difference enter the table entitled Table of proportional minutes for the intermediate days in almanac,[139] and look for the same number of degrees and minutes that you have in the difference. Fifth: consider how many days the difference is, and then under that number of days extract the degrees, minutes, seconds and tertias, and keep them in the table. Sixth: if in the difference of the two true motions there are beyond the degrees and minutes some seconds or tertias, then with changing the denominations[140] enter again the table under a similar number of days, as is the distance in that part of the table where you will find the numbers. Then the number has the denomination of a minute, and what follows will be the seconds and tertias, which will be added to the previous ones. You will enter in the same way with the tertias. Seventh: add all these together, and the whole sum shows how much such a planet moves in one day and thus it will correspond to one intermediate day. Eighth: add this motion, which corresponds to one day, to the previous motion, and you will have the true motion of the planet of the second day. And then add the same to the motion of the second day, and you will have the true motion for the third day, and so on until the day on which you had the true motion sought, and thus you will complete the almanac of the true motions in all the planets. **First caution:** pay a careful attention to the ends of the months, to how many final days each of them has, because according to this, you will search for the distances of true motions and afterwards it will be entered in the Table of proportional minutes under so many days. **Second caution:** it must be carefully noted whether the planet has been direct or retrograde, because if it is direct, that which corresponds to the whole of one day

139 I.e., the additional Table A-03.

140 This means that the entry and the result will correspond to smaller sexagesimal orders.

must be added to the obtained true motions. If, however, it is retrograde, the same must be subtracted. **Third caution:** regarding the three upper planets, when they are approaching retrogradation or direction, and in the whole retrogradation, it would be prudent to act in five by five. Otherwise, when they are direct, there is a certainty to work in ten by ten. In the Moon, if one does not pay attention to great precision, the operation could be done in three by three, and you would easily complete the almanac with the true motions. **For instance:** we want to continue with the creation of an almanac of true motions for the current year of the Lord 1487. First: we must have the true motion of the Sun for the beginning of that year,[141] this motion is called the radix and is $9^{s}$ 19;23,47,19°. Second: I must have the true motion of the Sun for the fifth day[142] and this is $9^{s}$ 24;30,2,29°. Third: I must subtract the previous true motion from the true motion for the fifth day and the difference remains: 5 degrees, 6 minutes, 15 seconds, 10 tertias. Fourth: with this difference, I must enter the Table of proportional minutes[143] under the five days and obtain the proportion, and the complete entry will be: 1 degree, 1 minute, 15 seconds, 2 tertias. If we add this to the true motion of the Sun on the last day of December, which is the radix, this true motion for the first day of January is: $9^{s}$ 20;25,2,21°. Fifth: if you add what you found in the table, to the motion of the first day of January, this true motion of the Sun for the second day is: $9^{s}$ 21;26,17,23°. And so you will continue with the true motions of the Sun throughout the year.

<27>

### Rule 27 on the true motion of the dragon's head, for easily placing it in an almanac

If you want to easily continue with the true motion of the dragon's head, proceed in this order: from the end of the year to its beginning at first find the mean motion of the dragon's head at the end of your year. Second: subtract it from 12 signs, and what results after the subtraction, will be the true position of the dragon's head at the end of your current year, that is, at the last day of December. Third: if to this true motion of the head of the dragon, found on the last day of December, you add the motion of the head in one day, it will be found for the penultimate day of December, and so you will continue continuously through the days and months in reverse order. **For example:** I want to continue with the true motion of the dragon's head for the current year of the Lord 1487. First, I must know the motion of the dragon's head at the end of the same year, that is, at the last day of December,[144] and this is: $1^{s}$ 22;45,54,23°. Second: I must subtract the same motion

141 According to the example in Rule 15, the radix of mean motion of the Sun for the year 1487 is $9^{s}$ 18;42,56,39°, radix of apogee $3^{s}$ 1;3,23,59° and radix of anomaly of the Sun (which is their difference) $6^{s}$ 17;39,32,40°. Interpolating in Table TRW-13, we find for this anomaly the equation of the Sun equal to $0^{s}$ 0;40,51,1°, which added to the mean motion gives the true motion $9^{s}$ 19;23,47,40°. However, other sets of the *Tabule resolute* (e.g. Prague, NK, X B 3, f. 96r) give the equation for $6^{s}$ 17° equal to $0^{s}$ 0;39,25° instead of $0^{s}$ 0;39,26°, which then leads to the true motion consistent with the present result within one tertia.

142 The mean motion, apogee and anomaly for the fifth day are calculated in the example in Rule 15 as $9^{s}$ 23;38,38,17°, $3^{s}$ 1;3,24,28°, and $6^{s}$ 22;35,13,49°, respectively. Interpolating in Table TRW-13 we find the corresponding equation $0^{s}$ 0;51,24,9,29° which agrees with the following result within 3 tertias.

143 See Table A-03. The same result can be obtained dividing the difference by the number (5) of days.

144 This can be found adding the cyclic radix for the year 1468 in the column G of the Table TRC-01 and its increment for 19 years in Table TRW-02.

from 12 signs, and after subtraction, the true motion of the dragon's head will remain for the end of the year, on the last day of December, and this will be: $10^{s}$ 7;14,5,37°. Third: to the same motion of the dragon's head found on the last day of December, I must add the motion of the dragon's head in one day,[145] namely $0^{s}$ 0;3,10,38°, and the true motion of the dragon's head to the penultimate December will be: $10^{s}$ 7;17,16,15°. If I add to that motion the motion of the dragon's head in one day, the true motion of the head for the previous day will be: $10^{s}$ 7;20,26,53°. And so you will continue with the motion of the dragon's head through the days and months in reverse order throughout the whole year.[146]

<28>

**Rule 28 on calculating the mean conjunctions and oppositions throughout the whole year**

If you want to find the first conjunction of the year, I will say the first, that is the one that occurs in the usual days of January and is called the 'radix of the conjunctions', first enter with the current years of Christ in the Tables of the conjunctions in collected years. If you would find it precisely, you would find immediately beside these collected years the first conjunction of that year in days of January, which belongs to February, in days, hours, minutes, seconds, and tertias. And beside this conjunction you will find the motion of the Sun and Moon, the motion of the anomaly of the Moon, and the motion of the argument of latitude of the Moon. Second: if you do not find your current year precisely, then you enter with the nearest year that you find in the tables of conjunctions and take the time of conjunction beside it. Similarly, take the mean motions of the Sun and Moon, the mean anomaly of the Moon and the mean argument of the latitude of the Moon. Third: with the remaining years that differ from your current years, you enter the Tables of expanded years, and find beside the years the time in days, hours, and minutes, likewise the motions of the Sun and Moon, the anomaly of the Moon, and the argument of latitude of the Moon, and add all these to its kind, as it has been described in the first rule. Fourth: subtract 31 days from the days resulting from this addition and you will have the first conjunction of the year in the days of January. It is called the 'radix of conjunctions' with its motions to the same time. Fifth: if you cannot subtract 31 days from the days resulting from such an addition, add again to it one lunation, namely 29 days, 12 hours and 44 minutes, and add also the motion found beside one lunation to the motions previously found, all to its kind. Sixth: subtract 31 days from this whole, and what remains in days, hours and minutes will be the time of the first conjunction in the days of January with its motions. And so this conjunction will be the radix of all the conjunctions of that year. **For instance:** we would like to find the first conjunction to the current year of the Lord 1487. We will first enter with the same current year in the Tables of conjunctions in the collected years, and if we do not find the current year precisely you enter with a closer one, that is, 1469, and beside

145 This can be found in Table TRW-05, column G.

146 Due to the rounded value of the daily motion, an error would accumulate by this procedure of up to 44 tertias in one year.

it we find[147] twelve in days, 18 in hours, 13 in minutes, 31 seconds, 21 tertias, 38 quartas, and in the mean motion of the Sun and Moon I find $10^{s}$ 1;39,9,11°. In the anomaly of the Moon, $1^{s}$ 11;53,6,19°. In the argument of latitude of the Moon, $11^{s}$ 17;38,37,19°. Third: with the remaining years, namely 18, I enter the Tables of expanded years,[148] and beside the same year 18, I get 12 in days, 18 hours, 59 minutes, 17 seconds, 1 tertia and 46 quartas. In the motion of the Sun I find $11^{s}$ 11;41,31,2°. In the anomaly of the Moon I get $11^{s}$ 1;20,9,20°. In the argument of latitude I find $10^{s}$ 28;50,56,33°. I must add all this together, each to its own kind, and when the addition is made, there will result in time 25 days, 13 hours, 12 minutes, 48 seconds, 23 tertias, and 25 quartas. In the motion of the Sun, it is $9^{s}$ 13;20,40,15°. In the anomaly of the Moon $0^{s}$ 13;13,15,39°. In the argument of latitude of the Moon $10^{s}$ 16;29,33,52°. Fourth: since I cannot subtract 31 days from the days resulting from such an addition, therefore I must add the time of one lunation with the motions placed beside it,[149] in time 29 days, 12 hours, 44 minutes, 3 seconds, 2 tertias, 59 quartas. In the motion of the Sun $0^{s}$ 29;6,24,12°. In the anomaly of the Moon $0^{s}$ 25;49,0,35°. In the argument of latitude of the Moon $1^{s}$ 0;40,13,46°. After I add each to its kind, there will result in time 24 days, 1 hour, 56 minutes, 51 seconds, 26 tertias, 24 quartas. In the motion of the Sun $10^{s}$ 12;27,4,27°. In the anomaly of the Moon $1^{s}$ 9;2,16,14°. In the argument of latitude of the Moon $11^{s}$ 17;9,47,38°. Therefore this conjunction with its motions will be the radix of all conjunctions to the current year of the Lord 1487. **First caution:** when in practice the conjunctions or oppositions have exceeded more than 24 hours, then subtract 24 hours and the remainder will be the time of the conjunction or opposition in hours, and for these 24 hours subtracted, add one to the day. **Second caution:** a conjunction found in a certain month is not of that month in which it is found, but of the following, as the conjunction of February is found in the days of January.

<29>

## Rule 29, how other mean conjunctions of the Sun and Moon are to be found in the circle of the year

If you want to find the other mean conjunction of the Sun and Moon in some year, first write in the table the day, the hours, minutes, seconds, and tertias of the first conjunction which you have as the radix. Second: enter the Table of mean conjunctions of the Sun and Moon and take the day, hours, minutes, seconds and tertias. Similarly, take the motion directly in the month in whose days you are looking for the conjunction. And add all this to the first conjunction, namely, time to time, that is, days to days, hours to hours, etc. Third: if more days come out of such a conjunction than the month contains, in whose days you seek the conjunction, then subtract the solar days from the lunar days thus derived, if you can, namely the days of the Sun placed directly in the line of the month in which you seek the conjunction, and count the remaining days to the beginning of that month and the hours

147 See Table TRC-09.

148 See Table TRW-11. The values of time of the conjunctions and oppositions in MS Toruń, BU, 74/III are slightly higher in tertias and quartas than in the later sets of the *Tabule resolute*.

149 These values are given in the first line of Table TRW-12.

from noon. Fourth: if the indirect solar days of the month with which you are working cannot be subtracted, which is the case when a month has two conjunctions, then subtract the days of the previous month and count the remaining days of that month. Similarly you will continue with the motion of the Sun and Moon, the motion of the anomaly of the Moon, and the argument of latitude of the Moon, and other conjunctions throughout whole the year. **First caution:** in a leap year you must be careful, because if it is a leap year and the leap day (intercalary day, *bissextus*) has passed, then add one more day to the given solar days[150] and the days found in the table of months, and after such addition subtract the solar days from the lunar days and you will have the time of conjunction. The lunar days are those which come by practice,[151] while those which are beside the month are solar. **Second caution:** if two conjunctions occur in the usual days of January, the first will be embolismal[152] and the second will be called February. However, the true embolism will be indicated by true and not by mean conjunctions. **Third caution:** if two conjunctions occur in the usual days of any month, one near the beginning and the other near the end, the first will be embolismal, the second will be of the following month. **For example:** for the year of the Lord 1487, I want to know the conjunction of March, which will be in the days of February. First I must have the first conjunction of that year, which is the radix with its motions. The radix of the conjunction is[153] 24 days, 1;56,51,26,24 hours. Second: I enter the Table of the mean conjunction of the Sun and Moon[154] beside the month in whose days I am looking for the day of the conjunction, and beside February I find 29 days, 12 hours, 44 minutes, 3 seconds, 2 tertias, 59 quartas. I find that the motion of the Sun and Moon is $0^{s}$ 29;6,24,12°. The motion of the anomaly of the Moon is $0^{s}$ 25;49,0,35°. The argument of latitude of the Moon is $1^{s}$ 0;40,13,46°. And after I add everything to its kind, that March conjunction results with its motions. First, however, I must subtract from the lunar days the solar days placed directly in this month in which I seek the conjunction, that is 31,[155] and the resulting time is 22 days, 14;40,54,29,23 hours. The mean motion of the Sun and Moon at that time is $11^{s}$ 11;33,28,39°, the mean anomaly of the Moon $2^{s}$ 4;51,16,49°, the mean argument of latitude of the Moon $0^{s}$ 17;50,1,24°.

<30>

**Rule 30 on the continuing mean conjunctions and oppositions of the Sun and Moon**

If you have the first conjunction of the Sun and Moon in a year, which is the radix, and you would like to have its opposition immediately after the time of the conjunction found,

150 Placed in Table TRW-12 in column S. This value is the total number of days from the beginning of the year till the beginning of the month.

151 The lunar days given in column L in Table TRW-12 are the total number of days till the beginning of the lunation. They are calculated together with the time in hours and their fractions in the column A by successive addition of the duration of one lunation given in the first line of the table.

152 That is an intercalary month.

153 See the result of example in the previous Rule 28.

154 I.e., Table TRW-12.

155 I.e., 24 days of the radix + 29 days of the first lunation = 53 days from the beginning of the year; minus 31 days of January gives February 22.

add the time from the conjunction to the opposition, namely 14 days, 18 hours and 22 minutes. Similarly in the motions and you will have the opposition of the month whose conjunction you have with its motions. Second: when you have oppositions and you want to have the conjunction of the immediately following month, add the same time, namely 14 days and 18 hours, to the found time of the opposition and similarly the motion to the motion and you will have the immediately following conjunction with the motion of the Sun and Moon, the mean anomaly of the Moon and the argument of latitude of the Moon. **Caution to be considered:** in such a continuation, when you wish to examine these continuations of oppositions and conjunctions, investigate the motion of the Sun for the time of conjunctions which you have found from the previous rule, because if it agrees with the motion of the Sun found by the continuation, you will be sure that the continuation is regular and you will understand that you have acted correctly according to the rule.[156]

<31>

### Rule 31 on finding the true conjunctions and oppositions of the Sun and Moon

If you want to know the time of true conjunction or opposition, first take the time of the mean conjunction or opposition according to the aforesaid rule. Similarly, the motion of the Sun and Moon, the anomaly of the Moon and the argument of latitude of the Moon. Second: find the true motion of the Sun at this time of the mean conjunction or opposition. Third: find the true motion of the Moon with the true anomaly of the Moon, with which the equation of the anomaly is found, which is added to or subtracted from the motion of the Moon. You don't need to look for the proportional part here, but you do need to work with the anomaly of the Moon, which you set aside. Fourth: when you have the true motion of the Sun and Moon, subtract the lesser from the greater in terms of degrees and other fractions and not in terms of signs. And consider in the difference which of them is greater in motion, and if it is the longitude of the Sun, the sign of the Sun is written at the longitude. But if it is the longitude of the Moon, the sign of the Moon is written at it. Fifth: if it is the longitude of the Sun, the true conjunction is to come after the mean conjunction. If, however, it is the longitude of the Moon, the true conjunction precedes the mean conjunction. In opposition, however, if the Moon has not yet come to the nadir of the Sun, the opposition is forthcoming.

But if it has passed, the opposition is already past. Sixth: convert this longitude – whether of the Sun or of the Moon – into seconds by multiplying by sixty. Seventh: after the longitude has been converted into seconds, divide the longitude thus converted into seconds by twelve, and if anything remains above half of twelve, add one to the number of the quotient, and this number of the quotient is a twelfth of the longitude. Eighth: this number of the quotient is added to the longitude converted into seconds, and its half obtained. Ninth: this half is converted into minutes by dividing it by sixty. Tenth: this half of the longitude with its twelfth reduced to minutes is added to the anomaly of the Moon at the time of the mean

156 Both procedures would be consistent, if the increments of the time and the mean motions between the conjunction and opposition will be precisely half of those for the whole lunation.

conjunction or opposition, if it was the longitude of the Sun, or subtracted from this anomaly if it was the longitude of the Moon, and there will be an anomaly of hour.[157] Eleventh: with this anomaly of hour search for the motion of the Moon in one hour with the double entry, that is one degree being held in order to have the difference.[158] Twelfth: with the anomaly of the Sun at the time of the mean opposition or conjunction, look for the motion of the Sun in one hour using the same table, also with double entry if you have minutes over a degree. Thirteenth: subtract the motion of the Sun in one hour from the motion of the Moon in one hour and the residuum is called *superlacio* or *superacio* according to how much the Moon moves more than the Sun in one hour. Fourteenth: convert this *superlacio* into the same units into which the simple longitude was converted, that is, to seconds. Fifteenth: divide the simple longitude by the *superlacio* and in the number of the quotient will come out the hours. Sixteenth: multiply the remainder by 60 and divide by the *superlacio* and in the number of quotient will come out the minutes. Seventeenth: multiply the remainder by 60 and divide by the *superlacio* and the seconds will come out. Eighteenth: if the remainder exceeds beyond half of the *superlacio*, then add one second to the seconds found and you will have extracted the time. Nineteenth and last: add this extracted time to the time of the mean conjunction or opposition, if it is the longitude of the Moon, and you will have the time of the true conjunction of the Sun with the Moon or the true opposition of the Sun with the Moon. **For example:** for the current year of the Lord 1487, I want to know the first conjunction, which is the conjunction of February in the days of January. First I must know the time of the mean conjunction,[159] and it is: 24 in days, 1;56,51,26,24 in hours. The mean motion at the same time is $10^s$ 12;27,4,27°. The mean anomaly of the Moon is $1^s$ 9;2,16,14°. The mean argument of latitude of the Moon is $11^s$ 17;9,47,38°. Second: at this time of the mean conjunction I must know the true motion of the Sun. The apogee of the Sun[160] is $3^s$ 1;3,26,16°. The anomaly of the Sun[161] is $7^s$ 11;23,38,11°. The true motion of the Sun[162] is $10^s$ 13;55,35,1°. Third: I must know the true motion of the Moon[163] and it is this: $10^s$ 9;32,41,38°. Fourth: when I have subtracted the lesser from the greater, this longitude of the Sun will remain: 4 degrees, 22 minutes, 53 seconds, 23 tertias. Fifth: I know that the true conjunction is yet to occur, because it is the longitude of the Sun. Sixth: I must convert the longitude of the Sun to seconds and it will come out 15773. Seventh: after I divide the longitude converted to seconds by 12, the twelfth of that longitude will come out: 1314. Eighth: when I add twelfth of

157 Anomaly of hour (*argumentum horar/i/um*) is an anomaly equated for finding the true motion of the Moon in one hour between the mean and true syzygies.

158 It means that the motion in one hour is to be interpolated between the two neighbouring values for anomalies which are tabulated at steps of one degree.

159 See the results (for the time as well as the corresponding mean motions) of example in Rule 28.

160 The radix of apogee of the Sun for the year 1487 was found $3^s$ 1;3,23,59° in the Rule 15. The increment in 24 days given in Table TRC-07c is $0^s$ 0;0,2,17,48°, which truncated (and neglecting the additional nearly two hours) gives the value presented here.

161 I.e., the difference of mean motion minus the apogee.

162 The interpolation in Table TRW-13 of the equation of the Sun (or better in its version in MS Cracow, BJ, 611, fol. 18v, which gives equation $0^s$ 1;29,33° for anomaly $7^s$ 12°) results in equation $0^s$ 1;28,30,35° and then the following true motion with an error of 1 tertia.

163 As noted in Rule 5, the centre of the Moon is zero in the conjunction. Following Rule 19, the equation of the Moon is then given by the anomaly of the Moon. Interpolating in Table TRW-15 we find the equation $0^s$ 2;54,22,49°, which subtracted from the mean motion results in the true motion of the Moon.

the longitude to the longitude in seconds, this longitude with its own twelfth in seconds will come out: 17087. Ninth: I take half of the longitude with its twelfth and it will come to 8543. Tenth: I convert this half to minutes and it is:[164] 16 minutes, 23 seconds. Eleventh: add this half converted to minutes to the anomaly of the Moon at the time of the mean conjunction, and the anomaly of the hour will be this: $1^s$ 9;18,39,14°. Twelfth: with this anomaly of hour search for the motion of the Moon in one hour in the proper table[165] with a double entry. First entry:[166] 30 minutes, 14 seconds, second entry: 30 minutes, 16 seconds. Difference: 2 seconds. And the complete motion of the Moon in one hour is this: 30 minutes, 14 seconds, 37 tertias. Thirteenth: with the anomaly of the Sun at the time of the mean conjunction search for the motion of the Sun in one hour and 2 minutes, 32 seconds will result.[167] Fourteenth: when I subtract the motion of the Sun in one hour from the motion of the Moon in one hour, there will remain this *superacio* of the Moon: 27 minutes, 42 seconds, 37 tertias. Fifteenth: I convert this *superacio* to the second and it will come out to 1662. Sixteenth: when I divide the simple longitude reduced to seconds[168] by the *superacio* also reduced to the second, in the quotient 9 hours come out and the residuum remains 815. Seventeenth: I multiply that residuum by 60 and divide again by the *superacio* in seconds and 29 minutes result as the quotient and the residuum 702 will remain. Eighteenth: if I multiply that residuum by 60 and divide by the *superacio* in seconds, 25 seconds will arrive in the quotient and the residuum will remain 570. There will therefore be extracted the time 9 hours, 29 minutes, 25 seconds, by which the time of the mean conjunction was distant from the true one. Finally: I add this extracted time to the mean conjunction and this true conjunction will come out: 24 days, 11 hours, 26 minutes, 17 seconds.[169]

<32>

**Rule 32 on finding the degree of ascendant and other houses of heaven**

If you want to find the degree of the ascendant at the time of conjunction or opposition or another time, whether this be hours counted from noon, or horological hours,[170] then first equate the time for which you want to know the ascendant of the hour in this way: enter the Table of equations of days with nights with the true degree of the Sun at that time, and find beside the degree of the Sun and under that sign in which the Sun is, the equation of days with nights, degrees and minutes. Second: for each degree take four

---

164 Correctly it should be $0^s$ 2;22,23°.

165 The Table of the true motion of both the Sun and Moon in one hour is included in the *Tabule resolute* as the Table TRW-14 in MS Toruń, BU, 74/III, f. 259r. This table is missing in later sets of *Tabule resolute*, but it often appears in the same manuscript in context of other astronomical tables.

166 Here are quoted values of the true motion of the Moon in one hour found in the table for entries $1^s$ 9° and $1^s$ 10°, which we can find, for instance in MS Prague, NK, X B 3, f. 117r, while in Table TRW-14 are given values $0^s$ 0;30,47° and $0^s$ 0;30,48° – see the commentary on Table TRW-14 in Chapter 22.

167 This value is present in MS Prague, NK, X B 3, f. 117v, while Table TRW-14 gives $0^s$ 0;2,31° for both entries $7^s$ 11° and $7^s$ 12°.

168 This means the longitude 15773 seconds between the Sun and Moon at the mean conjunction.

169 Just to illustrate the precision of the whole procedure of calculation, let us note that according to a modern ephemeris the conjunction occurred at 11 hours 41.6 minutes of the local Cracovian time.

170 I.e., hours counted from sunset.

minutes of an hour, that is, multiply the degrees you find by 4, and you will have minutes of hours. Third: for every minute, take four seconds of the hour, multiply the minutes by 4. Fourth: if more than 60 results, divide them by 60 and in the quotient the minutes will be found. Add them to the previous minutes and the remainder will be the seconds of hour. Fifth: add the minutes and seconds that come from the equation to your intended time. Sixth: search for the the degree of ascendant in this way: enter those tables with the true motion of the Sun under that sign in which the Sun is. If you have unequal hours in the intended time, take them. But if you have the equal hours at your intended time, take them also. Seventh: add what you find beside the degree of the Sun, whether in the unequal or equal hours, to your intended time. Eighth: search for that what results after the addition under the equal hours, and where you find such or a close number of hours and minutes, there beside it you will find the degree of ascendant with the other five houses, namely, the second, third, fourth, fifth, sixth. Ninth: when you have the first six houses, beside them in a similar degree you will calculate the other six houses. **For example:** I want to know the degree of the ascendant at the time of the January conjunction found in the previous rule. First: I must equate the intended time, namely 11 hours, 26 minutes, 17 seconds, by entering the Table of equations of days and nights with the true motion of the Sun at the time of the conjunction, namely the 13$^{th}$ of Aquarius,[171] and find 0 degrees, 6 minutes.[172] Second: when you multiply 6 minutes by 4, you get 24 seconds.[173] Third: I will add these 24 seconds to the seconds in the intended time and come up with this equated time: 11 hours, 26 minutes, 41 seconds. Fourth: with the true motion of the Sun, that is the 13$^{th}$ of Aquarius, I enter the Tables of the ascendant[174] and find 3 hours, 2 minutes under the hours of noon. Fifth: I will add this to my intended time, minutes to minutes and hours to hours, and it will come to 14 hours, 28 minutes, 41 seconds. Sixth: I search for all this under the equal hours, and where I find this or a closer number, beside it I find the six houses of the sky, that is, the ascendant, the second, the third, the fourth and the fifth, the sixth. Seventh: I enter the Tables of ascensions with 14 hours, 28 minutes or closer to find the ascendant. And if both numbers are equal, the next 27$^{th}$ of Libra is to be taken with the other five houses.[175] And the other six houses are saved opposite to those six. *Figura celi*[176] with the 12 houses is thus the following (see Fig. 24). **Caution:** Regarding this rule, whenever the number of hours becomes more than 24, 24 hours must be subtracted.

171 Using Tables TRC-01, TRW-02, TRW-05, and TRW-06 we can find the mean motion of the Sun at the true conjunction to be 10$^{s}$ 12;49,24,34°. The apogee of the Sun at the true conjunction is practically the same as at the mean conjunction found in the Rule 31, that is 3$^{s}$ 1;3,26,16°. The anomaly of the Sun thus reads 7$^{s}$ 11;45,58,18°. Interpolating in Table TRW-13 we find the equation of the Sun 0$^{s}$ 1;29,12° and hence the true conjunction occurring at 10$^{s}$ 14;18,37°, that is one degree more.

172 See Table A-05.

173 The minutes of a degree of right ascension are multiplied by 4 to get seconds of an hour of time shift.

174 The table is Bianchini's table of equations of houses; see, e.g., Chabás and Goldstein, *The Astronomical Tables of Giovanni Bianchini*, pp. 112–14; Hadrava and Hadravová, 'Nová akvizice Národní knihovny a její význam pro dějiny astronomie', *Dějiny věd a techniky*, 50 (2017), 192–208. It can be found, e.g., in MSS Cracow, BJ, 600, ff. 107v–108v; Cracow, BJ, 617, ff. 3r–5v and 6v–12r; or Prague, NK, 40 E 47 adl. 1, ff. 1r–6v.

175 This table is, for example, in MS Cracow, BJ, 600, f. 107v, where for Aries 27 we find equal hour 14 hours, 31 minutes.

176 *Figura celi* is the horoscope in the standard form of a square celestial chart. In manuscript **B**, it is inscribed in the centre: true conjunction of February in equated day 24$^{d}$ 11$^{h}$ 26$^{m}$ 17$^{s}$. The Sun and Moon at Aqr 13 are written in the 4$^{th}$ house, beginnings of the houses are: ascendent Lib 27, Sgr 2, Cap 3, Aqr 5, Psc 1, Psc 29, Ari 27, Gem 2, Cnc 3, Leo 5, Vir 1, and Vir 29.

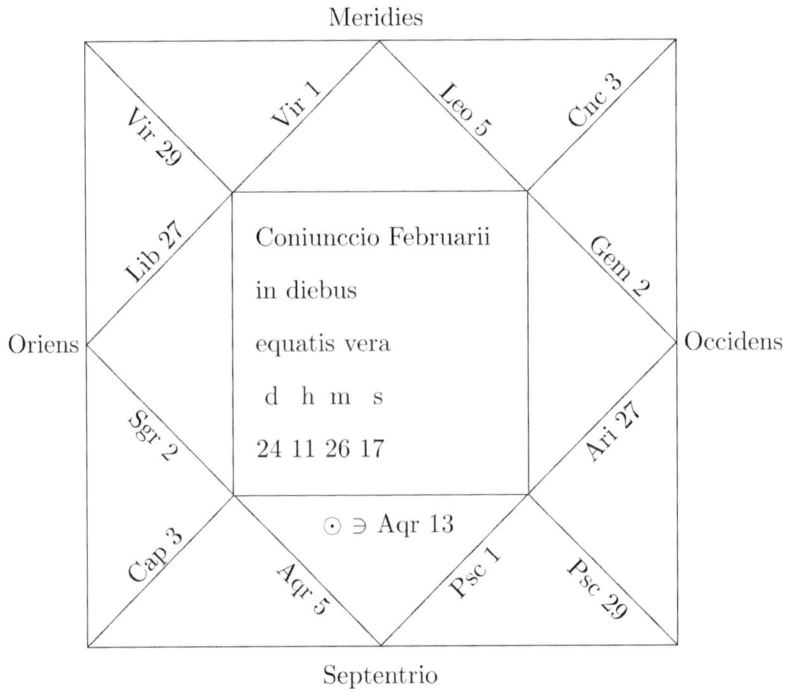


**Figure 24.** *Figura celi* for February conjunction 1487 found in MS Cracow, BJ, 600, f. 157r. Redrawn by the authors.

<33>

## Rule 33 on converting conjunctions and oppositions from noon hours to sunset hours

If you want to convert any conjunction or opposition, that is, the time of conjunction or opposition, to sunset hours and most completely, you must first know the time from noon to your intended time. You will know the time from noon from the table written below, which Table of noon shows this time for each day of the year.[177] Second: add the time of noon in your intended day to the time you have found from practice,[178] and what results after the addition will be the time of true conjunction or opposition in horological hours.[179] Third: if you wish to prove this, search according to the aforesaid rule for the ascendant with the other houses at the resulting complete horological hours. Fourth: if the same degree of ascendant with the other houses results, you will know that

177 This table can be found in the MS Cracow, BJ, 600, f. 110r. It is inscribed *Tabula, per quam scitur meridies omni die et per eandem reducuntur hore meridiei ad horas horalogii per solam addicionem.* It gives the syllable of cisiojanus and duration in hours and minutes from a sunset to next noon for each day of the year. It varies between 15 hours 55 minutes around June 12, and 20 hours 5 minutes around December 12. The equinoctial value of 18 hours is for March 11 and September 14.

178 I.e., from the computation.

179 I.e., equal hours counted from sunset.

you have converted it correctly. **Caution:** It is better to search for the ascendant with non-converted noon hours than with the converted horological hours. However, it is easy to recognise this **with the noon clock.**[180] And if in your calculated time the excess has not exceeded 12, then it is the same hour in the afternoon, because the noon clock touches one at one hour after the noon. But if more than 12 hours have elapsed in your intended time, subtract 12 and you will easily calculate the remainder. **For instance:** I want to convert the conjunction of February, which will be in the current year of the Lord 1487 on January 24, 11 hours, 26 minutes, 17 seconds. First I must know the time from noon on the same day 24 and in the table I find 19 hours, 23 minutes.[181] Second: when I add these hours of noon to the intended time, it will come to 6 hours, 49 minutes, 17 seconds, which is the time of conjunction in the complete horological hours. When I look for the ascendant at this time, 27 Libra results with other houses, as was found in the previous rule.

<34>

**Rule 34 on the entry of the Sun into Aries and into the other cardinal signs of the sky**

If you seek the entry of the Sun into Aries, or into any other cardinal sign[182] or any other sign of the heavens, first look for the true motion of the Sun at a time when the Sun, in your estimation, is not far from Aries or any other cardinal sign. Second: having the true motion of the Sun, subtract it from 12 signs if you want to have the entry of the Sun into Aries. However, if it is into Cancer, then subtract the true motion of the Sun from three signs. If into Libra, then from 6 signs. If it is into Capricorn, then from nine signs. And what remains is the distance of the Sun from Aries. Third: convert the remaining distance into the same unit, namely seconds or tertias, but it is more accurate to convert it into tertias. Fourth: find the true motion of the Sun after one hour by searching for the mean and true Sun after one hour. You will find it in this way: when you have the mean or true motion of the Sun, then go to the Loricated tables[183] and take that which stands beside one hour and add it to the true or mean motion and thus you will have one hour. Fifth: subtract the first found true motion of the Sun from the second found true motion of the Sun and the residuum will give the true motion of the Sun in one hour. Sixth: convert the motion of the Sun in one hour to the same units to which you converted the distance of the Sun from Aries, as to seconds or the tertias, but it is more precise to tertias. Seventh: divide the distance of the Sun from Aries in tertias by the motion of the Sun in one hour also in tertias and from the residuum[184] will come the hours. Eighth: multiply that residuum[185] by 60 and divide by the

180 *In horologio autem medio* refers to counting the equal hours from the noon and midnight as measured by a clock.

181 See the table inscribed *Tabula, per quam scitur meridies omni die…* in MS Cracow, BJ, 600, f. 110r.

182 Cardinal signs are the four zodiac signs of the year, i.e., Aries, Cancer, Libra, and Capricorn. See Helena Avelar de Carvalho, and Luis Ribeiro, *On the Heavenly Spheres. A Treatise on Traditional Astrology* (Tempe: American Federation of Astrologers, 2010), p. 45.

183 I.e., to Table TRW-06 – see the first caution to Rule 2. It gives the increment of the position of the mean Sun, which is approximately 2.5′ in one hour, while the increment of the equation of the Sun is at maximum about 6′′.

184 The hours will result in the quotient.

185 The remainder from the previous division should be used to get the minutes.

motion of the Sun in one hour in tertias and you will have minutes in the residuum. Ninth: multiply that residuum by 60 and in the number of the quotient will come the seconds. Tenth: multiply that residuum by 60 and divide by the same as before, and in the quotient will come the tertias. Eleventh: add this extracted time in hours, minutes, seconds, and tertias to your estimated time and you will arrive at the entry of the Sun into Aries. Twelfth: at that time, that is, of the entry of the Sun into Aries, search for the true motion of the Sun and if you find 0 sign, 0 degrees, 0 minutes, 0 seconds, 0 tertias. The entry of the Sun must be correctly calculated. Thirteenth: for this time make the *figura celi* and the twelve houses of the sky, as stated in rule 32, and similarly place and order the planets in such a figure, and you will have the *figura celi* and the constellations for the beginning and revolution of the year, which is when the Sun enters the first point of Aries. **For example:** I want to know the entry of the Sun into Aries for the current year of the Lord 1487. First I must to know the estimated time when the Sun is not far from Aries, and I take this: 11 days, 0 hours, 59 minutes, 31 seconds, 25 tertias. Second: I must know the true motion of the Sun at that estimated time. The mean motion[186] of the Sun is $11^s$ 27;45,6,13°. The apogee[187] of the Sun $3^s$ 1;3,30,38°. The anomaly of the Sun $8^s$ 26;41,31,35°. The equation of the Sun[188] is 2 degrees, 10 minutes, 0 seconds. The true motion of the Sun at that estimated time is $11^s$ 29;55,6,13°. Third: I must know the distance of the Sun from Aries and it is this: 4 minutes, 53 seconds, 47 tertias. Fourth: when I convert this distance to tertias, it results 17627. Fifth: when I look for the true motion of the Sun after one hour, the mean motion of the Sun is $11^s$ 27;47,34,4°. The true motion of the Sun after one hour: $11^s$ 29;57,34,4°. Sixth: when I subtract the true motion of the Sun, the lesser from the greater, the motion of the Sun in one hour will be 2 minutes, 27 seconds, 51 tertias.[189] Seventh: when I convert the motion of the Sun in one hour to tertias, it will come to 8871. Eighth: divide the distance of the Sun from Aries in tertias by the motion of the Sun in one hour also in tertias, in the quotient will come 1 hour. And the residuum will remain 8756. Ninth: if I multiply that residuum by 60 and divide by the motion of the Sun in one hour in tertias, in the quotient will come 59 minutes. Residuum 1971. Tenth: when I multiply that residuum by 60 and divide by the same as before, in the quotient will come 13 seconds, the remainder 2937. Eleventh: when I multiply that residuum by 60 and divide by the same as before, in the quotient will come 19 tertias and the residuum will remain 7671. For this residuum add one to the tertias and there are 20 tertias. Twelfth: I add that extracted time, that is, 1 hour, 59 minutes, 13 seconds, 20 tertias, to the estimated time and the entry of the Sun into Aries will occur on the 11th of March, at the second hour, 58 minutes, 44 seconds, 45 tertias. Thirteenth: I search for the true motion of the Sun at that time. The motion of the Sun will be $11^s$ 27;50,0,0°. To the mean motion I must add the equation found before, namely 2 degrees, 10 minutes, and the true motion of the Sun to

186 The date is 11th of March, so that the mean motion can be obtained adding the cyclical radix for 1468 from Table TRC-01, mean motions for 18 years from Table TRW-02, February from Table TRW-03, 11 days from Table TRW-05 and the fractions of hour from Table TRW-06.

187 This values of the apogee can be obtained from Tables TRP-07, TR-07a, and the first variants of Tables TRC-07b and TRC-07c.

188 See Table TRW-13.

189 This value could be found directly as the first one in column C of Table TRW-06. The equation of the Sun is in this example close to its maximum, so that it is constant. In a general case, the proportional part of its increment could be added to the mean motion in one hour to reach higher precision.

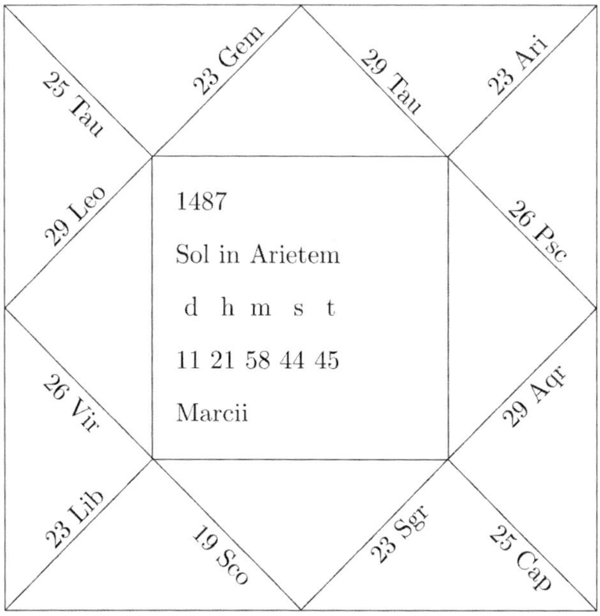


**Figure 25.** *Figura celi* for the entry of the Sun into Aries in 1487, as found in MS Cracow, BJ, 600, f. 157v. Redrawn by the authors.

the entry of the Sun into Aries will be $0^s$ 0;0,0,0°. Fourteenth: I look for the ascendant with the other 12 houses of the sky and this configuration of the sky will result (see Fig. 25):[190]

**Caution:** if you want to know the estimated time of the entry of the Sun into Aries or another cardinal sign, take first the radix of the entry of the Sun into Aries or another sign. For the year of the Lord 1468, you have the radix 10 in days, 12 in hours, 22 minutes, 41 seconds, 49 tertias.[191] Second: enter the table below, starting with one year and ending with 800 years, and enter with the distance from the year of radix to your current year, so that for the current year of the Lord 1487 you will enter with 19 years and beside it take the hours, minutes, seconds and tertias and add to the radix.[192] And if the hours exceed 24, subtract 24 and add one to the days. In a leap year, however, you will calculate the entry at the same time of the previous day.

190 Here we give *figura celi* according to witness **B**; the pattern is empty in **A** and entirely missing in **C**. The beginning of the tenth house should be 19 Tau instead of 29 Tau. It can be verified, e.g., in the table in MS Cracow, BJ, 600, f. 108v, where these values are given for the entry 29 Aqr. The value 23 Sgr is corrected in this table to 22 Sgr.

191 The table of radices of entries of the Sun into the four cardinal signs can be found in MS Cracow, BJ, 600, f. 110v (transcribed below). The values are given up to sextas for the meridian of Cracow in the year 1468.

192 The table of increments in years with a step of 1 till 20, step of 20 till 100, and step 100 till 800 years is also copied in MS Cracow, BJ, 600, f. 110v (transcribed below). This table suffers from various errors. For 19 years it gives value 14;36,49,$36^h$. Added to the above given radix in the year 1468 gives $11^d$ 2;59,31,$25^h$ which is used in the estimate of the conjunction in an example to Rule 34 (while two hours were omitted there). This shows why the author in the estimate also treated the fractions of an hour, which cause unnecessary complications of the subsequent calculation.

Cracow, BJ, 600, f. 110v:

*Introitus Solis in Arietem et alia signa cardinalia Anno Domini 1468 ad meridianum Cracoviensem, ubi est elevacio poli 50 graduum*

| | **D** | **H** | **M** | **2** | **3** | **4** | **5** | **6** |
|---|---|---|---|---|---|---|---|---|
| Ari | 10 | 12 | 22 | 41 | 49 | 22 | 57 | 36 |
| Cnc | 11 | 23 | 47 | 38 | 13 | 46 | 1 | 26 |
| Vir | 13 | 12 | 47 | 52 | 2 | 48 | 0 | 0 |
| Cap | 11 | 16 | 21 | 16 | 9 | 31 | 26 | 0 |

*Per hanc tabulam invenitur introitus Solis in Arietem*

| | **H** | **M** | **2** | **3** |
|---|---|---|---|---|
| 1 | 5 | 49 | 15 | 59 |
| 2 | 11 | 38 | 31 | 58 |
| 3 | 17 | 27 | 47 | 56 |
| 4 | 23 | 17 | 3 | 55 |
| 5 | 5 | 6 | 19 | 54 |
| 6 | 10 | 55 | 35 | 53 |
| 7 | 16 | 44 | 51 | 52 |
| 8 | 22 | 34 | 7 | 51 |
| 9 | 4 | 23 | 23 | 49 |
| 10 | 10 | 12 | 39 | 48 |
| 11 | 16 | 1 | 53 | 47 |
| 12 | 21 | 51 | 19 | 46 |
| 13 | 3 | 40 | 24 | 45 |
| 14 | 9 | 29 | 49 | 44 |
| 15 | 15 | 19 | 46 | 42 |
| 16 | 21 | 8 | 12 | 39 |
| 17 | 2 | 57 | 27 | 38 |
| 18 | 8 | 46 | 43 | 37 |
| 19 | 14 | 36 | 49 | 36 |
| 20 | 20 | 25 | 15 | 35 |
| 40 | 16 | 50 | 30 | 8 |
| 60 | 13 | 10 | 0 | 42 |
| 80 | 9 | 47 | 28 | 16 |
| 100 | 6 | 6 | 40 | 20 |
| 200 | 12 | 13 | 20 | 40 |
| 300 | 18 | 20 | 0 | 30 |
| 400 | 0 | 26 | 40 | 20 |
| 500 | 16 | 33 | 20 | 10 |
| 600 | 12 | 40 | 0 | 0 |
| 700 | 18 | 40 | 40 | 20 |
| 800 | 0 | 53 | 20 | 40 |

<35>

**Rule 35 on the conjunction of the other five planets**

If you wish to know the conjunction of certain planets, first find the true position of both planets at your estimated time of the conjunction. Second: subtract the lesser from the greater and note the remaining longitude. Third: convert this length to the same denomination, namely to seconds. Fourth: search for the motion of both after one day. Fifth: subtract the motion of any one found first from the motion found second, if both have been direct. Or, conversely, <subtract> the second found motion from the first found motion, if both have been retrograde, and what remains will be the motion of any day. Sixth: subtract the slower daily motion from the faster daily motion, so that in the motion of Jupiter and Mars, subtract the motion of Jupiter in one day from the motion of Mars in one day, and what remains is the *superlacio*. Seventh: convert this *superlacio* to the same denomination to which the longitude was converted. Eighth: divide the longitude by the *superlacio*, if you can, and if you can't, then write 0 days and in the quotient the days are found. Ninth: if anything remains after the division, it will be preserved in its proper unconverted form. The *superlacio* of a day is divided by 24 and you will have the *superlacio* of an hour. Tenth: divide the remainder, which you previously kept in its proper form, by *superlacio* of an hour, and in the number of quotient will come the hours. But if they cannot be divided, then put zero in the place of the hours and it indicates no hour. Eleventh: multiply the remainder by 60 and divide by the *superlacio* of an hour and in quotient the minutes of the hour come out. Twelfth: multiply the remainder by 60 and divide by the *superlacio* of an hour and in the quotient the seconds are found. Thirteenth: add this extracted time to the estimated time and you will have the true conjunction of the planets, if both planets have been direct or retrograde. Fourteenth: search for the motion of both planets at the same time and they agree in signs, degrees, minutes, seconds, and tertias. **Caution:** But if one the planets, whose conjunction you are seeking for, was direct and the other retrograde, then search for the motion of the direct in one day, that is, after the first day, as has been said. Similarly, search for the retrograde motion in one day, that is, after the first day. Then the second motion will be smaller than the first, that is, the retrograde motion on the second day is smaller than its motion on the first day. For when a planet is retrograde, its motion is always decreasing. Subtract, therefore, the second motion of this retrograde planet from the motion of the planet at the preceding day and what remains will be how much it goes back in one day. Then join those motions together, that is, the forward and backward motions, and divide the longitude by that result. And first convert it into seconds, just as you converted the longitude, and extract the time with its fractions, as was noted before. Add this extracted time to the estimated time, if the lighter one did not precede the slower one, or subtract if the lighter one preceded the slower one.[193] **For example:** for the current year of the Lord 1486 I want to know the conjunction of Saturn and Mercury, which will be on the 13th of December. First I need to know the true

193 The lighter (*levior*) planet means here the forward one, and the slower (*tardius*) is the retrograding planet. The planet with the higher ecliptic longitude 'precedes' the other one.

motion of both of them at the estimated time. The motion of Saturn is: $8^{s}$ 15;38,29,38°. And the true motion of Mercury is: $8^{s}$ 15;22,55,33°. And the longitude[194] is 15 minutes, 34 seconds, 5 tertias. When this longitude is converted to seconds, it will be 934. And the motion of Saturn after one day is: $8^{s}$ 15;45,8,35°. And the motion of Mercury after one day is: $8^{s}$ 16;33,24,1°. When I therefore subtract the motion of Saturn on the first day from the motion of Saturn on the second day, the motion of Saturn in one day will be: 6 minutes, 38 seconds, 57 tertias. When I subtract the motion of Mercury found on the first day from the motion of Mercury found on the second day, the motion of Mercury in one day will be 1 degree, 10 minutes, 28 seconds, 28 tertias. Then when I subtract the motion of Saturn in one day from the motion of Mercury in one day, the *superlacio* remains, namely, the amount by which the motion of Mercury is greater than that of Saturn in one day: 1 degree, 3 minutes, 49 seconds, 31 tertias. I convert the *superlacio* to the seconds by multiplying it by 60 and find 3829. Then I must divide the longitude by the *superlacio*, but I cannot, so it is a sign that nothing will be added in days. Then I divide the *superlacio* of the day by 24 and in the quotient number will come the *superlacio* in an hour 159. And the remainder is 13. Then by the hour *superlacio* I divide that which I have previously kept in its proper form, and in the quotient find 5 hours. The remainder is 139. Then I multiply this remainder by 60 and divide by the hour *superlacio*, and in the quotient number we get 52 minutes. The remainder is 72. Then I multiply that remainder by 60 and divide by the hour *superlacio*, and in the quotient number 27 seconds will come. The remainder will be 27. Therefore, the extracted time to be added to the estimated time of the conjunction is 0 in days, 5 hours, 52 minutes, 27 seconds.[195] The true conjunction of Saturn and Mercury will therefore be on December 13, at five o'clock in the afternoon, 52 minutes, 27 seconds, etc.[196]

<36>

**Rule 36 on the adaptation of radices to any other meridian**

Experience teaches us that cities are built at a considerable distance from each other. For some are closer to the east, and some are closer to the west. Therefore, the eastern ones have earlier sunrise, noon, and sunset than the western ones. It is therefore necessary to show the method of adapting tables made for a certain meridian to another meridian. When, therefore, you intend to do this, that is to adapt the radices of one city to the radices of another city, you must first know the longitudes of those cities, which you will be able to know from the Table of longitudes of cities, which is common. The longitude of a city is the distance from the zenith of the city to the east or west. In the eastern cities, the longitude is counted from the east, while in the western cities from the Algerian Gades

194 I.e., the elongation or the difference of ecliptical longitudes of both planets.

195 Because of the truncating of both the elongation and the daily superlacio to seconds and especially due to truncation of the hourly superlacio, this division has an error already in minutes; the correct value should be $5^{h}\ 51^{m}\ 14^{s}$.

196 A computation with modern ephemerides shows that the conjunction of Saturn and Mercury occured on December 11 at 6 hours UT in ecliptical longitude 255.78°. On December 13, at noon of the Cracovian time, the longitude of Saturn was 256.03°, compared to 255.64° given in this rule, but the longitude of Mercury was 259.10° instead of 255.38°.

(Pillars of Hercules), like with us, the longitude of the city is counted from the west, that is from the Pillars of Hercules. Second: having obtained the longitudes of the two, subtract the lesser from the greater and the difference of longitude will remain, so that one city is more distant from the east or the west than the other. Third: convert this distance or difference to fractions in this way. If there are more than 15 degrees in the difference, then always take one hour for every 15 degrees, because in one hour 15 equinoctial degrees rise. And for each degree remaining below 15 minutes, take 4 minutes of an hour. This is the degrees below 15; multiply by 4 and you will have minutes of an hour. Fourth: if there are minutes of degree in such a distance, for each minute of degree take 4 seconds of an hour, multiplying them by 4. Fifth: if the minutes or seconds exceed 60, divide by 60 and in the quotient grows a previous larger fraction. But if there is no excess in minutes or seconds beyond 60, let it be in its proper form. Sixth: with the fractions corresponding to such a distance, the mean motion of the planet is searched for, the radix which you wish to verify. If you have hours, enter with hours, if not, similarly enter with minutes and seconds. Seventh: Subtract the found motion corresponding to such a distance from the radix of that planet in the table where you found it if your city was more eastern, or add it if it was more western. **For instance:** if you want to adapt the radices of Prague to the meridian of Cracow, you must first know the longitudes of both cities. The longitude of the city of Cracow is 32 degrees, 36 minutes, the longitude of the city of Prague is 29 degrees, 30 minutes. Second: subtract the lesser from the greater and there will remain the difference of one city from the other, that is, the equinoctial degrees intercepted between Cracow and Prague, and there will remain a difference of 3 degrees, 6 minutes. Third: because these degrees do not exceed 15, then the distance between the meridian of Cracow and Prague will be zero hours. Fourth: multiply 3 degrees by 4 and there will be 12 minutes of an hour. I will also multiply the minutes of degree by 4 and there will be 24 seconds of an hour. The distance between Cracow and Prague in time reaches 0 hours, 12 minutes, 24 seconds. Although others, working with a higher resolution, say that there is zero hours, 12 minutes, 23 seconds, 59 tertias, 35 quartas.[197] If you add this to the Prague radix of the conjunctions, the Cracow radix will come out of the same. Hence the radix of Prague is: 5 in days, 2 hours, 35 minutes, 57 seconds, 46 tertias, 30 quartas.[198] Having therefore added the aforesaid distance, namely 12 minutes, 23 seconds, with other fractions, the radix of Cracow will be found for this mean conjunction: 5 days, 2 hours, 48 minutes, 21 seconds, 46 tertias, 5 quartas.[199] However, to have the radix of Cracow in the motions in the collected years, enter with 12 minutes and 24 seconds and not with other fractions in the table of the planet whose motion you want to adapt, as you will find in the table of the Sun: 0 minutes, 30 seconds, 33 tertias.[200] And subtract that from the radix of Prague,

197 The resolution to quartas of an hour in zonal time corresponds to a resolution in east-west direction about 12 km on the equator and 8 km at the geographical latitude 50° of Prague and Cracow. The true difference of geographical longitudes between Karolinum in Prague and Collegium Maius in Cracow is 5;30,37°, which corresponds to 22 minutes, 2.4 seconds in the zonal time.

198 See the time of conjunction of the Sun and Moon in the year 1429 in Table TRP-09.

199 See Table TRC-09.

200 The mean motion of the Sun is an example how to adapt the cyclical radices in collected years for Prague to the meridian of Cracow. In column C of Table TRW-06, we can find the mean motion of the Sun in 12 minutes equal to 0;0,29,34,10° and in 24 seconds 0;0,0,59,8,20°. Their sum rounded to tertias reads 0;0,30,33°.

which is 9 signs, 18 degrees, 47 minutes, 27 seconds, 18 tertias,[201] and this radis of Cracow will be: 9 signs, 18 degrees, 46 minutes, 56 seconds, 45 tertias.[202] **There is also another way of searching for Cracovian radices by means of distance from Prague.** Consider how many miles there are between Cracow and Prague as precisely as you can. And since it is commonly said that there are 60 miles, and as the road is often winding because of valleys and hills, I therefore accept 50 common miles. So if there were a direct route from Cracow to Prague, it would be that many miles. So for every 16 miles you will take 1 degree and you will have 3 degrees in the quotient after dividing by 16 and 2 miles will remain in the residuum. And since 3 minutes of a degree and 45 seconds correspond to each mile, then I take 3 minutes for each mile and you will have 6 minutes. You don't need to worry about seconds. Or if you wish to make the operation more precise, take 7 in minutes, but it is not necessary, because then there is no excess in degrees beyond 15, nor in minutes beyond 16. Multiply both of these by four, according to what has already been noted, and you will have 12 minutes of the hour and 24 seconds of the hour, as you already found.

<37>

### Rule 37 on finding the latitude of the Moon

If you want to find the latitude of the Moon, first find the argument for the latitude of the Moon. You can get to know this argument of the latitude of the Moon in two ways. If you subtract the true motion of the dragon's head from the true motion of the Moon, the remainder will be the equated argument of the latitude of the Moon; or if you add the true motion of the Moon to the motion of the head of the dragon, you will also have the true argument of the latitude of the Moon.[203] Second: with the argument of the latitude of the Moon enter the Table of the latitude of the Moon by double entry and get the latitude of the Moon placed in the direction of that sign. Third: subtract the lesser from the greater to have the difference. Fourth: multiply[204] the difference by minutes, seconds and tertias of the argument of the latitude of the Moon and search for the proportional part. Fifth: add or subtract this proportional part from the first entry and you will have the latitude of the Moon. And this latitude will be northern and ascending if the argument is less than three signs. But if it be from three signs to 6 signs, it will be northern descending. And if it be from the 6th sign to the 9th, it will be southern descending. But if it be from the 9th sign to the 12th sign, it will be southern ascending. **For instance:** I want to know

201 See the line for the year 1428, column C in Table TRP-01.
202 See Table TRC-01.
203 In the second possibility, there is added the mean motion of the Moon. See Rule 20.
204 The Latin verb *ducere* (to lead) means here (i.e. in the expression *differenciam duc in minuta, secunda et tertia in argumento latitudinis Lune*) to multiply. We can find it in this meaning, for instance, also in the chapter *De multiplicatione* by Schöner, 1536, where he writes on f. A2v: *Et scias hoc generale totius operationis, quod ex tali ductura unius per aliam minutiam semper productum in subtiliori denominatione proveniet… Verbi gratia: minuta per minuta ducta secunda producunt.* – 'And notice the general principle of the whole operation, that multiplying one fraction by another always yields a result in the lower order… For example: minutes multiplied by minutes produce seconds.' The next rules also use a variant *reducere*. This terminology is probably based on an implicit understanding of multiplication as a form of transforming/converting values to other units.

the latitude of the Moon on the day of the Assumption of Mary in the current year of the Lord 1486. First, you should know the argument of the latitude of the Moon,[205] and it is: 11$^{s}$ 28;6,49,32°. Second: with this argument, enter the Table of the latitude of the Moon with a double entry for the sake of precision. First entry: 0 degrees, 10 minutes, 27 seconds. The second entry is 0 degrees, 5 minutes, 33 seconds.[206] Third: subtract one from the other and the difference will remain 4 minutes, 54 seconds. Fourth: multiply this difference by the remainder of the argument of the latitude of the Moon,[207] searching for the proportional part, and find 33 seconds, 26 tertias. When I subtract the proportional part from the first entry, the latitude of the Moon will be 0 degrees, 9 minutes, 53 seconds, 34 tertias. And it is clear that the latitude of the Moon is southern ascending, because it is from 9 to 12 signs, etc.

<38>

**Rule 38 on finding the latitude of the three upper planets, Saturn, Jupiter, and Mars**

If you want to find the latitude of the three upper planets, first find the true centre of each of them. Second: for the true centre, whose latitude you are seeking, convert its signs into degrees, and add also your superfluous degrees, which you have.[208] Third: after converting, if you wish to have the latitude of Saturn, convert its centre to degrees, add 50 degrees. If you are searching for latitude of Jupiter, subtract 20 degrees from the converted centre. But if you are searching for the latitude of Mars, keep the converted centre in its proper form. Fourth: with the converted centre enter the Table of the latitude of the five planets,[209] where, if you find the converted centre in its proper form, beside it in the last column under the title of 'proportional minutes' take proportional minutes with double entry to have the difference. Fifth: if you do not find the degree or the converted centre in its proper form, enter with the nearest smaller degree and consider the degrees that remain, with which it was not entered, and similarly in the last column under the heading 'proportional minutes' enter with a double entry, and consider both entries. which was greater or lesser. Sixth: subtract the lesser proportional minutes from the greater and you will have the difference. Seventh: multiply the difference by the remainder of degrees, minutes, seconds and tertias, if you have to search for the proportional part. Eighth: divide the proportional part by six

205 See the example in Rule 21.

206 The Table of the lunar latitude is part of John of Lignères' tables (see Chabás and Saby, *The Tables of 1322 by John of Lignères*, p. 80–83). We can find such a table, e.g., in MSS Prague, NK, X B 3, f. 120r; Prague, NK, X A 23, f. 32r; Cracow, BJ, 553, f. 165v. In these tables, however, the value of the second entry is 0;5,13°. The value 0;5,33° is obviously incorrect, because it should correspond to the argument 1° and should thus be about one half of the first entry for the argument 2°. This incorrect value can be found, e.g., in MS Cracow, BJ, 600, f. 109r.

207 I.e., calculate the product 0;4,54 × 0;6,49,32 = 0;33,26,42,48.

208 The 'superfluous' or additional degrees are the corrections for angular distances between the apogees and nodes of the planets. They are +50° for Saturn, −20° for Jupiter and 0 for Mars according, e.g., to Rule 21 of canons *Priores astrologi*, see Saby, 'Les canons de Jean de Lignères', p. 207.

209 This table is part of John of Lignères' tables (see Chabás and Saby, *The Tables of 1322 by John of Lignères*, pp. 76–79). We can find it, e.g., in MS Cracow, BJ, 600, f. 109v. The headings of subtables for individual planets state the required shifts of the centres: *Latitudo Saturni per augmentacionem 50 graduum super cuspidem*, *Latitudo Iovis per remocionem 20 graduum a cuspide*, and *Latitudo Martis absque augmentacione et diminucione*.

and I will have a part proportional to six degrees.[210] Ninth: add the proportional part to the proportional minutes previously found, or subtract it according to the common method. If the first entry is greater than the second, subtract a proportional part. If the first is smaller, add it and you will have the equated proportional minutes. Note for them 'superior' (upper) or 'inferior' (lower) according to the inscription found in the table. Tenth: with the true anomaly converted to degrees and other degrees added, enter into the Table of latitudes, or with the nearest smaller. And if it is marked 'superior' at minutes, take the latitude of the planet under the title 'northern region'. But if it is marked 'inferior' at minutes, take the latitude under the title 'southern' with the double entry and consider which entry is greater. Eleventh: when you have entered with the double entry, take the difference and multiply it by the remaining degrees, if anything remains in the minutes, seconds, and tertias, if you have them, and search for a proportional part. Twelfth: divide the found proportional part by 6 degrees and you will have a proportional part to 6 degrees.[211] Thirteenth: add or subtract this proportional part from the first entry, according to what the inscription indicates, and there will be equated degrees and minutes. Fourteenth: multiply that which will come out last by the previously found minutes, which have been sought with the degrees of the true centre, and there will result the latitude of the planet sought by you. **First caution:** when you add 50 degrees in proportional minutes to the converted degrees of the centre, if there are more than 360 degrees that you do not have in the table, subtract 360 degrees and you will enter with the remainder or the closest, as stated. **Second caution:** when you search for the latitude for the following day and you find a greater latitude than the previous one, that latitude is called ascending latitude. But if you find a smaller one, it will be said to be descending. And it is so when both of them, that is either the greater and the lesser, are northern. But if both are southern and the following be greater, the planet must be said to be descending; and conversely, it should be named ascending when the southern latitude of the following day decreases or diminishes. **For instance:** I want to know the latitude of Saturn in the current year of the Lord 1486, for the day of the Assumption of Mary. First, I must know the true and equated centre of Saturn[212] and it is this: $11^s$ 27;28,38,21°. Second: when I convert the signs to degrees by adding other degrees in the same way, add 50 degrees, they come to 407.[213] And because they exceed 360, I therefore subtract 360 and 47 degrees will remain. Third: with the remaining degrees or with those nearer I will enter the Table of latitudes under the minutes in the last column and in the first entry I find 44 minutes, 24 seconds.[214] In the second entry I find 40 minutes, 0 seconds. The first is greater. Then when I subtract the lesser from the greater, there remains a difference 4 minutes, 24 seconds. Fourth: when I multiply the proportional minutes by the remainder of the degrees and minutes, that is by 5 degrees, 28 minutes, 38 seconds, 21 tertias, this proportional part will

210 The step of the table is six degrees. It is in a tradition which dates back to Ptolemy's *Almagest* XIII,5 (see Toomer, *Ptolemy's Almagest*, pp. 632–34).

211 Because the step of this table is 6 degrees, the proportional part of the increment is given by its product with the ratio of the remaining degrees and their fractions to this step. The procedure used here, that is to calculate first the product and then the division, is more precise than the reversed procedure with an intermediate rounding used in Rule 35.

212 See Rule 22.

213 I.e., $11 \times 30 + 27 + 50$.

214 See Chabás and Saby, *The Tables of 1322 by John of Lignères*, p. 79, the entry for 42°.

come out:[215] 24 minutes, 6 seconds, 0 tertias. Then when I divide this proportional part by six, I will have this proportional part at 6 degrees: 4 minutes, 1 second. That proportional part must be subtracted from the first entry and this equated proportional minutes will remain: 40 minutes, 23 seconds. That I have to keep aside and to denote it 'superior' because proportional minutes are taken in the upper part of the table. Then I must know the true anomaly of Saturn for the day of the Assumption of Mary,[216] and it is: $8^s$ 22;11,38,19°. And when I convert it to degrees, 262 degrees will result, and the closest smaller ones are 252.[217] I will enter the Table of latitude, which is at minutes marked by the inscription 'superior'. I take the latitude of Saturn under the inscription 'northern', and at the first entry I find 2 degrees, 39 minutes. Second entry is 2 degrees, 42 minutes.[218] The first one is lesser. Then when I subtract the lesser from the greater, there will be a difference of 3 minutes. Then I must multiply the difference by the remainder of degrees and other fractions, namely 10 degrees, 11 minutes, 38 seconds, 19 tertias, and this proportional part will result: 30 minutes, 34 seconds, 55 tertias. After this, divide the proportional part by six, and the proportional part to six degrees will come out: 5 minutes, 5 seconds,[219] which I add to the first entry, and the equated degrees and minutes will result.[220] Finally, when I multiply what finally results in equated degrees and minutes by the minutes previously found, which were sought, the degree of true centre of Saturn results in the latitude of Saturn on the day of the Assumption of Mary: 1 degree, 50 minutes, 26 seconds, 14 tertias, latitude northern.[221]

<39>

**Rule 39 on finding the latitude of Venus**

If you want to know the latitude of Venus, you will find three partial latitudes of it, of which you will finally have one complete. The first of these is called 'declination', the second 'reflection', the third 'deviation'. You will find the latitude of Venus, which is called the declination, in this way: first, convert the signs in the anomaly of Venus into degrees by adding the remaining degrees, and with these or the closest lesser ones enter the Table of Venus and take the declination with a double entry because of the difference and consider which entry is greater or lesser. Second: multiply the difference by the remainder of the degrees, the minutes, the seconds, and the tertias, as was noted above in searching for the proportional part. Third: add or subtract that proportional part found from the first entry according to the common method, and you will have the declination of Venus. Fourth: add 60 degrees to the true centre

215 I.e., 0;4,24 × 5;28,38,21 = 0;24,6,0,44,24.

216 In the eighth step of example to Rule 22, the value $8^s$ 22;11,38,19° was derived and it also agrees with the value 262 converted into degrees. In manuscripts **A** and **B**, however, a value $8^s$ 29;11,38,19° is written by an error.

217 Closest is 258.

218 See Chabás and Saby, *The Tables of 1322 by John of Lignères*, p. 77, where the value 2;39° is found for the entry 252° and 2;42° for 246°.

219 Precisely 0;5,5,49,9,30° which should be rounded to 0;5,6°.

220 This interpolation is completely wrong, because it should be done between the values 2;36° for 258° and 2;34° for 264°, and it should result in approximately 2;34,36,7°.

221 This value results from the incorrect interpolation of degrees and minutes. The correct interpolation would yield 1;44,3,21° (the true latitude according to modern ephemeris was 1;34°).

of Venus,[222] and if it exceeds 360, subtract 360, and with the remainder or something closer, enter the table and beside it take the proportional minutes with a double entry as before. Fifth: subtract the lesser minutes from the greater ones and take the difference. Sixth: multiply this difference by the remainder of degrees, minutes, seconds, and tertias, and take a proportional part. Seventh: divide the proportional part by 6 and you will have a proportional part to 6 degrees. Eighth: add or subtract the proportional part from the first entry according to the common method. Ninth: these two findings, namely the declination of Venus and the proportional minutes, multiply one by the other, and what results after the multiplication is the latitude of Venus, which is called the declination, which you keep aside with its title. **First caution:** if you have entered the table in the same direction both in the centre and in the anomaly, the declination will be southern. If, however, you enter with the anomaly in one part of the table, and with the centre in another, the latitude will be northern. **Second caution:** if you had 270 degrees in the centre with which to enter or closer to 360 degrees, you will find nothing in the declination, for it will be the latitude of the declination, so now you must enter with the centre adding 60 degrees.[223] **For instance:** I want to find the latitude of Venus, which is called declination, for the current year of the Lord 1486, on the day of the Nativity of Mary.[224] First I need to know the anomaly of Venus[225] and it is: $8^s\,1;53,49,59°$. And after I convert the signs into degrees there will be 241 degrees. Second: when I enter with double entry in the Table of Venus declination with these degrees or near greater or lesser, the first entry is 0 degrees, 59 minutes. Second entry is 1 degree, 13 minutes.[226] The first is lesser, the difference is 14 minutes. Then, when I multiply the difference by the remainder of the degrees,[227] this proportional part will result: 26 minutes, 33 seconds, 39 tertias. When I divide the proportional part by 6, this proportional part to 6 degrees will result: 4 minutes, 25 seconds,[228] which must be added to the first entry and the declination of Venus will result: 1 degree, 3 minutes, 25 seconds.[229] Then I must know the true centre of Venus[230] on the day of the Nativity of Mary and it is: $2^s\,23;8,53,14°$, which I also convert to degrees and add to it 60 degrees and find 143, with which or closer I enter the Table of proportional minutes with double entry. The first entry is 44 minutes, 24 seconds. The second is 48 minutes, 24 seconds.[231] The first is smaller. The difference is 4 minutes, 0 seconds. When I multiply the

222 The text on declination and reflection of Venus in Rule 39 is erroneous. For this problem see the end of the Chapter 23, 'Latitudes of planets'.

223 The meaning of the second caution is unclear.

224 It is the $8^{th}$ of September.

225 From Tables TRC-01, TRW-02, TRW-03, and TRW-05 we can find the mean motion $5^s\,26;21,7,21°$, mean anomaly $7^s\,29;44,50,0°$, from Tables TRP-07 and TRP-07a, TRC-07b, and TRC-07c the apogee $3^s\,1;3,14,7°$, the mean centre is thus $2^s\,25;17,53,14°$ and according to Table TRW-19 the equation of the centre is 2;9°, which added to the mean anomaly gives the true anomaly.

226 The first entry is for argument 240°, the second, however, for 234°. See Chabás and Saby, *The Tables of 1322 by John of Lignères*, p. 78.

227 $1;53,49,59 \times 0;14 = 0;26,33,39,46$.

228 More precisely $0;26,33,39,46/6 = 0;4,25,36,30$.

229 The proportional part should be subtracted, because the declination (which is decreasing here) is extrapolated from two smaller values of the argument. The next entry for 246° gives declination 0;45°, which means also an increment −0;14°. The correct interpolation should thus lead to the value of declination 0;54,34°.

230 It is the mean centre $2^s\,25;17,53,14°$ minus the equation of centre 2;9°.

231 The first entry is for the argument 138°, the second for 144°; see Chabás and Saby, *The Tables of 1322 by John of Lignères*, p. 79.

difference by the remainder of the degrees, that is, 5 degrees, 8 minutes, 53 seconds, 14 tertias, this proportional part will result: 20 minutes, 35 seconds, 33 tertias.[232] If I divide by 6, this proportional part to 6 degrees will result: 3 minutes, 26 seconds,[233] which must be added to the first entry, that is, 44 minutes, 24 seconds, and these equated proportional minutes will come out: 47 minutes, 50 seconds. Finally, I must multiply the declination of Venus by proportional minutes, and the latitude of Venus will result, which is called that declination, that is, 0 degrees, 50 minutes, 33 seconds, 26 tertias, north.[234]

**You can find the latitude of Venus, which is called 'reflection', as follows.** Enter first with the degrees of the true anomaly, signs converted to degrees with the remaining added, if there are any, or with the closest smaller ones, the Table of Venus and take the reflection with a double entry. Second: subtract the lesser entry from the greater and there will be the difference. Third: multiply the difference by the remainder of degrees, minutes, seconds and tertias and you will have a proportional part. Fourth: divide this proportional part by six and you will have a proportional part to six degrees. Fifth: add or subtract this proportional part from the first entry in the common way, and you will have the reflection of Venus. Sixth: subtract 60 from the degrees of the centre of Venus,[235] to which, if you could not subtract, add 360 and then subtract 60. Seventh: with these degrees of the true centre enter with double entry, searching for the proportional minutes and subtract the lesser from greater searching for the difference, which is multiplied by remainder of the degrees, minutes, seconds and tertias, and the proportional part will result. Divide this proportional part by six and you will have a proportional part to 6 degrees. Add or subtract this proportional part from the first entry in the common way, and you will have equated proportional minutes, which you will set aside. Finally: by these proportional minutes multiply the found reflection and the second latitude of Venus will result, which is called the equated latitude of reflection. **First caution:** if the centre for which you have now received the proportional minutes falls in the upper part of the table and the true anomaly is less than 180 degrees, the found reflection is northern. But if the anomaly is more than 180 degrees, it must be southern. **Second caution:** if the true centre falls in the lower part, the true anomaly is less than 180 degrees, the reflection is southern. If the true anomaly is more, it is northern. So set aside the latitude found in this way. **For example:** if I want to find the latitude of the reflection of Venus for the current year of the Lord 1486, for the day of the Nativity of Mary, first enter with the degrees of the true anomaly of Venus converted to degrees or with the closest smaller ones taking the reflection of Venus with double entry. And the true anomaly of Venus is: $8^s$ 1;53,49,59°, which when converted to degrees comes to 241. The first entry 2 degrees, 25 minutes, the second entry 2 degrees, 28 minutes.[236] The

232 5;8,53,14 × 0;4 = 0;20,35,32,56.

233 0;20,35,33/6 = 0;3,25,55,30.

234 1;3,25 × 0;47,50 = 0;50,33,25,50. However, with the properly interpolated declination 0;54,34 × 0;47,50 = 0;43,30,6,20.

235 The erroneous text *a gradibus centri Veneris deme 60* – 'subtract 60 from the degrees of the centre of Venus's is identical in all three copies with John of Głogów's canons we had available (**A** = Vatican, BAV, Pal. lat. 1439, f. 192r; **B** = Cracow, BJ, 600, f. 159r; **C** = Wrocław, BU, IV F 19, f. 263r) as well as in the canons *Astrorum* (in the MS Cracow, BJ, 600, f. 146v) which can be considered the inspiration for John of Głogów's rules. See the the Chapter 23, 'Latitudes of planets'.

236 The first entry is for the argument 240°. The second for 234°, which is thus again taken by a mistake instead of reflection 2;20° for argument 246°; see Chabás and Saby, *The Tables of 1322 by John of Lignères*, p. 78.

first is minor. Difference 3. The proportional part will be: 5 minutes, 41 seconds, 30 tertias.[237] Proportional minutes to six degrees, 0 minutes, 52 seconds,[238] which must be added to the first entry, and the latitude of the reflection of Venus will be: 2 degrees, 25 minutes, 52 seconds.[239] Then, when I subtract 60 degrees from the degrees of the centre there remain 23 degrees, with which, or with the closest lesser ones, I search for proportional minutes with double entry. First entry 57 minutes, 0 seconds, second entry 54 minutes, 36 seconds.[240] The first is major. The difference is 2 minutes, 24 seconds. When I multiply it by the remaining degrees and minutes, this proportional part will result: 12 minutes, 21 seconds, 19 tertias.[241] When I divide by six, there results a proportional part to these six degrees: 2 minutes, 3 seconds, 33 terrias,[242] which must be subtracted from the first entry, and these equated minutes will be: 54 minutes, 56 seconds, 27 tertias. Finally, when I have multiplied the reflection by the proportional minutes, the second latitude of Venus, which is called the equated latitude of reflection, will be: 2 degrees, 13 minutes, 34 seconds, 1 tertia,[243] south.

**To find the latitude of Venus, which is called the 'deviation'.** When you want to know the third latitude of Venus, which is called the deviation, divide by 6 the proportional minutes kept aside in the previous latitude of reflection, and you will get the third latitude, namely the latitude of the deviation, and it is always northern. Second: consider, whether these three latitudes are in the same title; join them together and the complete latitude of Venus is found. If, however, they are of different titles, then add those latitudes, which are of the same title, and, thirdly, subtract the lesser from the greater, and the remainder with the title of the greater will be the sought latitude of Venus. **For instance:** the proportional minutes kept aside in the latitude of the reflection are 54 minutes, 56 seconds, 27 tertias. When I divide by six, the latitude of the deviation of Venus will be: 0 degrees, 9 minutes, 9 seconds, 24 tertias, north. And because the latitudes are of different titles, I add together the two northern ones, and afterwards, when I have subtracted the smaller one from the larger one, the complete latitude of Venus will come out: 1 degree, 13 minutes, 51 seconds, 11 tertias, south.[244]

<40>

### Rule 40 and the last on finding the latitude of Mercury

Similarly, in the latitude of Mercury three latitudes are to be found, the first of which is called 'declination', the second 'reflection', the third 'deviation'. And if you would like to find the latitude of Mercury, which is called the declination, first convert the signs and degrees in the anomaly of Mercury. Similarly, with the nearest smaller degrees enter

---

237 1;53,49,59 × 0;3 = 0;5,41,29,57.

238 A correct division would yield 0;5,41,30/6 = 0;56,55.

239 A correct interpolation between the values 2;25° at 240° and 2;20° at 246° would result in 2;23,25,8° at 241;53,49,59°.

240 The first entry is for 18°, the second one for 24°. See Chabás and Saby, *The Tables of 1322 by John of Lignères*, p. 79.

241 5;8,53,14 × 0;2,24 = 0;12,21,19,45,36.

242 0;12,21,19/6 = 0;2,3,33,10.

243 The product 2;25,52 × 0;54,56,27 = 2;13,34,2,10,24. The correctly interpolated reflection would, however, result in 2;23,25,8 × 0;54,56,27 ≅ 2;11,19,34.

244 With the correctly interpolated declination and reflection, the result would be 1;30,45,38°, southern. Calculation with a modern ephemeris gives the ecliptical latitude −0;49°.

the table and find the declination of Mercury with double entry. Similarly convert the difference in the remainder of the degrees, minutes, seconds, and tertias, taking a proportional part, which you divide by 6, and you will have a proportional part to 6 degrees, which you can add or subtract, according to whether the first entry was greater or lesser, and you will have the declination of Mercury. Second: when the true centre of Mercury is converted to degrees, add 270 to those degrees, and if more degrees result than 360, subtract this, and with the remainder enter into the Table of minutes with double entry, and when you have the difference, convert it into the remainder of the degrees, minutes, seconds, and tertias, looking for the proportional part which divide by six, and you will have proportional minutes to 6 degrees, which add or subtract from the first entry in the common way, and you will have equated minutes. Third: multiply the declination by the equated proportional minutes and you will have the first latitude of Mercury, which is called the latitude of the declination of Mercury. **Caution:** if both the anomaly and the centre of Mercury, in which you found the proportional minutes, fall in the same part of the table, the found declination is southern. If they are in different parts, it is northern. So set that latitude aside with its title. **For instance:** I want to know the latitude of the declination of Mercury for the current year of the Lord 1486, the day of Saint Lucy.[245] First I must know the true anomaly of Mercury and it is: $7^s$ 2;44,53,23°. When I reduce it to degrees, there are 212 degrees; with this or the closest smaller one, namely 210, enter and the first entry will be 3 degrees, 7 minutes. Second 3 degrees, 26 minutes. The first is minor. The difference is 19 minutes. The proportional part will come out as follows: 0 degrees, 52 minutes, 12 seconds, 54 tertias, which when I divide by 6, this will come out: the proportional part must to be added to the first entry: 8 minutes, 42 seconds, and it will arrive at the declination of Mercury 3 degrees, 15 minutes, 42 seconds, north. Then I must know the true centre of Mercury and it is: 1 sign, 28 degrees, 42 minutes, 41 seconds, 42 tertias. When I convert it into degrees, after adding 270 degrees, there will be 328 degrees, with which or the smaller I will search for minutes with a double entry. First entry is 48 minutes, 24 seconds. Second is 44 minutes, 24 seconds.[246] The first is major. Difference is 4 minutes. The proportional part will be 18 minutes, 50 seconds, 46 tertias,[247] which when I divide by six, the proportional minutes come to 3 minutes, 8 seconds,[248] which must be subtracted from the first entry and these equated minutes are: 45 minutes, 16 seconds.[249] Finally, by multiplying the declination by the equated proportional minutes, this first latitude and declination of Mercury will be: 2 degrees, 27 minutes, 38 seconds, 41 tertias,[250] north.

**To find the latitude of Mercury, which is called reflection.** First, convert the signs and degrees of the anomaly of Mercury into one,[251] and with them or minor ones enter

245 It is the $13^{th}$ of December 1486.

246 The first entry is for the centre equal to 324°, the second for 318° instead of 0;52,0° for 330°. See Chabás and Saby, *The Tables of 1322 by John of Lignères*, p. 79.

247 This value is truncated from 0;4,0 × 4;42,41,42 = 0;18,50,46,48.

248 Again truncated from 0;18,50,46/6 = 0;3,8,27,40.

249 In a correct interpolation should be added 0;3,36 × 4;42,41,42/6 = 0;2,49,37,1,12, so that rounded to seconds, the proportional minutes should read 0;51,14.

250 0;45,16 × 3;15,42° = 2;27,38,41,12°.

251 I.e., into the same units, namely degrees.

the Table of Mercury taking its reflection with a double entry. Multiply the difference by the remainder searching for the proportional part, which you divide by six, and add or subtract the result from the first entry according in the common way, and you will have the reflection of Mercury. Second: with the centre of Mercury enter into the Table of minutes, first adding 180 degrees, and if it rises to more than 360 degrees, subtract that and enter with the remainder or the nearest smaller with double entry. Multiply the difference by the remainder and look for proportional minutes. Divide these by six and add or subtract the proportional part from the first entry in the common way. Set the minutes aside, as the third latitude is to be found from them. Finally, multiply the reflection by the equated minutes and you will get the second latitude of Mercury, which is called the latitude of reflection. **First caution:** if the centre with which I found the minutes falls at the upper part of the table, the true anomaly is less than 180 degrees and the reflection found will be northern. But if the true anomaly will be larger, it will be southern. **Second caution:** when the centre with which you found the minutes falls at the lower part of the table and the anomaly is less than 180 degrees, the reflection is southern. If more, it is northern. **For example:** I want to know the latitude of reflection of Mercury for the aforesaid day. First, I convert the signs and degrees of the anomaly into one and they come to 212, with that or the closer enter the Table of reflection of Mercury with double entry. The first entry will be: 1 degree, 45 minutes. The second 1 degree, 49 minutes.[252] The first is minor. Difference 4. The proportional part is: 10 minutes, 59 seconds, 34 tertias. When I divide this by 6, I will have this proportional part to 6 degrees: 1 minute, 50 seconds.[253] Add it and this equated reflection will be: 1 degree, 46 minutes, 50 seconds.[254] Then I have to convert the centre of Mercury in degrees by adding 180 degrees to it and 238 degrees will result, with which or the closest smaller ones enter into the Table of minutes and this first entry will be: 35 minutes, 12 seconds. Second entry 40 minutes, 0 seconds.[255] The difference is 4 minutes, 48 seconds. The proportional part will be: 22 minutes, 36 seconds, 55 tertias,[256] which when I divide by 6, the proportional part will be: 3 minutes, 46 seconds. By adding to the first entry, therefore, the equated minutes will be: 38 minutes, 58 seconds. Third: when the reflection is multiplied by minutes, this second latitude of Mercury, which is called the reflection, will come out: 1 degree, 9 minutes, 22 seconds, 56 tertias, south.

**To find the latitude of Mercury, which is called the deviation.** If you want to have the third latitude of Mercury, then take a quarter and a half from the proportional minutes kept, that is, divide those minutes by four, and in the quotient will be a quarter of these, therefore a fourth part. Add half of it, and thus the third latitude will be found, which is for Mercury always southern. Second: if all the latitudes are of the same title, join them together and the complete latitude of Mercury is found. If, however, they were of different titles, subtract the lesser from the greater, and the remainder of the greater with its title

252 The second entry corresponds to the argument 210°. However, the neighbouring entries of the reflection give 2;0° for the argument 216°, or 1;29° for 204°. See Chabás and Saby, *The Tables of 1322 by John of Lignères*, p. 79.

253 With correct entries it should be 0;6,52,13°.

254 Should be 1;51,51°.

255 The first entry is for the centre 234°, the second one for 228°, instead of 0;30,0 for 240°. See Chabás and Saby, *The Tables of 1322 by John of Lignères*, p. 79.

256 0;4,48 × 4;42,41,42 = 0;22,36,56,9,36.

will be the latitude of Mercury sought. **For instance:** if I want to know the latitude of Mercury, which is called the deviation, I must divide the previously found minutes by 4 and in the quotient that fourth part will be: 9 minutes, 44 seconds,[257] and in the residuum 2 will remain in excess. When we take a half of this fourth part, it will be: 5 minutes, 22 seconds,[258] which half, when I add to the fourth part, the latitude of the deviation of Mercury will be: 0 degrees, 15 minutes, 6 seconds, south. Finally, I add the two southern latitudes of Mercury, that is, the deviation and the reflection, and then subtract the smaller one from the larger one, and the final latitude of Mercury will remain: 1 degree, 2 minutes, 39 seconds, 45 tertias, north.[259] The end. Cracow.

## 8.4 Notes on Canons *Pro (introduccione) generali*

Just as Nicolaus Wodka in his Rule 36 gives instructions for the composition of a Table of oblique ascensions, John of Głogów in his Rule 8 advises how to check or recreate the Tables of mean motions. This explanation, however, is simplified and it is not consistent with the numbers which are actually found in the Tables of mean motions. John of Głogów suggests to take always the first value in the table, e.g. for one day, and by successive adding of this value to get its subsequent multiples. However, the first value is rounded to the whole tertias and the error due to neglecting lower fractions of the mean motion would soon increase to the tertias and to the higher orders. Tabulated values clearly reveal that they were calculated using more precise values and rounded only afterwards. It seems that John of Głogów simplified this problem for his pupils. If there are any simplifications in his text, it is not because he does not understand the material, but because he has simplified it for didactic purposes.

At first sight, it may appear that John of Głogów did not add much new content to the canons, but rather elaborated them in considerable detail. On the other hand, in his canons he has explained e.g. additional tables and some astrological matters to a much greater extent than his predecessors. He has numbered the individual steps to be followed by users of the tables ('first', 'second'..., up to the 'twenty-ninth and last' /*primo, secundo..., decimo nono et ultimo*/, as we find in his Rule 31) and he also added 'warnings' or 'cautions' (*cautelas*) on what to watch out for when working.

Another typical expression of the effort to make the explanation understandable for students is that John of Głogów added specific examples (in a few cases even two or more) to most of the rules in his canons. The only exceptions are rules 16, 23, 25, 30 and 36, where no example can be found. At the same time, John of Głogów often increased the precision of the numbers in his examples. He related the examples to two dates, 1486 (rules 1–8, 18–22, 24, 35, 37–40), and 1487 (rules 9–15, 17, 26–29, 31–34). The examples refer either to the beginning of the years in question or – most often – to the day of the Assumption of the Virgin Mary (15$^{th}$ of August), rarely also to the day of the Birth of the

257 The proportional minutes found in the calculation of the reflection 0;38,58/4 = 0;9,44.

258 Correctly should be 0;4,52.

259 I.e., 2;27,8,41° – (1;24,28,56° + 0;15,6°), however, these numbers are erroneous. For a comparison, the true latitude of Mercury according to modern ephemeris was southern, about –0;28°.

Virgin Mary (8th of September), the day of St Mary Magdalene (22nd of July), or the day of St Lucia (13th of December). He may have been motivated to choose the Assumption of the Virgin Mary day because of his relationship and reminiscence of the church of the same name in his hometown of Głogów. The years 1486 and 1487 are found in the examples in MSS Vatican, BAV, Pal. lat. 1439 and Cracow, BJ, 600. On the contrary, in MS Wrocław, BU, IV F 19, the examples were recalculated and updated by an anonymous scribe to the years 1493, 1494 and 1495. Several attached schematic figures of horoscopes remain blank in MS Vatican, BAV, Pal. lat. 1439, while in MS Cracow, BJ, 600 the figures are filled with data. In MS Wrocław, BU, IV F 19 the figures are completely missing.

John of Głogów's canons are above all evidence of a didactically presented work, reducing tabular computation to a series of steps for students to follow.

John of Głogów is also the author of the interpretation of the TR calculated to the Cracow meridian and glossed in MS Cracow, BJ, 2855, ff. 1r–24r from 1524.[260] Already by the title of the work (*Tabularum resolutarum compendiosa pro iuniorum institucione interpretacio per magistrum Iohannem* /manuscript per err.: Ihoannem/ *Glogovienssem recolecta*) we can see that the author again had at heart the benefit of the youth. Incipit and explicit: *Ad Tabularum resolutarum intelligendum aliqua generalia in tabula motuum planetarum, quas resolutas dicimus, introductorio sunt premitenda. Consideracio prima: Tabule, quarum praxis intendimus, resolute dicuntur* … × … *et figura celi ad hoc tempus verifficata est, ut patet in figura sequenti. Finis per me, Iohannem de Wijslijchow, anno Domini 1524*.

John of Głogów also often writes about how to use tables to make almanacs. According to Markowski he actually made almanacs, so he may be describing his own practice.

Already in the Introduction of this book (Chapter 1) we used John of Głogów's explanation of what the term TR means.

It is, of course, not impossible that in the future other canons or fragments of canons to the *Tabule resolute* will be found among the texts in the manuscripts. Markowski,[261] for example, mentions several other manuscripts written in Cracow and peripherally more or less related to the TR. But several of them (e.g. Cracow, BJ, 459, ff. 5r–7r[262] from *c.* 1429 or partly similar MS Cracow, BJ, 602, ff. 82r–85r,[263] also dated to 1429) are more theoretical treatises with no particular attention to the specifics of the TR and often are not divided into the usual canons.

260 Markowski, *Astronomica*, pp. 76–77 (item 112).

261 Markowski, *Astronomica*, pp. 22–23. Some other minor and marginal texts on TR are recorded by him on the pp. 275–76.

262 *Catalogus BJ*, III, pp. 24–30.

263 *Catalogus BJ*, III, pp. 196–97.

# 9 Early Modern Prints with Canons to the *Tabule resolute*

Three authors are known from the sixteenth century who published the TR with their canons. They are Ambrosius Lacher in 1511, Johannes Schöner in 1536 (and posthumously in 1551, 1587, and 1588) and Johannes Virdung also posthumously in 1542.

## 9.1 Ambrosius Lacher

The first words of the Lacher's edition (f. A1v–A2ra): *Tabule resolute de motibus planetarum aliorumque supercelestium mobilium iam de novo extracte, castigate atque revise per magistrum Ambrosium Lacher de Merspurg, mathematicum in studio novo Frankfordiano.* || *<T>anto revera digniorem primam causam iudicare debemus…*

Ambrosius Lacher was born in Meersburg am Bodensee 1470 and died in Frankfurt am Oder 1540.[1] He studied at the universities of Basel, Leipzig and Wittenberg and from 1506 worked at the University of Frankfurt am Oder. There he became the first professor of mathematics, astronomy, and music theory. In 1516 he served as rector of the university. Between 1506 and 1511, he published eight titles in his own printing house and on his own expenses, which he then used in his lectures to compensate the lack of textbooks. Ambrosius Lacher was thus one of the early humanist scholars who used the invention of printing directly for the dissemination of scientific knowledge itself. Among his prints were Euclid's Elementa and Alcabitius's Introduction to Astrology, but also the *Tabulae resolutae de motibus planetarum* (1511). This was the first printed edition of *Tabule resolute.* Lacher's tables were calculated for Leipzig, with the radices from 1388 to 2008.

The *Tabule resolute* with canons by Ambrosius Lacher were transcribed by hand in the sixteenth century; they can be found in MS Mainz, Wissenschaftliche Stadtbibliothek, HS I 536, ff. 60r–121r.[2] The manuscript was written in Antwerp in 1530, 1532, and 1533. The TR in the manuscript correspond to the 1511 edition. F. 60r: *Tabule resolute de motibus planetarum aliorumque mobilium supercelestium iam de novo extracte et castigate per magistrum Ambrosium Lacher de Merspurg, mathematicum in studio novo Franckfordiano.* On f. 102r the text of Rule 1 begins by these words: *Medium motum cuiuslibet planetarum per ipsius tabulas ad tempus quodcumque propositum invenire.*[3]

1 See Heinrich Grimm, 'Lacher, Ambrosius', in *Neue Deutsche Biographie*, 13 (1982), p. 367–68.

2 We are grateful to José Chabás for informing us about the manuscript.

3 See description of the whole manuscript: https://resolver.staatsbibliothek-berlin.de/HSP0006532500000000 [accessed: 12. 8. 2024]. See also Ernst Zinner, *Verzeichnis der astronomischen Handschriften des deutschen Kulturgebietes* (München: C. H. Beck, 1925), p. 344 (item 10863).

The first Rule, called *Medium motum cuiuslibet planetarum… invenire*, gives an example relating to the year 1502 on f. A6va–B1ra. We present it as an example of Lacher's contribution to the TR:

> *Quo facto omnes numeratores cuiusque denominacionis simul adde, unumquodque ad suum genus, ut in addicione fieri solet, et aggregatum ostendit motum medium, quem quesieras ad Lipsensium meridianum, ad quem has tabulas principaliter reduximus ob celebriorem prefati opidi famam, ut usui communi reddantur accomodaciores, nec aliis locis ab huius meridiano notabiliter discrepantibus, nisi reductive inservire possunt.*
>
> *Quod si forte loco habitacionis consideracionisve exigente ad alium meridianum motus supercelestium corporum infestigare (sic) proposueris, necesse est, ut tempus distantie meridianorum tue consideracionis locum et Lypsigk interiacens addas tempori proposito, ad quod motus invenire desideras, si tue consideracionis locus fuerit occidentalior Lypsigk, aut subtrahas, si fuerit orientalior, prout tabula huius rei gracia facta et ab unde informabit. Et tempus illud, quod facta addicione conflatur vel post subtractionem relinquitur, inserviet exacte tuo proposito, ad quod si motum alicuius planete vel mobilis inveneris, non aliter eveniet, quam si tabule nostre ad meridianum consideracionis tue vel habitacionis essent composite.*
>
> *Sed in hac re || tali fruamur exemplo: iubeor invenire medium motum Solis Anno Domini currente 1502 ad 4. diem mensis Aprilis, hora 17, minuto 34, secundo 10, transactis ad meridianum opidi Lypsensis. Quero ergo hunc annum, scilicet 1501 completum, in Tabula collectorum annorum, in linea numeri, et eum nusquam invenio precise propinquiorem. Ergo illo minorem ibi repertum 1488 suscipio motumque ei correspondente in loco Solis 9 signa, 19 gradus, 13 minuta, 23 secunda, 42 tercia extra scribo. Quo facto annos collectos ab annis propositis subtraho, residuabuntur 13, qui dicuntur expansi, penes quos in tabula eiusdem tituli in loco Solis 11 signa, 29 gradus, 50 minuta, 56, 45 inveniuntur, que prius notatis adiicio, ut radix Solis anni concrescat currentis 9 signa, 19 gradus, 4 minuta, 20, 27. Sic radix cuiuslibet mobilis ad principium anni currentis vel tamquam currentis constituatur. Cum etiam tempus propositum ultra annos integros includat menses, dies et cetera, quare cum transactis mensibus, scilicet Marcio, Tabulam mensium anni communis intrabo et motum Solis penes Marcium repertum 2 signa, 28 gradus, 42 minuta, 29, 26 radici subscribo. Deinde 4 dierum motum Solis 3 gradus, 56 minuta, 33, 18 in Tabula dierum inventum prescriptis subiungo. Haud secus cum 17 horis ex horarum tabula de columna Solis 0 gradus, 41 minuta, 53, 24 extrahendo cum supra positis scribo. Sed quia cum 34 minuta horarum tabulam simul ingredi non possum, numeri enim communis linea eiusdem tabule ultra 30 non extenditur, igitur || primo cum 30 minuta invenio 1 minutum, 13, 55, 25. Deinde cum residuo minutorum numero, scilicet 4, ex eadem tabula 0 minuta, 9, 51, 23 secundario introitu reperta cum prenotatis subscribo. Ultimo vero cum 10 secundis eandem tabulam ingrediendo in prima eius linea ocurrit secundorum numerus propter conformem denominacionem predicti numeri in tabulam missi. Hoc modo 0 in secundis, 24, 38, 28, que prius scriptis subiungo. Quo facto omnes differencias motus invicem addo debito ordine et prodibit motus Solis medius ad tempus propositum 0 signa, 22 gradus, 26 minuta, 40, 46, 26. Sic te in omnibus operari oportet quorumcumque ex nostris tabulis medios motus infestigare* (sic) *pretendis.*

> 'When this is done, then add all terms of each denomination together, each to its own kind, as is usually done in addition, and the sum will show the mean motion

you were seeking for the meridian of Leipzig, to which we have reduced these tables because of the famous reputation of that city, in order to make them more accommodating to common use. However, they cannot serve other places that deviate significantly from this meridian unless they are reduced.

If, however, you would by chance wish instead to find the motions of the heavenly bodies at your place of dwelling, or of necessity to a different meridian, it is necessary that you should add to the proposed time to which you wish to find out the motion, the temporal distance between the meridian of your place of consideration and Leipzig, if your place of consideration is more western than Leipzig, or subtract it, if it is more eastern, as indicated in the table that has been created for this matter.[4] And that time, which is fused by making the addition, or remains after the subtraction, will serve exactly your purpose, so that if you find the motion of any planet or mobile, that will not differ from that found if our tables had been composed for the meridian of your consideration or dwelling.

But in this matter let us enjoy such an example: I have to find the mean motion of the Sun in the current year 1502 on the fourth day of the month of April, at 17 hours, 34 minutes, 10 seconds having passed the meridian of the town of Leipzig. I therefore seek this year, namely 1501 completed, in the Table of collected years, in the line of numbers, and I do not find it anywhere closer precisely. Therefore I accept the smaller one found there, for 1488, and set aside the motion corresponding to it in the place of the Sun: 9 signs, 19 degrees, 13 minutes, 23 seconds, 42 tertias. When this is done, I subtract the collected years from the proposed years, and there are 13 years left, called expanded, for which in the table of the same title are found 11 signs, 29 degrees, 50 minutes, 56, 45 in place of the Sun.[5] This I add to the previously noted, so that the radix of the Sun of the current year is 9 signs, 19 degrees, 4 minutes, 20, 27. This is how the radix of any motion is determined at the beginning of the current year, as it is current. When the proposed time also includes whole months, days, etc., in addition to years, then I go with the completed months, namely March, into the Table of months in the common year and find for March the motion of the Sun 2 signs, 28 degrees, 42 minutes, 29, 26 and write it under the radix. Then I add the motion of the Sun for the four days found in the Table of days, i.e. 3 signs, 56 minutes, 33, 18.[6] The same is the case with the 17 hours from the Table of hours, from the column of the Sun, where I find 0 degrees, 41 minutes, 53, 24, and write that to the previous entries. But since I cannot enter the table together with 34 minutes of hours, for the common line of numbers in the same table does not extend beyond 30, so first with 30 minutes I find 1 minute, 13, 55, 25. Then I enter a second time with the remaining number of minutes, namely four, and in the same table I find 0 minutes, 9, 51, 23, and I write it to the previous one. Finally, after entering the same table with 10 seconds, the first column of the table shows the number of seconds due to the corresponding denomination of the

4 Ambrosius Lacher here gives the much simpler and more practical advice of converting local time to time on the reference meridian instead of adjusting the radices of all mean motions to use tables for another meridian.

5 See Table TRW-02, column C.

6 See Tables TRW-03 and TRW-05, columns C.

said number entered to the table. In this way it is 0 seconds, 24, 38, 28, which I add to the previously written.[7] When this is done, I add all the differences of motion to each other in due order and the mean motion of the Sun will come out at the intended time of 0 signs, 22 degrees, 26 minutes, 40, 46, 26. This way you must work in all of mean motions you intend to investigate from our tables.'[8]

### 9.1.1 *List of Rules*

**Ambrosius Lacher:**
(print from 1511 = **L**, according to the copy held at l'Observatoire de Paris, Rés. 20490)

On addition (*De addicione*, **L**, f. A2vb)
On subtraction (*De subtraccione*, **L**, f. A3rb)
On multiplication (*De multiplicacione*, **L**, f. A3va)
The rules on multiplication follow (*Sequuntur regule multiplicacionis*, **L**, f. A3vb)
On division (*De divisione*, **L**, f. A4vb)
On mean motions (*De motibus mediis*, **L**, f. A5va)

1. Finding the mean motion of any planet for a given time using tables (*Medium motum cuiuslibet planetarum per ipsius tabulas ad tempus quodcunque propositum invenire. <Proposicio 1>*, **L**, f. A5vb)
2. On the apogees of planets (*De planetarum augibus. Proposicio 2*, **L**, f. B1ra)
3. Finding the anomaly of the Sun (*Argumentum Solis elicere. Proposicio 3*, **L**, f. B1rb)
4. On mean anomalies of the three upper planets (*De mediis argumentis trium superiorum. Proposicium IIII*, **L**, f. B1va)
5. On the mean centre of the Moon (*De centro Lune medio. Proposicio V*, **L**, f. B1vb)
6. On the mean centre of five planets (*De medio centro quinque planetarum. Proposicio VI*, **L**, f. B1vb)
7. On the true motion of the Sun (*De vero motu Solis. Proposicio IX*, **L**, f. B3rb)[9]
8. On the true motion of the Moon (*De vero motu Lune. Proposicio X*, **L**, f. B3va)
9. On the explanation of the Tables of equations (*De Tabularum exposicione equacionum. Proposicio VII*, **L**, f. B2ra)
10. On the true motions (*De veris motibus. Proposicio VIII*, **L**, f. B2va)
11. On the true motion of five planets (*De vero motu quinque planetarum. <Proposicio> XI*, **L**, f. B4ra)
12. On warnings to keep in mind when working with tables (*De cautelis observandis in operacione tabularum. <Proposicio> XII*, **L**, f. C1vb)
13. On the head of dragon (*De genzahar Lune. <Proposicio> XIII*, **L**, f. C2ra)
14. On the properties of the planets (*De planetarum passionibus. <Proposicio> XIIII*, **L**, f. C2rb)

---

7 See Table TRW-06, column C.

8 It is interesting to note, that Lacher does not use the value $9^s$ 19;14,3,7° of the radix in 1488, which he gives in his tables for Leipzig (at 17;30° east of Toledo), but the value, which can be found in the tables for Cracow – see Table TRC-01, column C.

9 The folios with Propositions VII and VIII are bound only after Propositions IX and X in the copy at l'Observatoire de Paris.

15. On the latitude of the three upper planets (*De latitudine trium superiorum. <Proposicio> XV*, **L**, f. C2vb)
16. On the latitude of Venus (*De latitudine Veneris. <Proposicio> XVI*, **L**, f. C3va)
17. On the latitude of Mercury (*De latitudine Mercurii. <Proposicio> XVII*, **L**, f. C4va)
18. On the direction, retrogradation, and station of the planets (*De dirreccione, retrogradacione stacioneque planetarum. <Proposicio> XVIII*, **L**, f. D1vb)
19. To find the rise and set, appearance and hiding of planets under the Sun's rays (*Ortum et occasum, apparicionem occultacionemque planetarum sub radiis Solis cognoscere. <Proposicio> XIX*, **L**, f. D2va)
20. Finding the velocity, slowness, and equality of the planets[10] (*Planetarum velocitatem, tarditatem ac equalitatem invenire. <Proposicio> 20*, **L**, f. D3rb)
21. To know whether a planet is ascending or descending (*Ascensum et descensum planete scire. <Proposicio> XXI*, **L**, f. D3rb)
22. On the mean conjunctions and oppositions of luminaries (*De coniunccionibus opposicionibusque mediis luminarium. Proposicio XXII*, **L**, f. D3va)
23. Finding the mutual conjunctions of the planets (*Planetarum inter se coniuncciones invenire. <Proposicio> XXIIII*,[11] **L**, f. D5ra)
24. On the true conjunctions and oppositions of luminaries (*De coniunccionibus et opposicionibus veris luminarium. <Proposicio XXV>*, **L**, f. D5va)
25. On the conjunction of the planets with the stars of the eighth sphere (*De congressu planetarum cum stellis octave spere. <Proposicio XXVI>*, **L**, f. D6va)

### 9.1.2 Notes on Canons by Ambrosius Lacher: Visibility of Planets

Ambrosius Lacher, and after him Johannes Schöner, enriched the *Tabule resolute* collection with several tables from other sets of tables. The use of these newly added tables is described by both authors in their canons. For example, in Rule 19, Lacher and Schöner treat the use of Tables of visibility of planets, which they included into their editions with the title *Tabula apparitionis occultationisque planetarum*.[12] The tradition of this table dates back to Ptolemy's *Almagest* XIII,10.[13] The problem in question is that if a planet is close to conjunction with the Sun, it either rises shortly before sunrise or sets shortly after sunset, and can hardly be discerned in twilight by a naked eye. The visibility of the planet in a given night thus depends on its brightness and on the altitude (i.e. the depth) of the Sun below the horizon at the moment when the planet rises or sets. This altitude is given by the distance of the planet from the Sun in ecliptical longitude, by the inclination of the ecliptic to the horizon, and also by the ecliptical latitude of the planet.[14] The inclination depends

10 The meaning of the title of the rule emerges more clearly from Lacher's explanation on f. D3rb: *Quociescunque verus motus planete alicuius maior est suo motu medio, dicitur cursu suo velox. Et si minor, dicitur suo cursu tardus. Si autem verus motus et medius adequantur, tunc cursu suo equalis vocabitur.* – 'If the true motion of a planet is greater than its mean motion, it is said to be faster in its motion. If it is smaller, it is said to be slower in its motion. If the true and mean motions are equal, then its motion is said to be the same.'

11 Proposition XXIII is missing.

12 See Lacher, 1511, f. O6r; Schöner, 1536, f. S4v.

13 See Toomer, *Ptolemy's Almagest*, p. 647.

14 See Chapter 26.

on the geographical latitude of the observer, but it also varies at sunrise and sunset during the year according to the ecliptical longitude of the Sun. The Table of visibility thus gives for each planet in each zodiacal sign a limiting distance from the Sun in degrees of ecliptical longitude at which the planet appears or disappears at the background of the solar light scattered in the atmosphere. This threshold is to be compared with the actual difference of the ecliptical longitudes of the planet and the Sun and if the former is greater, the planet is invisible that night. In the case of the inner planets, Venus and Mercury, their upper and lower conjunction must be distinguished because they are brighter in the former and can thus be seen closer to the Sun.[15] Lacher and Schöner changed the layout of the table: they split it into two tables separately for the appearance and disappearance of always all planets, while their predecessors gave first all the phenomena for the outer planets and in the next table for the inner planets. However, they did not change the tabulated values, which are (up to scribal errors) the same as those found in al-Battānī's tables, in the Toledan Tables, or in the John of Lignères' Tables of 1322.[16] Like their sources, neither Lacher nor Schöner mentioned for which geographical latitude this table should correspond.

## 9.2 Johannes Schöner

Lacher's edition was followed in 1536 by a printing of the *Tabule resolute* and their canons composed by Johannes Schöner (*Tabulae astromonicae, quas vulgo… resolutas vocant*, Nuremberg, Johannes Petreius, ff. A1r–F1r) with the incipit: *Documenta generalia ex regulis arithmeticae de minutiis physicalibus pro usu harum tabularum necessaria praemittere. – Tanto revera digniorem primam causam iudicare debemus…*

After Schöner's death in 1547, his work, essentially unchanged, was reprinted in 1551, 1587, and 1588. It is practically the same text as Lacher's, only more carefully edited and with fewer errors. The edition differs most in the fact that Schöner has recalculated the tables as well as all examples in his canons for Nuremberg, where he worked, replacing Lacher's Leipzig.

Johannes Schöner[17] was born in Karlstadt am Main, Bavaria, in 1477, that is why he was sometimes called Carolostadius. He studied mathematics, medicine and theology at the University of Erfurt, became a bachelor in 1498 and was ordained a priest in 1500. Several years he lived in Nuremberg and then he served for about twenty years as canon of St James Abbey in Bamberg. At the same time he was interested in geography and made terrestrial globes. In 1525 he returned to Nuremberg where he taught mathematics at the Lutheran Gymnasium. For the next twenty years or so, he devoted himself to astronomy

15 The upper or lower parts of the motion on epicycle are recognised according to the values of equated anomaly, whether it is around zero or 180 degrees, respectively. The anomaly can also indicate if the planet is seen in the evening or in the morning, as described in the beginning of Rule 19.

16 See Carlo Alfonso Nallino, *Al-Battānī sive Albatenii Opus astronomicum I–III* (Milan: Ulrichus Hoeplius, 1899–1907), *II*, pp. 142–43; Pedersen, *The Toledan Tables*, pp. 1530–37; Chabás and Saby, *The Tables of 1322 by John of Lignères*, pp. 92–93.

17 See Felix Schmeidler, 'Schöner, Johannes', in *Neue Deutsche Biographie*, 23 (2007), pp. 405–06; Derek Jensen, 'Schöner, Johannes', in *The Biographical Encyclopedia of Astronomers*, ed. by Thomas Hockey (New York: Springer, 2007), pp. 1027–28.

and astrology in Nuremberg and led a printing workshop, which he founded. He edited and printed here many works of contemporary leading astronomers, including Johannes Regiomontanus and Bernard Walther. Georg Joachim Rheticus dedicated to Johannes Schöner his *Narratio prima*, published in Danzig 1540, which is a brief summary of Copernican heliocentric theory. Johannes Schöner died in Nuremberg in 1547.

To illustrate the method of Schöner's work, we transcribe here for comparison the passage from his canons which corresponds to Lacher's exposition and example of calculations of mean motions given above. The text coincides with Lacher's but Schöner mentions the Nuremberg meridian instead of Leipzig meridian given by Lacher. The example is derived from Schöner's edition 1536, f. B1r–v: *Quo facto omnes numeratores cuiusque denominationis*[18] *simul adde, unumquodque ad suum genus, ut in additione fieri solet, et aggregatum ostendit motum medium, quem quaesieras, ad Norimbergensem*[19] *meridianum, ad quem has tabulas principaliter reduximus, ob celebriorem praefati oppidi famam, ut usui communi reddantur accommodatiores, nec aliis locis ab huius meridiano notabiliter discrepantibus, nisi reductive inservire possunt.*

*Quod si forte loco habitationis considerationisve exigente ad alium meridianum motus supercoelestium corporum investigare proposueris, necesse est, ut tempus distantiae meridianorum tuae considerationis locum et Norimbergam*[20] *interiacens addas tempori proposito, ad quod motus invenire desideras, si tuae considerationis locus fuerit occidentalior Norimbergae,*[21] *aut subtrahas, si fuerit orientalior, prout tabula huius rei gratia facta et ab unde informabit. Et tempus illud, quod facta additione conflatur, vel post subtractionem relinquitur, inserviet exacte tuo proposito, ad quod si motum alicuius planetae vel mobi||lis inveneris, non aliter eveniet, quam si tabulae nostrae ad meridianum considerationis tuae vel habitationis essent compositae.*

*Sed in hac re tali fruamur exemplo: iubeor invenire medium motum Solis Anno Domini currente 1502 ad 4. diem mensis Aprilis, hora 17, minuto 34, secundo 10, transactis ad meridianum Norimbergensem.*[22] *Quaero ergo hunc annum, scilicet 1501 completum, in Tabula collectorum annorum, in linea numeri, et eum nusquam invenio praecise propinquiorem. Ergo illo minorem ibi repertum 1488 suscipio motumque ei correspondentem in loco Solis 9 signa, 19 gradus, 15 minuta, 1 secunda extra scribo. Quo facto annos collectos ab annis propositis subtraho, residuabuntur 13, qui dicuntur expansi, penes quos in tabula eiusdem tituli in loco Solis 11 signa, 29 gradus, 50 minuta, 57 secunda*[23] *inveniuntur, quae prius notatis adiicio, ut radix Solis anni concrescat currentis 9 signa, 19 gradus, 5 minuta, 58 secunda.*[24] *Sic radix cuiuslibet mobilis ad principium anni currentis vel tanquam currentis constituatur. Cum etiam tempus propositum ultra annos integros includat menses, dies et cetera, quare cum transactis mensibus, scilicet Marcio, Tabulam mensium anni communis intrabo et motum Solis penes Marcium repertum 2 signa, 28 gradus,*

18 Schöner uses classical orthography, albeit inconsistently. In the transcription, we unify it to the classical standard.

19 Here is Schöner's first substantive modification of Lacher's printing, changing Lacher's Leipzig meridian (*Lipsensium meridianum*) to the meridian of the city of Nuremberg (*Norimbergensem meridianum*).

20 Lacher: *Lypsigk*, Schöner: *Norimbergam.*

21 Lacher: *Lypsigk*, Schöner: *Norimbergae.*

22 Lacher: *opidi Lypsensis*, Schöner: *Norimbergensem.*

23 Lacher's numbers are slightly different: *11 signa, 29 gradus, 50 minuta, 56, 45 inveniuntur.*

24 Here, too, Lacher differs slightly in numbers: it varies from minutes to lower fractions. It is the difference between Leipzig and Nuremberg.

*42 minuta, 29 secunda radici subscribo. Deinde 4. dierum motum Solis 3 gradus, 56 minuta, 33, in Tabula dierum inventum praescriptis subiungo. Haud secus cum 17 horis ex horarum tabula de columna Solis 0 gradus, 41 minuta, 53 secunda extrahendo cum supra positis scribo. Sed quia cum 34 minuta horarum tabulam simul ingredi non possum, numeri enim communis linea eiusdem tabulae ultra 30 non extenditur, igitur primo cum 30 minuta invenio 1 minutum, 14 secunda. Deinde cum residuo minutorum numero, scilicet 4, ex eadem tabula 0 minuta, 10 secunda secundario introitu reperta cum praenotatis subscribo. Ultimo vero cum 10 secundis eandem tabulam ingrediendo in prima eius linea occurrit secundorum numerus propter conformem denominationem praedicti numeri in tabulam missi. Hoc modo 0 in secundis. Quo facto omnes differentias motus invicem addo debito ordine et prodibit motus Solis medius ad tempus propositum 0 signa, 22 gradus, 28 minuta, 17 secunda. Sic te in omnibus operari oportet quorumcunque ex nostris tabulis medios motus investigare praetendis.*

It can be seen that the text is taken literally from Lacher; Schöner even praises Nuremberg with the same words that Lacher chose for Leipzig. On the other hand, Schöner in the example consistently used the numbers from his tables recalculated for Nuremberg, while Lacher used the radix for Cracow in the example for Leipzig. Schöner also rounded the increments in fractions of an hour to seconds to be consistent with the precision of the radix and of other increments.

Schöner's edition further differs from Lacher's by adding an extra rule (*Vera loca stellarum fixarum ad quodcunque tempus invenire. Propositio IX*); the numbering of propositions (i.e. rules) is therefore shifted by one from this point. However, because later (in the seventeenth rule) he combined in one rule the explanations of the latitude of Venus and Mercury (*De latitudine Veneris et Mercurii. Propositio XVII*), which had been separated in Lacher, the numbering of the rules again coincides. Lacher's edition then ends with the 25th rule (*De congressu planetarum cum stellis octavae sphaerae. Propositio XXV*), whereas Schöner continues further to Rule 32, where his explanation ends.

### 9.2.1 *List of Rules*

**Johannes Schöner:**

(print from 1536 = **S1**; print from 1551 = **S2**; print from 1587 = **S3**; print from 1588 = **S4**):
On addition (*De additione*, **S1**, f. A1v; **S2**, p. 480; **S3**, f. C1v; **S4**, p. 2)
On subtraction (*De subtractione*, **S1**, f. A2r; **S2**, p. 480; **S3**, f. C2r; **S4**, p. 3)
On multiplication (*De multiplicatione*, **S1**, f. A1v; **S2**, p. 481; **S3**, f. C2v; **S4**, p. 4)
The rules of multiplication (*Regule multiplicationis*, **S1**, f. A2v; **S2**, p. 481; **S3**, f. C2v; **S4**, p. 4)
On division (*De divisione*, **S1**, f. A3v; **S2**, p. 483; **S3**, f. C3v; **S4**, p. 6)
On mean motions (*De motibus mediis*, **S1**, f. A4v; **S2**, p. 484; **S3**, f. D1r; **S4**, p. 9)

1. Finding the mean motion of any planet for a given time using tables (*Medium motum cuiuslibet planetarum per ipsius tabulas ad tempus quodcunque propositum invenire. Propositio I*, **S1**, f. A4v; **S2**, p. 485; **S3**, f. D1r; **S4**, p. 9)
2. On apogees of planets (*De planetarum augibus. Propositio II*, **S1**, f. B1v; **S2**, p. 486; **S3**, f. D2r; **S4**, p. 11)
3. Finding the anomaly of the Sun (*Argumentum Solis elicere. Propositio III*, **S1**, f. B2r; **S2**, p. 487; **S3**, f. D2v; **S4**, p. 12)
4. On the mean anomalies of the three upper planets (*De mediis argumentis trium superiorum. Propositium IIII*, **S1**, f. B2r; **S2**, p. 487; **S3**, f. D3r; **S4**, p. 13)

5. On the mean centre of the Moon (*De centro Lunae medio. Propositio V*, **S1**, f. B2v; **S2**, p. 488; **S3**, f. D3r; **S4**, p. 13)
6. On the mean centre of the five planets (*De medio centro quinque planetarum. Propositio VI*, **S1**, f. B2v; **S2**, p. 488; **S3**, f. D3v; **S4**, p. 14)
7. On the explanation of the Tables of equations (*De Tabularum aequationum expositione. Propositio VII*, **S1**, f. B2v; **S2**, p. 488; **S3**, f. D3v; **S4**, p. 14)
8. On the true motions (*De veris motibus. Propositio VIII*, **S1**, f. B3r; **S2**, p. 489; **S3**, f. D4v; **S4**, p. 16)
9. Finding the true positions of fixed stars for any time (*Vera loca stellarum fixarum ad quodcunque tempus invenire. Propositio IX*, **S1**, f. B4r; **S2**, p. 491; **S3**, f. E1r; **S4**, p. 17)
10. On the true motion of the Sun (*De vero motu Solis. Propositio X*, **S1**, f. B4r; **S2**, p. 491; **S3**, f. E1v; **S4**, p. 18)
11. On the true motion of the Moon (*De vero motu Lunae. Propositio XI*, **S1**, f. B4r; **S2**, p. 492; **S3**, f. E2r; **S4**, p. 19)
12. On the true motion of the five planets (*De vero motu quinque planetarum. Propositio XII*, **S1**, f. C1v; **S2**, p. 493; **S3**, f. E3r; **S4**, p. 21)
13. On warnings to keep in mind when working with tables (*De cautelis observandis in operatione tabularum. Propositio XIII*, **S1**, f. C2v; **S2**, p. 495; **S3**, f. E4r; **S4**, p. 23)
14. On the head of dragon (*De genzahar Lunae. Propositio XIIII*, **S1**, f. C3r; **S2**, p. 496; **S3**, f. E4v; **S4**, p. 24)
15. On the properties of the planets (*De planetarum passionibus. Propositio XV*, **S1**, f. C3r; **S2**, p. 496; **S3**, f. F1r; **S4**, p. 25)
16. On the latitude of the three upper planets (*De latitudine trium superiorum. Propositio XVI*, **S1**, f. C4r; **S2**, p. 497; **S3**, f. F1v; **S4**, p. 26)
17. On the latitude of Venus and Mercury (*De latitudine Veneris et Mercurii. Propositio XVII*, **S1**, f. C4v; **S2**, p. 498; **S3**, f. F2v; **S4**, p. 28)
18. On direction, retrogradation and station of planets (*De dirrectione, retrogradatione stationeque planetarum. Propositio XVIII*, **S1**, f. D1v; **S2**, p. 500; **S3**, f. F3v; **S4**, p. 30)
19. Finding the rise and set, the appearance and hiding of planets under the Sun's rays (*Ortum et occasum, apparitionem occultationemque planetarum sub radiis Solis cognoscere. Propositio XIX*, **S1**, f. D2r; **S2**, p. 501; **S3**, f. F4r; **S4**, p. 31)
20. Finding the velocity, slowness, and equality of planets[25] (*Planetarum velocitatem, tarditatem ac aequalitatem invenire. Propositio XX*, **S1**, f. D3r; **S2**, p. 502; **S3**, f. G1r; **S4**, p. 33)
21. To know whether a planet is ascending or descending (*Ascensum et descensum planetae scire. Propositio XXI*, **S1**, f. D3r; **S2**, p. 503; **S3**, f. G1r; **S4**, p. 33)
22. On the mean conjunctions and oppositions of the luminaries (*De coniunctionibus oppositionibusque mediis luminarium. Propositio XXII*, **S1**, f. D3r; **S2**, p. 504; **S3**, f. G1v; **S4**, p. 34)
23. Finding the mutual conjunctions of planets (*Planetarum inter se coniunctiones invenire. Propositio XXIII*, **S1**, f. D4v; **S2**, p. 507; **S3**, f. G3r; **S4**, p. 37)

25 See Rule 20 by Ambrosius Lacher, above.

24. On the true conjunctions and oppositions of the luminaries (*De coniunctionibus et opposicionibus veris luminarium. Propositio XXIIII*, **S1**, f. E1r; **S2**, p. 508; **S3**, f. G4r; **S4**, p. 39)
25. On the conjunction of the planets with the stars of the eighth sphere (*De congressu planetarum cum stellis octavae sphaerae. Propositio XXV*, **S1**, f. E2r; **S2**, p. 510; **S3**, f. H1r; **S4**, p. 41)
26. Finding the true motion of the Sun in one day, one hour, or several hours and their fractions (*Verum motum Solis in uno die, hora aut pluribus horis et horarum fractionibus colligere. Propositio XXVI*, **S1**, f. E2v; **S2**, p. 511; **S3**, f. H1v; **S4**, p. 42)
27. Finding the true motion of the Moon in one hour, or several hours and their fractions (*Verum motum Lunae in una hora vel pluribus et horarum fractionibus extrahere. Propositio XXVII*, **S1**, f. E2v; **S2**, p. 511; **S3**, f. H1v; **S4**, p. 42)
28. Finding the velocity of the three upper planets as well as Venus and Mercury, that is the various equated motion of the planet in one, two or several days using tables (*Buth trium superiorum ac Veneris et Mercurii, hoc est motum diversum aequatum planetae in uno die, duobus vel pluribus, per tabulas ad hoc ordinatas investigare. Propositio XXVIII*, **S1**, f. E3r; **S2**, p. 512; **S3**, f. H2r; **S4**, p. 43)
29. To create the tables of mean motions of planets (*Mediorum motuum planetarum tabulas formare. Propositio XXIX*, **S1**, f. E3v; **S2**, p. 513; **S3**, f. H2v; **S4**, p. 44)
30. Compilation and correction of the radices of planets (*Radices planetarum conscribere et emendare. Propositio XXX*, **S1**, f. E4r; **S2**, p. 515; **S3**, f. H3r; **S4**, p. 45)
31. Correction of the tables of mean motions (*Tabulas mediorum motuum corrigere. Propositio XXXI*, **S1**, f. E4v; **S2**, p. 516; **S3**, f. H3v; **S4**, p. 46)
32. Finding the revolution of the years[26] (*Habito principio annos mundi, nativitatum vel aliorum principiorum revolvere. Propositio XXXII*, **S1**, f. F1r; **S2**, p. 517; **S3**, f. H4r; **S4**, p. 47)

### 9.2.2 *Notes on Canons by Johannes Schöner*

Schöner's 1536 print contains, in addition to the *Tabule resolute*, an edition of Regiomontanus's work on astrological houses entitled *Ratio sive apodeixis duodecim domorum coeli*. The author of the preface to the book was the German humanist and professor at the University of Wittenberg Philipp Melanchthon (1497–1560), the well-known Protestant theologian, collaborator of Martin Luther and reformer of education. Although Melanchthon was proscribed by the Catholic Church in 1559 and his name was included in the list of prohibited books (*Index librorum prohibitorum*), his preface and his name are given in all reprints of Schöner's book published after that year.[27] The high evaluation and recommendation given to Schöner's book by the preface of such an influential scholar as Melanchthon was undoubtedly helpful in bringing the late medieval TR into Humanist circle. In the preface Melanchthon addresses Schöner and writes in flowery, rhetorical language and a refined humanistic style (f. iiiiir–v):

---

26 See Rules 53 and 54 of the canons *Quoniam plerique* by Nicolaus Wodka of Kwidzyn or Rule 34 of canons *Pro (introduccione) generali* by John of Głogów.

27 Philipp Melanchthon wrote a new preface to Schöner's *Opera omnia*, published in 1551.

*Quod igitur hoc tempore edidisti et expoliisti tua diligentia atque industria Tabulas resolutas, egregiam operam navasti, quia in scholis praecipue opus est his tabulis cum theoricas, ut vocant, motuum coelestium ad usum transferant et monstrent facilem et compendiariam rationem computandi motus coelestes. Quare sine eis theoricae prorsus luce carent et cognitis his tabulis iam magnum fructum capere discentes ex hoc studio possunt. Nam quod extremum est in arte ac finis doctrinae de motibus, id compendiaria via exhibet hic libellus, videlicet ut possimus motus planetarum statim computare aliquot seculorum, seu praeteritos, seu venturos. Haec meta est artis, quam avidissime expetunt studiosi. || Interdum desperatione abiiciunt aliqui studia, cum non prospiciunt metam, cum non capiunt fructum aliquem ex suis laboribus. At cognitis his tabulis non solum eximitur animis desperatio, sed animadversa tam facili computatione spes et magnitudo utilitatis generosa ingenia addiscendum accendunt. ... Quis non miretur tam brevi libello tot annorum descriptionem includi posse? Mirentur alii ligneas columbas aut alia opera automata. Hae tabulae multo magis sunt dignae admiratione, quae omnium siderum positus ostendunt nec unius tantum anni, sed multorum seculorum. Haec ea praefatus sum, ut admonerem invenes, qui haec lecturi sunt, et quantum debeant Schonero iuvanti publica studia editione talium librorum, quibus in scholis opus est, deinde ut utilitate harum tabularum excitati ad discendum invitentur.*

> 'By publishing and refining the *Tabule resolute* at this time with your diligence and effort, you have done an excellent service, because especially in schools such tables are needed to bring the doctrine of celestial motions into practical use and to show an easy and compact procedure for calculating celestial motions. Without them, the existing theories seem to lack light, but by knowing these tables the disciples can gain many fruits through study. This book gives in abbreviated form what is at the limits and at the end of this art and science, so that we may immediately calculate the motions of the planets for several centuries, whether past or future. It is the goal of this science that students eagerly seek. From time to time some of them become disillusioned and give up their studies when they do not see the goal before them, when they do not reap the fruits of their efforts. However, by understanding these tables, not only does hopelessness disappear, but by observing such an easy calculation, hope and a high degree of usefulness ignite noble minds to learn. ... Who would not admire that a description of so many years can be contained in so little book? Some admire wooden pigeons or other automata.[28] But these tables are worthy of much more admiration because they show the positions of all the stars, not just in one year, but over many centuries. I am writing a preface about them in order to point out to the young people who will read this how much they owe to Schöner for his help in public study by publishing such books, and finally to invite to study those who are interested in the usefulness of these tables.'

It should be mentioned that Johannes Schöner in his edition of the TR with their canons does not name the author, although he obviously and with only slight revisions took

28 A reference to traditional narratives of automata constructed by ancient inventors. Automata included, for example, statues that walked, winked, shook a spear or spoke, birds that made a human voice, or a flying wooden pigeon, probably powered by steam, constructed by the mechanic Archytas of Tarentum (5th–4th century BC). Evidence of this model comes from the work *Noctes Atticae* (10,12) by Aulus Gellius (2nd century AD).

over the work of Ambrosius Lacher. Similarly, the actual author of the treatise on the instrument intended for calculating solar and lunar eclipses, which is included under the title *Aequatorium astronomicum Ioannis Schoneri, mathematici Norici, cum novis canonibus et figuris sphaerarum singularum* in the commentary on equatoria published posthumously in 1551 in Schöner's *Opera omnia*, is not given. The author in this case was the master of the Prague University of the first half of the fifteenth century, John Šindel (Jan Ondřejův, called Šindel), who derived this instrument, or rather nomogram, from the universal multifunctional instrument called the albion, created by Richard of Wallingford in the fourteenth century. One copy of Šindel's work is preserved in Nuremberg, where this astronomer also worked for some time.[29]

## 9.3 Johannes Virdung

Johannes Virdung[30] was born around 1463 in Hassfurt, Bavaria, and died around 1538/39 in Heidelberg. Very little is known about his life, basically only what he wrote about himself in his own writings. The manuscript Vatican, BAV, Pal. lat. 1439 contains on ff. 331v–335r Virdung's handwritten report on his studies (*Relacio Johannis Virdung de studiis suis*).[31] In it we read, among other things, of his first studies in Leipzig (*ego, Johannes Virdung de Hasfurt, intitulatus in alma universitate Lipczensi anno 1481 sub rectoratu venerabilis et eximii viri, magistri Jacobi Sueci de Upsalia*) and then at the University of Cracow (*et post hoc 1484 Cracoviam versus me contuli ibidemque intitulatus eodem anno*) where he heard lectures by Albertus of Brudzewo. In 1486 he attended here the university lectures of John of Głogów and together they prepared astrological forecasts. He produced calendars, annual astrological prognostications (which were very popular and often printed) and he devoted himself to the interpretations of dramatic celestial phenomena, such as eclipses, conjunctions or the appearance of a comet. Virdung lived in Germany, but also travelled in France and Italy and he stayed at the court of King Christian II in Copenhagen.

His most important work, however, are his *Tabule resolute*, which he recalculated for the meridian of Heidelberg (*ad meridianum Edelbergensem*). Johannes Virdung's tables were prepared for publication by his student, the physician Jacob Curio from Hof (born 1497),[32] and printed posthumously in 1542 by the same Johannes Petreius who six years earlier had printed Schöner's TR. Jacob Curio authored the preface to Virdung's tables, addressed to Albertus, Archbishop of Mainz and Brandenburg, in whose service, at the Electoral Palatinate court, Virdung had worked. Curio writes that he and Virdung had enjoyed a long lasting friendship. Therefore, although himself is a physician, he prepared for publication Virdung's tables, which he believes to have been derived by Regiomontanus from Alfonsine tables: *statueram initio de motibus in numeros resolutis Tabulas, quas ex*

29 See Hadravová and Hadrava, 'The Eclipse Instrument by John Šindel', pp. 305–40.

30 Jürgen Hamel, 'Virdung, Johann', in *The Biographical Encyclopedia of Astronomers*, ed. by Thomas Hockey (New York: Springer, 2007), p. 1180; Lynn Thorndike, 'Johann Virdung of Hassfurt Again', *Isis* 25 (1936), 363–71; Siegmund Günther, 'Virdung, Johann', in *Allgemeine Deutsche Biographie*, 40 (1896), pp. 9–10.

31 See a description of this manuscript in Chapter 8.1.

32 August Hirsch, 'Curio, Jakob', in *Allgemeine Deutsche Biographie* 4 (1876), pp. 646–47.

*Alfonsinis, opinor, Regiomontanus derivavit* (the Preface, f. B1v). In the canons, Virdung similarly mistakenly believes that Regiomontanus had authored the TR: *in Tabulis resolutis, quas Regiomontanus, opinor, ex Alfonsinis, digessit* (the Rule 12, f. E4r). Virdung's canons are first introduced by a brief explanation of the theory of planets. Then all the other TR rules follow, and notes to the star catalogue are not omitted (see the table *Stellae ad annum 1526 ex Alfonso verificatae*, f. L2r–L4v). Examples which are related to the Heidelberg meridian in his canons help to resolve the basic problem that the numbers in his Tables of mean motions (starting on f. M1v) seemed to be inconsistent with the indicated starting year of 1510.[33] The book concludes with a celebratory poem from the pen of the German humanist Johannes Sinapius (1505–60), active in Ferrara.

An example from Virdung's canons, Rule 9 (f. C1r–v):

*De planetarum mediis supputandis motibus. Ut planetarum motus ad annos quoslibet, seu praeteritos, seu futuros, supputes, ... tabulas proposui in hoc opere tibi, e quibus facili via consequi potes doctrinam de motibus... Ut autem habeas, unde computare incipias. Constituere oportet radicem, quae hoc quidem loco haud alium est quam arcus medii sive aequalis motus inter duo et certa quaedam tempora depraehensus. Radices huiusmodi videbis infra, singulis mediorum motuum tabulis esse descriptas. Si libeat itaque indagare medium planetae alicuius motum, ut exempli gratia Martis ad annum 1525 incompletum, pete tabulam, cui titulus est Tabula medii motus Martis. Ubi, quia inter radices annum vicesimum quintum supra mille et quingentos non vides signatum, necesse erit motum ex<s>cribere correspondentem radici, quae inferior et vicinior est. In praesenti exemplo proxime inferior radix annos continet 1510. Cui in mo||tu respondent signa 5, gradus 17, minuta 43, secunda 0. Isti iam radici seu motui addendum est, quod correspondet annis 15. Siquidem motum quaeris ad 1525 non completum, in annorum columna numerum ergo quaere 15 et offerent se e regione[34] signa 11, gradus 20, minuta 50, secunda 38. Motum hunc adde radici et provenient post additionem signa 5, gradus 8, minuta 33, secunda 38. ... Sic medius Martis motus est ad annum 1525, sed non completum. Est autem annus, ut obiter dicam, completus, die semper ultimo Decembris, hora meridiei.*

'About calculating the mean motions of the planets. To calculate the motions of the planets for some years, past or future, ... I have given you in this volume tables by which you may easily learn the science of the motions... To get the motions, you must begin to count them. You must find out the radix, which here is nothing other than the angle of mean or equal motion, described between two definite times. You will see such radices below, they are written in tables of individual mean motions. So if you want to look up the mean motion of a planet, such as Mars in the unfinished year 1525, look for the table called the Table of mean motion of Mars. And as you will not find there recorded among the radices the year twenty-five over one thousand and five hundred, it will be necessary to write out the motion corresponding to the radix which is nearest and the smaller. In the example given, the nearest minor radix is the year 1510. It corresponds in motion to 5 signs, 17 degrees, 43 minutes

33 See Chapter 15.6.

34 The term *e regione* here does not refer to the geographical table, but to the Table of the expanded years.

> and 0 seconds. To this radix or motion must be added what corresponds to fifteen years. So if you are looking for a motion to the unfinished year 1525, look for the number 15 in the years column and it will give you: 11 signs, 20 degrees, 50 minutes and 38 seconds. Add this motion to the radix and the result is 5 signs, 8 degrees, 33 minutes and 38 seconds. … This is the mean motion of Mars for the year 1525, but unfinished. However, there is also a completed year, as I say in passing, which always ends on the last day of December at noon.'

The example continues in Rule 10 (f. C2r–v):

> *Id significantius exemplo cape. Quaesiturus medium motum Martis anno supra memorato 1525 ad decimum diem Aprilis, horam 4., minutum 25. post meridiem. Scribe ex superiori capite in abacum motum, quem ad annum 1525 incompletum sic comperisti: signa 5, gradus 8, minuta 33, secunda 38. Hic autem annus, cum sit communis, ingredi convenit pro mensurnis* (sic) *excerpendis motibus in Tabulam mensium anni communis. Cum Martio igitur completo, si ingressus fueris, offerent se signum 1, gradus 17, minuta 9, secunda 58. Ultra Marcium decem habes dies Aprilis, quare in Tabella dierum ingredi{t}or cum numero decimo et colliges in motu gradus 5, minuta 14, secunda 26. Post dies restant horae 4 et minuta 25. Quibus ut correspondentem motum prioribus addas, Tabellam pete horarum et aliarum fractionum temporis, ubi per 4 horas in motu reperies minuta 5, secunda 14. Per 25 minuta temporis colliges secunda 32, tertia 45. His omnibus ad se invicem additis medium Martis motum nondum recte collectum habes, nisi prius motum unius diei, scilicet minuta 31, secunda 26, toti summae adieceris. Decimus quintus enim annus, quo cum ingressus es, cruce signatur. Et antea dictum est, quod huiusmodi annus additionem motus unius diei requirat. Iam ergo in calculo, si quodlibet suo generi addideris, proveniet medius Martis motus ad praefatum tempus. Pariter et ad nobilem Edelbergam hoc modo:* || *signa 7, gradus 1, minuta 35, secunda 14, tertia 45.*
>
> 'You can understand it better from the example. It is to find the mean motion of Mars in the above-mentioned year 1525, on the tenth day of April, at four o'clock and 25 minutes after noon. According to the previous chapter, record on the abacus[35] the motion you found for the incomplete year 1525: 5 signs, 8 degrees, 33 minutes and 38 seconds. If the year is a non leap year to find the motions in months you need to enter the Table of months of the non leap year. So if you entered with March completed, you would be offered one sign, 17 degrees, 9 minutes and 58 seconds. You still have ten April days over March, so I'll enter the Table of days with the number ten and you will find 5 degrees, 14 minutes and 26 seconds in motion. There are 4 hours and 25 minutes left in the days. In order to add the corresponding motion to the previous one, look in the Table of hours and other fractions of time, and there, for the motion of four hours, you will find 5 minutes and 14 seconds. For the 25 minutes of time you will find 32 seconds and 45 tertias. Add it all together and you have the mean motion of Mars, but not quite right yet, until you add the motion

35 In this place (and also in Schöner, 1536, Rule 30) the term *abacus* is rarely mentioned to record numerical values.

of one day, 31 minutes and 26 seconds, to the total, because the fifteenth year you have entered is marked with a cross. And as has been said before, a year so marked requires the addition of the motion of one day. If you add everything correctly to its order, you will find the mean motion of Mars at the time mentioned. The same procedure is followed with the noble Heidelberg: 7 signs, 1 degree, 35 minutes, 14 seconds and 45 tertias.'

### 9.3.1 *List of Rules*

**Johannes Virdung:**
(print from 1542 = **V1**)

1. On astronomy and astrology and on their distinguishing, which the custom introduced between these terms (*De astronomia pariter et astrologia deque discrimine, quod inter voces has consuetudo induxit,* **V1**, f. B4r)
2. About some suitable aids for counting the motions of the Sun (*De nonnullis ad motus solaris supputationem pertinentibus adminiculis,* **V1**, f. C1v)
3. On what is necessary to compute the motion of the Moon (*De his, quae ad lunaris cursus supputationem cognitu sunt necessaria,* **V1**, f. C2v)
4. On what is involved in the theory, understanding and computation of the upper planets, i.e. Saturn, Jupiter, and Mars (*De his, quae ad superiorum planetarum, Saturni, Iovis et Martis, theorias, intellectum et calculum pertinent,* **V1**, f. C4v)
5. Theory on Venus and Mercury (*De Veneris et Mercurii theoria,* **V1**, f. D1v)
6. On the head and tail of dragon (*De capite et cauda draconis,* **V1**, f. D2v)
7. On progress, station and regress of the five planets (*De progressu, statione et regressu quinque planetarum,* **V1**, f. D3v)
8. On the velocity and slowness of motions (*De motuum velocitate tarditateque,* **V1**, f. D4v)
9. On computations the mean motions of planets (*De planetarum mediis supputandis motibus,* **V1**, f. E1r)
10. On caution in some rules for computations of the motions (*De observandis quibusdam in calculo motuum canonibus,* **V1**, f. E1v)
11. Beginnings or radices, on which our computation in the tables depends (*A quibus principiis sive radicibus pendeat nostra haec in tabulis supputatio,* **V1**, f. E2v)
12. On computation the true motion of the Sun (*De supputando vero motu Solis,* **V1**, f. E3v)[36]
13. On finding the true motion of the Moon (*De vero motu Lunae inquirendo,* **V1**, f. E4v)
14. On computations the true motions of the five planets (*De supputandis veris quinque planetarum motibus,* **V1**, f. F1v)
15. On investigation of the true motion of dragon (*De vero draconis motu vestigando,* **V1**, f. F4v)
16. On finding the true and equated argument of the latitude of the Moon (*De vero et aequato latitudinis lunaris argumento reperiendo,* **V1**, f. G1r)

---

36 Here are inserted *radices ad meridianum Edelbergensem* (= for Heidelberg).

17. On computations the apogees of the planets (*De augibus planetarum computandis,* **V1**, f. G1r)
18. On regress, progress, and stations of the planets (*De regressu, progressu et stationibus planetarum,* **V1**, f. G1v)
19. On finding the latitudes of the planets (*De planetarum inveniendis latitudinibus,* **V1**, f. G2r)
20. On investigating the motion of the Moon in latitude (*De motu Lunae in latum vestigando,* **V1**, f. G4r)
21. On finding the time of the conjunction or opposition of the Moon with Sun according to the mean motion (*De tempore inveniendo, quo secundum cursum medium Luna Soli congreditur, vel eidem opponitur,* **V1**, f. H1r)
22. On investigating the time of the true conjunctions and oppostitions of luminaries (*De tempore verarum coniunctionum et oppositionum in luminaribus vestigando,* **V1**, f. H1v)
23. On finding next times of the true conjunctions and oppostitions of luminaries (*De tempore iterum ad veras luminarium coniunctiones vel oppositiones definiendo,* **V1**, f. H2v)
24. On continuing and extending the times of the mean conjunctions and oppositions (*De mediarum coniunctionum et oppositionum temporibus continuandis extendendisque,* **V1**, f. H4v)
25. On investigating the mean conjunctions and oppositions of luminaries before the birth of Christ (*De mediis huiusmodi luminarium coniunctionibus et oppositionibus ante Natalem Christi vestigandis,* **V1**, f. H4v)
26. On finding the motions of the Sun and Moon in an hour (*De reperiendis Solis et Lunae motibus horariis,* **V1**, f. I1r)
27. On equating the time of conjunctions, oppositions, and other motions (*De tempore coniunctionum, oppositionum et motuum aliorum aequando,* **V1**, f. I1v)
28. On signs according to which we recognise if there can be any eclipse in a future conjunction and opposition (*De signis, per quae cognoscimus, an aliqua coniunctio oppositioque futura sit ecliptica,* **V1**, f. I2r)
29. On fixed stars, i.e. attached to the sky (*De fixis sive adhaerentibus caelo stellis,* **V1**, f. I2v)
30. On images or constellations of the northern sky (*De imaginibus sive constellationibus caeli septentrionalis,* **V1**, f. I3r)
31. On star images in the southern sky (*De stellarum imaginibus in austrum vergentibus,* **V1**, f. I4v)
32. On constellations on the zodiacal circle (*De constellationibus in orbe zodiaco,* **V1**, f. K1r)
33. On finding both the motion of apogees and fixed stars, and the eighth sphere (*De motu tum augium et fixorum syderum, tum octavi etiam orbis inquirendo,* **V1**, f. K3r)
34. On computing the motions of fixed stars from Alfonso to any time (*De fixorum syderum ab Alfonso computatis motibus ad quaevis tempora accommodandis,* **V1**, f. K4r)
35. Several stars verified from Alfonso to the year 1526 (*Stellae aliquot ad annum 1526 ex Alfonso verificatae,* **V1**, f. L2r)
36. List of several places with different meridians (*Enumeratio locorum aliquot, quorum diversi sunt meridiani,* **V1**, f. L3v)

### 9.3.2 Notes on Canons by Johannes Virdung

Virdung's canons contain rules which, in their form, are outside the conventions of canons not only for the TR but for tables in general. In fact, they include an explanation of the

Ptolemaic model of planetary motion itself, accompanied by drawings more usually found in theoretical works such as the *Theorica planetarum*.

Other rules are devoted to the verbal description of constellations and the stars that form them. These rules are concluded with a Table of coordinates of 93 selected stars ordered by increasing ecliptical longitude and corrected to the year 1526. Interestingly, due to a printing error, the lower part of f. 41v (L1v), which should have been blank between the heading of the Table of stars and the full-page table on the following page, contains the upper part of f. 40r (K4r) of Pomponius Gauricus's treatise *De sculptura*, published by Johannes Petreius in the same year as Johannes Virdung's edition, that is, in 1542.[37]

37 The humanist poet and philologist Pomponius Gauricus (Pomponio Gaurico, *c.* 1481–1530) was a professor at the University of Naples. See *Pomponii Gaurici Neapolitani De sculptura* (Nuremberg: Johannes Petreius, 1542).

# Overview of the Rules in the Canons to the *Tabule resolute*

The editions of selected canons to the TR and lists of rules in all other cases, given in the preceding chapters, make it possible to trace the interrelations and the gradual development of these texts. A first insight into the comparison of this material should be facilitated by the Table of rules contained in the various canons, presented below. This table gives an approximate correspondence between individual rules in the various sets of canons according to their content. The table is based on the Lists of rules that we have prepared for all the canons to the TR that are known to us. In the case of the canons *Girum* and *Nemo* (C5 and C6), which have more than one textual variant, we proceed from the first variant. To place the TR into the wider context of the Alfonsine tables, we have added to the list the canons by John of Lignères (*Quia ad inveniendum*), John of Saxony (*Tempus est mensura motus*) and John of Gmunden (*Quedam notabilia ad practicam tabularum deserviencia premittere*) as comparative material. The canons by John of Lignères and John of Saxony precede in time the *Tabule resolute*,[1] but the TR did not adopt some of their rules, especially those rules dealing with the conversion of time in the various calendars to the sexagesimal system used in the Parisian Alfonsine Tables. In our numerical table, we do not record those rules that were not adopted by the TR. And finally, the canons by John of Gmunden represent a Central European text, contemporary with the TR. Their theoretical description is briefly presented and described by José Chabás and edited by Porres de Mateo.[2] The topics of the rules of all three late medieval astronomers are characterised by original Latin titles which can be found in many manuscript copies of canons. We complemented them by brief English titles which characterise their content. The numbering of the rules in these three canons added for comparison draws from the following lists:

1 A more extensive list of titles, incipits, and explicits of various early Alfonsine canons and their rules is being prepared for print by Marie-Madeleine Saby and José Chabás. The list should help scholars to identify the different rules scattered individually in medieval manuscripts.

2 See Chabás, 'The Diffusion of the Alfonsine Tables: The Case of the Tabulae resolutae'; Porres de Mateo, 'Les tables astronomiques de Jean de Gmunden', pp. 405–504.

## 10.1 Comparison Material

**John of Lignères, *Quia ad inveniendum loca planetarum* (according to a critical edition based on ten manuscripts):**[3]

1. If you wish to convert any number of years, starting from the beginning of any known era, to quartas, tertias, seconds, and primes (*Cum ergo volueris reducere ad quarta, tertia, secunda, prima aliquem numerum annorum incipiencium a principio alicuius ere tibi note*)
2. On converting the years of any era to quartas, tertias, seconds, and primes[4] (*Item de reduccione annorum cuiuslibet here ad quartas, tertias, secundas et primas*)
3. On finding the years of any era from quartas (*De sciencia annorum cuiuslibet here per quartas*)
4. On finding 'nota' of any year, month, and day (*De nota cuiuslibet anni vel mensis vel diei invenienda*)
5. On converting the hours to fractions of days (*De reduccione horarum ad fracciones dierum*)
6. On converting minutes and other fractions of hours to fractions of days (*De reduccione minutorum et ceterarum fraccionum horarum ad fracciones dierum*)
7. On converting fractions of days to fractions of hours (*De reduccione fraccionum dierum ad fracciones horarum*)
8. On finding the mean motion of planets (*De medio motu planetarum inveniendo*)
9. On finding the mean conjunction and opposition of the Sun and Moon (*De media coniunccione et opposicione Solis et Lune invenienda*)
10. On finding the position of apogee of any planet (*De loco augis cuiuslibet planete inveniendo*)
11. On finding the true position of the Sun (*De vero loco Solis inveniendo*)
12. On finding the true position of the Moon (*De vero loco Lune inveniendo*)
13. On the position of the head of dragon (*De loco capitis draconis*)
14. On finding the true positions of the three upper planets (*De veris locis trium superiorum planetarum inveniendis*)
15. On finding the true position of Venus and Mercury (*De vero loco Veneris et Mercurii inveniendo*)
16. On finding the possibility of an eclipse of the Sun and Moon (*De possibilitate eclipsis Solis vel Lune invenienda*)

**John of Saxony, *Tempus est mensura motus* (canons to the PAT, according to edition based on Ratdolt's edition from 1483):**[5]

1. Time is a measure of the motion of the first mover… so if we want to know the motion (*Tempus est mensura motus primi mobilis… Cum igitur motum scire desideramus*)
2. Converting the numbers of years, months and days from the beginning of any era known to us to quartas, tertias, seconds, and primes without tables (*Numerum annorum,*

3 See Hadravová and Hadrava, 'John of Lignères, Quia ad inveniendum', pp. 270–98. Canons were probably created between 1322 and 1327.

4 This title does not correspond to the content of the rule, which does not deal with the converting of years of a certain era into quartas etc., but with the converting of the number of days (given in quartas, tertias, seconds, and primes) from one era to another.

5 See Poulle, *Les tables Alphonsines*, pp. 30–105 (Ratdolt 1483, ff. A2r–B8v). These canons originated in 1327.

*mensium et dierum a principio alicuius ere nobis note ad quarta, tercia, secunda et prima sine tabulis reducere*)

3. Converting the numbers of years, months and days starting from the beginning of any era known to us to quartas, tertias, seconds, and primes using tables (*Numerum annorum, mensium et dierum incipiencium a principio alicuius ere nobis note propositorum ad quarta, tercia, secunda et prima per tabulas ad hoc factas reducere*)
4. If we know quartas, tertias, seconds, and units from the beginning of any era known to us, you can find quartas, tertias, seconds, and primes starting from the beginning of any era unknown to us (*Cognitis quartis, terciis, secundis et primis a principio alicuius ere nobis note, per illa quarta, tercia, secunda et prima incipiencia a principio alicuius ere ignote invenire*)
5. If you know a date expressed in quartas, tertias, seconds, and primes, you can express it in years, months and days: this is the opposite of what is done in Chapter 3 (*Cognitis quartis, terciis, secundis et primis numerum annorum, mensium et dierum in eis contentorum invenire; et est conversum tercii capituli*)
6. Finding the 'note' for a year, a month and a day using the table prepared for this purpose (*Notam anni, mensis vel diei cuiusvis per tabulam ad hoc factam invenire*)
7. Transform, by means of the table established for this purpose, a given number of hours, minutes and other fractions of an hour into a number of minutes of a day and other fractions of a day (*Numerum horarum propositarum ac minutorum hore et aliarum fraccionum ad minuta et alias fracciones dierum per tabulas ad hoc factas invenire*)
8. Reduce minutes, seconds, tertias, etc. of the day to hours, minutes of hours, etc.: this is the opposite of the previous chapter (*Minuta dierum et secunda et tercia etc. ad horas et minuta horarum etc. reducere; istud est conversum capituli precedentis*)
9. What is meant by the term 'radix' when it is said 'radix of such or such motion' (*Quid intelligatur per hoc nomen: 'radix', cum dicitur 'radix talis vel talis motus' declarare*)
10. Find, using the tables prepared for this purpose, the revolution of the world in the years or in a nativity or in any other beginning expressed in hours and fractions of an hour (*Revolucionem annorum mundi vel nativitatis alicuius vel eciam cuiuscumque rei habentis inicium in horis et fraccionibus horarum per tabulas ad hoc factas invenire*)
11. Find the revolution of a year in terms of the ascending degree (*Revolucionem anni cuiuslibet quantum ad gradum ascendentem invenire*)
12. Find using the tables the mean motion of the Sun and Moon and other planets, the mean anomalies of the Moon, Venus and Mercury, the motion of the access and recess of the eighth sphere, the motion of apogees and fixed stars and argument of the latitude of the Moon, and the motion of the head of dragon (*Medium motum Solis et Lune et aliorum planetarum et eciam argumenta media Lune, Veneris et Mercurii, similiter motum accessus et recessus octave sphere et motum augium et stellarum fixarum et argumentum latitudinis Lune, similiter motum capitis draconis per tabulas invenire*)
13. Finding the time of the mean conjunction and opposition of the Sun and Moon using the table prepared for this purpose (*Tempus medie coniunccionis et opposicionis Solis et Lune per tabulam ad hoc factam invenire*)
14. Finding whether a solar eclipse in any conjunction or lunar eclipse in any opposition is possible (*Utrum eclypsis <Solis> in aliqua coniunccione vel eclypsis Lune in aliqua opposicione sit possibilis, invenire*)

15. How to correct the Tables of equal motions or rework them, if you have the motions in any day (*Modum corrigendi Tabulas motuum equalium et ipsas faciendi de novo, dum tamen habeantur motus cuiuslibet diei, subiungere*)
16. Finding the true position of the apogee of any planet using tables (*Verum locum augis cuiuslibet planete per tabulas invenire*)
17. Finding the true position of the Sun using tables (*Verum locum Solis per tabulas invenire*)
18. Finding the true position of the Moon using tables (*Verum locum Lune per tabulas invenire*)
19. Finding the true position of the head of dragon using tables (*Verum locum capitis draconis per tabulas invenire*)
20. Finding the true position of Saturn, Jupiter, and Mars using tables (*Verum locum Saturni Iovis et Martis per tabulas invenire*)
21. Finding the true position of Venus and Mercury using tables (*Verum locum Veneris et Mercurii per tabulas invenire*)
22. Finding the time of the true conjunction and opposition of the Sun and Moon (*Tempus vere coniunccionis et opposicionis Solis et Lune invenire*)
23. Finding the entry of the Sun into Aries or into any other sign (*Introitum Solis in Arietem vel in quodcumque signum volueris invenire*)
24. Finding the time of the quartile aspect of the Sun and Moon (*Tempus quarti aspectus Solis et Lune invenire*)
25. Finding the conjunction of Jupiter and Saturn and any other planet (*Tempus coniunccionis Iovis et Saturni et quorumlibet aliorum planetarum invenire*)
26. Finding the positions of fixed stars at any time (*Loca stellarum fixarum ad quodcumque tempus volueris invenire*)
27. Finding the conjuncton of any planet with a fixed star (*Coniunccionem cuiuslibet planete cum qualibet stellarum fixarum invenire*)

**John of Gmunden, incipit *Quedam notabilia* (canons to the *Tabule maiores* from 1430–1437 according to the Latin edition)**[6]

Prologue (*Prologus*)

1. Introductory note on the use of tables (*Quedam notabilia ad practicam tabularum deserviencia*)
2. Find out mean motion of the Sun and other planets (*Medium motum Solis et omnium planetarum etc.*)
3. Finding mean conjunction and opposition of the Sun and Moon (*Coniunccionem vel opposicionem mediam Solis et Lune… invenire*)
4. Finding the time of the first January conjunction of any year or the last conjunction in December of the previous year (*Tempus prime coniunccionis medie Ianuarii anni cuiuscumque volueris, aut ultime Decembris anni proximi precedentis etc.*)
5. Finding the time of the mean conjunction or opposition of the Sun and Moon following the first January conjunction or opposition etc. (*Quamcumque coniunccionem vel*

6 See Porres de Mateo, 'Les tables astronomiques de Jean de Gmunden', pp. 405–504; Chabás, 'The Diffusion of the Alfonsine Tables: The Case of the Tabulae resolutae'.

*opposicionem mediam Solis et Lune sequentem coniunccionem vel opposicionem primam Ianuarii… invenire*)

6. Finding possible eclipses of the Sun and Moon (*Possibilitatem eclipsis Solis et Lune invenire*)
7. Method of creating and correcting said tables (*Modum componendi et corrigendi tabulas predictas suiungere*)
8. Finding the radix of a city whose longitude is known using the radix of another city (*Radicem cuiuscumque civitatis, cuius longitudo fuerit nota, per radicem alterius civitatis… constituere*)
9. Finding the motion of the apogee of a planet and finding the anomaly of the Sun and the mean centre of the Saturn, Jupiter, Mars, Venus, and Mercury (*Motum augis cuiuslibet planete cognoscere… argumentum Solis et centrum medium Saturni, Iovis, Martis, Veneris et Mercurii invenire*)
10. Finding the proportional part of any difference of equations by the proportion of the remainder of the centre or the remainder of the anomaly to one degree (*Partem proporcionalem cuiuscumque differencie equacionum secundum proporcionem residui centri vel residui argumenti ad unum gradum… invenire*)
11. Finding the proportional part of some difference by the proportion of the remainder of the centre or the remainder of the anomaly to two, 3, 5, 6, 10, 15, or 30 degrees (*Partem proporcionalem cuiuscumque differencie secundum proporcionem residui centri vel residui argumenti ad 2, vel ad 3, vel ad 5, vel ad 6, vel ad 10, vel ad 15, vel ad 30 gradus invenire*)
12. Finding the Sun's true position at any time using the table prepared for this purpose (*Verum locum Solis ad quodcumque tempus propositum per tabulam ad hoc factam invenire*)
13. Finding the true motion of the Sun using the table of equations (*Verum motum Solis per tabulam equacionum ipsius invenire*)
14. Finding the true motion of the Moon using the table of equations with increment of one degree (*Verum locum Lune per tabulam equacionum ipsius auctam per unum gradum invenire*)
15. Finding the true motion of the Moon using the table of equations with increment of three degrees (*Verum motum Lune per tabulam equacionum auctam per 3 gradus invenire*)
16. Finding the true motion of the head of dragon and the true argument of the latitude of the Moon (*Verum motum capitis draconis et verum argumentum latitudinis Lune invenire*)
17. Finding the true motion of the Saturn, Jupiter, Mars, Venus, and Mercury using the tables of equations with increment of one degree or using the tables with increment of three degrees (*Verum motum Saturni, Iovis, Veneris et Mercurii per tabulas equacionum ipsorum auctas per unum gradum aut per tabulas eorum auctas per 3 gradus invenire*)
18. Finding the true motion of the Moon, Saturn, Jupiter, Mars, Venus, and Mercury using the tables of equations with increment of six degrees (*Verum motum Lune, Saturni, Iovis, Martis, Veneris et Mercurii per tabulas equacionum ipsorum auctas per 6 gradus invenire*)
19. Finding the time of the true conjunction and opposition of the Sun and Moon (*Tempus vere coniunccionis et opposicionis Solis et Lune invenire*)
20. Finding the time of the distance between the true and mean conjunction and the opposition using the tables prepared for this purpose (*Tempus distancie inter coniunccionem aut opposicionem veram et mediam per tabulas ad hoc factas invenire*)
21. Finding the motion of a planet in one day or one hour (*Motum cuiuscumque planete in uno die aut in una hora invenire*)

22. Finding the time of the true conjunction of a planet with any other (*Tempus coniunccionis vere cuiusvis planete cum quocumque alio invenire*)
23. Finding the entry of the Sun into Aries or some other sign (*Introitum Solis in Arietem vel in quodcumque signum volueris invenire*)
24. Finding of the fourth aspect of the Sun and Moon (*Tempus quarti aspectus Solis et Lune invenire*)
25. Finding if the planet is stationary, retrograde, or direct (*Utrum planeta sit stacionarius, retrogradus aut directus etc.*)
26. Finding the declination of the Sun or any degree of the zodiac (*Declinacionem Solis aut cuiuscumque gradus zodiaci invenire*)
27. Finding of the latitude of the Moon using tables (*Latitudinem Lune per tabulas invenire*)
28. Finding of the true latitude of Saturn, Jupiter, Venus and Mercury using tables (*Latitudinem verum Saturni, Iovis, mertis, Veneris et Mercurii per tabulas invenire*)
29. Finding declination of any planet using its degree and using the latitude (*Declinacionem cuiuslibet planete per gradum ipsius et per latitudinem invenire*)
30. To know, if the planet is ascending or descending in part of its latitude (*Utrum planeta sit ascendens vel descendens in parte latitudinis, in qua est, cognoscere*)
31. To know, if the planet is ascending or descending in its eccentre or epicycle (*Utrum planeta sit ascendens vel descendens in suo ecentrico vel epiciclo, cognoscere*)
32. Finding of heliacal rise and set of five planets (*Ortum et occasum elyacum 5 planetarum invenire*)
33. Finding the right sine of a given angle (*Cuiuslibet arcus propositi sinum rectum invenire*)
34. Finding the versed sine of a given angle (*Arcus propositi sinum versum invenire*)
35. Finding the angle of a given right sine (*Sinus recti propositi arcum invenire*)
36. Finding the angle of a given versed sine (*Sinus versi propositi arcum invenire*)
37. Finding the complete chord of a given angle (*Cuiuslibet arcus propositi cordam perfectam invenire*)
38. Finding the angle of a given complete chord (*Cuiuslibet corde perfecte proposite arcum invenire*)
39. Finding the direct shadow of an angle of a given altitude using tables prepared for this purpose (*Cuiuslibet arcus proposite altitudinis umbram rectam pet tabulas ad hoc factas invenire*)
40. Finding the reversed shadow of an angle of a given altitude using tables (*Cuiuslibet arcus proposite altitudinis umbram versam per tabulas invenire*)
41. Finding the angle of a given direct shadow using tables (*Umbre recte proposite arcum per tabulas invenire*)
42. Finding the angle of a given reversed shadow (*Umbre verse proposite arcum invenire*)
43. Finding the ascendant of an angle of the zodiac on the right sphere using the tables prepared for this purpose (*Cuiuslibet arcus zodiaci ascensionem in spera recta per tabulas ad hoc factas invenire*)
44. Converting given degrees of ascendants of the right horizon to equal degrees of the zodiac (*Gradus ascensionum propositos in orizonte recto ad gradus equales zodiaci reducere*)
45. Finding the latitude of a region and the altitude of the head of Aries etc. (*Latitudinem regionis et altitudinem capitis Arietis… invenire*)

46. Finding the meridian altitude of the Sun and stars using their declination (*Altitudinem Solis et stellarum meridianam per earum declinacionem invenire*)
47. Finding the ascendant of a given angle of the zodiac beginning from the equator on the right and oblique sphere (*Cuiuslibet arcus zodiaci ab equinocciali incipientis ascensionem in spera recta et obliqua invenire*)
48. Finding the ascensions of an angle of the zodiac on the oblique sphere using the table prepared for this purpose (*Ascensiones cuiuscumque arcus zodiaci in spera obliqua propositi per tabulam ad hoc factam invenire*)
49. Convert the given degrees of ascensions on the oblique horizon to equal degrees of the zodiac (*Gradus ascensionum propositos in orizonte obliquo ad gradus equales zodiaci reducere*)
50. Finding the quantity of the angle of any day and night (*Arcus cuiuslibet diei et eciam noctis quantitatem invenire*)
51. Finding the numbers of the degrees at one unequal hour of the day or night using the diurnal angle (*Numerum graduum in una hora inequali diei vel noctis per arcum diurnum invenire*)
52. Finding the numbers of equal hours of the day and night (*Numerum horarum equalium diei et eciam noctis invenire*)
53. Convert the unequal hours to equal hours and conversly (*Horas diei inequales ad equales reducere et econverso*)
54. Finding the ascendant of any hour (*Ascendens cuiuslibet hore invenire*)
55. Equate the twelve heavenly houses (*Duodecim domos celi equare*)
56. Finding the passed hours of the day using degrees of the ascendant (*Horas diei transactas per gradum ascendentem invenire*)
57. Finding longitudes and latitudes of fixed stars (*Loca stellarum fixarum in longitudine et latitudine… invenire*)
58. Finding the conjunction of any planet with any of fixed stars (*Coniunccionem cuiuslibet planete cum quacumque stellarum fixarum… invenire*)
59. Finding the radius of the Sun and the circle of the shadow at the point of the Moon's transit (*Semidyametrum Solis et Lune ac circuli umbre in loco transitus Lune invenire*)
60. Finding the difference of the Moon in longitude and latitude at the hour of the true conjunction of the Sun and Moon etc. (*Diversitatem aspectus Lune in longitudine et latitudine hora vere coniunccionis Solis et Lune et per eam horam coniunccionis visibilis, que est medium eclipsis Solis, invenire*)
61. Finding the quantity and duration of the eclipse of the Sun (*Eclipsis Solis quantitatem et duracionem invenire*)
62. Draw a picture of the eclipse of the Sun (*Figuram eclipsis Solis depingere*)
63. Finding the time of the eclipse of the Sun (*Tempus eclipsis Solis invenire*)
64. Finding the quantity and duration of the eclipse of the Moon (*Eclipsis Lune quantitatem et duracionem invenire*)
65. Draw a picture of the eclipse of the Moon (*Figuram eclipsis Lune depingere*)
66. Recognise the colours of eclipses before they occur (*Colores eclipsium, antequam eveniant, cognoscere*)

**Table of rules contained in various canons**

| | c1 | c2 | c3 | C4 | C5 | C6 | C7 | C8 | C9 | C10 | C11 | C12 | C13 |
|---|---|---|---|---|---|---|---|---|---|---|---|---|---|
| Ra | - | - | - | - | - | - | + | - | + | - | + | + | - |
| Rb | - | - | - | - | - | - | + | - | + | - | + | + | - |
| Rc | - | - | - | - | - | - | + | - | + | - | + | + | - |
| Rd | - | - | - | - | - | - | + | - | + | - | + | + | - |
| R1 | | | | 1 | 1 | | 1 | | | 1 | | | 11 |
| R2 | 8 | 12 | | 1 | 1 | 1 | 4 | 1 | 1 | 2 | 1 | 1 | 9,10 |
| R3 | 10 | 16 | 2 | 2 | 5 | | 3 | | | 3 | 2 | 2 | |
| R4 | | | | | | | 6 | | | 4 | 3 | 3 | |
| R5 | | | | | | | 7 | | | 5 | 5 | 5 | |
| R6 | | | | | | | 8 | | | 6 | 4 | 4 | |
| R7 | | | | | | | 9 | | | 7 | 6 | 6 | |
| R8 | | 15 | 7 | | 3 | | 10 | 3 | | 8 | | | |
| R9 | | | | | | | 11 | | | 9 | | | |
| R10 | | | | | | 16 | 12 | | | 10 | | | |
| R11 | | | | | | | 13 | | | 11 | | | |
| R12 | | | | | | | 14 | | | 12 | | | |
| R13 | | | | | | | 15 | | | 13 | | | |
| R14 | | | | | | | 16 | | | 14 | | | |
| R15 | | | | | | | 17 | | | 15 | | | |
| R16 | | | | | 6 | | 18 | 6 | 6 | 16 | 9 | 7 | |
| R17 | 10 | 16 | 9 | 2 | 5 | | 5, 19 | 5 | 2 | 17 | | | 17 |
| R18 | 11 | 17 | 12,13 | 4 | 7 | 2 | 20 | 7 | 7 | 18 | 7 | 10 | 12 |
| R19 | 12 | 18 | 14,15,18 | 5 | 8 | 3 | 21 | 8 | 8 | 19 | 8 | 11 | 13 |
| R20 | 13 | 19 | 16 | | 9 | 7,8 | | 9 | 12 | 20 | 13 | 14 | 6,15 |
| R21 | | | 16 | | | | | | 13 | 21 | | | |
| R22 | 14 | 20 | 17,18 | 6 | 10 | 4 | | 10 | 9 | 22 | 11 | 12 | 14 |
| R23 | 15 | 21 | 17,18 | 6 | 11 | 5 | 22 | 11 | | 23 | 11 | 12 | 14 |
| R24 | | | 25 | 6 | 12 | 6 | 24 | 12 | 10 | 24 | 18 | 18 | 7,18 |
| R25 | | | 30 | | 13 | | | 13 | | 25 | | | |
| R26 | | | | | 17 | 15 | 27 | 18 | 14-17 | 26 | | | |
| R27 | | | | | | 17 | 25,28 | | 19 | 27 | | | |
| R28 | | | 4 | 3 | 2 | 10 | 29 | 2 | 3 | 28 | 22 | 22 | 21 |
| R29 | 9 | 13 | 3,5 | 3 | 2 | 11 | 30 | 2 | 3 | 29,30 | 22 | 22 | 21 |
| R30 | 16 | 14 | | 3 | 2 | 12 | | 2 | 23 | | | | 28 |
| R31 | | 22 | 19,20 | 7 | 14 | 13 | 31 | 14 | 23,24 | 31 | 24 | 24 | 22 |
| R32 | | | 48,54,55 | | | | 38 | 17 | 26,42 | 32 | | | |
| R33 | | | 53 | | | | | | | 33 | | | |
| R34 | | 23 | 23 | 8 | 16 | 14 | 37 | 16 | 37 | 34 | | | |
| R35 | | 25 | 22 | | 15 | 9 | | 15 | 25 | 35 | 23 | 23 | 27 |
| R36 | | | 8 | Prologue | 4 | 18 | 2 | 4 | 5 | 36 | | | 11 |
| R37 | | | 27 | | | | 33 | | 4,22 | 37 | | | 20 |
| R38 | | | 28 | | | | 34 | | 11 | 38 | 15 | 16 | 19 |
| R39 | | | 28 | | | | 35 | | 11 | 39 | 16 | 17 | 19 |
| R40 | | | 28 | | | | 36 | | 11 | 40 | 17 | 17 | 19 |

**Explanatory notes to the columns:**

c1–c3 = auxiliary comparative canons; C4–C13 = canons to the *Tabule resolute*:

c1 = John of Lignères, *Quia ad inveniendum*; c2 = John of Saxony, *Tempus est mensura motus*; c3 = John of Gmunden, *Quedam notabilia* (i.e. canons to his *Tabule maiores*);

C4 = Petrus Cruciferus, *Mirabilis in altis Dominus*; C5 = Andrzej Grzymała, *Girum recensendo*, variant 1; C6 = Anonymous author, *Nemo integre sapit*, variant 1; C7 = Albertus of Brudzewo, *Astrorum (observatores) sapientes*; C8 = Nicolaus of Tuchów, *Quoniam Tabularum Alfonsi*; C9 = Nicolaus Wodka of Kwidzyn, *Quoniam plerique nostri temporis physici*; C10 = John of Głogów, *Pro (introduccione) generali*; C11 = Lacher's edition 1511; C12 = Schöner's edition 1536 etc.; C13 = Virdung's edition 1542

**Explanatory notes to the lines:**

Ra = Rule on addition; Rb = Rule on subtraction; Rc = Rule on multiplication; Rd = Rule on division; R1 = Radix of a year; R2 = Mean motions to any time; R3 = Radix of apogees; R4 = Anomaly of the Sun at any time; R5 = The centre of the Moon at any time; R6 = Anomalies of upper planets; R7 = Mean centres of the five planets; R8 = Creating of tables of mean motions; R9 = Mean motions of the Sun, Moon, and the anomaly of the Moon in the course of the year; R10 = Centre of the Moon for creating an almanac; R11 = Mean motions of the upper planets for creating an almanac; R12 = Equated anomaly of the upper planets for creating an almanac; R13 = Mean centres of the upper planets for creating an almanac; R14 = Mean anomaly of the lower planets for creating an almanac; R15 = Mean centres of the lower planets for creating an almanac; R16 = Search in the Tables of planetary equations; R17 = Common apogees; R18 = True motion of the Sun; R19 = True motion of the Moon; R20 = True position (motion) of the head of dragon; R21 = An argument of the latitude of the Moon; R22 = True positions (motions) of the upper planets; R23 = True positions (motions) of the lower planets; R24 = Retrogradation of planets; R25 = Ascending and descending planet; R26 = True motions for creating an almanac; R27 = True motions of the head of dragon for creating an almanac; R28 = First mean conjunction and opposition of the Sun and Moon in a year; R29 = Mean conjunctions and oppositions of the Sun and Moon in the course of a year; R30 = Possibility of eclipses; R31 = True conjunctions an oppositions of the Sun and Moon; R32 = The degree of ascendant and other houses of the sky; R33 = Reducing of the unequal to the equal hours; R34 = The entry of the Sun into Aries and other signs; R35 = Conjunctions of the five planets; R36 = Converting of radices to other meridian; R37 = Latitude of the Moon; R38 = Latitude of the upper planets; R39 = Latitude of Venus; R40 = Latitude of Mercury

**Notes to the table:**

The correspondence of the rules given in one line of the table cannot be perfect, because the medieval authors of the canons chose various arrangements of their explanations, and placed different emphasis on particular questions and on their grouping in the rules. The table can thus only be approximate. As a basic reference numbering of the rules in the table we chose their sequence in the canons *Pro (introduccione) generali* by John of Głogów, which are the most detailed in the main topic of *Tabule resolute*.

From the review of the rules in the various canons, several conclusions and tendencies emerge, some of which we previously noted:

- The canons for Salamanca by Nicolaus of Tuchów (C8) are dependent on the canons *Girum recensendo* by Andrzej Grzymała (C5).
- The canons by John of Głogów (C10) are obviously dependent on the canons *Astrorum (observatores) sapientes* by Albertus de Brudzewo (C7).
- The prints of the TR by Johannes Schöner (C12) depend on Lacher's edition (C11).
- John of Głogów and after him all prints completely omitted rules about eclipses and the search for eclipse possibilities (R30).
- It is conspicuous that rules R11–R15 are only contained in the canons by Albertus de Brudzewo (C7) and by John of Głogów (C10): both authors provide details about computation of mean motions for the construction of almanacs.
- Nicolaus Wodka in his canons *Quoniam plerique* (C9) devoted more space to a detailed explanation of *tacuinum* (i.e. one form of almanac) by computation of true motions of planets.

## 10.2 Summarizing Notes on Canons

We have collected and described a set of seven different types of canons to the TR, preserved in manuscripts (*Mirabilis, Girum, Nemo, Astrorum, Quoniam Tabularum Alfonsi, Quoniam plerique, Pro /introduccione/ generali*), some of which (at least *Girum* and *Nemo*) have several versions. Three other sets of rules, i.e. Lacher's, Schöner's and Virdung's canons, were printed in the sixteenth century. These canons are preserved in sixty manuscript copies. The numbers of individual surviving copies of canons to the TR provide us with valuable information. The first and fundamental TR accompanied by canons *Mirabilis* exist in seven copies and were followed by a large wave of TR recalculated for other meridians (Prague, Cracow and others) in the second half of the fifteenth century. The first canons in this wave were the *Girum*, preserved in a great number of copies (sixteen). Equally numerous copies are the canons *Astrorum* (sixteen), which are more extensive and partly more innovative than previous canons, since they add introductory chapters on addition, subtraction, multiplication and division, which were later adopted by Nicolaus Wodka of Kwidzyn and the authors of the printed TR, Ambrosius Lacher and Johannes Schöner. The overall structure of the canons *Astrorum* was also an inspiration for John of Głogów, whose canons are the most extensive and the youngest of those that have survived in manuscripts. They exist in eleven copies. The other canons are documented in fewer copies: *Nemo* in five manuscripts, *Quoniam Tabularum Alfonsi* for Salamanca in three copies and the version of *Quoniam plerique* for Bologna in two copies. It should be emphasized here, however, that despite of low number of copies of the TR and their canons prepared for Salamanca and Bologna, these witnesses point to the spreading of the TR beyond Central Europe to university centres in Hispania and Italy.

It is not a rule that canons to the TR are always accompanied in the manuscripts by the corresponding tables (and viceversa). Sometimes in codices we find only canons without tables or tables without canons. In the second part of our work we will deal with *Tabule resolute* themselves. The earliest are tables calculated for Wrocław with the radix in 1424 (MS Toruń, BU, 74/III). This was followed by tables for Prague, Cracow (e.g. Prague, NK, VIII G 24; Cracow, BJ, 600 and many others) and sporadically also for Bologna (Milan BA, C 278 b Inf.; London, BL, Add 22317) with the radix in 1428 – these tables are the largest in number. However, there are also tables computed for Salamanca with the radix beginning in 1348 and preserved in two copies (MSS Lisbon, ANTT, ML, 2115; Oxford, Bodl., Canon. misc. 27), for Prague starting from 1388 (MS Vatican, BAV, Pal. lat. 1381), or for Leipzig and Cracow starting from 1408 (MSS Leipzig, UB, 1473; Munich, BSB, Clm 51). There exist also tables which, although they may contain cyclical radices, are not the TR.[7] The printed TR were then calculated for the meridian of Leipzig (see Lacher 1511), Nuremberg (see Schöner 1536, 1551, 1578, 1588) and Heidelberg (see Virdung 1542). The following pages tell about all these types of *Tabule resolute.*

Interestingly, the surviving copies of the various canons to the TR are noticeably more numerous (60 copies in 49 manuscripts) than the surviving copies of the TR (45 copies in 44 manuscripts). The disproportion between the number of copies of canons and the copies of tables may be due to the influence of the university environment of the second half of the fifteenth century, which apparently demanded that students be able to use tables, know how to describe them and how to work with the finished tables. In studying, it was necessary to learn rules for using the tables and not so much how to compile them. Perhaps that is why students, recent graduates and teachers mostly copied the rules for praxis rather than the tables themselves.

7 See Chapter 18 for examples of these.

II

# Tables

The edition of the tables from the *Tabule resolute* follows in the Part II, after the editions of canons on their use in the preceding Part I. The tables are preserved in more than forty manuscript copies listed in Chapter 11 and in several printed editions. Tables belonging to the TR are extant in many variants compiled for different meridians. Some of the copies deviate substantially also in their layout. It would thus be practically impossible, or at least cumbersome, to include them all into one critical edition with variant readings indicating all the individual deviations for each tabulated number. Instead, therefore, we present transcriptions of entire tables of chosen representatives of each variant of the TR sets.

The oldest original version of the TR, composed by Petrus Cruciferus for Wrocław, is preserved in a single manuscript (MS Toruń, BU, 74/III). We have thus chosen it as the basic one and have transcribed this complete set of tables in Chapter 12. Transcriptions of tables which substantially differ, depending on the reference meridians (Prague, Cracow, Leipzig, Salamanca, Bologna etc.), are given in the subsequent Chapters 13–15. Selected additional tables are presented in Chapter 16. These tables were not a part of the original set of the TR and only some their subsets were usually included in the later reworkings of the TR. Newertheless some canons of the TR (e.g. by John of Głógow) describe how these tables are to be used in computations. In Chapter 17, the reader will find an overview of the folios on which each table appears in the various codices.

The copies of the tables naturally suffer from many scribal as well as computational errors. These can be more easily recognised by comparison with recomputations of the tables than by comparison against each other in the critical apparatus. The results of the recomputations and their comparison with the edited tables are presented in Part III.

11

# List of Manuscripts with the *Tabule resolute*

In the following list of manuscripts we give the folia or page numbers on which the basic tables TR-01 – TR-20 are written. Folia of individual tables and of additional tables A-01 – A-07 can be found in Chapter 17. If the listed manuscripts contain canons in addition to the basic tables of the TR, we list them as well; their detailed description can be found in the characteristics of the canons and their individual manuscripts in Part I. For those manuscripts that contain only the *Tabule resolute* (without canons) and have therefore not been discussed in previous interpretations, we also include, for the sake of context, a brief account of the other works of Alfonsine astronomy contained in them.

Cracow, BJ, 553, ff. 68r–103r (TR for Prague).

The incipit of the tables states that they were calculated for the meridian of Klosterneuburg[1] (*<I>ncipiunt Tabule de mediis et veris motibus planetarum super meridianum Cl<austroneoburgensem>*), but in fact their numerical values correspond to Prague.

As for other important Alfonsine works in this manuscript, Parisian Alfonsine Tables are contained on ff. 1r–20r, Oxford Tables of 1348 by William Batecombe are on ff. 25r – 65v, and Tables of 1322 by John of Lignères are on ff. 105r–108r, 147r–164v, 165v.

The manuscript may have been written in Wrocław – but it is not certain – around 1469. According to the *ex libris*, its later owner was Jacobus de Iłża senior, a master at the University of Cracow (*Liber magistri Iacobi de Yszlza*).[2]

Cracow, BJ, 560, pp. 9–99 (TR for Prague and Cracow).

The manuscript originated in Prague and Cracow, *c.* 1448. The pages relating to Prague (pp. 13–24, 31–44, 49–50, 53–56, 61–99) are written in calligraphy, the parts relating to the Cracow meridian (pp. 9–12, 25–29, 45–48, 51–52) are written in a slightly more cursory hand and it seems that they were probably inserted into the Prague tables secondarily. Among the known owners of this manuscript in Poland were Michael Falkener de Wratislavia (Wrocław), Matthias de Szydłów, Martinus Biem de Ołkusz and Johannes Broscius.

The TR are calculated both for Prague and Cracow meridians regardless of the fact that the incipit on p. 13 mentions Klosterneuburg:[3] *Incipiunt thabule de mediis et veris motibus*

1 The Klosterneuburg Augustinian monastery is located in Lower Austria. In the first half of the fifteenth century, especially in the time of prior Georg Müstinger, it flourished as a centre where mathematics, astronomy, and geography were cultivated. Georg Müstinger worked there together with Master of Vienna University John of Gmunden, Master of Prague University John Šindel (Johannes Andreae dictus Schindel, Jan Ondřejův known as Šindel) and many others. See e.g. Dana Bennett Durand, *The Vienna-Klosterneuburg Map Corpus of the Fifteenth Century. A Study in the Transition from Medieval to Modern Science* (Leiden: Brill, 1952), pp. 56–61 and passim.

2 *Catalogus BJ*, III, pp. 351–56.

3 *Catalogus BJ*, III, pp. 385–90.

*planetarum super meridianum Cl<austroneoburgensem>. Et primo in annis collectis. Anni collecti: 1428. Tabula augium.* The Klosterneuburg monastery is referred also in the TR in the MS Cracow, BJ, 553 (see above).

Cracow, BJ, 600, ff. 86r–100r (TR for Cracow)

The canons *Astrorum* are on ff. 137r–147v; *Pro* (*introduccione*) *generali* on ff. 149r–159v.

Cracow, BJ, 605, ff. 16r–53r (TR for Cracow)

The canons *Astrorum* are on ff. 97r–119v.

Cracow, BJ, 611, ff. 1r–36v[4] (TR for Cracow).

One of owners of the manuscript was Albertus of Brudzewo, among others the author of canons *Astrorum*.

Cracow, BJ, 614, ff. 1r–45v (TR for Cracow)

The canons *Nemo* are on ff. 67ra–73vb.

Cracow, BJ, 616, pp. 1–47[5] (TR for Cracow)

Following TR in this manuscript from the end of the fifteenth century are the *Tabule eclipsium* by Georg Peurbach (pp. 52–142). They are preserved in more then twenty copies.

The manuscript originated in Cracow, in the second half of the fifteenth century. One of its owners was Jacobus de Sierpc, teacher and dean at the University of Cracow at the beginning of the sixteenth century.

Cracow, BJ, 617, ff. 13r–37r[6] (TR for Prague and Cracow)

In this manuscript, the values in the Table of equation of the Sun differ from the standard Parisian Alfonsine Tables (f. 18v–19r).[7] Apogees of the Sun are given for years 1410–1600 at steps of ten years (f. 18r). Tables of mean conjunctions of the Sun and Moon are given in collected years 1400–3040 at steps of forty years (f. 97r).

As in the previous manuscript, in this codex we also find the *Tabule eclipsium* by Georg Peurbach (ff. 97r–133v).

Between the TR and Peurbach's *Tabule eclipsium*, Regiomontanus's *Tabule direccionum profeccionumque* are incorporated on ff. 37v–92r.

In this manuscript, we also find Giovanni Bianchini's Tables for the planets on ff. 134r–141v.

The manuscript originated in Cracow 1486 and was at one time owned by the Cracovian astronomer Martinus Biem de Ołkusz (*c.* 1470–1540). This manuscript is also remarkable

---

4 *Catalogus BJ*, IV, pp. 274–81.

5 *Catalogus BJ*, IV, pp. 308–12.

6 *Catalogus BJ*, IV, pp. 312–18.

7 See Chapter 21. This table was identified by R. L. Kremer to originate from the Peurbach's *Tabule eclipsium*, see Richard L. Kremer, 'The Parisian Alfonsine Tables in Manuscript, 1351–1525', in *Alfonsine Astronomy: Expanding the Scene*, ed. by José Chabás, Matthieu Husson, and Richard L. Kremer (Turnhout: Brepols, forthcoming).

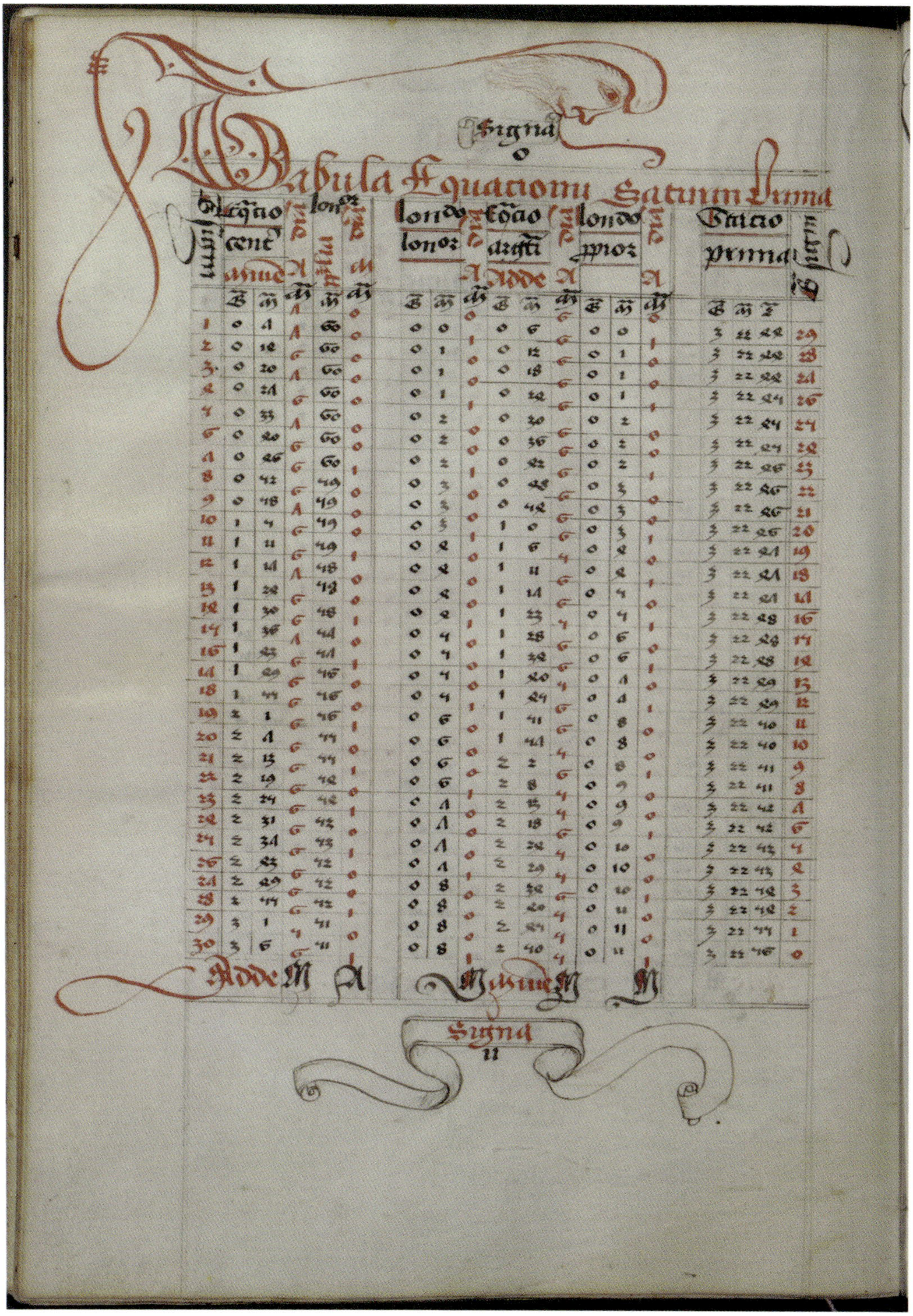


**Figure 26.** MS Cracow, BJ, 617, f. 22v: *Tabula equacionum Saturni prima* from the TR calculated for Prague and Cracow.

for the bordures and ribbons drawn on the edges of the folios. An example is the decorative initial with the head of a personified old and bearded Saturn (see Fig. 26). Its mostly negative astrological influence is based on the interpretation and reworking of ancient mythological ideas.

Cracow, BJ, 1846, pp. 1–65 (TR for Cracow)

The canons *Astrorum* are on pp. 157–196; the canons *Pro (introduccione) generali* on pp. 203–227.

Cracow, BJ, 1852, pp. 7–103 (TR for Prague and Cracow)

The canons *Girum* are on pp. 361–379.

Cracow, BJ, 1858, pp. 121–191 (TR for Cracow)

The canons *Astrorum* are on pp. 73–113.

Cracow, BJ, 1864, pp. 71–141 (TR for Cracow)

The canons *Girum* are on pp. 47–70; the canons *Astrorum* on pp. 32–38, 46 and 220.

Cracow, BJ, 1865, ff. 72r–111r (TR for Cracow)

The canons *Girum* are on ff. 58r–66v.

Cracow, BJ, 1916, pp. 57–128 (TR for Cracow)

The fragment of canons *Girum* are on pp. 47–48; the canons *Pro (introduccione) generali* on pp. 1–45.

Cracow, BJ, 1917, ff. 1r–34v (TR for Cracow)

Additional and other tables follow up to f. 68v.

The canons *Pro (introduccione) generali* are on ff. 84r–97r.

Cracow, BJ, 1920, pp. 25–98 (TR for Cracow)

The canons *Girum* are on pp. 183–201; the canons *Astrorum* on pp. 5–20; the canons *Pro (introduccione) generali* on pp. 3a–3b, 207–235.

Cracow, BJ, 1966, pp. 13–87 (TR for Cracow)

The canons *Astrorum* are on pp. 157–182.

Cracow, BJ, 2252, pp. 47–120 (TR for Cracow)

The canons *Girum* are on pp. 21–44.

Cracow, BJ, 2478, ff. 1r–38r (TR for Cracow)

The canons *Astrorum* are on ff. 73r–94v.

Cracow, BJ, 2480, pp. 1–75 (TR for Cracow)

The canons *Astrorum* are on pp. 181–215.

Cracow, BJ, 2650, pp. 75–154 (TR for Cracow)

The canons *Pro (introduccione) generali* are on pp. 3–64.

Cracow, BJ, 2664, ff. 1r–11v[8] (TR for Cracow, Incomplete).

Heidelberg, UB, 2503, ff. 32r–97r (TR for Cracow)

The canons *Astrorum* are on ff. 1r–23r.

Leipzig, UB, 1473, ff. 122r–126v[9] (TR for Leipzig)

The TR are calculated with the starting point already in the year 1408 and they are incomplete; tables of equations are missing.

The treatise on the Composition of the astrolabe by Cristannus of Prachatice can be found on ff. 134r–146v. The manuscript is dated to 1487.

Leipzig, UB, 1481, ff. 1r–38r[10] (TR for Cracow and partly for Prague)

Tables of mean motions in collected years are given for the meridian of Cracow and Tables of syzygies are first for Prague, followed by those for Cracow.

Lisbon, ANTT, ML, 2115, ff. 19r–56v (TR for Salamanca)

On ff. 59v–103r there are scattered different additional tables also for Salamanca.

The canons *Quoniam Tabularum Alfonsi* are on ff. 8r–18v.

London, BL, Add 22317, ff. 25r–89v (TR for Bologna)

Unlike the standard TR which give first all mean motions in collected years, then in expanded years, months, days and hours, this manuscript is one of a few exceptions which give mean motions in all time-scales separately for each planet on one folio – collected and expanded years and months on rectos, days and hours on versos of folios. This manuscript is thus not included in the Overview of the tables in the TR in Chapter 17. Some tables are not completed.

The canons *Quoniam plerique* are on ff. 1r–18r.

Mikulov, Mik 6361, ff. 47r–81r (TR for Prague)

The canons *Mirabilis* are on ff. 20v–22v.

---

8 Wisłocki, *Catalogus*, p. 631; Rosińska, *Scientific Writings*, p. 509–517 (item 1–18).

9 See <https://handschriftenportal.de/workspace>.

10 Markowski, *Astronomica*, p. 314. – According to Markowski, *Astronomica*, p. 280 and 314, there are also the TR in MS Leipzig, UB, 1491, ff. 1r–70r, but this is incorrect; the manuscript does not contain TR at all. The similar mistake is contained in Markowski, *Astronomica*, p. 282a: he claims that MS Cracow, BJ, 1480, pp. 1–138 contains the TR, but this manuscript does not include any astronomical treatises or tables.

### Mikulov, Mik 6361, ff. 135r–171v (TR for Prague)

The canons *Mirabilis* are on ff. 91v–92v.

### Milan, BA, C 278 b Inf., ff. 1r–31v (TR for Bologna)

Similarly as in London, BL, Add 22317, mean motions of the planets are given at all time-scales on both sides of one folio in MS Milan, BA, C 278 b Inf., ff. 1–9. Tables for the mean motions of fixed stars, access and recess, and the argument of latitude of the Moon are missing. Tables of apogees in collected and expanded years on ff. 10r–11r are followed by tables of conjunctions of the Sun and Moon on ff. 11v–12r and by the Table of geographical coordinates on f. 12v. The Tables of equations on the subsequent folios do not include the Table of access and recess. There are added additional tables on ff. 32r–64r.

The canons *Quoniam plerique* are on ff. 65r–76v.

### Munich, BSB, Clm 51, ff. 124r–135r (TR for Cracow)

The TR contain only mean motions and planetary equations. The cyclical radices are for the years 1408–1588. Similarly to Cracow, BJ, 617, the equations of the Sun differ from those common in PAT (see f. 126r), and the apogees of the Sun are given for the years 1410–1600 at steps of ten years on f. 125v.

The canons *Astrorum* are on ff. 49ra–58vb.

### Munich, BSB, Clm 214,[11] ff. 18r–32r (TR for Cracow)

Next folios – up to f. 39v – are devoted to the additional tables.

Besides of the TR, there is a Table of fixed stars (*Tabula stellarum fixarum verificatarum anno 1478*) on f. 4va, written probably by Albertus of Brudzewo. It is dated to 1478.

On ff. 40r–100v there are included *Tabule direccionum profeccionumque* from 1467 composed by Johannes Regiomontanus.

*Tabule eclipsium* by Georg Peurbach can be found on ff. 115r–166r.

### Munich, BSB, Clm 25011, ff. 1r–74r[12] (TR for Cracow)

This manuscript from the fifteenth century has the standard TR entered with the words *Tabule resolute de mediis et veris motibus planetarum super meridianum Cracoviensem ad annos 1428–1808*.

### Oxford, Bodl., Canon. misc. 27, ff. 33r–62r (TR for Salamanca)

The Tables of mean motions are written in a very compact form in this manuscript, starting with collected years up to hours on a single page separately for each planet. Hence this manuscript is not included in the Overview of the tables in the TR in Chapter 17.

The canons *Quoniam Tabularum Alfonsi* are on ff. 122v–129r.

---

11 Halm, *Catalogus codicum Latinorum… Vol. 1.1*, p. 47; Markowski, *Astronomica*, p. 316.

12 Karl Felix Halm, *Catalogus codicum Latinorum Bibliothecae regiae Monacensis. Vol. 2.4: Clm 21406–27268* (Munich: sumptibus Bibliothecae regiae, 1881), p. 156; Markowski, *Astronomica*, p. 317.

Oxford, Bodl., Canon. misc. 499, ff. 72r/62r–105r/95r[13] (TR for Prague)

The canons *Mirabilis* are on ff. 122ra/112ra–125vb/115vb.

Prague, NK, VIII G 24, ff. 2r–37r (TR for Prague)

The canons *Girum* are on ff. 60v–64v.

Prague, NK, X B 3, ff. 79r–114r (TR for Prague)

The canons *Mirabilis* are on ff. 73ra–75va.

Toruń, BU, 74/III, ff. 251r–268r (TR for Wrocław with Cyclical Radices Starting from 1424); ff. 283v–284v (Two Tables from another Set of the TR Related to Prague with Cyclical Radices from 1428).

The canons *Mirabilis* are on ff. 268va–273ra.

Vatican, BAV, Pal. lat. 1375, ff. 21r–34v[14] (for Cracow)

On f. 1r of this astronomical codex we can read a note on the ownership of the codex written by the hand of Johannes Virdung and dated to the year 1488: *Liber magistri Johannis Virdung de Hasfurt, scriptus manu propria Cracovie.*

The first text of the manuscript are Canons to the Tables for the planets by Bianchini (ff. 1r–8v), followed by First treatise on the equatorium by John of Lignères (ff. 8v–10v); Bianchini's Tables for the planets are found on ff. 45r–52v and 185r–262v, *Tabule eclipsium* by Georg Peurbach are on ff. 55r–106r, Tabule direccionum profeccionumque by Johannes Regiomontanus are on ff. 106v–171v, and *Almagest*-based star catalogue is on ff. 175r–176r.

Vatican, BAV, Pal. lat. 1381, ff. 113r–120v (for Prague)

An incomplete set of the TR includes only mean motions. Tables of equations can be found in separate parts of the manuscript: equations of Saturn, Jupiter and Venus – at steps of one degree – are on ff. 72r–76r; equations of all planets – at steps of five degrees – are on ff. 101v–103r, etc.

The canons *Nemo* are on ff. 121r–123r.

Vienna, ÖNB, Cod. 5227, ff. 37r–77v (for Prague and Cracow)

Tables of mean motions in collected years (denoted in our edition as TRC-01) and Tables of mean conjunctions (TRC-09) and mean oppositions (TRC-10) of luminaries for Cracow precede the TR for Prague.

The canons *Girum* are on ff. 26r–36v; the canons *Astrorum* on ff. 1r–22r.

---

13 upper/lower numbering.

14 Markowski, *Astronomica*, p. 287; Schuba, *Die Quadriviums-Handschriften*, pp. 88–93; Thorndike, 'Johann Virdung of Hassfurt Again'.

Vienna, ÖNB, Cod. 5240, ff. 1r–36r (for Prague)

The canons *Girum* are on ff. 67r–76r.

Vienna, ÖNB, Cod. 5245, ff. 62r–98r (for Cracow and partly for Prague)

At the beginning of the TR, there are mean motions for Cracow, but on f. 126r is their extension for years 1348–1408 for Prague. The mean syzygies on ff. 72v–74r are for Prague while for Cracow they are only on ff. 126v–128r, after the additional tables.

The canons *Girum* are on ff. 27r–34v.

Vienna, ÖNB, Cod. 5275, ff. 88r–123r (for Cracow and partly for Prague)

Similarly to the MS Vienna, ÖNB, Cod. 5245 (see above), MS Cod. 5275 also contains mean motions for Cracow and on ff. 97v–99r syzygies for Prague while syzygies for Cracow are added on ff. 151r–152v only.

The canons *Girum* are on ff. 36r–42r.

Wrocław, BU, IV F 19, ff. 47r–75v (for Prague)

The canons *Pro* (*introduccione*) *generali* are on ff. 247v–264v.

**Set of tables used for comparison:**

Leipzig, UB, 1482, ff. 13r–25v (Tables for Leipzig)

The tables in this manuscript are strangely biased: they have some features of the TR, but do not contain all the tables. Their radix is only for the year 1400. They contain mean motions (ff. 13r–24v), that is, from f. 13r to 15v are collected years (*anni collecti*) and months, from f. 16r to 18v are expanded years (*anni expansi*), from 19r to 21v the same is repeated in days, on ff. 22r–24v everything is in hours. On f. 25r are the conjunctions of the Sun and Moon, and on f. 25v the geographical table is present. We have included these tables outside the TR and use them only for comparison.

The canons *Nemo* are on ff. 2r–9r.

# *Tabule resolute* for the Meridian of Wrocław

Table TRW-01 of mean motions in collected years (Toruń, BU, 74/III, f. 251r)[1]

| Tabula mediorum motuum planetarum ad annos Christi collectos super meridianum Wratisslaviensem | | | | | | | | | | | | | | | | | | | | |
|---|---|---|---|---|---|---|---|---|---|---|---|---|---|---|---|---|---|---|---|---|
| Numerus annorum Christi collectorum | Tabula medii motus augium et stellarum fixarum | | | | | Tabula medii motus accessus et recessus octave spere | | | | | Tabula medii motus Solis, Veneris et Mercurii | | | | | Tabula medii motus Lune | | | | |
| | S | G | M | 2 | 3 | S | G | M | 2 | 3 | S | G | M | 2 | 3 | S | G | M | 2 | 3 |
| 1424 | 0 | 10 | 27 | 43 | 21 | 2 | 12 | 26 | 37 | 24 | 9 | 18 | 45 | 7 | 1 | 1 | 25 | 40 | 33 | 54 |
| 1444 | 0 | 10 | 36 | 32 | 20 | 2 | 13 | 28 | 20 | 16 | 9 | 18 | 53 | 56 | 0 | 6 | 9 | 14 | 21 | 27 |
| 1464 | 0 | 10 | 45 | 21 | 19 | 2 | 14 | 30 | 3 | 7 | 9 | 19 | 2 | 44 | 58 | 10 | 22 | 48 | 9 | 1 |
| 1484 | 0 | 10 | 54 | 10 | 17 | 2 | 15 | 31 | 45 | 58 | 9 | 19 | 11 | 33 | 57 | 3 | 6 | 21 | 56 | 35 |
| 1504 | 0 | 11 | 2 | 59 | 16 | 2 | 16 | 33 | 28 | 50 | 9 | 19 | 20 | 22 | 56 | 7 | 19 | 55 | 44 | 8 |
| 1524 | 0 | 11 | 11 | 48 | 15 | 2 | 17 | 35 | 11 | 41 | 9 | 19 | 29 | 11 | 55 | 0 | 3 | 29 | 31 | 42 |
| 1544 | 0 | 11 | 20 | 37 | 14 | 2 | 18 | 36 | 54 | 33 | 9 | 19 | 38 | 0 | 53 | 4 | 17 | 3 | 19 | 16 |
| 1564 | 0 | 11 | 29 | 26 | 12 | 2 | 19 | 38 | 37 | 24 | 9 | 19 | 46 | 49 | 52 | 9 | 0 | 37 | 6 | 50 |
| 1584 | 0 | 11 | 38 | 15 | 11 | 2 | 20 | 40 | 20 | 16 | 9 | 19 | 55 | 38 | 51 | 1 | 14 | 10 | 54 | 23 |
| 1604 | 0 | 11 | 47 | 4 | 10 | 2 | 21 | 42 | 3 | 7 | 9 | 20 | 4 | 27 | 50 | 5 | 27 | 44 | 41 | 57 |
| 1624 | 0 | 11 | 55 | 53 | 9 | 2 | 22 | 43 | 45 | 58 | 9 | 20 | 13 | 16 | 48 | 10 | 11 | 18 | 29 | 31 |
| 1644 | 0 | 12 | 4 | 42 | 8 | 2 | 23 | 45 | 28 | 50 | 9 | 20 | 22 | 5 | 47 | 2 | 24 | 52 | 17 | 4 |
| 1664 | 0 | 12 | 13 | 31 | 6 | 2 | 24 | 47 | 11 | 41 | 9 | 20 | 30 | 54 | 46 | 7 | 8 | 26 | 4 | 38 |
| 1684 | 0 | 12 | 22 | 20 | 5 | 2 | 25 | 48 | 54 | 33 | 9 | 20 | 39 | 43 | 45 | 11 | 21 | 59 | 52 | 12 |
| 1704 | 0 | 12 | 31 | 9 | 4 | 2 | 26 | 50 | 37 | 24 | 9 | 20 | 48 | 32 | 44 | 4 | 5 | 33 | 39 | 46 |
| adde et habebis radicem Pragensem | | | | | | | | | | | | | | 34 | 30 | | | 7 | 41 | 11 |
| | | | A | | | | | B | | | | | C | | | | | D | | |

1 See Fig. 1.

Continuation of the Table TRW-01 (Toruń, BU, 74/III, f. 251v)

| Residua pars Tabule mediorum motuum planetarum ad annos Christi collectos super <meridianum> Wratisslaviensem | | | | | | | | | | | | | | | | | | | | |
|---|---|---|---|---|---|---|---|---|---|---|---|---|---|---|---|---|---|---|---|---|
| Numerus annorum Christi collectorum | Tabula medii motus argumenti Lune | | | | | Tabula medii motus argumenti latitudinis Lune | | | | | Tabula medii motus capitis draconis | | | | | Tabula medii motus Saturni | | | | |
| | **S** | **G** | **M** | **2** | **3** | **S** | **G** | **M** | **2** | **3** | **S** | **G** | **M** | **2** | **3** | **S** | **G** | **M** | **2** | **3** |
| 1424 | 4 | 27 | 44 | 23 | 28 | 10 | 29 | 43 | 48 | 26 | 9 | 4 | 17 | 14 | 51 | 7 | 6 | 16 | 27 | 6 |
| 1444 | 6 | 7 | 28 | 49 | 48 | 4 | 10 | 7 | 27 | 12 | 10 | 1 | 7 | 6 | 4 | 3 | 10 | 58 | 4 | 12 |
| 1464 | 7 | 17 | 13 | 16 | 8 | 9 | 20 | 31 | 5 | 58 | 10 | 27 | 56 | 57 | 16 | 11 | 15 | 39 | 41 | 19 |
| 1484 | 8 | 26 | 57 | 42 | 28 | 3 | 0 | 54 | 44 | 44 | 11 | 24 | 46 | 48 | 28 | 7 | 20 | 21 | 18 | 27 |
| 1504 | 10 | 6 | 42 | 8 | 49 | 8 | 11 | 18 | 23 | 30 | 0 | 21 | 36 | 39 | 41 | 3 | 25 | 2 | 55 | 32 |
| 1524 | 11 | 16 | 26 | 35 | 9 | 1 | 21 | 42 | 2 | 16 | 1 | 18 | 26 | 30 | 53 | 11 | 29 | 44 | 32 | 39 |
| 1544 | 0 | 26 | 11 | 1 | 29 | 7 | 2 | 5 | 41 | 2 | 2 | 15 | 16 | 22 | 5 | 8 | 4 | 26 | 9 | 45 |
| 1564 | 2 | 5 | 55 | 27 | 49 | 0 | 12 | 29 | 19 | 48 | 3 | 12 | 6 | 13 | 18 | 4 | 9 | 7 | 46 | 52 |
| 1584 | 3 | 15 | 39 | 54 | 10 | 5 | 22 | 52 | 58 | 34 | 4 | 8 | 56 | 4 | 30 | 0 | 13 | 49 | 23 | 59 |
| 1604 | 4 | 25 | 24 | 20 | 30 | 11 | 3 | 16 | 37 | 20 | 5 | 5 | 45 | 55 | 42 | 8 | 18 | 31 | 1 | 5 |
| 1624 | 6 | 5 | 8 | 46 | 50 | 4 | 13 | 40 | 16 | 6 | 6 | 2 | 35 | 46 | 55 | 4 | 23 | 12 | 38 | 12 |
| 1644 | 7 | 14 | 53 | 13 | 10 | 9 | 24 | 3 | 54 | 52 | 6 | 29 | 25 | 38 | 7 | 0 | 27 | 54 | 15 | 18 |
| 1664 | 8 | 24 | 37 | 39 | 31 | 3 | 4 | 27 | 33 | 38 | 7 | 26 | 15 | 29 | 19 | 9 | 2 | 35 | 52 | 25 |
| 1684 | 10 | 4 | 22 | 5 | 51 | 8 | 14 | 51 | 12 | 28 | 8 | 23 | 5 | 20 | 31 | 5 | 7 | 17 | 29 | 32 |
| 1704 | 11 | 14 | 6 | 32 | 11 | 1 | 25 | 14 | 51 | 10 | 9 | 19 | 55 | 11 | 44 | 1 | 11 | 59 | 6 | 38 |
| adde et habebis radicem Pragensem | | | 7 | 37 | 17 | | | 7 | 23 | 2 | | | | 1 | 51 | | | | 1 | 11 |
| | E | | | | | F | | | | | G | | | | | H | | | | |

Remaining part of the Table TRW-01 (Toruń, BU, 74/III, f. 252r)

| Residua pars Tabule mediorum motuum planetarum ad annos Christi collectos super <meridianum> Wratisslaviensem | | | | | | | | | | | | | | | | | | | | |
|---|---|---|---|---|---|---|---|---|---|---|---|---|---|---|---|---|---|---|---|---|
| Numerus annorum Christi collectorum | Tabula medii motus Iovis | | | | | Tabula medii motus Martis | | | | | Tabula medii motus argumenti Veneris | | | | | Tabula medii motus argumenti Mercurii | | | | |
| | S | G | M | 2 | 3 | S | G | M | 2 | 3 | S | G | M | 2 | 3 | S | G | M | 2 | 3 |
| 1424 | 7 | 6 | 20 | 54 | 8 | 3 | 7 | 29 | 48 | 20 | 0 | 29 | 1 | 49 | 47 | 7 | 0 | 49 | 1 | 9 |
| 1444 | 3 | 13 | 35 | 30 | 25 | 10 | 25 | 48 | 46 | 10 | 7 | 2 | 40 | 40 | 43 | 7 | 15 | 16 | 40 | 0 |
| 1464 | 11 | 20 | 50 | 6 | 42 | 6 | 14 | 7 | 44 | 0 | 1 | 6 | 19 | 31 | 39 | 7 | 29 | 44 | 18 | 52 |
| 1484 | 7 | 28 | 14 | 42 | 59 | 2 | 2 | 26 | 41 | 50 | 7 | 9 | 58 | 22 | 35 | 8 | 14 | 11 | 57 | 43 |
| 1504 | 4 | 5 | 29 | 19 | 17 | 9 | 20 | 45 | 39 | 40 | 1 | 13 | 37 | 13 | 31 | 8 | 28 | 39 | 36 | 35 |
| 1524 | 0 | 12 | 43 | 55 | 34 | 5 | 9 | 4 | 37 | 30 | 7 | 17 | 16 | 4 | 27 | 9 | 13 | 7 | 15 | 26 |
| 1544 | 8 | 19 | 58 | 31 | 51 | 0 | 27 | 23 | 35 | 21 | 1 | 20 | 54 | 55 | 23 | 9 | 27 | 34 | 54 | 18 |
| 1564 | 4 | 27 | 13 | 8 | 9 | 8 | 15 | 42 | 33 | 11 | 7 | 24 | 33 | 46 | 19 | 10 | 12 | 2 | 33 | 9 |
| 1584 | 1 | 4 | 27 | 44 | 26 | 4 | 4 | 1 | 31 | 1 | 1 | 28 | 12 | 37 | 15 | 10 | 26 | 30 | 12 | 0 |
| 1604 | 9 | 11 | 42 | 20 | 43 | 11 | 22 | 20 | 28 | 51 | 8 | 1 | 51 | 28 | 11 | 11 | 10 | 57 | 50 | 52 |
| 1624 | 5 | 18 | 56 | 57 | 0 | 7 | 10 | 39 | 26 | 41 | 2 | 5 | 30 | 19 | 7 | 11 | 25 | 25 | 29 | 43 |
| 1644 | 1 | 26 | 11 | 33 | 18 | 2 | 28 | 58 | 24 | 31 | 8 | 9 | 9 | 10 | 3 | 0 | 9 | 53 | 8 | 35 |
| 1664 | 10 | 3 | 26 | 9 | 35 | 10 | 17 | 17 | 22 | 21 | 2 | 12 | 48 | 0 | 59 | 0 | 24 | 20 | 47 | 26 |
| 1684 | 6 | 10 | 40 | 45 | 52 | 6 | 5 | 36 | 20 | 12 | 8 | 16 | 26 | 51 | 55 | 1 | 8 | 48 | 26 | 18 |
| 1704 | 2 | 17 | 55 | 22 | 9 | 1 | 23 | 55 | 18 | 2 | 2 | 20 | 5 | 42 | 51 | 1 | 23 | 16 | 5 | 9 |
| | | | | 2 | 55 | | | | 18 | 21 | | | | 21 | 35 | | | 1 | 48 | 45 |
| | | | I | | | | | J | | | | | K | | | | | L | | |

Table TRW-02 of mean motions in expanded years (Toruń, BU, 74/III, f. 252v)

| Tabula mediorum motuum planetarum ad annos Christi expansos | | | | | | | | | | | | | | | | | | | | |
|---|---|---|---|---|---|---|---|---|---|---|---|---|---|---|---|---|---|---|---|---|
| Numerus annorum expansorum | Tabula motus augium et stellarum | | | | | Tabula medii motus accessus et recessus octave spere | | | | | Tabula motus Solis, Veneris et Mercurii | | | | | Tabula medii motus Lune | | | | |
| | S | G | M | 2 | 3 | S | G | M | 2 | 3 | S | G | M | 2 | 3 | S | G | M | 2 | 3 |
| 1 | 0 | 0 | 0 | 26 | 26 | 0 | 0 | 3 | 5 | 1 | 11 | 29 | 45 | 39 | 22 | 4 | 9 | 23 | 2 | 37 |
| 2 | 0 | 0 | 0 | 52 | 52 | 0 | 0 | 6 | 10 | 2 | 11 | 29 | 31 | 18 | 44 | 8 | 18 | 46 | 5 | 15 |
| 3 | 0 | 0 | 1 | 19 | 18 | 0 | 0 | 9 | 15 | 3 | 11 | 29 | 16 | 58 | 6 | 0 | 28 | 9 | 7 | 52 |
| 4 | 0 | 0 | 1 | 45 | 48 | 0 | 0 | 12 | 20 | 34 | 0 | 0 | 1 | 45 | 48 | 5 | 20 | 42 | 45 | 31 |
| 5 | 0 | 0 | 2 | 12 | 14 | 0 | 0 | 15 | 25 | 35 | 11 | 29 | 47 | 25 | 10 | 10 | 0 | 5 | 48 | 8 |
| 6 | 0 | 0 | 2 | 38 | 39 | 0 | 0 | 18 | 30 | 36 | 11 | 29 | 33 | 4 | 32 | 2 | 9 | 28 | 50 | 45 |
| 7 | 0 | 0 | 3 | 5 | 5 | 0 | 0 | 21 | 35 | 37 | 11 | 29 | 18 | 43 | 54 | 6 | 18 | 51 | 53 | 23 |
| 8 | 0 | 0 | 3 | 31 | 35 | 0 | 0 | 24 | 41 | 9 | 0 | 0 | 3 | 31 | 36 | 11 | 11 | 25 | 31 | 1 |
| 9 | 0 | 0 | 3 | 58 | 1 | 0 | 0 | 27 | 46 | 10 | 11 | 29 | 49 | 10 | 58 | 3 | 20 | 44 | 33 | 39 |
| 10 | 0 | 0 | 4 | 24 | 27 | 0 | 0 | 30 | 51 | 11 | 11 | 29 | 34 | 50 | 20 | 8 | 0 | 11 | 36 | 16 |
| 11 | 0 | 0 | 4 | 50 | 53 | 0 | 0 | 33 | 56 | 11 | 11 | 29 | 20 | 29 | 42 | 0 | 9 | 34 | 38 | 54 |
| 12 | 0 | 0 | 5 | 17 | 23 | 0 | 0 | 37 | 1 | 43 | 0 | 0 | 5 | 17 | 23 | 5 | 2 | 8 | 16 | 32 |
| 13 | 0 | 0 | 5 | 43 | 49 | 0 | 0 | 40 | 6 | 44 | 11 | 29 | 50 | 56 | 45 | 9 | 11 | 31 | 19 | 10 |
| 14 | 0 | 0 | 6 | 10 | 15 | 0 | 0 | 43 | 11 | 45 | 11 | 29 | 36 | 36 | 7 | 1 | 20 | 54 | 21 | 47 |
| 15 | 0 | 0 | 6 | 36 | 41 | 0 | 0 | 46 | 16 | 46 | 11 | 29 | 22 | 15 | 29 | 6 | 0 | 17 | 24 | 24 |
| 16 | 0 | 0 | 7 | 3 | 11 | 0 | 0 | 49 | 22 | 17 | 0 | 0 | 7 | 3 | 11 | 10 | 22 | 51 | 2 | 3 |
| 17 | 0 | 0 | 7 | 29 | 37 | 0 | 0 | 52 | 27 | 18 | 11 | 29 | 52 | 42 | 33 | 3 | 2 | 14 | 4 | 40 |
| 18 | 0 | 0 | 7 | 56 | 3 | 0 | 0 | 55 | 32 | 19 | 11 | 29 | 38 | 21 | 55 | 7 | 11 | 37 | 7 | 18 |
| 19 | 0 | 0 | 8 | 22 | 28 | 0 | 0 | 58 | 37 | 20 | 11 | 29 | 24 | 1 | 17 | 11 | 21 | 0 | 9 | 55 |
| 20 | 0 | 0 | 8 | 48 | 59 | 0 | 1 | 1 | 42 | 51 | 0 | 0 | 8 | 48 | 59 | 4 | 13 | 33 | 47 | 34 |
| | | | A | | | | | B | | | | | C | | | | | D | | |

Continuation of the Table TRW-02 (Toruń, BU, 74/III, f. 253r)

| Residua pars Tabule mediorum motuum planetarum ad annos Christi expansos | | | | | | | | | | | | | | | | | | | | |
|---|---|---|---|---|---|---|---|---|---|---|---|---|---|---|---|---|---|---|---|---|
| Numerus annorum expansorum | Tabula motus argumenti Lune | | | | | Tabula medii motus argumenti latitudinis Lune | | | | | Tabula medii motus capitis draconis | | | | | Tabula medii motus Saturni | | | | |
| | S | G | M | 2 | 3 | S | G | M | 2 | 3 | S | G | M | 2 | 3 | S | G | M | 2 | 3 |
| 1 | 2 | 28 | 43 | 14 | 50 | 4 | 28 | 42 | 44 | 31 | 0 | 19 | 19 | 41 | 54 | 0 | 12 | 13 | 34 | 43 |
| 2 | 5 | 27 | 26 | 29 | 39 | 9 | 27 | 25 | 29 | 3 | 1 | 8 | 39 | 23 | 48 | 0 | 24 | 27 | 9 | 25 |
| 3 | 8 | 26 | 9 | 44 | 29 | 2 | 26 | 8 | 13 | 34 | 1 | 27 | 59 | 5 | 42 | 1 | 6 | 40 | 44 | 8 |
| 4 | 0 | 7 | 56 | 53 | 16 | 8 | 8 | 4 | 43 | 45 | 2 | 17 | 21 | 58 | 14 | 1 | 18 | 56 | 19 | 25 |
| 5 | 3 | 6 | 40 | 8 | 6 | 1 | 6 | 47 | 28 | 17 | 3 | 6 | 41 | 40 | 9 | 2 | 1 | 9 | 54 | 8 |
| 6 | 6 | 5 | 23 | 22 | 55 | 6 | 5 | 30 | 12 | 48 | 3 | 26 | 1 | 22 | 3 | 2 | 13 | 23 | 28 | 50 |
| 7 | 9 | 4 | 6 | 37 | 45 | 11 | 4 | 12 | 57 | 20 | 4 | 15 | 21 | 3 | 57 | 2 | 25 | 37 | 3 | 33 |
| 8 | 0 | 15 | 53 | 46 | 32 | 4 | 16 | 9 | 27 | 30 | 5 | 4 | 43 | 56 | 29 | 3 | 7 | 52 | 38 | 50 |
| 9 | 3 | 14 | 37 | 1 | 22 | 9 | 14 | 52 | 12 | 2 | 5 | 24 | 3 | 38 | 23 | 3 | 20 | 6 | 13 | 33 |
| 10 | 6 | 13 | 20 | 16 | 11 | 2 | 13 | 34 | 56 | 33 | 6 | 13 | 23 | 20 | 17 | 4 | 2 | 19 | 48 | 16 |
| 11 | 9 | 12 | 3 | 31 | 1 | 7 | 12 | 17 | 41 | 5 | 7 | 2 | 43 | 2 | 11 | 4 | 14 | 33 | 22 | 58 |
| 12 | 0 | 23 | 50 | 39 | 48 | 0 | 24 | 14 | 11 | 16 | 7 | 22 | 5 | 54 | 43 | 4 | 26 | 48 | 58 | 15 |
| 13 | 3 | 22 | 33 | 54 | 38 | 5 | 22 | 56 | 55 | 47 | 8 | 11 | 25 | 36 | 37 | 5 | 9 | 2 | 32 | 58 |
| 14 | 6 | 21 | 17 | 9 | 27 | 10 | 21 | 39 | 40 | 18 | 9 | 0 | 45 | 18 | 32 | 5 | 21 | 16 | 7 | 41 |
| 15 | 9 | 20 | 0 | 24 | 17 | 3 | 20 | 22 | 24 | 50 | 9 | 20 | 5 | 0 | 26 | 6 | 3 | 29 | 42 | 24 |
| 16 | 1 | 1 | 47 | 33 | 4 | 9 | 2 | 18 | 55 | 1 | 10 | 9 | 27 | 52 | 58 | 6 | 15 | 45 | 17 | 41 |
| 17 | 4 | 0 | 30 | 47 | 54 | 2 | 1 | 1 | 39 | 32 | 10 | 28 | 47 | 34 | 52 | 6 | 27 | 58 | 52 | 23 |
| 18 | 6 | 29 | 14 | 2 | 43 | 6 | 29 | 44 | 24 | 4 | 11 | 18 | 7 | 16 | 46 | 7 | 10 | 12 | 27 | 6 |
| 19 | 9 | 27 | 57 | 17 | 33 | 11 | 28 | 27 | 8 | 35 | 0 | 7 | 26 | 58 | 40 | 7 | 22 | 26 | 1 | 49 |
| 20 | 1 | 9 | 44 | 26 | 20 | 5 | 10 | 23 | 38 | 46 | 0 | 26 | 49 | 51 | 12 | 8 | 4 | 41 | 37 | 6 |
| | E | | | | | F | | | | | G | | | | | H | | | | |

Remaining part of the Table TRW-02 (Toruń, BU, 74/III, f. 253v)

| Residua pars Tabule mediorum motuum planetarum ad annos Christi expansos | | | | | | | | | | | | | | | | | | | | |
|---|---|---|---|---|---|---|---|---|---|---|---|---|---|---|---|---|---|---|---|---|
| Numerus annorum expansorum | Tabula motus Iovis | | | | | Tabula medii motus Martis | | | | | Tabula medii motus argumenti Veneris | | | | | Tabula medii motus argumenti Mercurii | | | | |
| | **S** | **G** | **M** | **2** | **3** | **S** | **G** | **M** | **2** | **3** | **S** | **G** | **M** | **2** | **3** | **S** | **G** | **M** | **2** | **3** |
| 1 | 1 | 0 | 20 | 29 | 0 | 6 | 11 | 17 | 5 | 14 | 7 | 15 | 1 | 41 | 41 | 1 | 23 | 56 | 46 | 55 |
| 2 | 2 | 0 | 40 | 58 | 0 | 0 | 22 | 34 | 10 | 28 | 3 | 0 | 3 | 23 | 23 | 3 | 17 | 53 | 33 | 49 |
| 3 | 3 | 1 | 1 | 27 | 0 | 7 | 3 | 51 | 15 | 42 | 10 | 15 | 5 | 5 | 3 | 5 | 11 | 50 | 20 | 44 |
| 4 | 4 | 1 | 26 | 55 | 16 | 1 | 15 | 39 | 47 | 34 | 6 | 0 | 43 | 46 | 11 | 7 | 8 | 53 | 31 | 46 |
| 5 | 5 | 1 | 47 | 24 | 16 | 7 | 26 | 56 | 52 | 48 | 1 | 15 | 45 | 27 | 52 | 9 | 2 | 50 | 18 | 41 |
| 6 | 6 | 2 | 7 | 53 | 16 | 2 | 8 | 13 | 58 | 2 | 9 | 0 | 47 | 9 | 33 | 10 | 26 | 47 | 5 | 36 |
| 7 | 7 | 2 | 28 | 22 | 16 | 8 | 19 | 31 | 3 | 16 | 4 | 15 | 48 | 51 | 14 | 0 | 20 | 43 | 52 | 30 |
| 8 | 8 | 2 | 53 | 50 | 31 | 3 | 1 | 19 | 35 | 8 | 0 | 1 | 27 | 32 | 22 | 2 | 17 | 47 | 3 | 33 |
| 9 | 9 | 3 | 14 | 19 | 31 | 9 | 12 | 36 | 40 | 22 | 7 | 16 | 29 | 14 | 3 | 4 | 11 | 43 | 50 | 27 |
| 10 | 10 | 3 | 34 | 48 | 31 | 3 | 23 | 53 | 45 | 36 | 3 | 1 | 30 | 55 | 44 | 6 | 5 | 40 | 37 | 22 |
| 11 | 11 | 3 | 55 | 17 | 31 | 10 | 5 | 10 | 50 | 50 | 10 | 16 | 32 | 37 | 25 | 7 | 29 | 37 | 24 | 16 |
| 12 | 0 | 4 | 20 | 45 | 47 | 4 | 16 | 59 | 22 | 42 | 6 | 2 | 11 | 18 | 34 | 9 | 26 | 34 | 35 | 19 |
| 13 | 1 | 4 | 41 | 14 | 47 | 10 | 28 | 16 | 27 | 56 | 1 | 17 | 13 | 0 | 15 | 11 | 20 | 37 | 22 | 13 |
| 14 | 2 | 5 | 1 | 43 | 47 | 5 | 9 | 33 | 33 | 10 | 9 | 2 | 14 | 41 | 56 | 1 | 14 | 34 | 9 | 8 |
| 15 | 3 | 5 | 22 | 12 | 47 | 10 | 20 | 50 | 38 | 24 | 4 | 17 | 16 | 23 | 36 | 3 | 8 | 30 | 56 | 3 |
| 16 | 4 | 5 | 47 | 41 | 2 | 6 | 2 | 39 | 10 | 16 | 0 | 2 | 55 | 4 | 45 | 5 | 5 | 34 | 7 | 5 |
| 17 | 5 | 6 | 8 | 10 | 2 | 0 | 13 | 56 | 15 | 30 | 7 | 17 | 56 | 46 | 26 | 6 | 29 | 30 | 54 | 0 |
| 18 | 6 | 6 | 28 | 39 | 2 | 6 | 25 | 13 | 20 | 44 | 3 | 2 | 58 | 28 | 7 | 8 | 23 | 27 | 40 | 54 |
| 19 | 7 | 6 | 49 | 8 | 2 | 1 | 6 | 30 | 25 | 58 | 10 | 18 | 0 | 9 | 48 | 10 | 17 | 24 | 27 | 49 |
| 20 | 8 | 7 | 14 | 36 | 17 | 7 | 18 | 18 | 57 | 50 | 6 | 3 | 38 | 50 | 56 | 0 | 14 | 27 | 38 | 51 |

I J K L

Table TRW-03 of mean motions in months in non-leap years (Toruń, BU, 74/III, f. 251r)[2]

| Tabula mediorum motuum planetarum in mensibus et hoc in anno communi | | | | | | | | | | | | | | | | | | | | |
|---|---|---|---|---|---|---|---|---|---|---|---|---|---|---|---|---|---|---|---|---|
| | Tabula medii motus augium etc. | | | | | Tabula medii motus octave spere | | | | | Tabula medii motus Solis, Veneris et Mercurii | | | | | Tabula medii motus Lune | | | | |
| | **S** | **G** | **M** | **2** | **3** | **S** | **G** | **M** | **2** | **3** | **S** | **G** | **M** | **2** | **3** | **S** | **G** | **M** | **2** | **3** |
| Ian | 0 | 0 | 0 | 2 | 15 | 0 | 0 | 0 | 15 | 43 | 1 | 0 | 33 | 18 | 8 | 1 | 18 | 28 | 5 | 39 |
| Feb | 0 | 0 | 0 | 4 | 16 | 0 | 0 | 0 | 29 | 54 | 1 | 28 | 9 | 11 | 18 | 1 | 27 | 24 | 26 | 14 |
| Mar | 0 | 0 | 0 | 6 | 31 | 0 | 0 | 0 | 45 | 37 | 2 | 28 | 42 | 29 | 26 | 3 | 15 | 52 | 31 | 53 |
| Apr | 0 | 0 | 0 | 8 | 41 | 0 | 0 | 1 | 0 | 50 | 3 | 28 | 16 | 39 | 15 | 4 | 21 | 10 | 2 | 30 |
| May | 0 | 0 | 0 | 10 | 56 | 0 | 0 | 1 | 16 | 32 | 4 | 28 | 49 | 57 | 23 | 6 | 9 | 38 | 8 | 9 |
| Iun | 0 | 0 | 0 | 13 | 6 | 0 | 0 | 1 | 31 | 45 | 5 | 28 | 24 | 7 | 12 | 7 | 14 | 55 | 38 | 47 |
| Iul | 0 | 0 | 0 | 15 | 21 | 0 | 0 | 1 | 47 | 28 | 6 | 28 | 57 | 25 | 20 | 9 | 3 | 23 | 44 | 26 |
| Aug | 0 | 0 | 0 | 17 | 36 | 0 | 0 | 2 | 3 | 10 | 7 | 29 | 30 | 43 | 28 | 10 | 21 | 51 | 50 | 5 |
| Sep | 0 | 0 | 0 | 19 | 46 | 0 | 0 | 2 | 18 | 23 | 8 | 29 | 4 | 53 | 17 | 11 | 27 | 9 | 20 | 42 |
| Oct | 0 | 0 | 0 | 22 | 1 | 0 | 0 | 2 | 34 | 6 | 9 | 29 | 38 | 11 | 25 | 1 | 15 | 37 | 26 | 21 |
| Nov | 0 | 0 | 0 | 24 | 11 | 0 | 0 | 2 | 49 | 18 | 10 | 29 | 12 | 21 | 14 | 2 | 20 | 54 | 56 | 59 |
| Dec | 0 | 0 | 0 | 26 | 26 | 0 | 0 | 3 | 5 | 1 | 11 | 29 | 45 | 39 | 22 | 4 | 9 | 23 | 2 | 37 |

A B C D

2 See Fig. 1.

Continuation of the Table TRW-03 (Toruń, BU, 74/III, f. 251v)

| Residua pars Tabule mediorum motuum planetarum in mensibus et hoc in anno communi | | | | | | | | | | | | | | | | | | | | |
|---|---|---|---|---|---|---|---|---|---|---|---|---|---|---|---|---|---|---|---|---|
| | Tabula medii motus argumenti Lune | | | | | Tabula medii motus argumenti latitudinis Lune | | | | | Tabula medii motus capitis draconis | | | | | Tabula medii motus Saturni | | | | |
| | **S** | **G** | **M** | **2** | **3** | **S** | **G** | **M** | **2** | **3** | **S** | **G** | **M** | **2** | **3** | **S** | **G** | **M** | **2** | **3** |
| Ian | 1 | 15 | 0 | 52 | 43 | 1 | 20 | 6 | 35 | 21 | 0 | 1 | 38 | 29 | 42 | 0 | 1 | 2 | 18 | 14 |
| Feb | 1 | 20 | 50 | 3 | 33 | 2 | 0 | 31 | 53 | 43 | 0 | 3 | 7 | 27 | 29 | 0 | 1 | 58 | 34 | 42 |
| Mar | 3 | 5 | 50 | 56 | 15 | 3 | 20 | 38 | 29 | 4 | 0 | 4 | 45 | 57 | 11 | 0 | 3 | 0 | 52 | 57 |
| Apr | 4 | 7 | 47 | 55 | 1 | 4 | 27 | 31 | 18 | 45 | 0 | 6 | 21 | 16 | 15 | 0 | 4 | 1 | 10 | 35 |
| May | 5 | 22 | 48 | 47 | 43 | 6 | 17 | 37 | 54 | 5 | 0 | 7 | 59 | 45 | 56 | 0 | 5 | 3 | 28 | 49 |
| Iun | 6 | 24 | 45 | 46 | 29 | 7 | 24 | 30 | 43 | 47 | 0 | 9 | 35 | 5 | 0 | 0 | 6 | 3 | 46 | 28 |
| Iul | 8 | 9 | 46 | 39 | 11 | 9 | 14 | 37 | 19 | 7 | 0 | 11 | 13 | 34 | 42 | 0 | 7 | 6 | 4 | 42 |
| Aug | 9 | 24 | 47 | 31 | 54 | 11 | 4 | 43 | 54 | 28 | 0 | 12 | 52 | 4 | 23 | 0 | 8 | 8 | 22 | 57 |
| Sep | 10 | 26 | 44 | 30 | 39 | 0 | 11 | 36 | 44 | 9 | 0 | 14 | 27 | 23 | 27 | 0 | 9 | 8 | 40 | 35 |
| Oct | 0 | 11 | 45 | 23 | 22 | 2 | 1 | 43 | 19 | 29 | 0 | 16 | 5 | 53 | 9 | 0 | 10 | 10 | 58 | 50 |
| Nov | 1 | 13 | 42 | 22 | 7 | 3 | 8 | 36 | 9 | 11 | 0 | 17 | 41 | 12 | 12 | 0 | 11 | 11 | 16 | 28 |
| Dec | 2 | 28 | 43 | 14 | 50 | 4 | 28 | 42 | 44 | 31 | 0 | 19 | 19 | 41 | 54 | 0 | 12 | 13 | 34 | 43 |
| | E | | | | | F | | | | | G | | | | | H | | | | |

Remaining part of the Table TRW-03 (Toruń, BU, 74/III, f. 252r)

| Residua pars Tabule mediorum motuum planetarum ad menses in anno communi | | | | | | | | | | | | | | | | | | | | |
|---|---|---|---|---|---|---|---|---|---|---|---|---|---|---|---|---|---|---|---|---|
| | Tabula medii motus Iovis | | | | | Tabula medii motus Martis | | | | | Tabula medii motus argumenti Veneris | | | | | Tabula medii motus argumenti Mercurii | | | | |
| | **S** | **G** | **M** | **2** | **3** | **S** | **G** | **M** | **2** | **3** | **S** | **G** | **M** | **2** | **3** | **S** | **G** | **M** | **2** | **3** |
| Ian | 0 | 2 | 34 | 36 | 59 | 0 | 16 | 14 | 45 | 59 | 0 | 19 | 6 | 43 | 9 | 3 | 6 | 18 | 27 | 59 |
| Feb | 0 | 4 | 54 | 16 | 12 | 1 | 0 | 55 | 12 | 1 | 1 | 6 | 22 | 27 | 57 | 6 | 3 | 17 | 43 | 35 |
| Mar | 0 | 7 | 28 | 53 | 11 | 1 | 17 | 9 | 58 | 0 | 1 | 25 | 29 | 11 | 6 | 9 | 9 | 36 | 11 | 34 |
| Apr | 0 | 9 | 58 | 30 | 54 | 2 | 2 | 53 | 17 | 20 | 2 | 13 | 58 | 54 | 48 | 0 | 12 | 48 | 15 | 25 |
| May | 0 | 12 | 33 | 7 | 53 | 2 | 19 | 8 | 3 | 19 | 3 | 3 | 5 | 37 | 57 | 3 | 19 | 6 | 43 | 24 |
| Iun | 0 | 15 | 2 | 45 | 37 | 3 | 4 | 51 | 22 | 39 | 3 | 21 | 35 | 21 | 39 | 6 | 22 | 18 | 47 | 16 |
| Iul | 0 | 17 | 37 | 22 | 36 | 3 | 21 | 6 | 8 | 38 | 4 | 10 | 42 | 4 | 49 | 9 | 28 | 37 | 15 | 15 |
| Aug | 0 | 20 | 11 | 59 | 35 | 4 | 7 | 20 | 54 | 36 | 4 | 29 | 48 | 47 | 58 | 1 | 4 | 55 | 43 | 14 |
| Sep | 0 | 22 | 41 | 37 | 18 | 4 | 23 | 4 | 13 | 56 | 5 | 18 | 18 | 31 | 40 | 4 | 8 | 7 | 47 | 51 |
| Oct | 0 | 25 | 16 | 14 | 17 | 5 | 9 | 18 | 59 | 55 | 6 | 7 | 25 | 14 | 50 | 7 | 14 | 26 | 15 | 4 |
| Nov | 0 | 27 | 45 | 52 | 1 | 5 | 25 | 2 | 19 | 15 | 6 | 25 | 54 | 58 | 32 | 10 | 17 | 38 | 18 | 56 |
| Dec | 1 | 0 | 20 | 29 | 0 | 6 | 11 | 17 | 5 | 14 | 7 | 15 | 1 | 41 | 41 | 1 | 23 | 56 | 46 | 55 |
| | I | | | | | J | | | | | K | | | | | L | | | | |

Table TRW-04 of mean motions in months in leap years (Toruń, BU, 74/III, f. 252v)

| Tabula mediorum motuum planetarum in mensibus et hoc in anno bisextili | | | | | | | | | | | | | | | | | | | | |
|---|---|---|---|---|---|---|---|---|---|---|---|---|---|---|---|---|---|---|---|---|
| | Motus augium et stellarum | | | | | Motus octave spere | | | | | Motus Solis, Veneris et Mercurii | | | | | Motus Lune | | | | |
| | S | G | M | 2 | 3 | S | G | M | 2 | 3 | S | G | M | 2 | 3 | S | G | M | 2 | 3 |
| Ian | 0 | 0 | 0 | 2 | 15 | 0 | 0 | 0 | 15 | 43 | 1 | 0 | 33 | 18 | 8 | 1 | 18 | 28 | 5 | 39 |
| Feb | 0 | 0 | 0 | 4 | 21 | 0 | 0 | 0 | 30 | 25 | 1 | 29 | 8 | 19 | 37 | 2 | 10 | 35 | 1 | 15 |
| Mar | 0 | 0 | 0 | 6 | 35 | 0 | 0 | 0 | 46 | 8 | 2 | 29 | 41 | 37 | 46 | 3 | 29 | 3 | 6 | 54 |
| Apr | 0 | 0 | 0 | 8 | 46 | 0 | 0 | 1 | 1 | 20 | 3 | 29 | 15 | 47 | 34 | 5 | 4 | 20 | 37 | 32 |
| May | 0 | 0 | 0 | 11 | 0 | 0 | 0 | 1 | 17 | 3 | 4 | 29 | 49 | 5 | 43 | 6 | 22 | 48 | 43 | 10 |
| Iun | 0 | 0 | 0 | 13 | 11 | 0 | 0 | 1 | 32 | 15 | 5 | 29 | 23 | 15 | 31 | 7 | 28 | 6 | 13 | 48 |
| Iul | 0 | 0 | 0 | 15 | 25 | 0 | 0 | 1 | 47 | 58 | 6 | 29 | 56 | 33 | 40 | 9 | 16 | 34 | 19 | 27 |
| Aug | 0 | 0 | 0 | 17 | 40 | 0 | 0 | 2 | 3 | 41 | 8 | 0 | 29 | 51 | 48 | 11 | 5 | 2 | 25 | 6 |
| Sep | 0 | 0 | 0 | 19 | 50 | 0 | 0 | 2 | 18 | 53 | 9 | 0 | 4 | 1 | 36 | 0 | 10 | 19 | 55 | 43 |
| Oct | 0 | 0 | 0 | 22 | 5 | 0 | 0 | 2 | 34 | 36 | 10 | 0 | 37 | 19 | 45 | 1 | 28 | 48 | 1 | 22 |
| Nov | 0 | 0 | 0 | 24 | 16 | 0 | 0 | 2 | 49 | 49 | 11 | 0 | 11 | 29 | 33 | 3 | 4 | 5 | 32 | 0 |
| Dec | 0 | 0 | 0 | 26 | 30 | 0 | 0 | 3 | 5 | 31 | 0 | 0 | 44 | 47 | 42 | 4 | 22 | 33 | 37 | 38 |

A B C D

Continuation of the Table TRW-04 (Toruń, BU, 74/III, f. 253r)

| Residua pars Tabule mediorum motuum planetarum ad menses in anno bisextili | | | | | | | | | | | | | | | | | | | | |
|---|---|---|---|---|---|---|---|---|---|---|---|---|---|---|---|---|---|---|---|---|
| | Motus argumenti Lune | | | | | Argumenti latitudinis Lune | | | | | Capitis draconis | | | | | Motus Saturni | | | | |
| | S | G | M | 2 | 3 | S | G | M | 2 | 3 | S | G | M | 2 | 3 | S | G | M | 2 | 3 |
| Ian | 1 | 15 | 0 | 52 | 43 | 1 | 20 | 6 | 35 | 21 | 0 | 1 | 38 | 29 | 42 | 0 | 1 | 2 | 18 | 14 |
| Feb | 2 | 3 | 53 | 57 | 30 | 2 | 13 | 45 | 39 | 22 | 0 | 3 | 10 | 38 | 7 | 0 | 2 | 0 | 35 | 18 |
| Mar | 3 | 18 | 54 | 50 | 13 | 4 | 3 | 52 | 14 | 43 | 0 | 4 | 49 | 7 | 49 | 0 | 3 | 2 | 53 | 32 |
| Apr | 4 | 20 | 51 | 48 | 58 | 5 | 10 | 45 | 4 | 24 | 0 | 6 | 24 | 26 | 53 | 0 | 4 | 3 | 11 | 11 |
| May | 6 | 5 | 52 | 41 | 41 | 7 | 0 | 51 | 39 | 45 | 0 | 8 | 2 | 56 | 34 | 0 | 5 | 5 | 29 | 25 |
| Iun | 7 | 7 | 49 | 40 | 26 | 8 | 7 | 44 | 29 | 26 | 0 | 9 | 38 | 15 | 38 | 0 | 6 | 5 | 18 | 4 |
| Iul | 8 | 22 | 50 | 33 | 9 | 9 | 27 | 51 | 4 | 47 | 0 | 11 | 16 | 45 | 20 | 0 | 7 | 8 | 5 | 19 |
| Aug | 10 | 7 | 51 | 25 | 51 | 11 | 17 | 57 | 40 | 7 | 0 | 12 | 55 | 15 | 1 | 0 | 8 | 10 | 23 | 32 |
| Sep | 11 | 9 | 48 | 24 | 37 | 0 | 24 | 50 | 29 | 48 | 0 | 14 | 30 | 34 | 5 | 0 | 9 | 10 | 41 | 12 |
| Oct | 0 | 24 | 49 | 17 | 19 | 2 | 14 | 57 | 5 | 9 | 0 | 16 | 9 | 3 | 47 | 0 | 10 | 12 | 59 | 27 |
| Nov | 1 | 26 | 46 | 16 | 5 | 3 | 21 | 49 | 54 | 50 | 0 | 17 | 44 | 22 | 50 | 0 | 11 | 13 | 17 | 5 |
| Dec | 3 | 11 | 47 | 8 | 47 | 5 | 11 | 56 | 30 | 11 | 0 | 19 | 22 | 52 | 32 | 0 | 12 | 15 | 35 | 18 |

E F G H

Remaining part of the Table TRW-04 (Toruń, BU, 74/III, f. 253v)

| Residua pars <Tabule> mediorum motuum planetarum in mensibus et hoc in anno bisextili | | | | | | | | | | | | | | | | | | | | |
|---|---|---|---|---|---|---|---|---|---|---|---|---|---|---|---|---|---|---|---|---|
| | Motus Iovis | | | | | Motus Martis | | | | | Motus argumenti Veneris | | | | | Motus argumenti Mercurii | | | | |
| | S | G | M | 2 | 3 | S | G | M | 2 | 3 | S | G | M | 2 | 3 | S | G | M | 2 | 3 |
| Ian | 0 | 2 | 34 | 36 | 59 | 0 | 16 | 14 | 45 | 59 | 0 | 19 | 6 | 43 | 9 | 3 | 6 | 18 | 27 | 59 |
| Feb | 0 | 4 | 59 | 15 | 27 | 1 | 1 | 26 | 38 | 41 | 1 | 6 | 59 | 27 | 24 | 6 | 6 | 24 | 7 | 43 |
| Mar | 0 | 7 | 33 | 52 | 26 | 1 | 17 | 41 | 24 | 39 | 1 | 26 | 6 | 10 | 33 | 9 | 12 | 42 | 35 | 42 |
| Apr | 0 | 10 | 3 | 30 | 10 | 2 | 3 | 24 | 43 | 59 | 2 | 14 | 35 | 54 | 15 | 0 | 15 | 54 | 39 | 33 |
| May | 0 | 12 | 38 | 7 | 9 | 2 | 19 | 39 | 29 | 58 | 3 | 3 | 42 | 37 | 25 | 3 | 22 | 13 | 7 | 32 |
| Iun | 0 | 15 | 7 | 44 | 52 | 3 | 5 | 22 | 49 | 18 | 3 | 22 | 12 | 21 | 7 | 6 | 25 | 25 | 11 | 23 |
| Iul | 0 | 17 | 42 | 21 | 51 | 3 | 21 | 37 | 35 | 16 | 4 | 11 | 19 | 4 | 16 | 10 | 1 | 43 | 39 | 23 |
| Aug | 0 | 20 | 16 | 58 | 50 | 4 | 7 | 52 | 21 | 15 | 5 | 0 | 25 | 47 | 26 | 1 | 8 | 2 | 7 | 22 |
| Sep | 0 | 22 | 46 | 36 | 34 | 4 | 23 | 35 | 40 | 35 | 5 | 18 | 55 | 31 | 8 | 4 | 11 | 14 | 11 | 13 |
| Oct | 0 | 25 | 21 | 13 | 33 | 5 | 9 | 50 | 26 | 34 | 6 | 8 | 2 | 14 | 17 | 7 | 17 | 32 | 39 | 12 |
| Nov | 0 | 27 | 50 | 51 | 16 | 5 | 25 | 33 | 45 | 54 | 6 | 26 | 31 | 57 | 59 | 10 | 20 | 44 | 43 | 3 |
| Dec | 1 | 0 | 25 | 28 | 15 | 6 | 11 | 48 | 31 | 53 | 7 | 15 | 38 | 41 | 8 | 1 | 27 | 3 | 11 | 2 |
| | I | | | | | J | | | | | K | | | | | L | | | | |

Table TRW-05 of mean motions in days (Toruń, BU, 74/III, f. 254r)

| Tabula mediorum motuum planetarum ad dies mensium | | | | | | | | | | | | | | | | | | | | | |
|---|---|---|---|---|---|---|---|---|---|---|---|---|---|---|---|---|---|---|---|---|
| Numerus dierum | Tabula medii motus augium et stellarum fixarum | | | | | Tabula medii motus accessus et recessus octave spere | | | | | Tabula medii motus Solis, Veneris et Mercurii | | | | | Tabula medii motus Lune | | | | |
| | S | G | M | 2 | 3 | S | G | M | 2 | 3 | S | G | M | 2 | 3 | S | G | M | 2 | 3 |
| 1 | 0 | 0 | 0 | 0 | 4 | 0 | 0 | 0 | 0 | 30 | 0 | 0 | 59 | 8 | 20 | 0 | 13 | 10 | 35 | 1 |
| 2 | 0 | 0 | 0 | 0 | 9 | 0 | 0 | 0 | 1 | 1 | 0 | 1 | 58 | 16 | 39 | 0 | 26 | 21 | 10 | 3 |
| 3 | 0 | 0 | 0 | 0 | 13 | 0 | 0 | 0 | 1 | 31 | 0 | 2 | 57 | 24 | 59 | 1 | 9 | 31 | 45 | 4 |
| 4 | 0 | 0 | 0 | 0 | 17 | 0 | 0 | 0 | 2 | 2 | 0 | 3 | 56 | 33 | 18 | 1 | 22 | 42 | 20 | 5 |
| 5 | 0 | 0 | 0 | 0 | 22 | 0 | 0 | 0 | 2 | 32 | 0 | 4 | 55 | 41 | 38 | 2 | 5 | 52 | 55 | 6 |
| 6 | 0 | 0 | 0 | 0 | 26 | 0 | 0 | 0 | 3 | 2 | 0 | 5 | 54 | 49 | 58 | 2 | 19 | 3 | 30 | 8 |
| 7 | 0 | 0 | 0 | 0 | 30 | 0 | 0 | 0 | 3 | 33 | 0 | 6 | 53 | 58 | 17 | 3 | 2 | 14 | 5 | 9 |
| 8 | 0 | 0 | 0 | 0 | 35 | 0 | 0 | 0 | 4 | 3 | 0 | 7 | 53 | 6 | 37 | 3 | 15 | 24 | 40 | 10 |
| 9 | 0 | 0 | 0 | 0 | 39 | 0 | 0 | 0 | 4 | 34 | 0 | 8 | 52 | 14 | 57 | 3 | 28 | 35 | 15 | 11 |
| 10 | 0 | 0 | 0 | 0 | 43 | 0 | 0 | 0 | 5 | 4 | 0 | 9 | 51 | 23 | 16 | 4 | 11 | 45 | 50 | 13 |
| 11 | 0 | 0 | 0 | 0 | 48 | 0 | 0 | 0 | 5 | 35 | 0 | 10 | 50 | 31 | 36 | 4 | 24 | 56 | 25 | 14 |
| 12 | 0 | 0 | 0 | 0 | 52 | 0 | 0 | 0 | 6 | 5 | 0 | 11 | 49 | 39 | 55 | 5 | 8 | 7 | 0 | 15 |
| 13 | 0 | 0 | 0 | 0 | 56 | 0 | 0 | 0 | 6 | 35 | 0 | 12 | 48 | 48 | 15 | 5 | 21 | 17 | 35 | 16 |
| 14 | 0 | 0 | 0 | 1 | 1 | 0 | 0 | 0 | 7 | 6 | 0 | 13 | 47 | 56 | 35 | 6 | 4 | 28 | 10 | 18 |
| 15 | 0 | 0 | 0 | 1 | 5 | 0 | 0 | 0 | 7 | 36 | 0 | 14 | 47 | 4 | 54 | 6 | 17 | 38 | 45 | 19 |
| 16 | 0 | 0 | 0 | 1 | 10 | 0 | 0 | 0 | 8 | 7 | 0 | 15 | 46 | 13 | 14 | 7 | 0 | 49 | 20 | 20 |
| 17 | 0 | 0 | 0 | 1 | 14 | 0 | 0 | 0 | 8 | 37 | 0 | 16 | 45 | 21 | 34 | 7 | 13 | 59 | 55 | 21 |
| 18 | 0 | 0 | 0 | 1 | 18 | 0 | 0 | 0 | 9 | 7 | 0 | 17 | 44 | 29 | 53 | 7 | 27 | 10 | 30 | 23 |
| 19 | 0 | 0 | 0 | 1 | 23 | 0 | 0 | 0 | 9 | 38 | 0 | 18 | 43 | 38 | 13 | 8 | 10 | 21 | 5 | 24 |
| 20 | 0 | 0 | 0 | 1 | 27 | 0 | 0 | 0 | 10 | 8 | 0 | 19 | 42 | 46 | 32 | 8 | 23 | 31 | 40 | 25 |
| 21 | 0 | 0 | 0 | 1 | 31 | 0 | 0 | 0 | 10 | 39 | 0 | 20 | 41 | 54 | 54 | 9 | 6 | 42 | 15 | 26 |
| 22 | 0 | 0 | 0 | 1 | 36 | 0 | 0 | 0 | 11 | 9 | 0 | 21 | 41 | 3 | 12 | 9 | 19 | 52 | 50 | 28 |

| Tabula mediorum motuum planetarum ad dies mensium | | | | | | | | | | | | | | | | | | | | |
|---|---|---|---|---|---|---|---|---|---|---|---|---|---|---|---|---|---|---|---|---|
| Numerus dierum | Tabula medii motus augium et stellarum fixarum | | | | | Tabula medii motus accessus et recessus octave spere | | | | | Tabula medii motus Solis, Veneris et Mercurii | | | | | Tabula medii motus Lune | | | | |
| | S | G | M | 2 | 3 | S | G | M | 2 | 3 | S | G | M | 2 | 3 | S | G | M | 2 | 3 |
| 23 | 0 | 0 | 0 | 1 | 40 | 0 | 0 | 0 | 11 | 40 | 0 | 22 | 40 | 11 | 31 | 10 | 3 | 3 | 25 | 29 |
| 24 | 0 | 0 | 0 | 1 | 44 | 0 | 0 | 0 | 12 | 10 | 0 | 23 | 39 | 19 | 51 | 10 | 16 | 14 | 0 | 30 |
| 25 | 0 | 0 | 0 | 1 | 49 | 0 | 0 | 0 | 12 | 40 | 0 | 24 | 38 | 28 | 11 | 10 | 29 | 24 | 35 | 31 |
| 26 | 0 | 0 | 0 | 1 | 53 | 0 | 0 | 0 | 13 | 11 | 0 | 25 | 37 | 36 | 30 | 11 | 12 | 35 | 10 | 33 |
| 27 | 0 | 0 | 0 | 1 | 57 | 0 | 0 | 0 | 13 | 41 | 0 | 26 | 36 | 44 | 50 | 11 | 25 | 45 | 45 | 34 |
| 28 | 0 | 0 | 0 | 2 | 2 | 0 | 0 | 0 | 14 | 12 | 0 | 27 | 35 | 53 | 9 | 0 | 8 | 56 | 20 | 35 |
| 29 | 0 | 0 | 0 | 2 | 6 | 0 | 0 | 0 | 14 | 42 | 0 | 28 | 35 | 1 | 29 | 0 | 22 | 6 | 55 | 36 |
| 30 | 0 | 0 | 0 | 2 | 10 | 0 | 0 | 0 | 15 | 12 | 0 | 29 | 34 | 9 | 49 | 1 | 5 | 17 | 30 | 38 |
| 31 | 0 | 0 | 0 | 2 | 15 | 0 | 0 | 0 | 15 | 43 | 1 | 0 | 33 | 18 | 8 | 1 | 18 | 28 | 5 | 39 |

A B C D

Continuation of the Table TRW-05 (Toruń, BU, 74/III, f. 254v)

| Residua pars Tabule mediorum motuum planetarum ad dies mensium | | | | | | | | | | | | | | | | | | | | |
|---|---|---|---|---|---|---|---|---|---|---|---|---|---|---|---|---|---|---|---|---|
| Numerus dierum | Tabula medii motus argumenti Lune | | | | | Tabula medii motus argumenti latitudinis Lune | | | | | Tabula medii motus capitis draconis | | | | | Tabula medii motus Saturni | | | | |
| | S | G | M | 2 | 3 | S | G | M | 2 | 3 | S | G | M | 2 | 3 | S | G | M | 2 | 3 |
| 1 | 0 | 13 | 3 | 53 | 58 | 0 | 13 | 13 | 45 | 39 | 0 | 0 | 3 | 10 | 38 | 0 | 0 | 2 | 0 | 35 |
| 2 | 0 | 26 | 7 | 47 | 55 | 0 | 26 | 27 | 31 | 19 | 0 | 0 | 6 | 21 | 16 | 0 | 0 | 4 | 1 | 11 |
| 3 | 1 | 9 | 11 | 41 | 53 | 1 | 9 | 41 | 16 | 58 | 0 | 0 | 9 | 31 | 54 | 0 | 0 | 6 | 1 | 46 |
| 4 | 1 | 22 | 15 | 35 | 50 | 1 | 22 | 55 | 2 | 37 | 0 | 0 | 12 | 42 | 32 | 0 | 0 | 8 | 2 | 21 |
| 5 | 2 | 5 | 19 | 29 | 48 | 2 | 6 | 8 | 48 | 17 | 0 | 0 | 15 | 53 | 11 | 0 | 0 | 10 | 2 | 56 |
| 6 | 2 | 18 | 23 | 23 | 45 | 2 | 19 | 22 | 33 | 56 | 0 | 0 | 19 | 3 | 49 | 0 | 0 | 12 | 3 | 32 |
| 7 | 3 | 1 | 27 | 17 | 43 | 3 | 2 | 36 | 19 | 36 | 0 | 0 | 22 | 14 | 27 | 0 | 0 | 14 | 4 | 7 |
| 8 | 3 | 14 | 31 | 11 | 40 | 3 | 15 | 50 | 5 | 15 | 0 | 0 | 25 | 25 | 5 | 0 | 0 | 16 | 4 | 42 |
| 9 | 3 | 27 | 35 | 5 | 38 | 3 | 29 | 3 | 50 | 54 | 0 | 0 | 28 | 35 | 43 | 0 | 0 | 18 | 5 | 18 |
| 10 | 4 | 10 | 38 | 59 | 35 | 4 | 12 | 17 | 36 | 34 | 0 | 0 | 31 | 46 | 21 | 0 | 0 | 20 | 5 | 53 |
| 11 | 4 | 23 | 42 | 53 | 33 | 4 | 25 | 31 | 22 | 13 | 0 | 0 | 34 | 56 | 59 | 0 | 0 | 22 | 6 | 28 |
| 12 | 5 | 6 | 46 | 47 | 30 | 5 | 8 | 45 | 7 | 52 | 0 | 0 | 38 | 7 | 37 | 0 | 0 | 24 | 7 | 4 |
| 13 | 5 | 19 | 50 | 41 | 28 | 5 | 21 | 58 | 53 | 32 | 0 | 0 | 41 | 18 | 16 | 0 | 0 | 26 | 7 | 39 |
| 14 | 6 | 2 | 54 | 35 | 25 | 6 | 5 | 12 | 39 | 11 | 0 | 0 | 44 | 28 | 54 | 0 | 0 | 28 | 8 | 14 |
| 15 | 6 | 15 | 58 | 29 | 23 | 6 | 18 | 26 | 24 | 51 | 0 | 0 | 47 | 39 | 32 | 0 | 0 | 30 | 8 | 49 |
| 16 | 6 | 29 | 2 | 23 | 20 | 7 | 1 | 40 | 10 | 30 | 0 | 0 | 50 | 50 | 10 | 0 | 0 | 32 | 9 | 25 |
| 17 | 7 | 12 | 6 | 17 | 18 | 7 | 14 | 53 | 56 | 9 | 0 | 0 | 54 | 0 | 48 | 0 | 0 | 34 | 10 | 0 |
| 18 | 7 | 25 | 10 | 11 | 15 | 7 | 28 | 7 | 41 | 49 | 0 | 0 | 57 | 11 | 26 | 0 | 0 | 36 | 10 | 35 |
| 19 | 8 | 8 | 14 | 5 | 13 | 8 | 11 | 21 | 27 | 28 | 0 | 1 | 0 | 22 | 4 | 0 | 0 | 38 | 11 | 11 |
| 20 | 8 | 21 | 17 | 59 | 10 | 8 | 24 | 35 | 13 | 7 | 0 | 1 | 3 | 32 | 42 | 0 | 0 | 40 | 11 | 46 |
| 21 | 9 | 4 | 21 | 53 | 8 | 9 | 7 | 48 | 58 | 47 | 0 | 1 | 6 | 43 | 21 | 0 | 0 | 42 | 12 | 21 |
| 22 | 9 | 17 | 25 | 47 | 5 | 9 | 21 | 2 | 44 | 26 | 0 | 1 | 9 | 53 | 59 | 0 | 0 | 44 | 12 | 56 |

| Residua pars Tabule mediorum motuum planetarum ad dies mensium | | | | | | | | | | | | | | | | | | | | |
|---|---|---|---|---|---|---|---|---|---|---|---|---|---|---|---|---|---|---|---|---|
| Numerus dierum | Tabula medii motus argumenti Lune | | | | | Tabula medii motus argumenti latitudinis Lune | | | | | Tabula medii motus capitis draconis | | | | | Tabula medii motus Saturni | | | | |
| | S | G | M | 2 | 3 | S | G | M | 2 | 3 | S | G | M | 2 | 3 | S | G | M | 2 | 3 |
| 23 | 10 | 0 | 29 | 41 | 3 | 10 | 4 | 16 | 30 | 6 | 0 | 1 | 13 | 4 | 37 | 0 | 0 | 46 | 13 | 32 |
| 24 | 10 | 13 | 33 | 35 | 0 | 10 | 17 | 30 | 15 | 45 | 0 | 1 | 16 | 15 | 15 | 0 | 0 | 48 | 14 | 7 |
| 25 | 10 | 26 | 37 | 28 | 58 | 11 | 0 | 44 | 1 | 24 | 0 | 1 | 19 | 25 | 53 | 0 | 0 | 50 | 14 | 42 |
| 26 | 11 | 9 | 41 | 22 | 55 | 11 | 13 | 57 | 47 | 4 | 0 | 1 | 22 | 36 | 31 | 0 | 0 | 52 | 15 | 18 |
| 27 | 11 | 22 | 45 | 16 | 53 | 11 | 27 | 11 | 32 | 43 | 0 | 1 | 25 | 47 | 9 | 0 | 0 | 54 | 15 | 53 |
| 28 | 0 | 5 | 49 | 10 | 50 | 0 | 10 | 25 | 18 | 22 | 0 | 1 | 28 | 57 | 47 | 0 | 0 | 56 | 16 | 28 |
| 29 | 0 | 18 | 53 | 4 | 48 | 0 | 23 | 39 | 4 | 2 | 0 | 1 | 32 | 8 | 26 | 0 | 0 | 58 | 17 | 4 |
| 30 | 1 | 1 | 56 | 58 | 45 | 1 | 6 | 52 | 49 | 41 | 0 | 1 | 35 | 19 | 4 | 0 | 1 | 0 | 17 | 39 |
| 31 | 1 | 15 | 0 | 52 | 43 | 1 | 20 | 6 | 35 | 21 | 0 | 1 | 38 | 29 | 42 | 0 | 1 | 2 | 18 | 14 |

E F G H

Remaining part of the Table TRW-05 (Toruń, BU, 74/III, f. 255r)

| Residua pars Tabule mediorum motuum planetarum ad dies mensium | | | | | | | | | | | | | | | | | | | | |
|---|---|---|---|---|---|---|---|---|---|---|---|---|---|---|---|---|---|---|---|---|
| Numerus dierum | Tabula medii motus Iovis | | | | | Tabula medii motus Martis | | | | | Tabula medii argumenti Veneris | | | | | Tabula medii argumenti Mercurii | | | | |
| | S | G | M | 2 | 3 | S | G | M | 2 | 3 | S | G | M | 2 | 3 | S | G | M | 2 | 3 |
| 1 | 0 | 0 | 4 | 59 | 15 | 0 | 0 | 31 | 26 | 39 | 0 | 0 | 36 | 59 | 27 | 0 | 3 | 6 | 24 | 8 |
| 2 | 0 | 0 | 9 | 58 | 31 | 0 | 1 | 2 | 53 | 17 | 0 | 1 | 13 | 58 | 55 | 0 | 6 | 12 | 48 | 15 |
| 3 | 0 | 0 | 14 | 57 | 46 | 0 | 1 | 34 | 19 | 56 | 0 | 1 | 50 | 58 | 22 | 0 | 9 | 19 | 12 | 23 |
| 4 | 0 | 0 | 19 | 57 | 2 | 0 | 2 | 5 | 46 | 35 | 0 | 2 | 27 | 57 | 50 | 0 | 12 | 25 | 36 | 31 |
| 5 | 0 | 0 | 24 | 56 | 17 | 0 | 2 | 37 | 13 | 13 | 0 | 3 | 4 | 57 | 17 | 0 | 15 | 32 | 0 | 39 |
| 6 | 0 | 0 | 29 | 55 | 33 | 0 | 3 | 8 | 39 | 52 | 0 | 3 | 41 | 56 | 44 | 0 | 18 | 38 | 24 | 46 |
| 7 | 0 | 0 | 34 | 54 | 48 | 0 | 3 | 40 | 6 | 31 | 0 | 4 | 18 | 56 | 12 | 0 | 21 | 44 | 48 | 54 |
| 8 | 0 | 0 | 39 | 54 | 4 | 0 | 4 | 11 | 33 | 9 | 0 | 4 | 55 | 55 | 39 | 0 | 24 | 51 | 13 | 2 |
| 9 | 0 | 0 | 44 | 53 | 19 | 0 | 4 | 42 | 59 | 48 | 0 | 5 | 32 | 55 | 7 | 0 | 27 | 57 | 37 | 9 |
| 10 | 0 | 0 | 49 | 52 | 35 | 0 | 5 | 14 | 26 | 27 | 0 | 6 | 9 | 54 | 34 | 1 | 1 | 4 | 1 | 17 |
| 11 | 0 | 0 | 54 | 51 | 50 | 0 | 5 | 45 | 53 | 5 | 0 | 6 | 46 | 54 | 1 | 1 | 4 | 10 | 25 | 25 |
| 12 | 0 | 0 | 59 | 51 | 5 | 0 | 6 | 17 | 19 | 44 | 0 | 7 | 23 | 53 | 29 | 1 | 7 | 16 | 49 | 33 |
| 13 | 0 | 1 | 4 | 50 | 21 | 0 | 6 | 48 | 46 | 23 | 0 | 8 | 0 | 52 | 56 | 1 | 10 | 23 | 13 | 40 |
| 14 | 0 | 1 | 9 | 49 | 36 | 0 | 7 | 20 | 13 | 1 | 0 | 8 | 37 | 52 | 24 | 1 | 13 | 29 | 37 | 48 |
| 15 | 0 | 1 | 14 | 48 | 52 | 0 | 7 | 51 | 39 | 40 | 0 | 9 | 14 | 51 | 51 | 1 | 16 | 36 | 1 | 56 |
| 16 | 0 | 1 | 19 | 48 | 7 | 0 | 8 | 23 | 6 | 19 | 0 | 9 | 51 | 51 | 18 | 1 | 19 | 42 | 26 | 3 |
| 17 | 0 | 1 | 24 | 47 | 23 | 0 | 8 | 54 | 32 | 57 | 0 | 10 | 28 | 50 | 46 | 1 | 22 | 48 | 50 | 11 |
| 18 | 0 | 1 | 29 | 46 | 38 | 0 | 9 | 25 | 59 | 36 | 0 | 11 | 5 | 50 | 13 | 1 | 25 | 55 | 14 | 19 |
| 19 | 0 | 1 | 34 | 45 | 54 | 0 | 9 | 57 | 26 | 15 | 0 | 11 | 42 | 49 | 41 | 1 | 29 | 1 | 38 | 27 |
| 20 | 0 | 1 | 39 | 45 | 9 | 0 | 10 | 28 | 52 | 53 | 0 | 12 | 19 | 49 | 8 | 2 | 2 | 8 | 2 | 34 |
| 21 | 0 | 1 | 44 | 44 | 24 | 0 | 11 | 0 | 19 | 32 | 0 | 12 | 56 | 48 | 35 | 2 | 5 | 14 | 26 | 42 |
| 22 | 0 | 1 | 49 | 43 | 40 | 0 | 11 | 31 | 46 | 11 | 0 | 13 | 33 | 48 | 3 | 2 | 8 | 20 | 50 | 50 |

| Residua pars Tabule mediorum motuum planetarum ad dies mensium | | | | | | | | | | | | | | | | | | | | |
|---|---|---|---|---|---|---|---|---|---|---|---|---|---|---|---|---|---|---|---|---|
| Numerus dierum | Tabula medii motus Iovis | | | | | Tabula medii motus Martis | | | | | Tabula medii argumenti Veneris | | | | | Tabula medii argumenti Mercurii | | | | |
| | **S** | **G** | **M** | **2** | **3** | **S** | **G** | **M** | **2** | **3** | **S** | **G** | **M** | **2** | **3** | **S** | **G** | **M** | **2** | **3** |
| 23 | 0 | 1 | 54 | 42 | 55 | 0 | 12 | 3 | 12 | 49 | 0 | 14 | 10 | 47 | 30 | 2 | 11 | 27 | 14 | 57 |
| 24 | 0 | 1 | 59 | 42 | 11 | 0 | 12 | 34 | 39 | 28 | 0 | 14 | 47 | 46 | 58 | 2 | 14 | 33 | 39 | 5 |
| 25 | 0 | 2 | 4 | 41 | 26 | 0 | 13 | 6 | 6 | 7 | 0 | 15 | 24 | 46 | 25 | 2 | 17 | 40 | 3 | 13 |
| 26 | 0 | 2 | 9 | 40 | 42 | 0 | 13 | 37 | 32 | 45 | 0 | 16 | 1 | 45 | 52 | 2 | 20 | 46 | 27 | 20 |
| 27 | 0 | 2 | 14 | 39 | 57 | 0 | 14 | 8 | 59 | 24 | 0 | 16 | 38 | 45 | 20 | 2 | 23 | 52 | 51 | 28 |
| 28 | 0 | 2 | 19 | 39 | 13 | 0 | 14 | 40 | 26 | 3 | 0 | 17 | 15 | 44 | 47 | 2 | 26 | 59 | 15 | 36 |
| 29 | 0 | 2 | 24 | 38 | 28 | 0 | 15 | 11 | 52 | 41 | 0 | 17 | 52 | 44 | 15 | 3 | 0 | 5 | 39 | 44 |
| 30 | 0 | 2 | 29 | 37 | 44 | 0 | 15 | 43 | 19 | 20 | 0 | 18 | 29 | 43 | 42 | 3 | 3 | 12 | 3 | 51 |
| 31 | 0 | 2 | 34 | 36 | 59 | 0 | 16 | 14 | 45 | 59 | 0 | 19 | 6 | 43 | 9 | 3 | 6 | 18 | 27 | 59 |

I J K L

Table TRW-06 of mean motions in hours and their fractions (Toruń, BU, 74/III, f. 255v)

| Tabula mediorum motuum planetarum ad horas et fracciones horarum | | | | | | | | | | | | | | | | | | | | | | | | | | | | |
|---|---|---|---|---|---|---|---|---|---|---|---|---|---|---|---|---|---|---|---|---|---|---|---|---|---|---|---|---|
| Numerus horarum | Motus augium et stellarum fixarum | | | | Motus accessus et recessus octave spere | | | | Motus Solis, Veneris et Mercurii | | | | Motus medii Lune | | | | Motus medii argumenti Lune | | | | Motus argumenti latitudinis Lune | | | |
| | G | M | 2 | 3 | G | M | 2 | 3 | G | M | 2 | 3 | G | M | 2 | 3 | G | M | 2 | 3 | G | M | 2 | 3 |
| 1 | 0 | 0 | 0 | 0 | 0 | 0 | 0 | 1 | 0 | 2 | 27 | 51 | 0 | 32 | 56 | 28 | 0 | 32 | 39 | 45 | 0 | 33 | 4 | 24 |
| 2 | 0 | 0 | 0 | 0 | 0 | 0 | 0 | 3 | 0 | 4 | 55 | 42 | 1 | 5 | 52 | 55 | 1 | 5 | 19 | 29 | 1 | 6 | 8 | 48 |
| 3 | 0 | 0 | 0 | 1 | 0 | 0 | 0 | 4 | 0 | 7 | 23 | 32 | 1 | 38 | 49 | 23 | 1 | 37 | 59 | 15 | 1 | 39 | 13 | 12 |
| 4 | 0 | 0 | 0 | 1 | 0 | 0 | 0 | 5 | 0 | 9 | 51 | 23 | 2 | 11 | 45 | 50 | 2 | 10 | 39 | 0 | 2 | 12 | 17 | 37 |
| 5 | 0 | 0 | 0 | 1 | 0 | 0 | 0 | 6 | 0 | 12 | 19 | 14 | 2 | 44 | 42 | 18 | 2 | 43 | 18 | 44 | 2 | 45 | 22 | 1 |
| 6 | 0 | 0 | 0 | 1 | 0 | 0 | 0 | 8 | 0 | 14 | 47 | 5 | 3 | 17 | 38 | 45 | 3 | 15 | 58 | 29 | 3 | 18 | 26 | 25 |
| 7 | 0 | 0 | 0 | 1 | 0 | 0 | 0 | 9 | 0 | 17 | 14 | 56 | 3 | 50 | 35 | 13 | 3 | 48 | 38 | 14 | 3 | 51 | 30 | 49 |
| 8 | 0 | 0 | 0 | 1 | 0 | 0 | 0 | 10 | 0 | 19 | 42 | 47 | 4 | 23 | 31 | 40 | 4 | 21 | 17 | 59 | 4 | 24 | 35 | 13 |
| 9 | 0 | 0 | 0 | 2 | 0 | 0 | 0 | 11 | 0 | 22 | 10 | 37 | 4 | 56 | 28 | 8 | 4 | 53 | 57 | 44 | 4 | 57 | 39 | 37 |
| 10 | 0 | 0 | 0 | 2 | 0 | 0 | 0 | 13 | 0 | 24 | 38 | 28 | 5 | 29 | 24 | 36 | 5 | 26 | 37 | 29 | 5 | 30 | 44 | 1 |
| 11 | 0 | 0 | 0 | 2 | 0 | 0 | 0 | 14 | 0 | 27 | 6 | 19 | 6 | 2 | 21 | 3 | 5 | 59 | 17 | 14 | 6 | 3 | 48 | 25 |
| 12 | 0 | 0 | 0 | 2 | 0 | 0 | 0 | 15 | 0 | 29 | 34 | 10 | 6 | 31 | 17 | 31 | 6 | 31 | 56 | 59 | 6 | 36 | 52 | 50 |
| 13 | 0 | 0 | 0 | 2 | 0 | 0 | 0 | 16 | 0 | 32 | 2 | 1 | 7 | 8 | 13 | 58 | 7 | 4 | 36 | 44 | 7 | 9 | 57 | 14 |
| 14 | 0 | 0 | 0 | 3 | 0 | 0 | 0 | 18 | 0 | 34 | 29 | 51 | 7 | 41 | 10 | 26 | 7 | 37 | 16 | 29 | 7 | 43 | 1 | 38 |
| 15 | 0 | 0 | 0 | 3 | 0 | 0 | 0 | 19 | 0 | 36 | 57 | 42 | 8 | 14 | 6 | 53 | 8 | 9 | 56 | 13 | 8 | 16 | 6 | 2 |
| 16 | 0 | 0 | 0 | 3 | 0 | 0 | 0 | 20 | 0 | 39 | 25 | 33 | 8 | 47 | 3 | 21 | 8 | 42 | 35 | 58 | 8 | 49 | 10 | 26 |
| 17 | 0 | 0 | 0 | 3 | 0 | 0 | 0 | 22 | 0 | 41 | 53 | 24 | 9 | 19 | 59 | 48 | 9 | 15 | 15 | 43 | 9 | 22 | 14 | 50 |
| 18 | 0 | 0 | 0 | 3 | 0 | 0 | 0 | 23 | 0 | 44 | 21 | 15 | 9 | 52 | 56 | 16 | 9 | 47 | 55 | 28 | 9 | 55 | 19 | 14 |
| 19 | 0 | 0 | 0 | 3 | 0 | 0 | 0 | 24 | 0 | 46 | 49 | 6 | 10 | 25 | 52 | 43 | 10 | 20 | 35 | 13 | 10 | 28 | 23 | 39 |
| 20 | 0 | 0 | 0 | 4 | 0 | 0 | 0 | 25 | 0 | 49 | 16 | 56 | 10 | 58 | 49 | 11 | 10 | 53 | 14 | 58 | 11 | 1 | 28 | 3 |
| 21 | 0 | 0 | 0 | 4 | 0 | 0 | 0 | 27 | 0 | 51 | 44 | 47 | 11 | 31 | 45 | 39 | 11 | 25 | 54 | 43 | 11 | 34 | 32 | 27 |
| 22 | 0 | 0 | 0 | 4 | 0 | 0 | 0 | 28 | 0 | 54 | 12 | 38 | 12 | 4 | 42 | 6 | 11 | 58 | 34 | 28 | 12 | 7 | 36 | 51 |

| Tabula mediorum motuum planetarum ad horas et fracciones horarum | | | | | | | | | | | | | | | | | | | | | | | | |
|---|---|---|---|---|---|---|---|---|---|---|---|---|---|---|---|---|---|---|---|---|---|---|---|---|
| Numerus horarum | Motus augium et stellarum fixarum | | | | Motus accessus et recessus octave spere | | | | Motus Solis, Veneris et Mercurii | | | | Motus medii Lune | | | | Motus medii argumenti Lune | | | | Motus argumenti latitudinis Lune | | | |
| | **G** | **M** | **2** | **3** | **G** | **M** | **2** | **3** | **G** | **M** | **2** | **3** | **G** | **M** | **2** | **3** | **G** | **M** | **2** | **3** | **G** | **M** | **2** | **3** |
| 23 | 0 | 0 | 0 | 4 | 0 | 0 | 0 | 29 | 0 | 56 | 40 | 29 | 12 | 37 | 38 | 34 | 12 | 31 | 14 | 13 | 12 | 40 | 41 | 15 |
| 24 | 0 | 0 | 0 | 4 | 0 | 0 | 0 | 30 | 0 | 59 | 8 | 20 | 13 | 10 | 35 | 1 | 13 | 3 | 53 | 58 | 13 | 13 | 45 | 39 |
| 25 | 0 | 0 | 0 | 5 | 0 | 0 | 0 | 32 | 1 | 1 | 36 | 10 | 13 | 43 | 31 | 29 | 13 | 36 | 33 | 42 | 13 | 46 | 50 | 4 |
| 26 | 0 | 0 | 0 | 5 | 0 | 0 | 0 | 33 | 1 | 4 | 4 | 1 | 14 | 16 | 27 | 56 | 14 | 9 | 13 | 27 | 14 | 19 | 54 | 28 |
| 27 | 0 | 0 | 0 | 5 | 0 | 0 | 0 | 34 | 1 | 6 | 31 | 52 | 14 | 49 | 24 | 24 | 14 | 41 | 53 | 12 | 14 | 52 | 58 | 52 |
| 28 | 0 | 0 | 0 | 5 | 0 | 0 | 0 | 36 | 1 | 8 | 59 | 43 | 15 | 22 | 20 | 52 | 15 | 14 | 32 | 57 | 15 | 26 | 3 | 16 |
| 29 | 0 | 0 | 0 | 5 | 0 | 0 | 0 | 37 | 1 | 11 | 27 | 34 | 15 | 55 | 17 | 19 | 15 | 47 | 12 | 42 | 15 | 59 | 7 | 40 |
| 30 | 0 | 0 | 0 | 5 | 0 | 0 | 0 | 38 | 1 | 13 | 55 | 25 | 16 | 28 | 13 | 47 | 16 | 19 | 52 | 27 | 16 | 32 | 12 | 4 |
| m | m | 2 | 3 | 4 | m | 2 | 3 | 4 | m | 2 | 3 | 4 | m | 2 | 3 | 4 | m | 2 | 3 | 4 | m | 2 | 3 | 4 |
| 2 | 2 | 3 | 4 | | 2 | 3 | 4 | | 2 | 3 | 4 | | 2 | 3 | 4 | | 2 | 3 | 4 | | 2 | 3 | 4 | |
| 3 | 3 | 4 | | | 3 | 4 | | | 3 | 4 | | | 3 | 4 | | | 3 | 4 | | | 3 | 4 | | |
| 4 | 4 | | | | 4 | | | | 4 | | | | 4 | | | | 4 | | | | 4 | | | |

A B C D E F

Remaining part of the Table TRW-06 (Toruń, BU, 74/III, f. 256r)

| Residua pars Tabule motuum planetarum ad horas et fracciones horarum | | | | | | | | | | | | | | | | | | | | | | | | |
|---|---|---|---|---|---|---|---|---|---|---|---|---|---|---|---|---|---|---|---|---|---|---|---|---|
| Numerus horarum | Motus medius capitis draconis | | | | Motus medius Saturni | | | | Motus medius Iovis | | | | Motus medius Martis | | | | Motus argumenti Veneris | | | | Motus argumenti Mercurii | | | |
| | G | M | 2 | 3 | G | M | 2 | 3 | G | M | 2 | 3 | G | M | 2 | 3 | G | M | 2 | 3 | G | M | 2 | 3 |
| 1 | 0 | 0 | 7 | 57 | 0 | 0 | 5 | 1 | 0 | 0 | 12 | 28 | 0 | 1 | 18 | 37 | 0 | 1 | 32 | 29 | 0 | 7 | 46 | 0 |
| 2 | 0 | 0 | 15 | 53 | 0 | 0 | 10 | 3 | 0 | 0 | 24 | 56 | 0 | 2 | 37 | 13 | 0 | 3 | 4 | 57 | 0 | 15 | 32 | 1 |
| 3 | 0 | 0 | 23 | 50 | 0 | 0 | 15 | 4 | 0 | 0 | 37 | 24 | 0 | 3 | 55 | 50 | 0 | 4 | 37 | 26 | 0 | 23 | 18 | 1 |
| 4 | 0 | 0 | 31 | 46 | 0 | 0 | 20 | 6 | 0 | 0 | 49 | 53 | 0 | 5 | 14 | 26 | 0 | 6 | 9 | 55 | 0 | 31 | 4 | 1 |
| 5 | 0 | 0 | 39 | 43 | 0 | 0 | 25 | 7 | 0 | 1 | 2 | 21 | 0 | 6 | 33 | 3 | 0 | 7 | 42 | 23 | 0 | 38 | 50 | 2 |
| 6 | 0 | 0 | 47 | 40 | 0 | 0 | 30 | 9 | 0 | 1 | 14 | 49 | 0 | 7 | 51 | 40 | 0 | 9 | 14 | 52 | 0 | 46 | 36 | 2 |
| 7 | 0 | 0 | 55 | 36 | 0 | 0 | 35 | 10 | 0 | 1 | 27 | 17 | 0 | 9 | 10 | 16 | 0 | 10 | 47 | 21 | 0 | 54 | 22 | 2 |
| 8 | 0 | 1 | 3 | 33 | 0 | 0 | 40 | 12 | 0 | 1 | 39 | 45 | 0 | 10 | 28 | 53 | 0 | 12 | 19 | 49 | 1 | 2 | 8 | 3 |
| 9 | 0 | 1 | 11 | 29 | 0 | 0 | 45 | 13 | 0 | 1 | 52 | 13 | 0 | 11 | 47 | 29 | 0 | 13 | 52 | 18 | 1 | 9 | 54 | 3 |
| 10 | 0 | 1 | 19 | 26 | 0 | 0 | 50 | 15 | 0 | 2 | 4 | 41 | 0 | 13 | 6 | 6 | 0 | 15 | 24 | 47 | 1 | 17 | 40 | 3 |
| 11 | 0 | 1 | 27 | 22 | 0 | 0 | 55 | 16 | 0 | 2 | 17 | 9 | 0 | 14 | 24 | 42 | 0 | 16 | 57 | 15 | 1 | 25 | 26 | 3 |
| 12 | 0 | 1 | 35 | 19 | 0 | 1 | 0 | 18 | 0 | 2 | 29 | 38 | 0 | 15 | 43 | 19 | 0 | 18 | 29 | 44 | 1 | 33 | 12 | 4 |
| 13 | 0 | 1 | 43 | 16 | 0 | 1 | 5 | 19 | 0 | 2 | 42 | 6 | 0 | 17 | 1 | 56 | 0 | 20 | 2 | 12 | 1 | 40 | 58 | 4 |
| 14 | 0 | 1 | 51 | 12 | 0 | 1 | 10 | 21 | 0 | 2 | 54 | 34 | 0 | 18 | 20 | 32 | 0 | 21 | 34 | 41 | 1 | 48 | 44 | 4 |
| 15 | 0 | 1 | 59 | 9 | 0 | 1 | 15 | 22 | 0 | 3 | 7 | 2 | 0 | 19 | 39 | 9 | 0 | 23 | 7 | 10 | 1 | 56 | 30 | 5 |
| 16 | 0 | 2 | 7 | 5 | 0 | 1 | 20 | 23 | 0 | 3 | 19 | 30 | 0 | 20 | 57 | 46 | 0 | 24 | 39 | 38 | 2 | 4 | 16 | 5 |
| 17 | 0 | 2 | 15 | 2 | 0 | 1 | 25 | 25 | 0 | 3 | 31 | 58 | 0 | 22 | 16 | 22 | 0 | 26 | 12 | 6 | 2 | 12 | 2 | 5 |
| 18 | 0 | 2 | 22 | 59 | 0 | 1 | 30 | 27 | 0 | 3 | 44 | 26 | 0 | 23 | 34 | 59 | 0 | 27 | 44 | 36 | 2 | 19 | 48 | 6 |
| 19 | 0 | 2 | 30 | 55 | 0 | 1 | 35 | 28 | 0 | 3 | 56 | 55 | 0 | 24 | 53 | 35 | 0 | 29 | 17 | 4 | 2 | 27 | 34 | 6 |
| 20 | 0 | 2 | 38 | 52 | 0 | 1 | 40 | 29 | 0 | 4 | 9 | 23 | 0 | 26 | 12 | 12 | 0 | 30 | 49 | 33 | 2 | 35 | 20 | 6 |
| 21 | 0 | 2 | 46 | 48 | 0 | 1 | 45 | 31 | 0 | 4 | 21 | 51 | 0 | 27 | 30 | 49 | 0 | 32 | 22 | 1 | 2 | 43 | 6 | 7 |
| 22 | 0 | 2 | 54 | 45 | 0 | 1 | 50 | 32 | 0 | 4 | 34 | 19 | 0 | 28 | 49 | 25 | 0 | 33 | 54 | 30 | 2 | 50 | 52 | 7 |

| Residua pars Tabule motuum planetarum ad horas et fracciones horarum | | | | | | | | | | | | | | | | | | | | | | | | |
|---|---|---|---|---|---|---|---|---|---|---|---|---|---|---|---|---|---|---|---|---|---|---|---|---|
| Numerus horarum | Motus medius capitis draconis | | | | Motus medius Saturni | | | | Motus medius Iovis | | | | Motus medius Martis | | | | Motus argumenti Veneris | | | | Motus argumenti Mercurii | | | |
| | G | M | 2 | 3 | G | M | 2 | 3 | G | M | 2 | 3 | G | M | 2 | 3 | G | M | 2 | 3 | G | M | 2 | 3 |
| 23 | 0 | 3 | 2 | 41 | 0 | 1 | 55 | 34 | 0 | 4 | 46 | 47 | 0 | 30 | 8 | 2 | 0 | 35 | 26 | 59 | 2 | 58 | 38 | 7 |
| 24 | 0 | 3 | 10 | 38 | 0 | 2 | 0 | 35 | 0 | 4 | 59 | 15 | 0 | 31 | 26 | 39 | 0 | 36 | 59 | 27 | 3 | 6 | 24 | 8 |
| 25 | 0 | 3 | 18 | 35 | 0 | 2 | 5 | 37 | 0 | 5 | 11 | 44 | 0 | 32 | 45 | 15 | 0 | 38 | 31 | 56 | 3 | 14 | 10 | 8 |
| 26 | 0 | 3 | 26 | 31 | 0 | 2 | 10 | 38 | 0 | 5 | 24 | 12 | 0 | 34 | 3 | 52 | 0 | 40 | 4 | 25 | 3 | 21 | 56 | 8 |
| 27 | 0 | 3 | 34 | 28 | 0 | 2 | 15 | 40 | 0 | 5 | 36 | 40 | 0 | 35 | 22 | 29 | 0 | 41 | 36 | 53 | 3 | 29 | 42 | 9 |
| 28 | 0 | 3 | 42 | 24 | 0 | 2 | 20 | 41 | 0 | 5 | 49 | 8 | 0 | 36 | 41 | 5 | 0 | 43 | 9 | 22 | 3 | 37 | 28 | 9 |
| 29 | 0 | 3 | 50 | 21 | 0 | 2 | 25 | 43 | 0 | 6 | 1 | 36 | 0 | 37 | 59 | 42 | 0 | 44 | 41 | 51 | 3 | 45 | 14 | 9 |
| 30 | 0 | 3 | 58 | 18 | 0 | 2 | 30 | 44 | 0 | 6 | 14 | 4 | 0 | 39 | 18 | 18 | 0 | 46 | 14 | 19 | 3 | 53 | 0 | 10 |
| m | m | 2 | 3 | 4 | m | 2 | 3 | 4 | m | 2 | 3 | 4 | m | 2 | 3 | 4 | m | 2 | 3 | 4 | m | 2 | 3 | 4 |
| 2 | 2 | 3 | 4 | | 2 | 3 | 4 | | 2 | 3 | 4 | | 2 | 3 | 4 | | 2 | 3 | 4 | | 2 | 3 | 4 | |
| 3 | 3 | 4 | | | 3 | 4 | | | 3 | 4 | | | 3 | 4 | | | 3 | 4 | | | 3 | 4 | | |
| 4 | 4 | | | | 4 | | | | 4 | | | | 4 | | | | 4 | | | | 4 | | | |

G H I J K L

Table TRW-07 of true planetary apogees in collected years (Toruń, BU, 74/III, f. 256v)

| Tabula augium verarum omnium planetarum ad annos Christi collectos | | | | | | | | | | | | | | | | | | | | |
|---|---|---|---|---|---|---|---|---|---|---|---|---|---|---|---|---|---|---|---|---|
| Numerus annorum collectorum | Aux in communis | | | | | Aux Solis et Veneris | | | | | Aux Saturni | | | | | Differencia | | | Differencia differencie | |
| | S | G | M | 2 | 3 | S | G | M | 2 | 3 | S | G | M | 2 | 3 | M | 2 | 3 | 2 | 3 |
| 1424 | 0 | 19 | 2 | 23 | 14 | 3 | 0 | 27 | 46 | 18 | 8 | 12 | 26 | 5 | 18 | 11 | 40 | 8 | 35 | 0 |
| 1444 | 0 | 19 | 14 | 3 | 22 | 3 | 0 | 39 | 26 | 26 | 8 | 12 | 37 | 45 | 26 | 11 | 30 | 4 | 34 | 30 |
| 1464 | 0 | 19 | 24 | 33 | 26 | 3 | 0 | 50 | 56 | 30 | 8 | 12 | 49 | 15 | 30 | 11 | 20 | 2 | 34 | 0 |
| 1484 | 0 | 19 | 36 | 53 | 28 | 3 | 1 | 2 | 16 | 32 | 8 | 13 | 0 | 35 | 32 | 11 | 9 | 27 | 33 | 28 |
| 1504 | 0 | 19 | 48 | 2 | 55 | 3 | 1 | 13 | 25 | 59 | 8 | 13 | 11 | 44 | 59 | 10 | 59 | 29 | 32 | 58 |
| 1524 | 0 | 19 | 59 | 2 | 24 | 3 | 1 | 24 | 25 | 28 | 8 | 13 | 22 | 44 | 28 | 10 | 49 | 57 | 32 | 30 |
| 1544 | 0 | 20 | 9 | 52 | 21 | 3 | 1 | 35 | 15 | 25 | 8 | 13 | 33 | 34 | 25 | 10 | 40 | 27 | 32 | 1 |
| 1564 | 0 | 20 | 20 | 32 | 48 | 3 | 1 | 45 | 55 | 52 | 8 | 13 | 44 | 14 | 52 | 10 | 30 | 15 | 31 | 31 |
| 1584 | 0 | 20 | 31 | 3 | 3 | 3 | 1 | 56 | 26 | 7 | 8 | 13 | 54 | 45 | 7 | 10 | 17 | 35 | 30 | 53 |
| 1604 | 0 | 20 | 41 | 20 | 38 | 3 | 2 | 6 | 43 | 42 | 8 | 14 | 5 | 2 | 42 | 10 | 7 | 19 | 30 | 22 |
| 1624 | 0 | 20 | 51 | 27 | 57 | 3 | 2 | 16 | 51 | 1 | 8 | 14 | 15 | 10 | 1 | 9 | 58 | 27 | 29 | 55 |
| 1644 | 0 | 21 | 1 | 26 | 24 | 3 | 2 | 26 | 49 | 28 | 8 | 14 | 25 | 8 | 28 | 9 | 47 | 58 | 29 | 24 |
| 1664 | 0 | 21 | 11 | 14 | 22 | 3 | 2 | 36 | 37 | 26 | 8 | 14 | 34 | 56 | 26 | 9 | 37 | 24 | 28 | 52 |
| 1684 | 0 | 21 | 20 | 51 | 46 | 3 | 2 | 46 | 14 | 50 | 8 | 14 | 44 | 33 | 50 | 9 | 26 | 50 | 28 | 20 |
| 1704 | 0 | 21 | 30 | 18 | 36 | 3 | 2 | 55 | 41 | 40 | 8 | 14 | 54 | 0 | 40 | 0 | 0 | 0 | 0 | 0 |
| Radices Incarnacionis Christi | | | | | | 2 | 11 | 25 | 23 | 4 | 7 | 23 | 23 | 42 | 4 | | | | | |
| | | | A | | | | | B | | | | | C | | | | D | | E | |

| | Aux Iovis | | | | | Aux Martis | | | | | Aux Mercurii | | | | | Differencia | | | Differencia differencie | |
|---|---|---|---|---|---|---|---|---|---|---|---|---|---|---|---|---|---|---|---|---|
| | **S** | **G** | **M** | **2** | **3** | **S** | **G** | **M** | **2** | **3** | **S** | **G** | **M** | **2** | **h** | **M** | **2** | **3** | **2** | **3** |
| 1424 | 5 | 22 | 39 | 23 | 18 | 4 | 14 | 14 | 36 | 18 | 6 | 29 | 41 | 56 | 18 | 11 | 40 | 8 | 35 | 0 |
| 1444 | 5 | 22 | 51 | 3 | 26 | 4 | 14 | 26 | 16 | 26 | 6 | 29 | 53 | 36 | 26 | 11 | 30 | 4 | 34 | 30 |
| 1464 | 5 | 23 | 2 | 33 | 30 | 4 | 14 | 37 | 46 | 30 | 7 | 0 | 5 | 6 | 30 | 11 | 20 | 2 | 34 | 0 |
| 1484 | 5 | 23 | 13 | 53 | 32 | 4 | 14 | 49 | 6 | 32 | 7 | 0 | 16 | 26 | 32 | 11 | 9 | 27 | 33 | 28 |
| 1504 | 5 | 23 | 25 | 2 | 59 | 4 | 15 | 0 | 15 | 59 | 7 | 0 | 27 | 35 | 59 | 10 | 59 | 29 | 32 | 58 |
| 1524 | 5 | 23 | 36 | 2 | 28 | 4 | 15 | 11 | 15 | 28 | 7 | 0 | 38 | 35 | 28 | 10 | 40 | 57 | 32 | 30 |
| 1544 | 5 | 23 | 46 | 52 | 25 | 4 | 15 | 22 | 5 | 25 | 7 | 0 | 49 | 25 | 25 | 10 | 40 | 27 | 32 | 1 |
| 1564 | 5 | 23 | 57 | 32 | 52 | 4 | 15 | 32 | 45 | 52 | 7 | 1 | 0 | 5 | 52 | 10 | 30 | 15 | 31 | 31 |
| 1584 | 5 | 24 | 8 | 3 | 7 | 4 | 15 | 43 | 16 | 7 | 7 | 1 | 10 | 36 | 7 | 10 | 17 | 35 | 30 | 53 |
| 1604 | 5 | 24 | 18 | 20 | 42 | 4 | 15 | 53 | 33 | 42 | 7 | 1 | 20 | 53 | 42 | 10 | 7 | 19 | 30 | 22 |
| 1624 | 5 | 24 | 28 | 28 | 1 | 4 | 16 | 3 | 41 | 1 | 7 | 1 | 31 | 1 | 1 | 9 | 58 | 27 | 29 | 55 |
| 1644 | 5 | 24 | 38 | 26 | 28 | 4 | 16 | 13 | 39 | 28 | 7 | 1 | 40 | 59 | 28 | 9 | 47 | 58 | 29 | 24 |
| 1664 | 5 | 24 | 48 | 14 | 26 | 4 | 16 | 23 | 27 | 26 | 7 | 1 | 50 | 47 | 26 | 9 | 37 | 24 | 28 | 52 |
| 1684 | 5 | 24 | 57 | 51 | 50 | 4 | 16 | 33 | 4 | 50 | 7 | 2 | 0 | 24 | 50 | 9 | 26 | 50 | 28 | 20 |
| 1704 | 5 | 25 | 7 | 18 | 40 | 4 | 16 | 42 | 31 | 40 | 7 | 2 | 9 | 51 | 40 | 0 | 0 | 0 | 0 | 0 |
| Radices | 5 | 3 | 37 | 0 | 4 | 3 | 25 | 12 | 13 | 4 | 6 | 10 | 39 | 33 | 4 | | | | | |

1 addatur, cum praxis sit ad meridianum Pragensem in omnibus augium

F G H D E

Table TRW-08 of equation of access and recess (Toruń, BU, 74/III, f. 257r)

| Tabula equacionis accessus et recessus octave spere | | | | | | | | | | |
|---|---|---|---|---|---|---|---|---|---|---|
| | 0 6 | | | 1 7 | | | 2 8 | | | |
| Gradus octave spere | Equacio | | | Equacio | | | Equacio | | | Gradus octave spere |
| | G | M | 2 | G | M | 2 | G | M | 2 | |
| 1 | 0 | 9 | 25 | 4 | 37 | 17 | 7 | 51 | 49 | 29 |
| 2 | 0 | 18 | 49 | 4 | 45 | 18 | 7 | 57 | 19 | 28 |
| 3 | 0 | 28 | 11 | 4 | 53 | 14 | 8 | 0 | 41 | 27 |
| 4 | 0 | 37 | 32 | 5 | 1 | 5 | 8 | 4 | 56 | 26 |
| 5 | 0 | 46 | 52 | 5 | 8 | 51 | 8 | 9 | 2 | 25 |
| 6 | 0 | 56 | 12 | 5 | 16 | 30 | 8 | 12 | 58 | 24 |
| 7 | 1 | 5 | 31 | 5 | 24 | 4 | 8 | 16 | 45 | 23 |
| 8 | 1 | 14 | 28 | 5 | 31 | 33 | 8 | 20 | 23 | 22 |
| 9 | 1 | 24 | 4 | 5 | 38 | 57 | 8 | 23 | 52 | 21 |
| 10 | 1 | 33 | 20 | 5 | 46 | 16 | 8 | 27 | 11 | 20 |
| 11 | 1 | 42 | 34 | 5 | 53 | 26 | 8 | 30 | 23 | 19 |
| 12 | 1 | 51 | 46 | 6 | 0 | 29 | 8 | 33 | 24 | 18 |
| 13 | 2 | 0 | 57 | 6 | 7 | 26 | 8 | 36 | 15 | 17 |
| 14 | 2 | 10 | 6 | 6 | 14 | 17 | 8 | 38 | 56 | 16 |
| 15 | 2 | 19 | 13 | 6 | 21 | 2 | 8 | 41 | 28 | 15 |
| 16 | 2 | 28 | 17 | 6 | 27 | 40 | 8 | 43 | 50 | 14 |
| 17 | 2 | 37 | 16 | 6 | 34 | 10 | 8 | 46 | 2 | 13 |
| 18 | 2 | 46 | 11 | 6 | 40 | 33 | 8 | 48 | 5 | 12 |
| 19 | 2 | 55 | 2 | 6 | 46 | 49 | 8 | 49 | 59 | 11 |
| 20 | 3 | 3 | 49 | 6 | 52 | 58 | 8 | 51 | 44 | 10 |
| 21 | 3 | 12 | 47 | 6 | 59 | 0 | 8 | 53 | 19 | 9 |
| 22 | 3 | 21 | 36 | 7 | 4 | 53 | 8 | 54 | 41 | 8 |
| 23 | 3 | 30 | 20 | 7 | 10 | 38 | 8 | 55 | 55 | 7 |
| 24 | 3 | 38 | 57 | 7 | 16 | 15 | 8 | 57 | 0 | 6 |
| 25 | 3 | 47 | 27 | 7 | 21 | 44 | 8 | 57 | 55 | 5 |
| 26 | 3 | 55 | 54 | 7 | 27 | 7 | 8 | 58 | 40 | 4 |
| 27 | 4 | 4 | 17 | 7 | 32 | 21 | 8 | 59 | 15 | 3 |
| 28 | 4 | 12 | 38 | 7 | 37 | 27 | 8 | 59 | 40 | 2 |
| 29 | 4 | 20 | 55 | 7 | 42 | 23 | 8 | 59 | 55 | 1 |
| 30 | 4 | 29 | 10 | 7 | 47 | 10 | 9 | 0 | 0 | 0 |
| | 11 | 5 | | 10 | 4 | | 9 | 3 | | |
| | | A | | | B | | | C | | |

Table TRW-09 of mean conjunctions of the Sun and Moon in collected years (Toruń, BU, 74/III, f. 257v)

| Tabula medie coniunccionis Solis et Lune in annis Christi collectis super meridianum Wratislaviensem | | | | | | | | | | | | | | | | | | | | | |
|---|---|---|---|---|---|---|---|---|---|---|---|---|---|---|---|---|---|---|---|---|---|
| Numerus annorum collectorum | Tempus coniunccionis medie | | | | | | Motus Solis et Lune | | | | | Motus argumenti Lune | | | | | Argumentum latitudinis Lune | | | | |
| | D | H | M | 2 | 3 | 4 | S | G | M | 2 | 3 | S | G | M | 2 | 3 | S | G | M | 2 | 3 |
| 1425 | 19 | 2 | 51 | 28 | 16 | 5 | 10 | 7 | 35 | 47 | 45 | 1 | 7 | 31 | 49 | 21 | 7 | 12 | 39 | 47 | 2 |
| 1445 | 8 | 4 | 12 | 1 | 36 | 19 | 9 | 26 | 57 | 23 | 38 | 9 | 24 | 17 | 13 | 17 | 7 | 28 | 16 | 27 | 49 |
| 1465 | 26 | 18 | 16 | 37 | 55 | 26 | 10 | 15 | 25 | 23 | 42 | 7 | 6 | 51 | 37 | 46 | 9 | 14 | 33 | 22 | 21 |
| 1485 | 15 | 19 | 37 | 11 | 11 | 52 | 10 | 4 | 46 | 59 | 35 | 3 | 23 | 37 | 1 | 40 | 10 | 0 | 10 | 3 | 7 |
| 1505 | 4 | 20 | 57 | 44 | 28 | 9 | 9 | 24 | 8 | 35 | 28 | 0 | 10 | 22 | 25 | 35 | 10 | 15 | 46 | 43 | 52 |
| 1525 | 23 | 11 | 2 | 20 | 47 | 25 | 10 | 12 | 36 | 35 | 32 | 9 | 22 | 56 | 50 | 4 | 0 | 2 | 3 | 38 | 24 |
| 1545 | 12 | 12 | 22 | 54 | 3 | 43 | 10 | 1 | 58 | 11 | 24 | 6 | 9 | 42 | 13 | 58 | 0 | 17 | 40 | 19 | 9 |
| 1565 | 1 | 13 | 43 | 27 | 20 | 0 | 9 | 21 | 19 | 47 | 17 | 2 | 26 | 27 | 37 | 52 | 1 | 3 | 16 | 59 | 55 |
| 1585 | 20 | 3 | 48 | 3 | 39 | 17 | 10 | 9 | 47 | 47 | 22 | 0 | 9 | 2 | 2 | 22 | 2 | 19 | 33 | 54 | 26 |
| 1605 | 9 | 5 | 8 | 36 | 55 | 34 | 9 | 29 | 9 | 23 | 14 | 8 | 25 | 47 | 26 | 16 | 3 | 5 | 10 | 35 | 12 |
| 1625 | 27 | 19 | 13 | 13 | 14 | 50 | 10 | 17 | 37 | 23 | 18 | 6 | 8 | 21 | 52 | 45 | 4 | 21 | 27 | 29 | 43 |
| 1645 | 16 | 20 | 33 | 46 | 31 | 8 | 10 | 6 | 58 | 59 | 11 | 2 | 25 | 7 | 14 | 39 | 5 | 7 | 4 | 10 | 29 |
| 1665 | 5 | 21 | 54 | 19 | 47 | 26 | 9 | 26 | 20 | 35 | 4 | 11 | 11 | 52 | 38 | 34 | 5 | 22 | 40 | 51 | 14 |
| 1685 | 24 | 11 | 58 | 56 | 6 | 42 | 10 | 14 | 48 | 35 | 8 | 8 | 24 | 27 | 3 | 3 | 7 | 8 | 57 | 45 | 42 |
| 1705 | 13 | 13 | 19 | 29 | 22 | 59 | 10 | 4 | 10 | 11 | 1 | 5 | 11 | 12 | 26 | 57 | 7 | 24 | 34 | 26 | 31 |
| | A | | | | | | B | | | | | C | | | | | D | | | | |

Table TRW-10 of mean oppositions of the Sun and Moon in collected years (Toruń, BU, 74/III, f. 257v)

| Tabula medie opposicionis Solis et Lune in annis Christi collectis | | | | | | | | | | | | | | | | | | | | | |
|---|---|---|---|---|---|---|---|---|---|---|---|---|---|---|---|---|---|---|---|---|---|
| Numerus annorum collectorum | Tempus opposicionis medie | | | | | | Motus Solis et Lune | | | | | Motus argumenti Lune | | | | | Argumentum latitudinis Lune | | | | |
| | D | H | M | 2 | 3 | 4 | S | G | M | 2 | 3 | S | G | M | 2 | 3 | S | G | M | 2 | 3 |
| 1425 | 4 | 8 | 29 | 26 | 44 | 36 | 9 | 23 | 2 | 35 | 39 | 6 | 24 | 37 | 19 | 4 | 0 | 27 | 19 | 40 | 9 |
| 1445 | 22 | 22 | 34 | 3 | 7 | 48 | 10 | 11 | 30 | 35 | 44 | 4 | 7 | 11 | 43 | 34 | 2 | 13 | 36 | 34 | 42 |
| 1465 | 11 | 23 | 54 | 36 | 23 | 57 | 10 | 0 | 52 | 11 | 36 | 0 | 23 | 57 | 7 | 29 | 2 | 29 | 13 | 15 | 28 |
| 1485 | 1 | 1 | 15 | 9 | 40 | 23 | 9 | 20 | 13 | 47 | 29 | 9 | 10 | 42 | 31 | 23 | 3 | 14 | 49 | 56 | 14 |
| 1505 | 19 | 15 | 19 | 45 | 59 | 38 | 10 | 8 | 41 | 47 | 34 | 6 | 23 | 16 | 55 | 52 | 5 | 1 | 6 | 50 | 45 |
| 1525 | 8 | 16 | 40 | 19 | 15 | 56 | 9 | 28 | 3 | 23 | 26 | 3 | 10 | 2 | 19 | 47 | 5 | 16 | 43 | 31 | 31 |
| 1545 | 27 | 6 | 44 | 55 | 35 | 12 | 10 | 16 | 31 | 23 | 30 | 0 | 22 | 36 | 44 | 15 | 7 | 3 | 0 | 26 | 2 |
| 1565 | 16 | 8 | 5 | 28 | 51 | 29 | 10 | 5 | 52 | 59 | 23 | 9 | 9 | 22 | 8 | 9 | 7 | 18 | 37 | 6 | 48 |
| 1585 | 5 | 9 | 26 | 2 | 7 | 48 | 9 | 25 | 14 | 35 | 16 | 5 | 26 | 7 | 32 | 5 | 8 | 4 | 13 | 47 | 33 |
| 1605 | 23 | 23 | 30 | 38 | 27 | 3 | 10 | 13 | 42 | 35 | 20 | 3 | 8 | 41 | 56 | 33 | 9 | 20 | 30 | 42 | 5 |
| 1625 | 13 | 0 | 51 | 11 | 43 | 21 | 10 | 3 | 4 | 11 | 12 | 11 | 25 | 27 | 22 | 28 | 10 | 6 | 7 | 22 | 50 |
| 1645 | 2 | 2 | 11 | 44 | 59 | 39 | 9 | 22 | 25 | 47 | 5 | 8 | 12 | 12 | 44 | 22 | 10 | 21 | 44 | 3 | 36 |
| 1665 | 20 | 16 | 16 | 21 | 18 | 55 | 10 | 10 | 53 | 47 | 10 | 5 | 24 | 47 | 8 | 51 | 0 | 8 | 0 | 58 | 7 |
| 1685 | 9 | 17 | 36 | 54 | 35 | 13 | 10 | 0 | 15 | 23 | 2 | 2 | 11 | 32 | 32 | 46 | 0 | 23 | 37 | 38 | 49 |
| 1705 | 28 | 7 | 41 | 30 | 54 | 28 | 10 | 18 | 43 | 23 | 7 | 11 | 24 | 6 | 57 | 14 | 2 | 9 | 54 | 33 | 24 |
| | A | | | | | | B | | | | | C | | | | | D | | | | |

Table TRW-11 of mean conjunctions and oppositions of the Sun and Moon in expanded years (Toruń, BU, 74/III, f. 258r)

| Tabula medie coniunccionis et opposicionis Solis et Lune ad annos Christi expansos | | | | | | | | | | | | | | | | | | | | | | | |
|---|---|---|---|---|---|---|---|---|---|---|---|---|---|---|---|---|---|---|---|---|---|---|---|
| Anni expansi | | Tempus coniunccionis et opposicionis | | | | | | | Motus Solis et Lune | | | | | Motus argumenti Lune | | | | | Argumentum latitudinis Lune | | | | |
| | | **D** | **H** | **M** | **2** | **3** | **4** | | **S** | **G** | **M** | **2** | **3** | **S** | **G** | **M** | **2** | **3** | **S** | **G** | **M** | **2** | **3** |
| | 1 | 20 | 8 | 48 | 36 | 35 | 48 | | 11 | 19 | 16 | 50 | 20 | 10 | 9 | 48 | 6 | 59 | 0 | 8 | 2 | 45 | 13 |
| | 2 | 9 | 17 | 37 | 13 | 11 | 36 | | 11 | 8 | 33 | 40 | 39 | 8 | 19 | 36 | 13 | 59 | 0 | 16 | 5 | 30 | 26 |
| B | 3 | 28 | 15 | 9 | 52 | 50 | 23 | E | 11 | 26 | 56 | 55 | 10 | 7 | 25 | 13 | 21 | 33 | 1 | 24 | 48 | 29 | 26 |
| | 4 | 16 | 23 | 58 | 29 | 26 | 11 | | 11 | 16 | 13 | 45 | 30 | 6 | 5 | 1 | 28 | 33 | 2 | 2 | 51 | 14 | 39 |
| | 5 | 6 | 8 | 47 | 6 | 1 | 59 | | 11 | 5 | 30 | 35 | 49 | 4 | 14 | 49 | 35 | 32 | 2 | 10 | 53 | 59 | 52 |
| | 6 | 25 | 6 | 19 | 45 | 40 | 46 | E | 11 | 23 | 53 | 50 | 31 | 3 | 20 | 26 | 43 | 6 | 3 | 19 | 36 | 58 | 51 |
| B | 7 | 14 | 15 | 8 | 22 | 16 | 34 | | 11 | 13 | 10 | 40 | 40 | 2 | 0 | 14 | 50 | 6 | 3 | 27 | 39 | 44 | 5 |
| | 8 | 2 | 23 | 56 | 58 | 52 | 22 | | 11 | 2 | 27 | 31 | 0 | 0 | 10 | 2 | 57 | 5 | 4 | 5 | 42 | 29 | 18 |
| | 9 | 12 | 21 | 29 | 38 | 31 | 9 | E | 11 | 20 | 50 | 45 | 31 | 11 | 15 | 40 | 4 | 39 | 5 | 14 | 25 | 28 | 17 |
| | 10 | 11 | 6 | 18 | 15 | 6 | 57 | | 11 | 10 | 7 | 35 | 51 | 9 | 25 | 28 | 11 | 39 | 5 | 22 | 28 | 13 | 30 |
| B | 11 | 30 | 3 | 50 | 54 | 45 | 44 | E | 11 | 28 | 30 | 50 | 22 | 9 | 1 | 5 | 19 | 13 | 7 | 1 | 11 | 12 | 30 |
| | 12 | 18 | 12 | 39 | 31 | 21 | 32 | | 11 | 17 | 47 | 40 | 41 | 7 | 10 | 53 | 26 | 13 | 7 | 9 | 13 | 57 | 43 |
| | 13 | 7 | 21 | 28 | 7 | 57 | 20 | | 11 | 7 | 4 | 30 | 1 | 5 | 20 | 41 | 33 | 12 | 7 | 17 | 16 | 42 | 56 |
| | 14 | 26 | 19 | 0 | 47 | 36 | 7 | E | 11 | 25 | 27 | 45 | 32 | 4 | 26 | 18 | 40 | 46 | 8 | 25 | 59 | 41 | 55 |
| B | 15 | 16 | 3 | 49 | 24 | 11 | 55 | | 11 | 14 | 44 | 35 | 52 | 3 | 6 | 6 | 47 | 46 | 9 | 4 | 2 | 27 | 9 |
| | 16 | 4 | 12 | 37 | 0 | 47 | 43 | | 11 | 4 | 1 | 26 | 11 | 1 | 15 | 54 | 54 | 45 | 9 | 12 | 5 | 12 | 21 |
| | 17 | 23 | 10 | 10 | 40 | 26 | 30 | E | 11 | 22 | 24 | 40 | 42 | 0 | 21 | 32 | 2 | 20 | 10 | 20 | 48 | 11 | 21 |
| | 18 | 12 | 18 | 59 | 17 | 2 | 18 | | 11 | 11 | 41 | 31 | 2 | 11 | 1 | 20 | 9 | 19 | 10 | 28 | 50 | 56 | 34 |
| B | 19 | 2 | 3 | 47 | 53 | 38 | 6 | | 11 | 0 | 58 | 21 | 22 | 9 | 11 | 8 | 16 | 18 | 11 | 6 | 53 | 41 | 48 |
| | 20 | 20 | 1 | 20 | 33 | 18 | 53 | E | 11 | 19 | 21 | 35 | 53 | 8 | 16 | 45 | 23 | 53 | 0 | 15 | 36 | 40 | 47 |

A B C D

Table TRW-12 of mean conjunctions and oppositions of the Sun and Moon in months (Toruń, BU, 74/III, f. 258r)

| Tabula medie coniunccionis et opposicionis Solis et Lune in mensibus | | | | | | | | | | | | | | | | | | | | | | |
|---|---|---|---|---|---|---|---|---|---|---|---|---|---|---|---|---|---|---|---|---|---|---|
| | Dies Lune | Coniuncciones et opposiciones | | | | | Dies Solis | Motus Solis et Lune | | | | | Motus argumenti Lune | | | | | Argumentum latitudinis Lune | | | | |
| | | H | M | 2 | 3 | 4 | | S | G | M | 2 | 3 | S | G | M | 2 | 3 | S | G | M | 2 | 3 |
| Feb | 29 | 12 | 44 | 3 | 2 | 59 | 31 | 0 | 29 | 6 | 24 | 12 | 0 | 25 | 49 | 0 | 35 | 1 | 0 | 40 | 13 | 46 |
| Mar | 59 | 1 | 28 | 6 | 5 | 58 | 59 | 1 | 28 | 12 | 48 | 23 | 1 | 21 | 38 | 1 | 10 | 2 | 1 | 20 | 27 | 32 |
| Apr | 88 | 14 | 12 | 9 | 8 | 57 | 90 | 2 | 27 | 19 | 12 | 35 | 2 | 17 | 27 | 1 | 45 | 3 | 2 | 0 | 41 | 18 |
| May | 118 | 2 | 56 | 12 | 11 | 56 | 120 | 3 | 26 | 25 | 36 | 46 | 3 | 13 | 16 | 2 | 20 | 4 | 2 | 40 | 55 | 4 |
| Iun | 147 | 15 | 40 | 15 | 14 | 55 | 151 | 4 | 25 | 32 | 0 | 58 | 4 | 9 | 5 | 2 | 55 | 5 | 3 | 21 | 8 | 50 |
| Iul | 177 | 4 | 24 | 18 | 17 | 54 | 181 | 5 | 24 | 38 | 25 | 10 | 5 | 4 | 54 | 3 | 30 | 6 | 4 | 1 | 22 | 37 |
| Aug | 206 | 17 | 8 | 21 | 20 | 53 | 212 | 6 | 23 | 44 | 49 | 21 | 6 | 0 | 43 | 4 | 5 | 7 | 4 | 41 | 36 | 23 |
| Sep | 236 | 5 | 52 | 24 | 23 | 52 | 243 | 7 | 22 | 51 | 13 | 33 | 6 | 26 | 32 | 4 | 40 | 8 | 5 | 21 | 50 | 9 |
| Oct | 265 | 18 | 36 | 27 | 26 | 51 | 273 | 8 | 21 | 57 | 37 | 45 | 7 | 22 | 21 | 5 | 15 | 9 | 6 | 2 | 3 | 55 |
| Nov | 295 | 7 | 20 | 30 | 29 | 50 | 304 | 9 | 21 | 4 | 1 | 56 | 8 | 18 | 10 | 5 | 50 | 10 | 6 | 42 | 17 | 41 |
| Dec | 324 | 20 | 4 | 33 | 32 | 49 | 334 | 10 | 20 | 10 | 26 | 8 | 9 | 13 | 59 | 6 | 24 | 11 | 7 | 22 | 31 | 27 |
| Ian | 354 | 8 | 48 | 36 | 35 | 48 | 365 | 11 | 19 | 16 | 50 | 20 | 10 | 9 | 48 | 6 | 59 | 0 | 8 | 2 | 45 | 13 |
| | | A | | | | | | B | | | | | C | | | | | D | | | | |

Table TRW-13 of equation of the Sun (Toruń, BU, 74/III, f. 258v)

| Tabula equacionis Solis | | | | | | | | | | | | | | | | | | | |
|---|---|---|---|---|---|---|---|---|---|---|---|---|---|---|---|---|---|---|---|
| minue | | 0 | | | 1 | | | 2 | | | 3 | | | 4 | | | 5 | | |
| Gradus | Equacio | | | Equacio | | | Equacio | | | Equacio | | | Equacio | | | Equacio | | | Gradus |
| | G | M | 2 | G | M | 2 | G | M | 2 | G | M | 2 | G | M | 2 | G | M | 2 | |
| 1 | 0 | 2 | 10 | 1 | 4 | 46 | 1 | 51 | 51 | 2 | 9 | 59 | 1 | 53 | 46 | 1 | 5 | 1 | 29 |
| 2 | 0 | 4 | 19 | 1 | 6 | 37 | 1 | 52 | 56 | 2 | 10 | 0 | 1 | 52 | 35 | 1 | 2 | 54 | 28 |
| 3 | 0 | 6 | 27 | 1 | 8 | 28 | 1 | 54 | 0 | 2 | 10 | 0 | 1 | 51 | 24 | 1 | 0 | 47 | 27 |
| 4 | 0 | 8 | 36 | 1 | 10 | 19 | 1 | 55 | 6 | 2 | 10 | 0 | 1 | 50 | 12 | 0 | 58 | 40 | 26 |
| 5 | 0 | 10 | 44 | 1 | 12 | 9 | 1 | 56 | 9 | 2 | 9 | 57 | 1 | 48 | 59 | 0 | 56 | 33 | 25 |
| 6 | 0 | 12 | 52 | 1 | 13 | 58 | 1 | 57 | 11 | 2 | 9 | 51 | 1 | 47 | 46 | 0 | 54 | 25 | 24 |
| 7 | 0 | 15 | 2 | 1 | 15 | 41 | 1 | 58 | 2 | 2 | 9 | 36 | 1 | 46 | 20 | 0 | 52 | 17 | 23 |
| 8 | 0 | 17 | 10 | 1 | 17 | 24 | 1 | 58 | 52 | 2 | 9 | 20 | 1 | 44 | 53 | 0 | 50 | 9 | 22 |
| 9 | 0 | 19 | 19 | 1 | 19 | 6 | 1 | 59 | 41 | 2 | 9 | 2 | 1 | 43 | 26 | 0 | 48 | 1 | 21 |
| 10 | 0 | 21 | 28 | 1 | 20 | 48 | 2 | 0 | 26 | 2 | 8 | 45 | 1 | 41 | 57 | 0 | 45 | 53 | 20 |
| 11 | 0 | 23 | 36 | 1 | 22 | 29 | 2 | 1 | 16 | 2 | 8 | 25 | 1 | 40 | 27 | 0 | 43 | 44 | 19 |
| 12 | 0 | 25 | 45 | 1 | 24 | 10 | 2 | 2 | 2 | 2 | 8 | 6 | 1 | 38 | 57 | 0 | 41 | 35 | 18 |
| 13 | 0 | 27 | 53 | 1 | 25 | 50 | 2 | 2 | 42 | 2 | 7 | 41 | 1 | 37 | 25 | 0 | 39 | 26 | 17 |
| 14 | 0 | 30 | 1 | 1 | 27 | 29 | 2 | 3 | 21 | 2 | 7 | 14 | 1 | 35 | 53 | 0 | 37 | 16 | 16 |
| 15 | 0 | 32 | 8 | 1 | 29 | 8 | 2 | 3 | 59 | 2 | 6 | 46 | 1 | 34 | 20 | 0 | 35 | 6 | 15 |
| 16 | 0 | 34 | 16 | 1 | 30 | 46 | 2 | 4 | 36 | 2 | 6 | 18 | 1 | 32 | 46 | 0 | 32 | 51 | 14 |
| 17 | 0 | 36 | 23 | 1 | 32 | 23 | 2 | 5 | 16 | 2 | 5 | 48 | 1 | 31 | 12 | 0 | 30 | 35 | 13 |
| 18 | 0 | 38 | 30 | 1 | 33 | 59 | 2 | 5 | 48 | 2 | 5 | 18 | 1 | 29 | 37 | 0 | 28 | 19 | 12 |
| 19 | 0 | 40 | 37 | 1 | 35 | 30 | 2 | 6 | 17 | 2 | 4 | 42 | 1 | 27 | 50 | 0 | 26 | 1 | 11 |
| 20 | 0 | 42 | 43 | 1 | 37 | 0 | 2 | 6 | 45 | 2 | 4 | 5 | 1 | 26 | 3 | 0 | 23 | 42 | 10 |
| 21 | 0 | 44 | 49 | 1 | 38 | 30 | 2 | 7 | 12 | 2 | 3 | 27 | 1 | 24 | 16 | 0 | 21 | 22 | 9 |
| 22 | 0 | 46 | 55 | 1 | 39 | 58 | 2 | 7 | 37 | 2 | 2 | 37 | 1 | 22 | 28 | 0 | 19 | 1 | 8 |

| Tabula equacionis Solis | | | | | | | | | | | | | | | | | | |
|---|---|---|---|---|---|---|---|---|---|---|---|---|---|---|---|---|---|---|
| minue | | 0 | | | 1 | | | 2 | | | 3 | | | 4 | | | 5 | | |
| **Gradus** | **Equacio** | | | **Equacio** | | | **Equacio** | | | **Equacio** | | | **Equacio** | | | **Equacio** | | | **Gradus** |
| | **G** | **M** | **2** | **G** | **M** | **2** | **G** | **M** | **2** | **G** | **M** | **2** | **G** | **M** | **2** | **G** | **M** | **2** | |
| 23 | 0 | 48 | 59 | 1 | 41 | 27 | 2 | 8 | 2 | 2 | 1 | 45 | 1 | 20 | 40 | 0 | 16 | 40 | 7 |
| 24 | 0 | 51 | 4 | 1 | 42 | 54 | 2 | 8 | 27 | 2 | 0 | 51 | 1 | 18 | 51 | 0 | 14 | 19 | 6 |
| 25 | 0 | 53 | 4 | 1 | 44 | 14 | 2 | 8 | 45 | 1 | 59 | 53 | 1 | 17 | 0 | 0 | 11 | 58 | 5 |
| 26 | 0 | 55 | 2 | 1 | 45 | 34 | 2 | 9 | 1 | 1 | 58 | 55 | 1 | 15 | 8 | 0 | 9 | 36 | 4 |
| 27 | 0 | 57 | 1 | 1 | 46 | 53 | 2 | 9 | 17 | 1 | 57 | 57 | 1 | 13 | 13 | 0 | 7 | 12 | 3 |
| 28 | 0 | 58 | 59 | 1 | 48 | 10 | 2 | 9 | 32 | 1 | 56 | 57 | 1 | 11 | 16 | 0 | 4 | 48 | 2 |
| 29 | 1 | 0 | 57 | 1 | 49 | 28 | 2 | 9 | 45 | 1 | 55 | 57 | 1 | 9 | 10 | 0 | 2 | 24 | 1 |
| 30 | 1 | 2 | 54 | 1 | 50 | 54 | 2 | 9 | 57 | 1 | 54 | 57 | 1 | 7 | 7 | 0 | 0 | 0 | 0 |
| adde | | 11 | | | 10 | | | 9 | | | 8 | | | 7 | | | 6 | | adde |
| | | A | | | B | | | C | | | D | | | E | | | F | | |

Table TRW-14 of true motion of the Sun and Moon in one hour (Toruń, BU, 74/III, f. 259r)

| Tabula veri motus Solis et Lune in una hora | | | | | | | | | | | | | | | | | | | | | | | | | |
|---|---|---|---|---|---|---|---|---|---|---|---|---|---|---|---|---|---|---|---|---|---|---|---|---|---|
| | 0 | | | | 1 | | | | 2 | | | | 3 | | | | 4 | | | | 5 | | | | |
| | Motus Solis | | Motus Lune | | Motus Solis | | Motus Lune | | Motus Solis | | Motus Lune | | Motus Solis | | Motus Lune | | Motus Solis | | Motus Lune | | Motus Solis | | Motus Lune | | |
| G | M | 2 | M | 2 | M | 2 | M | 2 | M | 2 | M | 2 | M | 2 | M | 2 | M | 2 | M | 2 | M | 2 | M | 2 | G |
| 1 | 2 | 23 | 30 | 18 | 2 | 24 | 30 | 37 | 2 | 25 | 31 | 26 | 2 | 28 | 32 | 45 | 2 | 30 | 34 | 17 | 2 | 32 | 35 | 33 | 29 |
| 2 | 2 | 23 | 30 | 18 | 2 | 24 | 30 | 38 | 2 | 25 | 31 | 29 | 2 | 28 | 32 | 48 | 2 | 30 | 34 | 20 | 2 | 32 | 35 | 35 | 28 |
| 3 | 2 | 23 | 30 | 18 | 2 | 24 | 30 | 40 | 2 | 26 | 31 | 31 | 2 | 28 | 32 | 51 | 2 | 30 | 34 | 23 | 2 | 32 | 35 | 37 | 27 |
| 4 | 2 | 23 | 30 | 19 | 2 | 24 | 30 | 42 | 2 | 26 | 31 | 33 | 2 | 28 | 32 | 53 | 2 | 30 | 34 | 26 | 2 | 32 | 35 | 39 | 26 |
| 5 | 2 | 23 | 30 | 19 | 2 | 24 | 30 | 43 | 2 | 26 | 31 | 36 | 2 | 28 | 32 | 56 | 2 | 30 | 34 | 29 | 2 | 32 | 35 | 41 | 25 |
| 6 | 2 | 23 | 30 | 19 | 2 | 24 | 30 | 44 | 2 | 26 | 31 | 38 | 2 | 28 | 32 | 59 | 2 | 30 | 34 | 32 | 2 | 32 | 35 | 43 | 24 |
| 7 | 2 | 23 | 30 | 20 | 2 | 24 | 30 | 45 | 2 | 26 | 31 | 41 | 2 | 28 | 33 | 4 | 2 | 30 | 34 | 35 | 2 | 32 | 35 | 45 | 23 |
| 8 | 2 | 23 | 30 | 20 | 2 | 24 | 30 | 46 | 2 | 26 | 31 | 43 | 2 | 28 | 33 | 8 | 2 | 30 | 34 | 38 | 2 | 32 | 35 | 47 | 22 |
| 9 | 2 | 23 | 30 | 20 | 2 | 24 | 30 | 47 | 2 | 26 | 31 | 46 | 2 | 29 | 33 | 13 | 2 | 30 | 34 | 41 | 2 | 33 | 35 | 48 | 21 |
| 10 | 2 | 23 | 30 | 21 | 2 | 24 | 30 | 48 | 2 | 26 | 31 | 48 | 2 | 29 | 33 | 18 | 2 | 30 | 34 | 43 | 2 | 33 | 35 | 49 | 20 |
| 11 | 2 | 23 | 30 | 21 | 2 | 24 | 30 | 50 | 2 | 26 | 31 | 51 | 2 | 29 | 33 | 22 | 2 | 30 | 34 | 46 | 2 | 33 | 35 | 51 | 19 |
| 12 | 2 | 23 | 30 | 21 | 2 | 24 | 30 | 51 | 2 | 26 | 31 | 53 | 2 | 29 | 33 | 27 | 2 | 30 | 34 | 49 | 2 | 33 | 35 | 52 | 18 |
| 13 | 2 | 23 | 30 | 22 | 2 | 24 | 30 | 53 | 2 | 26 | 31 | 56 | 2 | 29 | 33 | 29 | 2 | 30 | 34 | 52 | 2 | 33 | 35 | 52 | 17 |
| 14 | 2 | 23 | 30 | 22 | 2 | 24 | 30 | 54 | 2 | 26 | 31 | 58 | 2 | 29 | 33 | 30 | 2 | 30 | 34 | 54 | 2 | 33 | 35 | 53 | 16 |
| 15 | 2 | 23 | 30 | 23 | 2 | 25 | 30 | 58 | 2 | 27 | 32 | 1 | 2 | 29 | 33 | 32 | 2 | 30 | 34 | 57 | 2 | 33 | 35 | 53 | 15 |
| 16 | 2 | 23 | 30 | 23 | 2 | 25 | 31 | 0 | 2 | 27 | 32 | 3 | 2 | 29 | 33 | 33 | 2 | 30 | 34 | 59 | 2 | 33 | 35 | 54 | 14 |
| 17 | 2 | 23 | 30 | 24 | 2 | 25 | 31 | 0 | 2 | 27 | 32 | 6 | 2 | 29 | 33 | 35 | 2 | 30 | 35 | 2 | 2 | 33 | 35 | 54 | 13 |
| 18 | 2 | 23 | 30 | 24 | 2 | 25 | 31 | 1 | 2 | 27 | 32 | 8 | 2 | 29 | 33 | 36 | 2 | 31 | 35 | 4 | 2 | 33 | 35 | 54 | 12 |
| 19 | 2 | 23 | 30 | 25 | 2 | 25 | 31 | 3 | 2 | 27 | 32 | 11 | 2 | 29 | 33 | 39 | 2 | 31 | 35 | 7 | 2 | 33 | 35 | 55 | 11 |
| 20 | 2 | 23 | 30 | 25 | 2 | 25 | 31 | 5 | 2 | 27 | 32 | 14 | 2 | 29 | 33 | 42 | 2 | 31 | 35 | 9 | 2 | 33 | 35 | 57 | 10 |
| 21 | 2 | 23 | 30 | 26 | 2 | 25 | 31 | 7 | 2 | 27 | 32 | 17 | 2 | 29 | 33 | 46 | 2 | 31 | 35 | 11 | 2 | 33 | 35 | 58 | 9 |

| Tabula veri motus Solis et Lune in una hora | | | | | | | | | | | | | | | | | | | | | | | | | |
|---|---|---|---|---|---|---|---|---|---|---|---|---|---|---|---|---|---|---|---|---|---|---|---|---|---|
| | 0 | | | | 1 | | | | 2 | | | | 3 | | | | 4 | | | | 5 | | | | |
| | Motus Solis | | Motus Lune | | Motus Solis | | Motus Lune | | Motus Solis | | Motus Lune | | Motus Solis | | Motus Lune | | Motus Solis | | Motus Lune | | Motus Solis | | Motus Lune | | |
| G | M | 2 | M | 2 | M | 2 | M | 2 | M | 2 | M | 2 | M | 2 | M | 2 | M | 2 | M | 2 | M | 2 | M | 2 | G |
| 22 | 2 | 23 | 30 | 27 | 2 | 25 | 31 | 8 | 2 | 27 | 32 | 19 | 2 | 29 | 33 | 49 | 2 | 31 | 35 | 13 | 2 | 33 | 35 | 59 | 8 |
| 23 | 2 | 23 | 30 | 27 | 2 | 25 | 31 | 10 | 2 | 27 | 32 | 23 | 2 | 29 | 33 | 54 | 2 | 31 | 35 | 16 | 2 | 33 | 36 | 1 | 7 |
| 24 | 2 | 23 | 30 | 28 | 2 | 25 | 31 | 12 | 2 | 27 | 32 | 25 | 2 | 29 | 33 | 55 | 2 | 31 | 35 | 18 | 2 | 33 | 36 | 2 | 6 |
| 25 | 2 | 23 | 30 | 29 | 2 | 25 | 31 | 14 | 2 | 27 | 32 | 26 | 2 | 29 | 33 | 58 | 2 | 31 | 35 | 20 | 2 | 33 | 36 | 2 | 5 |
| 26 | 2 | 23 | 30 | 31 | 2 | 25 | 31 | 16 | 2 | 27 | 32 | 31 | 2 | 29 | 34 | 1 | 2 | 31 | 35 | 23 | 2 | 33 | 36 | 3 | 4 |
| 27 | 2 | 24 | 30 | 32 | 2 | 25 | 31 | 18 | 2 | 28 | 32 | 34 | 2 | 29 | 34 | 5 | 2 | 31 | 35 | 25 | 2 | 33 | 36 | 3 | 3 |
| 28 | 2 | 24 | 30 | 33 | 2 | 25 | 31 | 20 | 2 | 28 | 32 | 36 | 2 | 29 | 34 | 8 | 2 | 31 | 35 | 27 | 2 | 33 | 36 | 3 | 2 |
| 29 | 2 | 24 | 30 | 35 | 2 | 25 | 31 | 22 | 2 | 28 | 32 | 38 | 2 | 29 | 34 | 11 | 2 | 31 | 35 | 29 | 2 | 33 | 36 | 4 | 1 |
| 30 | 2 | 24 | 30 | 36 | 2 | 25 | 31 | 24 | 2 | 28 | 32 | 42 | 2 | 29 | 34 | 14 | 2 | 31 | 35 | 31 | 2 | 33 | 36 | 4 | 0 |
| | 11 | | | | 10 | | | | 9 | | | | 8 | | | | 7 | | | | 6 | | | | |
| | A | | B | | C | | D | | E | | F | | G | | H | | I | | J | | K | | L | | |

Table TRW-15 of equations of the Moon (Toruń, BU, 74/III, f. 259v)

| Tabula equacionis Lune | | | | | | | | | | | | | | | | | |
|---|---|---|---|---|---|---|---|---|---|---|---|---|---|---|---|---|---|
| | 0 | | | | | | | | 1 | | | | | | | | |
| Gradus | Equacio centri | | Minuta propor-cionalia | Diversitas dyametri | | Equacio argumenti | | | Equacio centri | | Minuta propor-cionalia | Diversitas dyametri | | Equacio argumenti | | | Gradus |
| | G | M | | G | M | G | M | 2 | G | M | | G | M | G | M | 2 | |
| 1 | 0 | 9 | 0 | 0 | 3 | 0 | 4 | 46 | 4 | 32 | 3 | 1 | 12 | 2 | 21 | 43 | 29 |
| 2 | 0 | 18 | 0 | 0 | 5 | 0 | 9 | 31 | 4 | 41 | 3 | 1 | 14 | 2 | 25 | 55 | 28 |
| 3 | 0 | 27 | 0 | 0 | 7 | 0 | 14 | 15 | 4 | 49 | 4 | 1 | 16 | 2 | 30 | 5 | 27 |
| 4 | 0 | 36 | 0 | 0 | 10 | 0 | 19 | 0 | 4 | 58 | 4 | 1 | 19 | 2 | 34 | 12 | 26 |
| 5 | 0 | 45 | 0 | 0 | 12 | 0 | 23 | 44 | 5 | 7 | 4 | 1 | 21 | 2 | 38 | 17 | 25 |
| 6 | 0 | 53 | 0 | 0 | 14 | 0 | 28 | 28 | 5 | 15 | 4 | 1 | 23 | 2 | 42 | 21 | 24 |
| 7 | 1 | 2 | 0 | 0 | 17 | 0 | 33 | 11 | 5 | 24 | 5 | 1 | 25 | 2 | 46 | 22 | 23 |
| 8 | 1 | 11 | 0 | 0 | 19 | 0 | 37 | 54 | 5 | 33 | 5 | 1 | 27 | 2 | 50 | 19 | 22 |
| 9 | 1 | 20 | 0 | 0 | 21 | 0 | 42 | 37 | 5 | 41 | 5 | 1 | 29 | 2 | 54 | 14 | 21 |
| 10 | 1 | 29 | 0 | 0 | 24 | 0 | 47 | 19 | 5 | 50 | 5 | 1 | 31 | 2 | 58 | 7 | 20 |
| 11 | 1 | 38 | 0 | 0 | 26 | 0 | 52 | 0 | 5 | 59 | 6 | 1 | 33 | 3 | 1 | 58 | 19 |
| 12 | 1 | 46 | 1 | 0 | 28 | 0 | 56 | 41 | 6 | 7 | 6 | 1 | 35 | 3 | 5 | 46 | 18 |
| 13 | 1 | 55 | 1 | 0 | 31 | 1 | 1 | 20 | 6 | 16 | 6 | 1 | 37 | 3 | 9 | 31 | 17 |
| 14 | 2 | 4 | 1 | 0 | 33 | 1 | 5 | 59 | 6 | 25 | 7 | 1 | 39 | 3 | 13 | 13 | 16 |
| 15 | 2 | 13 | 1 | 0 | 35 | 1 | 10 | 38 | 6 | 33 | 7 | 1 | 40 | 3 | 16 | 51 | 15 |
| 16 | 2 | 22 | 1 | 0 | 38 | 1 | 15 | 15 | 6 | 42 | 7 | 1 | 42 | 3 | 20 | 26 | 14 |
| 17 | 2 | 31 | 1 | 0 | 40 | 1 | 19 | 51 | 6 | 50 | 8 | 1 | 44 | 3 | 23 | 59 | 13 |
| 18 | 2 | 39 | 1 | 0 | 42 | 1 | 24 | 27 | 6 | 58 | 8 | 1 | 45 | 3 | 27 | 30 | 12 |
| 19 | 2 | 48 | 1 | 0 | 45 | 1 | 29 | 0 | 7 | 7 | 8 | 1 | 47 | 3 | 30 | 57 | 11 |
| 20 | 2 | 57 | 2 | 0 | 47 | 1 | 33 | 32 | 7 | 15 | 9 | 1 | 48 | 3 | 34 | 20 | 10 |

| Tabula equacionis Lune | | | | | | | | | | | | | | | | | |
|---|---|---|---|---|---|---|---|---|---|---|---|---|---|---|---|---|---|
| | 0 | | | | | | | | 1 | | | | | | | | |
| Gradus | Equacio centri | | Minuta propor-cionalia | Diversitas dyametri | | Equacio argumenti | | | Equacio centri | | Minuta propor-cionalia | Diversitas dyametri | | Equacio argumenti | | | Gradus |
| | G | M | | G | M | G | M | 2 | G | M | | G | M | G | M | 2 | |
| 21 | 3 | 5 | 2 | 0 | 49 | 1 | 38 | 3 | 7 | 23 | 9 | 1 | 49 | 3 | 37 | 40 | 9 |
| 22 | 3 | 14 | 2 | 0 | 52 | 1 | 42 | 33 | 7 | 32 | 9 | 1 | 51 | 3 | 40 | 57 | 8 |
| 23 | 3 | 23 | 2 | 0 | 54 | 1 | 47 | 1 | 7 | 40 | 10 | 1 | 53 | 3 | 44 | 10 | 7 |
| 24 | 3 | 31 | 2 | 0 | 57 | 1 | 51 | 27 | 7 | 48 | 10 | 1 | 54 | 3 | 47 | 20 | 6 |
| 25 | 3 | 40 | 2 | 0 | 59 | 1 | 55 | 52 | 7 | 56 | 10 | 1 | 56 | 3 | 50 | 26 | 5 |
| 26 | 3 | 49 | 2 | 1 | 1 | 2 | 0 | 15 | 8 | 4 | 11 | 1 | 58 | 3 | 53 | 29 | 4 |
| 27 | 3 | 57 | 3 | 1 | 3 | 2 | 4 | 37 | 8 | 12 | 11 | 1 | 59 | 3 | 56 | 30 | 3 |
| 28 | 4 | 6 | 3 | 1 | 6 | 2 | 8 | 57 | 8 | 20 | 11 | 2 | 1 | 3 | 59 | 26 | 2 |
| 29 | 4 | 15 | 3 | 1 | 8 | 2 | 13 | 14 | 8 | 28 | 12 | 2 | 2 | 4 | 2 | 17 | 1 |
| 30 | 4 | 23 | 3 | 1 | 10 | 2 | 17 | 29 | 8 | 36 | 12 | 2 | 3 | 4 | 5 | 4 | 0 |
| | 11 | | | | | | | | 10 | | | | | | | | |

C M D A C M D A

Continuation of the Table TRW-15 (Toruń, BU, 74/III, f. 260r)

| Tabula equacionis Lune | | | | | | | | | | | | | | | | | |
|---|---|---|---|---|---|---|---|---|---|---|---|---|---|---|---|---|---|
| | 2 | | | | | | | | 3 | | | | | | | | |
| Gradus | Equacio centri | | Minuta propor-cionalia | Diversitas dyametri | | Equacio argumenti | | | Equacio centri | | Minuta propor-cionalia | Diversitas dyametri | | Equacio argumenti | | | Gradus |
| | G | M | | G | M | G | M | 2 | G | M | | G | M | G | M | 2 | |
| 1 | 8 | 44 | 13 | 2 | 5 | 4 | 7 | 47 | 12 | 5 | 27 | 2 | 37 | 4 | 55 | 18 | 29 |
| 2 | 8 | 52 | 13 | 2 | 6 | 4 | 10 | 27 | 12 | 10 | 27 | 2 | 37 | 4 | 55 | 37 | 28 |
| 3 | 8 | 59 | 14 | 2 | 7 | 4 | 13 | 3 | 12 | 15 | 28 | 2 | 38 | 4 | 55 | 49 | 27 |
| 4 | 9 | 7 | 14 | 2 | 9 | 4 | 15 | 35 | 12 | 20 | 28 | 2 | 38 | 4 | 55 | 55 | 26 |
| 5 | 9 | 15 | 15 | 2 | 10 | 4 | 18 | 3 | 12 | 24 | 29 | 2 | 38 | 4 | 55 | 0 | 25 |
| 6 | 9 | 22 | 15 | 2 | 12 | 4 | 20 | 27 | 12 | 28 | 30 | 2 | 38 | 4 | 55 | 56 | 24 |
| 7 | 9 | 30 | 15 | 2 | 13 | 4 | 22 | 47 | 12 | 32 | 30 | 2 | 38 | 4 | 55 | 43 | 23 |
| 8 | 9 | 37 | 16 | 2 | 14 | 4 | 25 | 2 | 12 | 36 | 31 | 2 | 39 | 4 | 55 | 25 | 22 |
| 9 | 9 | 44 | 16 | 2 | 15 | 4 | 27 | 12 | 12 | 39 | 31 | 2 | 39 | 4 | 55 | 4 | 21 |
| 10 | 9 | 52 | 17 | 2 | 16 | 4 | 29 | 18 | 12 | 42 | 32 | 2 | 39 | 4 | 54 | 41 | 20 |
| 11 | 9 | 59 | 17 | 2 | 17 | 4 | 31 | 20 | 12 | 45 | 32 | 2 | 39 | 4 | 54 | 12 | 19 |
| 12 | 10 | 6 | 18 | 2 | 18 | 4 | 33 | 18 | 12 | 48 | 33 | 2 | 39 | 4 | 53 | 38 | 18 |
| 13 | 10 | 13 | 18 | 2 | 19 | 4 | 35 | 11 | 12 | 51 | 33 | 2 | 40 | 4 | 52 | 59 | 17 |
| 14 | 10 | 20 | 19 | 2 | 20 | 4 | 36 | 59 | 12 | 54 | 34 | 2 | 40 | 4 | 52 | 14 | 16 |
| 15 | 10 | 27 | 19 | 2 | 21 | 4 | 38 | 43 | 12 | 56 | 35 | 2 | 40 | 4 | 51 | 22 | 15 |
| 16 | 10 | 34 | 20 | 2 | 22 | 4 | 40 | 23 | 12 | 58 | 35 | 2 | 40 | 4 | 50 | 22 | 14 |
| 17 | 10 | 41 | 20 | 2 | 23 | 4 | 41 | 58 | 13 | 0 | 36 | 2 | 40 | 4 | 49 | 17 | 13 |
| 18 | 10 | 48 | 21 | 2 | 24 | 4 | 43 | 28 | 13 | 2 | 36 | 2 | 40 | 4 | 48 | 10 | 12 |
| 19 | 10 | 55 | 21 | 2 | 25 | 4 | 44 | 53 | 13 | 4 | 37 | 2 | 40 | 4 | 46 | 54 | 11 |
| 20 | 11 | 2 | 22 | 2 | 26 | 4 | 46 | 13 | 13 | 5 | 37 | 2 | 39 | 4 | 45 | 33 | 10 |

| Tabula equacionis Lune | | | | | | | | | | | | | | | | | |
|---|---|---|---|---|---|---|---|---|---|---|---|---|---|---|---|---|---|
| | 2 | | | | | | | | 3 | | | | | | | | |
| **Gradus** | **Equacio centri** | | **Minuta propor-cionalia** | **Diversitas dyametri** | | **Equacio argumenti** | | | **Equacio centri** | | **Minuta propor-cionalia** | **Diversitas dyametri** | | **Equacio argumenti** | | | **Gradus** |
| | **G** | **M** | | **G** | **M** | **G** | **M** | **2** | **G** | **M** | | **G** | **M** | **G** | **M** | **2** | |
| 21 | 11 | 8 | 22 | 2 | 27 | 4 | 47 | 26 | 13 | 6 | 38 | 2 | 39 | 4 | 44 | 7 | 9 |
| 22 | 11 | 15 | 23 | 2 | 28 | 4 | 48 | 35 | 13 | 7 | 38 | 2 | 38 | 4 | 42 | 34 | 8 |
| 23 | 11 | 21 | 23 | 2 | 29 | 4 | 49 | 38 | 13 | 8 | 39 | 2 | 38 | 4 | 40 | 56 | 7 |
| 24 | 11 | 27 | 23 | 2 | 30 | 4 | 50 | 41 | 13 | 9 | 39 | 2 | 37 | 4 | 39 | 15 | 6 |
| 25 | 11 | 33 | 24 | 2 | 31 | 4 | 51 | 38 | 13 | 9 | 40 | 2 | 36 | 4 | 37 | 29 | 5 |
| 26 | 11 | 39 | 24 | 2 | 32 | 4 | 52 | 28 | 13 | 8 | 40 | 2 | 36 | 4 | 35 | 37 | 4 |
| 27 | 11 | 44 | 25 | 2 | 33 | 4 | 53 | 11 | 13 | 7 | 41 | 2 | 35 | 4 | 33 | 41 | 3 |
| 28 | 11 | 50 | 25 | 2 | 34 | 4 | 53 | 50 | 13 | 6 | 41 | 2 | 34 | 4 | 31 | 44 | 2 |
| 29 | 11 | 55 | 26 | 2 | 35 | 4 | 54 | 25 | 13 | 5 | 42 | 2 | 33 | 4 | 29 | 20 | 1 |
| 30 | 12 | 0 | 26 | 2 | 36 | 4 | 54 | 54 | 13 | 4 | 42 | 2 | 31 | 4 | 27 | 0 | 0 |
| | 9 | | | | | | | | 8 | | | | | | | | |
| | C | | M | D | | A | | | C | | M | D | | A | | | |

Remaining part of the Table TRW-15 (Toruń, BU, 74/III, f. 260v)

| Tabula equacionis Lune | | | | | | | | | | | | | | | | | |
|---|---|---|---|---|---|---|---|---|---|---|---|---|---|---|---|---|---|
| | 4 | | | | | | | | 5 | | | | | | | | |
| Gradus | Equacio centri | | Minuta propor-cionalia | Diversitas dyametri | | Equacio argumenti | | | Equacio centri | | Minuta propor-cionalia | Diversitas dyametri | | Equacio argumenti | | | Gradus |
| | G | M | | G | M | G | M | 2 | G | M | | G | M | G | M | 2 | |
| 1 | 13 | 3 | 43 | 2 | 30 | 4 | 24 | 38 | 9 | 8 | 56 | 1 | 32 | 2 | 34 | 52 | 29 |
| 2 | 13 | 1 | 44 | 2 | 29 | 4 | 22 | 11 | 8 | 53 | 56 | 1 | 29 | 2 | 30 | 6 | 28 |
| 3 | 12 | 59 | 44 | 2 | 27 | 4 | 19 | 38 | 8 | 38 | 56 | 1 | 26 | 2 | 25 | 16 | 27 |
| 4 | 12 | 56 | 45 | 2 | 26 | 4 | 16 | 58 | 8 | 22 | 56 | 1 | 24 | 2 | 20 | 23 | 26 |
| 5 | 12 | 53 | 45 | 2 | 25 | 4 | 14 | 13 | 8 | 5 | 57 | 1 | 21 | 2 | 15 | 26 | 25 |
| 6 | 12 | 50 | 45 | 2 | 23 | 4 | 11 | 23 | 7 | 48 | 57 | 1 | 18 | 2 | 10 | 26 | 24 |
| 7 | 12 | 46 | 46 | 2 | 22 | 4 | 8 | 28 | 7 | 31 | 57 | 1 | 16 | 2 | 5 | 22 | 23 |
| 8 | 12 | 41 | 46 | 2 | 21 | 4 | 5 | 31 | 7 | 14 | 57 | 1 | 13 | 2 | 0 | 17 | 22 |
| 9 | 12 | 36 | 47 | 2 | 19 | 4 | 2 | 30 | 6 | 56 | 57 | 1 | 10 | 1 | 55 | 9 | 21 |
| 10 | 12 | 30 | 47 | 2 | 18 | 3 | 59 | 20 | 6 | 39 | 58 | 1 | 8 | 1 | 49 | 58 | 20 |
| 11 | 12 | 23 | 47 | 2 | 17 | 3 | 56 | 5 | 6 | 21 | 58 | 1 | 5 | 1 | 44 | 44 | 19 |
| 12 | 12 | 16 | 48 | 2 | 15 | 3 | 52 | 47 | 6 | 3 | 58 | 1 | 2 | 1 | 39 | 27 | 18 |
| 13 | 12 | 9 | 48 | 2 | 14 | 3 | 49 | 23 | 5 | 45 | 58 | 0 | 59 | 1 | 34 | 9 | 17 |
| 14 | 12 | 2 | 49 | 2 | 12 | 3 | 45 | 52 | 5 | 27 | 58 | 0 | 56 | 1 | 28 | 49 | 16 |
| 15 | 11 | 54 | 49 | 2 | 10 | 3 | 42 | 17 | 5 | 8 | 58 | 0 | 52 | 1 | 23 | 26 | 15 |
| 16 | 11 | 46 | 50 | 2 | 9 | 3 | 38 | 37 | 4 | 49 | 59 | 0 | 49 | 1 | 18 | 1 | 14 |
| 17 | 11 | 38 | 50 | 2 | 7 | 3 | 34 | 53 | 4 | 30 | 59 | 0 | 46 | 1 | 12 | 34 | 13 |
| 18 | 11 | 29 | 51 | 2 | 5 | 3 | 31 | 3 | 4 | 11 | 59 | 0 | 42 | 1 | 7 | 6 | 12 |
| 19 | 11 | 20 | 51 | 2 | 3 | 3 | 27 | 10 | 3 | 52 | 59 | 0 | 39 | 1 | 1 | 36 | 11 |
| 20 | 11 | 11 | 52 | 2 | 1 | 3 | 23 | 12 | 3 | 32 | 59 | 0 | 36 | 0 | 56 | 5 | 10 |

| Tabula equacionis Lune | | | | | | | | | | | | | | | | | |
|---|---|---|---|---|---|---|---|---|---|---|---|---|---|---|---|---|---|
| | 4 | | | | | | | | 5 | | | | | | | | |
| Gradus | Equacio centri | | Minuta propor-cionalia | Diversitas dyametri | | Equacio argumenti | | | Equacio centri | | Minuta propor-cionalia | Diversitas dyametri | | Equacio argumenti | | | Gradus |
| | G | M | | G | M | G | M | 2 | G | M | | G | M | G | M | 2 | |
| 21 | 11 | 2 | 52 | 1 | 58 | 3 | 19 | 9 | 3 | 12 | 59 | 0 | 32 | 0 | 50 | 32 | 9 |
| 22 | 10 | 52 | 53 | 1 | 56 | 3 | 15 | 2 | 2 | 52 | 59 | 0 | 29 | 0 | 44 | 58 | 8 |
| 23 | 10 | 43 | 53 | 1 | 54 | 3 | 10 | 50 | 2 | 32 | 60 | 0 | 25 | 0 | 39 | 23 | 7 |
| 24 | 10 | 33 | 53 | 1 | 51 | 3 | 6 | 35 | 2 | 11 | 60 | 0 | 21 | 0 | 33 | 47 | 6 |
| 25 | 10 | 22 | 54 | 1 | 49 | 3 | 2 | 15 | 1 | 50 | 60 | 0 | 18 | 0 | 28 | 10 | 5 |
| 26 | 10 | 11 | 54 | 1 | 46 | 2 | 57 | 51 | 1 | 29 | 60 | 0 | 15 | 0 | 22 | 33 | 4 |
| 27 | 10 | 0 | 54 | 1 | 43 | 2 | 53 | 23 | 1 | 7 | 60 | 0 | 11 | 0 | 16 | 56 | 3 |
| 28 | 9 | 48 | 55 | 1 | 41 | 2 | 48 | 51 | 0 | 45 | 60 | 0 | 8 | 0 | 11 | 18 | 2 |
| 29 | 9 | 35 | 55 | 1 | 38 | 2 | 44 | 15 | 0 | 23 | 60 | 0 | 4 | 0 | 5 | 40 | 1 |
| 30 | 9 | 22 | 55 | 1 | 35 | 2 | 39 | 35 | 0 | 0 | 60 | 0 | 0 | 0 | 0 | 0 | 0 |
| | 7 | | | | | | | | 6 | | | | | | | | |
| | C | | M | D | | A | | | C | | M | D | | A | | | |

Table TRW-16 of equations of Saturn (Toruń, BU, 74/III, f. 261r)

| Tabula equacionis Saturni | | | | | | | | | | | | | | | | | | | | | | | | | |
|---|---|---|---|---|---|---|---|---|---|---|---|---|---|---|---|---|---|---|---|---|---|---|---|---|---|
| | 0 | | | | | | | | | | | | 1 | | | | | | | | | | | | |
| Gradus | Equacio centri | | Minuta propor-cionalia | Longitudo longior | | Equacio argumenti | | Longitudo propior | | Stacio prima | | | Equacio centri | | Minuta propor-cionalia | Longitudo longior | | Equacio argumenti | | Longitudo propior | | Stacio prima | | | Gradus |
| | G | M | Longior | G | M | G | M | G | M | S | G | M | G | M | Longior | G | M | G | M | G | M | S | G | M | |
| 1 | 0 | 7 | 60 | 0 | 0 | 0 | 6 | 0 | 0 | 3 | 22 | 44 | 3 | 12 | 50 | 0 | 9 | 2 | 55 | 0 | 12 | 3 | 22 | 57 | 29 |
| 2 | 0 | 14 | 60 | 0 | 1 | 0 | 12 | 0 | 1 | 3 | 22 | 44 | 3 | 18 | 50 | 0 | 9 | 3 | 0 | 0 | 12 | 3 | 22 | 58 | 28 |
| 3 | 0 | 20 | 60 | 0 | 1 | 0 | 18 | 0 | 1 | 3 | 22 | 44 | 3 | 23 | 49 | 0 | 9 | 3 | 5 | 0 | 12 | 3 | 22 | 59 | 27 |
| 4 | 0 | 27 | 60 | 0 | 1 | 0 | 24 | 0 | 1 | 3 | 22 | 45 | 3 | 29 | 49 | 0 | 9 | 3 | 10 | 0 | 12 | 3 | 23 | 0 | 26 |
| 5 | 0 | 33 | 60 | 0 | 2 | 0 | 30 | 0 | 2 | 3 | 22 | 45 | 3 | 34 | 48 | 0 | 9 | 3 | 15 | 0 | 13 | 3 | 23 | 1 | 25 |
| 6 | 0 | 40 | 60 | 0 | 2 | 0 | 36 | 0 | 2 | 3 | 22 | 45 | 3 | 39 | 48 | 0 | 10 | 3 | 20 | 0 | 13 | 3 | 23 | 2 | 24 |
| 7 | 0 | 46 | 60 | 0 | 2 | 0 | 42 | 0 | 2 | 3 | 22 | 46 | 3 | 45 | 47 | 0 | 10 | 3 | 25 | 0 | 13 | 3 | 23 | 3 | 23 |
| 8 | 0 | 52 | 60 | 0 | 3 | 0 | 48 | 0 | 3 | 3 | 22 | 46 | 3 | 50 | 46 | 0 | 10 | 3 | 30 | 0 | 14 | 3 | 23 | 4 | 22 |
| 9 | 0 | 58 | 59 | 0 | 3 | 0 | 54 | 0 | 3 | 3 | 22 | 46 | 3 | 55 | 46 | 0 | 10 | 3 | 35 | 0 | 14 | 3 | 23 | 5 | 21 |
| 10 | 1 | 5 | 59 | 0 | 3 | 1 | 0 | 0 | 3 | 3 | 22 | 46 | 4 | 0 | 45 | 0 | 10 | 3 | 40 | 0 | 14 | 3 | 23 | 6 | 20 |
| 11 | 1 | 11 | 59 | 0 | 4 | 1 | 6 | 0 | 4 | 3 | 22 | 47 | 4 | 5 | 45 | 0 | 11 | 3 | 45 | 0 | 15 | 3 | 23 | 7 | 19 |
| 12 | 1 | 17 | 59 | 0 | 4 | 1 | 11 | 0 | 4 | 3 | 22 | 47 | 4 | 10 | 44 | 0 | 11 | 3 | 49 | 0 | 15 | 3 | 23 | 8 | 18 |
| 13 | 1 | 24 | 58 | 0 | 4 | 1 | 17 | 0 | 5 | 3 | 22 | 47 | 4 | 15 | 44 | 0 | 11 | 3 | 54 | 0 | 15 | 3 | 23 | 9 | 17 |
| 14 | 1 | 30 | 58 | 0 | 4 | 1 | 23 | 0 | 5 | 3 | 22 | 48 | 4 | 20 | 43 | 0 | 11 | 3 | 59 | 0 | 16 | 3 | 23 | 10 | 16 |
| 15 | 1 | 36 | 58 | 0 | 5 | 1 | 28 | 0 | 6 | 3 | 22 | 48 | 4 | 25 | 42 | 0 | 11 | 4 | 3 | 0 | 16 | 3 | 23 | 11 | 15 |
| 16 | 1 | 43 | 57 | 0 | 5 | 1 | 33 | 0 | 6 | 3 | 22 | 48 | 4 | 30 | 42 | 0 | 12 | 4 | 8 | 0 | 16 | 3 | 23 | 12 | 14 |
| 17 | 1 | 49 | 57 | 0 | 5 | 1 | 40 | 0 | 7 | 3 | 22 | 49 | 4 | 35 | 41 | 0 | 12 | 4 | 13 | 0 | 17 | 3 | 23 | 13 | 13 |
| 18 | 1 | 55 | 56 | 0 | 5 | 1 | 45 | 0 | 7 | 3 | 22 | 49 | 4 | 39 | 40 | 0 | 12 | 4 | 17 | 0 | 17 | 3 | 23 | 14 | 12 |
| 19 | 2 | 1 | 56 | 0 | 6 | 1 | 51 | 0 | 8 | 3 | 22 | 50 | 4 | 44 | 40 | 0 | 12 | 4 | 22 | 0 | 17 | 3 | 23 | 15 | 11 |
| 20 | 2 | 7 | 56 | 0 | 6 | 1 | 57 | 0 | 8 | 3 | 22 | 50 | 4 | 48 | 39 | 0 | 13 | 4 | 26 | 0 | 18 | 3 | 23 | 16 | 10 |
| 21 | 2 | 13 | 55 | 0 | 6 | 2 | 2 | 0 | 8 | 3 | 22 | 51 | 4 | 52 | 38 | 0 | 13 | 4 | 30 | 0 | 18 | 3 | 23 | 17 | 9 |
| 22 | 2 | 19 | 55 | 0 | 6 | 2 | 8 | 0 | 9 | 3 | 22 | 51 | 4 | 56 | 37 | 0 | 13 | 4 | 34 | 0 | 18 | 3 | 23 | 18 | 8 |

| Tabula equacionis Saturni | | | | | | | | | | | | | | | | | | | | | | | | | | |
|---|---|---|---|---|---|---|---|---|---|---|---|---|---|---|---|---|---|---|---|---|---|---|---|---|---|---|
| | 0 | | | | | | | | | | | | 1 | | | | | | | | | | | | | |
| Gradus | Equacio centri | | Minuta propor-cionalia | Longitudo longior | | Equacio argumenti | | Longitudo propior | | Stacio prima | | | Equacio centri | | Minuta propor-cionalia | Longitudo longior | | Equacio argumenti | | Longitudo propior | | Stacio prima | | | Gradus |
| | G | M | Longior | G | M | G | M | G | M | S | G | M | G | M | Longior | G | M | G | M | G | M | S | G | M | |
| 23 | 2 | 25 | 54 | 0 | 7 | 2 | 13 | 0 | 9 | 3 | 22 | 52 | 5 | 1 | 36 | 0 | 13 | 4 | 38 | 0 | 19 | 3 | 23 | 19 | 7 |
| 24 | 2 | 31 | 54 | 0 | 7 | 2 | 18 | 0 | 9 | 3 | 22 | 52 | 5 | 5 | 35 | 0 | 14 | 4 | 42 | 0 | 19 | 3 | 23 | 20 | 6 |
| 25 | 2 | 37 | 53 | 0 | 7 | 2 | 24 | 0 | 10 | 3 | 22 | 53 | 5 | 9 | 35 | 0 | 14 | 4 | 46 | 0 | 19 | 3 | 23 | 21 | 5 |
| 26 | 2 | 43 | 53 | 0 | 7 | 2 | 29 | 0 | 10 | 3 | 22 | 53 | 5 | 13 | 34 | 0 | 14 | 4 | 50 | 0 | 19 | 3 | 23 | 22 | 4 |
| 27 | 2 | 49 | 52 | 0 | 8 | 2 | 34 | 0 | 10 | 3 | 22 | 54 | 5 | 17 | 33 | 0 | 14 | 4 | 53 | 0 | 19 | 3 | 23 | 24 | 3 |
| 28 | 2 | 55 | 52 | 0 | 8 | 2 | 40 | 0 | 11 | 3 | 22 | 54 | 5 | 21 | 32 | 0 | 15 | 4 | 57 | 0 | 19 | 3 | 23 | 25 | 2 |
| 29 | 3 | 1 | 51 | 0 | 8 | 2 | 45 | 0 | 11 | 3 | 22 | 55 | 5 | 25 | 31 | 0 | 15 | 5 | 1 | 0 | 20 | 3 | 23 | 26 | 1 |
| 30 | 3 | 6 | 51 | 0 | 8 | 2 | 50 | 0 | 11 | 3 | 22 | 56 | 5 | 29 | 30 | 0 | 15 | 5 | 4 | 0 | 20 | 3 | 23 | 28 | 0 |
| | 11 | | | | | | | | | | | | 10 | | | | | | | | | | | | | |
| | A | | B | C | | D | | E | | F | | | A | | B | C | | D | | E | | F | | | |

Continuation of the Table TRW-16 (Toruń, BU, 74/III, f. 261v)

| Tabula equacionis Saturni | | | | | | | | | | | | | | | | | | | | | | | | | |
|---|---|---|---|---|---|---|---|---|---|---|---|---|---|---|---|---|---|---|---|---|---|---|---|---|---|
| | 2 | | | | | | | | | | | | 3 | | | | | | | | | | | | |
| Gradus | Equacio centri | | Minuta propor-cionalia | Longitudo longior | | Equacio argumenti | | Longitudo propior | | Stacio prima | | | Equacio centri | | Minuta propor-cionalia | Longitudo longior | | Equacio argumenti | | Longitudo propior | | Stacio prima | | | Gradus |
| | G | M | Longior | G | M | G | M | G | M | S | G | M | G | M | Propior | G | M | G | M | G | M | S | G | M | |
| 1 | 5 | 33 | 29 | 0 | 16 | 5 | 8 | 0 | 20 | 3 | 23 | 29 | 6 | 31 | 4 | 0 | 19 | 6 | 12 | 0 | 23 | 3 | 24 | 12 | 29 |
| 2 | 5 | 37 | 28 | 0 | 16 | 5 | 12 | 0 | 20 | 3 | 23 | 30 | 6 | 31 | 5 | 0 | 20 | 6 | 12 | 0 | 23 | 3 | 24 | 13 | 28 |
| 3 | 5 | 41 | 27 | 0 | 16 | 5 | 15 | 0 | 20 | 3 | 23 | 32 | 6 | 31 | 6 | 0 | 20 | 6 | 12 | 0 | 24 | 3 | 24 | 15 | 27 |
| 4 | 5 | 44 | 26 | 0 | 16 | 5 | 19 | 0 | 20 | 3 | 23 | 33 | 6 | 31 | 7 | 0 | 20 | 6 | 13 | 0 | 24 | 3 | 24 | 16 | 26 |
| 5 | 5 | 47 | 25 | 0 | 16 | 5 | 22 | 0 | 20 | 3 | 23 | 34 | 6 | 30 | 8 | 0 | 20 | 6 | 13 | 0 | 24 | 3 | 24 | 18 | 25 |
| 6 | 5 | 50 | 24 | 0 | 17 | 5 | 25 | 0 | 20 | 3 | 23 | 36 | 6 | 30 | 9 | 0 | 20 | 6 | 13 | 0 | 24 | 3 | 24 | 19 | 24 |
| 7 | 5 | 53 | 23 | 0 | 17 | 5 | 28 | 0 | 21 | 3 | 23 | 37 | 6 | 29 | 10 | 0 | 20 | 6 | 13 | 0 | 24 | 3 | 24 | 20 | 23 |
| 8 | 5 | 56 | 22 | 0 | 17 | 5 | 31 | 0 | 21 | 3 | 23 | 38 | 6 | 29 | 11 | 0 | 20 | 6 | 13 | 0 | 24 | 3 | 24 | 22 | 22 |
| 9 | 5 | 59 | 21 | 0 | 17 | 5 | 34 | 0 | 21 | 3 | 23 | 39 | 6 | 28 | 12 | 0 | 20 | 6 | 13 | 0 | 24 | 3 | 24 | 23 | 21 |
| 10 | 6 | 2 | 20 | 0 | 17 | 5 | 37 | 0 | 21 | 3 | 23 | 41 | 6 | 28 | 13 | 0 | 21 | 6 | 12 | 0 | 24 | 3 | 24 | 24 | 20 |
| 11 | 6 | 5 | 19 | 0 | 17 | 5 | 40 | 0 | 21 | 3 | 23 | 42 | 6 | 27 | 14 | 0 | 21 | 6 | 12 | 0 | 25 | 3 | 24 | 26 | 19 |
| 12 | 6 | 7 | 18 | 0 | 18 | 5 | 42 | 0 | 21 | 3 | 23 | 44 | 6 | 26 | 15 | 0 | 21 | 6 | 12 | 0 | 25 | 3 | 24 | 27 | 18 |
| 13 | 6 | 9 | 16 | 0 | 18 | 5 | 45 | 0 | 21 | 3 | 23 | 45 | 6 | 25 | 15 | 0 | 21 | 6 | 11 | 0 | 25 | 3 | 24 | 28 | 17 |
| 14 | 6 | 12 | 15 | 0 | 18 | 5 | 47 | 0 | 21 | 3 | 23 | 47 | 6 | 24 | 16 | 0 | 21 | 6 | 10 | 0 | 25 | 3 | 24 | 30 | 16 |
| 15 | 6 | 14 | 14 | 0 | 18 | 5 | 49 | 0 | 21 | 3 | 23 | 49 | 6 | 22 | 17 | 0 | 21 | 6 | 9 | 0 | 25 | 3 | 24 | 31 | 15 |
| 16 | 6 | 16 | 13 | 0 | 18 | 5 | 51 | 0 | 21 | 3 | 23 | 51 | 6 | 21 | 18 | 0 | 21 | 6 | 8 | 0 | 25 | 3 | 24 | 32 | 14 |
| 17 | 6 | 18 | 12 | 0 | 18 | 5 | 53 | 0 | 21 | 3 | 23 | 53 | 6 | 19 | 19 | 0 | 21 | 6 | 7 | 0 | 25 | 3 | 24 | 34 | 13 |
| 18 | 6 | 19 | 11 | 0 | 18 | 5 | 55 | 0 | 21 | 3 | 23 | 55 | 6 | 17 | 20 | 0 | 20 | 6 | 5 | 0 | 25 | 3 | 24 | 36 | 12 |
| 19 | 6 | 21 | 9 | 0 | 18 | 5 | 57 | 0 | 22 | 3 | 23 | 57 | 6 | 16 | 20 | 0 | 20 | 6 | 4 | 0 | 25 | 3 | 24 | 37 | 11 |
| 20 | 6 | 22 | 8 | 0 | 18 | 5 | 59 | 0 | 22 | 3 | 23 | 58 | 6 | 14 | 21 | 0 | 20 | 6 | 2 | 0 | 25 | 3 | 24 | 39 | 10 |
| 21 | 6 | 23 | 7 | 0 | 18 | 6 | 0 | 0 | 22 | 3 | 23 | 59 | 6 | 12 | 22 | 0 | 20 | 6 | 0 | 0 | 25 | 3 | 24 | 40 | 9 |
| 22 | 6 | 25 | 6 | 0 | 19 | 6 | 2 | 0 | 22 | 3 | 24 | 1 | 6 | 10 | 23 | 0 | 20 | 5 | 59 | 0 | 24 | 3 | 24 | 41 | 8 |

| Tabula equacionis Saturni | | | | | | | | | | | | | | | | | | | | | | | | | |
|---|---|---|---|---|---|---|---|---|---|---|---|---|---|---|---|---|---|---|---|---|---|---|---|---|---|
| | 2 | | | | | | | | | | | | 3 | | | | | | | | | | | | |
| Gradus | Equacio centri | | Minuta propor-cionalia | Longitudo longior | | Equacio argumenti | | Longitudo propior | | Stacio prima | | | Equacio centri | | Minuta propor-cionalia | Longitudo longior | | Equacio argumenti | | Longitudo propior | | Stacio prima | | | Gradus |
| | G | M | Longior | G | M | G | M | G | M | S | G | M | G | M | Propior | G | M | G | M | G | M | S | G | M | |
| 23 | 6 | 26 | 5 | 0 | 19 | 6 | 4 | 0 | 22 | 3 | 24 | 3 | 6 | 8 | 24 | 0 | 20 | 5 | 57 | 0 | 24 | 3 | 24 | 42 | 7 |
| 24 | 6 | 27 | 4 | 0 | 19 | 6 | 5 | 0 | 22 | 3 | 24 | 4 | 6 | 6 | 25 | 0 | 19 | 5 | 55 | 0 | 24 | 3 | 24 | 44 | 6 |
| 25 | 6 | 25 | 3 | 0 | 19 | 6 | 7 | 0 | 22 | 3 | 24 | 6 | 6 | 4 | 26 | 0 | 19 | 5 | 53 | 0 | 24 | 3 | 24 | 45 | 5 |
| 26 | 6 | 29 | 2 | 0 | 19 | 6 | 8 | 0 | 23 | 3 | 24 | 7 | 6 | 1 | 26 | 0 | 19 | 5 | 51 | 0 | 24 | 3 | 24 | 47 | 4 |
| 27 | 6 | 27 | 1 | 0 | 19 | 6 | 9 | 0 | 23 | 3 | 24 | 8 | 5 | 58 | 27 | 0 | 19 | 5 | 48 | 0 | 24 | 3 | 24 | 48 | 3 |
| 28 | 6 | 30 | Propior 1 | 0 | 19 | 6 | 10 | 0 | 23 | 3 | 24 | 9 | 5 | 55 | 28 | 0 | 19 | 5 | 46 | 0 | 23 | 3 | 24 | 48 | 2 |
| 29 | 6 | 30 | 2 | 0 | 19 | 6 | 11 | 0 | 23 | 3 | 24 | 10 | 5 | 52 | 29 | 0 | 19 | 5 | 43 | 0 | 23 | 3 | 24 | 49 | 1 |
| 30 | 6 | 31 | 3 | 0 | 19 | 6 | 11 | 0 | 23 | 3 | 24 | 11 | 5 | 49 | 30 | 0 | 19 | 5 | 41 | 0 | 23 | 3 | 24 | 50 | 0 |
| | 9 | | | | | | | | | | | | 8 | | | | | | | | | | | | |
| | A | | B | C | | D | | E | | F | | | A | | B | C | | D | | E | | F | | | |

Remaining part of the Table TRW-16 (Toruń, BU, 74/III, f. 262r)

| Tabula equacionis Saturni | | | | | | | | | | | | | | | | | | | | | | | | | |
|---|---|---|---|---|---|---|---|---|---|---|---|---|---|---|---|---|---|---|---|---|---|---|---|---|---|
| | 4 | | | | | | | | | | | | 5 | | | | | | | | | | | | |
| Gradus | Equacio centri | | Minuta proporcionalia | Longitudo longior | | Equacio argumenti | | Longitudo propior | | Stacio prima | | | Equacio centri | | Minuta proporcionalia | Longitudo longior | | Equacio argumenti | | Longitudo propior | | Stacio prima | | | Gradus |
| | G | M | Propior | G | M | G | M | G | M | S | G | M | G | M | Propior | G | M | G | M | G | M | S | G | M | |
| 1 | 5 | 46 | 30 | 0 | 19 | 5 | 37 | 0 | 23 | 3 | 24 | 51 | 3 | 18 | 52 | 0 | 12 | 3 | 19 | 0 | 13 | 3 | 25 | 20 | 29 |
| 2 | 5 | 43 | 31 | 0 | 19 | 5 | 34 | 0 | 23 | 3 | 24 | 52 | 3 | 12 | 53 | 0 | 11 | 3 | 13 | 0 | 13 | 3 | 25 | 20 | 28 |
| 3 | 5 | 40 | 32 | 0 | 19 | 5 | 31 | 0 | 23 | 3 | 24 | 53 | 3 | 6 | 53 | 0 | 11 | 3 | 7 | 0 | 13 | 3 | 25 | 21 | 27 |
| 4 | 5 | 36 | 33 | 0 | 19 | 5 | 28 | 0 | 23 | 3 | 24 | 55 | 2 | 59 | 54 | 0 | 11 | 3 | 1 | 0 | 12 | 3 | 25 | 21 | 26 |
| 5 | 5 | 32 | 33 | 0 | 18 | 5 | 24 | 0 | 22 | 3 | 24 | 56 | 2 | 53 | 54 | 0 | 10 | 2 | 54 | 0 | 12 | 3 | 25 | 22 | 25 |
| 6 | 5 | 28 | 34 | 0 | 18 | 5 | 21 | 0 | 22 | 3 | 24 | 57 | 2 | 46 | 55 | 0 | 10 | 2 | 48 | 0 | 12 | 3 | 25 | 22 | 24 |
| 7 | 5 | 24 | 35 | 0 | 18 | 5 | 18 | 0 | 22 | 3 | 24 | 58 | 2 | 40 | 55 | 0 | 9 | 2 | 42 | 0 | 11 | 3 | 25 | 23 | 23 |
| 8 | 5 | 20 | 36 | 0 | 18 | 5 | 14 | 0 | 22 | 3 | 24 | 59 | 2 | 34 | 56 | 0 | 9 | 2 | 36 | 0 | 11 | 3 | 25 | 23 | 22 |
| 9 | 5 | 16 | 37 | 0 | 18 | 5 | 10 | 0 | 21 | 3 | 25 | 0 | 2 | 27 | 56 | 0 | 8 | 2 | 29 | 0 | 11 | 3 | 25 | 24 | 21 |
| 10 | 5 | 12 | 37 | 0 | 17 | 5 | 6 | 0 | 21 | 3 | 25 | 1 | 2 | 21 | 57 | 0 | 8 | 2 | 23 | 0 | 10 | 3 | 25 | 24 | 20 |
| 11 | 5 | 8 | 38 | 0 | 17 | 5 | 2 | 0 | 21 | 3 | 25 | 2 | 2 | 14 | 57 | 0 | 7 | 2 | 16 | 0 | 10 | 3 | 25 | 25 | 19 |
| 12 | 5 | 3 | 39 | 0 | 17 | 4 | 58 | 0 | 21 | 3 | 25 | 3 | 2 | 7 | 58 | 0 | 7 | 2 | 9 | 0 | 10 | 3 | 25 | 25 | 18 |
| 13 | 4 | 58 | 40 | 0 | 17 | 4 | 54 | 0 | 20 | 3 | 25 | 4 | 2 | 0 | 58 | 0 | 6 | 2 | 2 | 0 | 9 | 3 | 25 | 25 | 17 |
| 14 | 4 | 53 | 41 | 0 | 17 | 4 | 50 | 0 | 20 | 3 | 25 | 5 | 1 | 53 | 59 | 0 | 6 | 1 | 55 | 0 | 9 | 3 | 25 | 26 | 16 |
| 15 | 4 | 48 | 42 | 0 | 16 | 4 | 45 | 0 | 20 | 3 | 25 | 6 | 1 | 46 | 59 | 0 | 6 | 1 | 48 | 0 | 8 | 3 | 25 | 26 | 15 |
| 16 | 4 | 43 | 42 | 0 | 16 | 4 | 41 | 0 | 19 | 3 | 25 | 7 | 1 | 39 | 59 | 0 | 5 | 1 | 41 | 0 | 8 | 3 | 25 | 26 | 14 |
| 17 | 4 | 38 | 43 | 0 | 16 | 4 | 36 | 0 | 19 | 3 | 25 | 8 | 1 | 32 | 59 | 0 | 5 | 1 | 34 | 0 | 7 | 3 | 25 | 26 | 13 |
| 18 | 4 | 33 | 44 | 0 | 16 | 4 | 31 | 0 | 19 | 3 | 25 | 9 | 1 | 25 | 60 | 0 | 5 | 1 | 27 | 0 | 7 | 3 | 25 | 27 | 12 |
| 19 | 4 | 28 | 44 | 0 | 16 | 4 | 26 | 0 | 18 | 3 | 25 | 10 | 1 | 18 | 60 | 0 | 5 | 1 | 20 | 0 | 6 | 3 | 25 | 27 | 11 |
| 20 | 4 | 23 | 45 | 0 | 15 | 4 | 21 | 0 | 18 | 3 | 25 | 11 | 1 | 11 | 60 | 0 | 4 | 1 | 13 | 0 | 6 | 3 | 25 | 27 | 10 |
| 21 | 4 | 17 | 46 | 0 | 15 | 4 | 16 | 0 | 18 | 3 | 25 | 12 | 1 | 4 | 60 | 0 | 4 | 1 | 6 | 0 | 5 | 3 | 25 | 28 | 9 |
| 22 | 4 | 12 | 46 | 0 | 15 | 4 | 11 | 0 | 17 | 3 | 25 | 13 | 0 | 57 | 60 | 0 | 4 | 0 | 59 | 0 | 5 | 3 | 25 | 28 | 8 |

**Tabula equacionis Saturni**

| | 4 | | | | | | | | | | | | 5 | | | | | | | | | | | | |
|---|---|---|---|---|---|---|---|---|---|---|---|---|---|---|---|---|---|---|---|---|---|---|---|---|---|
| Gradus | Equacio centri | | Minuta propor-cionalia | Longitudo longior | | Equacio argumenti | | Longitudo propior | | Stacio prima | | | Equacio centri | | Minuta propor-cionalia | Longitudo longior | | Equacio argumenti | | Longitudo propior | | Stacio prima | | | Gradus |
| | G | M | Propior | G | M | G | M | G | M | S | G | M | G | M | Propior | G | M | G | M | G | M | S | G | M | |
| 23 | 4 | 6 | 47 | 0 | 15 | 4 | 6 | 0 | 17 | 3 | 25 | 14 | 0 | 50 | 60 | 0 | 3 | 0 | 52 | 0 | 4 | 3 | 25 | 28 | 7 |
| 24 | 4 | 0 | 48 | 0 | 14 | 4 | 0 | 0 | 17 | 3 | 25 | 15 | 0 | 42 | 60 | 0 | 3 | 0 | 45 | 0 | 4 | 3 | 25 | 28 | 6 |
| 25 | 3 | 54 | 49 | 0 | 14 | 3 | 55 | 0 | 16 | 3 | 25 | 16 | 0 | 36 | 60 | 0 | 3 | 0 | 38 | 0 | 3 | 3 | 25 | 29 | 5 |
| 26 | 3 | 48 | 49 | 0 | 14 | 3 | 49 | 0 | 16 | 3 | 25 | 17 | 0 | 29 | 60 | 0 | 2 | 0 | 31 | 0 | 3 | 3 | 25 | 29 | 4 |
| 27 | 3 | 42 | 50 | 0 | 13 | 3 | 43 | 0 | 15 | 3 | 25 | 18 | 0 | 22 | 60 | 0 | 2 | 0 | 23 | 0 | 2 | 3 | 25 | 29 | 3 |
| 28 | 3 | 36 | 50 | 0 | 13 | 3 | 37 | 0 | 15 | 3 | 25 | 18 | 0 | 15 | 60 | 0 | 1 | 0 | 16 | 0 | 2 | 3 | 25 | 30 | 2 |
| 29 | 3 | 30 | 51 | 0 | 12 | 3 | 31 | 0 | 14 | 3 | 25 | 19 | 0 | 8 | 60 | 0 | 1 | 0 | 8 | 0 | 1 | 3 | 25 | 30 | 1 |
| 30 | 3 | 24 | 51 | 0 | 12 | 3 | 25 | 0 | 14 | 3 | 25 | 19 | 0 | 0 | 60 | 0 | 0 | 0 | 0 | 0 | 0 | 3 | 25 | 30 | 0 |
| | 7 | | | | | | | | | | | | 6 | | | | | | | | | | | | |
| | A | | B | C | | D | | E | | F | | | A | | B | C | | D | | E | | F | | | |

Table TRW-17 of equations of Jupiter (Toruń, BU, 74/III, f. 262v)

| Tabula equacionis Iovis | | | | | | | | | | | | | | | | | | | | | | | | | |
|---|---|---|---|---|---|---|---|---|---|---|---|---|---|---|---|---|---|---|---|---|---|---|---|---|---|
| | 0 | | | | | | | | | | | | 1 | | | | | | | | | | | | |
| Gradus | Equacio centri | | Minuta proporcionalia | Longitudo longior | | Equacio argumenti | | Longitudo propior | | Stacio prima | | | Equacio centri | | Minuta proporcionalia | Longitudo longior | | Equacio argumenti | | Longitudo propior | | Stacio prima | | | Gradus |
| | G | M | Longior | G | M | G | M | G | M | S | G | M | G | M | Longior | G | M | G | M | G | M | S | G | M | |
| 1 | 0 | 6 | 60 | 0 | 0 | 0 | 10 | 0 | 0 | 4 | 4 | 5 | 2 | 56 | 52 | 0 | 10 | 4 | 51 | 0 | 11 | 4 | 4 | 18 | 29 |
| 2 | 0 | 12 | 60 | 0 | 1 | 0 | 20 | 0 | 1 | 4 | 4 | 5 | 3 | 1 | 51 | 0 | 11 | 5 | 0 | 0 | 12 | 4 | 4 | 19 | 28 |
| 3 | 0 | 18 | 60 | 0 | 1 | 0 | 29 | 0 | 1 | 4 | 4 | 5 | 3 | 6 | 51 | 0 | 11 | 5 | 8 | 0 | 12 | 4 | 4 | 20 | 27 |
| 4 | 0 | 24 | 60 | 0 | 1 | 0 | 39 | 0 | 1 | 4 | 4 | 5 | 3 | 11 | 50 | 0 | 11 | 5 | 16 | 0 | 12 | 4 | 4 | 21 | 26 |
| 5 | 0 | 30 | 60 | 0 | 2 | 0 | 49 | 0 | 2 | 4 | 4 | 5 | 3 | 17 | 50 | 0 | 12 | 5 | 25 | 0 | 13 | 4 | 4 | 22 | 25 |
| 6 | 0 | 36 | 60 | 0 | 2 | 0 | 58 | 0 | 2 | 4 | 4 | 5 | 3 | 21 | 49 | 0 | 12 | 5 | 34 | 0 | 13 | 4 | 4 | 23 | 24 |
| 7 | 0 | 42 | 60 | 0 | 2 | 1 | 8 | 0 | 2 | 4 | 4 | 6 | 3 | 26 | 49 | 0 | 12 | 5 | 43 | 0 | 13 | 4 | 4 | 24 | 23 |
| 8 | 0 | 48 | 60 | 0 | 3 | 1 | 18 | 0 | 3 | 4 | 4 | 6 | 3 | 31 | 48 | 0 | 13 | 5 | 52 | 0 | 14 | 4 | 4 | 25 | 22 |
| 9 | 0 | 53 | 60 | 0 | 3 | 1 | 28 | 0 | 3 | 4 | 4 | 6 | 3 | 36 | 47 | 0 | 13 | 6 | 0 | 0 | 14 | 4 | 4 | 26 | 21 |
| 10 | 0 | 59 | 60 | 0 | 3 | 1 | 37 | 0 | 3 | 4 | 4 | 6 | 3 | 41 | 47 | 0 | 13 | 6 | 9 | 0 | 14 | 4 | 4 | 27 | 20 |
| 11 | 1 | 5 | 60 | 0 | 4 | 1 | 47 | 0 | 4 | 4 | 4 | 6 | 3 | 45 | 46 | 0 | 14 | 6 | 17 | 0 | 15 | 4 | 4 | 28 | 19 |
| 12 | 1 | 11 | 60 | 0 | 4 | 1 | 57 | 0 | 4 | 4 | 4 | 6 | 3 | 49 | 46 | 0 | 14 | 6 | 25 | 0 | 15 | 4 | 4 | 29 | 18 |
| 13 | 1 | 17 | 59 | 0 | 4 | 2 | 6 | 0 | 5 | 4 | 4 | 7 | 3 | 54 | 45 | 0 | 14 | 6 | 33 | 0 | 15 | 4 | 4 | 30 | 17 |
| 14 | 1 | 23 | 59 | 0 | 5 | 2 | 15 | 0 | 5 | 4 | 4 | 7 | 3 | 59 | 44 | 0 | 15 | 6 | 41 | 0 | 16 | 4 | 4 | 31 | 16 |
| 15 | 1 | 28 | 59 | 0 | 5 | 2 | 24 | 0 | 5 | 4 | 4 | 7 | 4 | 3 | 43 | 0 | 15 | 6 | 49 | 0 | 16 | 4 | 4 | 32 | 15 |
| 16 | 1 | 34 | 59 | 0 | 5 | 2 | 34 | 0 | 6 | 4 | 4 | 8 | 4 | 8 | 43 | 0 | 15 | 6 | 57 | 0 | 17 | 4 | 4 | 33 | 14 |
| 17 | 1 | 40 | 58 | 0 | 6 | 2 | 43 | 0 | 6 | 4 | 4 | 8 | 4 | 12 | 42 | 0 | 16 | 7 | 5 | 0 | 17 | 4 | 4 | 34 | 13 |
| 18 | 1 | 45 | 58 | 0 | 6 | 2 | 52 | 0 | 7 | 4 | 4 | 8 | 4 | 16 | 41 | 0 | 16 | 7 | 12 | 0 | 18 | 4 | 4 | 35 | 12 |
| 19 | 1 | 51 | 58 | 0 | 6 | 3 | 2 | 0 | 7 | 4 | 4 | 9 | 4 | 20 | 40 | 0 | 16 | 7 | 20 | 0 | 18 | 4 | 4 | 37 | 11 |
| 20 | 1 | 57 | 57 | 0 | 7 | 3 | 11 | 0 | 8 | 4 | 4 | 9 | 4 | 24 | 39 | 0 | 17 | 7 | 28 | 0 | 19 | 4 | 4 | 38 | 10 |
| 21 | 2 | 2 | 57 | 0 | 7 | 3 | 20 | 0 | 8 | 4 | 4 | 9 | 4 | 28 | 38 | 0 | 17 | 7 | 35 | 0 | 19 | 4 | 4 | 40 | 9 |
| 22 | 2 | 8 | 57 | 0 | 7 | 3 | 30 | 0 | 8 | 4 | 4 | 10 | 4 | 32 | 37 | 0 | 17 | 7 | 43 | 0 | 19 | 4 | 4 | 42 | 8 |

| Tabula equacionis Iovis | | | | | | | | | | | | | | | | | | | | | | | | | |
|---|---|---|---|---|---|---|---|---|---|---|---|---|---|---|---|---|---|---|---|---|---|---|---|---|---|
| | 0 | | | | | | | | | | | | 1 | | | | | | | | | | | | |
| Gradus | Equacio centri | | Minuta propor-cionalia | Longitudo longior | | Equacio argumenti | | Longitudo propior | | Stacio prima | | | Equacio centri | | Minuta propor-cionalia | Longitudo longior | | Equacio argumenti | | Longitudo propior | | Stacio prima | | | Gradus |
| | G | M | Longior | G | M | G | M | G | M | S | G | M | G | M | Longior | G | M | G | M | G | M | S | G | M | |
| 23 | 2 | 13 | 56 | 0 | 8 | 3 | 39 | 0 | 9 | 4 | 4 | 10 | 4 | 36 | 36 | 0 | 18 | 7 | 50 | 0 | 20 | 4 | 4 | 43 | 7 |
| 24 | 2 | 18 | 56 | 0 | 8 | 3 | 48 | 0 | 9 | 4 | 4 | 11 | 4 | 39 | 35 | 0 | 18 | 7 | 57 | 0 | 20 | 4 | 4 | 44 | 6 |
| 25 | 2 | 24 | 55 | 0 | 8 | 3 | 57 | 0 | 9 | 4 | 4 | 12 | 4 | 43 | 34 | 0 | 18 | 8 | 4 | 0 | 20 | 4 | 4 | 46 | 5 |
| 26 | 2 | 30 | 55 | 0 | 9 | 4 | 6 | 0 | 10 | 4 | 4 | 13 | 4 | 47 | 33 | 0 | 19 | 8 | 11 | 0 | 21 | 4 | 4 | 48 | 4 |
| 27 | 2 | 35 | 54 | 0 | 9 | 4 | 15 | 0 | 10 | 4 | 4 | 14 | 4 | 50 | 33 | 0 | 19 | 8 | 17 | 0 | 21 | 4 | 4 | 49 | 3 |
| 28 | 2 | 41 | 54 | 0 | 9 | 4 | 24 | 0 | 10 | 4 | 4 | 15 | 4 | 54 | 32 | 0 | 19 | 8 | 24 | 0 | 21 | 4 | 4 | 51 | 2 |
| 29 | 2 | 46 | 53 | 0 | 10 | 4 | 33 | 0 | 11 | 4 | 4 | 16 | 4 | 58 | 31 | 0 | 20 | 8 | 31 | 0 | 22 | 4 | 4 | 52 | 1 |
| 30 | 2 | 51 | 53 | 0 | 10 | 4 | 44 | 0 | 11 | 4 | 4 | 17 | 5 | 1 | 30 | 0 | 20 | 8 | 37 | 0 | 22 | 4 | 4 | 53 | 0 |
| | 11 | | | | | | | | | | | | 10 | | | | | | | | | | | | |
| | A | | B | C | | D | | E | | F | | | A | | B | C | | D | | E | | F | | | |

Continuation of the Table TRW-17 (Toruń, BU, 74/III, f. 263r)

| Tabula equacionis Iovis | | | | | | | | | | | | | | | | | | | | | | | | | |
|---|---|---|---|---|---|---|---|---|---|---|---|---|---|---|---|---|---|---|---|---|---|---|---|---|---|
| | 2 | | | | | | | | | | | | 3 | | | | | | | | | | | | |
| Gradus | Equacio centri | | Minuta propor-cionalia | Longitudo longior | | Equacio argumenti | | Longitudo propior | | Stacio prima | | | Equacio centri | | Minuta propor-cionalia | Longitudo longior | | Equacio argumenti | | Longitudo propior | | Stacio prima | | | Gradus |
| | G | M | Longior | G | M | G | M | G | M | S | G | M | G | M | Propior | G | M | G | M | G | M | S | G | M | |
| 1 | 5 | 5 | 29 | 0 | 20 | 8 | 44 | 0 | 22 | 4 | 4 | 55 | 5 | 57 | 3 | 0 | 27 | 10 | 53 | 0 | 31 | 4 | 5 | 43 | 29 |
| 2 | 5 | 9 | 28 | 0 | 21 | 8 | 50 | 0 | 23 | 4 | 4 | 56 | 5 | 57 | 4 | 0 | 28 | 10 | 55 | 0 | 31 | 4 | 5 | 44 | 28 |
| 3 | 5 | 12 | 27 | 0 | 21 | 8 | 56 | 0 | 23 | 4 | 4 | 57 | 5 | 57 | 5 | 0 | 28 | 10 | 57 | 0 | 31 | 4 | 5 | 46 | 27 |
| 4 | 5 | 15 | 26 | 0 | 21 | 9 | 2 | 0 | 23 | 4 | 4 | 59 | 5 | 57 | 5 | 0 | 28 | 10 | 59 | 0 | 31 | 4 | 5 | 47 | 26 |
| 5 | 5 | 18 | 25 | 0 | 22 | 9 | 8 | 0 | 24 | 4 | 5 | 1 | 5 | 57 | 6 | 0 | 28 | 11 | 0 | 0 | 31 | 4 | 5 | 48 | 25 |
| 6 | 5 | 20 | 24 | 0 | 22 | 9 | 14 | 0 | 24 | 4 | 5 | 2 | 5 | 57 | 7 | 0 | 28 | 11 | 1 | 0 | 31 | 4 | 5 | 50 | 24 |
| 7 | 5 | 23 | 23 | 0 | 22 | 9 | 20 | 0 | 24 | 4 | 5 | 4 | 5 | 56 | 8 | 0 | 28 | 11 | 2 | 0 | 31 | 4 | 5 | 51 | 23 |
| 8 | 5 | 26 | 22 | 0 | 23 | 9 | 26 | 0 | 25 | 4 | 5 | 6 | 5 | 56 | 9 | 0 | 28 | 11 | 2 | 0 | 32 | 4 | 5 | 52 | 22 |
| 9 | 5 | 28 | 21 | 0 | 23 | 9 | 31 | 0 | 25 | 4 | 5 | 7 | 5 | 55 | 10 | 0 | 29 | 11 | 3 | 0 | 32 | 4 | 5 | 54 | 21 |
| 10 | 5 | 31 | 20 | 0 | 23 | 9 | 36 | 0 | 25 | 4 | 5 | 9 | 5 | 55 | 11 | 0 | 29 | 11 | 3 | 0 | 32 | 4 | 5 | 55 | 20 |
| 11 | 5 | 33 | 19 | 0 | 24 | 9 | 41 | 0 | 26 | 4 | 5 | 11 | 5 | 54 | 12 | 0 | 29 | 11 | 3 | 0 | 32 | 4 | 5 | 56 | 19 |
| 12 | 5 | 35 | 18 | 0 | 24 | 9 | 47 | 0 | 26 | 4 | 5 | 12 | 5 | 53 | 13 | 0 | 29 | 11 | 3 | 0 | 32 | 4 | 5 | 58 | 18 |
| 13 | 5 | 37 | 17 | 0 | 24 | 9 | 51 | 0 | 26 | 4 | 5 | 14 | 5 | 52 | 14 | 0 | 29 | 11 | 2 | 0 | 32 | 4 | 5 | 59 | 17 |
| 14 | 5 | 39 | 16 | 0 | 25 | 9 | 56 | 0 | 27 | 4 | 5 | 15 | 5 | 51 | 15 | 0 | 29 | 11 | 2 | 0 | 32 | 4 | 6 | 0 | 16 |
| 15 | 5 | 41 | 15 | 0 | 25 | 10 | 0 | 0 | 27 | 4 | 5 | 17 | 5 | 49 | 16 | 0 | 29 | 11 | 2 | 0 | 32 | 4 | 6 | 1 | 15 |
| 16 | 5 | 43 | 14 | 0 | 25 | 10 | 5 | 0 | 27 | 4 | 5 | 19 | 5 | 48 | 17 | 0 | 29 | 11 | 1 | 0 | 32 | 4 | 6 | 3 | 14 |
| 17 | 5 | 45 | 13 | 0 | 25 | 10 | 9 | 0 | 28 | 4 | 5 | 21 | 5 | 46 | 18 | 0 | 30 | 11 | 0 | 0 | 33 | 4 | 6 | 4 | 13 |
| 18 | 5 | 46 | 12 | 0 | 26 | 10 | 13 | 0 | 28 | 4 | 5 | 22 | 5 | 44 | 19 | 0 | 30 | 10 | 59 | 0 | 33 | 4 | 6 | 6 | 12 |
| 19 | 5 | 48 | 11 | 0 | 26 | 10 | 17 | 0 | 28 | 4 | 5 | 24 | 5 | 43 | 20 | 0 | 30 | 10 | 57 | 0 | 33 | 4 | 6 | 7 | 11 |
| 20 | 5 | 49 | 10 | 0 | 26 | 10 | 21 | 0 | 29 | 4 | 5 | 26 | 5 | 41 | 21 | 0 | 30 | 10 | 55 | 0 | 33 | 4 | 6 | 8 | 10 |
| 21 | 5 | 50 | 9 | 0 | 26 | 10 | 25 | 0 | 29 | 4 | 5 | 27 | 5 | 39 | 22 | 0 | 30 | 10 | 53 | 0 | 33 | 4 | 6 | 10 | 9 |
| 22 | 5 | 51 | 8 | 0 | 26 | 10 | 29 | 0 | 29 | 4 | 5 | 29 | 5 | 37 | 22 | 0 | 30 | 10 | 51 | 0 | 33 | 4 | 6 | 11 | 8 |

**Tabula equacionis Iovis**

| | 2 | | | | | | | | | | | | | 3 | | | | | | | | | | | | | |
|---|---|---|---|---|---|---|---|---|---|---|---|---|---|---|---|---|---|---|---|---|---|---|---|---|---|---|
| **Gradus** | **Equacio centri** | | **Minuta propor-cionalia** | **Longitudo longior** | | **Equacio argumenti** | | **Longitudo propior** | | **Stacio prima** | | | **Equacio centri** | | **Minuta propor-cionalia** | **Longitudo longior** | | **Equacio argumenti** | | **Longitudo propior** | | **Stacio prima** | | | **Gradus** |
| | **G** | **M** | **Longior** | **G** | **M** | **G** | **M** | **G** | **M** | **S** | **G** | **M** | **G** | **M** | **Propior** | **G** | **M** | **G** | **M** | **G** | **M** | **S** | **G** | **M** | |
| 23 | 5 | 52 | 7 | 0 | 26 | 10 | 32 | 0 | 29 | 4 | 5 | 31 | 5 | 35 | 23 | 0 | 30 | 10 | 48 | 0 | 33 | 4 | 6 | 12 | 7 |
| 24 | 5 | 53 | 6 | 0 | 27 | 10 | 35 | 0 | 30 | 4 | 5 | 32 | 5 | 33 | 24 | 0 | 30 | 10 | 45 | 0 | 33 | 4 | 6 | 14 | 6 |
| 25 | 5 | 54 | 5 | 0 | 27 | 10 | 38 | 0 | 30 | 4 | 5 | 34 | 5 | 31 | 25 | 0 | 30 | 10 | 42 | 0 | 33 | 4 | 6 | 15 | 5 |
| 26 | 5 | 55 | 4 | 0 | 27 | 10 | 41 | 0 | 30 | 4 | 5 | 36 | 5 | 29 | 26 | 0 | 30 | 10 | 39 | 0 | 33 | 4 | 6 | 17 | 4 |
| 27 | 5 | 55 | 3 | 0 | 27 | 10 | 44 | 0 | 30 | 4 | 5 | 37 | 5 | 27 | 27 | 0 | 30 | 10 | 35 | 0 | 33 | 4 | 6 | 19 | 3 |
| 28 | 5 | 56 | 2 | 0 | 27 | 10 | 47 | 0 | 30 | 4 | 5 | 39 | 5 | 25 | 28 | 0 | 30 | 10 | 31 | 0 | 33 | 4 | 6 | 20 | 2 |
| 29 | 5 | 56 | **Propior** 1 | 0 | 27 | 10 | 49 | 0 | 30 | 4 | 5 | 41 | 5 | 22 | 29 | 0 | 29 | 10 | 27 | 0 | 33 | 4 | 6 | 22 | 1 |
| 30 | 5 | 56 | 2 | 0 | 27 | 10 | 51 | 0 | 30 | 4 | 5 | 42 | 5 | 19 | 30 | 0 | 29 | 10 | 23 | 0 | 33 | 4 | 6 | 24 | 0 |
| | 9 | | | | | | | | | | | | 8 | | | | | | | | | | | | |
| | A | | B | C | | D | | E | | F | | | A | | B | C | | D | | E | | F | | | |

Remaining part of the Table TRW-17 (Toruń, BU, 74/III, f. 263v)

| Tabula equacionis Iovis | | | | | | | | | | | | | | | | | | | | | | | | | |
|---|---|---|---|---|---|---|---|---|---|---|---|---|---|---|---|---|---|---|---|---|---|---|---|---|---|
| | 4 | | | | | | | | | | | | 5 | | | | | | | | | | | | |
| Gradus | Equacio centri | | Minuta propor-cionalia | Longitudo longior | | Equacio argumenti | | Longitudo propior | | Stacio prima | | | Equacio centri | | Minuta propor-cionalia | Longitudo longior | | Equacio argumenti | | Longitudo propior | | Stacio prima | | | Gradus |
| | G | M | Propior | G | M | G | M | G | M | S | G | M | G | M | Propior | G | M | G | M | G | M | S | G | M | |
| 1 | 5 | 16 | 31 | 0 | 29 | 10 | 19 | 0 | 32 | 4 | 6 | 25 | 3 | 1 | 52 | 0 | 21 | 6 | 23 | 0 | 22 | 4 | 7 | 0 | 29 |
| 2 | 5 | 13 | 32 | 0 | 29 | 10 | 15 | 0 | 32 | 4 | 6 | 27 | 2 | 55 | 53 | 0 | 20 | 6 | 12 | 0 | 21 | 4 | 7 | 1 | 28 |
| 3 | 5 | 10 | 33 | 0 | 29 | 10 | 10 | 0 | 32 | 4 | 6 | 29 | 2 | 49 | 53 | 0 | 19 | 6 | 0 | 0 | 20 | 4 | 7 | 2 | 27 |
| 4 | 5 | 6 | 34 | 0 | 29 | 10 | 5 | 0 | 32 | 4 | 6 | 30 | 2 | 44 | 53 | 0 | 19 | 5 | 48 | 0 | 20 | 4 | 7 | 2 | 26 |
| 5 | 5 | 3 | 35 | 0 | 29 | 10 | 0 | 0 | 32 | 4 | 6 | 32 | 2 | 38 | 54 | 0 | 18 | 5 | 36 | 0 | 19 | 4 | 7 | 3 | 25 |
| 6 | 4 | 59 | 36 | 0 | 29 | 9 | 54 | 0 | 32 | 4 | 6 | 34 | 2 | 32 | 54 | 0 | 17 | 5 | 24 | 0 | 18 | 4 | 7 | 4 | 24 |
| 7 | 4 | 55 | 37 | 0 | 28 | 9 | 48 | 0 | 32 | 4 | 6 | 35 | 2 | 26 | 55 | 0 | 17 | 5 | 12 | 0 | 18 | 4 | 7 | 5 | 23 |
| 8 | 4 | 51 | 38 | 0 | 28 | 9 | 42 | 0 | 31 | 4 | 6 | 36 | 2 | 20 | 55 | 0 | 16 | 5 | 0 | 0 | 17 | 4 | 7 | 5 | 22 |
| 9 | 4 | 47 | 39 | 0 | 28 | 9 | 36 | 0 | 31 | 4 | 6 | 37 | 2 | 14 | 56 | 0 | 15 | 4 | 47 | 0 | 16 | 4 | 7 | 6 | 21 |
| 10 | 4 | 43 | 40 | 0 | 28 | 9 | 30 | 0 | 31 | 4 | 6 | 38 | 2 | 8 | 56 | 0 | 15 | 4 | 34 | 0 | 16 | 4 | 7 | 6 | 20 |
| 11 | 4 | 39 | 41 | 0 | 28 | 9 | 24 | 0 | 31 | 4 | 6 | 39 | 2 | 2 | 57 | 0 | 14 | 4 | 22 | 0 | 15 | 4 | 7 | 7 | 19 |
| 12 | 4 | 35 | 41 | 0 | 28 | 9 | 17 | 0 | 31 | 4 | 6 | 40 | 1 | 56 | 57 | 0 | 13 | 4 | 9 | 0 | 14 | 4 | 7 | 7 | 18 |
| 13 | 4 | 31 | 42 | 0 | 27 | 9 | 10 | 0 | 31 | 4 | 6 | 41 | 1 | 50 | 57 | 0 | 13 | 3 | 56 | 0 | 14 | 4 | 7 | 8 | 17 |
| 14 | 4 | 27 | 43 | 0 | 27 | 9 | 2 | 0 | 30 | 4 | 6 | 42 | 1 | 43 | 58 | 0 | 12 | 3 | 42 | 0 | 13 | 4 | 7 | 8 | 16 |
| 15 | 4 | 22 | 44 | 0 | 27 | 8 | 54 | 0 | 30 | 4 | 6 | 43 | 1 | 37 | 58 | 0 | 11 | 3 | 29 | 0 | 12 | 4 | 7 | 9 | 15 |
| 16 | 4 | 18 | 44 | 0 | 26 | 8 | 47 | 0 | 30 | 4 | 6 | 44 | 1 | 30 | 58 | 0 | 11 | 3 | 16 | 0 | 12 | 4 | 7 | 9 | 14 |
| 17 | 4 | 13 | 45 | 0 | 26 | 8 | 39 | 0 | 30 | 4 | 6 | 45 | 1 | 24 | 59 | 0 | 10 | 3 | 3 | 0 | 11 | 4 | 7 | 9 | 13 |
| 18 | 4 | 8 | 46 | 0 | 26 | 8 | 31 | 0 | 29 | 4 | 6 | 46 | 1 | 18 | 59 | 0 | 9 | 2 | 49 | 0 | 10 | 4 | 7 | 10 | 12 |
| 19 | 4 | 4 | 46 | 0 | 25 | 8 | 22 | 0 | 29 | 4 | 6 | 47 | 1 | 12 | 59 | 0 | 9 | 2 | 35 | 0 | 10 | 4 | 7 | 10 | 11 |
| 20 | 3 | 59 | 47 | 0 | 25 | 8 | 13 | 0 | 28 | 4 | 6 | 48 | 1 | 5 | 59 | 0 | 8 | 2 | 21 | 0 | 9 | 4 | 7 | 10 | 10 |
| 21 | 3 | 54 | 47 | 0 | 25 | 8 | 4 | 0 | 28 | 4 | 6 | 49 | 0 | 59 | 60 | 0 | 7 | 2 | 7 | 0 | 8 | 4 | 7 | 10 | 9 |
| 22 | 3 | 49 | 48 | 0 | 24 | 7 | 55 | 0 | 27 | 4 | 6 | 50 | 0 | 52 | 60 | 0 | 6 | 1 | 53 | 0 | 7 | 4 | 7 | 11 | 8 |

**Tabula equacionis Iovis**

| | 4 | | | | | | | | | | | | 5 | | | | | | | | | | | | |
|---|---|---|---|---|---|---|---|---|---|---|---|---|---|---|---|---|---|---|---|---|---|---|---|---|---|
| **Gradus** | **Equacio centri** | | **Minuta propor-cionalia** | **Longitudo longior** | | **Equacio argumenti** | | **Longitudo propior** | | **Stacio prima** | | | **Equacio centri** | | **Minuta propor-cionalia** | **Longitudo longior** | | **Equacio argumenti** | | **Longitudo propior** | | **Stacio prima** | | | **Gradus** |
| | **G** | **M** | **Propior** | **G** | **M** | **G** | **M** | **G** | **M** | **S** | **G** | **M** | **G** | **M** | **Propior** | **G** | **M** | **G** | **M** | **G** | **M** | **S** | **G** | **M** | |
| 23 | 3 | 44 | 48 | 0 | 24 | 7 | 45 | 0 | 27 | 4 | 6 | 52 | 0 | 46 | 60 | 0 | 6 | 1 | 39 | 0 | 7 | 4 | 7 | 11 | 7 |
| 24 | 3 | 38 | 49 | 0 | 24 | 7 | 36 | 0 | 26 | 4 | 6 | 53 | 0 | 39 | 60 | 0 | 5 | 1 | 25 | 0 | 6 | 4 | 7 | 11 | 6 |
| 25 | 3 | 33 | 49 | 0 | 23 | 7 | 26 | 0 | 26 | 4 | 6 | 54 | 0 | 33 | 60 | 0 | 5 | 1 | 11 | 0 | 5 | 4 | 7 | 11 | 5 |
| 26 | 3 | 28 | 50 | 0 | 23 | 7 | 16 | 0 | 25 | 4 | 6 | 55 | 0 | 27 | 60 | 0 | 4 | 0 | 57 | 0 | 4 | 4 | 7 | 11 | 4 |
| 27 | 3 | 22 | 50 | 0 | 23 | 7 | 6 | 0 | 24 | 4 | 6 | 56 | 0 | 20 | 60 | 0 | 3 | 0 | 43 | 0 | 3 | 4 | 7 | 11 | 3 |
| 28 | 3 | 17 | 51 | 0 | 22 | 6 | 55 | 0 | 24 | 4 | 6 | 57 | 0 | 14 | 60 | 0 | 2 | 0 | 29 | 0 | 2 | 4 | 7 | 11 | 2 |
| 29 | 3 | 12 | 51 | 0 | 22 | 6 | 45 | 0 | 23 | 4 | 6 | 58 | 0 | 7 | 60 | 0 | 1 | 0 | 15 | 0 | 1 | 4 | 7 | 11 | 1 |
| 30 | 3 | 6 | 52 | 0 | 21 | 6 | 34 | 0 | 22 | 4 | 6 | 59 | 0 | 0 | 60 | 0 | 0 | 0 | 0 | 0 | 0 | 4 | 7 | 11 | 0 |
| | 7 | | | | | | | | | | | | 6 | | | | | | | | | | | | |
| | A | | B | C | | D | | E | | F | | | A | | B | C | | D | | E | | F | | | |

Table TRW-18 of equations of Mars (Toruń, BU, 74/III, f. 264r)

| Tabula equacionis Martis | | | | | | | | | | | | | | | | | | | | | | | | | |
|---|---|---|---|---|---|---|---|---|---|---|---|---|---|---|---|---|---|---|---|---|---|---|---|---|---|
| | 0 | | | | | | | | | | | | 1 | | | | | | | | | | | | |
| Gradus | Equacio centri | | Minuta propor-cionalia | Longitudo longior | | Equacio argumenti | | Longitudo propior | | Stacio prima | | | Equacio centri | | Minuta propor-cionalia | Longitudo longior | | Equacio argumenti | | Longitudo propior | | Stacio prima | | | Gradus |
| | G | M | Longior | G | M | G | M | G | M | S | G | M | G | M | Longior | G | M | G | M | G | M | S | G | M | |
| 1 | 0 | 11 | 60 | 0 | 2 | 0 | 24 | 0 | 2 | 5 | 7 | 28 | 5 | 26 | 51 | 0 | 44 | 12 | 15 | 0 | 48 | 5 | 8 | 18 | 29 |
| 2 | 0 | 22 | 60 | 0 | 3 | 0 | 48 | 0 | 3 | 5 | 7 | 29 | 5 | 36 | 51 | 0 | 45 | 12 | 38 | 0 | 50 | 5 | 8 | 21 | 28 |
| 3 | 0 | 33 | 60 | 0 | 4 | 1 | 12 | 0 | 4 | 5 | 7 | 30 | 5 | 45 | 50 | 0 | 47 | 13 | 1 | 0 | 51 | 5 | 8 | 24 | 27 |
| 4 | 0 | 44 | 60 | 0 | 6 | 1 | 36 | 0 | 6 | 5 | 7 | 31 | 5 | 55 | 50 | 0 | 48 | 13 | 25 | 0 | 53 | 5 | 8 | 27 | 26 |
| 5 | 0 | 55 | 60 | 0 | 7 | 2 | 0 | 0 | 7 | 5 | 7 | 32 | 6 | 4 | 49 | 0 | 50 | 13 | 48 | 0 | 55 | 5 | 8 | 30 | 25 |
| 6 | 1 | 5 | 60 | 0 | 8 | 2 | 24 | 0 | 9 | 5 | 7 | 33 | 6 | 13 | 49 | 0 | 51 | 14 | 11 | 0 | 56 | 5 | 8 | 33 | 24 |
| 7 | 1 | 16 | 59 | 0 | 10 | 2 | 48 | 0 | 10 | 5 | 7 | 34 | 6 | 22 | 48 | 0 | 52 | 14 | 34 | 0 | 58 | 5 | 8 | 37 | 23 |
| 8 | 1 | 27 | 59 | 0 | 11 | 3 | 12 | 0 | 12 | 5 | 7 | 34 | 6 | 31 | 47 | 0 | 54 | 14 | 57 | 1 | 0 | 5 | 8 | 41 | 22 |
| 9 | 1 | 38 | 59 | 0 | 12 | 3 | 36 | 0 | 13 | 5 | 7 | 35 | 6 | 40 | 47 | 0 | 56 | 15 | 20 | 1 | 1 | 5 | 8 | 45 | 21 |
| 10 | 1 | 49 | 59 | 0 | 14 | 3 | 59 | 0 | 15 | 5 | 7 | 36 | 6 | 49 | 46 | 0 | 57 | 15 | 43 | 1 | 2 | 5 | 8 | 49 | 20 |
| 11 | 2 | 0 | 59 | 0 | 15 | 4 | 23 | 0 | 16 | 5 | 7 | 37 | 6 | 58 | 45 | 0 | 59 | 16 | 6 | 1 | 5 | 5 | 8 | 53 | 19 |
| 12 | 2 | 10 | 59 | 0 | 16 | 4 | 46 | 0 | 18 | 5 | 7 | 38 | 7 | 7 | 45 | 1 | 0 | 16 | 29 | 1 | 6 | 5 | 8 | 57 | 18 |
| 13 | 2 | 21 | 58 | 0 | 18 | 5 | 10 | 0 | 20 | 5 | 7 | 39 | 7 | 16 | 44 | 1 | 2 | 16 | 52 | 1 | 8 | 5 | 9 | 2 | 17 |
| 14 | 2 | 32 | 58 | 0 | 19 | 5 | 34 | 0 | 21 | 5 | 7 | 40 | 7 | 24 | 44 | 1 | 3 | 17 | 15 | 1 | 10 | 5 | 9 | 6 | 16 |
| 15 | 2 | 42 | 58 | 0 | 20 | 5 | 57 | 0 | 23 | 5 | 7 | 42 | 7 | 32 | 43 | 1 | 5 | 17 | 38 | 1 | 11 | 5 | 9 | 6 | 15 |
| 16 | 2 | 53 | 57 | 0 | 22 | 6 | 21 | 0 | 24 | 5 | 7 | 44 | 7 | 41 | 42 | 1 | 6 | 18 | 1 | 1 | 13 | 5 | 9 | 13 | 14 |
| 17 | 3 | 3 | 57 | 0 | 23 | 6 | 44 | 0 | 26 | 5 | 7 | 45 | 7 | 49 | 41 | 1 | 8 | 18 | 24 | 1 | 15 | 5 | 9 | 17 | 13 |
| 18 | 3 | 13 | 57 | 0 | 24 | 7 | 8 | 0 | 28 | 5 | 7 | 47 | 7 | 57 | 40 | 1 | 9 | 18 | 46 | 1 | 16 | 5 | 9 | 21 | 12 |
| 19 | 3 | 24 | 56 | 0 | 26 | 7 | 32 | 0 | 29 | 5 | 7 | 49 | 8 | 5 | 40 | 1 | 11 | 19 | 9 | 1 | 18 | 5 | 9 | 26 | 11 |
| 20 | 3 | 35 | 56 | 0 | 27 | 7 | 56 | 0 | 31 | 5 | 7 | 51 | 8 | 13 | 39 | 1 | 12 | 19 | 31 | 1 | 20 | 5 | 9 | 31 | 10 |
| 21 | 3 | 45 | 56 | 0 | 28 | 8 | 19 | 0 | 32 | 5 | 7 | 52 | 8 | 20 | 38 | 1 | 14 | 19 | 53 | 1 | 22 | 5 | 9 | 36 | 9 |
| 22 | 3 | 56 | 55 | 0 | 30 | 8 | 43 | 0 | 34 | 5 | 7 | 54 | 8 | 27 | 37 | 1 | 15 | 20 | 16 | 1 | 24 | 5 | 9 | 41 | 8 |

| Tabula equacionis Martis | | | | | | | | | | | | | | | | | | | | | | | | |
|---|---|---|---|---|---|---|---|---|---|---|---|---|---|---|---|---|---|---|---|---|---|---|---|---|
| | 0 | | | | | | | | | | | | 1 | | | | | | | | | | | |
| Gradus | Equacio centri | | Minuta propor-cionalia | Longitudo longior | | Equacio argumenti | | Longitudo propior | | Stacio prima | | | Equacio centri | | Minuta propor-cionalia | Longitudo longior | | Equacio argumenti | | Longitudo propior | | Stacio prima | | | Gradus |
| | G | M | Longior | G | M | G | M | G | M | S | G | M | G | M | Longior | G | M | G | M | G | M | S | G | M | |
| 23 | 4 | 6 | 55 | 0 | 32 | 9 | 6 | 0 | 35 | 5 | 7 | 56 | 8 | 35 | 36 | 1 | 17 | 20 | 38 | 1 | 26 | 5 | 9 | 46 | 7 |
| 24 | 4 | 16 | 55 | 0 | 33 | 9 | 30 | 0 | 37 | 5 | 7 | 59 | 8 | 42 | 35 | 1 | 18 | 21 | 0 | 1 | 28 | 5 | 9 | 51 | 6 |
| 25 | 4 | 26 | 54 | 0 | 35 | 9 | 54 | 0 | 38 | 5 | 8 | 2 | 8 | 50 | 34 | 1 | 20 | 21 | 23 | 1 | 30 | 5 | 9 | 56 | 5 |
| 26 | 4 | 36 | 54 | 0 | 37 | 10 | 18 | 0 | 40 | 5 | 8 | 4 | 8 | 57 | 33 | 1 | 21 | 21 | 45 | 1 | 32 | 5 | 10 | 1 | 4 |
| 27 | 4 | 46 | 53 | 0 | 38 | 10 | 41 | 0 | 41 | 5 | 8 | 6 | 9 | 4 | 32 | 1 | 23 | 22 | 7 | 1 | 34 | 5 | 10 | 6 | 3 |
| 28 | 4 | 56 | 53 | 0 | 40 | 11 | 5 | 0 | 43 | 5 | 8 | 9 | 9 | 11 | 31 | 1 | 24 | 22 | 29 | 1 | 36 | 5 | 10 | 11 | 2 |
| 29 | 5 | 6 | 52 | 0 | 41 | 11 | 28 | 0 | 44 | 5 | 8 | 12 | 9 | 18 | 30 | 1 | 26 | 22 | 51 | 1 | 38 | 5 | 10 | 16 | 1 |
| 30 | 5 | 16 | 52 | 0 | 42 | 11 | 51 | 0 | 46 | 5 | 8 | 15 | 9 | 24 | 30 | 1 | 27 | 23 | 13 | 1 | 40 | 5 | 10 | 21 | 0 |
| | 11 | | | | | | | | | | | | 10 | | | | | | | | | | | | |
| | A | | B | C | | D | | E | | F | | | A | | B | C | | D | | E | | F | | | |

Continuation of the Table TRW-18 (Toruń, BU, 74/III, f. 264v)

| Tabula equacionis Martis | | | | | | | | | | | | | | | | | | | | | | | | | |
|---|---|---|---|---|---|---|---|---|---|---|---|---|---|---|---|---|---|---|---|---|---|---|---|---|---|
| | 2 | | | | | | | | | | | | 3 | | | | | | | | | | | | |
| Gradus | Equacio centri | | Minuta propor-cionalia | Longitudo longior | | Equacio argumenti | | Longitudo propior | | Stacio prima | | | Equacio centri | | Minuta propor-cionalia | Longitudo longior | | Equacio argumenti | | Longitudo propior | | Stacio prima | | | Gradus |
| | G | M | Longior | G | M | G | M | G | M | S | G | M | G | M | Propior | G | M | G | M | G | M | S | G | M | |
| 1 | 9 | 31 | 29 | 1 | 29 | 23 | 35 | 1 | 42 | 5 | 10 | 27 | 11 | 23 | 4 | 2 | 30 | 33 | 40 | 2 | 52 | 5 | 13 | 31 | 29 |
| 2 | 9 | 37 | 28 | 1 | 30 | 23 | 57 | 1 | 44 | 5 | 10 | 33 | 11 | 24 | 5 | 2 | 32 | 33 | 58 | 2 | 55 | 5 | 13 | 36 | 28 |
| 3 | 9 | 43 | 27 | 1 | 32 | 24 | 18 | 1 | 46 | 5 | 10 | 39 | 11 | 24 | 6 | 2 | 35 | 34 | 15 | 2 | 57 | 5 | 13 | 41 | 27 |
| 4 | 9 | 49 | 26 | 1 | 34 | 24 | 40 | 1 | 48 | 5 | 10 | 44 | 11 | 24 | 7 | 2 | 37 | 34 | 32 | 3 | 0 | 5 | 13 | 46 | 26 |
| 5 | 9 | 55 | 25 | 1 | 36 | 25 | 1 | 1 | 50 | 5 | 10 | 50 | 11 | 24 | 8 | 2 | 39 | 34 | 49 | 3 | 3 | 5 | 13 | 51 | 25 |
| 6 | 10 | 0 | 24 | 1 | 37 | 25 | 22 | 1 | 53 | 5 | 10 | 56 | 11 | 24 | 9 | 2 | 42 | 35 | 6 | 3 | 6 | 5 | 13 | 57 | 24 |
| 7 | 10 | 5 | 23 | 1 | 39 | 25 | 44 | 1 | 55 | 5 | 11 | 2 | 11 | 23 | 10 | 2 | 44 | 35 | 23 | 3 | 9 | 5 | 14 | 3 | 23 |
| 8 | 10 | 10 | 22 | 1 | 41 | 26 | 5 | 1 | 57 | 5 | 11 | 8 | 11 | 23 | 11 | 2 | 47 | 35 | 40 | 3 | 12 | 5 | 14 | 9 | 22 |
| 9 | 10 | 15 | 21 | 1 | 43 | 26 | 26 | 1 | 59 | 5 | 11 | 14 | 11 | 22 | 12 | 2 | 49 | 35 | 56 | 3 | 15 | 5 | 14 | 14 | 21 |
| 10 | 10 | 20 | 20 | 1 | 45 | 26 | 47 | 2 | 1 | 5 | 11 | 19 | 11 | 21 | 13 | 2 | 51 | 36 | 12 | 3 | 19 | 5 | 14 | 20 | 20 |
| 11 | 10 | 25 | 19 | 1 | 47 | 27 | 8 | 2 | 3 | 5 | 11 | 25 | 11 | 20 | 14 | 2 | 54 | 36 | 28 | 3 | 22 | 5 | 14 | 26 | 19 |
| 12 | 10 | 29 | 17 | 1 | 49 | 27 | 29 | 2 | 6 | 5 | 11 | 31 | 11 | 19 | 15 | 2 | 56 | 36 | 43 | 3 | 25 | 5 | 14 | 31 | 18 |
| 13 | 10 | 34 | 16 | 1 | 51 | 27 | 50 | 2 | 8 | 5 | 11 | 37 | 11 | 17 | 16 | 2 | 59 | 36 | 58 | 3 | 28 | 5 | 14 | 38 | 17 |
| 14 | 10 | 38 | 15 | 1 | 53 | 28 | 11 | 2 | 10 | 5 | 11 | 43 | 11 | 15 | 16 | 3 | 1 | 37 | 13 | 3 | 32 | 5 | 14 | 42 | 16 |
| 15 | 10 | 42 | 14 | 1 | 55 | 28 | 31 | 2 | 12 | 5 | 11 | 50 | 11 | 13 | 17 | 3 | 4 | 37 | 27 | 3 | 36 | 5 | 14 | 48 | 15 |
| 16 | 10 | 46 | 13 | 1 | 57 | 28 | 52 | 2 | 14 | 5 | 11 | 56 | 11 | 11 | 18 | 3 | 7 | 37 | 41 | 3 | 39 | 5 | 14 | 53 | 14 |
| 17 | 10 | 50 | 12 | 1 | 59 | 29 | 11 | 2 | 16 | 5 | 12 | 2 | 11 | 9 | 19 | 3 | 10 | 37 | 55 | 3 | 43 | 5 | 14 | 59 | 13 |
| 18 | 10 | 53 | 11 | 2 | 1 | 29 | 32 | 2 | 19 | 5 | 12 | 9 | 11 | 6 | 20 | 3 | 13 | 38 | 9 | 3 | 47 | 5 | 15 | 5 | 12 |
| 19 | 10 | 57 | 10 | 2 | 3 | 29 | 52 | 2 | 21 | 5 | 12 | 15 | 11 | 3 | 21 | 3 | 16 | 38 | 23 | 3 | 50 | 5 | 15 | 11 | 11 |
| 20 | 11 | 0 | 9 | 2 | 5 | 30 | 12 | 2 | 23 | 5 | 12 | 21 | 11 | 0 | 22 | 3 | 19 | 38 | 36 | 3 | 54 | 5 | 15 | 19 | 10 |
| 21 | 11 | 3 | 8 | 2 | 8 | 30 | 32 | 2 | 26 | 5 | 12 | 28 | 10 | 57 | 22 | 3 | 22 | 38 | 48 | 3 | 58 | 5 | 15 | 22 | 9 |
| 22 | 11 | 6 | 7 | 2 | 10 | 30 | 52 | 2 | 28 | 5 | 12 | 34 | 10 | 53 | 23 | 3 | 25 | 39 | 1 | 4 | 1 | 5 | 15 | 28 | 8 |

**Tabula equacionis Martis**

| | 2 | | | | | | | | | | | | | 3 | | | | | | | | | | | | |
|---|---|---|---|---|---|---|---|---|---|---|---|---|---|---|---|---|---|---|---|---|---|---|---|---|---|
| **Gradus** | **Equacio centri** | | **Minuta propor-cionalia** | **Longitudo longior** | | **Equacio argumenti** | | **Longitudo propior** | | **Stacio prima** | | | **Equacio centri** | | **Minuta propor-cionalia** | **Longitudo longior** | | **Equacio argumenti** | | **Longitudo propior** | | **Stacio prima** | | | **Gradus** |
| | **G** | **M** | **Longior** | **G** | **M** | **G** | **M** | **G** | **M** | **S** | **G** | **M** | **G** | **M** | **Propior** | **G** | **M** | **G** | **M** | **G** | **M** | **S** | **G** | **M** | |
| 23 | 11 | 9 | 5 | 2 | 12 | 31 | 11 | 2 | 30 | 5 | 12 | 40 | 10 | 49 | 24 | 3 | 28 | 39 | 13 | 4 | 5 | 5 | 15 | 34 | 7 |
| 24 | 11 | 12 | 4 | 2 | 14 | 31 | 30 | 2 | 33 | 5 | 12 | 47 | 10 | 45 | 25 | 3 | 32 | 39 | 24 | 4 | 9 | 5 | 15 | 39 | 6 |
| 25 | 11 | 15 | 3 | 2 | 16 | 31 | 49 | 2 | 35 | 5 | 12 | 53 | 10 | 41 | 26 | 3 | 35 | 39 | 35 | 4 | 13 | 5 | 15 | 44 | 5 |
| 26 | 11 | 17 | 2 | 2 | 18 | 32 | 8 | 2 | 38 | 5 | 12 | 59 | 10 | 37 | 27 | 3 | 39 | 39 | 45 | 4 | 17 | 5 | 15 | 49 | 4 |
| 27 | 11 | 19 | 1 | 2 | 20 | 32 | 27 | 2 | 41 | 5 | 13 | 6 | 10 | 33 | 27 | 3 | 43 | 39 | 56 | 4 | 21 | 5 | 15 | 55 | 3 |
| 28 | 11 | 21 | **Propior** 1 | 2 | 23 | 32 | 46 | 2 | 43 | 5 | 13 | 12 | 10 | 29 | 28 | 3 | 46 | 40 | 5 | 4 | 26 | 5 | 16 | 0 | 2 |
| 29 | 11 | 22 | 2 | 2 | 25 | 33 | 4 | 2 | 46 | 5 | 13 | 18 | 10 | 25 | 29 | 3 | 50 | 40 | 14 | 4 | 30 | 5 | 16 | 6 | 1 |
| 30 | 11 | 23 | 3 | 2 | 27 | 33 | 22 | 2 | 49 | 5 | 13 | 25 | 10 | 21 | 30 | 3 | 54 | 40 | 23 | 4 | 35 | 5 | 16 | 11 | 0 |
| | 9 | | | | | | | | | | | | 8 | | | | | | | | | | | | |
| | A | | B | C | | D | | E | | F | | | A | | B | C | | D | | E | | F | | | |

Remaining part of the Table TRW-18 (Toruń, BU, 74/III, f. 265r)

| Tabula equacionis Martis | | | | | | | | | | | | | | | | | | | | | | | | | |
|---|---|---|---|---|---|---|---|---|---|---|---|---|---|---|---|---|---|---|---|---|---|---|---|---|---|
| | 4 | | | | | | | | | | | | 5 | | | | | | | | | | | | |
| Gradus | Equacio centri | | Minuta propor-cionalia | Longitudo longior | | Equacio argumenti | | Longitudo propior | | Stacio prima | | | Equacio centri | | Minuta propor-cionalia | Longitudo longior | | Equacio argumenti | | Longitudo propior | | Stacio prima | | | Gradus |
| | G | M | Propior | G | M | G | M | G | M | S | G | M | G | M | Propior | G | M | G | M | G | M | S | G | M | |
| 1 | 10 | 17 | 30 | 3 | 57 | 40 | 30 | 4 | 40 | 5 | 16 | 16 | 6 | 5 | 52 | 5 | 36 | 36 | 57 | 7 | 35 | 5 | 18 | 25 | 29 |
| 2 | 10 | 12 | 31 | 4 | 0 | 40 | 37 | 4 | 45 | 5 | 16 | 21 | 5 | 54 | 52 | 5 | 37 | 36 | 25 | 7 | 40 | 5 | 18 | 28 | 28 |
| 3 | 10 | 6 | 32 | 4 | 4 | 40 | 44 | 4 | 50 | 5 | 16 | 26 | 5 | 43 | 53 | 5 | 38 | 35 | 52 | 7 | 45 | 5 | 18 | 31 | 27 |
| 4 | 10 | 0 | 33 | 4 | 7 | 40 | 49 | 4 | 55 | 5 | 16 | 31 | 5 | 32 | 53 | 5 | 38 | 35 | 15 | 7 | 50 | 5 | 18 | 35 | 26 |
| 5 | 9 | 54 | 34 | 4 | 10 | 40 | 54 | 5 | 0 | 5 | 16 | 36 | 5 | 21 | 54 | 5 | 38 | 34 | 35 | 7 | 54 | 5 | 18 | 38 | 25 |
| 6 | 9 | 48 | 35 | 4 | 14 | 40 | 59 | 5 | 5 | 5 | 16 | 41 | 5 | 9 | 55 | 5 | 38 | 33 | 52 | 7 | 58 | 5 | 18 | 41 | 24 |
| 7 | 9 | 41 | 36 | 4 | 17 | 41 | 2 | 5 | 10 | 5 | 16 | 46 | 4 | 57 | 55 | 5 | 37 | 33 | 7 | 8 | 0 | 5 | 18 | 43 | 23 |
| 8 | 9 | 34 | 37 | 4 | 21 | 41 | 5 | 5 | 15 | 5 | 16 | 51 | 4 | 45 | 56 | 5 | 36 | 32 | 20 | 8 | 2 | 5 | 18 | 45 | 22 |
| 9 | 9 | 27 | 38 | 4 | 24 | 41 | 8 | 5 | 21 | 5 | 16 | 56 | 4 | 32 | 57 | 5 | 34 | 31 | 30 | 8 | 3 | 5 | 18 | 47 | 21 |
| 10 | 9 | 20 | 39 | 4 | 28 | 41 | 9 | 5 | 26 | 5 | 17 | 1 | 4 | 20 | 57 | 5 | 30 | 30 | 36 | 8 | 2 | 5 | 18 | 48 | 20 |
| 11 | 9 | 13 | 40 | 4 | 31 | 41 | 10 | 5 | 31 | 5 | 17 | 6 | 4 | 8 | 57 | 5 | 25 | 29 | 38 | 8 | 0 | 5 | 18 | 51 | 19 |
| 12 | 9 | 5 | 41 | 4 | 35 | 41 | 10 | 5 | 37 | 5 | 17 | 11 | 3 | 55 | 58 | 5 | 18 | 28 | 35 | 7 | 58 | 5 | 18 | 53 | 18 |
| 13 | 8 | 57 | 41 | 4 | 38 | 41 | 7 | 5 | 43 | 5 | 17 | 16 | 3 | 43 | 58 | 5 | 10 | 27 | 28 | 7 | 55 | 5 | 18 | 55 | 17 |
| 14 | 8 | 49 | 42 | 4 | 41 | 41 | 4 | 5 | 49 | 5 | 17 | 20 | 3 | 31 | 58 | 5 | 1 | 26 | 16 | 7 | 51 | 5 | 18 | 57 | 16 |
| 15 | 8 | 41 | 42 | 4 | 45 | 41 | 0 | 5 | 55 | 5 | 17 | 24 | 3 | 18 | 58 | 4 | 52 | 25 | 3 | 7 | 47 | 5 | 18 | 58 | 15 |
| 16 | 8 | 32 | 43 | 4 | 48 | 40 | 55 | 6 | 1 | 5 | 17 | 29 | 3 | 5 | 58 | 4 | 41 | 23 | 45 | 7 | 34 | 5 | 19 | 1 | 14 |
| 17 | 8 | 23 | 44 | 4 | 52 | 40 | 50 | 6 | 8 | 5 | 17 | 33 | 2 | 52 | 59 | 4 | 30 | 22 | 24 | 7 | 26 | 5 | 19 | 2 | 13 |
| 18 | 8 | 14 | 45 | 4 | 56 | 40 | 45 | 6 | 15 | 5 | 17 | 37 | 2 | 39 | 59 | 4 | 18 | 21 | 0 | 7 | 6 | 5 | 19 | 5 | 12 |
| 19 | 8 | 5 | 46 | 4 | 59 | 40 | 39 | 6 | 21 | 5 | 17 | 41 | 2 | 26 | 59 | 4 | 4 | 19 | 29 | 6 | 42 | 5 | 19 | 6 | 11 |
| 20 | 7 | 56 | 47 | 5 | 3 | 40 | 31 | 6 | 27 | 5 | 17 | 45 | 2 | 13 | 59 | 3 | 48 | 17 | 58 | 6 | 16 | 5 | 19 | 7 | 10 |
| 21 | 7 | 47 | 47 | 5 | 7 | 40 | 21 | 6 | 34 | 5 | 17 | 49 | 1 | 59 | 59 | 3 | 32 | 16 | 26 | 5 | 49 | 5 | 19 | 8 | 9 |
| 22 | 7 | 37 | 48 | 5 | 11 | 40 | 8 | 6 | 41 | 5 | 17 | 53 | 1 | 46 | 59 | 3 | 12 | 14 | 45 | 5 | 22 | 5 | 19 | 9 | 8 |

**Tabula equacionis Martis**

| | 4 | | | | | | | | | | | | 5 | | | | | | | | | | | | |
|---|---|---|---|---|---|---|---|---|---|---|---|---|---|---|---|---|---|---|---|---|---|---|---|---|---|
| **Gradus** | **Equacio centri** | | **Minuta propor-cionalia** | **Longitudo longior** | | **Equacio argumenti** | | **Longitudo propior** | | **Stacio prima** | | | **Equacio centri** | | **Minuta propor-cionalia** | **Longitudo longior** | | **Equacio argumenti** | | **Longitudo propior** | | **Stacio prima** | | | **Gradus** |
| | **G** | **M** | **Propior** | **G** | **M** | **G** | **M** | **G** | **M** | **S** | **G** | **M** | **G** | **M** | **Propior** | **G** | **M** | **G** | **M** | **G** | **M** | **S** | **G** | **M** | |
| 23 | 7 | 27 | 48 | 5 | 15 | 39 | 53 | 6 | 47 | 5 | 17 | 57 | 1 | 33 | 59 | 2 | 50 | 13 | 1 | 4 | 54 | 5 | 19 | 10 | 7 |
| 24 | 7 | 17 | 49 | 5 | 18 | 39 | 37 | 6 | 53 | 5 | 18 | 1 | 1 | 20 | 59 | 2 | 27 | 11 | 15 | 4 | 26 | 5 | 19 | 11 | 6 |
| 25 | 7 | 7 | 49 | 5 | 22 | 39 | 20 | 6 | 59 | 5 | 18 | 5 | 1 | 7 | 60 | 2 | 4 | 9 | 27 | 3 | 46 | 5 | 19 | 12 | 5 |
| 26 | 6 | 57 | 50 | 5 | 25 | 39 | 1 | 7 | 6 | 5 | 18 | 8 | 0 | 54 | 60 | 1 | 40 | 7 | 37 | 3 | 3 | 5 | 19 | 13 | 4 |
| 27 | 6 | 47 | 50 | 5 | 28 | 38 | 40 | 7 | 12 | 5 | 18 | 11 | 0 | 40 | 60 | 1 | 16 | 5 | 45 | 2 | 20 | 5 | 19 | 13 | 3 |
| 28 | 6 | 37 | 50 | 5 | 30 | 38 | 16 | 7 | 19 | 5 | 18 | 15 | 0 | 27 | 60 | 0 | 51 | 3 | 52 | 1 | 34 | 5 | 19 | 14 | 2 |
| 29 | 6 | 26 | 51 | 5 | 32 | 37 | 51 | 7 | 24 | 5 | 18 | 18 | 0 | 14 | 60 | 0 | 26 | 1 | 57 | 0 | 47 | 5 | 19 | 14 | 1 |
| 30 | 6 | 16 | 51 | 5 | 34 | 37 | 25 | 7 | 30 | 5 | 18 | 21 | 0 | 0 | 60 | 0 | 0 | 0 | 0 | 0 | 0 | 5 | 19 | 15 | 0 |
| | 7 | | | | | | | | | | | | 6 | | | | | | | | | | | | |
| | A | | B | C | | D | | E | | F | | | A | | B | C | | D | | E | | F | | | |

Table TRW-19 of equations of Venus (Toruń, BU, 74/III, f. 265v)

| Tabula equacionis Veneris | | | | | | | | | | | | | | | | | | | | | | | | | | | |
|---|---|---|---|---|---|---|---|---|---|---|---|---|---|---|---|---|---|---|---|---|---|---|---|---|---|---|---|
| | 0 | | | | | | | | | | | | | 1 | | | | | | | | | | | | | |
| Gradus | Equacio centri | | Minuta propor-cionalia | Longitudo longior | | Equacio argumenti | | Longitudo propior | | Stacio prima | | | Equacio centri | | Minuta propor-cionalia | Longitudo longior | | Equacio argumenti | | Longitudo propior | | Stacio prima | | | Gradus |
| | G | M | Longior | G | M | G | M | G | M | S | G | M | G | M | Longior | G | M | G | M | G | M | S | G | M | |
| 1 | 0 | 2 | 60 | 0 | 0 | 0 | 26 | 0 | 0 | 5 | 15 | 51 | 1 | 5 | 51 | 0 | 9 | 12 | 55 | 0 | 10 | 5 | 16 | 1 | 29 |
| 2 | 0 | 4 | 60 | 0 | 1 | 0 | 51 | 0 | 1 | 5 | 15 | 51 | 1 | 7 | 51 | 0 | 10 | 13 | 20 | 0 | 11 | 5 | 16 | 2 | 28 |
| 3 | 0 | 6 | 60 | 0 | 1 | 1 | 16 | 0 | 1 | 5 | 15 | 51 | 1 | 9 | 50 | 0 | 10 | 13 | 44 | 0 | 11 | 5 | 16 | 3 | 27 |
| 4 | 0 | 9 | 60 | 0 | 1 | 1 | 41 | 0 | 1 | 5 | 15 | 52 | 1 | 10 | 50 | 0 | 10 | 14 | 9 | 0 | 11 | 5 | 16 | 4 | 26 |
| 5 | 0 | 11 | 60 | 0 | 1 | 2 | 6 | 0 | 2 | 5 | 15 | 52 | 1 | 12 | 49 | 0 | 11 | 14 | 34 | 0 | 11 | 5 | 16 | 5 | 25 |
| 6 | 0 | 13 | 60 | 0 | 1 | 2 | 31 | 0 | 2 | 5 | 15 | 52 | 1 | 14 | 49 | 0 | 11 | 14 | 58 | 0 | 12 | 5 | 16 | 6 | 24 |
| 7 | 0 | 15 | 60 | 0 | 1 | 2 | 56 | 0 | 2 | 5 | 15 | 54 | 1 | 16 | 48 | 0 | 11 | 15 | 23 | 0 | 12 | 5 | 16 | 7 | 23 |
| 8 | 0 | 17 | 59 | 0 | 1 | 3 | 21 | 0 | 3 | 5 | 15 | 54 | 1 | 17 | 48 | 0 | 12 | 15 | 48 | 0 | 12 | 5 | 16 | 8 | 22 |
| 9 | 0 | 19 | 59 | 0 | 2 | 3 | 46 | 0 | 3 | 5 | 15 | 54 | 1 | 19 | 47 | 0 | 12 | 16 | 12 | 0 | 12 | 5 | 16 | 9 | 21 |
| 10 | 0 | 21 | 59 | 0 | 2 | 4 | 11 | 0 | 3 | 5 | 15 | 54 | 1 | 21 | 47 | 0 | 12 | 16 | 37 | 0 | 13 | 5 | 16 | 10 | 20 |
| 11 | 0 | 24 | 59 | 0 | 2 | 4 | 36 | 0 | 4 | 5 | 15 | 54 | 1 | 22 | 46 | 0 | 13 | 17 | 1 | 0 | 13 | 5 | 16 | 11 | 19 |
| 12 | 0 | 26 | 59 | 0 | 3 | 5 | 1 | 0 | 4 | 5 | 15 | 54 | 1 | 24 | 45 | 0 | 13 | 17 | 25 | 0 | 13 | 5 | 16 | 12 | 18 |
| 13 | 0 | 28 | 58 | 0 | 3 | 5 | 26 | 0 | 4 | 5 | 15 | 55 | 1 | 26 | 45 | 0 | 13 | 17 | 50 | 0 | 13 | 5 | 16 | 13 | 17 |
| 14 | 0 | 30 | 58 | 0 | 3 | 5 | 51 | 0 | 5 | 5 | 15 | 55 | 1 | 27 | 44 | 0 | 14 | 18 | 14 | 0 | 14 | 5 | 16 | 14 | 16 |
| 15 | 0 | 32 | 58 | 0 | 4 | 6 | 16 | 0 | 5 | 5 | 15 | 56 | 1 | 29 | 43 | 0 | 14 | 18 | 38 | 0 | 14 | 5 | 16 | 15 | 15 |
| 16 | 0 | 34 | 57 | 0 | 4 | 6 | 41 | 0 | 5 | 5 | 15 | 56 | 1 | 32 | 42 | 0 | 14 | 19 | 3 | 0 | 14 | 5 | 16 | 16 | 14 |
| 17 | 0 | 36 | 57 | 0 | 5 | 7 | 6 | 0 | 6 | 5 | 15 | 56 | 1 | 33 | 42 | 0 | 15 | 19 | 27 | 0 | 14 | 5 | 16 | 17 | 13 |
| 18 | 0 | 38 | 57 | 0 | 5 | 7 | 31 | 0 | 6 | 5 | 15 | 57 | 1 | 34 | 41 | 0 | 15 | 19 | 51 | 0 | 15 | 5 | 16 | 18 | 12 |
| 19 | 0 | 41 | 56 | 0 | 5 | 7 | 56 | 0 | 6 | 5 | 15 | 57 | 1 | 36 | 40 | 0 | 15 | 20 | 15 | 0 | 15 | 5 | 16 | 19 | 11 |
| 20 | 0 | 43 | 56 | 0 | 6 | 8 | 21 | 0 | 7 | 5 | 15 | 57 | 1 | 37 | 40 | 0 | 16 | 20 | 39 | 0 | 15 | 5 | 16 | 20 | 10 |
| 21 | 0 | 45 | 56 | 0 | 6 | 8 | 46 | 0 | 7 | 5 | 15 | 58 | 1 | 39 | 39 | 0 | 16 | 21 | 3 | 0 | 16 | 5 | 16 | 21 | 9 |
| 22 | 0 | 47 | 55 | 0 | 6 | 9 | 11 | 0 | 7 | 5 | 15 | 58 | 1 | 40 | 38 | 0 | 16 | 21 | 27 | 0 | 16 | 5 | 16 | 22 | 8 |

| Tabula equacionis Veneris | | | | | | | | | | | | | | | | | | | | | | | | | |
|---|---|---|---|---|---|---|---|---|---|---|---|---|---|---|---|---|---|---|---|---|---|---|---|---|---|
| | 0 | | | | | | | | | | | | 1 | | | | | | | | | | | | |
| Gradus | Equacio centri | | Minuta propor-cionalia | Longitudo longior | | Equacio argumenti | | Longitudo propior | | Stacio prima | | | Equacio centri | | Minuta propor-cionalia | Longitudo longior | | Equacio argumenti | | Longitudo propior | | Stacio prima | | | Gradus |
| | G | M | Longior | G | M | G | M | G | M | S | G | M | G | M | Longior | G | M | G | M | G | M | S | G | M | |
| 23 | 0 | 49 | 55 | 0 | 7 | 9 | 36 | 0 | 8 | 5 | 15 | 58 | 1 | 42 | 37 | 0 | 17 | 21 | 51 | 0 | 16 | 5 | 16 | 23 | 7 |
| 24 | 0 | 51 | 55 | 0 | 7 | 10 | 1 | 0 | 8 | 5 | 15 | 59 | 1 | 43 | 36 | 0 | 17 | 22 | 15 | 0 | 17 | 5 | 16 | 24 | 6 |
| 25 | 0 | 53 | 54 | 0 | 7 | 10 | 26 | 0 | 8 | 5 | 15 | 59 | 1 | 44 | 35 | 0 | 17 | 22 | 39 | 0 | 17 | 5 | 16 | 25 | 5 |
| 26 | 0 | 55 | 54 | 0 | 8 | 10 | 51 | 0 | 9 | 5 | 15 | 59 | 1 | 46 | 34 | 0 | 18 | 23 | 3 | 0 | 17 | 5 | 16 | 26 | 4 |
| 27 | 0 | 57 | 53 | 0 | 8 | 11 | 16 | 0 | 9 | 5 | 16 | 0 | 1 | 47 | 33 | 0 | 18 | 23 | 27 | 0 | 18 | 5 | 16 | 27 | 3 |
| 28 | 0 | 59 | 53 | 0 | 8 | 11 | 41 | 0 | 9 | 5 | 16 | 0 | 1 | 48 | 32 | 0 | 18 | 23 | 51 | 0 | 18 | 5 | 16 | 28 | 2 |
| 29 | 1 | 1 | 52 | 0 | 9 | 12 | 6 | 0 | 10 | 5 | 16 | 0 | 1 | 50 | 31 | 0 | 19 | 24 | 15 | 0 | 19 | 5 | 16 | 29 | 1 |
| 30 | 1 | 3 | 52 | 0 | 9 | 12 | 30 | 0 | 10 | 5 | 16 | 0 | 1 | 51 | 30 | 0 | 19 | 24 | 38 | 0 | 19 | 5 | 16 | 30 | 0 |
| | 11 | | | | | | | | | | | | 10 | | | | | | | | | | | | |
| | A | | B | C | | D | | E | | F | | | A | | B | C | | D | | E | | F | | | |

Continuation of the Table TRW-19 (Toruń, BU, 74/III, f. 266r)

| Tabula equacionis Veneris | | | | | | | | | | | | | | | | | | | | | | | | | | |
|---|---|---|---|---|---|---|---|---|---|---|---|---|---|---|---|---|---|---|---|---|---|---|---|---|---|---|
| | 2 | | | | | | | | | | | | 3 | | | | | | | | | | | | | |
| Gradus | Equacio centri | | Minuta propor-cionalia | Longitudo longior | | Equacio argumenti | | Longitudo propior | | Stacio prima | | | Equacio centri | | Minuta propor-cionalia | Longitudo longior | | Equacio argumenti | | Longitudo propior | | Stacio prima | | | Gradus |
| | G | M | Longior | G | M | G | M | G | M | S | G | M | G | M | Propior | G | M | G | M | G | M | S | G | M | |
| 1 | 1 | 52 | 29 | 0 | 19 | 25 | 2 | 0 | 19 | 5 | 16 | 31 | 2 | 10 | 3 | 0 | 34 | 36 | 4 | 0 | 35 | 5 | 17 | 12 | 29 |
| 2 | 1 | 53 | 28 | 0 | 20 | 25 | 25 | 0 | 20 | 5 | 16 | 32 | 2 | 10 | 4 | 0 | 35 | 36 | 24 | 0 | 36 | 5 | 17 | 13 | 28 |
| 3 | 1 | 54 | 27 | 0 | 20 | 25 | 48 | 0 | 20 | 5 | 16 | 33 | 2 | 10 | 5 | 0 | 35 | 36 | 44 | 0 | 37 | 5 | 17 | 14 | 27 |
| 4 | 1 | 55 | 26 | 0 | 20 | 26 | 11 | 0 | 21 | 5 | 16 | 34 | 2 | 10 | 6 | 0 | 36 | 37 | 4 | 0 | 37 | 5 | 17 | 16 | 26 |
| 5 | 1 | 56 | 26 | 0 | 21 | 26 | 34 | 0 | 21 | 5 | 16 | 35 | 2 | 10 | 7 | 0 | 36 | 37 | 23 | 0 | 38 | 5 | 17 | 15 | 25 |
| 6 | 1 | 57 | 25 | 0 | 21 | 26 | 57 | 0 | 22 | 5 | 16 | 36 | 2 | 10 | 8 | 0 | 37 | 37 | 43 | 0 | 38 | 5 | 17 | 17 | 24 |
| 7 | 1 | 58 | 24 | 0 | 22 | 27 | 20 | 0 | 22 | 5 | 16 | 37 | 2 | 10 | 9 | 0 | 37 | 38 | 2 | 0 | 39 | 5 | 17 | 18 | 23 |
| 8 | 1 | 59 | 23 | 0 | 22 | 27 | 43 | 0 | 23 | 5 | 16 | 38 | 2 | 10 | 10 | 0 | 38 | 38 | 21 | 0 | 40 | 5 | 17 | 19 | 22 |
| 9 | 2 | 0 | 22 | 0 | 23 | 28 | 6 | 0 | 23 | 5 | 16 | 39 | 2 | 9 | 11 | 0 | 38 | 38 | 40 | 0 | 40 | 5 | 17 | 20 | 21 |
| 10 | 2 | 1 | 21 | 0 | 23 | 28 | 29 | 0 | 24 | 5 | 16 | 40 | 2 | 9 | 12 | 0 | 39 | 38 | 59 | 0 | 41 | 5 | 17 | 21 | 20 |
| 11 | 2 | 1 | 20 | 0 | 24 | 28 | 52 | 0 | 24 | 5 | 16 | 41 | 2 | 9 | 13 | 0 | 39 | 39 | 17 | 0 | 42 | 5 | 17 | 22 | 19 |
| 12 | 2 | 2 | 19 | 0 | 24 | 29 | 14 | 0 | 25 | 5 | 16 | 43 | 2 | 9 | 14 | 0 | 40 | 39 | 35 | 0 | 43 | 5 | 17 | 23 | 18 |
| 13 | 2 | 3 | 18 | 0 | 25 | 29 | 37 | 0 | 25 | 5 | 16 | 45 | 2 | 8 | 15 | 0 | 41 | 39 | 53 | 0 | 43 | 5 | 17 | 24 | 17 |
| 14 | 2 | 3 | 16 | 0 | 25 | 29 | 59 | 0 | 26 | 5 | 16 | 47 | 2 | 8 | 16 | 0 | 41 | 40 | 11 | 0 | 44 | 5 | 17 | 25 | 16 |
| 15 | 2 | 4 | 15 | 0 | 25 | 30 | 21 | 0 | 27 | 5 | 16 | 49 | 2 | 7 | 17 | 0 | 42 | 40 | 29 | 0 | 45 | 5 | 17 | 26 | 15 |
| 16 | 2 | 5 | 14 | 0 | 26 | 30 | 43 | 0 | 27 | 5 | 16 | 50 | 2 | 7 | 18 | 0 | 43 | 40 | 46 | 0 | 46 | 5 | 17 | 27 | 14 |
| 17 | 2 | 5 | 13 | 0 | 26 | 31 | 5 | 0 | 28 | 5 | 16 | 52 | 2 | 6 | 19 | 0 | 43 | 41 | 3 | 0 | 47 | 5 | 17 | 28 | 13 |
| 18 | 2 | 6 | 12 | 0 | 26 | 31 | 27 | 0 | 28 | 5 | 16 | 53 | 2 | 5 | 20 | 0 | 44 | 41 | 20 | 0 | 47 | 5 | 17 | 29 | 12 |
| 19 | 2 | 6 | 11 | 0 | 27 | 31 | 49 | 0 | 29 | 5 | 16 | 54 | 2 | 5 | 21 | 0 | 45 | 41 | 37 | 0 | 48 | 5 | 17 | 30 | 11 |
| 20 | 2 | 7 | 10 | 0 | 27 | 32 | 11 | 0 | 29 | 5 | 16 | 56 | 2 | 4 | 21 | 0 | 46 | 41 | 53 | 0 | 49 | 5 | 17 | 32 | 10 |
| 21 | 2 | 7 | 9 | 0 | 28 | 32 | 33 | 0 | 30 | 5 | 16 | 58 | 2 | 4 | 22 | 0 | 47 | 42 | 9 | 0 | 50 | 5 | 17 | 33 | 9 |
| 22 | 2 | 8 | 8 | 0 | 28 | 32 | 55 | 0 | 30 | 5 | 16 | 59 | 2 | 3 | 23 | 0 | 48 | 42 | 24 | 0 | 51 | 5 | 17 | 34 | 8 |

**Tabula equacionis Veneris**

| | 2 | | | | | | | | | | | | 3 | | | | | | | | | | | | |
|---|---|---|---|---|---|---|---|---|---|---|---|---|---|---|---|---|---|---|---|---|---|---|---|---|---|
| **Gradus** | **Equacio centri** | | **Minuta propor-cionalia** | **Longitudo longior** | | **Equacio argumenti** | | **Longitudo propior** | | **Stacio prima** | | | **Equacio centri** | | **Minuta propor-cionalia** | **Longitudo longior** | | **Equacio argumenti** | | **Longitudo propior** | | **Stacio prima** | | | **Gradus** |
| | **G** | **M** | Longior | **G** | **M** | **G** | **M** | **G** | **M** | **S** | **G** | **M** | **G** | **M** | Propior | **G** | **M** | **G** | **M** | **G** | **M** | **S** | **G** | **M** | |
| 23 | 2 | 9 | 7 | 0 | 29 | 33 | 17 | 0 | 31 | 5 | 17 | 0 | 2 | 2 | 24 | 0 | 49 | 42 | 39 | 0 | 52 | 5 | 17 | 36 | 7 |
| 24 | 2 | 9 | 5 | 0 | 30 | 33 | 38 | 0 | 31 | 5 | 17 | 2 | 2 | 1 | 25 | 0 | 50 | 42 | 54 | 0 | 52 | 5 | 17 | 37 | 6 |
| 25 | 2 | 9 | 4 | 0 | 30 | 34 | 0 | 0 | 32 | 5 | 17 | 3 | 2 | 0 | 26 | 0 | 51 | 43 | 7 | 0 | 53 | 5 | 17 | 38 | 5 |
| 26 | 2 | 9 | 3 | 0 | 31 | 34 | 21 | 0 | 32 | 5 | 17 | 5 | 1 | 59 | 27 | 0 | 51 | 43 | 21 | 0 | 54 | 5 | 17 | 39 | 4 |
| 27 | 2 | 9 | 2 | 0 | 32 | 34 | 42 | 0 | 33 | 5 | 17 | 7 | 1 | 58 | 28 | 0 | 52 | 43 | 35 | 0 | 55 | 5 | 17 | 41 | 3 |
| 28 | 2 | 10 | 1 | 0 | 32 | 35 | 3 | 0 | 33 | 5 | 17 | 8 | 1 | 57 | 29 | 0 | 53 | 43 | 48 | 0 | 56 | 5 | 17 | 42 | 2 |
| 29 | 2 | 10 | Propior 1 | 0 | 33 | 35 | 24 | 0 | 34 | 5 | 17 | 10 | 1 | 56 | 30 | 0 | 54 | 44 | 1 | 0 | 56 | 5 | 17 | 43 | 1 |
| 30 | 2 | 10 | 2 | 0 | 33 | 35 | 44 | 0 | 34 | 5 | 17 | 11 | 1 | 55 | 31 | 0 | 54 | 44 | 13 | 0 | 57 | 5 | 17 | 45 | 0 |
| | 9 | | | | | | | | | | | | 8 | | | | | | | | | | | | |

A B C D E F A B C D E F

Remaining part of the Table TRW-19 (Toruń, BU, 74/III, f. 266v)

| Tabula equacionis Veneris | | | | | | | | | | | | | | | | | | | | | | | | | |
|---|---|---|---|---|---|---|---|---|---|---|---|---|---|---|---|---|---|---|---|---|---|---|---|---|---|
| | 4 | | | | | | | | | | | | 5 | | | | | | | | | | | | |
| Gradus | Equacio centri | | Minuta propor-cionalia | Longitudo longior | | Equacio argumenti | | Longitudo propior | | Stacio prima | | | Equacio centri | | Minuta propor-cionalia | Longitudo longior | | Equacio argumenti | | Longitudo propior | | Stacio prima | | | Gradus |
| | G | M | Propior | G | M | G | M | G | M | S | G | M | G | M | Propior | G | M | G | M | G | M | S | G | M | |
| 1 | 1 | 54 | 32 | 0 | 55 | 44 | 25 | 0 | 58 | 5 | 17 | 46 | 1 | 5 | 53 | 1 | 35 | 43 | 15 | 1 | 40 | 5 | 18 | 11 | 29 |
| 2 | 1 | 53 | 33 | 0 | 56 | 44 | 36 | 0 | 59 | 5 | 17 | 48 | 1 | 3 | 54 | 1 | 36 | 42 | 48 | 1 | 42 | 5 | 18 | 11 | 28 |
| 3 | 1 | 52 | 34 | 0 | 57 | 44 | 47 | 1 | 0 | 5 | 17 | 49 | 1 | 1 | 54 | 1 | 37 | 42 | 18 | 1 | 44 | 5 | 18 | 12 | 27 |
| 4 | 1 | 50 | 35 | 0 | 58 | 44 | 57 | 1 | 1 | 5 | 17 | 50 | 0 | 59 | 55 | 1 | 38 | 41 | 45 | 1 | 46 | 5 | 18 | 12 | 26 |
| 5 | 1 | 49 | 36 | 0 | 59 | 45 | 6 | 1 | 2 | 5 | 17 | 51 | 0 | 57 | 55 | 1 | 39 | 41 | 8 | 1 | 47 | 5 | 18 | 13 | 25 |
| 6 | 1 | 48 | 37 | 1 | 0 | 45 | 14 | 1 | 3 | 5 | 17 | 52 | 0 | 55 | 56 | 1 | 39 | 40 | 28 | 1 | 48 | 5 | 18 | 14 | 24 |
| 7 | 1 | 46 | 38 | 1 | 1 | 45 | 21 | 1 | 5 | 5 | 17 | 53 | 0 | 52 | 56 | 1 | 40 | 39 | 46 | 1 | 49 | 5 | 18 | 14 | 23 |
| 8 | 1 | 45 | 39 | 1 | 2 | 45 | 27 | 1 | 5 | 5 | 17 | 54 | 0 | 50 | 56 | 1 | 40 | 38 | 58 | 1 | 50 | 5 | 18 | 15 | 22 |
| 9 | 1 | 44 | 40 | 1 | 3 | 45 | 33 | 1 | 7 | 5 | 17 | 54 | 0 | 48 | 57 | 1 | 41 | 38 | 7 | 1 | 51 | 5 | 18 | 16 | 21 |
| 10 | 1 | 42 | 40 | 1 | 4 | 45 | 39 | 1 | 9 | 5 | 17 | 55 | 0 | 46 | 57 | 1 | 41 | 37 | 12 | 1 | 51 | 5 | 18 | 16 | 20 |
| 11 | 1 | 41 | 41 | 1 | 5 | 45 | 45 | 1 | 10 | 5 | 17 | 56 | 0 | 44 | 57 | 1 | 42 | 36 | 12 | 1 | 52 | 5 | 18 | 17 | 19 |
| 12 | 1 | 39 | 42 | 1 | 6 | 45 | 50 | 1 | 11 | 5 | 17 | 57 | 0 | 42 | 57 | 1 | 42 | 35 | 7 | 1 | 52 | 5 | 18 | 17 | 18 |
| 13 | 1 | 38 | 43 | 1 | 7 | 45 | 54 | 1 | 12 | 5 | 17 | 57 | 0 | 40 | 58 | 1 | 41 | 33 | 57 | 1 | 52 | 5 | 18 | 17 | 17 |
| 14 | 1 | 36 | 44 | 1 | 9 | 45 | 57 | 1 | 13 | 5 | 17 | 58 | 0 | 37 | 58 | 1 | 40 | 32 | 44 | 1 | 51 | 5 | 18 | 18 | 16 |
| 15 | 1 | 34 | 44 | 1 | 10 | 45 | 59 | 1 | 15 | 5 | 17 | 59 | 0 | 35 | 58 | 1 | 38 | 31 | 24 | 1 | 50 | 5 | 18 | 18 | 15 |
| 16 | 1 | 33 | 45 | 1 | 11 | 45 | 59 | 1 | 16 | 5 | 18 | 0 | 0 | 33 | 58 | 1 | 36 | 29 | 58 | 1 | 49 | 5 | 18 | 18 | 14 |
| 17 | 1 | 31 | 46 | 1 | 12 | 45 | 58 | 1 | 17 | 5 | 18 | 1 | 0 | 31 | 58 | 1 | 34 | 28 | 25 | 1 | 46 | 5 | 18 | 18 | 13 |
| 18 | 1 | 30 | 47 | 1 | 14 | 45 | 57 | 1 | 18 | 5 | 18 | 2 | 0 | 28 | 59 | 1 | 31 | 26 | 46 | 1 | 42 | 5 | 18 | 19 | 12 |
| 19 | 1 | 28 | 47 | 1 | 15 | 45 | 55 | 1 | 20 | 5 | 18 | 3 | 0 | 26 | 59 | 1 | 28 | 25 | 2 | 1 | 38 | 5 | 18 | 19 | 11 |
| 20 | 1 | 26 | 48 | 1 | 17 | 45 | 51 | 1 | 21 | 5 | 18 | 3 | 0 | 24 | 59 | 1 | 24 | 23 | 12 | 1 | 33 | 5 | 18 | 19 | 10 |
| 21 | 1 | 24 | 48 | 1 | 19 | 45 | 46 | 1 | 23 | 5 | 18 | 4 | 0 | 21 | 59 | 1 | 19 | 21 | 15 | 1 | 27 | 5 | 18 | 19 | 9 |
| 22 | 1 | 23 | 49 | 1 | 21 | 45 | 39 | 1 | 24 | 5 | 18 | 5 | 0 | 19 | 59 | 1 | 12 | 19 | 11 | 1 | 21 | 5 | 18 | 20 | 8 |

**Tabula equacionis Veneris**

| | 4 | | | | | | | | | | | | 5 | | | | | | | | | | | | |
|---|---|---|---|---|---|---|---|---|---|---|---|---|---|---|---|---|---|---|---|---|---|---|---|---|---|
| **Gradus** | **Equacio centri** | | **Minuta propor-cionalia** | **Longitudo longior** | | **Equacio argumenti** | | **Longitudo propior** | | **Stacio prima** | | | **Equacio centri** | | **Minuta propor-cionalia** | **Longitudo longior** | | **Equacio argumenti** | | **Longitudo propior** | | **Stacio prima** | | | **Gradus** |
| | **G** | **M** | **Propior** | **G** | **M** | **G** | **M** | **G** | **M** | **S** | **G** | **M** | **G** | **M** | **Propior** | **G** | **M** | **G** | **M** | **G** | **M** | **S** | **G** | **M** | |
| 23 | 1 | 21 | 49 | 1 | 23 | 45 | 31 | 1 | 25 | 5 | 18 | 5 | 0 | 17 | 59 | 1 | 4 | 17 | 2 | 1 | 14 | 5 | 18 | 20 | 7 |
| 24 | 1 | 19 | 50 | 1 | 24 | 45 | 21 | 1 | 27 | 5 | 18 | 6 | 0 | 14 | 60 | 0 | 57 | 14 | 47 | 1 | 6 | 5 | 18 | 20 | 6 |
| 25 | 1 | 17 | 50 | 1 | 26 | 45 | 9 | 1 | 28 | 5 | 18 | 7 | 0 | 12 | 60 | 0 | 48 | 12 | 46 | 0 | 57 | 5 | 18 | 21 | 5 |
| 26 | 1 | 15 | 51 | 1 | 28 | 44 | 55 | 1 | 30 | 5 | 18 | 7 | 0 | 10 | 60 | 0 | 40 | 10 | 4 | 0 | 46 | 5 | 18 | 21 | 4 |
| 27 | 1 | 13 | 51 | 1 | 29 | 44 | 39 | 1 | 32 | 5 | 18 | 8 | 0 | 7 | 60 | 0 | 31 | 7 | 38 | 0 | 35 | 5 | 18 | 21 | 3 |
| 28 | 1 | 11 | 52 | 1 | 31 | 44 | 21 | 1 | 34 | 5 | 18 | 9 | 0 | 5 | 60 | 0 | 21 | 5 | 9 | 0 | 24 | 5 | 18 | 21 | 2 |
| 29 | 1 | 9 | 52 | 1 | 32 | 44 | 1 | 1 | 36 | 5 | 18 | 9 | 0 | 3 | 60 | 0 | 11 | 2 | 34 | 0 | 12 | 5 | 18 | 21 | 1 |
| 30 | 1 | 7 | 53 | 1 | 33 | 43 | 39 | 1 | 38 | 5 | 18 | 10 | 0 | 0 | 60 | 0 | 0 | 0 | 0 | 0 | 0 | 5 | 18 | 21 | 0 |
| | 7 | | | | | | | | | | | | 6 | | | | | | | | | | | | |
| | A | | B | C | | D | | E | | F | | | A | | B | C | | D | | E | | F | | | |

Table TRW-20 of equations of Mercury (Toruń, BU, 74/III, f. 267r)

| Tabula equacionis Mercurii | | | | | | | | | | | | | | | | | | | | | | | | | |
|---|---|---|---|---|---|---|---|---|---|---|---|---|---|---|---|---|---|---|---|---|---|---|---|---|---|
| | 0 | | | | | | | | | | | | 1 | | | | | | | | | | | | |
| Gradus | Equacio centri | | Minuta propor-cionalia | Longitudo longior | | Equacio argumenti | | Longitudo propior | | Stacio prima | | | Equacio centri | | Minuta propor-cionalia | Longitudo longior | | Equacio argumenti | | Longitudo propior | | Stacio prima | | | Gradus |
| | G | M | Minuatur | G | M | G | M | G | M | S | G | M | G | M | Minuatur | G | M | G | M | G | M | S | G | M | |
| 1 | 0 | 3 | 60 | 0 | 2 | 0 | 17 | 0 | 1 | 4 | 27 | 14 | 1 | 20 | 43 | 0 | 51 | 8 | 20 | 0 | 29 | 4 | 26 | 34 | 29 |
| 2 | 0 | 6 | 60 | 0 | 4 | 0 | 33 | 0 | 2 | 4 | 27 | 14 | 1 | 23 | 42 | 0 | 53 | 8 | 35 | 0 | 30 | 4 | 26 | 32 | 28 |
| 3 | 0 | 9 | 60 | 0 | 5 | 0 | 49 | 0 | 3 | 4 | 27 | 14 | 1 | 25 | 41 | 0 | 54 | 8 | 50 | 0 | 31 | 4 | 26 | 29 | 27 |
| 4 | 0 | 12 | 59 | 0 | 7 | 1 | 5 | 0 | 4 | 4 | 27 | 13 | 1 | 28 | 40 | 0 | 56 | 9 | 6 | 0 | 32 | 4 | 26 | 27 | 26 |
| 5 | 0 | 15 | 59 | 0 | 9 | 1 | 22 | 0 | 4 | 4 | 27 | 13 | 1 | 30 | 39 | 0 | 58 | 9 | 21 | 0 | 33 | 4 | 26 | 24 | 25 |
| 6 | 0 | 17 | 59 | 0 | 10 | 1 | 38 | 0 | 5 | 4 | 27 | 12 | 1 | 33 | 38 | 0 | 59 | 9 | 36 | 0 | 34 | 4 | 26 | 21 | 24 |
| 7 | 0 | 20 | 58 | 0 | 12 | 1 | 55 | 0 | 6 | 4 | 27 | 12 | 1 | 36 | 36 | 1 | 1 | 9 | 51 | 0 | 35 | 4 | 26 | 19 | 23 |
| 8 | 0 | 23 | 58 | 0 | 14 | 2 | 11 | 0 | 7 | 4 | 27 | 11 | 1 | 38 | 35 | 1 | 2 | 10 | 6 | 0 | 36 | 4 | 26 | 17 | 22 |
| 9 | 0 | 25 | 58 | 0 | 15 | 2 | 27 | 0 | 8 | 4 | 27 | 11 | 1 | 40 | 34 | 1 | 4 | 10 | 21 | 0 | 37 | 4 | 26 | 15 | 21 |
| 10 | 0 | 28 | 57 | 0 | 17 | 2 | 44 | 0 | 9 | 4 | 27 | 10 | 1 | 43 | 33 | 1 | 5 | 10 | 36 | 0 | 38 | 4 | 26 | 12 | 20 |
| 11 | 0 | 30 | 57 | 0 | 19 | 3 | 0 | 0 | 10 | 4 | 27 | 9 | 1 | 45 | 32 | 1 | 7 | 10 | 51 | 0 | 39 | 4 | 26 | 9 | 19 |
| 12 | 0 | 33 | 57 | 0 | 20 | 3 | 16 | 0 | 11 | 4 | 27 | 8 | 1 | 47 | 31 | 1 | 8 | 11 | 6 | 0 | 40 | 4 | 26 | 6 | 18 |
| 13 | 0 | 35 | 56 | 0 | 22 | 3 | 32 | 0 | 12 | 4 | 27 | 7 | 1 | 50 | 29 | 1 | 10 | 11 | 21 | 0 | 41 | 4 | 26 | 4 | 17 |
| 14 | 0 | 38 | 56 | 0 | 23 | 3 | 48 | 0 | 13 | 4 | 27 | 6 | 1 | 52 | 28 | 1 | 12 | 11 | 36 | 0 | 42 | 4 | 26 | 1 | 16 |
| 15 | 0 | 40 | 55 | 0 | 24 | 4 | 5 | 0 | 14 | 4 | 27 | 5 | 1 | 54 | 27 | 1 | 13 | 11 | 50 | 0 | 43 | 4 | 25 | 58 | 15 |
| 16 | 0 | 43 | 55 | 0 | 26 | 4 | 21 | 0 | 15 | 4 | 27 | 3 | 1 | 57 | 25 | 1 | 15 | 12 | 5 | 0 | 44 | 4 | 25 | 56 | 14 |
| 17 | 0 | 45 | 54 | 0 | 28 | 4 | 37 | 0 | 16 | 4 | 27 | 2 | 1 | 59 | 24 | 1 | 16 | 12 | 19 | 0 | 45 | 4 | 25 | 53 | 13 |
| 18 | 0 | 48 | 54 | 0 | 29 | 4 | 53 | 0 | 17 | 4 | 27 | 0 | 2 | 1 | 23 | 1 | 18 | 12 | 34 | 0 | 46 | 4 | 25 | 50 | 12 |
| 19 | 0 | 50 | 53 | 0 | 31 | 5 | 9 | 0 | 18 | 4 | 26 | 59 | 2 | 4 | 21 | 1 | 20 | 12 | 48 | 0 | 47 | 4 | 25 | 48 | 11 |
| 20 | 0 | 53 | 53 | 0 | 33 | 5 | 25 | 0 | 19 | 4 | 26 | 56 | 2 | 6 | 20 | 1 | 22 | 13 | 2 | 0 | 48 | 4 | 25 | 45 | 10 |
| 21 | 0 | 55 | 52 | 0 | 34 | 5 | 41 | 0 | 20 | 4 | 26 | 54 | 2 | 8 | 19 | 1 | 23 | 13 | 16 | 0 | 48 | 4 | 25 | 42 | 9 |
| 22 | 0 | 58 | 51 | 0 | 36 | 5 | 57 | 0 | 21 | 4 | 26 | 54 | 2 | 10 | 17 | 1 | 25 | 13 | 30 | 0 | 49 | 4 | 25 | 40 | 8 |

**Tabula equacionis Mercurii**

| | 0 | | | | | | | | | | | 1 | | | | | | | | | | | |
|---|---|---|---|---|---|---|---|---|---|---|---|---|---|---|---|---|---|---|---|---|---|---|---|
| Gradus | Equacio centri | | Minuta propor-cionalia | Longitudo longior | | Equacio argumenti | | Longitudo propior | | Stacio prima | | | Equacio centri | | Minuta propor-cionalia | Longitudo longior | | Equacio argumenti | | Longitudo propior | | Stacio prima | | | Gradus |
| | G | M | Minuatur | G | M | G | M | G | M | S | G | M | G | M | Minuatur | G | M | G | M | G | M | S | G | M | |
| 23 | 1 | 0 | 51 | 0 | 38 | 6 | 13 | 0 | 22 | 4 | 26 | 50 | 2 | 12 | 16 | 1 | 27 | 13 | 44 | 0 | 50 | 4 | 25 | 37 | 7 |
| 24 | 1 | 2 | 50 | 0 | 39 | 6 | 29 | 0 | 23 | 4 | 26 | 48 | 2 | 14 | 15 | 1 | 28 | 13 | 58 | 0 | 51 | 4 | 25 | 34 | 6 |
| 25 | 1 | 5 | 49 | 0 | 41 | 6 | 45 | 0 | 24 | 4 | 26 | 46 | 2 | 16 | 13 | 1 | 30 | 14 | 12 | 0 | 52 | 4 | 25 | 30 | 5 |
| 26 | 1 | 8 | 48 | 0 | 43 | 7 | 1 | 0 | 24 | 4 | 26 | 44 | 2 | 18 | 12 | 1 | 32 | 14 | 26 | 0 | 53 | 4 | 25 | 27 | 4 |
| 27 | 1 | 10 | 47 | 0 | 44 | 7 | 17 | 0 | 25 | 4 | 26 | 42 | 2 | 19 | 11 | 1 | 34 | 14 | 38 | 0 | 53 | 4 | 25 | 24 | 3 |
| 28 | 1 | 13 | 46 | 0 | 46 | 7 | 33 | 0 | 26 | 4 | 26 | 40 | 2 | 21 | 9 | 1 | 36 | 14 | 52 | 0 | 54 | 4 | 25 | 22 | 2 |
| 29 | 1 | 15 | 45 | 0 | 48 | 7 | 49 | 0 | 27 | 4 | 26 | 38 | 2 | 23 | 8 | 1 | 38 | 15 | 5 | 0 | 55 | 4 | 25 | 20 | 1 |
| 30 | 1 | 17 | 44 | 0 | 49 | 8 | 4 | 0 | 28 | 4 | 26 | 36 | 2 | 25 | 7 | 1 | 39 | 15 | 18 | 0 | 56 | 4 | 25 | 18 | 0 |
| | 11 | | | | | | | | | | | 10 | | | | | | | | | | | |
| | A | | B | C | | D | | E | | F | | | A | | B | C | | D | | E | | F | | | |

Continuation of the Table TRW-20 (Toruń, BU, 74/III, f. 267v)

| Tabula equacionis Mercurii | | | | | | | | | | | | | | | | | | | | | | | | | |
|---|---|---|---|---|---|---|---|---|---|---|---|---|---|---|---|---|---|---|---|---|---|---|---|---|---|
| | 2 | | | | | | | | | | | | 3 | | | | | | | | | | | | |
| Gradus | Equacio centri | | Minuta propor-cionalia | Longitudo longior | | Equacio argumenti | | Longitudo propior | | Stacio prima | | | Equacio centri | | Minuta propor-cionalia | Longitudo longior | | Equacio argumenti | | Longitudo propior | | Stacio prima | | | Gradus |
| | G | M | Minuatur | G | M | G | M | G | M | S | G | M | G | M | Addatur | G | M | G | M | G | M | S | G | M | |
| 1 | 2 | 27 | 5 | 1 | 41 | 15 | 31 | 0 | 57 | 4 | 25 | 16 | 3 | 1 | 46 | 2 | 31 | 20 | 40 | 1 | 30 | 4 | 24 | 37 | 29 |
| 2 | 2 | 29 | 4 | 1 | 43 | 15 | 44 | 0 | 58 | 4 | 25 | 16 | 3 | 1 | 47 | 2 | 33 | 20 | 47 | 1 | 31 | 4 | 24 | 36 | 28 |
| 3 | 2 | 31 | 2 | 1 | 44 | 15 | 56 | 1 | 0 | 4 | 25 | 11 | 3 | 2 | 48 | 2 | 34 | 20 | 54 | 1 | 32 | 4 | 24 | 36 | 27 |
| 4 | 2 | 33 | 1 | 1 | 46 | 16 | 9 | 1 | 1 | 4 | 25 | 9 | 3 | 2 | 49 | 2 | 36 | 21 | 1 | 1 | 33 | 4 | 24 | 35 | 26 |
| 5 | 2 | 34 | Addatur 1 | 1 | 48 | 16 | 21 | 1 | 2 | 4 | 25 | 7 | 3 | 2 | 50 | 2 | 38 | 21 | 7 | 1 | 34 | 4 | 24 | 35 | 25 |
| 6 | 2 | 36 | 2 | 1 | 49 | 16 | 33 | 1 | 4 | 4 | 25 | 3 | 3 | 2 | 50 | 2 | 39 | 21 | 13 | 1 | 35 | 4 | 24 | 34 | 24 |
| 7 | 2 | 38 | 4 | 1 | 51 | 16 | 45 | 1 | 5 | 4 | 25 | 1 | 3 | 2 | 51 | 2 | 41 | 21 | 19 | 1 | 36 | 4 | 24 | 34 | 23 |
| 8 | 2 | 40 | 6 | 1 | 53 | 16 | 57 | 1 | 6 | 4 | 25 | 0 | 3 | 1 | 52 | 2 | 43 | 21 | 24 | 1 | 37 | 4 | 24 | 33 | 22 |
| 9 | 2 | 41 | 8 | 1 | 54 | 17 | 9 | 1 | 7 | 4 | 24 | 58 | 3 | 1 | 53 | 2 | 44 | 21 | 29 | 1 | 38 | 4 | 24 | 33 | 21 |
| 10 | 2 | 43 | 10 | 1 | 56 | 17 | 21 | 1 | 8 | 4 | 24 | 56 | 3 | 1 | 54 | 2 | 46 | 21 | 34 | 1 | 39 | 4 | 24 | 32 | 20 |
| 11 | 2 | 44 | 12 | 1 | 58 | 17 | 32 | 1 | 9 | 4 | 24 | 55 | 3 | 0 | 55 | 2 | 48 | 21 | 38 | 1 | 40 | 4 | 24 | 32 | 19 |
| 12 | 2 | 45 | 14 | 1 | 59 | 17 | 43 | 1 | 11 | 4 | 24 | 54 | 3 | 0 | 56 | 2 | 49 | 21 | 42 | 1 | 41 | 4 | 24 | 31 | 18 |
| 13 | 2 | 47 | 16 | 2 | 1 | 17 | 54 | 1 | 12 | 4 | 24 | 53 | 2 | 59 | 56 | 2 | 50 | 21 | 46 | 1 | 42 | 4 | 24 | 31 | 17 |
| 14 | 2 | 48 | 18 | 2 | 3 | 18 | 5 | 1 | 13 | 4 | 24 | 52 | 2 | 59 | 57 | 2 | 52 | 21 | 49 | 1 | 43 | 4 | 24 | 31 | 16 |
| 15 | 2 | 49 | 20 | 2 | 4 | 18 | 16 | 1 | 14 | 4 | 24 | 51 | 2 | 58 | 57 | 2 | 53 | 21 | 52 | 1 | 44 | 4 | 24 | 30 | 15 |
| 16 | 2 | 50 | 22 | 2 | 6 | 18 | 27 | 1 | 15 | 4 | 24 | 50 | 2 | 58 | 58 | 2 | 55 | 21 | 55 | 1 | 45 | 4 | 24 | 30 | 14 |
| 17 | 2 | 51 | 24 | 2 | 8 | 18 | 37 | 1 | 16 | 4 | 24 | 49 | 2 | 57 | 58 | 2 | 57 | 21 | 57 | 1 | 46 | 4 | 24 | 30 | 13 |
| 18 | 2 | 52 | 25 | 2 | 9 | 18 | 47 | 1 | 17 | 4 | 24 | 48 | 2 | 56 | 58 | 2 | 58 | 21 | 59 | 1 | 47 | 4 | 24 | 30 | 12 |
| 19 | 2 | 53 | 27 | 2 | 11 | 18 | 57 | 1 | 18 | 4 | 24 | 47 | 2 | 55 | 59 | 3 | 0 | 22 | 0 | 1 | 48 | 4 | 24 | 30 | 11 |
| 20 | 2 | 54 | 29 | 2 | 13 | 19 | 7 | 1 | 19 | 4 | 24 | 46 | 2 | 54 | 59 | 3 | 1 | 22 | 1 | 1 | 49 | 4 | 24 | 30 | 10 |
| 21 | 2 | 55 | 30 | 2 | 14 | 19 | 16 | 1 | 20 | 4 | 24 | 45 | 2 | 53 | 59 | 3 | 2 | 22 | 2 | 1 | 50 | 4 | 24 | 30 | 9 |
| 22 | 2 | 56 | 32 | 2 | 16 | 19 | 25 | 1 | 21 | 4 | 24 | 44 | 2 | 52 | 59 | 3 | 3 | 22 | 2 | 1 | 51 | 4 | 24 | 30 | 8 |

| Tabula equacionis Mercurii | | | | | | | | | | | | | | | | | | | | | | | | | |
|---|---|---|---|---|---|---|---|---|---|---|---|---|---|---|---|---|---|---|---|---|---|---|---|---|---|
| | 2 | | | | | | | | | | | | 3 | | | | | | | | | | | | |
| **Gradus** | **Equacio centri** | | **Minuta propor-cionalia** | **Longitudo longior** | | **Equacio argumenti** | | **Longitudo propior** | | **Stacio prima** | | | **Equacio centri** | | **Minuta propor-cionalia** | **Longitudo longior** | | **Equacio argumenti** | | **Longitudo propior** | | **Stacio prima** | | | **Gradus** |
| | **G** | **M** | **Addatur** | **G** | **M** | **G** | **M** | **G** | **M** | **S** | **G** | **M** | **G** | **M** | **Addatur** | **G** | **M** | **G** | **M** | **G** | **M** | **S** | **G** | **M** | |
| 23 | 2 | 57 | 34 | 2 | 18 | 19 | 34 | 1 | 22 | 4 | 24 | 43 | 2 | 51 | 60 | 3 | 3 | 22 | 1 | 1 | 52 | 4 | 24 | 30 | 7 |
| 24 | 2 | 58 | 35 | 2 | 19 | 19 | 44 | 1 | 23 | 4 | 24 | 42 | 2 | 50 | 60 | 3 | 4 | 22 | 0 | 1 | 53 | 4 | 24 | 30 | 6 |
| 25 | 2 | 58 | 37 | 2 | 21 | 19 | 53 | 1 | 24 | 4 | 24 | 41 | 2 | 49 | 60 | 3 | 4 | 21 | 59 | 1 | 54 | 4 | 24 | 29 | 5 |
| 26 | 2 | 59 | 38 | 2 | 23 | 20 | 2 | 1 | 25 | 4 | 24 | 40 | 2 | 48 | 60 | 3 | 5 | 21 | 58 | 1 | 55 | 4 | 24 | 29 | 4 |
| 27 | 2 | 59 | 40 | 2 | 24 | 20 | 10 | 1 | 26 | 4 | 24 | 39 | 2 | 46 | 60 | 3 | 6 | 21 | 56 | 1 | 56 | 4 | 24 | 29 | 3 |
| 28 | 3 | 0 | 41 | 2 | 26 | 20 | 18 | 1 | 27 | 4 | 24 | 38 | 2 | 45 | 60 | 3 | 6 | 21 | 53 | 1 | 56 | 4 | 24 | 29 | 2 |
| 29 | 3 | 0 | 43 | 2 | 28 | 20 | 25 | 1 | 28 | 4 | 24 | 37 | 2 | 43 | 60 | 3 | 7 | 21 | 50 | 1 | 57 | 4 | 24 | 29 | 1 |
| 30 | 3 | 1 | 44 | 2 | 2 | 20 | 33 | 1 | 29 | 4 | 24 | 37 | 2 | 41 | 60 | 3 | 8 | 21 | 47 | 1 | 57 | 4 | 24 | 29 | 0 |
| | 9 | | | | | | | | | | | | 8 | | | | | | | | | | | | |
| | A | | B | C | | D | | E | | F | | | A | | B | C | | D | | E | | F | | | |

Remaining part of the Table TRW-20 (Toruń, BU, 74/III, f. 268r)

| Tabula equacionis Mercurii | | | | | | | | | | | | | | | | | | | | | | | | | |
|---|---|---|---|---|---|---|---|---|---|---|---|---|---|---|---|---|---|---|---|---|---|---|---|---|---|
| | 4 | | | | | | | | | | | | 5 | | | | | | | | | | | | |
| Gradus | Equacio centri | | Minuta propor-cionalia | Longitudo longior | | Equacio argumenti | | Longitudo propior | | Stacio prima | | | Equacio centri | | Minuta propor-cionalia | Longitudo longior | | Equacio argumenti | | Longitudo propior | | Stacio prima | | | Gradus |
| | G | M | Addatur | G | M | G | M | G | M | S | G | M | G | M | Addatur | G | M | G | M | G | M | S | G | M | |
| 1 | 2 | 39 | 60 | 3 | 8 | 21 | 43 | 1 | 58 | 4 | 24 | 29 | 1 | 30 | 48 | 2 | 39 | 15 | 8 | 1 | 45 | 4 | 24 | 36 | 29 |
| 2 | 2 | 37 | 60 | 3 | 9 | 21 | 38 | 1 | 58 | 4 | 24 | 29 | 1 | 27 | 47 | 2 | 36 | 14 | 44 | 1 | 43 | 4 | 24 | 36 | 28 |
| 3 | 2 | 35 | 59 | 3 | 9 | 21 | 33 | 1 | 58 | 4 | 24 | 29 | 1 | 24 | 47 | 2 | 32 | 14 | 20 | 1 | 41 | 4 | 24 | 37 | 27 |
| 4 | 2 | 34 | 59 | 3 | 9 | 21 | 27 | 1 | 59 | 4 | 24 | 29 | 1 | 21 | 46 | 2 | 29 | 13 | 55 | 1 | 39 | 4 | 24 | 37 | 26 |
| 5 | 2 | 32 | 59 | 3 | 10 | 21 | 21 | 1 | 59 | 4 | 24 | 29 | 1 | 18 | 46 | 2 | 25 | 13 | 29 | 1 | 37 | 4 | 24 | 37 | 25 |
| 6 | 2 | 30 | 59 | 3 | 10 | 21 | 15 | 1 | 59 | 4 | 24 | 29 | 1 | 15 | 45 | 2 | 21 | 13 | 3 | 1 | 34 | 4 | 24 | 38 | 24 |
| 7 | 2 | 28 | 59 | 3 | 11 | 21 | 8 | 2 | 0 | 4 | 24 | 29 | 1 | 12 | 45 | 2 | 17 | 12 | 36 | 1 | 32 | 4 | 24 | 38 | 23 |
| 8 | 2 | 26 | 58 | 3 | 11 | 21 | 1 | 2 | 0 | 4 | 24 | 29 | 1 | 9 | 44 | 2 | 13 | 12 | 9 | 1 | 29 | 4 | 24 | 38 | 22 |
| 9 | 2 | 24 | 58 | 3 | 12 | 20 | 53 | 2 | 0 | 4 | 24 | 30 | 1 | 6 | 44 | 2 | 9 | 11 | 41 | 1 | 26 | 4 | 24 | 39 | 21 |
| 10 | 2 | 22 | 58 | 3 | 12 | 20 | 44 | 2 | 0 | 4 | 24 | 30 | 1 | 3 | 43 | 2 | 5 | 11 | 12 | 1 | 23 | 4 | 24 | 39 | 20 |
| 11 | 2 | 20 | 57 | 3 | 12 | 20 | 35 | 2 | 1 | 4 | 24 | 30 | 1 | 0 | 43 | 2 | 0 | 10 | 43 | 1 | 20 | 4 | 24 | 39 | 19 |
| 12 | 2 | 18 | 57 | 3 | 11 | 20 | 25 | 2 | 1 | 4 | 24 | 30 | 0 | 57 | 43 | 1 | 55 | 10 | 13 | 1 | 17 | 4 | 24 | 39 | 18 |
| 13 | 2 | 16 | 57 | 3 | 11 | 20 | 14 | 2 | 1 | 4 | 24 | 30 | 0 | 54 | 42 | 1 | 50 | 9 | 43 | 1 | 14 | 4 | 24 | 39 | 17 |
| 14 | 2 | 14 | 56 | 3 | 10 | 20 | 2 | 2 | 1 | 4 | 24 | 30 | 0 | 51 | 42 | 1 | 44 | 9 | 13 | 1 | 11 | 4 | 24 | 39 | 16 |
| 15 | 2 | 11 | 56 | 3 | 9 | 19 | 50 | 2 | 1 | 4 | 24 | 31 | 0 | 48 | 42 | 1 | 38 | 8 | 40 | 1 | 7 | 4 | 24 | 40 | 15 |
| 16 | 2 | 9 | 56 | 3 | 8 | 19 | 37 | 2 | 1 | 4 | 24 | 31 | 0 | 45 | 42 | 1 | 32 | 8 | 7 | 1 | 4 | 4 | 24 | 40 | 14 |
| 17 | 2 | 7 | 55 | 3 | 7 | 19 | 24 | 2 | 0 | 4 | 24 | 31 | 0 | 42 | 41 | 1 | 26 | 7 | 34 | 1 | 0 | 4 | 24 | 40 | 13 |
| 18 | 2 | 4 | 55 | 3 | 6 | 19 | 10 | 2 | 0 | 4 | 24 | 32 | 0 | 39 | 41 | 1 | 19 | 7 | 1 | 0 | 56 | 4 | 24 | 40 | 12 |
| 19 | 2 | 2 | 54 | 3 | 5 | 18 | 55 | 2 | 0 | 4 | 24 | 32 | 0 | 35 | 41 | 1 | 13 | 6 | 27 | 0 | 52 | 4 | 24 | 41 | 11 |
| 20 | 2 | 0 | 54 | 3 | 4 | 18 | 40 | 2 | 0 | 4 | 24 | 32 | 0 | 32 | 41 | 1 | 7 | 5 | 53 | 0 | 47 | 4 | 24 | 41 | 10 |
| 21 | 1 | 57 | 53 | 3 | 2 | 18 | 24 | 2 | 0 | 4 | 24 | 33 | 0 | 28 | 41 | 1 | 1 | 5 | 19 | 0 | 43 | 4 | 24 | 41 | 9 |
| 22 | 1 | 55 | 53 | 3 | 1 | 18 | 7 | 1 | 59 | 4 | 24 | 33 | 0 | 25 | 41 | 0 | 55 | 4 | 44 | 0 | 38 | 4 | 24 | 41 | 8 |

| Tabula equacionis Mercurii | | | | | | | | | | | | | | | | | | | | | | | | | |
|---|---|---|---|---|---|---|---|---|---|---|---|---|---|---|---|---|---|---|---|---|---|---|---|---|---|
| | 4 | | | | | | | | | | | | 5 | | | | | | | | | | | | |
| Gradus | Equacio centri | | Minuta propor-cionalia | Longitudo longior | | Equacio argumenti | | Longitudo propior | | Stacio prima | | | Equacio centri | | Minuta propor-cionalia | Longitudo longior | | Equacio argumenti | | Longitudo propior | | Stacio prima | | | Gradus |
| | G | M | Addatur | G | M | G | M | G | M | S | G | M | G | M | Addatur | G | M | G | M | G | M | S | G | M | |
| 23 | 1 | 52 | 52 | 2 | 59 | 17 | 50 | 1 | 59 | 4 | 24 | 33 | 0 | 22 | 40 | 0 | 48 | 4 | 10 | 0 | 33 | 4 | 24 | 41 | 7 |
| 24 | 1 | 49 | 52 | 2 | 57 | 17 | 32 | 1 | 58 | 4 | 24 | 34 | 0 | 19 | 40 | 0 | 42 | 3 | 35 | 0 | 28 | 4 | 24 | 41 | 6 |
| 25 | 1 | 47 | 51 | 2 | 55 | 17 | 14 | 1 | 57 | 4 | 24 | 34 | 0 | 16 | 40 | 0 | 35 | 3 | 0 | 0 | 24 | 4 | 24 | 41 | 5 |
| 26 | 1 | 44 | 51 | 2 | 53 | 16 | 55 | 1 | 55 | 4 | 24 | 34 | 0 | 13 | 40 | 0 | 28 | 2 | 24 | 0 | 19 | 4 | 24 | 41 | 4 |
| 27 | 1 | 41 | 50 | 2 | 51 | 16 | 35 | 1 | 53 | 4 | 24 | 35 | 0 | 9 | 40 | 0 | 21 | 1 | 48 | 0 | 14 | 4 | 24 | 41 | 3 |
| 28 | 1 | 38 | 49 | 2 | 48 | 16 | 14 | 1 | 51 | 4 | 24 | 35 | 0 | 6 | 40 | 0 | 14 | 1 | 12 | 0 | 10 | 4 | 24 | 42 | 2 |
| 29 | 1 | 35 | 49 | 2 | 45 | 15 | 53 | 1 | 49 | 4 | 24 | 35 | 0 | 3 | 40 | 0 | 7 | 0 | 36 | 0 | 5 | 4 | 24 | 42 | 1 |
| 30 | 1 | 32 | 48 | 2 | 42 | 15 | 31 | 1 | 47 | 4 | 24 | 36 | 0 | 0 | 40 | 0 | 0 | 0 | 0 | 0 | 0 | 4 | 24 | 42 | 0 |
| | 7 | | | | | | | | | | | | 6 | | | | | | | | | | | | |
| | A | | B | C | | D | | E | | F | | | A | | B | C | | D | | E | | F | | | |

# 13

# *Tabule resolute* for the Meridian of Prague

The original *Tabule resolute* compiled by Petrus Cruciferus for the meridian of Wrocław start the cyclical radices for the year 1424; most of the later versions start for the year 1428 and continue at steps of 20 years. There were also some changes in the layout of some of these tables and in the order of the tables, which we shall document in Chapter 17.

The first such 1428 tables were computed for the meridian of Prague. They are preserved in several copies in different manuscripts. We can take MS Prague, NK, X B 3 as a typical example of the *Tabule resolute* for Prague (see Fig. 27). This copy is written more sparsely on the folios of the convolute, so it can keep the natural order of the Tables of mean motions from the collected years to the shortest time steps of hour and their fractions (like the Tables TRW-01 – TRW-06). We transcribe the Table of mean motions in collected years. The other Tables of mean motions are not influenced by the change of the reference meridian and the shifted year of the first radix. Up to the scribal errors and the more sparse layout, they are thus identical with Tables TRW-02 – TRW-06 and hence we do not transcribe them here. The only noticeable change is that the Tables of mean motions in expanded years are enriched by a letter 'B' in the leftmost column for the years 4, 8, 12, 16, and 20, denoting that these years are leap (*bisextiles*).

There is a change in the later versions of the *Tabule resolute* with respect to the original tables for Wrocław; after the Tables of mean motions there follow the Tables of syzygies (that is of conjunctions and oppositions of the Sun and Moon) in collected and expanded years and in months (i.e. the Tables TRW-09 – TRW-12). The Tables of syzygies in collected years start for the year 1429 and are again dependent on the chosen meridian. In MS Prague, NK, X B 3 they have the form reproduced here in Table TRP-09. It is interesting to note several scribal errors in this manuscript: '45' tertias instead of '5' are written for the year 1449 in the column A. The digit '4' is protruding and appears to have been inserted additionally in this manuscript, but the number '45' is regularly written in both copies of the TR in MS Mikulov, RM, Mik 6331, ff. 56v and 144v, while the correct value '5' tertias is written, e.g., in MSS Oxford, Bodl., Canon. misc. 499, f. 80r/70r; Prague, NK, VIII G 24, f. 11v; Wrocław, BU, IV F 19, f. 53v, etc. Similarly, in column C, '1' degree for the year 1429 and '24' seconds for the year 1449 are written in the rubric and corrected in black ink to '12' degrees or '42' seconds, respectively. The erroneous values without any correction are written again in both copies in the MS Mikulov, RM, Mik 6331, while the correct values are in the other manuscripts.

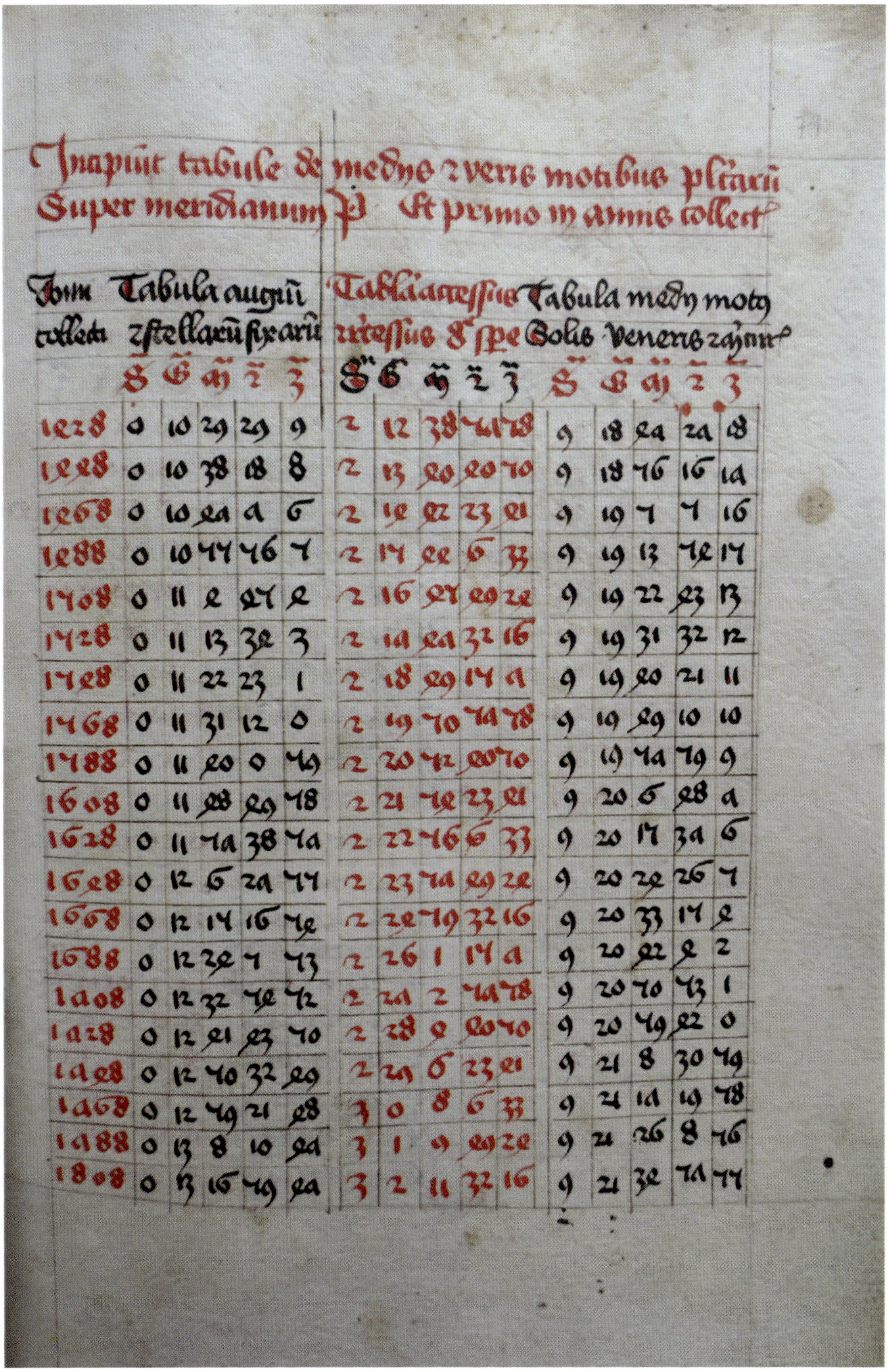

**Figure 27.** Incipit of the *Tabule resolute* in the MS Prague, NK, X B 3, f. 79r.

Table TRP-01 of mean motions in collected years for the meridian of Prague (Prague, NK, X B 3, f. 79r)

| Incipiunt Tabule de mediis et veris motibus planetarum super meridianum Pragensem, et primo in annis collectis | | | | | | | | | | | | | | | |
|---|---|---|---|---|---|---|---|---|---|---|---|---|---|---|---|
| **Anni collecti** | **Tabula augium et stellarum fixarum** | | | | | **Tabula accessus et recessus octave spere** | | | | | **Tabula medii motus Solis, Veneris et Mercurii** | | | | |
| | **S** | **G** | **M** | **2** | **3** | **S** | **G** | **M** | **2** | **3** | **S** | **G** | **M** | **2** | **3** |
| 1428 | 0 | 10 | 29 | 29 | 9 | 2 | 12 | 38 | 57 | 58 | 9 | 18 | 47 | 27 | 18 |
| 1448 | 0 | 10 | 38 | 18 | 8 | 2 | 13 | 40 | 40 | 50 | 9 | 18 | 56 | 16 | 17 |
| 1468 | 0 | 10 | 47 | 7 | 6 | 2 | 14 | 42 | 23 | 41 | 9 | 19 | 5 | 5 | 16 |
| 1488 | 0 | 10 | 55 | 56 | 5 | 2 | 15 | 44 | 6 | 33 | 9 | 19 | 13 | 54 | 15 |
| 1508 | 0 | 11 | 4 | 45 | 4 | 2 | 16 | 45 | 49 | 24 | 9 | 19 | 22 | 43 | 13 |
| 1528 | 0 | 11 | 13 | 34 | 3 | 2 | 17 | 47 | 32 | 16 | 9 | 19 | 31 | 32 | 12 |
| 1548 | 0 | 11 | 22 | 23 | 1 | 2 | 18 | 49 | 15 | 7 | 9 | 19 | 40 | 21 | 11 |
| 1568 | 0 | 11 | 31 | 12 | 0 | 2 | 19 | 50 | 57 | 58 | 9 | 19 | 49 | 10 | 10 |
| 1588 | 0 | 11 | 40 | 0 | 59 | 2 | 20 | 52 | 40 | 50 | 9 | 19 | 57 | 59 | 9 |
| 1608 | 0 | 11 | 48 | 49 | 58 | 2 | 21 | 54 | 23 | 41 | 9 | 20 | 6 | 48 | 7 |
| 1628 | 0 | 11 | 57 | 38 | 57 | 2 | 22 | 56 | 6 | 33 | 9 | 20 | 15 | 37 | 6 |
| 1648 | 0 | 12 | 6 | 27 | 55 | 2 | 23 | 57 | 49 | 24 | 9 | 20 | 24 | 26 | 7 |
| 1668 | 0 | 12 | 15 | 16 | 54 | 2 | 24 | 59 | 32 | 16 | 9 | 20 | 33 | 15 | 4 |
| 1688 | 0 | 12 | 24 | 5 | 43 | 2 | 26 | 1 | 15 | 7 | 9 | 20 | 42 | 4 | 2 |
| 1708 | 0 | 12 | 32 | 54 | 52 | 2 | 27 | 2 | 57 | 58 | 9 | 20 | 50 | 53 | 1 |
| 1728 | 0 | 12 | 41 | 43 | 50 | 2 | 28 | 4 | 40 | 50 | 9 | 20 | 59 | 42 | 0 |
| 1748 | 0 | 12 | 50 | 32 | 49 | 2 | 29 | 6 | 23 | 41 | 9 | 21 | 8 | 30 | 59 |
| 1768 | 0 | 12 | 59 | 21 | 48 | 3 | 0 | 8 | 6 | 33 | 9 | 21 | 17 | 19 | 58 |
| 1788 | 0 | 13 | 8 | 10 | 47 | 3 | 1 | 9 | 49 | 24 | 9 | 21 | 26 | 8 | 56 |
| 1808 | 0 | 13 | 16 | 59 | 47 | 3 | 2 | 11 | 32 | 16 | 9 | 21 | 34 | 57 | 55 |

A B C

Continuation of the Table TRP-01 (Prague, NK, X B 3, f. 79v)

| Residua pars Tabule de mediis motibus planetarum in annis collectis | | | | | | | | | | | | | | | |
|---|---|---|---|---|---|---|---|---|---|---|---|---|---|---|---|
| Anni collecti | Tabula motus medii Lune | | | | | Tabula medii argumenti Lune | | | | | Tabula argumenti latitudinis Lune | | | | |
| | S | G | M | 2 | 3 | S | G | M | 2 | 3 | S | G | M | 2 | 3 |
| 1428 | 7 | 16 | 31 | 0 | 35 | 5 | 5 | 48 | 54 | 0 | 7 | 7 | 56 | 15 | 12 |
| 1448 | 0 | 0 | 4 | 48 | 8 | 6 | 15 | 33 | 20 | 20 | 0 | 18 | 19 | 53 | 58 |
| 1468 | 4 | 13 | 38 | 35 | 42 | 7 | 25 | 17 | 46 | 41 | 7 | 28 | 43 | 32 | 45 |
| 1488 | 8 | 27 | 12 | 22 | 16 | 9 | 5 | 2 | 13 | 1 | 11 | 9 | 7 | 11 | 31 |
| 1508 | 1 | 10 | 46 | 10 | 50 | 10 | 14 | 46 | 39 | 21 | 4 | 19 | 30 | 50 | 17 |
| 1528 | 5 | 24 | 19 | 58 | 23 | 11 | 34 | 31 | 5 | 41 | 9 | 29 | 54 | 29 | 3 |
| 1548 | 10 | 7 | 53 | 45 | 57 | 1 | 4 | 15 | 32 | 2 | 3 | 10 | 18 | 7 | 49 |
| 1568 | 2 | 21 | 27 | 33 | 31 | 2 | 13 | 59 | 58 | 22 | 8 | 20 | 41 | 46 | 35 |
| 1588 | 7 | 5 | 1 | 21 | 4 | 3 | 23 | 44 | 24 | 42 | 2 | 1 | 5 | 25 | 21 |
| 1608 | 11 | 18 | 35 | 8 | 38 | 5 | 3 | 28 | 51 | 2 | 7 | 11 | 29 | 4 | 7 |
| 1628 | 4 | 2 | 8 | 56 | 12 | 6 | 13 | 13 | 17 | 23 | 0 | 21 | 52 | 42 | 53 |
| 1648 | 8 | 15 | 42 | 43 | 46 | 7 | 22 | 57 | 43 | 43 | 6 | 2 | 16 | 21 | 39 |
| 1668 | 0 | 29 | 16 | 31 | 19 | 9 | 2 | 42 | 10 | 3 | 11 | 12 | 40 | 0 | 25 |
| 1688 | 7 | 12 | 50 | 18 | 43 | 10 | 12 | 26 | 36 | 23 | 4 | 23 | 3 | 39 | 11 |
| 1708 | 9 | 26 | 24 | 6 | 27 | 11 | 22 | 11 | 2 | 44 | 10 | 3 | 27 | 17 | 57 |
| 1728 | 2 | 9 | 57 | 54 | 0 | 1 | 1 | 55 | 29 | 4 | 3 | 13 | 50 | 56 | 43 |
| 1748 | 6 | 25 | 31 | 41 | 34 | 2 | 11 | 39 | 55 | 24 | 8 | 24 | 14 | 35 | 29 |
| 1768 | 11 | 7 | 5 | 29 | 8 | 3 | 21 | 24 | 21 | 44 | 2 | 4 | 38 | 14 | 15 |
| 1788 | 3 | 20 | 39 | 16 | 42 | 5 | 1 | 8 | 48 | 5 | 7 | 15 | 1 | 53 | 1 |
| 1808 | 8 | 4 | 13 | 4 | 16 | 6 | 10 | 53 | 14 | 25 | 0 | 25 | 25 | 31 | 47 |

D E F

Continuation of the Table TRP-01 (Prague, NK, X B 3, f. 80r)

| Residua pars Tabule de mediis motibus planetarum in annis collectis | | | | | | | | | | | | | | | |
|---|---|---|---|---|---|---|---|---|---|---|---|---|---|---|---|
| Anni collecti | Tabula motus capitis draconis | | | | | Tabula medii motus Saturni | | | | | Tabula medii motus Iovis | | | | |
| | S | G | M | 2 | 3 | S | G | M | 2 | 3 | S | G | M | 2 | 3 |
| 1428 | 11 | 21 | 39 | 14 | 57 | 8 | 25 | 12 | 47 | 41 | 11 | 7 | 47 | 52 | 18 |
| 1448 | 0 | 18 | 29 | 6 | 10 | 4 | 29 | 54 | 24 | 48 | 7 | 15 | 2 | 28 | 35 |
| 1468 | 1 | 15 | 18 | 57 | 22 | 1 | 4 | 36 | 1 | 55 | 3 | 22 | 17 | 4 | 52 |
| 1488 | 2 | 12 | 8 | 48 | 34 | 9 | 9 | 17 | 39 | 1 | 11 | 29 | 31 | 41 | 9 |
| 1508 | 3 | 8 | 58 | 39 | 47 | 5 | 13 | 59 | 16 | 8 | 8 | 6 | 46 | 17 | 27 |
| 1528 | 4 | 5 | 48 | 30 | 59 | 1 | 18 | 40 | 53 | 14 | 4 | 14 | 0 | 53 | 44 |
| 1548 | 5 | 2 | 38 | 22 | 11 | 9 | 23 | 22 | 30 | 21 | 0 | 21 | 15 | 30 | 1 |
| 1568 | 5 | 29 | 28 | 13 | 23 | 5 | 28 | 4 | 7 | 28 | 8 | 28 | 30 | 6 | 19 |
| 1588 | 6 | 26 | 18 | 4 | 36 | 2 | 2 | 45 | 44 | 34 | 5 | 5 | 44 | 42 | 36 |
| 1608 | 7 | 23 | 7 | 55 | 48 | 10 | 7 | 27 | 21 | 41 | 1 | 12 | 59 | 18 | 53 |
| 1628 | 8 | 19 | 57 | 47 | 0 | 6 | 12 | 8 | 58 | 48 | 9 | 20 | 13 | 55 | 10 |
| 1648 | 9 | 16 | 47 | 38 | 13 | 2 | 16 | 50 | 35 | 54 | 5 | 27 | 28 | 31 | 28 |
| 1668 | 10 | 13 | 37 | 29 | 25 | 10 | 21 | 32 | 13 | 1 | 2 | 4 | 43 | 7 | 45 |
| 1688 | 11 | 10 | 27 | 20 | 37 | 6 | 26 | 13 | 50 | 7 | 10 | 11 | 57 | 44 | 2 |
| 1708 | 0 | 7 | 17 | 11 | 50 | 3 | 0 | 55 | 27 | 14 | 6 | 19 | 12 | 20 | 19 |
| 1728 | 1 | 4 | 7 | 3 | 2 | 11 | 5 | 37 | 4 | 21 | 2 | 26 | 26 | 56 | 37 |
| 1748 | 2 | 0 | 56 | 54 | 14 | 7 | 10 | 18 | 41 | 27 | 11 | 3 | 41 | 32 | 54 |
| 1768 | 2 | 27 | 46 | 45 | 27 | 3 | 15 | 0 | 18 | 34 | 7 | 10 | 56 | 9 | 1 |
| 1788 | 3 | 24 | 36 | 36 | 39 | 11 | 19 | 41 | 55 | 40 | 3 | 18 | 10 | 45 | 28 |
| 1808 | 4 | 21 | 26 | 27 | 51 | 7 | 24 | 23 | 32 | 47 | 11 | 25 | 25 | 21 | 46 |

G H I

Remaining part of the Table TRP-01 (Prague, NK, X B 3, f. 80v)

| Residua pars Tabule de mediis motibus planetarum in annis collectis | | | | | | | | | | | | | | | |
|---|---|---|---|---|---|---|---|---|---|---|---|---|---|---|---|
| Anni collecti | Tabula motus Martis | | | | | Tabula medii argumenti Veneris | | | | | Tabula medii argumenti Mercurii | | | | |
| | S | G | M | 2 | 3 | S | G | M | 2 | 3 | S | G | M | 2 | 3 |
| 1428 | 4 | 23 | 9 | 54 | 15 | 6 | 29 | 45 | 57 | 2 | 2 | 9 | 44 | 25 | 38 |
| 1448 | 0 | 11 | 28 | 52 | 5 | 1 | 3 | 24 | 48 | 7 | 2 | 24 | 12 | 4 | 30 |
| 1468 | 7 | 29 | 47 | 49 | 56 | 7 | 7 | 3 | 39 | 4 | 3 | 8 | 39 | 43 | 21 |
| 1488 | 3 | 18 | 6 | 47 | 46 | 1 | 10 | 24 | 30 | 0 | 3 | 23 | 7 | 22 | 13 |
| 1508 | 11 | 6 | 25 | 45 | 36 | 7 | 14 | 21 | 20 | 56 | 4 | 7 | 35 | 1 | 4 |
| 1528 | 6 | 24 | 44 | 43 | 26 | 1 | 18 | 0 | 11 | 52 | 4 | 22 | 2 | 39 | 56 |
| 1548 | 2 | 13 | 3 | 41 | 16 | 7 | 21 | 39 | 2 | 48 | 5 | 6 | 30 | 18 | 47 |
| 1568 | 10 | 1 | 22 | 39 | 6 | 1 | 25 | 17 | 53 | 44 | 5 | 20 | 57 | 57 | 38 |
| 1588 | 5 | 19 | 41 | 36 | 56 | 7 | 28 | 56 | 44 | 40 | 6 | 5 | 25 | 36 | 30 |
| 1608 | 1 | 8 | 0 | 34 | 47 | 2 | 2 | 35 | 35 | 36 | 6 | 19 | 53 | 15 | 21 |
| 1628 | 8 | 26 | 19 | 32 | 37 | 8 | 6 | 14 | 26 | 32 | 7 | 4 | 20 | 54 | 13 |
| 1648 | 4 | 14 | 38 | 30 | 27 | 2 | 9 | 53 | 17 | 28 | 7 | 18 | 48 | 33 | 4 |
| 1668 | 0 | 2 | 57 | 28 | 17 | 8 | 13 | 32 | 8 | 24 | 8 | 3 | 16 | 11 | 56 |
| 1688 | 7 | 21 | 16 | 26 | 7 | 2 | 17 | 10 | 59 | 20 | 8 | 17 | 43 | 50 | 47 |
| 1708 | 3 | 9 | 35 | 23 | 57 | 8 | 20 | 49 | 50 | 16 | 9 | 2 | 11 | 29 | 38 |
| 1728 | 10 | 27 | 54 | 21 | 48 | 2 | 24 | 28 | 41 | 12 | 9 | 16 | 39 | 8 | 30 |
| 1748 | 6 | 16 | 13 | 19 | 38 | 8 | 28 | 7 | 32 | 8 | 10 | 1 | 6 | 47 | 21 |
| 1768 | 2 | 4 | 32 | 17 | 28 | 3 | 1 | 46 | 32 | 4 | 10 | 15 | 34 | 26 | 13 |
| 1788 | 9 | 22 | 51 | 15 | 18 | 9 | 5 | 25 | 14 | 0 | 11 | 0 | 2 | 5 | 4 |
| 1808 | 5 | 11 | 10 | 13 | 8 | 3 | 9 | 4 | 4 | 56 | 11 | 14 | 29 | 43 | 55 |

J K L

Table TRP-09 of mean conjunctions of the Sun and Moon in collected years for the meridian of Prague (Prague, NK, X B 3, ff. 88v–89r)

| Tabula medie coniunccionis Solis et Lune cum motibus Solis et Lune in annis Cristi collectis | | | | | | | | | | | | | | | | | | | | | | |
|---|---|---|---|---|---|---|---|---|---|---|---|---|---|---|---|---|---|---|---|---|---|---|
| Anni collecti | Tempus medie coniunccionis Solis et Lune | | | | | | Medius motus Solis et Lune | | | | | Anni collecti | Medius motus argumenti Lune | | | | | Medium argumentum latitudinis Lune | | | | |
| | D | H | M | 2 | 3 | 4 | S | G | M | 2 | 3 | | S | G | M | 2 | 3 | S | G | M | 2 | 3 |
| 1429 | 5 | 2 | 35 | 57 | 46 | 30 | 9 | 23 | 49 | 33 | 15 | 1429 | 7 | 12 | 33 | 17 | 56 | 9 | 15 | 31 | 1 | 43 |
| 1449 | 23 | 16 | 40 | 34 | 45 | 46 | 10 | 12 | 17 | 33 | 19 | 1449 | 4 | 25 | 7 | 42 | 25 | 11 | 1 | 47 | 56 | 14 |
| 1469 | 12 | 18 | 1 | 7 | 22 | 3 | 10 | 1 | 39 | 9 | 12 | 1469 | 1 | 11 | 53 | 6 | 19 | 11 | 17 | 24 | 37 | 0 |
| 1489 | 1 | 19 | 21 | 40 | 38 | 21 | 9 | 21 | 0 | 45 | 5 | 1489 | 9 | 28 | 38 | 30 | 13 | 0 | 3 | 1 | 17 | 45 |
| 1509 | 20 | 9 | 26 | 16 | 57 | 37 | 10 | 9 | 28 | 45 | 9 | 1509 | 7 | 11 | 12 | 54 | 43 | 1 | 19 | 18 | 12 | 17 |
| 1529 | 9 | 10 | 46 | 50 | 13 | 55 | 9 | 28 | 50 | 21 | 2 | 1529 | 3 | 27 | 58 | 18 | 37 | 2 | 4 | 54 | 53 | 2 |
| 1549 | 28 | 0 | 51 | 26 | 33 | 11 | 10 | 17 | 18 | 21 | 6 | 1549 | 1 | 10 | 32 | 43 | 6 | 3 | 21 | 11 | 47 | 34 |
| 1569 | 17 | 2 | 11 | 59 | 49 | 29 | 10 | 6 | 39 | 56 | 59 | 1569 | 9 | 27 | 18 | 7 | 1 | 4 | 6 | 48 | 28 | 19 |
| 1589 | 6 | 3 | 32 | 33 | 5 | 46 | 9 | 26 | 1 | 32 | 55 | 1589 | 6 | 14 | 3 | 30 | 55 | 4 | 22 | 25 | 9 | 5 |
| 1609 | 24 | 17 | 37 | 9 | 27 | 1 | 10 | 14 | 29 | 32 | 56 | 1609 | 3 | 26 | 37 | 55 | 25 | 6 | 8 | 42 | 3 | 38 |
| 1629 | 13 | 18 | 57 | 42 | 41 | 20 | 10 | 3 | 51 | 8 | 48 | 1629 | 0 | 13 | 23 | 19 | 18 | 6 | 24 | 18 | 44 | 22 |
| 1649 | 2 | 20 | 18 | 15 | 57 | 37 | 9 | 23 | 12 | 44 | 41 | 1649 | 9 | 0 | 8 | 43 | 13 | 7 | 9 | 55 | 25 | 8 |
| 1669 | 21 | 10 | 22 | 52 | 16 | 54 | 10 | 11 | 40 | 44 | 45 | 1669 | 6 | 12 | 43 | 7 | 42 | 8 | 26 | 12 | 19 | 39 |
| 1689 | 10 | 11 | 43 | 25 | 33 | 11 | 10 | 1 | 2 | 20 | 38 | 1689 | 2 | 29 | 28 | 31 | 36 | 9 | 11 | 49 | 0 | 25 |
| 1709 | 29 | 1 | 48 | 1 | 52 | 28 | 10 | 19 | 30 | 20 | 42 | 1709 | 0 | 12 | 2 | 56 | 5 | 10 | 28 | 5 | 54 | 56 |
| 1729 | 18 | 3 | 8 | 35 | 8 | 45 | 10 | 8 | 51 | 56 | 35 | 1729 | 8 | 28 | 48 | 20 | 0 | 11 | 13 | 42 | 35 | 42 |
| 1749 | 7 | 4 | 29 | 8 | 25 | 3 | 9 | 28 | 13 | 32 | 28 | 1749 | 5 | 15 | 33 | 43 | 54 | 11 | 29 | 19 | 16 | 27 |
| 1769 | 25 | 18 | 33 | 44 | 44 | 19 | 10 | 16 | 41 | 32 | 32 | 1769 | 2 | 28 | 8 | 8 | 23 | 1 | 15 | 36 | 10 | 59 |
| 1789 | 14 | 19 | 54 | 18 | 0 | 36 | 10 | 6 | 3 | 8 | 25 | 1789 | 11 | 14 | 53 | 32 | 17 | 2 | 1 | 12 | 51 | 44 |
| 1809 | 3 | 21 | 14 | 51 | 16 | 54 | 9 | 25 | 24 | 44 | 17 | 1809 | 8 | 1 | 38 | 56 | 12 | 2 | 16 | 49 | 32 | 30 |

A B C D

Table TRP-10 of mean oppositions of the Sun and Moon in collected years for the meridian of Prague (Prague, NK, X B 3, ff. 89v–90r)

| Tabula medie opposicionis Solis et Lune cum motibus Solis et Lune in annis Cristi collectis | | | | | | | | | | | | | | | | | | | | | | |
|---|---|---|---|---|---|---|---|---|---|---|---|---|---|---|---|---|---|---|---|---|---|---|
| Anni collecti | Tempus medie opposicionis Solis et Lune | | | | | | Medius motus Solis et Lune | | | | | Anni collecti | Medius motus argumenti Lune | | | | | Medium argumentum latitudinis Lune | | | | |
| | D | H | M | 2 | 3 | 4 | S | G | M | 2 | 3 | | S | G | M | 2 | 3 | S | G | M | 2 | 3 |
| 1429 | 19 | 20 | 57 | 59 | 17 | 59 | 10 | 8 | 22 | 45 | 21 | 1429 | 1 | 25 | 27 | 48 | 13 | 4 | 0 | 51 | 8 | 36 |
| 1449 | 8 | 22 | 18 | 32 | 34 | 7 | 9 | 27 | 44 | 21 | 14 | 1449 | 10 | 12 | 13 | 12 | 7 | 4 | 16 | 27 | 49 | 21 |
| 1469 | 27 | 12 | 23 | 8 | 53 | 33 | 10 | 16 | 12 | 21 | 18 | 1469 | 7 | 24 | 47 | 36 | 37 | 6 | 2 | 44 | 43 | 53 |
| 1489 | 16 | 13 | 43 | 42 | 9 | 20 | 10 | 5 | 33 | 57 | 11 | 1489 | 4 | 11 | 33 | 0 | 31 | 6 | 18 | 21 | 24 | 38 |
| 1509 | 5 | 15 | 4 | 15 | 26 | 8 | 9 | 24 | 55 | 33 | 3 | 1509 | 0 | 28 | 18 | 24 | 25 | 7 | 3 | 58 | 5 | 24 |
| 1529 | 24 | 5 | 8 | 51 | 45 | 24 | 10 | 13 | 23 | 33 | 8 | 1529 | 10 | 10 | 52 | 48 | 54 | 8 | 20 | 14 | 59 | 55 |
| 1549 | 13 | 6 | 29 | 25 | 1 | 42 | 10 | 2 | 45 | 9 | 0 | 1549 | 6 | 27 | 38 | 12 | 49 | 9 | 5 | 51 | 40 | 41 |
| 1569 | 2 | 7 | 49 | 58 | 17 | 59 | 9 | 22 | 6 | 44 | 53 | 1569 | 3 | 14 | 23 | 36 | 43 | 9 | 21 | 28 | 21 | 26 |
| 1589 | 20 | 21 | 54 | 34 | 37 | 16 | 10 | 10 | 34 | 44 | 57 | 1589 | 0 | 26 | 58 | 1 | 12 | 11 | 7 | 45 | 15 | 58 |
| 1609 | 9 | 23 | 15 | 7 | 55 | 31 | 9 | 29 | 56 | 20 | 50 | 1609 | 9 | 13 | 43 | 25 | 8 | 11 | 23 | 21 | 56 | 45 |
| 1629 | 28 | 13 | 19 | 44 | 12 | 49 | 10 | 18 | 24 | 20 | 54 | 1629 | 6 | 26 | 17 | 49 | 36 | 1 | 9 | 38 | 51 | 15 |
| 1649 | 17 | 14 | 40 | 17 | 29 | 7 | 10 | 7 | 45 | 56 | 47 | 1649 | 3 | 13 | 3 | 13 | 30 | 1 | 25 | 15 | 32 | 1 |
| 1669 | 6 | 16 | 0 | 50 | 45 | 24 | 9 | 27 | 7 | 32 | 40 | 1669 | 11 | 29 | 48 | 37 | 24 | 2 | 10 | 52 | 12 | 46 |
| 1689 | 25 | 6 | 5 | 27 | 4 | 41 | 10 | 15 | 35 | 32 | 44 | 1689 | 9 | 12 | 23 | 1 | 54 | 3 | 27 | 9 | 7 | 18 |
| 1709 | 14 | 7 | 26 | 0 | 20 | 58 | 10 | 4 | 57 | 8 | 37 | 1709 | 5 | 29 | 8 | 25 | 48 | 4 | 12 | 45 | 48 | 3 |
| 1729 | 3 | 8 | 46 | 33 | 37 | 16 | 9 | 24 | 18 | 44 | 29 | 1729 | 2 | 15 | 53 | 49 | 42 | 4 | 28 | 22 | 28 | 49 |
| 1749 | 21 | 22 | 51 | 9 | 56 | 32 | 10 | 12 | 46 | 44 | 34 | 1749 | 11 | 28 | 28 | 14 | 11 | 6 | 14 | 39 | 23 | 20 |
| 1769 | 11 | 0 | 11 | 43 | 12 | 49 | 10 | 2 | 8 | 20 | 26 | 1769 | 8 | 15 | 13 | 38 | 6 | 7 | 0 | 16 | 4 | 6 |
| 1789 | 0 | 1 | 32 | 16 | 29 | 7 | 9 | 21 | 29 | 56 | 19 | 1789 | 5 | 1 | 49 | 2 | 0 | 7 | 15 | 52 | 44 | 51 |
| 1809 | 18 | 15 | 36 | 52 | 48 | 23 | 10 | 9 | 57 | 56 | 23 | 1809 | 2 | 14 | 33 | 26 | 29 | 9 | 2 | 9 | 39 | 23 |

A B C D

The Tables of syzygies in expanded years and in months in the *Tabule resolute* for Prague are basically the same as in Tables TRW-11 and TRW-12 for Wrocław, so we shall not reproduce them here. It is interesting to note that Table TRW-14 of true motion of luminaries in an hour is omitted from the corpus of the *Tabule resolute* in most copies of their later versions for Prague and Cracow; it is, however, often present in many of the manuscripts in the context of other astronomical tables.

After the Tables of syzygies follow the tables of apogees only in MS Prague, NK, X B 3. Like the Tables of mean motions, the Table of apogees in collected years also starts for the year 1428. Moreover, it has a slightly different layout from Table TRW-07. The columns A (the 'common' apogee), D (the 'difference'), and E (inscribed as 'difference of difference' but in reality it gives values D/20) are missing. Instead of these columns, there is added an extended table (TRP-07a) entitled 'Table of apogees in expanded years', which gives annual increments of longitudes of the apogees. For the users of the tables it resembles the calculation of mean motions; however because of the non-linearity of the trepidation, the increments in the expanded years must be given differently for each twenty-year period. In Oxford, Bodl., Canon. misc. 499, ff. 85v/75v and 86r/76r glosses are added by a later hand as are Tables of apogees in months (unfinished), days, and hours, which are often found in the later *Tabule resolute* for Cracow and which are also described in the canons written by authors from Cracow. The Tables of apogees in months and days are also given, for instance, in Cracow, BJ, 1852, p. 60, which contains Tables for meridians of both Prague and Cracow. The Tables of apogees in months, days, and hours are, however, inconsistent with the assumption of trepidation.[1]

A significant change in the layout of the *Tabule resolute* from the initial set compiled for Wrocław is the addition of columns with differences between the subsequent lines in the Tables of equations. The values of differences are vertically shifted for half of the line height to make clear to which pair of values they correspond. The differences are given in the Tables of equations of the motion of access and recess, of the Sun and Moon, as well as of the planets, but not for the first stations of the planets (see Fig. 28).

The *Tabule resolute* for the meridian of Prague are also included in MS Oxford, Bodl., Canon. misc. 499, ff. 72r/62r–105r/95r. Their incipit is identical to that in MS Prague, NK, X B 3. The mean motions in collected and expanded years, in days and hours are written more densely in four columns per page. On ff. 85v/75v and 86r/76r we can find on the margins small Tables of apogees in months, days, and hours written by a later hand. Such tables are characteristic for the later tables for the meridian of Cracow.

An interesting case is the manuscript Mikulov, RM, Mik 6361, which contains two sets of the TR for Prague, equally as two copies of the canons *Mirabilis in altis Dominus*. The first denoted here as 'A' can be found on folios 47r–81r, the second ('B') on folios 135r–171v. Interestingly, f. 82, which follows after the folios with the copy of text 'A', has a vacat on recto and a Table of mean motions of the Moon, centre of the Moon, and anomaly of the Moon, which are given up to quartas. The folios in the copy 'B' are bound in an incorrect sequence; ff. 159–171 should be correctly placed between f. 146 and f. 147.

The MS Prague, NK, VIII G 24 contains the TR for Prague on ff. 2r–37r.

1 See Chapter 19 for details.

Table TRP-07 of apogees of planets in collected years for the meridian of Prague (Prague, NK, X B 3, ff. 92v–93r)

| Tabula augium planetarum ad annos Cristi collectos | | | | | | | | | | | | | | | | | | | | | | | | | | | |
|---|---|---|---|---|---|---|---|---|---|---|---|---|---|---|---|---|---|---|---|---|---|---|---|---|---|---|---|
| Anni collecti | Auges Solis et Veneris | | | | | Auges Saturni | | | | | Auges Iovis | | | | | Anni collecti | Auges Martis | | | | | Auges Mercurii | | | | | |
| | S | G | M | 2 | 3 | S | G | M | 2 | 3 | S | G | M | 2 | 3 | | S | G | M | 2 | 3 | S | G | M | 2 | 3 |
| 1428 | 3 | 0 | 30 | 7 | 12 | 8 | 12 | 28 | 26 | 16 | 5 | 22 | 41 | 44 | 16 | 1428 | 4 | 14 | 16 | 57 | 16 | 6 | 29 | 44 | 17 | 16 |
| 1448 | 3 | 0 | 41 | 45 | 21 | 8 | 12 | 40 | 4 | 21 | 5 | 22 | 53 | 22 | 21 | 1448 | 4 | 14 | 28 | 35 | 21 | 6 | 29 | 55 | 55 | 21 |
| 1468 | 3 | 0 | 53 | 13 | 34 | 8 | 12 | 51 | 32 | 34 | 5 | 23 | 4 | 50 | 34 | 1468 | 4 | 14 | 40 | 3 | 34 | 7 | 0 | 7 | 23 | 34 |
| 1488 | 3 | 1 | 4 | 31 | 33 | 8 | 13 | 2 | 50 | 33 | 5 | 23 | 16 | 8 | 33 | 1488 | 4 | 14 | 51 | 21 | 33 | 7 | 0 | 18 | 41 | 33 |
| 1508 | 3 | 1 | 15 | 38 | 57 | 8 | 13 | 13 | 57 | 57 | 5 | 23 | 27 | 15 | 57 | 1508 | 4 | 15 | 2 | 28 | 57 | 7 | 0 | 29 | 48 | 57 |
| 1528 | 3 | 1 | 26 | 36 | 34 | 8 | 13 | 24 | 55 | 34 | 5 | 23 | 38 | 13 | 34 | 1528 | 4 | 15 | 13 | 26 | 34 | 7 | 0 | 40 | 46 | 34 |
| 1548 | 3 | 1 | 37 | 24 | 40 | 8 | 13 | 35 | 43 | 40 | 5 | 23 | 49 | 1 | 40 | 1548 | 4 | 15 | 24 | 14 | 40 | 7 | 0 | 51 | 34 | 40 |
| 1568 | 3 | 1 | 48 | 3 | 16 | 8 | 13 | 46 | 22 | 16 | 5 | 23 | 59 | 40 | 16 | 1568 | 4 | 15 | 34 | 53 | 16 | 7 | 1 | 2 | 13 | 16 |
| 1588 | 3 | 1 | 58 | 31 | 28 | 8 | 13 | 56 | 50 | 28 | 5 | 24 | 10 | 8 | 28 | 1588 | 4 | 15 | 45 | 21 | 28 | 7 | 1 | 12 | 41 | 28 |
| 1608 | 3 | 2 | 8 | 46 | 22 | 8 | 14 | 7 | 5 | 22 | 5 | 24 | 20 | 23 | 22 | 1608 | 4 | 15 | 55 | 36 | 22 | 7 | 1 | 22 | 56 | 22 |
| 1628 | 3 | 2 | 18 | 52 | 12 | 8 | 14 | 17 | 11 | 12 | 5 | 24 | 30 | 29 | 12 | 1628 | 4 | 16 | 5 | 42 | 12 | 7 | 1 | 33 | 2 | 12 |
| 1648 | 3 | 2 | 28 | 48 | 38 | 8 | 14 | 27 | 7 | 38 | 5 | 24 | 40 | 25 | 38 | 1648 | 4 | 16 | 15 | 38 | 38 | 7 | 1 | 42 | 58 | 38 |
| 1668 | 3 | 2 | 38 | 34 | 33 | 8 | 14 | 36 | 53 | 33 | 5 | 24 | 50 | 11 | 33 | 1668 | 4 | 16 | 25 | 24 | 33 | 7 | 1 | 52 | 44 | 33 |
| 1688 | 3 | 2 | 48 | 9 | 41 | 8 | 14 | 46 | 28 | 41 | 5 | 24 | 59 | 46 | 41 | 1688 | 4 | 16 | 34 | 59 | 41 | 7 | 2 | 2 | 19 | 41 |
| 1708 | 3 | 2 | 57 | 34 | 10 | 8 | 14 | 55 | 53 | 10 | 5 | 25 | 9 | 11 | 10 | 1708 | 4 | 16 | 44 | 24 | 10 | 7 | 2 | 11 | 44 | 10 |
| 1728 | 3 | 3 | 6 | 48 | 5 | 8 | 15 | 5 | 7 | 5 | 5 | 25 | 18 | 25 | 5 | 1728 | 4 | 16 | 53 | 38 | 5 | 7 | 2 | 20 | 58 | 5 |
| 1748 | 3 | 3 | 15 | 51 | 25 | 8 | 15 | 14 | 10 | 25 | 5 | 25 | 27 | 28 | 25 | 1748 | 4 | 17 | 2 | 41 | 25 | 7 | 2 | 30 | 1 | 25 |
| 1768 | 3 | 3 | 24 | 44 | 11 | 8 | 15 | 23 | 3 | 11 | 5 | 25 | 36 | 21 | 11 | 1768 | 4 | 17 | 11 | 34 | 11 | 7 | 2 | 38 | 54 | 11 |
| 1788 | 3 | 3 | 33 | 26 | 23 | 8 | 15 | 31 | 45 | 23 | 5 | 25 | 45 | 3 | 23 | 1788 | 4 | 17 | 20 | 16 | 23 | 7 | 2 | 47 | 36 | 23 |
| 1808 | 3 | 3 | 41 | 58 | 1 | 8 | 15 | 40 | 17 | 1 | 5 | 25 | 53 | 25 | 1 | 1808 | 4 | 17 | 28 | 48 | 1 | 7 | 2 | 56 | 8 | 1 |
| Radix | 2 | 11 | 25 | 23 | 4 | 7 | 23 | 23 | 42 | 4 | 5 | 3 | 37 | 0 | 4 | Radix | 3 | 25 | 12 | 13 | 4 | 6 | 10 | 39 | 33 | 4 |
| | B | | | | | C | | | | | F | | | | | | G | | | | | H | | | | |

Table TRP-07a of apogees of planets in expanded years (Prague, NK, X B 3, ff. 93r–93v)

| Tabula augium planetarum ad annos Christi expansos | | | | | | | | | | | | | | | | | | | | | | | |
|---|---|---|---|---|---|---|---|---|---|---|---|---|---|---|---|---|---|---|---|---|---|---|---|
| Anni expansi | 1428 | | | Anni expansi | 1448 | | | 1468 | | | 1488 | | | 1508 | | | 1528 | | | 1548 | | | |
| | M | 2 | 3 | | M | 2 | 3 | M | 2 | 3 | M | 2 | 3 | M | 2 | 3 | M | 2 | 3 | M | 2 | 3 | |
| 1 | 0 | 35 | 13 | 1 | 0 | 34 | 42 | 0 | 34 | 15 | 0 | 33 | 44 | 0 | 33 | 13 | 0 | 32 | 45 | 0 | 32 | 14 | |
| 2 | 1 | 10 | 26 | 2 | 1 | 9 | 25 | 1 | 8 | 29 | 1 | 7 | 27 | 1 | 6 | 26 | 1 | 5 | 30 | 1 | 4 | 31 | |
| 3 | 1 | 45 | 39 | 3 | 1 | 44 | 7 | 1 | 42 | 44 | 1 | 41 | 11 | 1 | 39 | 39 | 1 | 38 | 15 | 1 | 36 | 52 | |
| 4 | 2 | 20 | 58 | 4 | 2 | 18 | 55 | 2 | 17 | 4 | 2 | 15 | 0 | 2 | 12 | 57 | 2 | 11 | 6 | 2 | 9 | 0 | |
| 5 | 2 | 56 | 12 | 5 | 2 | 53 | 37 | 2 | 51 | 18 | 2 | 48 | 44 | 2 | 45 | 59 | 2 | 43 | 24 | 2 | 40 | 50 | |
| 6 | 3 | 31 | 25 | 6 | 3 | 28 | 20 | 3 | 25 | 24 | 3 | 22 | 2 | 3 | 18 | 44 | 3 | 15 | 42 | 3 | 12 | 40 | |
| 7 | 4 | 6 | 32 | 7 | 4 | 2 | 42 | 3 | 59 | 8 | 3 | 55 | 15 | 3 | 51 | 29 | 3 | 47 | 59 | 3 | 44 | 29 | |
| 8 | 4 | 41 | 21 | 8 | 4 | 37 | 2 | 4 | 32 | 57 | 4 | 28 | 33 | 4 | 24 | 19 | 4 | 20 | 22 | 4 | 16 | 24 | |
| 9 | 5 | 16 | 3 | 9 | 5 | 11 | 16 | 5 | 6 | 41 | 5 | 1 | 46 | 4 | 57 | 5 | 4 | 52 | 39 | 4 | 48 | 14 | |
| 10 | 5 | 50 | 45 | 10 | 5 | 45 | 31 | 5 | 40 | 24 | 5 | 34 | 59 | 5 | 29 | 50 | 5 | 24 | 57 | 5 | 20 | 4 | |
| 11 | 6 | 25 | 27 | 11 | 6 | 19 | 45 | 6 | 14 | 8 | 6 | 8 | 12 | 6 | 2 | 35 | 5 | 57 | 14 | 5 | 51 | 53 | |
| 12 | 7 | 0 | 15 | 12 | 6 | 54 | 5 | 6 | 47 | 57 | 6 | 41 | 30 | 6 | 35 | 25 | 6 | 29 | 37 | 6 | 23 | 48 | |
| 13 | 7 | 34 | 58 | 13 | 7 | 28 | 20 | 7 | 21 | 41 | 7 | 14 | 43 | 7 | 8 | 10 | 7 | 1 | 54 | 6 | 55 | 38 | |
| 14 | 8 | 9 | 40 | 14 | 8 | 2 | 35 | 7 | 55 | 25 | 7 | 47 | 56 | 7 | 40 | 56 | 7 | 34 | 11 | 7 | 27 | 27 | |
| 15 | 8 | 44 | 22 | 15 | 8 | 36 | 49 | 8 | 29 | 9 | 8 | 21 | 9 | 8 | 13 | 41 | 8 | 6 | 29 | 7 | 59 | 17 | |
| 16 | 9 | 19 | 10 | 16 | 9 | 11 | 9 | 9 | 2 | 58 | 8 | 54 | 27 | 8 | 46 | 31 | 8 | 38 | 52 | 8 | 31 | 12 | |
| 17 | 9 | 53 | 52 | 17 | 9 | 45 | 24 | 9 | 36 | 42 | 9 | 27 | 40 | 9 | 19 | 16 | 9 | 11 | 9 | 9 | 3 | 1 | |
| 18 | 10 | 28 | 35 | 18 | 10 | 19 | 38 | 10 | 10 | 25 | 10 | 0 | 53 | 9 | 52 | 2 | 9 | 43 | 26 | 9 | 34 | 51 | |
| 19 | 11 | 3 | 17 | 19 | 10 | 53 | 53 | 10 | 44 | 9 | 10 | 34 | 6 | 10 | 24 | 47 | 10 | 15 | 44 | 10 | 6 | 41 | |
| 20 | 11 | 38 | 5 | 20 | 11 | 28 | 13 | 11 | 17 | 58 | 11 | 7 | 24 | 10 | 57 | 37 | 10 | 48 | 6 | 10 | 38 | 36 | |
| | | A | | | | B | | | C | | | D | | | E | | | F | | | G | | |

Continuation of the Table TRP-07a (Prague, NK, X B 3, f. 94r)

| Residua pars Tabule augium planetarum in annis Christi expansis | | | | | | | | | | | | | | | | | | |
|---|---|---|---|---|---|---|---|---|---|---|---|---|---|---|---|---|---|---|
| **Anni expansi** | **1568** | | | **1588** | | | **1608** | | | **1628** | | | **1648** | | | **1668** | | |
| | **M** | **2** | **3** | **M** | **2** | **3** | **M** | **2** | **3** | **M** | **2** | **3** | **M** | **2** | **3** | **M** | **2** | **3** |
| 1 | 0 | 31 | 50 | 0 | 31 | 19 | 0 | 30 | 39 | 0 | 30 | 14 | 0 | 29 | 37 | 0 | 28 | 49 |
| 2 | 1 | 3 | 39 | 1 | 2 | 38 | 1 | 1 | 13 | 1 | 0 | 8 | 0 | 58 | 53 | 0 | 57 | 34 |
| 3 | 1 | 35 | 27 | 1 | 33 | 31 | 1 | 31 | 27 | 1 | 29 | 54 | 1 | 28 | 8 | 1 | 26 | 18 |
| 4 | 2 | 6 | 51 | 2 | 4 | 15 | 2 | 1 | 46 | 1 | 59 | 45 | 1 | 57 | 28 | 1 | 55 | 8 |
| 5 | 2 | 38 | 9 | 2 | 34 | 54 | 2 | 32 | 0 | 2 | 29 | 31 | 2 | 26 | 44 | 2 | 23 | 52 |
| 6 | 3 | 9 | 28 | 3 | 5 | 32 | 3 | 2 | 14 | 2 | 59 | 18 | 2 | 55 | 59 | 2 | 52 | 37 |
| 7 | 3 | 40 | 47 | 3 | 36 | 11 | 3 | 32 | 28 | 3 | 29 | 4 | 3 | 25 | 15 | 3 | 21 | 22 |
| 8 | 4 | 12 | 11 | 4 | 6 | 55 | 4 | 2 | 47 | 3 | 58 | 55 | 3 | 54 | 35 | 3 | 50 | 11 |
| 9 | 4 | 43 | 30 | 4 | 37 | 34 | 4 | 33 | 1 | 4 | 28 | 41 | 4 | 23 | 50 | 4 | 18 | 56 |
| 10 | 5 | 14 | 49 | 5 | 8 | 12 | 5 | 3 | 15 | 4 | 58 | 28 | 4 | 53 | 6 | 4 | 47 | 40 |
| 11 | 5 | 46 | 7 | 5 | 38 | 51 | 5 | 33 | 29 | 5 | 28 | 14 | 5 | 22 | 21 | 5 | 16 | 25 |
| 12 | 6 | 17 | 31 | 6 | 9 | 35 | 6 | 3 | 48 | 5 | 58 | 5 | 5 | 51 | 42 | 5 | 45 | 14 |
| 13 | 6 | 48 | 50 | 6 | 40 | 13 | 6 | 34 | 2 | 6 | 27 | 51 | 6 | 20 | 57 | 6 | 13 | 59 |
| 14 | 7 | 20 | 9 | 7 | 10 | 52 | 7 | 4 | 16 | 6 | 57 | 38 | 6 | 50 | 13 | 6 | 42 | 43 |
| 15 | 7 | 51 | 28 | 7 | 41 | 31 | 7 | 34 | 30 | 7 | 27 | 24 | 7 | 19 | 28 | 7 | 11 | 28 |
| 16 | 8 | 22 | 52 | 8 | 12 | 14 | 8 | 4 | 49 | 7 | 57 | 15 | 7 | 48 | 48 | 7 | 40 | 17 |
| 17 | 8 | 54 | 10 | 8 | 42 | 53 | 8 | 35 | 3 | 8 | 27 | 1 | 8 | 18 | 4 | 8 | 9 | 2 |
| 18 | 9 | 25 | 29 | 9 | 13 | 32 | 9 | 5 | 17 | 8 | 56 | 48 | 8 | 47 | 19 | 8 | 37 | 47 |
| 19 | 9 | 56 | 48 | 9 | 44 | 11 | 9 | 35 | 31 | 9 | 26 | 34 | 9 | 16 | 35 | 9 | 6 | 31 |
| 20 | 10 | 28 | 12 | 10 | 14 | 55 | 10 | 5 | 50 | 9 | 56 | 25 | 9 | 45 | 55 | 9 | 35 | 8 |

H I J K L M

Remaining part of the Table TRP-07a (Prague, NK, X B 3, f. 94v)

| Residua pars Tabule augium planetarum in annis Christi expansis | | | | | | | | | | | | | | | | | | |
|---|---|---|---|---|---|---|---|---|---|---|---|---|---|---|---|---|---|---|
| **Anni expansi** | **1688** | | | **1708** | | | **1728** | | | **1748** | | | **1768** | | | **1788** | | |
| | **M** | **2** | **3** | **M** | **2** | **3** | **M** | **2** | **3** | **M** | **2** | **3** | **M** | **2** | **3** | **M** | **2** | **3** |
| 1 | 0 | 28 | 14 | 0 | 27 | 43 | 0 | 27 | 12 | 0 | 26 | 41 | 0 | 26 | 10 | 0 | 25 | 40 |
| 2 | 0 | 56 | 28 | 0 | 55 | 26 | 0 | 54 | 24 | 0 | 53 | 23 | 0 | 52 | 21 | 0 | 51 | 19 |
| 3 | 1 | 24 | 41 | 1 | 23 | 9 | 1 | 21 | 36 | 1 | 20 | 4 | 1 | 18 | 31 | 1 | 16 | 59 |
| 4 | 1 | 53 | 0 | 1 | 50 | 56 | 1 | 48 | 53 | 1 | 46 | 49 | 1 | 44 | 46 | 1 | 42 | 43 |
| 5 | 2 | 21 | 14 | 2 | 18 | 39 | 2 | 16 | 5 | 2 | 13 | 31 | 2 | 10 | 56 | 2 | 8 | 22 |
| 6 | 2 | 49 | 27 | 2 | 46 | 22 | 2 | 43 | 17 | 2 | 40 | 12 | 2 | 37 | 7 | 2 | 34 | 2 |
| 7 | 3 | 17 | 41 | 3 | 14 | 5 | 3 | 10 | 29 | 3 | 6 | 53 | 3 | 3 | 17 | 2 | 59 | 41 |
| 8 | 3 | 45 | 59 | 3 | 41 | 53 | 3 | 37 | 46 | 3 | 33 | 39 | 3 | 29 | 32 | 3 | 25 | 25 |
| 9 | 4 | 14 | 13 | 4 | 9 | 36 | 4 | 4 | 58 | 4 | 0 | 20 | 3 | 55 | 43 | 3 | 51 | 5 |
| 10 | 4 | 42 | 27 | 4 | 37 | 19 | 4 | 32 | 10 | 4 | 27 | 1 | 4 | 21 | 53 | 4 | 16 | 44 |
| 11 | 5 | 10 | 41 | 5 | 5 | 1 | 4 | 59 | 22 | 4 | 53 | 43 | 4 | 48 | 3 | 4 | 42 | 24 |
| 12 | 5 | 38 | 59 | 5 | 32 | 49 | 5 | 26 | 39 | 5 | 20 | 28 | 5 | 14 | 18 | 5 | 8 | 8 |
| 13 | 6 | 7 | 13 | 6 | 0 | 32 | 5 | 53 | 51 | 5 | 47 | 10 | 5 | 40 | 29 | 5 | 33 | 47 |
| 14 | 6 | 35 | 27 | 6 | 28 | 15 | 6 | 21 | 3 | 6 | 13 | 51 | 6 | 6 | 39 | 5 | 59 | 27 |
| 15 | 7 | 3 | 41 | 6 | 55 | 58 | 6 | 48 | 15 | 6 | 40 | 32 | 6 | 32 | 49 | 6 | 25 | 7 |
| 16 | 7 | 31 | 59 | 7 | 23 | 45 | 7 | 15 | 32 | 7 | 7 | 18 | 6 | 59 | 4 | 6 | 50 | 50 |
| 17 | 8 | 0 | 13 | 7 | 51 | 28 | 7 | 42 | 44 | 7 | 33 | 59 | 7 | 25 | 9 | 7 | 16 | 7 |
| 18 | 8 | 28 | 27 | 8 | 19 | 11 | 8 | 9 | 54 | 8 | 0 | 21 | 7 | 50 | 49 | 7 | 41 | 16 |
| 19 | 8 | 56 | 40 | 8 | 46 | 38 | 8 | 36 | 35 | 8 | 26 | 32 | 8 | 16 | 28 | 8 | 6 | 25 |
| 20 | 9 | 24 | 29 | 9 | 13 | 55 | 9 | 3 | 21 | 8 | 52 | 46 | 8 | 42 | 12 | 8 | 31 | 38 |

N O P Q R S

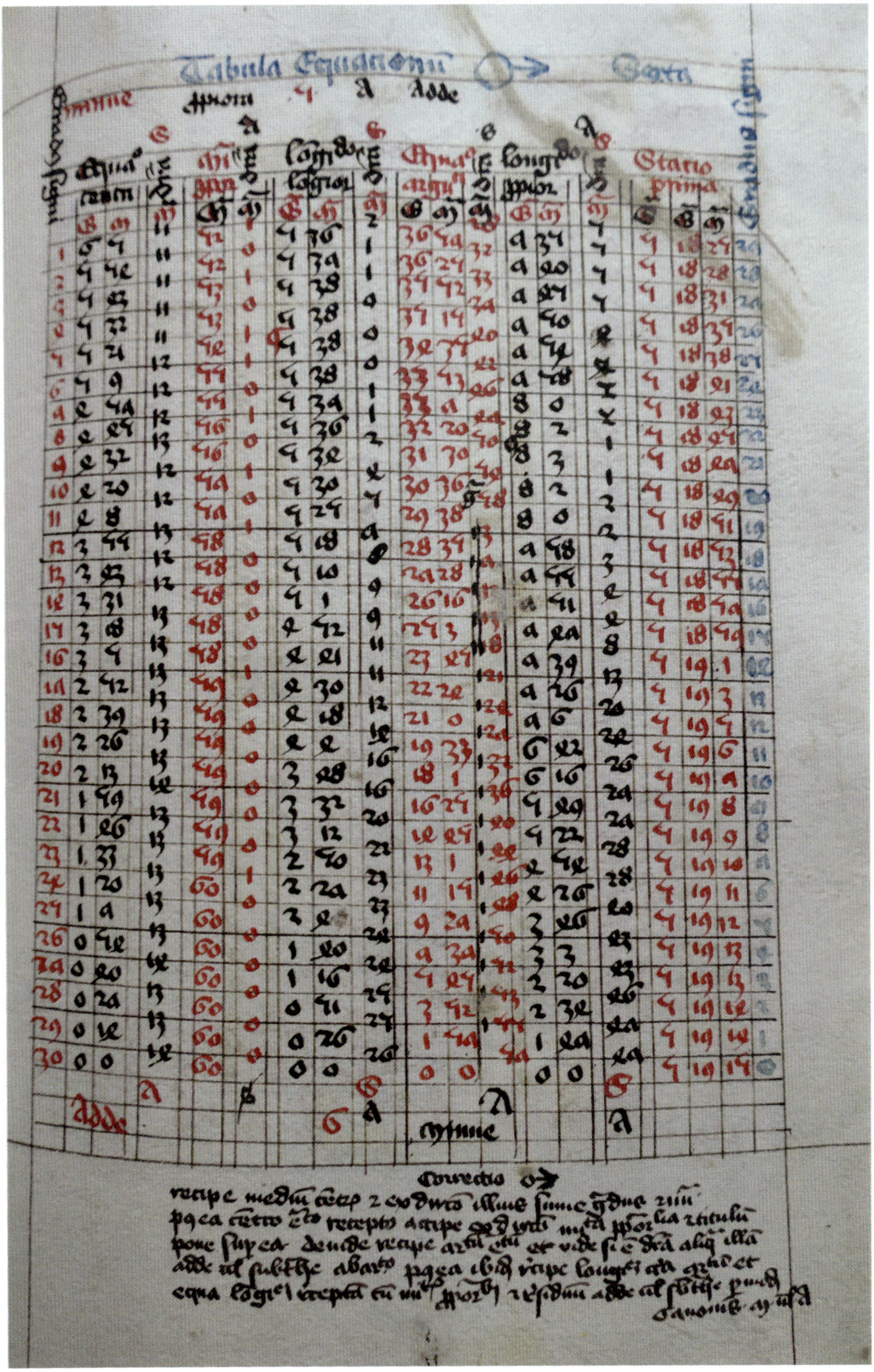

**Figure 28.** An example of the Table of equations of Mars in the *Tabule resolute*, Prague, NK, X B 3, f. 108r.

Another set of the *Tabule resolute* for the meridian of Prague can be found in MS Wrocław, BU, IV F 19, ff. 47r–75v. These tables are written more densely with mean motions in months below the collected years and four columns on page, like in the MS Toruń, BU, 74/III with the original tables for Wrocław. However, the same values with mean motions from f. 47 are repeated again on f. 48. The cyclical radices are extrapolated to the year 1408 in an empty line in this second copy as well as in the following last f. 49r with the collected years. The Tables of apogees follow again only after those of syzygies like in the other tables for Prague. F. 60 repeats f. 59 with the parts five and six of the Table of equations of the Moon.

The tables in MSS Cracow, BJ, 553, and Cracow, BJ, 560, are inscribed to be composed for the meridian of Klosterneuburg, but the values of cyclical radices of mean motions are again for the meridian of Prague.

# *Tabule resolute* for the Meridian of Cracow

Most of the preserved TR are computed for the meridian of Cracow, the geographical longitude of which is taken to be 21°36′ or $1^h\,26^m\,24^s$ to the east from the meridian of Toledo. The cyclical radices in these tables are usually computed for the years 1428 to 1808, as are the tables for Prague. We can take MS Cracow, BJ, 611 as a typical example of these tables for Cracow. In this manuscript we find the following Tables of mean motions in collected years:

Table TRC-01 of mean motions in collected years for the meridian of Cracow (Cracow, BJ, 611, f. 1r)

| Incipiunt Tabule de mediis et veris motibus planetarum super meridianum Cracoviensem verificate, et primo in annis Cristi collectis | | | | | | | | | | | | | | | |
|---|---|---|---|---|---|---|---|---|---|---|---|---|---|---|---|
| | Tabula augium et stellarum fixarum | | | | | Tabula accessus et recessus octave spere | | | | | Thabula medii motus Solis, Veneris et Mercurii | | | | |
| | S | G | M | 2 | 3 | S | G | M | 2 | 3 | S | G | M | 2 | 3 |
| 1428 | 0 | 10 | 29 | 29 | 9 | 2 | 12 | 38 | 57 | 58 | 9 | 18 | 46 | 56 | 45 |
| 1448 | 0 | 10 | 38 | 18 | 8 | 2 | 13 | 40 | 40 | 50 | 9 | 18 | 55 | 45 | 44 |
| 1468 | 0 | 10 | 47 | 7 | 6 | 2 | 14 | 42 | 23 | 41 | 9 | 19 | 4 | 34 | 43 |
| 1488 | 0 | 10 | 55 | 56 | 5 | 2 | 15 | 44 | 6 | 33 | 9 | 19 | 13 | 23 | 42 |
| 1508 | 0 | 11 | 4 | 45 | 4 | 2 | 16 | 45 | 49 | 24 | 9 | 19 | 22 | 12 | 40 |
| 1528 | 0 | 11 | 13 | 34 | 3 | 2 | 17 | 47 | 32 | 16 | 9 | 19 | 31 | 1 | 39 |
| 1548 | 0 | 11 | 22 | 23 | 1 | 2 | 18 | 49 | 15 | 7 | 9 | 19 | 39 | 50 | 38 |
| 1568 | 0 | 11 | 31 | 12 | 0 | 2 | 19 | 50 | 57 | 58 | 9 | 19 | 48 | 39 | 37 |
| 1588 | 0 | 11 | 40 | 0 | 59 | 2 | 20 | 52 | 40 | 50 | 9 | 19 | 57 | 28 | 35 |
| 1608 | 0 | 11 | 48 | 49 | 58 | 2 | 21 | 54 | 23 | 41 | 9 | 20 | 6 | 17 | 34 |
| 1628 | 0 | 11 | 57 | 38 | 57 | 2 | 22 | 56 | 6 | 33 | 9 | 20 | 15 | 6 | 33 |
| 1648 | 0 | 12 | 6 | 27 | 55 | 2 | 23 | 57 | 49 | 24 | 9 | 20 | 23 | 55 | 32 |
| 1668 | 0 | 12 | 15 | 16 | 54 | 2 | 24 | 59 | 32 | 16 | 9 | 20 | 32 | 44 | 30 |
| 1688 | 0 | 12 | 24 | 5 | 53 | 2 | 26 | 1 | 15 | 7 | 9 | 20 | 41 | 33 | 29 |
| 1708 | 0 | 12 | 32 | 54 | 52 | 2 | 27 | 2 | 57 | 58 | 9 | 20 | 50 | 22 | 28 |
| 1728 | 0 | 12 | 41 | 43 | 50 | 2 | 28 | 4 | 40 | 50 | 9 | 20 | 59 | 11 | 27 |
| 1748 | 0 | 12 | 50 | 32 | 49 | 2 | 29 | 6 | 23 | 41 | 9 | 21 | 8 | 0 | 25 |
| 1768 | 0 | 12 | 59 | 21 | 48 | 3 | 0 | 8 | 6 | 33 | 9 | 21 | 16 | 49 | 24 |
| 1788 | 0 | 13 | 8 | 10 | 47 | 3 | 1 | 9 | 49 | 24 | 9 | 21 | 25 | 38 | 23 |
| 1808 | 0 | 13 | 16 | 59 | 46 | 3 | 2 | 11 | 32 | 15 | 9 | 21 | 34 | 27 | 22 |

A B C

**Continuation of the Table TRC-01 (Cracow, BJ, 611, f. 1v)**

| Residua pars Thabule mediorum motuum planetarum in annis Christi collectis | | | | | | | | | | | | | | | |
|---|---|---|---|---|---|---|---|---|---|---|---|---|---|---|---|
| **Anni collecti** | **Tabula medii motus Lune** | | | | | **Tabula medii argumenti Lune** | | | | | **Tabula medii argumenti latitudinis Lune** | | | | |
| | **S** | **G** | **M** | **2** | **3** | **S** | **G** | **M** | **2** | **3** | **S** | **G** | **M** | **2** | **3** |
| 1428 | 7 | 16 | 24 | 12 | 7 | 5 | 5 | 42 | 8 | 59 | 7 | 7 | 49 | 25 | 6 |
| 1448 | 11 | 29 | 57 | 59 | 40 | 6 | 15 | 26 | 35 | 20 | 0 | 18 | 13 | 3 | 53 |
| 1468 | 4 | 13 | 31 | 47 | 14 | 7 | 25 | 11 | 1 | 40 | 5 | 28 | 36 | 42 | 39 |
| 1488 | 8 | 27 | 5 | 34 | 48 | 9 | 4 | 55 | 28 | 0 | 11 | 9 | 0 | 21 | 25 |
| 1508 | 1 | 10 | 39 | 22 | 22 | 10 | 14 | 39 | 54 | 20 | 4 | 19 | 24 | 0 | 11 |
| 1528 | 5 | 24 | 13 | 9 | 55 | 11 | 24 | 24 | 20 | 41 | 9 | 29 | 47 | 38 | 57 |
| 1548 | 10 | 7 | 46 | 57 | 29 | 1 | 4 | 8 | 47 | 1 | 3 | 10 | 11 | 17 | 43 |
| 1568 | 2 | 21 | 20 | 45 | 3 | 2 | 13 | 53 | 13 | 21 | 8 | 20 | 34 | 56 | 29 |
| 1588 | 7 | 4 | 54 | 32 | 36 | 3 | 23 | 37 | 39 | 41 | 2 | 0 | 58 | 35 | 15 |
| 1608 | 11 | 18 | 28 | 20 | 10 | 5 | 3 | 22 | 6 | 2 | 7 | 11 | 22 | 14 | 1 |
| 1628 | 4 | 2 | 2 | 7 | 44 | 6 | 13 | 6 | 32 | 22 | 0 | 21 | 45 | 52 | 47 |
| 1648 | 8 | 15 | 35 | 55 | 18 | 7 | 22 | 50 | 58 | 42 | 6 | 2 | 9 | 31 | 33 |
| 1668 | 0 | 29 | 9 | 42 | 55 | 9 | 2 | 35 | 25 | 2 | 11 | 12 | 33 | 10 | 19 |
| 1688 | 5 | 12 | 43 | 30 | 25 | 10 | 12 | 19 | 51 | 23 | 4 | 22 | 56 | 49 | 24 |
| 1708 | 9 | 26 | 17 | 17 | 59 | 11 | 22 | 4 | 17 | 43 | 10 | 3 | 20 | 28 | 10 |
| 1728 | 2 | 9 | 51 | 5 | 33 | 1 | 1 | 48 | 44 | 3 | 3 | 13 | 44 | 6 | 56 |
| 1748 | 6 | 23 | 24 | 53 | 6 | 2 | 11 | 33 | 10 | 23 | 8 | 24 | 7 | 45 | 42 |
| 1768 | 11 | 6 | 58 | 40 | 40 | 3 | 21 | 17 | 36 | 44 | 2 | 4 | 31 | 24 | 28 |
| 1788 | 3 | 20 | 32 | 28 | 14 | 5 | 1 | 2 | 3 | 4 | 7 | 14 | 55 | 3 | 14 |
| 1808 | 8 | 4 | 6 | 15 | 47 | 6 | 10 | 46 | 29 | 24 | 0 | 25 | 18 | 42 | 0 |

D E F

Continuation of the Table TRC-01 (Cracow, BJ, 611, f. 2r)

| Residua pars Thabule mediorum motuum planetarum in annis Christi collectis | | | | | | | | | | | | | | | |
|---|---|---|---|---|---|---|---|---|---|---|---|---|---|---|---|
| Anni collecti | Tabula medii motus capitis draconis | | | | | Tabula medii motus Saturni | | | | | Tabula medii motus Iovis | | | | |
| | S | G | M | 2 | 3 | S | G | M | 2 | 3 | S | G | M | 2 | 3 |
| 1428 | 11 | 21 | 39 | 13 | 18 | 8 | 25 | 12 | 46 | 39 | 11 | 7 | 47 | 49 | 43 |
| 1448 | 0 | 18 | 29 | 4 | 31 | 4 | 29 | 54 | 23 | 46 | 7 | 15 | 2 | 26 | 0 |
| 1468 | 1 | 15 | 18 | 55 | 43 | 1 | 4 | 36 | 0 | 52 | 3 | 22 | 17 | 2 | 18 |
| 1488 | 2 | 12 | 8 | 46 | 55 | 9 | 9 | 17 | 37 | 59 | 11 | 29 | 31 | 38 | 35 |
| 1508 | 3 | 8 | 58 | 38 | 8 | 5 | 13 | 59 | 15 | 6 | 8 | 6 | 46 | 14 | 52 |
| 1528 | 4 | 5 | 48 | 29 | 20 | 1 | 18 | 40 | 52 | 12 | 4 | 14 | 0 | 51 | 9 |
| 1548 | 5 | 2 | 38 | 20 | 32 | 9 | 23 | 22 | 29 | 19 | 0 | 21 | 15 | 27 | 27 |
| 1568 | 5 | 29 | 28 | 11 | 45 | 5 | 28 | 4 | 6 | 25 | 8 | 28 | 30 | 3 | 44 |
| 1588 | 6 | 26 | 18 | 2 | 57 | 2 | 2 | 45 | 43 | 35 | 5 | 5 | 44 | 40 | 1 |
| 1608 | 7 | 23 | 7 | 54 | 9 | 10 | 7 | 27 | 20 | 39 | 1 | 12 | 59 | 16 | 18 |
| 1628 | 8 | 19 | 57 | 45 | 22 | 6 | 12 | 8 | 57 | 45 | 9 | 20 | 13 | 52 | 36 |
| 1648 | 9 | 16 | 47 | 36 | 34 | 2 | 16 | 50 | 34 | 52 | 5 | 27 | 28 | 28 | 53 |
| 1668 | 10 | 13 | 37 | 27 | 46 | 10 | 21 | 32 | 11 | 59 | 2 | 4 | 43 | 5 | 10 |
| 1688 | 11 | 10 | 27 | 18 | 59 | 6 | 26 | 13 | 49 | 5 | 10 | 11 | 57 | 41 | 27 |
| 1708 | 0 | 7 | 17 | 10 | 11 | 3 | 0 | 55 | 26 | 12 | 6 | 19 | 12 | 17 | 45 |
| 1728 | 1 | 4 | 7 | 1 | 23 | 11 | 5 | 37 | 3 | 18 | 2 | 26 | 26 | 54 | 2 |
| 1748 | 2 | 0 | 56 | 52 | 36 | 7 | 10 | 18 | 40 | 25 | 11 | 3 | 41 | 30 | 19 |
| 1768 | 2 | 27 | 46 | 43 | 48 | 3 | 15 | 0 | 17 | 32 | 7 | 10 | 56 | 6 | 36 |
| 1788 | 3 | 24 | 36 | 35 | 0 | 11 | 19 | 41 | 54 | 38 | 3 | 18 | 10 | 42 | 54 |
| 1808 | 4 | 21 | 26 | 26 | 13 | 7 | 24 | 23 | 31 | 45 | 11 | 25 | 25 | 19 | 11 |

G H I

Remaining part of the Table TRC-01 (Cracow, BJ, 611, f. 2v)

| Residuum precedentis Tabule in annis collectis | | | | | | | | | | | | | | | |
|---|---|---|---|---|---|---|---|---|---|---|---|---|---|---|---|
| Anni collecti | Tabula medii motus Martis | | | | | Tabula medii argumenti Veneris | | | | | Tabula medii argumenti Mercurii | | | | |
| | S | G | M | 2 | 3 | S | G | M | 2 | 3 | S | G | M | 2 | 3 |
| 1428 | 4 | 23 | 9 | 38 | 1 | 6 | 29 | 45 | 38 | 5 | 2 | 9 | 42 | 49 | 20 |
| 1448 | 0 | 11 | 28 | 35 | 51 | 1 | 3 | 24 | 29 | 1 | 2 | 24 | 10 | 28 | 11 |
| 1468 | 7 | 29 | 47 | 33 | 41 | 7 | 7 | 3 | 19 | 57 | 3 | 8 | 38 | 7 | 3 |
| 1488 | 3 | 18 | 6 | 31 | 31 | 1 | 10 | 42 | 10 | 53 | 3 | 23 | 5 | 45 | 54 |
| 1508 | 11 | 6 | 25 | 29 | 21 | 7 | 14 | 21 | 1 | 49 | 4 | 7 | 33 | 24 | 46 |
| 1528 | 6 | 24 | 44 | 27 | 11 | 1 | 17 | 59 | 52 | 45 | 4 | 22 | 1 | 3 | 37 |
| 1548 | 2 | 13 | 3 | 25 | 1 | 7 | 21 | 38 | 43 | 41 | 5 | 6 | 28 | 42 | 28 |
| 1568 | 10 | 1 | 22 | 22 | 52 | 1 | 25 | 17 | 34 | 37 | 5 | 20 | 56 | 21 | 20 |
| 1588 | 5 | 19 | 41 | 20 | 42 | 7 | 28 | 56 | 25 | 33 | 6 | 5 | 24 | 0 | 11 |
| 1608 | 1 | 8 | 0 | 18 | 32 | 2 | 2 | 35 | 16 | 29 | 6 | 19 | 51 | 39 | 3 |
| 1628 | 8 | 26 | 19 | 16 | 22 | 8 | 6 | 14 | 7 | 25 | 7 | 4 | 19 | 17 | 54 |
| 1648 | 4 | 14 | 38 | 14 | 12 | 2 | 9 | 52 | 58 | 21 | 7 | 18 | 46 | 56 | 46 |
| 1668 | 0 | 2 | 57 | 12 | 2 | 8 | 13 | 31 | 49 | 27 | 8 | 3 | 14 | 35 | 37 |
| 1688 | 7 | 21 | 16 | 9 | 52 | 2 | 17 | 10 | 40 | 13 | 8 | 17 | 42 | 14 | 28 |
| 1708 | 3 | 9 | 35 | 7 | 42 | 8 | 20 | 49 | 31 | 9 | 9 | 2 | 9 | 53 | 20 |
| 1728 | 10 | 27 | 54 | 5 | 33 | 2 | 24 | 28 | 22 | 5 | 9 | 16 | 37 | 32 | 11 |
| 1748 | 6 | 16 | 13 | 3 | 23 | 8 | 28 | 7 | 13 | 1 | 10 | 1 | 5 | 11 | 3 |
| 1768 | 2 | 4 | 32 | 1 | 13 | 3 | 1 | 46 | 3 | 57 | 10 | 15 | 32 | 49 | 54 |
| 1788 | 9 | 22 | 50 | 59 | 3 | 9 | 5 | 24 | 54 | 53 | 11 | 0 | 0 | 28 | 46 |
| 1808 | 5 | 11 | 9 | 57 | 53 | 3 | 9 | 3 | 45 | 49 | 11 | 14 | 28 | 7 | 37 |

J K L

Table TRC-09 of mean conjunctions of the Sun and Moon in collected years for the meridian of Cracow (Cracow, BJ, 611, f. 11r-v)

| Tabula medie coniunccionis Solis et Lune cum motibus eorundem argumentorum Lune in annis collectis | | | | | | | | | | | | | | | | | Finis precedentis | | | | | |
|---|---|---|---|---|---|---|---|---|---|---|---|---|---|---|---|---|---|---|---|---|---|---|
| Anni collecti | Tempus medie coniunccionis Solis et Lune | | | | | | Medius motus Solis et Lune | | | | | Tabula motus medii argumenti Lune | | | | | | Tabula argumenti latitudinis Lune | | | | |
| | D | H | M | 2 | 3 | 4 | S | G | M | 2 | 3 | S | G | M | 2 | 3 | | S | G | M | 2 | 3 |
| 1429 | 5 | 2 | 48 | 21 | 46 | 5 | 9 | 23 | 49 | 33 | 15 | 7 | 12 | 33 | 17 | 55 | 1429 | 9 | 15 | 45 | 2 | 1 |
| 1449 | 23 | 16 | 52 | 58 | 5 | 22 | 10 | 12 | 17 | 33 | 19 | 4 | 25 | 7 | 42 | 25 | 1449 | 11 | 2 | 1 | 56 | 33 |
| 1469 | 12 | 18 | 13 | 31 | 21 | 39 | 10 | 1 | 39 | 9 | 12 | 1 | 11 | 53 | 6 | 19 | 1469 | 11 | 17 | 38 | 37 | 19 |
| 1489 | 1 | 19 | 34 | 4 | 37 | 57 | 9 | 21 | 0 | 45 | 4 | 9 | 28 | 38 | 30 | 13 | 1489 | 0 | 3 | 15 | 18 | 4 |
| 1509 | 20 | 9 | 38 | 40 | 57 | 13 | 10 | 9 | 28 | 45 | 9 | 7 | 11 | 12 | 54 | 43 | 1509 | 1 | 19 | 32 | 12 | 36 |
| 1529 | 9 | 10 | 59 | 14 | 13 | 30 | 9 | 28 | 50 | 21 | 2 | 3 | 27 | 58 | 18 | 37 | 1529 | 2 | 5 | 8 | 53 | 21 |
| 1549 | 28 | 1 | 3 | 50 | 32 | 47 | 10 | 17 | 18 | 21 | 6 | 1 | 10 | 32 | 43 | 6 | 1549 | 3 | 21 | 25 | 47 | 53 |
| 1569 | 17 | 2 | 24 | 23 | 49 | 4 | 10 | 6 | 39 | 56 | 59 | 9 | 27 | 18 | 7 | 0 | 1569 | 4 | 7 | 2 | 28 | 38 |
| 1589 | 6 | 3 | 44 | 57 | 5 | 22 | 9 | 26 | 1 | 32 | 51 | 6 | 14 | 3 | 30 | 55 | 1589 | 4 | 22 | 39 | 9 | 24 |
| 1609 | 24 | 17 | 49 | 33 | 24 | 38 | 10 | 14 | 29 | 32 | 56 | 3 | 26 | 37 | 55 | 24 | 1609 | 6 | 8 | 56 | 3 | 56 |
| 1629 | 13 | 19 | 10 | 6 | 40 | 55 | 10 | 3 | 51 | 8 | 48 | 0 | 13 | 23 | 19 | 18 | 1629 | 6 | 24 | 32 | 44 | 41 |
| 1649 | 2 | 20 | 30 | 39 | 57 | 13 | 9 | 23 | 12 | 44 | 41 | 9 | 0 | 8 | 43 | 12 | 1649 | 7 | 10 | 9 | 25 | 27 |
| 1669 | 21 | 10 | 35 | 16 | 16 | 29 | 10 | 11 | 40 | 44 | 45 | 6 | 12 | 43 | 7 | 42 | 1669 | 8 | 26 | 26 | 19 | 58 |
| 1689 | 10 | 11 | 55 | 49 | 32 | 47 | 10 | 1 | 2 | 20 | 35 | 2 | 29 | 28 | 31 | 36 | 1689 | 9 | 12 | 3 | 0 | 44 |
| 1709 | 29 | 2 | 10 | 25 | 52 | 3 | 10 | 19 | 30 | 20 | 42 | 0 | 12 | 2 | 56 | 5 | 1709 | 10 | 28 | 19 | 59 | 15 |
| 1729 | 18 | 3 | 20 | 59 | 8 | 20 | 10 | 8 | 51 | 56 | 35 | 8 | 28 | 48 | 19 | 59 | 1729 | 11 | 13 | 56 | 36 | 1 |
| 1749 | 7 | 4 | 41 | 32 | 24 | 38 | 9 | 28 | 13 | 32 | 28 | 5 | 15 | 33 | 43 | 54 | 1749 | 11 | 29 | 32 | 16 | 46 |
| 1769 | 25 | 18 | 46 | 8 | 43 | 54 | 10 | 16 | 41 | 32 | 32 | 2 | 28 | 8 | 8 | 23 | 1769 | 1 | 15 | 50 | 11 | 18 |
| 1789 | 14 | 20 | 6 | 42 | 0 | 12 | 10 | 6 | 3 | 8 | 25 | 11 | 14 | 53 | 32 | 17 | 1789 | 2 | 1 | 26 | 52 | 4 |
| 1809 | 3 | 21 | 27 | 15 | 16 | 29 | 9 | 25 | 24 | 44 | 17 | 8 | 1 | 38 | 56 | 12 | 1809 | 2 | 17 | 3 | 32 | 49 |
| | | | A | | | | | | B | | | | | | C | | | | | D | | |

The table of time of conjunctions continues by a later hand up to the year 2009.

The time of syzygies reflects the zonal time in geographical longitude of the city for which it is calculated, while the other quantities given in Tables TRW-09 and TRW-10 are independent of the choice of meridian. If we compare Table TRC-09 with TRP-09, and similarly TRC-10 with TRP-10, we find that the times differ mostly for $0^d$ 0;12,23,59,35 hours. This is exactly the value mentioned by John of Głogów in Rule 36 of his canons to the *Tabule resolute* as a precise distance between the Cracow and Prague. It is interesting to note that, compared to calculation with the standard values of parameters for the Parisian Alfonsine Tables,[2] the mean argument of latitude of the Moon for both conjunctions and oppositions of luminaries are systematically higher (a difference of 0;14,0,20°) in most *Tabule resolute* copies for the meridian of Cracow, but MS Vienna, ÖNB 5245 is an exception in this respect.

The Tables of true apogees are in principle also dependent on the meridian. However, the change of the longitude of apogees in one year decreases from 0;0,35,13° in 1428 to 0;0,25,40° in 1808, as can be seen in the first line of Table TRP-07a. The maximal change in a one day is thus less than 0;0,0,6° and 0;0,0,0,3° for the 12 minutes 24 seconds of the assumed time difference between Prague and Cracow. This is why John of Głogów wrote in Rule 3 of his canons that it is not necessary to work with a Table of apogees in hours and also why the Tables of apogees in collected and expanded years for Prague and Cracow are the same (again up to the scribal errors). Anyway, there is a difference between the *Tabule resolute* for Prague and some sets of the tables for Cracow; the latter also contain Tables of apogees in months and days. There are also references to these tables in canons by authors from Cracow.

Due to the nonlinearity of trepidation, the Tables of apogees in months (and days as well) should differ for each year. We can really find different tables in different manuscripts, sometimes even within the same manuscript. However, titles never specify for which year they are computed. We can only estimate the time of their validity comparing the last value (usually assigned to December) with the difference between the subsequent values of the longitudes of apogees in the expanded years. As an example, we shall present here the Tables of apogees in months from MS Cracow, BJ, 600, where there are two versions of these tables.

2 See Chapter 18, Tab. 1.

Table TRC-10 of mean oppositions of the Sun and Moon in collected years for the meridian of Cracow (Cracow, BJ, 611, f. 11v–12r)

| Tabula medie opposicionis Solis et Lune cum motibus eorum et argumentis Lune in annis Cristi collectis | | | | | | Residuum Tabule precedentis medie opposicionis Solis et Lune cum motibus | | | | | | | | | | | | | | | |
|---|---|---|---|---|---|---|---|---|---|---|---|---|---|---|---|---|---|---|---|---|---|
| | | | | | | | Medius motus Solis et Lune | | | | | Medium argumentum Lune | | | | | Medium argumentum latitudinis Lune | | | | |
| D | H | M | 2 | 3 | 4 | | S | G | M | 2 | 3 | S | G | M | 2 | 3 | S | G | M | 2 | 3 |
| 19 | 21 | 10 | 23 | 17 | 35 | 1429 | 10 | 8 | 22 | 45 | 21 | 1 | 25 | 27 | 48 | 13 | 4 | 1 | 5 | 8 | 54 |
| 8 | 22 | 30 | 56 | 33 | 52 | 1449 | 9 | 31 | 44 | 21 | 14 | 10 | 12 | 13 | 12 | 7 | 4 | 16 | 41 | 49 | 40 |
| 27 | 12 | 35 | 32 | 53 | 8 | 1469 | 10 | 16 | 13 | 21 | 18 | 7 | 24 | 47 | 36 | 36 | 6 | 2 | 58 | 44 | 12 |
| 16 | 13 | 56 | 6 | 9 | 26 | 1489 | 10 | 5 | 33 | 57 | 15 | 4 | 11 | 33 | 0 | 31 | 6 | 18 | 35 | 24 | 57 |
| 5 | 15 | 16 | 39 | 25 | 43 | 1509 | 9 | 24 | 55 | 33 | 3 | 0 | 28 | 18 | 24 | 25 | 7 | 4 | 12 | 5 | 43 |
| 24 | 5 | 21 | 15 | 45 | 0 | 1529 | 10 | 13 | 23 | 33 | 8 | 10 | 10 | 52 | 48 | 44 | 8 | 20 | 29 | 0 | 14 |
| 13 | 6 | 41 | 49 | 1 | 17 | 1549 | 10 | 2 | 45 | 9 | 0 | 6 | 27 | 38 | 12 | 49 | 9 | 6 | 5 | 41 | 0 |
| 2 | 8 | 2 | 22 | 17 | 35 | 1569 | 9 | 22 | 6 | 44 | 53 | 3 | 14 | 23 | 36 | 43 | 9 | 21 | 42 | 21 | 45 |
| 20 | 22 | 6 | 58 | 36 | 51 | 1589 | 10 | 10 | 34 | 44 | 57 | 0 | 26 | 58 | 1 | 12 | 11 | 7 | 59 | 16 | 17 |
| 9 | 23 | 27 | 31 | 53 | 8 | 1609 | 9 | 29 | 56 | 20 | 50 | 9 | 13 | 43 | 25 | 6 | 11 | 23 | 35 | 57 | 2 |
| 28 | 13 | 32 | 8 | 12 | 25 | 1629 | 10 | 18 | 24 | 20 | 54 | 6 | 26 | 17 | 49 | 36 | 1 | 9 | 52 | 51 | 34 |
| 17 | 14 | 52 | 41 | 28 | 42 | 1649 | 10 | 7 | 45 | 56 | 57 | 3 | 13 | 3 | 13 | 30 | 1 | 25 | 29 | 32 | 20 |
| 6 | 16 | 13 | 14 | 45 | 0 | 1669 | 9 | 27 | 7 | 32 | 40 | 11 | 29 | 48 | 37 | 24 | 2 | 11 | 6 | 13 | 5 |
| 25 | 6 | 17 | 51 | 4 | 16 | 1689 | 10 | 15 | 35 | 32 | 44 | 9 | 12 | 23 | 1 | 53 | 3 | 27 | 23 | 7 | 37 |
| 14 | 7 | 38 | 24 | 20 | 33 | 1709 | 10 | 4 | 57 | 8 | 37 | 5 | 29 | 8 | 25 | 48 | 4 | 12 | 59 | 48 | 22 |
| 3 | 8 | 58 | 57 | 36 | 51 | 1729 | 9 | 24 | 18 | 44 | 39 | 2 | 15 | 53 | 49 | 42 | 4 | 28 | 36 | 29 | 8 |
| 21 | 23 | 3 | 33 | 56 | 7 | 1749 | 10 | 13 | 46 | 44 | 33 | 11 | 28 | 28 | 14 | 11 | 6 | 14 | 53 | 23 | 39 |
| 11 | 0 | 24 | 7 | 12 | 25 | 1769 | 10 | 2 | 8 | 20 | 26 | 8 | 15 | 13 | 38 | 6 | 7 | 0 | 30 | 4 | 25 |
| 0 | 1 | 44 | 40 | 28 | 42 | 1789 | 9 | 21 | 29 | 56 | 19 | 5 | 1 | 59 | 2 | 0 | 7 | 16 | 6 | 45 | 10 |
| 18 | 15 | 49 | 16 | 47 | 58 | 1809 | 10 | 9 | 57 | 56 | 23 | 2 | 14 | 33 | 26 | 29 | 9 | 2 | 23 | 39 | 42 |

A B C D

Table TRC-07b of apogees in months (Cracow, BJ, 611, f. 90r–v)

| | Tabula augium planetarum in mensibus in anno conmuni | | | | Tabula augium in mensibus nova | | | |
|---|---|---|---|---|---|---|---|---|
| | M | 2 | 3 | 4 | M | 2 | 3 | 4 |
| Ianuarius | 0 | 2 | 56 | 43 | 0 | 2 | 54 | 31 |
| Februarius | 0 | 5 | 36 | 19 | 0 | 5 | 32 | 10 |
| Marcius | 0 | 8 | 33 | 2 | 0 | 8 | 26 | 42 |
| Apprilis | 0 | 11 | 24 | 2 | 0 | 11 | 15 | 36 |
| Maiius | 0 | 14 | 20 | 45 | 0 | 14 | 10 | 7 |
| Iunius | 0 | 17 | 11 | 45 | 0 | 16 | 59 | 1 |
| Iulius | 0 | 20 | 8 | 28 | 0 | 19 | 53 | 33 |
| Augustus | 0 | 23 | 5 | 11 | 0 | 22 | 48 | 5 |
| September | 0 | 25 | 56 | 11 | 0 | 25 | 36 | 59 |
| October | 0 | 28 | 52 | 54 | 0 | 28 | 31 | 32 |
| November | 0 | 31 | 43 | 54 | 0 | 31 | 20 | 25 |
| December | 0 | 34 | 40 | 37 | 0 | 34 | 14 | 57 |
| | A | | | | B | | | |

The same pair of Tables in the months is written in the margin of MS Cracow, BJ, 611, f. 16v.

The version A, truncated to tertias, can also be found in MS Cracow, BJ, 560, p. 51, and Cracow, BJ, 1852, p. 60.

In MS Cracow, BJ, 600, f. 90v, there are also two versions ('A' and 'B') of Tables of motion of apogees in days, which correspond to the above versions in months (the version 'A' is added here in the margin). They are, again in the margin, written also in MS Cracow, BJ, 611, f. 17r, and on f. 16v is added yet another version ('C'). The version 'A' is written in the margin by a later hand also in MS Oxford, Bodl., Can. misc. 499, f. 85v/75v (as well as the corresponding Table in hours on f. 86r/76r, while a prepared frame for the Table in months on f. 85v/75v remained empty).

Finally, a Table of apogees in hours is written also in the margin in MS Cracow, BJ, 611, f. 17r, which complements the Table in days (variant 'C') on the previous folio and has the same rate of the motion.

Table TRC-07c of apogees in days (Cracow, BJ, 600, f. 90v; Cracow, BJ, 611, 16v)

| Tabula augium aliqua in diebus | | | | Tabula nova augium in diebus | | | Tabula augium nova in diebus | | | |
|---|---|---|---|---|---|---|---|---|---|---|
| **dies** | **2** | **3** | **4** | **2** | **3** | **4** | **2** | **3** | **4** | **5** |
| 1 | 0 | 5 | 42 | 0 | 5 | 37 | 0 | 5 | 25 | 29 |
| 2 | 0 | 11 | 24 | 0 | 11 | 14 | 0 | 10 | 50 | 58 |
| 3 | 0 | 17 | 6 | 0 | 16 | 51 | 0 | 16 | 16 | 27 |
| 4 | 0 | 22 | 48 | 0 | 22 | 28 | 0 | 21 | 21 | 56 |
| 5 | 0 | 28 | 30 | 0 | 28 | 5 | 0 | 27 | 7 | 25 |
| 6 | 0 | 34 | 12 | 0 | 33 | 42 | 0 | 32 | 32 | 54 |
| 7 | 0 | 39 | 54 | 0 | 39 | 19 | 0 | 37 | 58 | 23 |
| 8 | 0 | 45 | 36 | 0 | 44 | 56 | 0 | 43 | 23 | 52 |
| 9 | 0 | 51 | 18 | 0 | 50 | 33 | 0 | 48 | 49 | 21 |
| 10 | 0 | 57 | 1 | 0 | 56 | 16 | 0 | 54 | 14 | 50 |
| 11 | 1 | 2 | 42 | 1 | 1 | 47 | 0 | 59 | 40 | 19 |
| 12 | 1 | 8 | 24 | 1 | 7 | 24 | 1 | 5 | 5 | 48 |
| 13 | 1 | 14 | 6 | 1 | 13 | 1 | 1 | 10 | 31 | 17 |
| 14 | 1 | 19 | 40 | 1 | 18 | 38 | 1 | 15 | 56 | 46 |
| 15 | 1 | 25 | 30 | 1 | 24 | 15 | 1 | 21 | 22 | 15 |
| 16 | 1 | 31 | 12 | 1 | 29 | 52 | 1 | 26 | 47 | 44 |
| 17 | 1 | 36 | 54 | 1 | 35 | 29 | 1 | 32 | 13 | 13 |
| 18 | 1 | 42 | 36 | 1 | 41 | 6 | 1 | 37 | 38 | 42 |
| 19 | 1 | 48 | 18 | 1 | 46 | 43 | 1 | 43 | 4 | 11 |
| 20 | 1 | 54 | 0 | 1 | 52 | 26 | 1 | 48 | 29 | 40 |
| 21 | 1 | 59 | 42 | 1 | 57 | 57 | 1 | 53 | 55 | 9 |
| 22 | 2 | 5 | 24 | 2 | 3 | 34 | 1 | 59 | 20 | 38 |
| 23 | 2 | 11 | 6 | 2 | 9 | 11 | 2 | 4 | 46 | 7 |
| 24 | 2 | 17 | 48 | 2 | 14 | 48 | 2 | 10 | 11 | 36 |
| 25 | 2 | 22 | 30 | 2 | 20 | 25 | 2 | 15 | 37 | 5 |
| 26 | 2 | 28 | 12 | 2 | 26 | 2 | 2 | 21 | 2 | 38 |
| 27 | 2 | 33 | 54 | 2 | 31 | 39 | 2 | 26 | 28 | 3 |
| 28 | 2 | 39 | 36 | 2 | 37 | 38 | 2 | 31 | 53 | 32 |
| 29 | 2 | 45 | 18 | 2 | 43 | 16 | 2 | 37 | 19 | 1 |
| 30 | 2 | 51 | 0 | 2 | 48 | 54 | 2 | 42 | 44 | 30 |
| 31 | 2 | 56 | 42 | 2 | 54 | 31 | 2 | 48 | 9 | 59 |

A B C

Table TRC-07d of apogees in hours (Cracow, BJ, 611, f. 17r)

| **Tabula augium nova in horis et minutis** | | | |
|---|---|---|---|
| **hora** | **2** | **3** | **4** |
| 1 | 0 | 13 | 34 |
| 2 | 0 | 27 | 7 |
| 3 | 0 | 40 | 41 |
| 4 | 0 | 54 | 15 |
| 5 | 1 | 7 | 48 |
| 6 | 1 | 21 | 22 |
| 7 | 1 | 34 | 56 |
| 8 | 1 | 48 | 30 |
| 9 | 2 | 2 | 3 |
| 10 | 2 | 15 | 37 |
| 11 | 2 | 29 | 11 |
| 12 | 2 | 42 | 44 |
| 13 | 2 | 56 | 18 |
| 14 | 3 | 9 | 52 |
| 15 | 3 | 23 | 26 |
| 16 | 3 | 36 | 59 |
| 17 | 3 | 50 | 33 |
| 18 | 4 | 4 | 7 |
| 19 | 4 | 17 | 40 |
| 20 | 4 | 31 | 17 |
| 21 | 4 | 44 | 48 |
| 22 | 4 | 58 | 22 |
| 23 | 5 | 11 | 55 |
| 24 | 5 | 25 | 29 |
| 25 | 5 | 39 | 3 |
| 26 | 5 | 52 | 36 |
| 27 | 6 | 6 | 10 |
| 28 | 6 | 19 | 44 |
| 29 | 6 | 33 | 18 |
| 30 | 6 | 46 | 51 |
| **M** | **4** | **5** | **6** |

C

15

# Peculiar Sets of the *Tabule resolute*

## 15.1 *Tabule resolute* for the Meridian of Leipzig

Some sets of the TR which we can find in manuscripts and later also in printed editions were compiled for other meridians. One example of such a set is contained in MS Leipzig, UB, 1473, ff. 122r–126v. The Tables of mean motions in this manuscript differ in several aspects from the majority of previous sets for Wrocław, Prague, and Cracow (see the Table TRL-01). The cyclical radices are extrapolated back to the year 1408, but their series ends at the year 1548. They are given for different quantities and in different order; there are missing mean motions of apogees and fixed stars, of access and recess, and of the argument of latitude of the Moon, but there are added anomalies of Saturn, Jupiter, Mars, and the Moon.[1] The most important difference is that the mean motions are given to a precision of minutes, instead of tertias as in the standard sets. Minutes are more realistic with respect to the precision of the whole theory; they make the tables more compact and simplify the calculations. Yet it also decreases the precision with which we can check the geographical longitude of the meridian for which the tables were compiled (to about half degree in the case of the Moon, which is the most indicative). This meridian is not given at the Table of mean motions in the collected years, but in the headings of the Table of mean conjunctions of the Sun and Moon on f. 126r; it is written that this table is compiled for Leipzig. The traditionally accepted distance of Leipzig from Toledo was 17;30° or $1^h\ 10^m$. This distance is one degree less than the distance of Prague from Toledo.[2] The radices computed for this longitude are in a good agreement with the values given in the Table TRL-01, so that we can accept that the whole set was compiled for Leipzig.

The values in Tables of mean motions in expanded years, non-leap and leap months, days, and hours in MS Lepzig, UB, 1473 agree (except for the scribal error, sometimes even in the signs) with the values in the standard *Tabule resolute* rounded to the tertias, or, in the case of the additional columns of the anomalies of planets, to the values obtained from them by the corresponding subtractions and/or doubling. We thus do not reproduce here these Tables of mean motions. Similarly, the Table of planetary apogees in collected years on f. 125v is equivalent to Table TRP-07 rounded to tertias and limited to the year 1548. Unlike the Tables of mean motions, the Table of apogees is not extrapolated to the year 1408. Due to the rounding to tertias, the motion of apogees in expanded years does not differ significantly for the different twenty-years cycles, so that Table TRP-07a, which

1 These anomalies enable to avoid the subtraction from the mean motion of the Sun, or in the case of the Moon subtraction of the motion of the Sun and doubling the remainder, but it comes at the cost of adding more numbers for collected and expanded years, months etc.

2 See Richard L. Kremer and Jerzy Dobrzycki, 'Alfonsine Meridians', 187–99.

Table TRL-01 of mean motions in collected years for the meridian of Leipzig (Leipzig, UB, 1473, f. 122r–v)

| Tabula annorum collectorum | | | | | | | | | | | | | | | | | | | | | | |
|---|---|---|---|---|---|---|---|---|---|---|---|---|---|---|---|---|---|---|---|---|---|---|
| | Capitis | | | Saturni | | | Argumenti Saturni | | | Iovis | | | Argumenti Iovis | | | Martis | | | Argumenti Martis | | |
| | S | G | M | S | G | M | S | G | M | S | G | M | S | G | M | S | G | M | S | G | M |
| 1408 | 10 | 24 | 49 | 0 | 20 | 32 | 8 | 28 | 6 | 3 | 0 | 33 | 6 | 18 | 5 | 9 | 4 | 51 | 0 | 13 | 47 |
| 1428 | 11 | 21 | 29 | 8 | 25 | 13 | 0 | 23 | 34 | 11 | 7 | 48 | 10 | 10 | 59 | 4 | 23 | 10 | 4 | 25 | 37 |
| 1448 | 0 | 18 | 29 | 4 | 29 | 54 | 4 | 19 | 2 | 7 | 15 | 2 | 2 | 3 | 54 | 0 | 11 | 29 | 9 | 7 | 27 |
| 1468 | 1 | 15 | 19 | 1 | 4 | 36 | 8 | 14 | 25 | 3 | 22 | 17 | 5 | 26 | 48 | 7 | 29 | 48 | 1 | 19 | 17 |
| 1488 | 2 | 12 | 9 | 9 | 9 | 18 | 0 | 9 | 55 | 11 | 29 | 32 | 9 | 19 | 41 | 3 | 18 | 7 | 6 | 1 | 7 |
| 1508 | 3 | 8 | 59 | 5 | 13 | 59 | 4 | 5 | 23 | 8 | 6 | 46 | 1 | 12 | 36 | 11 | 6 | 25 | 10 | 12 | 57 |
| 1528 | 4 | 5 | 49 | 1 | 18 | 41 | 8 | 0 | 50 | 4 | 14 | 1 | 5 | 5 | 30 | 6 | 24 | 44 | 2 | 24 | 47 |
| 1548 | 5 | 2 | 38 | 9 | 23 | 22 | 11 | 26 | 18 | 0 | 21 | 15 | 8 | 28 | 25 | 2 | 13 | 3 | 7 | 6 | 37 |

Y A B C D E F G

| Tabula annorum collectorum | | | | | | | | | | | | | | | | | | | | |
|---|---|---|---|---|---|---|---|---|---|---|---|---|---|---|---|---|---|---|---|---|
| Anni collecti | Solis, Veneris et Mercurii | | | Argumenti Veneris | | | Argumenti Mercurii | | | Anni collecti | Lune | | | Argumenti Lune | | | Centri Lune | | |
| | S | G | M | S | G | M | S | G | M | | S | G | M | S | G | M | S | G | M |
| 1408 | 9 | 18 | 38 | 0 | 26 | 8 | 1 | 25 | 17 | 1408 | 3 | 2 | 59 | 3 | 26 | 7 | 10 | 28 | 40 |
| 1428 | 9 | 18 | 48 | 6 | 29 | 46 | 2 | 9 | 45 | 1428 | 7 | 16 | 33 | 5 | 5 | 51 | 7 | 25 | 30 |
| 1448 | 9 | 18 | 57 | 1 | 3 | 24 | 2 | 24 | 13 | 1448 | 0 | 0 | 7 | 6 | 15 | 36 | 4 | 22 | 20 |
| 1468 | 9 | 19 | 6 | 7 | 7 | 3 | 3 | 8 | 40 | 1468 | 4 | 13 | 41 | 7 | 25 | 20 | 1 | 10 | 10 |
| 1488 | 9 | 19 | 14 | 1 | 10 | 42 | 3 | 23 | 8 | 1488 | 8 | 27 | 15 | 9 | 5 | 4 | 10 | 16 | 0 |
| 1508 | 9 | 19 | 23 | 7 | 14 | 21 | 4 | 7 | 36 | 1508 | 1 | 10 | 48 | 10 | 14 | 49 | 7 | 12 | 50 |
| 1528 | 9 | 19 | 32 | 1 | 18 | 0 | 4 | 22 | 3 | 1528 | 5 | 24 | 22 | 11 | 24 | 43 | 4 | 9 | 40 |
| 1548 | 9 | 19 | 41 | 7 | 21 | 39 | 5 | 6 | 31 | 1548 | 10 | 7 | 56 | 1 | 4 | 18 | 1 | 6 | 30 |

Y H I J Y K L M

Table TRL-09 of radices of mean conjunctions of the Sun and Moon for the meridian of Leipzig (Leipzig, UB, 1473, f. 126r)

| Tabula ad Liptzk radicum coniunccionum | | | | | | | | | | | | | | | | | | | |
|---|---|---|---|---|---|---|---|---|---|---|---|---|---|---|---|---|---|---|---|
| Anni | Media coniunccio | | | Motus Solis et Lune | | | Argumentum Lune | | | Anni | Media coniunccio | | | Motus Solis et Lune | | | Argumentum Lune | | |
| | D | H | M | S | G | M | S | G | M | | D | H | M | S | G | M | S | G | M |
| 1469 | 12 | 17 | 58 | 10 | 1 | 39 | 1 | 11 | 53 | 30 | 11 | 14 | 20 | 10 | 0 | 15 | 8 | 19 | 55 |
| 1 | 2 | 2 | 46 | 9 | 20 | 56 | 11 | 21 | 41 | 31 | 0 | 23 | 9 | 9 | 19 | 32 | 6 | 29 | 43 |
| 2 | 21 | 0 | 19 | 10 | 9 | 19 | 10 | 27 | 18 | 32 | 18 | 20 | 42 | 10 | 7 | 55 | 6 | 5 | 20 |
| 3 | 10 | 9 | 7 | 9 | 28 | 36 | 9 | 7 | 6 | 33 | 8 | 5 | 30 | 9 | 27 | 12 | 4 | 15 | 9 |
| 4 | 28 | 6 | 40 | 10 | 16 | 59 | 8 | 12 | 43 | 34 | 27 | 3 | 3 | 10 | 15 | 35 | 3 | 20 | 46 |
| 5 | 17 | 15 | 29 | 10 | 6 | 16 | 6 | 22 | 32 | 35 | 16 | 11 | 52 | 10 | 4 | 52 | 2 | 0 | 34 |
| 6 | 7 | 0 | 17 | 9 | 25 | 33 | 5 | 2 | 20 | 36 | 4 | 20 | 40 | 9 | 24 | 8 | 0 | 10 | 22 |
| 7 | 25 | 21 | 50 | 10 | 13 | 56 | 4 | 7 | 57 | 37 | 23 | 18 | 13 | 10 | 12 | 32 | 11 | 15 | 59 |
| 8 | 14 | 6 | 39 | 10 | 3 | 13 | 2 | 17 | 45 | 38 | 13 | 3 | 1 | 10 | 1 | 49 | 9 | 25 | 47 |
| 9 | 3 | 15 | 27 | 9 | 22 | 30 | 0 | 27 | 33 | 39 | 2 | 11 | 50 | 9 | 21 | 5 | 8 | 5 | 35 |
| 10 | 22 | 13 | 0 | 10 | 10 | 53 | 0 | 3 | 10 | 40 | 20 | 9 | 23 | 10 | 9 | 29 | 7 | 11 | 13 |
| 11 | 11 | 21 | 48 | 10 | 0 | 10 | 10 | 12 | 58 | 41 | 9 | 18 | 11 | 9 | 28 | 46 | 5 | 21 | 1 |
| 12 | 0 | 6 | 37 | 9 | 19 | 27 | 8 | 22 | 46 | 42 | 28 | 15 | 44 | 10 | 17 | 9 | 4 | 26 | 38 |
| 13 | 19 | 4 | 10 | 10 | 7 | 50 | 7 | 28 | 24 | 43 | 18 | 0 | 33 | 10 | 6 | 26 | 3 | 6 | 26 |
| 14 | 8 | 12 | 58 | 9 | 27 | 2 | 6 | 8 | 12 | 44 | 6 | 9 | 21 | 9 | 25 | 43 | 1 | 16 | 14 |
| 15 | 27 | 10 | 31 | 10 | 15 | 30 | 5 | 13 | 49 | 45 | 25 | 6 | 54 | 10 | 14 | 6 | 0 | 21 | 52 |
| 16 | 15 | 19 | 20 | 10 | 4 | 46 | 3 | 23 | 37 | 46 | 14 | 15 | 42 | 10 | 3 | 23 | 11 | 1 | 40 |
| 17 | 5 | 4 | 8 | 9 | 24 | 4 | 2 | 3 | 25 | 47 | 4 | 0 | 31 | 9 | 22 | 40 | 9 | 11 | 28 |
| 18 | 24 | 1 | 41 | 10 | 12 | 27 | 1 | 9 | 2 | 48 | 21 | 22 | 4 | 10 | 11 | 3 | 8 | 17 | 5 |
| 19 | 13 | 10 | 29 | 10 | 1 | 43 | 11 | 18 | 50 | 49 | 11 | 6 | 52 | 10 | 0 | 20 | 6 | 26 | 53 |
| 20 | 1 | 19 | 18 | 9 | 21 | 1 | 9 | 28 | 38 | 50 | 0 | 15 | 41 | 9 | 19 | 37 | 5 | 6 | 41 |
| 21 | 20 | 16 | 51 | 10 | 9 | 24 | 9 | 4 | 15 | 51 | 19 | 13 | 14 | 10 | 8 | 0 | 4 | 12 | 18 |
| 22 | 10 | 1 | 39 | 9 | 28 | 41 | 7 | 14 | 3 | 52 | 7 | 22 | 2 | 9 | 27 | 17 | 2 | 22 | 6 |
| 23 | 28 | 23 | 12 | 10 | 17 | 4 | 6 | 19 | 40 | 53 | 26 | 19 | 35 | 10 | 15 | 40 | 1 | 27 | 44 |
| 24 | 17 | 8 | 1 | 10 | 6 | 21 | 4 | 29 | 28 | 54 | 16 | 4 | 23 | 10 | 4 | 57 | 0 | 7 | 32 |
| 25 | 6 | 16 | 49 | 9 | 25 | 38 | 3 | 9 | 17 | 55 | 5 | 13 | 12 | 9 | 24 | 14 | 10 | 17 | 20 |
| 26 | 25 | 14 | 22 | 10 | 14 | 1 | 2 | 14 | 54 | 56 | 23 | 10 | 45 | 10 | 12 | 36 | 9 | 22 | 57 |
| 27 | 14 | 23 | 10 | 10 | 3 | 18 | 0 | 24 | 42 | 57 | 12 | 19 | 33 | 10 | 1 | 54 | 8 | 2 | 45 |
| 28 | 3 | 7 | 59 | 9 | 22 | 34 | 11 | 4 | 30 | 58 | 2 | 4 | 22 | 9 | 21 | 11 | 6 | 12 | 33 |
| 29 | 22 | 5 | 32 | 10 | 10 | 58 | 10 | 10 | 7 | 59 | 21 | 1 | 55 | 10 | 9 | 33 | 5 | 18 | 10 |
| Y | A | | | B | | | C | | | Y | A | | | B | | | C | | |

Remaining part of the Table TRL-09 (Leipzig, UB, 1473, f. 126v)

| Residua pars Tabelle radicum coniunccionum | | | | | | | | | |
|---|---|---|---|---|---|---|---|---|---|
| **Anni** | **Media coniunccio** | | | **Motus Solis et Lune** | | | **Argumentum Lune** | | |
| | **D** | **H** | **M** | **S** | **G** | **M** | **S** | **G** | **M** |
| 60 | 9 | 10 | 43 | 9 | 28 | 50 | 3 | 27 | 58 |
| 61 | 28 | 8 | 16 | 10 | 17 | 13 | 3 | 3 | 35 |
| 62 | 17 | 17 | 5 | 10 | 6 | 30 | 1 | 13 | 23 |
| 63 | 7 | 1 | 53 | 9 | 25 | 47 | 11 | 23 | 11 |
| 64 | 24 | 23 | 26 | 10 | 14 | 10 | 10 | 28 | 48 |
| 65 | 14 | 8 | 14 | 10 | 3 | 27 | 9 | 8 | 37 |
| 66 | 3 | 17 | 3 | 9 | 22 | 44 | 7 | 18 | 26 |
| 67 | 22 | 14 | 36 | 10 | 11 | 7 | 6 | 24 | 2 |
| 68 | 10 | 23 | 24 | 10 | 0 | 24 | 5 | 3 | 50 |
| 69 | 0 | 8 | 13 | 9 | 19 | 41 | 3 | 13 | 38 |
| 70 | 19 | 5 | 46 | 10 | 8 | 4 | 2 | 19 | 15 |
| 71 | 8 | 14 | 34 | 9 | 27 | 21 | 0 | 29 | 3 |
| 72 | 26 | 12 | 7 | 10 | 15 | 44 | 0 | 4 | 40 |
| 73 | 15 | 20 | 55 | 10 | 5 | 1 | 10 | 14 | 29 |
| 74 | 5 | 5 | 44 | 9 | 24 | 18 | 8 | 24 | 17 |
| 75 | 24 | 3 | 17 | 10 | 12 | 41 | 7 | 29 | 54 |

Y A B C

| Motus a coniunccione ad opposicionem | | | | | | | | |
|---|---|---|---|---|---|---|---|---|
| 14 | 18 | 22 | 0 | 14 | 33 | 6 | 12 | 54 |
| **D** | **H** | **M** | **S** | **G** | **M** | **S** | **G** | **M** |

| Sequitur Tabella pro renovacione radicum coniunccionum | | | | | | | | | | |
|---|---|---|---|---|---|---|---|---|---|---|
| | **D** | **H** | **M** | **S** | **G** | **M** | **S** | **G** | **M** | |
| 1317 | 0 | 11 | 44 | 11 | 29 | 22 | 2 | 4 | 22 | adde |
| 1393 | 0 | 5 | 52 | 11 | 29 | 41 | 7 | 2 | 11 | adde |
| 1460 | 0 | 0 | 0 | 0 | 0 | 0 | 0 | 0 | 0 | 0 0 |
| 1545 | 0 | 5 | 52 | 11 | 29 | 41 | 7 | 2 | 11 | minue |
| 1621 | 0 | 11 | 44 | 11 | 29 | 22 | 2 | 4 | 22 | minue |
| 1697 | 0 | 17 | 37 | 11 | 29 | 3 | 9 | 6 | 33 | minue |

takes several pages in tables for Prague and Cracow, simplifies to a small table of increments from 1 to 11 tertias in the course of the twenty years on f. 125v in the tables for Leipzig.

Instead of a Table of cyclical radices of conjunctions of the Sun and Moon in collected years as in Tables TRW-09, TRP-09, or TRC-09, the tables for Leipzig contain on f. 126r–v a Table TRL-09 of radices of conjunctions of the luminaries for the year 1469 and for each of the subsequent 75 years. While the first radix is denoted by the number 1469 of the year, the subsequent entries are denoted by numbers from 1 to 75, which are to be added to the first item to get the final year. This is because this Table of radices is complemented on f. 126v by a small table which gives the increments of time and the additional angular quantities of the conjunctions in three previous and following periods of 76 years.[3] The roles of the Tables in collected and expanded years are thus reversed in these tables compared to the other sets of the *Tabule resolute*.[4] Both these tables give the time of the first mean conjunction in a year rounded to minutes of an hour, and the ecliptic longitude of the Sun and Moon and anomaly of the Moon, rounded to minutes of a degree, while the argument of latitude of the Moon is skipped as is the case in the Tables of mean motions. The traditionally accepted distance of one degree in geographical longitude from Prague to the Leipzig should correspond to the local time of the conjunction in Leipzig, smaller by four minutes than in Prague. Comparing the Table TRL-09 with values truncated from TRP-09 we find differences of three minutes or four minutes with respect to the properly rounded values for Prague in the years 1489 and 1529 (numbers of years 20 and 60 in the Table TRL-09). These values are thus consistent with the inscription of the whole table that the radices are calculated for Leipzig and also with the assumption that this complete set of the *Tabule resolute* was modified to the meridian of Leipzig. On f. 126v is added a one-line table with increments of the time and angles between the mean conjunction and opposition. The Tables can thus be used for both conjunctions and oppositions. There is, also on f. 126v, a Table of mean conjunctions in months corresponding to Table TRW-12, but again rounded to minutes with omitted argument of the latitude of the Moon. The tables in MS Leipzig, BU, 1473 finish on f. 127r by a Table of differences of zonal time in other cities with respect to the Leipzig. The Tables of equations are missing in this manuscript.

## 15.2 *Tabule resolute* for the Meridian of Salamanca

Another interesting example of the *Tabule resolute* for a different meridian can be found in MS Lisbon, ANTT, ML, 2115. This manuscript contains the *Tabule resolute* for the meridian of Salamanca on ff. 19r–56v, as it is inscribed there (see the Table TRS-01). The cyclical radices of mean motions are extrapolated in these tables back to the year 1348 and they continue up to the year 1788. Consistent with this, the Tables of syzygies on folios 29v–31r are computed for collected years from 1349 to 1789, the Tables of apogees on folios 33v–34r for years 1348 to 1788 and also the Tables of apogees in expanded years on folios 34v–36v. On folio 37r we can find Tables of apogees in months and days, which resemble similar Tables TRC-07b and TRC-07c in the sets compiled for the meridian of Cracow. The values in both these tables correspond to the 'new' variant 'B', but they are rounded to tertias only. The link between MS Lisbon, ANTT, ML, 2115, and the Cracow school is also obvious from the presence on ff. 8r–18v of canons with the incipit *Quoniam Tabularum Alfonsi* (see Chapter 6) written by Nicolaus Polonus in Salamanca on the basis of canons

3 By a mistake, the central item in this table of increments is denoted 1460 instead of 1469 on f. 126v.

4 A similar change in the role of collected and expanded years can be found, e.g., in a version of the *Tabule magne* for Erfurt in MS Vatican, BAV, Pal. lat. 1412, f. 116r, but not in the original *Tabule magne* or in the Tables of 1322 by John of Lignères; see Chabás, *Computational Astronomy in the Middle Ages*, pp. 200–01, and Chabás and Saby, *The Tables of 1322 by John of Lignères*, pp. 94–97.

*Girum recensendo* (see Chapter 3) by Andrzej Grzymała. Polonus gives in his canons the longitude of Salamanca as 8;16° from the Pillars of Hercules or 25;46° from the 'true west' to the east. These values are mutually consistent and it follows from them that Salamanca is 2;44° to the west from Toledo (because Toledo is 11° east of the Pillars of Hercules). This value is also consistent with the values of mean motions in collected years given in the corresponding tables in MS Lisbon, ANTT, ML, 2115 transcribed here in Table TRS-01. The other Tables of mean motions in expanded years, months, days and hours are basically the same as in other sets of the *Tabule resolute*, so we shall not reproduce them here.

**Table TRS-01 of mean motions in collected years for the meridian of Salamanca (Lisbon, ANTT, ML, 2115, f. 19r)**

| **Tabule de mediis et veris motibus planetarum super meridianum Salamantinum ordinate** | | | | | | | | | | | | | | | |
|---|---|---|---|---|---|---|---|---|---|---|---|---|---|---|---|
| **Anni Christi collecti** | **Tabula augium et stellarum fixarum** | | | | | **Tabula accessus et recessus octave <spere>** | | | | | **Tabula medii motus Solis, Veneris et Mercurii** | | | | |
| | **S** | **G** | **M** | **2** | **3** | **S** | **G** | **M** | **2** | **3** | **S** | **G** | **M** | **2** | **3** |
| 1348 | 0 | 9 | 54 | 13 | 14 | 2 | 8 | 32 | 6 | 35 | 9 | 18 | 15 | 40 | 40 |
| 1368 | 0 | 10 | 3 | 2 | 12 | 2 | 9 | 33 | 49 | 26 | 9 | 18 | 24 | 29 | 39 |
| 1388 | 0 | 10 | 11 | 51 | 11 | 2 | 10 | 35 | 32 | 17 | 9 | 18 | 33 | 18 | 38 |
| 1408 | 0 | 10 | 20 | 40 | 10 | 2 | 11 | 37 | 15 | 9 | 9 | 18 | 42 | 7 | 37 |
| 1428 | 0 | 10 | 29 | 29 | 9 | 2 | 12 | 38 | 58 | 0 | 9 | 18 | 50 | 56 | 36 |
| 1448 | 0 | 10 | 38 | 18 | 8 | 2 | 13 | 40 | 40 | 52 | 9 | 18 | 59 | 45 | 34 |
| 1468 | 0 | 10 | 47 | 7 | 6 | 2 | 14 | 42 | 23 | 43 | 9 | 19 | 8 | 34 | 33 |
| 1488 | 0 | 10 | 55 | 56 | 5 | 2 | 15 | 44 | 6 | 35 | 9 | 19 | 17 | 23 | 32 |
| 1508 | 0 | 11 | 4 | 45 | 4 | 2 | 16 | 45 | 49 | 26 | 9 | 19 | 26 | 12 | 31 |
| 1528 | 0 | 11 | 13 | 34 | 3 | 2 | 17 | 47 | 32 | 17 | 9 | 19 | 35 | 1 | 29 |
| 1548 | 0 | 11 | 22 | 23 | 2 | 2 | 18 | 49 | 15 | 9 | 9 | 19 | 43 | 50 | 28 |
| 1568 | 0 | 11 | 31 | 12 | 0 | 2 | 19 | 50 | 58 | 0 | 9 | 19 | 52 | 39 | 27 |
| 1588 | 0 | 11 | 40 | 0 | 59 | 2 | 20 | 52 | 40 | 52 | 9 | 20 | 1 | 28 | 26 |
| 1608 | 0 | 11 | 48 | 49 | 58 | 2 | 21 | 54 | 23 | 43 | 9 | 20 | 10 | 17 | 24 |
| 1628 | 0 | 11 | 57 | 38 | 56 | 2 | 22 | 56 | 6 | 34 | 9 | 20 | 19 | 6 | 23 |
| 1648 | 0 | 12 | 6 | 27 | 55 | 2 | 23 | 57 | 49 | 26 | 9 | 20 | 27 | 55 | 22 |
| 1668 | 0 | 12 | 15 | 16 | 54 | 2 | 24 | 59 | 32 | 17 | 9 | 20 | 36 | 44 | 21 |
| 1688 | 0 | 12 | 24 | 5 | 53 | 2 | 26 | 1 | 15 | 9 | 9 | 20 | 45 | 33 | 19 |
| 1708 | 0 | 12 | 32 | 54 | 52 | 2 | 27 | 2 | 58 | 0 | 9 | 20 | 54 | 22 | 18 |
| 1728 | 0 | 12 | 41 | 43 | 50 | 2 | 28 | 4 | 40 | 52 | 9 | 21 | 3 | 11 | 17 |
| 1748 | 0 | 12 | 50 | 32 | 49 | 2 | 29 | 6 | 23 | 43 | 9 | 21 | 12 | 0 | 16 |
| 1768 | 0 | 12 | 59 | 21 | 48 | 3 | 0 | 8 | 6 | 34 | 9 | 21 | 20 | 49 | 17 |
| 1788 | 0 | 13 | 8 | 10 | 47 | 3 | 1 | 9 | 49 | 26 | 9 | 21 | 29 | 38 | 13 |
| | | | A | | | | | B | | | | | C | | |

**Continuation of the Table TRS-01 (Lisbon, ANTT, ML, 2115, f. 19v)**

| Residua pars Tabule de mediis motibus planetarum in annis Christi collectis | | | | | | | | | | | | | | | |
|---|---|---|---|---|---|---|---|---|---|---|---|---|---|---|---|
| **Anni Christi collecti** | **Tabula medii motus Lune** | | | | | **Tabula medii argumenti Lune** | | | | | **Tabula medii argumenti latitudinis Lune** | | | | |
| | **S** | **G** | **M** | **2** | **3** | **S** | **G** | **M** | **2** | **3** | **S** | **G** | **M** | **2** | **3** |
| 1348 | 1 | 23 | 2 | 28 | 7 | 11 | 27 | 37 | 22 | 47 | 9 | 27 | 8 | 29 | 10 |
| 1368 | 6 | 6 | 36 | 15 | 41 | 1 | 7 | 21 | 49 | 7 | 3 | 7 | 32 | 7 | 56 |
| 1388 | 10 | 20 | 10 | 3 | 15 | 2 | 17 | 6 | 15 | 28 | 8 | 17 | 55 | 46 | 42 |
| 1408 | 3 | 3 | 43 | 50 | 48 | 3 | 26 | 50 | 41 | 48 | 1 | 28 | 19 | 25 | 28 |
| 1428 | 7 | 17 | 17 | 38 | 26 | 5 | 6 | 35 | 8 | 8 | 7 | 8 | 43 | 4 | 14 |
| 1448 | 0 | 0 | 51 | 25 | 56 | 6 | 16 | 19 | 34 | 28 | 0 | 19 | 6 | 43 | 0 |
| 1468 | 4 | 14 | 25 | 13 | 29 | 7 | 26 | 4 | 0 | 49 | 5 | 29 | 30 | 21 | 46 |
| 1488 | 8 | 27 | 59 | 1 | 3 | 9 | 5 | 48 | 27 | 9 | 11 | 9 | 54 | 0 | 33 |
| 1508 | 1 | 11 | 32 | 48 | 36 | 10 | 15 | 32 | 53 | 29 | 4 | 20 | 17 | 39 | 19 |
| 1528 | 5 | 25 | 6 | 36 | 11 | 11 | 25 | 17 | 19 | 49 | 10 | 0 | 41 | 18 | 5 |
| 1548 | 10 | 8 | 40 | 23 | 44 | 1 | 5 | 1 | 46 | 10 | 3 | 11 | 4 | 56 | 51 |
| 1568 | 2 | 22 | 14 | 11 | 18 | 2 | 14 | 46 | 12 | 30 | 8 | 21 | 28 | 35 | 37 |
| 1588 | 7 | 5 | 47 | 58 | 52 | 3 | 24 | 30 | 38 | 50 | 2 | 1 | 52 | 14 | 23 |
| 1608 | 11 | 19 | 21 | 46 | 26 | 5 | 4 | 15 | 5 | 10 | 7 | 12 | 15 | 53 | 9 |
| 1628 | 4 | 2 | 55 | 33 | 59 | 6 | 13 | 59 | 31 | 31 | 0 | 22 | 39 | 31 | 55 |
| 1648 | 8 | 16 | 29 | 21 | 33 | 7 | 23 | 43 | 57 | 51 | 6 | 3 | 3 | 10 | 41 |
| 1668 | 1 | 0 | 3 | 9 | 7 | 9 | 3 | 28 | 24 | 11 | 11 | 13 | 26 | 49 | 27 |
| 1688 | 5 | 13 | 36 | 56 | 41 | 10 | 13 | 12 | 50 | 31 | 4 | 23 | 50 | 28 | 14 |
| 1708 | 9 | 27 | 10 | 44 | 14 | 11 | 22 | 57 | 16 | 52 | 10 | 4 | 14 | 7 | 0 |
| 1728 | 2 | 10 | 44 | 31 | 48 | 1 | 2 | 41 | 43 | 12 | 3 | 14 | 37 | 45 | 46 |
| 1748 | 6 | 24 | 18 | 19 | 22 | 2 | 12 | 26 | 9 | 32 | 8 | 25 | 1 | 24 | 32 |
| 1768 | 11 | 7 | 52 | 6 | 55 | 3 | 22 | 10 | 35 | 52 | 2 | 5 | 25 | 3 | 18 |
| 1788 | 3 | 21 | 25 | 54 | 29 | 5 | 1 | 55 | 2 | 13 | 7 | 15 | 48 | 42 | 4 |

D E F

Continuation of the Table TRS-01 (Lisbon, ANTT, ML, 2115, f. 20r)

| **Residua pars Tabule de mediis motibus planetarum in annis Christi collectis** | | | | | | | | | | | | | | | |
|---|---|---|---|---|---|---|---|---|---|---|---|---|---|---|---|
| **Anni Christi collecti** | **Tabula medii capitis draconis** | | | | | **Tabula medii Saturni** | | | | | **Tabula medii Iovis** | | | | |
| | **S** | **G** | **M** | **2** | **3** | **S** | **G** | **M** | **2** | **3** | **S** | **G** | **M** | **2** | **3** |
| 1348 | 8 | 4 | 20 | 1 | 22 | 0 | 6 | 26 | 26 | 22 | 2 | 8 | 49 | 44 | 48 |
| 1368 | 9 | 1 | 9 | 52 | 34 | 8 | 11 | 8 | 3 | 28 | 10 | 16 | 4 | 21 | 5 |
| 1388 | 9 | 27 | 59 | 43 | 47 | 4 | 15 | 49 | 40 | 35 | 6 | 23 | 18 | 57 | 22 |
| 1408 | 10 | 24 | 49 | 34 | 59 | 0 | 20 | 31 | 17 | 41 | 3 | 0 | 33 | 33 | 39 |
| 1428 | 11 | 21 | 39 | 26 | 11 | 8 | 25 | 12 | 54 | 48 | 11 | 7 | 48 | 9 | 57 |
| 1448 | 0 | 18 | 29 | 17 | 24 | 4 | 29 | 54 | 31 | 55 | 7 | 15 | 2 | 46 | 14 |
| 1468 | 1 | 15 | 19 | 8 | 36 | 1 | 4 | 36 | 9 | 1 | 3 | 22 | 17 | 22 | 31 |
| 1488 | 2 | 12 | 8 | 59 | 49 | 9 | 9 | 17 | 46 | 7 | 11 | 29 | 31 | 58 | 48 |
| 1508 | 3 | 8 | 58 | 51 | 1 | 5 | 13 | 59 | 23 | 14 | 8 | 6 | 46 | 35 | 6 |
| 1528 | 4 | 5 | 48 | 42 | 13 | 1 | 18 | 41 | 0 | 20 | 4 | 14 | 1 | 11 | 23 |
| 1548 | 5 | 2 | 38 | 33 | 25 | 9 | 23 | 22 | 37 | 27 | 0 | 21 | 15 | 47 | 40 |
| 1568 | 5 | 29 | 28 | 24 | 39 | 5 | 28 | 4 | 14 | 33 | 8 | 28 | 30 | 23 | 57 |
| 1588 | 6 | 26 | 18 | 15 | 50 | 2 | 2 | 45 | 51 | 40 | 5 | 5 | 45 | 0 | 15 |
| 1608 | 7 | 23 | 8 | 7 | 2 | 10 | 7 | 27 | 28 | 47 | 1 | 12 | 59 | 36 | 31 |
| 1628 | 8 | 19 | 57 | 58 | 15 | 6 | 12 | 9 | 5 | 53 | 9 | 20 | 14 | 12 | 49 |
| 1648 | 9 | 16 | 47 | 49 | 27 | 2 | 16 | 50 | 43 | 0 | 5 | 27 | 28 | 49 | 6 |
| 1668 | 10 | 13 | 37 | 40 | 39 | 10 | 21 | 32 | 20 | 6 | 2 | 4 | 43 | 25 | 24 |
| 1688 | 11 | 10 | 27 | 31 | 52 | 6 | 26 | 13 | 57 | 13 | 10 | 11 | 58 | 1 | 41 |
| 1708 | 0 | 7 | 17 | 23 | 4 | 3 | 0 | 55 | 34 | 20 | 6 | 19 | 12 | 37 | 58 |
| 1728 | 1 | 4 | 7 | 14 | 16 | 11 | 5 | 37 | 11 | 26 | 2 | 26 | 27 | 14 | 15 |
| 1748 | 2 | 0 | 57 | 5 | 29 | 7 | 10 | 18 | 48 | 33 | 11 | 3 | 41 | 50 | 33 |
| 1768 | 2 | 27 | 46 | 56 | 41 | 3 | 15 | 0 | 25 | 40 | 7 | 10 | 56 | 26 | 50 |
| 1788 | 3 | 24 | 36 | 47 | 53 | 11 | 19 | 42 | 2 | 46 | 3 | 18 | 11 | 3 | 7 |

G H I

MS Lisbon, ANTT, ML, 2115 also contains Tables of mean syzygies in collected years on ff. 29v–31r. They are extrapolated back to the year 1349. Comparing the time of mean conjunction in the year 1429 in Table TRS-09 for Salamanca with that in TRP-09 for Prague, we find the difference in zonal time to be 1;24,56 hour without one quarta. This means the distance in the geographical longitude 21;14°, which is consistent with the distance of Salamanca 2;44° to the west of Toledo and Prague 18;30° to the east of Toledo. In other years the difference varies within a few quartas. Similarly, in comparison with the corresponding Table TRC-09 for Cracow, we find the difference between the local times of the mean conjunctions to be nearly 1;37,20 hours, which corresponds to 24;20° in geographical longitude, and agrees with the commonly given distance 21;36° of Cracow from Toledo. More precisely, about 30 quartas of an hour are missing in the temporal difference in most of the collected years, which is close to the 25 quartas mentioned by John of Głogów in his Rule 36 as missing in the difference of 12 minutes and 24 seconds between Prague and Cracow.

**Remaining part of the Table TRS-01 (Lisbon, ANTT, ML, 2115, f. 20v)**

| Residua pars Tabule mediorum motuum planetarum in annis Christi collectis | | | | | | | | | | | | | | | |
|---|---|---|---|---|---|---|---|---|---|---|---|---|---|---|---|
| **Anni Christi collecti** | **Tabula medii motus Martis** | | | | | **Tabula medii argumenti Veneris** | | | | | **Tabula medii argumenti Mercurii** | | | | |
| | **S** | **G** | **M** | **2** | **3** | **S** | **G** | **M** | **2** | **3** | **S** | **G** | **M** | **2** | **3** |
| 1348 | 10 | 9 | 55 | 54 | 10 | 6 | 15 | 12 | 44 | 43 | 0 | 12 | 4 | 45 | 53 |
| 1368 | 5 | 28 | 14 | 52 | 0 | 0 | 18 | 51 | 35 | 39 | 0 | 26 | 32 | 24 | 44 |
| 1388 | 1 | 16 | 33 | 49 | 51 | 6 | 22 | 30 | 26 | 35 | 1 | 11 | 0 | 3 | 36 |
| 1408 | 9 | 4 | 52 | 47 | 41 | 0 | 26 | 9 | 17 | 31 | 1 | 25 | 27 | 42 | 27 |
| 1428 | 4 | 23 | 11 | 45 | 31 | 6 | 29 | 48 | 8 | 27 | 2 | 9 | 55 | 21 | 19 |
| 1448 | 0 | 11 | 30 | 43 | 21 | 1 | 3 | 26 | 59 | 23 | 2 | 24 | 23 | 0 | 10 |
| 1468 | 7 | 29 | 49 | 41 | 11 | 7 | 7 | 5 | 50 | 19 | 3 | 8 | 50 | 39 | 2 |
| 1488 | 3 | 18 | 8 | 39 | 1 | 1 | 10 | 44 | 41 | 15 | 3 | 23 | 18 | 17 | 53 |
| 1508 | 11 | 6 | 27 | 36 | 51 | 7 | 14 | 23 | 32 | 11 | 4 | 7 | 45 | 56 | 44 |
| 1528 | 6 | 24 | 46 | 34 | 42 | 1 | 18 | 2 | 23 | 7 | 4 | 22 | 13 | 35 | 36 |
| 1548 | 2 | 13 | 5 | 32 | 32 | 7 | 21 | 41 | 14 | 3 | 5 | 6 | 41 | 14 | 27 |
| 1568 | 10 | 1 | 24 | 30 | 22 | 1 | 25 | 20 | 4 | 59 | 5 | 21 | 8 | 53 | 19 |
| 1588 | 5 | 19 | 43 | 28 | 12 | 7 | 28 | 58 | 55 | 55 | 6 | 5 | 36 | 32 | 10 |
| 1608 | 1 | 8 | 2 | 26 | 2 | 2 | 2 | 37 | 46 | 51 | 6 | 20 | 4 | 11 | 1 |
| 1628 | 8 | 26 | 21 | 23 | 52 | 8 | 6 | 16 | 37 | 47 | 7 | 4 | 31 | 49 | 53 |
| 1648 | 4 | 14 | 40 | 21 | 42 | 2 | 9 | 55 | 28 | 43 | 7 | 18 | 59 | 28 | 44 |
| 1668 | 0 | 2 | 59 | 19 | 33 | 8 | 13 | 34 | 19 | 39 | 8 | 3 | 27 | 7 | 36 |
| 1688 | 7 | 21 | 18 | 17 | 23 | 2 | 17 | 13 | 10 | 35 | 8 | 17 | 54 | 46 | 27 |
| 1708 | 3 | 9 | 37 | 15 | 13 | 8 | 20 | 52 | 1 | 31 | 9 | 2 | 22 | 25 | 18 |
| 1728 | 10 | 27 | 56 | 13 | 3 | 2 | 24 | 30 | 52 | 27 | 9 | 16 | 50 | 4 | 10 |
| 1748 | 6 | 16 | 15 | 10 | 53 | 8 | 28 | 9 | 43 | 23 | 10 | 1 | 17 | 43 | 1 |
| 1768 | 2 | 4 | 34 | 8 | 43 | 3 | 1 | 48 | 34 | 19 | 10 | 15 | 45 | 21 | 53 |
| 1788 | 9 | 22 | 53 | 6 | 34 | 9 | 5 | 27 | 25 | 15 | 11 | 0 | 30 | 0 | 44 |

J K L

Table TRS-10 of mean oppositions is also adapted to the meridian of Salamanca and extrapolated to the year 1349. We do not reproduce here the Tables of syzygies in expanded years and months which have the standard form.

The Tables of apogees in collected and expanded years are calculated back to the year 1348, but otherwise they have the same form as the corresponding Tables TRP-07 and TRP-07a in the sets of tables for Prague and Cracow. There are also Tables of apogees in months and days, which correspond to variants B of Tables TRC-07b and TRC-07c, but they are rounded or truncated to tertias only. The Table of apogees in months is given in variants for both the non-leap and leap year. The Tables of equations follow in the same form as in the sets of tables for Prague and Cracow.

The *Tabule resolute* for the meridian of Salamanca can also be found in MS Oxford, Bodl., Canon. misc. 27, ff. 33r–62r. This copy has, however, very different layout from the other versions of the *Tabule resolute*. The Tables of mean motions on ff. 33r–38v are not

Table TRS-09 of mean conjunctions of the Sun and Moon in collected years for the meridian of Salamanca (Lisbon, ANTT, ML, 2115, f. 29v-30r)

| Tabula medie coniunccionis Solis et Lune cum motibus in annis Christi collectis | | | | | | | | | | | | Residua pars Tabule medie coniunccionis Solis et Lune cum motibus in annis Christi collectis | | | | | | | | | | |
|---|---|---|---|---|---|---|---|---|---|---|---|---|---|---|---|---|---|---|---|---|---|---|
| Anni collecti | Tempus medie coniunccionis Solis et Lune | | | | | | Medius motus Solis et Lune | | | | | Anni collecti | Medium argumentum Lune | | | | | Medium argumentum latitudinis Lune | | | | |
| | D | H | M | 2 | 3 | 4 | S | G | M | 2 | 3 | | S | G | M | 2 | 3 | S | G | M | 2 | 3 |
| 1349 | 19 | 7 | 4 | 45 | 38 | 20 | 10 | 7 | 16 | 45 | 32 | 1349 | 8 | 9 | 42 | 41 | 44 | 6 | 12 | 24 | 4 | 54 |
| 1369 | 8 | 8 | 25 | 18 | 54 | 38 | 9 | 26 | 38 | 21 | 25 | 1369 | 4 | 26 | 28 | 5 | 38 | 6 | 28 | 0 | 45 | 40 |
| E 1389 | 26 | 22 | 29 | 55 | 13 | 55 | 10 | 15 | 6 | 21 | 30 | 1389 | 2 | 9 | 2 | 30 | 7 | 8 | 14 | 17 | 40 | 12 |
| 1409 | 15 | 23 | 50 | 28 | 30 | 13 | 10 | 4 | 27 | 57 | 23 | 1409 | 10 | 25 | 47 | 54 | 1 | 8 | 29 | 54 | 20 | 58 |
| 1429 | 5 | 1 | 11 | 1 | 46 | 31 | 9 | 23 | 49 | 33 | 15 | 1429 | 7 | 12 | 33 | 17 | 55 | 9 | 15 | 31 | 1 | 44 |
| E 1449 | 23 | 15 | 15 | 38 | 5 | 47 | 10 | 12 | 17 | 33 | 19 | 1449 | 4 | 25 | 7 | 42 | 25 | 11 | 1 | 47 | 56 | 14 |
| 1469 | 12 | 16 | 36 | 11 | 22 | 8 | 10 | 1 | 39 | 9 | 12 | 1469 | 1 | 11 | 53 | 6 | 19 | 11 | 17 | 24 | 37 | 0 |
| 1489 | 1 | 17 | 56 | 44 | 38 | 24 | 9 | 21 | 0 | 45 | 5 | 1489 | 9 | 28 | 38 | 30 | 13 | 0 | 3 | 1 | 17 | 45 |
| E 1509 | 20 | 8 | 1 | 20 | 57 | 41 | 10 | 9 | 28 | 45 | 9 | 1509 | 7 | 11 | 12 | 54 | 43 | 1 | 19 | 18 | 12 | 17 |
| 1529 | 9 | 9 | 21 | 54 | 13 | 59 | 9 | 28 | 50 | 21 | 2 | 1529 | 3 | 27 | 58 | 18 | 37 | 2 | 4 | 54 | 52 | 2 |
| E 1549 | 27 | 23 | 26 | 30 | 33 | 15 | 10 | 17 | 18 | 21 | 6 | 1549 | 1 | 10 | 32 | 43 | 6 | 3 | 21 | 11 | 47 | 34 |
| 1569 | 17 | 0 | 47 | 3 | 49 | 33 | 10 | 6 | 39 | 56 | 59 | 1569 | 9 | 27 | 18 | 7 | 1 | 4 | 6 | 48 | 28 | 19 |
| 1589 | 6 | 2 | 7 | 37 | 5 | 52 | 9 | 26 | 1 | 32 | 51 | 1589 | 6 | 14 | 3 | 30 | 55 | 4 | 22 | 25 | 9 | 5 |
| E 1609 | 24 | 16 | 12 | 13 | 25 | 9 | 10 | 14 | 29 | 32 | 56 | 1609 | 3 | 26 | 37 | 55 | 24 | 6 | 8 | 42 | 3 | 36 |
| 1629 | 13 | 17 | 32 | 46 | 41 | 55 | 10 | 3 | 51 | 8 | 48 | 1629 | 0 | 13 | 23 | 19 | 18 | 6 | 24 | 18 | 44 | 22 |
| 1649 | 2 | 18 | 53 | 19 | 57 | 45 | 9 | 23 | 12 | 44 | 41 | 1649 | 9 | 0 | 8 | 43 | 12 | 7 | 9 | 55 | 25 | 8 |
| E 1669 | 21 | 8 | 57 | 56 | 17 | 2 | 10 | 11 | 40 | 44 | 45 | 1669 | 6 | 12 | 43 | 7 | 42 | 8 | 26 | 12 | 19 | 39 |
| 1689 | 10 | 10 | 18 | 29 | 33 | 19 | 10 | 1 | 2 | 20 | 38 | 1689 | 2 | 29 | 28 | 31 | 36 | 9 | 11 | 49 | 0 | 25 |
| E 1709 | 29 | 0 | 23 | 5 | 52 | 36 | 10 | 19 | 30 | 20 | 42 | 1709 | 0 | 12 | 2 | 56 | 5 | 10 | 28 | 5 | 54 | 56 |
| 1729 | 18 | 1 | 43 | 39 | 8 | 54 | 10 | 8 | 51 | 56 | 35 | 1729 | 8 | 28 | 48 | 19 | 59 | 11 | 13 | 42 | 35 | 42 |
| 1749 | 7 | 3 | 4 | 12 | 25 | 12 | 9 | 28 | 13 | 32 | 28 | 1749 | 5 | 15 | 33 | 43 | 54 | 11 | 29 | 19 | 16 | 27 |
| E 1769 | 25 | 17 | 8 | 48 | 44 | 29 | 10 | 16 | 41 | 32 | 31 | 1769 | 2 | 28 | 8 | 8 | 23 | 1 | 15 | 36 | 10 | 59 |
| 1789 | 14 | 18 | 29 | 22 | 0 | 47 | 10 | 6 | 3 | 8 | 25 | 1789 | 11 | 14 | 53 | 32 | 17 | 2 | 1 | 12 | 51 | 44 |

A B C D

Table TRS-10 of mean oppositions of the Sun and Moon in collected years for the meridian of Salamanca (Lisbon, ANTT, ML, 2115, f. 30v-31r)

| Tabula medie opposicionis Solis et Lune cum motibus in annis Christi collectis | | | | | | | | | | | | Residua pars Tabule medie opposicionis (*ms. per err.*: coniunccionis) Solis et Lune | | | | | | | | | | |
|---|---|---|---|---|---|---|---|---|---|---|---|---|---|---|---|---|---|---|---|---|---|---|
| Anni Christi collecti | Tempus medie opposicionis Solis et Lune | | | | | | Medius motus Solis et Lune | | | | | Anni Christi collecti | Medium argumentum Lune | | | | | Medium argumentum latitudinis Lune | | | | |
| | D | H | M | 2 | 3 | 4 | S | G | M | 2 | 3 | | S | G | M | 2 | 3 | S | G | M | 2 | 3 |
| E 1349 | 3 | 1 | 26 | 47 | 9 | 49 | 10 | 21 | 49 | 57 | 38 | 1349 | 2 | 22 | 37 | 12 | 1 | 0 | 27 | 44 | 11 | 47 |
| 1369 | 23 | 2 | 47 | 20 | 26 | 7 | 10 | 11 | 11 | 33 | 31 | 1369 | 11 | 9 | 22 | 35 | 55 | 1 | 13 | 20 | 52 | 33 |
| 1389 | 12 | 4 | 7 | 53 | 42 | 25 | 10 | 29 | 39 | 33 | 36 | 1389 | 18 | 21 | 57 | 0 | 24 | 2 | 29 | 37 | 47 | 5 |
| E 1409 | 30 | 18 | 12 | 30 | 1 | 42 | 10 | 19 | 1 | 9 | 29 | 1409 | 5 | 8 | 42 | 24 | 18 | 3 | 15 | 14 | 27 | 51 |
| 1429 | 19 | 19 | 33 | 3 | 18 | 0 | 10 | 8 | 22 | 45 | 21 | 1429 | 1 | 25 | 25 | 48 | 13 | 4 | 0 | 51 | 8 | 36 |
| 1449 | 8 | 20 | 53 | 36 | 34 | 18 | 9 | 27 | 44 | 21 | 14 | 1449 | 10 | 12 | 13 | 12 | 7 | 4 | 16 | 27 | 49 | 21 |
| E 1469 | 27 | 10 | 58 | 12 | 53 | 35 | 10 | 16 | 12 | 21 | 18 | 1469 | 7 | 24 | 47 | 36 | 36 | 6 | 2 | 44 | 43 | 53 |
| 1489 | 16 | 12 | 18 | 46 | 9 | 53 | 10 | 5 | 33 | 57 | 11 | 1489 | 4 | 11 | 33 | 0 | 31 | 6 | 18 | 21 | 24 | 38 |
| 1509 | 5 | 13 | 39 | 19 | 26 | 11 | 9 | 24 | 55 | 33 | 3 | 1509 | 0 | 28 | 18 | 24 | 25 | 7 | 3 | 58 | 5 | 24 |
| E 1529 | 24 | 3 | 43 | 55 | 45 | 28 | 10 | 13 | 23 | 33 | 8 | 1529 | 10 | 10 | 52 | 48 | 54 | 8 | 25 | 14 | 59 | 55 |
| 1549 | 13 | 5 | 4 | 29 | 1 | 46 | 10 | 2 | 45 | 9 | 0 | 1549 | 6 | 27 | 38 | 12 | 49 | 9 | 5 | 51 | 40 | 41 |
| 1569 | 2 | 6 | 25 | 2 | 18 | 4 | 9 | 22 | 6 | 44 | 53 | 1569 | 3 | 14 | 23 | 36 | 43 | 9 | 7 | 28 | 21 | 26 |
| E 1589 | 20 | 20 | 29 | 38 | 37 | 21 | 10 | 10 | 34 | 44 | 57 | 1589 | 0 | 26 | 58 | 1 | 12 | 11 | 23 | 45 | 15 | 58 |
| 1609 | 28 | 21 | 50 | 11 | 53 | 38 | 9 | 29 | 56 | 20 | 50 | 1609 | 9 | 13 | 43 | 25 | 6 | 11 | 9 | 21 | 56 | 45 |
| E 1629 | 17 | 11 | 54 | 48 | 12 | 55 | 10 | 18 | 24 | 20 | 54 | 1629 | 6 | 26 | 17 | 49 | 36 | 1 | 25 | 38 | 51 | 15 |
| 1649 | 6 | 13 | 15 | 21 | 29 | 13 | 10 | 7 | 45 | 56 | 47 | 1649 | 3 | 12 | 3 | 13 | 30 | 1 | 10 | 15 | 32 | 1 |
| 1669 | 25 | 14 | 35 | 54 | 45 | 31 | 9 | 27 | 7 | 32 | 40 | 1669 | 11 | 29 | 48 | 37 | 24 | 2 | 27 | 52 | 12 | 46 |
| E 1689 | 14 | 4 | 40 | 31 | 4 | 48 | 10 | 15 | 35 | 32 | 44 | 1689 | 9 | 15 | 23 | 1 | 23 | 3 | 12 | 9 | 7 | 18 |
| 1709 | 3 | 6 | 1 | 4 | 21 | 6 | 10 | 4 | 57 | 8 | 37 | 1709 | 5 | 29 | 8 | 25 | 48 | 4 | 28 | 45 | 48 | 3 |
| 1729 | 21 | 7 | 21 | 37 | 37 | 24 | 9 | 24 | 18 | 44 | 29 | 1729 | 2 | 15 | 53 | 49 | 41 | 4 | 14 | 22 | 28 | 49 |
| E 1749 | 10 | 21 | 26 | 13 | 56 | 51 | 10 | 12 | 46 | 44 | 34 | 1749 | 11 | 28 | 28 | 14 | 11 | 6 | 0 | 39 | 23 | 20 |
| 1769 | 0 | 22 | 46 | 47 | 12 | 59 | 10 | 2 | 8 | 20 | 26 | 1769 | 8 | 15 | 13 | 38 | 6 | 7 | 1 | 16 | 4 | 5 |
| 1789 | 0 | 0 | 7 | 20 | 29 | 17 | 9 | 21 | 29 | 56 | 19 | 1789 | 5 | 1 | 59 | 2 | 0 | 7 | | 52 | 44 | 51 |
| Y | A | | | | | | B | | | | | Y | C | | | | | D | | | | |

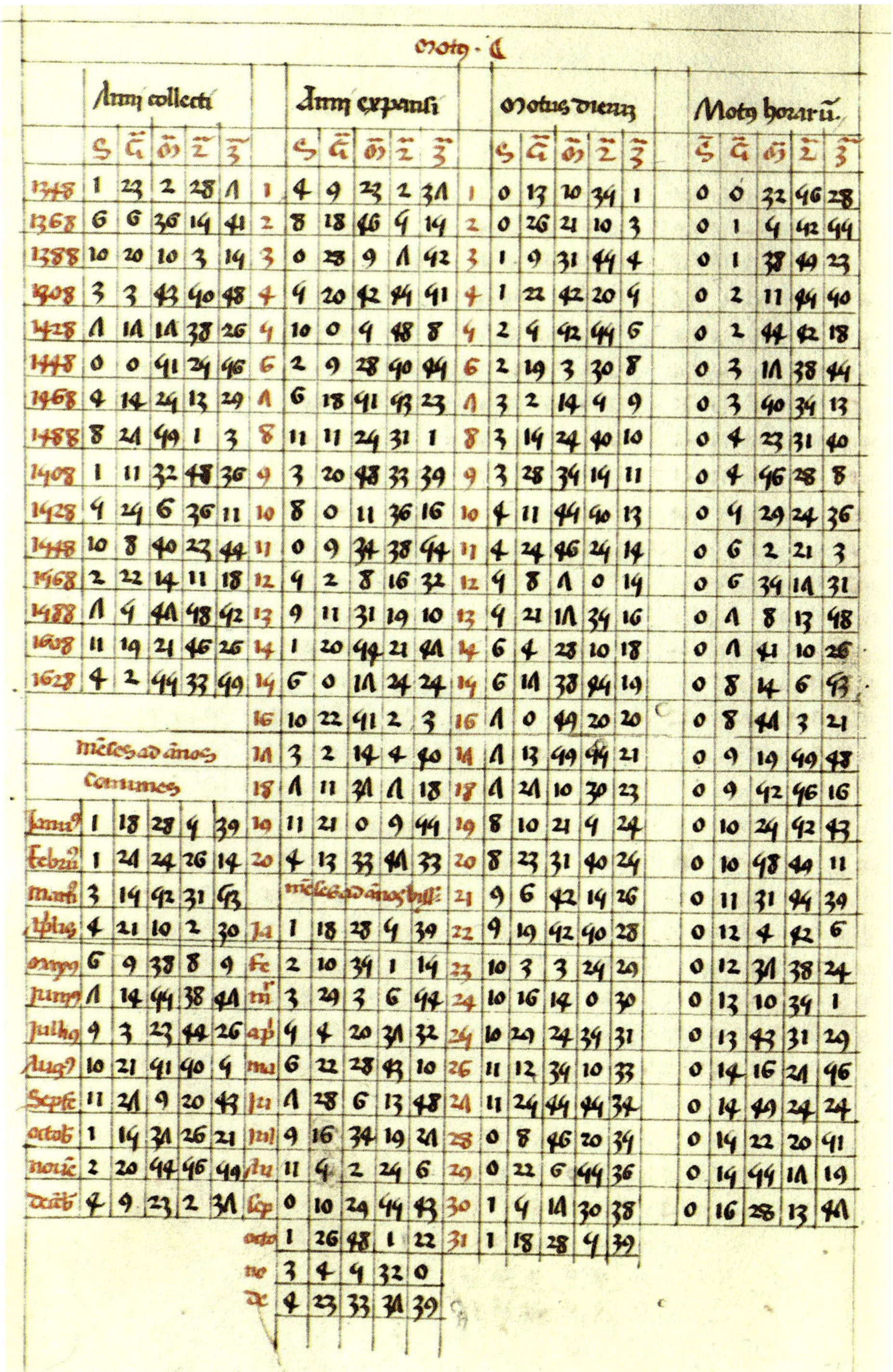


**Figure 29.** An example of the Table of mean motion of the Moon in longitude from the *Tabule resolute*, MS Oxford, Bodl., Canon. misc. 27, f. 34v.

organized according to subsequently decreasing intervals of time, that is, the collected years, expanded years, months, days, and hours like in the Tables TRW-01 – TRW-06, but instead, all these subtables are given on one page, which deals with one quantity, either, e.g., mean motion of apogees, the Sun, Moon, its anomalies, head of dragon, individual planets, or of their anomalies (as an example, see Fig. 29).

A similar arrangement can be found, e.g., in MS Trento, Fondazione Biblioteca San Bernardino, MS 313, ff. 34r-38v, where several mean motions are derived from *Tabule resolute* for Prague meridian by rounding to minutes. Another example is MS Prague, NK, I G 6, ff. 91v–97v. However, in this example, only one radix is given and collected years give increments of mean motions in twenties, hundreds, and thousands of years (as it is in the *Tabule magne* by John of Lignères), while in MS Oxford, Bodl., Canon. misc. 27, the subtables of collected years give the cyclical radices like in the *Tabule resolute*. The range of collected years starts again in 1348 as in the MS Lisbon, ANTT, ML, 2115, but it ends already at 1628 or 1648. These two manuscripts share also errors,[5] so that in spite of the different layout, they can be both classified as the *Tabule resolute*. The Tables of mean motions are followed by Tables of true apogees in collected years from 1348 to 1648 on f. 39r and expanded years from 1348 to 1628 on ff. 39v–40r. A blank space at bottom of folios 39r–40r is used to add mean motions in twenties and hundreds of years. The same Tables of apogees in months and days as in MS Lisbon, ANTT, ML, 2115, are written at the bottom of f. 40v, where at the top, the Table of mean conjunctions of the Sun and Moon in collected years 1349–1669 is given. The Table of syzygies in expanded year is written at f. 42r together with the increments from the conjunctions to opposition and to quarter of the Moon. The Table of syzygies in months is written on a shorter folio 41r and once more at bottom of f. 42v, with a difference that the former starts by January while the later by February. On the top of f. 42v the Table of oppositions of the Sun and Moon in collected years 1349–1789 is written, rounded to tertias. The set of the *Tabule resolute* is then completed on ff. 43r–62r by the Tables of equations in their standard form and is immediately followed by the additional Tables A-04 of sexagesimal multiplication, A-03 of proportional minutes for almanac, etc.

## 15.3 *Tabule resolute* for the Meridian of Bologna

A similar, though less compact, example of rearrangement of Tables of mean motions is presented by the *Tabule resolute* for the Bologna meridian. They are preserved in MS Milan, BA, C 278 b Inf., ff. 1r–31v and MS London, BL, Add 22317, ff. 25r–89v. In both these manuscripts are also copied the canones *Quoniam plerique* for their use (see Chapter 7). The Tables of mean motions of the Sun (with Venus and Mercury), the Moon, anomaly of the Moon, head of dragon, Saturn, Jupiter, Mars, anomaly of Venus, and anomaly of Mercury are given here, each on one folio. On its recto, in a top left table, are always written cyclical

5 For example, the mean motion of the Moon in 1428 is in both manuscripts greater for 4 tertias than it should be. See Alena Hadravová and Petr Hadrava, 'Cyclical Radices in the Tabulae resolutae and Their Adaptation to the Meridian of Another City', in *Alfonsine Astronomy: Computational Practices*, ed. by Matthieu Husson, José Chabás, and Richard L. Kremer (Turnhout: Brepols, forthcoming).

Table TRB-01 of mean motions for the meridian of Bologna (London, BL, Add 22317, f. 26r)

| **Tabula medii motus Lune ad meridianum Bononie Italie** | | | | | | | | | | |
|---|---|---|---|---|---|---|---|---|---|---|
| **Anni Christi collecti** | **Motus** | | | | **Anni Christi expansi** | | **Motus** | | | |
| | **S** | **G** | **M** | **2** | | | **S** | **G** | **M** | **2** |
| 1428 | 7 | 16 | 25 | 18 | | 1 | 4 | 9 | 23 | 3 |
| 1448 | 11 | 29 | 58 | 6 | | 2 | 8 | 18 | 46 | 5 |
| 1468 | 4 | 13 | 32 | 53 | | 3 | 0 | 28 | 9 | 8 |
| 1488 | 8 | 27 | 6 | 42 | B | 4 | 5 | 20 | 42 | 46 |
| 1508 | 1 | 10 | 40 | 28 | | 5 | 10 | 0 | 5 | 48 |
| 1528 | 5 | 24 | 14 | 16 | | 6 | 2 | 9 | 28 | 51 |
| 1548 | 10 | 7 | 48 | 4 | | 7 | 6 | 18 | 51 | 53 |
| 1568 | 2 | 21 | 21 | 51 | B | 8 | 11 | 11 | 25 | 31 |
| 1588 | 7 | 4 | 55 | 39 | | 9 | 3 | 20 | 48 | 34 |
| 1608 | 11 | 18 | 29 | 26 | | 10 | 8 | 0 | 11 | 36 |
| 1628 | 4 | 2 | 3 | 15 | | 11 | 0 | 9 | 34 | 39 |
| 1648 | 8 | 15 | 37 | 1 | B | 12 | 5 | 2 | 8 | 17 |
| 1668 | 0 | 29 | 10 | 49 | | 13 | 9 | 11 | 31 | 19 |
| 1688 | 5 | 12 | 44 | 36 | | 14 | 1 | 20 | 54 | 22 |
| 1708 | 9 | 26 | 18 | 24 | | 15 | 6 | 0 | 17 | 24 |
| 1728 | 2 | 9 | 52 | 12 | B | 16 | 10 | 22 | 51 | 2 |
| 1748 | 6 | 23 | 25 | 59 | | 17 | 3 | 2 | 14 | 5 |
| 1768 | 11 | 6 | 59 | 47 | | 18 | 7 | 11 | 37 | 7 |
| 1788 | 3 | 20 | 33 | 34 | | 19 | 11 | 21 | 0 | 0 |
| 1808 | 8 | 4 | 7 | 22 | B | 20 | 4 | 13 | 33 | 48 |

| **Motus mensium in anno communi** | | | | | **Motus mensium in anno bisextili** | | | | |
|---|---|---|---|---|---|---|---|---|---|
| | **S** | **G** | **M** | **2** | | **S** | **G** | **M** | **2** |
| Januarius | 1 | 18 | 28 | 6 | Januarius | 1 | 18 | 26 | 6 |
| Februarius | 1 | 27 | 24 | 27 | Februarius | 2 | 10 | 35 | 1 |
| Martius | 3 | 15 | 52 | 32 | Martius | 3 | 29 | 3 | 7 |
| Aprilis | 4 | 21 | 10 | 3 | Aprilis | 5 | 4 | 20 | 38 |
| Madius | 6 | 9 | 38 | 8 | Madius | 6 | 22 | 48 | 43 |
| Junius | 7 | 14 | 55 | 39 | Junius | 11 | 28 | 6 | 14 |
| Jullius | 9 | 3 | 23 | 44 | Jullius | 9 | 16 | 34 | 19 |
| Augustus | 10 | 21 | 51 | 50 | Augustus | 11 | 5 | 2 | 25 |
| Setember | 11 | 27 | 9 | 21 | Setember | 0 | 10 | 19 | 56 |
| Octuber | 1 | 15 | 37 | 26 | Octuber | 1 | 28 | 48 | 1 |
| November | 2 | 26 | 54 | 57 | November | 3 | 4 | 5 | 32 |
| December | 4 | 9 | 23 | 3 | December | 4 | 22 | 33 | 38 |

radices in collected years from 1428 to 1808, in a top right table are the mean motions in expanded years, and at the bottom of the page are the Tables of mean motions in months in non-leap and leap years. These tables are thus analogous to Tables TRW-01 – TRW-04. As an example, we present here Table TRB-01 with the mean motions of the Moon from MS London, BL, Add 22317, f. 26r. On the verso of each folio are then presented in two columns the Tables of mean motions in days and hours, that is, analogous to Tables TRW-05 and TRW-06. As can be seen, the mean motions are rounded to seconds of a degree. Comparing Tables TRB-01 and TRC-01 (column D), we find, e.g., for the year 1428 a difference of 0;1,6° = 66 seconds between the radices for Bologna and Cracow. Taking into account the speed of the Moon, this corresponds to the distance 0.5009° in geographical longitude between the meridians of Bologna and Cracow.

As to the tables in MS London, BL, Add 22317, ff. 25r, 26r etc., it may be remarked that their scribe was probably Italian, because he writes 'Setember' instead of 'September' and 'Madius' instead of 'Maius', which seems to be an influence of Italian ('Settembre', 'Maggio').

Like the Tables of mean motions, the Tables of apogees in both collected and expanded years are also rounded to seconds of a degree. In agreement with the Tables TRP-07 and TRP-07a, the collected years run from 1428 to 1808, and the expanded years from 1428 to 1788. A small tablet with differences between the apogees of the Sun and the planets Saturn, Jupiter, Mars, and Mercury is included, which enables calculation of the apogees of these planets by simply adding these differences to the apogee of the Sun.

The Table of mean conjunctions of the luminaries is rounded to seconds of an hour and the corresponding motions to seconds of a degree. The column with the argument of latitude of the Moon is missing, just as the argument of latitude is skipped from the Tables of mean motions. By comparison with Table TRC-09, we find that the times of the conjunctions for the meridian of Bologna are 2 minutes less than for the meridian of Cracow. This agrees precisely with the half-degree difference between the geographical longitudes of Cracow and Bologna (Bologna being to the west), which we have found from the differences of mean motions of the Moon. Similarly from a comparison with the Table TRP-09, Bologna should be 2;36° east of the meridian of Prague (while in reality it is about 3° west of Prague and 8;30° west of Cracow). An interesting change is that the corresponding values, which are inscribed by the collected years 1429 to 1809 in Tables TRC-09 or TRP-09, are attributed to the years 1428–1808 in Table TRB-09 for the meridian of Bologna; that is, the first conjunction occuring in the new year is denoted as a radix for the completed previous year, similar to the terminology for the radices of mean motions. A Table of mean oppositions of the Sun and Moon analogous to Tables TRP-10 and TRC-10 is missing. Instead, only the times and the corresponding increments of motions (the elongation of the luminaries, anomaly of the Moon, and also the motion of the head of dragon) are given between the conjunction and opposition, as well as between two subsequent conjunctions.

Between the Table of mean conjunctions and the subsequent Tables of equations, a Table of geographical coordinates of different cities is included. Nicolaus Wodka of Kwidzyn refers in his canons *Quoniam plerique* to the geographical longitudes from such a table if the tables for Bologna would be used for another city (see Chapter 7). However, the longitudes of Bologna (33;30° from the Pillars of Hercules)[6] and Cracow (44;45°) or Prague (37;0°) in the table copied in these manuscripts are not consistent with the values

6 In MS London, BL, Add 22317, f. 36v, the longitude of Bologna is corrected to 32;30°. This, however, violates the otherwise maintained order of the cities according to the increasing longitude, because the value 33;30° is kept for the preceding Florence as well as for the subsequent Venice.

Table TRB-09 of mean conjunctions of the Sun and Moon in collected years for the meridian of Bologna (London, BL, Add 22317, f. 35v)

| Tabula medie coniunccionis Solis et Lune cum suis motibus | | | | | | | | | | | | |
|---|---|---|---|---|---|---|---|---|---|---|---|---|
| Anni Christi collecti | Tempus medie coniunccionis Solis et Lune | | | | Medius motus Solis et Lune | | | | Medius motus argumenti Lune | | | |
| | D | H | M | 2 | S | G | M | 2 | S | G | M | 2 |
| 1428 | 5 | 2 | 46 | 22 | 9 | 23 | 49 | 33 | 7 | 12 | 33 | 18 |
| 1448 | 23 | 16 | 50 | 58 | 10 | 12 | 17 | 33 | 4 | 25 | 7 | 42 |
| 1468 | 12 | 18 | 11 | 31 | 10 | 1 | 39 | 9 | 1 | 11 | 53 | 6 |
| 1488 | 1 | 19 | 32 | 5 | 9 | 21 | 0 | 45 | 9 | 28 | 38 | 30 |
| 1508 | 20 | 9 | 36 | 41 | 10 | 9 | 28 | 45 | 7 | 11 | 12 | 55 |
| 1528 | 9 | 10 | 57 | 14 | 9 | 28 | 50 | 21 | 3 | 27 | 58 | 19 |
| 1548 | 28 | 1 | 1 | 51 | 10 | 17 | 18 | 21 | 1 | 10 | 32 | 43 |
| 1568 | 17 | 2 | 22 | 24 | 10 | 6 | 39 | 57 | 9 | 27 | 18 | 7 |
| 1588 | 6 | 3 | 42 | 57 | 9 | 26 | 1 | 33 | 6 | 14 | 3 | 31 |
| 1608 | 24 | 17 | 47 | 33 | 10 | 14 | 29 | 33 | 3 | 26 | 37 | 55 |
| 1628 | 23 | 19 | 8 | 7 | 10 | 3 | 51 | 9 | 0 | 13 | 23 | 19 |
| 1648 | 2 | 20 | 28 | 40 | 9 | 23 | 12 | 45 | 9 | 0 | 8 | 43 |
| 1668 | 21 | 10 | 33 | 16 | 10 | 11 | 40 | 45 | 6 | 12 | 43 | 8 |
| 1688 | 10 | 11 | 53 | 50 | 10 | 1 | 2 | 21 | 2 | 29 | 28 | 32 |
| 1708 | 29 | 2 | 8 | 56 | 10 | 19 | 30 | 21 | 0 | 12 | 2 | 56 |
| 1728 | 18 | 3 | 18 | 59 | 10 | 8 | 51 | 57 | 8 | 28 | 48 | 20 |
| 1748 | 7 | 4 | 39 | 32 | 9 | 29 | 13 | 32 | 5 | 15 | 33 | 44 |
| 1768 | 25 | 18 | 44 | 9 | 10 | 16 | 41 | 23 | 2 | 28 | 8 | 8 |
| 1788 | 14 | 20 | 4 | 42 | 10 | 6 | 3 | 8 | 11 | 14 | 53 | 32 |
| 1808 | 3 | 21 | 25 | 15 | 9 | 25 | 24 | 44 | 8 | 1 | 38 | 56 |

A B C

**Una lunacio est tempus cum motibus a coniunccione ad coniunccionem**

| D | H | M | 2 | S | G | M | 2 | S | G | M | 2 |
|---|---|---|---|---|---|---|---|---|---|---|---|
| 29 | 12 | 44 | 3 | 0 | 29 | 6 | 24 | 0 | 25 | 49 | 1 |

**Tempus cum motibus a coniunccione ad opposicionem**

| D | H | M | 2 | S | G | M | 2 | S | G | M | 2 |
|---|---|---|---|---|---|---|---|---|---|---|---|
| 14 | 18 | 22 | 2 | 0 | 14 | 33 | 12 | 6 | 12 | 54 | 30 |

**Medius cursus capitis draconis a coniunccione ad coniunccionem**

| S | G | M | 2 |
|---|---|---|---|
| 0 | 1 | 33 | 50 |

which we can uncover from the Tables of mean motions or mean conjunctions of the luminaries.[7] The Tables of equations of the Sun, Moon and the planets, which then follow, are basically equivalent to the corresponding tables in other sets of the *Tabule resolute*.

Various additional tables follow after the end of the *Tabule resolute*. Particularly, interesting are *Tabule moderantes prolixitates laboris tachuini* (Tables to mitigate the tediousness of *tacuinum* preparation) which start on f. 32r in MS Milan, BA, C 278 b Inf.

These auxiliary tables for a composition of the *tacuinum*, that is an almanac with daily true longitudes and aspects of the planets in the tradition established in Bologna and northern Italy, give mean motions at steps used throughout the year. The composition of the *tacuinum* by interpolation between these chosen days for which the planetary positions were calculated precisely is described in details by Nicolaus Wodka in his canons, see Chapter 7.

## 15.4 *Tabule resolute* by Ambrosius Lacher, published 1511

The tables in Lacher's edition start after his canons on f. E1r and are titled *Tabule resolute de motibus planetarum aliorumque mobilium supercelestium*, repeating the title of the entire edition. It is followed by the additional Table A-04 of sexagesimal multiplication, which seems to correspond to the fact that his canons also start with an explanation of arithmetic. Only after that do the Tables of mean motions in collected years start. The cyclical radices are given in the standard form up to tertias, but are unusually extended from 1388 until 2008.[8] Comparing the radix $7^s$ 16;32,59,10° of the mean motion of the Moon for the year 1428 in Lacher's tables with $7^s$ 17;11,38,14° computed for the same epoch at Toledo, we find a difference 0;38,39,4°. Dividing it by the daily motion of the Moon, the time shift between the meridian of Toledo and that taken by Lacher for Leipzig is 1;10,24 hours, which means that the Leipzig should be 17;36° east of Toledo. Similarly, from a comparison with the radix for Prague in Table TRP-01, we find at difference 0;3,36 hours or 0;0,54° west of Prague; and from a comparison with Cracow in Table TRC-01 we get 16 minutes of an hour or 4° west of Cracow.[9] The other Tables of mean motion in expanded years, months, days, and hours are basically similar to those in the older manuscript tradition. An insignificant but interesting change is that the upper planets are written in all these tables in a reversed order Mars, Jupiter, and Saturn, and only the Table of mean motions in days includes as its last column also the centre of the Moon.

The Table of apogees in collected years gives the values again for the years 1388 to 2008 in an order of the Sun with Venus, Mercury, Mars, Jupiter, and Saturn. Like Table TRP-07a the common apogees are given in all expanded years from 1388 to 2008. Table TRC-07b of apogees in (non-leap) months is given in the 'new' variant B up to quartas,

7 For a discussion of the frequent inconsistency of the geographical and astronomical data in medieval manuscripts see Kremer and Dobrzycki, 'Alfonsine Meridians'.

8 Taking into account the additional twenty expanded years, Lacher's tables are thus prepared for use until today.

9 Let us note that the *Tabula regionum et urbium* on f. O5v in Lacher's edition gives again completely different time shifts: 1 hour 27 minutes for Toledo, 14 minutes for Prague and 46 minutes for Cracow.

and similarly also the Table TRC-07c of apogees in days. It is interesting to note, that this Table of apogees in days has the same error in quartas for days from 28 to 31 in MSS Cracow, BJ, 600, f. 90v, Cracow, BJ, 611, f. 17r, and Munich, BSB, Clm 214, f. 22v but that error is corrected in Lacher's edition.

The Tables of mean syzygies (in the order oppositions and then conjunctions) are printed in the standard form up to quartas of an hour for cyclical radices from 1389 to 2009. Comparing Lacher's Table of mean conjunctions of the luminaries with the Table TRC-09 for Cracow, we find that the time at Leipzig should be precisely for 16 minutes of an hour less in the year 1429, while in comparison with Table TRP-09 for Prague it is smaller by 0;3,36,0,25 hours. The difference of the same 25 quartas from the integer seconds, which John of Głogów mentions in his Rule 36 to be a more precise distance between Prague and Cracow and which we also find in the difference between the Tables of mean syzygies of these cities, may indicate that Lacher derived his values from the tables for Cracow. There is, however a difference between the times of conjunctions for Cracow and Lacher's values, which gradually increases up to 11 quartas in the year 1809. After the Tables of syzygies in expanded years and in months, which have the form usual in the *Tabule resolute*, Lacher included the Table of true motions of luminaries, analogous to Table TRW-14 (with a slightly different values especially for the Moon) which was skipped in later manuscript copies of the corpus after that for Wrocław, but often appears in other sets of tables. There is also included the additional Table A-05 of equation of time and a Table of latitudes of the Moon and declinations of the five planets before the Tables of equations start. The Tables of equations have again a standard form, but they are ordered from Mercury to Saturn. The already mentioned Table of regions and cities follows after them.

## 15.5 *Tabule resolute* by Johannes Schöner, published first in 1536

As do the Lacher's tables, Schöner's tables start with the additional Table A-04 of sexagesimal multiplication. The Tables of mean motions in collected years then follow for the years from 1388 to 2008 (with a 20-year period, 1388, 1408, 1428 etc.) and in the same order of the subtables found in Lacher's tables. The cyclical radices, as well as all subsequent mean motions are, however, rounded to only seconds of a degree. By the title *Tabulae radicum mediorum motum in annis collectis et completis* Schöner emphasizes that the values of mean motions in a collected year correspond to the completed year, that is to the noon of the last day of December, so that it is a radix for the following year. Comparing Schöner's mean motion $7^s$ 16;41,16° of the Moon in the completed year 1428 with the value $7^s$ 17;11,38,14° for Toledo, we find from the difference 0;30,22° and from the daily motion 13.1764° of the Moon at difference in zonal time 0;55,18,40 hours or 13;49,40° in geographical longitude between Toledo and Nuremberg, for which Schöner adapted his version of the *Tabule resolute*. The Tables of mean motions in expanded years are continued after the year 20, again in the tradition of John of Lignères' *Tabule magne* by increments in twenties of years up to 100, in hundreds of years up to 1600, and the values for 2000 years are also listed. There are also radices of the mean motions for Nuremberg at the Incarnation. This radix for the Moon reads $4^s$ 2;16,39°, which is for 0;30,11° less than the Alfonsine radix for Toledo. This would correspond to the distance of Nuremberg from Toledo 0;54,58,38 hours in time and 13;44,39° in geographical longitude. The mean motions in months (in both common and leap years), days, and hours are organized as in the Lacher's tables (e.g., a column with anomaly of the Moon is added only in the Tables in days). Because they are rounded to seconds, all values for mean motions of apogees and fixed stars in hours are zero.

The Tables of apogees in both collected and expanded years are also given for the years from 1388 to 2008. They are rounded to seconds of a degree, as in the Tables of apogees in months and days.

The Tables of mean syzygies, which follow after the Tables of apogees, are also rounded to seconds of both an hour and a degree. The first is the Table of radices of mean oppositions, but it is incorrectly inscribed as a Table of conjunctions in the edition of 1536. This mistake is corrected in later editions of Schöner's tables in 1551, 1587, and 1588. The radices of mean oppositions and conjunctions are given for the collected years from 1389 to 2009. The time of opposition in 1429 is $19^{d}$ 20;39,3 hours, which is for 0;55,4 hours greater than the corresponding value for the Toledo. The same difference also appears in the radix of the conjunction in 1429. This corresponds to a difference in the zonal time 0;55,4 hours of 13;46° in geographical longitude between the Toledo and Nuremberg. This value is closer to the true distance 15;5° than is the value 1;11 hours or 17;45° in longitude given for the distance of Toledo from Nuremberg in the Schöner's Table of regions and cities.

The other tables in Schöner's edition are organized similarly to Lacher's edition. The main difference is in the Tables of planetary latitudes, which, however, do not belong to the set of the *Tabule resolute*. Lacher included a one-page Table of latitudes of the five planets at steps of six degrees, which is a part of Tables of 1322 by John of Lignères,[10] and which can be found, e.g., in MS Cracow, BJ, 600, f. 109v. Johannes Schöner replaced it by a set of Tables of latitudes of individual planets at steps of one degree.

## 15.6 *Tabule resolute* by Johannes Virdung, published posthumously in 1542

In their layout, the tables prepared by Virdung continue the tradition of the tables compiled by Nicolaus Wodka of Kwidzyn, which can also be seen in the copy of the TR by Nicolaus of Tuchów in MS Oxford, Bodl., Canon. misc. 27. They give the mean motions in collected and expanded years, months, days, and hours for each of the planets, always on one page. All these quantities are printed with a precision up to seconds of degree. The cyclical radices in Virdung's tables are calculated for Heidelberg. If we compare the mean of the printed values of radices for the Moon with the values computed for Toledo, we find a difference 0;23,23,35°, which would corresponds to the distance of Heidelberg from Toledo 10;39° in geographical longitude and 0;42,36 hours in the zonal time (again quite different from the value 1;10 hour given in the Table of coordinates of cities on f. L3v, but closer to the true distance 12;42°). These cyclical radices are given for the years from 1510 to 1590 with the standard step of twenty years. It means that unlike all previous *Tabule resolute*, the collected years are not chosen to be leap years (e.g. 1424 in the tables for Wrocław or

10 See Chabás and Saby, *The Tables of 1322 by John of Lignères*, pp. 77–79.

1428 for Prague, Cracow and others). They are also really radices for the given year, that is the values for 1510 are those at the noon of December 31, 1509. A consequence of this change is that already the third year in expanded years is leap instead of the fourth, and hence it should have increments of mean motions in 366 instead of 365 days. Nevertheless, Virdung's tables include the same mean motions in expanded years as found already in Petrus Cruciferus's tables for Wrocław, but every leap year is denoted by a cross, which means that the motion in one day must be added. Like Schöner's Tables of mean motions in expanded years, they are continued after the first twenty years by values in twenties of years up to 100 and in hundreds of years up to 1000. In Rule 11 of Virdung's canons, radices for the meridian of Heidelberg at the Incarnation are given,[11] so that the mean motions could be calculated for any year beyond the range of collected years. The radix of the mean motion of the Moon given there differs by 0;24,41,56° from the standard Parisian Alfonsine Tables for Toledo, which would imply the distance of Heidelberg from Toledo equal to approximately 11;15° in longitude or about 45 minutes in zonal time.

The Tables of mean motions are followed in Virdung's tables by Tables of equations. While the former are rearranged to give all quantities for one planet on the same page, the former are oppositely reordered to give equation of anomaly and the farther and nearer distance (*longitudo longior, propior*) for all planets in one table, and equation of centre and proportional minutes in another. It is possible to compress these Tables of equations for two or more planets onto one page because the values are tabulated only for every fifth degree. Instead of any Table on stations of planets for different angles, which is included as a subtable in the Tables of equations in the other *Tabule resolute*, there is included only the additional Table A-01.

The Tables of apogees, which follow the Tables of equations, give apogees of the Sun in collected years from 1508 to 1628, incremented in corresponding twenties of expanded years, in months, and in days. These tables thus continue in the tradition of Tables TRP-07, TRP-07a, TRC-07b, and TRC-07c instead of Virdung's tables of mean motions. All values are rounded to seconds and the Tables of apogees thus take less than a page. A small Table of distances of the apogees of Saturn, Jupiter, Mars, and Mercury, from the apogee of the Sun (which is identical with apogee of Venus) is contained in Rule 17 of Virdung's canons dealing with the computation of the apogees. The Tables of apogees are then followed by Tables of planetary latitudes, which are not a part of other *Tabule resolute*.

Virdung's tables then include Tables for finding syzygies, which differ significantly from the other *Tabule resolute*. They indicate the times in days, hours, minutes, and seconds remaining from the last conjunction of the Sun and Moon to the end of each of the collected years from 1500 to 2000, expanded years from 1 to 20 and months in a common or leap year. To find the time of a conjunction or opposition of the Sun and Moon, their sum is then to be subtracted from a semi-integer multiple of the lunacy, given in the *Tabula revolutionum*. Alternatively, there is also given a radix of 16 days 17;25,56 hours for Heidelberg and increments in collected years in twenties, hundreds and thousands of years, from which the radices at collected years could be calculated. Such a radix for Toledo at the Incarnation calculated with parameters of Parisian Alfonsine Tables should be 16 days 18;27,49 hours,

11 It is written there that they are included thanks to a suggestion by Jacop Curio.

which would correspond to a distance of Heidelberg from Toledo of 51 minutes 53 seconds in zonal time. However, the various values given in the Tables of conjunctions are not mutually consistent, so that the geographical longitude assumed for Heidelberg cannot be reliably derived from them. The Tables of syzygies do not contain the motions of the Sun and Moon like the Tables TRW-09 – TRW-12. To compute the true syzygies, there follows a Table of equations of the luminaries at mean syzygies.

There are also Tables of the motions of the Sun and Moon in one hour in the variant like in MS Prague, NK, X B 3[12] but in a layout different from Table TRW-14, separated for each of the luminaries on one page. Table A-05 of the equation of time follows. Then come Tables of mean motions of apogees and fixed stars, of the motion of access and recess and Table TRW-08 of its equation. For these two mean motions, annual radices are given for the years 1525–1604/07 and daily motions for every day in a year, but also increments in expanded years up to 20, and in twenties and hundreds years up to one thousand, and true motions of the trepidation for the collected years from 1520 to 1700. It should also be mentioned that the canons, which precede the Virdung's tables, are followed by tables of coordinates of selected stars and cities. The former contains 93 stars verified to the year 1526, ordered according to increasing ecliptical longitude. The cities in the later table are ordered alphabetically and the geographical longitudes are expressed in differences in zonal time from Heidelberg.

12 See Chapter 22, Tab. 2.

# Additional Tables

The set of tables comprising the *Tabule resolute* is not precisely specified. We have defined here the corpus of the *Tabule resolute* according to MS Toruń, BU, 74/III with the original tables for the meridian of Wrocław. However, for instance Table TRW-14 of the true motion of the Sun and Moon in an hour was either omitted or moved to other parts of the convolute in many other manuscripts. On the other hand, the corpus is enriched in some manuscripts by various additional tables, which we can also find in other sets of tables. G. Rosińska lists some of them as 'Tables which usually follow the *Tabule resolute* and the *Tabule Alphonsi*'.[1] She includes among these additional tables the Tables of equation of time, Tables of ascensions in the right sphere and the oblique sphere for the middle and for the end of the seventh climate, and the Table of proportional minutes in the intermediate days for almanacs. J. Chabás considers these tables and also the sexagesimal multiplication table and a small Table of stations and retrogradation to be a regular part of content of the *Tabule resolute*.[2] Regardless of how we define which tables belong to the *Tabule resolute* itself and which are supplementary, it is worthy to pay attention to the wider set of the tables. We thus give transcriptions of the additional tables, derived from selected manuscripts, which are most often found together with the above described basic sets of the *Tabule resolute*.

#### Table A-01 of stations and retrogradation

This Table of planetary retrogradation is taken from the set of Tables of 1322 by John of Lignères.[3] The 'beginning of the first station' and 'beginning of retrogradation' are the minimum and maximum values of the anomaly at which the first station may occur. These 'beginnings' – apart from having slightly different numerical values – can be found in column F of the first stations in the Tables of planetary equations TRW-16 – TRW-20. The table A-01 is thus superfluous in the context of the *Tabule resolute*. The ends of retrogradation and of the second stations are simple complements to twelve signs of the beginnings, so that this information is not very important.[4] This table is also found in manuscripts with other sets of tables such as Prague, KMK, M.CLXII (1524), f. 54v (Parisian Alfonsine Tables with radices to Erfurt). In some manuscripts with the *Tabule resolute* can also be found another Table of retrogradation from the set of John of Lignères' Tables of 1322, which in the tradition of Ptolemy's *Almagest* gives the stations at steps of six degrees in the anomaly[5] (see, e.g., Oxford, Bodl., Canon. misc. 499, f. 160r/150r, or Oxford, Bodl., Canon. misc. 27, f. 89v).

1 Rosińska, *Scientific Writings*, pp. 519–21.
2 Chabás, *Computational Astronomy in the Middle Ages*, pp. 316–17.
3 See Chabás and Saby, *The Tables of 1322 by John of Lignères*, pp. 88–89.
4 The wrong value of 7 signs for the end of the second station of Mars is a scribal mistake of the copy in MS Prague, NK, X B 3.
5 See Chabás and Saby, *The Tables of 1322 by John of Lignères*, pp. 90–91; Toomer, *Ptolemy's Almagest*, pp. 555–88.

Example of Table A-01 taken from Prague, NK, X B 3, f. 114v

| **Tabula retrogradacionis, dirrecionis, stacionis planetarum et intratur cum argumento vero eorum** | | | |
|---|---|---|---|
| **S** | **G** | **M** | **porcio equata Saturni** |
| | 24 | 44 | inicium prime stacionis |
| 3 | 25 | 30 | inicium retrogradacionis |
| 8 | 4 | 30 | finis retrogradacionis |
| 8 | 7 | 16 | finis stacionis secunde |
| | | | |
| **S** | **G** | **M** | **porcio equata Iovis** |
| 4 | 4 | 5 | inicium stacionis prime |
| 4 | 7 | 11 | inicium retrogradacionis |
| 7 | 22 | 49 | finis retrogradacionis |
| 7 | 25 | 55 | finis stacionis secunde |
| | | | |
| **S** | **G** | **M** | **porcio equata Martis** |
| 5 | 7 | 28 | inicium stacionis prime |
| 5 | 19 | 5 | inicium retrogradacionis |
| 6 | 10 | 45 | finis retrogradacionis |
| 7 | 22 | 32 | finis stacionis secunde |
| | | | |
| **S** | **G** | **M** | **porcio equata Veneris** |
| 5 | 15 | 51 | inicium stacionis prime |
| 5 | 18 | 21 | inicium retrogradacionis |
| 6 | 11 | 39 | finis retrogradacionis |
| 6 | 24 | 9 | finis stacionis secunde |
| | | | |
| **S** | **G** | **M** | **porcio equata Mercurii** |
| 4 | 24 | 42 | inicium stacionis prime |
| 4 | 27 | 14 | inicium retrogradacionis |
| 7 | 2 | 46 | finis retrogradacionis |
| 7 | 5 | 18 | finis stacionis secunde |

**Table A-02 of mean motions of planets in selected numbers of days**

Table A-02 gives mean motions of anomalies and centres of planets in several numbers of days which were usually chosen as steps for calculations of true motions of planets in almanacs. To calculate the centres, the uniform motion of apogees is linearised assuming the precession velocity 0;0,0,5,42° per day. This value is also given in the last subtable and agrees with the value in the column A of the Table TRC-07c. This velocity, which is used for the Sun with Mercury and Venus and also for Mars, gives a motion of 0;0,34,40,30° in a non-leap year. According to Table TRP-07a, it corresponds to precession after the year 1448. In the case of Saturn and Jupiter, Table A-02 uses the precession velocity of 0;0,0,5,48° per day, that is, 0;0,35,17,0° in a non-leap year. This value would correspond to years before 1428 according to Table TRP-07a.[6]

**Example of Table A-02 taken from Prague, NK, X B 3, f. 115r**

| centra Veneris et Mercurii | S | G | M | 2 | 3 | 4 |
|---|---|---|---|---|---|---|
| argumentum Solis unius diei | 0 | 0 | 59 | 8 | 13 | 55 |
| argumentum Solis duorum dierum | 0 | 1 | 58 | 16 | 27 | 51 |
| argumentum Solis trium dierum | 0 | 2 | 57 | 24 | 41 | 46 |
| argumentum Solis 4 dierum | 0 | 3 | 56 | 32 | 55 | 41 |
| argumentum Solis 5 dierum | 0 | 4 | 55 | 41 | 9 | 37 |
| argumentum Solis 6 dierum | 0 | 5 | 54 | 49 | 23 | 32 |
| argumentum Solis 28 dierum | 0 | 27 | 35 | 50 | 29 | 49 |
| argumentum Solis 30 dierum | 0 | 29 | 34 | 6 | 57 | 37 |
| argumentum Solis 31 dierum | 1 | 0 | 33 | 15 | 11 | 35 |
| | | | | | | |
| centrum Lune unius diei | 0 | 24 | 22 | 53 | 23 | 16 |
| centrum Lune 15 dierum | 0 | 5 | 43 | 20 | 48 | 56 |
| centrum Lune 28 dierum | 10 | 22 | 46 | 54 | 51 | 20 |
| centrum Lune 29 dierum | 11 | 17 | 3 | 48 | 14 | 36 |
| centrum Lune 30 dierum | 0 | 11 | 26 | 41 | 37 | 52 |
| centrum Lune 31 dierum | 1 | 5 | 49 | 35 | 1 | 8 |
| | | | | | | |
| centrum Saturni unius diei | 0 | 0 | 2 | 0 | 29 | 30 |
| centrum Saturni 8 dierum | 0 | 0 | 16 | 3 | 55 | 57 |
| centrum Saturni 10 dierum | 0 | 0 | 20 | 4 | 54 | 57 |
| centrum Saturni 11 dierum | 0 | 0 | 22 | 5 | 34 | 27 |
| centrum Saturni 28 dierum | 0 | 0 | 56 | 13 | 45 | 51 |
| centrum Saturni 30 dierum | 0 | 1 | 0 | 14 | 44 | 51 |
| centrum Saturni 31 dierum | 0 | 1 | 2 | 15 | 14 | 21 |
| | | | | | | |
| argumentum Saturni unius diei | 0 | 0 | 57 | 7 | 44 | 20 |
| argumentum Saturni 8 dierum | 0 | 7 | 37 | 1 | 54 | 37 |
| argumentum Saturni 10 dierum | 0 | 9 | 31 | 17 | 23 | 17 |
| argumentum Saturni 11 dierum | 0 | 10 | 28 | 24 | 7 | 36 |
| argumentum Saturni 28 dierum | 0 | 26 | 39 | 36 | 41 | 11 |
| argumentum Saturni 30 dierum | 0 | 28 | 33 | 52 | 9 | 49 |
| argumentum Saturni 31 dierum | 0 | 29 | 30 | 59 | 54 | 9 |
| | | | | | | |
| centrum Iovis unius diei | 0 | 0 | 4 | 59 | 9 | 39 |
| centrum Iovis 4 dierum | 0 | 0 | 19 | 56 | 38 | 36 |
| centrum Iovis 6 dierum | 0 | 0 | 29 | 54 | 57 | 55 |
| centrum Iovis 7 dierum | 0 | 0 | 34 | 54 | 7 | 34 |

6 See Kremer, 'Abbreviating the Alfonsine Tables in Cracow'.

| | S | G | M | 2 | 3 | 4 |
|---|---|---|---|---|---|---|
| centrum Iovis 8 dierum | 0 | 0 | 39 | 53 | 17 | 12 |
| centrum Iovis 28 dierum | 0 | 2 | 19 | 36 | 30 | 16 |
| centrum Iovis 30 dierum | 0 | 2 | 29 | 34 | 49 | 34 |
| centrum Iovis 31 dierum | 0 | 2 | 34 | 33 | 59 | 13 |
| argumentum Iovis unius diei | 0 | 0 | 54 | 9 | 4 | 10 |
| argumentum Iovis 4 dierum | 0 | 3 | 36 | 36 | 16 | 41 |
| argumentum Iovis 6 dierum | 0 | 5 | 24 | 54 | 25 | 1 |
| argumentum Iovis 7 dierum | 0 | 6 | 19 | 3 | 29 | 11 |
| argumentum Iovis 8 dierum | 0 | 7 | 13 | 12 | 33 | 21 |
| argumentum Iovis 28 dierum | 0 | 25 | 16 | 13 | 56 | 46 |
| argumentum Iovis 30 dierum | 0 | 27 | 4 | 32 | 5 | 16 |
| argumentum Iovis 31 dierum | 0 | 27 | 58 | 41 | 9 | 26 |
| centrum Martis unius diei | 0 | 0 | 31 | 26 | 32 | 58 |
| centrum Martis 4 dierum | 0 | 2 | 5 | 46 | 11 | 52 |
| centrum Martis 6 dierum | 0 | 3 | 8 | 39 | 17 | 48 |
| centrum Martis 7 dierum | 0 | 3 | 40 | 5 | 50 | 46 |
| centrum Martis 28 dierum | 0 | 14 | 40 | 36 | 30 | 16 |
| centrum Martis 30 dierum | 0 | 15 | 43 | 16 | 29 | 0 |
| centrum Martis 31 dierum | 0 | 16 | 14 | 33 | 59 | 13 |
| argumentum Martis unius diei | 0 | 0 | 27 | 41 | 40 | 57 |
| argumentum Martis 4 dierum | 0 | 1 | 50 | 46 | 43 | 49 |
| argumentum Martis 6 dierum | 0 | 2 | 46 | 10 | 5 | 43 |
| argumentum Martis 7 dierum | 0 | 3 | 13 | 51 | 46 | 41 |
| argumentum Martis 28 dierum | 0 | 12 | 55 | 27 | 6 | 43 |
| argumentum Martis 30 dierum | 0 | 13 | 50 | 50 | 28 | 37 |
| argumentum Martis 31 dierum | 0 | 14 | 18 | 32 | 9 | 34 |
| motus augium unius diei | | | | 0 | 5 | 42 |
| motus augium 5 dierum | | | | 0 | 28 | 30 |
| motus augium 30 dierum | | | | 2 | 51 | 0 |
| motus augium 10 dierum | | | | 0 | 57 | 0 |

**Table A-03 for interpolation to intermediate days for making an almanac**

This extended table gives parts 1/2, 1/3… 1/7 of angles from 0 to 9 degrees and parts 1/8… 1/11 of angles from 0 to 3 degrees at steps of one minute. Its purpose is explained in Rule 26 of canons by John of Głogów. We transcribe here the first and the last part of the table only.

**Example of Table A-03 taken from Oxford, Bodl., Canon. misc. 499, f. 106v/96v**

| Incipit Tabula minutorum proporcionalium pro diebus intermediis in almanach faciendis | | | | | | | | | | | | | | | | | | | | |
|---|---|---|---|---|---|---|---|---|---|---|---|---|---|---|---|---|---|---|---|---|
| Differencia | | 2 dies | | | 3 dies | | | 4 dies | | | 5 dies | | | 6 dies | | | 7 dies | | | |
| G | M | G | M | 2 | G | M | 2 | G | M | 2 | G | M | 2 | G | M | 2 | G | M | 2 | 3 |
| 0 | 1 | 0 | 0 | 30 | 0 | 0 | 20 | 0 | 0 | 15 | 0 | 0 | 12 | 0 | 0 | 10 | 0 | 0 | 8 | 34 |
| 0 | 2 | 0 | 1 | 0 | 0 | 0 | 40 | 0 | 0 | 30 | 0 | 0 | 24 | 0 | 0 | 20 | 0 | 0 | 17 | 9 |
| 0 | 3 | 0 | 1 | 30 | 0 | 1 | 0 | 0 | 0 | 45 | 0 | 0 | 36 | 0 | 0 | 30 | 0 | 0 | 25 | 43 |
| 0 | 4 | 0 | 2 | 0 | 0 | 1 | 20 | 0 | 1 | 0 | 0 | 0 | 48 | 0 | 0 | 40 | 0 | 0 | 34 | 17 |
| 0 | 5 | 0 | 2 | 30 | 0 | 1 | 40 | 0 | 1 | 15 | 0 | 1 | 0 | 0 | 0 | 50 | 0 | 0 | 42 | 51 |
| 0 | 6 | 0 | 3 | 0 | 0 | 2 | 0 | 0 | 1 | 30 | 0 | 1 | 12 | 0 | 1 | 0 | 0 | 0 | 51 | 26 |
| 0 | 7 | 0 | 3 | 30 | 0 | 2 | 20 | 0 | 1 | 45 | 0 | 1 | 24 | 0 | 1 | 10 | 0 | 1 | 0 | 0 |
| 0 | 8 | 0 | 4 | 0 | 0 | 2 | 40 | 0 | 2 | 0 | 0 | 1 | 36 | 0 | 1 | 20 | 0 | 1 | 8 | 34 |
| 0 | 9 | 0 | 4 | 30 | 0 | 3 | 0 | 0 | 2 | 15 | 0 | 1 | 48 | 0 | 1 | 30 | 0 | 1 | 17 | 9 |
| 0 | 10 | 0 | 5 | 0 | 0 | 3 | 20 | 0 | 2 | 30 | 0 | 2 | 0 | 0 | 1 | 40 | 0 | 1 | 25 | 43 |
| 0 | 11 | 0 | 5 | 30 | 0 | 3 | 40 | 0 | 2 | 45 | 0 | 2 | 12 | 0 | 1 | 50 | 0 | 1 | 34 | 17 |
| 0 | 12 | 0 | 6 | 0 | 0 | 4 | 0 | 0 | 3 | 0 | 0 | 2 | 24 | 0 | 2 | 0 | 0 | 1 | 42 | 51 |
| 0 | 13 | 0 | 6 | 30 | 0 | 4 | 20 | 0 | 3 | 15 | 0 | 2 | 36 | 0 | 2 | 10 | 0 | 1 | 51 | 26 |
| 0 | 14 | 0 | 7 | 0 | 0 | 4 | 40 | 0 | 3 | 30 | 0 | 2 | 48 | 0 | 2 | 20 | 0 | 2 | 0 | 0 |
| 0 | 15 | 0 | 7 | 30 | 0 | 5 | 0 | 0 | 3 | 45 | 0 | 3 | 0 | 0 | 2 | 30 | 0 | 2 | 8 | 34 |
| 0 | 16 | 0 | 8 | 0 | 0 | 5 | 20 | 0 | 4 | 0 | 0 | 3 | 12 | 0 | 2 | 40 | 0 | 2 | 17 | 9 |
| 0 | 17 | 0 | 8 | 30 | 0 | 5 | 40 | 0 | 4 | 15 | 0 | 3 | 24 | 0 | 2 | 50 | 0 | 2 | 25 | 43 |
| 0 | 18 | 0 | 9 | 0 | 0 | 6 | 0 | 0 | 4 | 30 | 0 | 3 | 36 | 0 | 3 | 0 | 0 | 2 | 34 | 17 |
| 0 | 19 | 0 | 9 | 30 | 0 | 6 | 20 | 0 | 4 | 45 | 0 | 3 | 48 | 0 | 3 | 10 | 0 | 2 | 42 | 51 |
| 0 | 20 | 0 | 10 | 0 | 0 | 6 | 40 | 0 | 5 | 0 | 0 | 4 | 0 | 0 | 3 | 20 | 0 | 2 | 51 | 26 |
| 0 | 21 | 0 | 10 | 30 | 0 | 7 | 0 | 0 | 5 | 15 | 0 | 4 | 12 | 0 | 3 | 30 | 0 | 3 | 0 | 0 |
| 0 | 22 | 0 | 11 | 0 | 0 | 7 | 20 | 0 | 5 | 30 | 0 | 4 | 24 | 0 | 3 | 40 | 0 | 3 | 8 | 34 |
| 0 | 23 | 0 | 11 | 30 | 0 | 7 | 40 | 0 | 5 | 45 | 0 | 4 | 36 | 0 | 3 | 50 | 0 | 3 | 17 | 9 |
| 0 | 24 | 0 | 12 | 0 | 0 | 8 | 0 | 0 | 6 | 0 | 0 | 4 | 48 | 0 | 4 | 0 | 0 | 3 | 25 | 43 |
| 0 | 25 | 0 | 12 | 30 | 0 | 8 | 20 | 0 | 6 | 15 | 0 | 5 | 0 | 0 | 4 | 10 | 0 | 3 | 34 | 17 |
| 0 | 26 | 0 | 13 | 0 | 0 | 8 | 40 | 0 | 6 | 30 | 0 | 5 | 12 | 0 | 4 | 20 | 0 | 3 | 42 | 51 |
| 0 | 27 | 0 | 13 | 30 | 0 | 9 | 0 | 0 | 6 | 45 | 0 | 5 | 24 | 0 | 4 | 30 | 0 | 3 | 51 | 26 |
| 0 | 28 | 0 | 14 | 0 | 0 | 9 | 20 | 0 | 7 | 0 | 0 | 5 | 36 | 0 | 4 | 40 | 0 | 4 | 0 | 0 |
| 0 | 29 | 0 | 14 | 30 | 0 | 9 | 40 | 0 | 7 | 15 | 0 | 5 | 48 | 0 | 4 | 50 | 0 | 4 | 8 | 34 |
| 0 | 30 | 0 | 15 | 0 | 0 | 10 | 0 | 0 | 7 | 30 | 0 | 6 | 0 | 0 | 5 | 0 | 0 | 4 | 17 | 9 |

Remaining part of Table A-03 (Oxford, Bodl., Canon. misc. 499, f. 118r/108r)

| Differencia | | Residua pars Tabule minutorum proporcionabilium pro diebus intermediis in almanach faciendis | | | | | | | | | | | | | | | |
|---|---|---|---|---|---|---|---|---|---|---|---|---|---|---|---|---|---|
| | | 8 dies | | | | 9 dies | | | | 10 dies | | | | 11 dies | | | |
| **G** | **M** | **G** | **M** | **2** | **3** | **G** | **M** | **2** | **3** | **G** | **M** | **2** | **3** | **G** | **M** | **2** | **3** |
| 2 | 31 | 0 | 18 | 52 | 30 | 0 | 16 | 46 | 40 | 0 | 15 | 6 | 0 | 0 | 13 | 43 | 38 |
| 2 | 32 | 0 | 19 | 0 | 0 | 0 | 16 | 53 | 20 | 0 | 15 | 12 | 0 | 0 | 13 | 49 | 5 |
| 2 | 33 | 0 | 19 | 7 | 30 | 0 | 17 | 0 | 0 | 0 | 15 | 18 | 0 | 0 | 13 | 54 | 33 |
| 2 | 34 | 0 | 19 | 15 | 0 | 0 | 17 | 6 | 40 | 0 | 15 | 24 | 0 | 0 | 14 | 0 | 0 |
| 2 | 35 | 0 | 19 | 22 | 30 | 0 | 17 | 13 | 20 | 0 | 15 | 30 | 0 | 0 | 14 | 5 | 27 |
| 2 | 36 | 0 | 19 | 30 | 0 | 0 | 17 | 20 | 0 | 0 | 15 | 36 | 0 | 0 | 14 | 10 | 54 |
| 2 | 37 | 0 | 19 | 37 | 30 | 0 | 17 | 26 | 40 | 0 | 15 | 42 | 0 | 0 | 14 | 16 | 22 |
| 2 | 38 | 0 | 19 | 45 | 0 | 0 | 17 | 33 | 20 | 0 | 15 | 48 | 0 | 0 | 14 | 21 | 49 |
| 2 | 39 | 0 | 19 | 52 | 30 | 0 | 17 | 40 | 0 | 0 | 15 | 54 | 0 | 0 | 14 | 27 | 16 |
| 2 | 40 | 0 | 20 | 0 | 0 | 0 | 17 | 46 | 40 | 0 | 16 | 0 | 0 | 0 | 14 | 32 | 44 |
| 2 | 41 | 0 | 20 | 7 | 30 | 0 | 17 | 53 | 20 | 0 | 16 | 6 | 0 | 0 | 14 | 38 | 11 |
| 2 | 42 | 0 | 20 | 15 | 0 | 0 | 18 | 0 | 0 | 0 | 16 | 12 | 0 | 0 | 14 | 43 | 38 |
| 2 | 43 | 0 | 20 | 22 | 30 | 0 | 18 | 6 | 40 | 0 | 16 | 18 | 0 | 0 | 14 | 49 | 5 |
| 2 | 44 | 0 | 20 | 30 | 0 | 0 | 18 | 13 | 20 | 0 | 16 | 24 | 0 | 0 | 14 | 54 | 33 |
| 2 | 45 | 0 | 20 | 37 | 30 | 0 | 18 | 20 | 0 | 0 | 16 | 30 | 0 | 0 | 15 | 0 | 0 |
| 2 | 46 | 0 | 20 | 45 | 0 | 0 | 18 | 26 | 40 | 0 | 16 | 36 | 0 | 0 | 15 | 5 | 27 |
| 2 | 47 | 0 | 20 | 52 | 30 | 0 | 18 | 33 | 20 | 0 | 16 | 42 | 0 | 0 | 15 | 10 | 55 |
| 2 | 48 | 0 | 21 | 0 | 0 | 0 | 18 | 40 | 0 | 0 | 16 | 48 | 0 | 0 | 15 | 16 | 22 |
| 2 | 49 | 0 | 21 | 7 | 30 | 0 | 18 | 46 | 40 | 0 | 16 | 54 | 0 | 0 | 15 | 21 | 49 |
| 2 | 50 | 0 | 21 | 15 | 0 | 0 | 18 | 53 | 20 | 0 | 17 | 0 | 0 | 0 | 15 | 27 | 16 |
| 2 | 51 | 0 | 21 | 22 | 30 | 0 | 19 | 0 | 0 | 0 | 17 | 6 | 0 | 0 | 15 | 32 | 44 |
| 2 | 52 | 0 | 21 | 30 | 0 | 0 | 19 | 6 | 40 | 0 | 17 | 12 | 0 | 0 | 15 | 38 | 11 |
| 2 | 53 | 0 | 21 | 37 | 30 | 0 | 19 | 13 | 20 | 0 | 17 | 18 | 0 | 0 | 15 | 43 | 38 |
| 2 | 54 | 0 | 21 | 45 | 0 | 0 | 19 | 20 | 0 | 0 | 17 | 24 | 0 | 0 | 15 | 49 | 5 |
| 2 | 55 | 0 | 21 | 52 | 30 | 0 | 19 | 26 | 40 | 0 | 17 | 30 | 0 | 0 | 15 | 54 | 33 |
| 2 | 56 | 0 | 22 | 0 | 0 | 0 | 19 | 33 | 20 | 0 | 17 | 36 | 0 | 0 | 16 | 0 | 0 |
| 2 | 57 | 0 | 22 | 7 | 30 | 0 | 19 | 40 | 0 | 0 | 17 | 42 | 0 | 0 | 16 | 5 | 27 |
| 2 | 58 | 0 | 22 | 15 | 0 | 0 | 19 | 46 | 40 | 0 | 17 | 48 | 0 | 0 | 16 | 10 | 55 |
| 2 | 59 | 0 | 22 | 22 | 30 | 0 | 19 | 53 | 20 | 0 | 17 | 54 | 0 | 0 | 16 | 16 | 22 |
| 3 | 0 | 0 | 22 | 30 | 0 | 0 | 20 | 0 | 0 | 0 | 18 | 0 | 0 | 0 | 16 | 21 | 49 |

**Table A-04 of sexagesimal multiplication**

This extended table gives in sexagesimal notation multiples of two numbers current from 1 to 60. We can find it, e.g., in Santritter's edition of the Parisian Alfonsine Tables (Venice: 1492, ff. G5v–H3r), where it is denoted as the *Tabula tabularum*.[7] Its usage for interpolation is explained in Rule 18 of canons by John of Głogów. We again transcribe here only the first and the last part of the table.

**Example of Table A-04 taken from Prague, NK, VIII G 24, f. 50v**

| Incipit thabula ad inveniendum partem proporcionalem ab uno usque ad (*del.*: 19) novem | | | | | | | | | | | | | | | | | | |
|---|---|---|---|---|---|---|---|---|---|---|---|---|---|---|---|---|---|---|
| | 1 | | 2 | | 3 | | 4 | | 5 | | 6 | | 7 | | 8 | | 9 | |
| | G | M | G | M | G | M | G | M | G | M | G | M | G | M | G | M | G | M |
| 1 | 0 | 1 | 0 | 2 | 0 | 3 | 0 | 4 | 0 | 5 | 0 | 6 | 0 | 7 | 0 | 8 | 0 | 9 |
| 2 | 0 | 2 | 0 | 4 | 0 | 6 | 0 | 8 | 0 | 10 | 0 | 12 | 0 | 14 | 0 | 16 | 0 | 18 |
| 3 | 0 | 3 | 0 | 6 | 0 | 9 | 0 | 12 | 0 | 15 | 0 | 18 | 0 | 21 | 0 | 24 | 0 | 27 |
| 4 | 0 | 4 | 0 | 8 | 0 | 12 | 0 | 16 | 0 | 20 | 0 | 24 | 0 | 28 | 0 | 32 | 0 | 36 |
| 5 | 0 | 5 | 0 | 10 | 0 | 15 | 0 | 20 | 0 | 25 | 0 | 30 | 0 | 35 | 0 | 40 | 0 | 45 |
| 6 | 0 | 6 | 0 | 12 | 0 | 18 | 0 | 24 | 0 | 30 | 0 | 36 | 0 | 42 | 0 | 48 | 0 | 54 |
| 7 | 0 | 7 | 0 | 14 | 0 | 21 | 0 | 28 | 0 | 35 | 0 | 42 | 0 | 49 | 0 | 56 | 1 | 3 |
| 8 | 0 | 8 | 0 | 16 | 0 | 24 | 0 | 32 | 0 | 40 | 0 | 48 | 0 | 56 | 1 | 4 | 1 | 12 |
| 9 | 0 | 9 | 0 | 18 | 0 | 27 | 0 | 36 | 0 | 45 | 0 | 54 | 1 | 3 | 1 | 12 | 1 | 21 |
| 10 | 0 | 10 | 0 | 20 | 0 | 30 | 0 | 40 | 0 | 50 | 1 | 0 | 1 | 10 | 1 | 20 | 1 | 30 |
| 11 | 0 | 11 | 0 | 22 | 0 | 33 | 0 | 44 | 0 | 55 | 1 | 6 | 1 | 17 | 1 | 28 | 1 | 39 |
| 12 | 0 | 12 | 0 | 24 | 0 | 36 | 0 | 48 | 1 | 0 | 1 | 12 | 1 | 24 | 1 | 36 | 1 | 48 |
| 13 | 0 | 13 | 0 | 26 | 0 | 39 | 0 | 52 | 1 | 5 | 1 | 18 | 1 | 31 | 1 | 44 | 1 | 57 |
| 14 | 0 | 14 | 0 | 28 | 0 | 42 | 0 | 56 | 1 | 10 | 1 | 24 | 1 | 38 | 1 | 52 | 2 | 6 |
| 15 | 0 | 15 | 0 | 30 | 0 | 45 | 1 | 0 | 1 | 15 | 1 | 30 | 1 | 45 | 2 | 0 | 2 | 15 |
| 16 | 0 | 16 | 0 | 32 | 0 | 48 | 1 | 4 | 1 | 20 | 1 | 36 | 1 | 52 | 2 | 8 | 2 | 24 |
| 17 | 0 | 17 | 0 | 34 | 0 | 51 | 1 | 8 | 1 | 25 | 1 | 42 | 1 | 59 | 2 | 16 | 2 | 33 |

7 See Matthieu Husson, 'La Tabula tabularum de Jean de Murs et les modèles de l'arithmétique médiévale', *Cahiers de recherches médiévales et humanistes*, 27 (2014), 97–122.

**Incipit thabula ad inveniendum partem proporcionalem ab uno usque ad *(del.:* 19*)* novem**

| | 1 | | 2 | | 3 | | 4 | | 5 | | 6 | | 7 | | 8 | | 9 | |
|---|---|---|---|---|---|---|---|---|---|---|---|---|---|---|---|---|---|---|
| | G | M | G | M | G | M | G | M | G | M | G | M | G | M | G | M | G | M |
| 18 | 0 | 18 | 0 | 36 | 0 | 54 | 1 | 12 | 1 | 30 | 1 | 48 | 2 | 6 | 2 | 24 | 2 | 42 |
| 19 | 0 | 19 | 0 | 38 | 0 | 57 | 1 | 16 | 1 | 35 | 1 | 54 | 2 | 13 | 2 | 32 | 2 | 51 |
| 20 | 0 | 20 | 0 | 40 | 1 | 0 | 1 | 20 | 1 | 40 | 2 | 0 | 2 | 20 | 2 | 40 | 3 | 0 |
| 21 | 0 | 21 | 0 | 42 | 1 | 3 | 1 | 24 | 1 | 45 | 2 | 6 | 2 | 27 | 2 | 48 | 3 | 9 |
| 22 | 0 | 22 | 0 | 44 | 1 | 6 | 1 | 28 | 1 | 50 | 2 | 12 | 2 | 34 | 2 | 56 | 3 | 18 |
| 23 | 0 | 23 | 0 | 46 | 1 | 9 | 1 | 32 | 1 | 55 | 2 | 18 | 2 | 41 | 3 | 4 | 3 | 27 |
| 24 | 0 | 24 | 0 | 48 | 1 | 12 | 1 | 36 | 2 | 0 | 2 | 24 | 2 | 48 | 3 | 12 | 3 | 36 |
| 25 | 0 | 25 | 0 | 50 | 1 | 15 | 1 | 40 | 2 | 5 | 2 | 30 | 2 | 55 | 3 | 20 | 3 | 45 |
| 26 | 0 | 26 | 0 | 52 | 1 | 18 | 1 | 44 | 2 | 10 | 2 | 36 | 3 | 2 | 3 | 28 | 3 | 54 |
| 27 | 0 | 27 | 0 | 54 | 1 | 21 | 1 | 48 | 2 | 15 | 2 | 42 | 3 | 9 | 3 | 36 | 4 | 3 |
| 28 | 0 | 28 | 0 | 56 | 1 | 24 | 1 | 52 | 2 | 20 | 2 | 48 | 3 | 16 | 3 | 44 | 4 | 12 |
| 29 | 0 | 29 | 0 | 58 | 1 | 27 | 1 | 56 | 2 | 25 | 2 | 54 | 3 | 23 | 3 | 52 | 4 | 21 |
| 30 | 0 | 30 | 1 | 0 | 1 | 30 | 2 | 0 | 2 | 30 | 3 | 0 | 3 | 30 | 4 | 0 | 4 | 30 |

**Remaining part of Table A-04 (Prague, NK, VIII G 24, f. 57r)**

| | 55 | | 56 | | 57 | | 58 | | 59 | | 60 | |
|---|---|---|---|---|---|---|---|---|---|---|---|---|
| | G | M | G | M | G | M | G | M | G | M | G | M |
| 31 | 28 | 25 | 28 | 56 | 29 | 27 | 29 | 58 | 30 | 29 | 31 | 0 |
| 32 | 29 | 20 | 29 | 52 | 30 | 24 | 30 | 56 | 31 | 28 | 32 | 0 |
| 33 | 30 | 15 | 30 | 48 | 31 | 21 | 31 | 54 | 32 | 27 | 33 | 0 |
| 34 | 31 | 10 | 31 | 44 | 32 | 18 | 32 | 52 | 33 | 26 | 34 | 0 |
| 35 | 32 | 5 | 32 | 40 | 33 | 15 | 33 | 50 | 34 | 25 | 35 | 0 |
| 36 | 33 | 0 | 33 | 36 | 34 | 12 | 34 | 48 | 35 | 24 | 36 | 0 |
| 37 | 33 | 55 | 34 | 32 | 35 | 9 | 35 | 46 | 36 | 23 | 37 | 0 |
| 38 | 34 | 50 | 35 | 28 | 36 | 6 | 36 | 44 | 37 | 22 | 38 | 0 |
| 39 | 35 | 45 | 36 | 24 | 37 | 3 | 37 | 42 | 38 | 21 | 39 | 0 |
| 40 | 36 | 40 | 37 | 20 | 38 | 0 | 38 | 40 | 39 | 20 | 40 | 0 |
| 41 | 37 | 35 | 38 | 16 | 38 | 57 | 39 | 38 | 40 | 19 | 41 | 0 |
| 42 | 38 | 30 | 39 | 12 | 39 | 54 | 40 | 36 | 41 | 18 | 42 | 0 |
| 43 | 39 | 25 | 40 | 8 | 40 | 51 | 41 | 34 | 42 | 17 | 43 | 0 |
| 44 | 40 | 20 | 41 | 4 | 41 | 48 | 42 | 32 | 43 | 16 | 44 | 0 |
| 45 | 41 | 15 | 42 | 0 | 42 | 45 | 43 | 30 | 44 | 15 | 45 | 0 |
| 46 | 42 | 10 | 42 | 56 | 43 | 42 | 44 | 28 | 45 | 14 | 46 | 0 |
| 47 | 43 | 5 | 43 | 52 | 44 | 39 | 45 | 26 | 46 | 13 | 47 | 0 |
| 48 | 44 | 0 | 44 | 48 | 45 | 36 | 46 | 24 | 47 | 12 | 48 | 0 |
| 49 | 44 | 55 | 45 | 44 | 46 | 33 | 47 | 22 | 48 | 11 | 49 | 0 |
| 50 | 45 | 50 | 46 | 40 | 47 | 30 | 48 | 20 | 49 | 10 | 50 | 0 |
| 51 | 46 | 45 | 47 | 36 | 48 | 27 | 49 | 18 | 50 | 9 | 51 | 0 |
| 52 | 47 | 40 | 48 | 32 | 49 | 24 | 50 | 16 | 51 | 8 | 52 | 0 |
| 53 | 48 | 35 | 49 | 28 | 50 | 21 | 51 | 14 | 52 | 7 | 53 | 0 |
| 54 | 49 | 30 | 50 | 24 | 51 | 18 | 52 | 12 | 53 | 6 | 54 | 0 |
| 55 | 50 | 25 | 51 | 20 | 52 | 15 | 53 | 10 | 54 | 5 | 55 | 0 |
| 56 | 51 | 20 | 52 | 16 | 53 | 12 | 54 | 8 | 55 | 4 | 56 | 0 |
| 57 | 52 | 15 | 53 | 12 | 54 | 9 | 55 | 6 | 56 | 3 | 57 | 0 |
| 58 | 53 | 10 | 54 | 8 | 55 | 6 | 56 | 4 | 57 | 2 | 58 | 0 |
| 59 | 54 | 5 | 55 | 4 | 56 | 3 | 57 | 2 | 58 | 1 | 59 | 0 |
| 60 | 55 | 0 | 56 | 0 | 57 | 0 | 58 | 0 | 59 | 0 | 60 | 0 |

**Table A-05 of equation of time**

The Table of equation of time (*Tabula equacionis dierum cum noctibus*) gives the difference between the apparent and mean solar time, see Chapter 25. Its use is described, e.g., by John of Głogów in Rule 32 of his canons *Pro (introduccione) generali*, see Chapter 8.

**Example of Table A-05 taken from Prague, NK, X B 3, f. 144v**

| Tabula equacionis dierum cum noctibus | | | | | | | | | | | | | | | | | | | | | | | | |
|---|---|---|---|---|---|---|---|---|---|---|---|---|---|---|---|---|---|---|---|---|---|---|---|---|
| Linea numeri | 9 | | 10 | | 11 | | 0 | | 1 | | 2 | | 3 | | 4 | | 5 | | 6 | | 7 | | 8 | |
| | Capricorni | | Aquarii | | Piscium | | Arietis | | Thauri | | Geminorum | | Cancri | | Leonis | | Virginis | | Libre | | Scorpii | | Sagittarii | |
| G | G | M | G | M | G | M | G | M | G | M | G | M | G | M | G | M | G | M | G | M | G | M | G | M |
| 1 | 3 | 46 | 0 | 38 | 0 | 9 | 2 | 7 | 4 | 33 | 5 | 19 | 4 | 4 | 2 | 51 | 3 | 43 | 6 | 5 | 7 | 53 | 7 | 5 |
| 2 | 3 | 37 | 0 | 34 | 0 | 11 | 2 | 12 | 4 | 36 | 5 | 17 | 4 | 1 | 2 | 50 | 3 | 47 | 6 | 10 | 7 | 54 | 7 | 0 |
| 3 | 3 | 29 | 0 | 31 | 0 | 13 | 2 | 17 | 4 | 39 | 5 | 15 | 3 | 57 | 2 | 50 | 3 | 51 | 6 | 15 | 7 | 54 | 6 | 55 |
| 4 | 3 | 22 | 0 | 27 | 0 | 15 | 2 | 22 | 4 | 43 | 5 | 14 | 3 | 54 | 2 | 50 | 3 | 55 | 6 | 20 | 7 | 55 | 6 | 50 |
| 5 | 3 | 14 | 0 | 23 | 0 | 16 | 2 | 27 | 4 | 46 | 5 | 13 | 3 | 59 | 2 | 49 | 3 | 59 | 6 | 25 | 7 | 55 | 6 | 45 |
| 6 | 3 | 7 | 0 | 21 | 0 | 21 | 2 | 33 | 4 | 49 | 5 | 12 | 3 | 47 | 2 | 50 | 4 | 4 | 6 | 30 | 7 | 56 | 6 | 40 |
| 7 | 3 | 0 | 0 | 19 | 0 | 23 | 2 | 38 | 4 | 53 | 5 | 10 | 3 | 44 | 2 | 51 | 4 | 8 | 6 | 35 | 7 | 56 | 6 | 44 |
| 8 | 2 | 55 | 0 | 16 | 0 | 26 | 2 | 42 | 4 | 56 | 5 | 8 | 3 | 41 | 2 | 51 | 4 | 12 | 6 | 40 | 7 | 57 | 6 | 28 |
| 9 | 2 | 45 | 0 | 14 | 0 | 29 | 2 | 48 | 4 | 58 | 5 | 6 | 3 | 38 | 2 | 52 | 4 | 17 | 6 | 44 | 7 | 57 | 6 | 24 |
| 10 | 2 | 38 | 0 | 12 | 0 | 33 | 2 | 53 | 5 | 1 | 5 | 4 | 3 | 35 | 2 | 52 | 4 | 22 | 6 | 48 | 7 | 56 | 6 | 16 |
| 11 | 2 | 31 | 0 | 10 | 0 | 37 | 2 | 59 | 5 | 3 | 4 | 1 | 3 | 32 | 2 | 54 | 4 | 26 | 6 | 53 | 7 | 56 | 6 | 10 |
| 12 | 2 | 24 | 0 | 8 | 0 | 40 | 3 | 4 | 5 | 6 | 4 | 59 | 3 | 28 | 2 | 56 | 4 | 31 | 6 | 57 | 7 | 55 | 6 | 2 |
| 13 | 2 | 17 | 0 | 6 | 0 | 44 | 3 | 9 | 5 | 9 | 4 | 57 | 3 | 25 | 2 | 58 | 4 | 36 | 7 | 1 | 7 | 55 | 5 | 58 |
| 14 | 2 | 16 | 0 | 4 | 0 | 48 | 3 | 15 | 5 | 10 | 4 | 55 | 3 | 22 | 2 | 58 | 4 | 40 | 7 | 5 | 7 | 54 | 5 | 52 |
| 15 | 2 | 3 | 0 | 3 | 0 | 52 | 3 | 21 | 5 | 12 | 4 | 52 | 3 | 19 | 3 | 0 | 4 | 46 | 7 | 8 | 7 | 53 | 5 | 45 |
| 16 | 1 | 57 | 0 | 2 | 0 | 56 | 3 | 27 | 5 | 14 | 4 | 42 | 3 | 16 | 3 | 2 | 4 | 50 | 7 | 12 | 7 | 51 | 5 | 38 |
| 17 | 1 | 51 | 0 | 1 | 1 | 1 | 3 | 31 | 5 | 15 | 4 | 46 | 3 | 13 | 3 | 5 | 4 | 56 | 7 | 16 | 7 | 50 | 5 | 30 |
| 18 | 1 | 45 | 0 | 0 | 1 | 5 | 3 | 36 | 5 | 17 | 4 | 43 | 3 | 12 | 3 | 7 | 5 | 1 | 7 | 20 | 7 | 48 | 5 | 23 |

| Tabula equacionis dierum cum noctibus | | | | | | | | | | | | | | | | | | | | | | | | |
|---|---|---|---|---|---|---|---|---|---|---|---|---|---|---|---|---|---|---|---|---|---|---|---|---|
| Linea numeri | 9 | | 10 | | 11 | | 0 | | 1 | | 2 | | 3 | | 4 | | 5 | | 6 | | 7 | | 8 | |
| | Capricorni | | Aquarii | | Piscium | | Arietis | | Thauri | | Geminorum | | Cancri | | Leonis | | Virginis | | Libre | | Scorpii | | Sagittarii | |
| G | G | M | G | M | G | M | G | M | G | M | G | M | G | M | G | M | G | M | G | M | G | M | G | M |
| 19 | 1 | 39 | 0 | 0 | 1 | 9 | 3 | 40 | 5 | 18 | 4 | 40 | 3 | 9 | 3 | 9 | 5 | 6 | 7 | 23 | 7 | 46 | 5 | 16 |
| 20 | 1 | 33 | 0 | 0 | 1 | 14 | 3 | 45 | 5 | 19 | 4 | 37 | 3 | 7 | 3 | 11 | 5 | 11 | 7 | 26 | 7 | 44 | 5 | 9 |
| 21 | 1 | 27 | 0 | 0 | 1 | 19 | 3 | 50 | 5 | 19 | 4 | 34 | 3 | 5 | 3 | 14 | 5 | 16 | 7 | 29 | 7 | 41 | 5 | 2 |
| 22 | 1 | 22 | 0 | 0 | 1 | 23 | 3 | 55 | 5 | 20 | 4 | 31 | 3 | 3 | 3 | 17 | 5 | 20 | 7 | 32 | 7 | 38 | 4 | 55 |
| 23 | 1 | 16 | 0 | 0 | 1 | 28 | 4 | 0 | 5 | 20 | 4 | 28 | 3 | 1 | 3 | 20 | 5 | 25 | 7 | 35 | 7 | 36 | 4 | 48 |
| 24 | 1 | 10 | 0 | 0 | 1 | 33 | 4 | 4 | 5 | 20 | 4 | 25 | 2 | 59 | 3 | 23 | 5 | 30 | 7 | 38 | 7 | 32 | 4 | 40 |
| 25 | 1 | 5 | 0 | 0 | 1 | 37 | 4 | 9 | 5 | 21 | 4 | 22 | 2 | 57 | 3 | 26 | 5 | 35 | 7 | 41 | 7 | 28 | 4 | 32 |
| 26 | 1 | 1 | 0 | 1 | 1 | 42 | 4 | 13 | 5 | 21 | 4 | 19 | 2 | 56 | 3 | 29 | 5 | 40 | 7 | 43 | 7 | 25 | 4 | 24 |
| 27 | 0 | 57 | 0 | 2 | 1 | 47 | 4 | 17 | 5 | 21 | 4 | 16 | 2 | 55 | 3 | 31 | 5 | 45 | 7 | 45 | 7 | 21 | 4 | 17 |
| 28 | 0 | 52 | 0 | 3 | 1 | 52 | 4 | 21 | 5 | 20 | 4 | 13 | 2 | 54 | 3 | 33 | 5 | 50 | 7 | 47 | 7 | 17 | 4 | 9 |
| 29 | 0 | 47 | 0 | 4 | 1 | 57 | 4 | 25 | 5 | 20 | 4 | 10 | 2 | 53 | 3 | 35 | 5 | 55 | 7 | 49 | 7 | 14 | 4 | 2 |
| 30 | 0 | 42 | 0 | 6 | 2 | 2 | 4 | 29 | 5 | 20 | 4 | 7 | 2 | 52 | 3 | 37 | 6 | 0 | 7 | 51 | 7 | 10 | 3 | 54 |

**Table A-06 of right ascensions**

The Table of right ascensions gives a projection of each degree of ecliptic on the celestial equator, as it could be seen by simultaneous rising above the horizon at the geographical equator, where the sphere is 'right' – see Chapter 25 for details. Such a table is contained in many different sets of tables.[8] A precursor of this table can be already found in the *Almagest*.[9]

**Example of Table A-06 taken from Cracow, BJ, 617, f. 6r**

| Tabula ascensionum signorum in circulo recto | | | | | | | | | | | | | | | | | | | | | | | | | |
|---|---|---|---|---|---|---|---|---|---|---|---|---|---|---|---|---|---|---|---|---|---|---|---|---|---|
| gradus signi | 9 | | 10 | | 11 | | 0 | | 1 | | 2 | | 3 | | 4 | | 5 | | 6 | | 7 | | 8 | | gradus signi |
| | Cap | | Aqr | | Psc | | Ari | | Tau | | Gem | | Cnc | | Leo | | Vir | | Lib | | Sco | | Sgr | | |
| | G | M | G | M | G | M | G | M | G | M | G | M | G | M | G | M | G | M | G | M | G | M | G | M | |
| 1 | 1 | 6 | 33 | 15 | 63 | 4 | 90 | 55 | 118 | 50 | 148 | 50 | 181 | 6 | 213 | 15 | 243 | 4 | 270 | 55 | 298 | 50 | 328 | 50 | 1 |
| 2 | 2 | 11 | 34 | 17 | 64 | 1 | 91 | 50 | 119 | 48 | 149 | 53 | 182 | 11 | 214 | 17 | 244 | 1 | 271 | 50 | 299 | 48 | 329 | 53 | 2 |
| 3 | 3 | 17 | 35 | 19 | 64 | 58 | 92 | 45 | 120 | 46 | 150 | 56 | 183 | 17 | 215 | 19 | 244 | 58 | 272 | 45 | 300 | 46 | 330 | 56 | 3 |
| 4 | 4 | 22 | 36 | 21 | 65 | 55 | 93 | 40 | 121 | 44 | 151 | 59 | 184 | 22 | 216 | 21 | 245 | 55 | 273 | 40 | 301 | 44 | 331 | 59 | 4 |
| 5 | 5 | 28 | 37 | 23 | 66 | 52 | 94 | 35 | 122 | 42 | 153 | 2 | 185 | 28 | 217 | 23 | 246 | 52 | 274 | 35 | 302 | 42 | 333 | 2 | 5 |
| 6 | 6 | 33 | 38 | 24 | 67 | 48 | 95 | 30 | 123 | 40 | 154 | 5 | 186 | 33 | 218 | 24 | 247 | 48 | 275 | 30 | 303 | 40 | 334 | 5 | 6 |
| 7 | 7 | 38 | 39 | 26 | 68 | 45 | 96 | 25 | 124 | 38 | 155 | 9 | 187 | 38 | 219 | 26 | 248 | 45 | 276 | 25 | 304 | 38 | 335 | 9 | 7 |
| 8 | 8 | 43 | 40 | 27 | 69 | 41 | 97 | 20 | 125 | 36 | 156 | 13 | 188 | 43 | 220 | 27 | 249 | 41 | 277 | 20 | 305 | 36 | 336 | 13 | 8 |
| 9 | 9 | 48 | 41 | 28 | 70 | 37 | 98 | 16 | 126 | 35 | 157 | 17 | 189 | 48 | 221 | 28 | 250 | 37 | 278 | 16 | 306 | 35 | 337 | 17 | 9 |
| 10 | 10 | 53 | 42 | 29 | 71 | 33 | 99 | 11 | 127 | 34 | 158 | 21 | 190 | 53 | 222 | 29 | 251 | 33 | 279 | 11 | 307 | 34 | 338 | 21 | 10 |
| 11 | 11 | 58 | 43 | 29 | 72 | 29 | 100 | 6 | 128 | 33 | 159 | 25 | 191 | 58 | 223 | 29 | 252 | 29 | 280 | 6 | 308 | 33 | 339 | 25 | 11 |
| 12 | 13 | 3 | 44 | 30 | 73 | 25 | 101 | 1 | 129 | 32 | 160 | 29 | 193 | 3 | 224 | 30 | 253 | 25 | 281 | 0 | 309 | 32 | 340 | 29 | 12 |
| 13 | 14 | 8 | 45 | 30 | 74 | 21 | 101 | 57 | 130 | 31 | 161 | 33 | 194 | 8 | 225 | 30 | 254 | 21 | 281 | 57 | 310 | 31 | 341 | 33 | 13 |
| 14 | 15 | 13 | 46 | 30 | 75 | 17 | 102 | 52 | 131 | 30 | 162 | 38 | 195 | 13 | 226 | 30 | 255 | 17 | 282 | 52 | 311 | 30 | 342 | 37 | 14 |

8 See, e.g., Chabás and Saby, *The Tables of 1322 by John of Lignères*, pp. 62–65.
9 *Almagest* II,8 (see Toomer, *Ptolemy's Almagest*, p. 100).

| gradus signi | 9 | | 10 | | 11 | | 0 | | 1 | | 2 | | 3 | | 4 | | 5 | | 6 | | 7 | | 8 | | gradus signi |
|---|---|---|---|---|---|---|---|---|---|---|---|---|---|---|---|---|---|---|---|---|---|---|---|---|---|
| | Cap | | Aqr | | Psc | | Ari | | Tau | | Gem | | Cnc | | Leo | | Vir | | Lib | | Sco | | Sgr | | |
| | G | M | G | M | G | M | G | M | G | M | G | M | G | M | G | M | G | M | G | M | G | M | G | M | |
| 15 | 16 | 18 | 47 | 30 | 76 | 12 | 103 | 47 | 132 | 30 | 163 | 42 | 196 | 18 | 227 | 30 | 256 | 12 | 283 | 47 | 312 | 30 | 343 | 42 | 15 |
| 16 | 17 | 22 | 48 | 30 | 77 | 8 | 104 | 43 | 133 | 30 | 164 | 47 | 197 | 22 | 228 | 30 | 257 | 8 | 284 | 43 | 313 | 30 | 344 | 47 | 16 |
| 17 | 18 | 26 | 49 | 29 | 78 | 3 | 105 | 39 | 134 | 30 | 165 | 52 | 198 | 27 | 229 | 29 | 258 | 3 | 285 | 39 | 314 | 30 | 345 | 52 | 17 |
| 18 | 19 | 31 | 50 | 28 | 78 | 59 | 106 | 35 | 135 | 30 | 166 | 57 | 199 | 31 | 230 | 28 | 258 | 59 | 286 | 35 | 315 | 30 | 346 | 57 | 18 |
| 19 | 20 | 35 | 51 | 27 | 79 | 54 | 107 | 31 | 136 | 31 | 168 | 2 | 200 | 35 | 231 | 27 | 259 | 54 | 287 | 31 | 316 | 31 | 348 | 2 | 19 |
| 20 | 21 | 39 | 52 | 26 | 80 | 49 | 108 | 27 | 137 | 31 | 169 | 7 | 201 | 39 | 232 | 26 | 260 | 49 | 288 | 27 | 317 | 31 | 349 | 7 | 20 |
| 21 | 22 | 43 | 53 | 25 | 81 | 44 | 109 | 23 | 138 | 32 | 170 | 12 | 202 | 43 | 233 | 25 | 261 | 44 | 289 | 23 | 318 | 32 | 350 | 12 | 21 |
| 22 | 23 | 47 | 54 | 24 | 82 | 40 | 110 | 19 | 139 | 33 | 171 | 17 | 203 | 47 | 234 | 24 | 262 | 40 | 290 | 19 | 319 | 33 | 351 | 17 | 22 |
| 23 | 24 | 51 | 55 | 22 | 83 | 35 | 111 | 15 | 140 | 34 | 172 | 22 | 204 | 51 | 235 | 22 | 263 | 35 | 291 | 15 | 320 | 34 | 352 | 22 | 23 |
| 24 | 25 | 55 | 56 | 20 | 84 | 30 | 112 | 12 | 141 | 36 | 173 | 27 | 205 | 55 | 236 | 20 | 264 | 30 | 292 | 12 | 321 | 36 | 353 | 27 | 24 |
| 25 | 26 | 58 | 57 | 18 | 85 | 25 | 113 | 8 | 142 | 37 | 174 | 33 | 206 | 58 | 237 | 18 | 265 | 25 | 293 | 8 | 322 | 37 | 354 | 33 | 25 |
| 26 | 28 | 1 | 58 | 16 | 86 | 20 | 114 | 5 | 143 | 39 | 175 | 38 | 208 | 1 | 238 | 16 | 266 | 20 | 294 | 5 | 323 | 39 | 355 | 38 | 26 |
| 27 | 29 | 4 | 59 | 14 | 87 | 15 | 115 | 2 | 144 | 41 | 176 | 43 | 209 | 4 | 239 | 14 | 267 | 15 | 295 | 2 | 324 | 41 | 356 | 43 | 27 |
| 28 | 30 | 7 | 60 | 12 | 88 | 10 | 115 | 59 | 145 | 43 | 177 | 49 | 210 | 7 | 240 | 12 | 268 | 10 | 295 | 59 | 325 | 43 | 357 | 49 | 28 |
| 29 | 31 | 10 | 61 | 10 | 89 | 5 | 116 | 56 | 146 | 45 | 178 | 54 | 211 | 10 | 241 | 10 | 269 | 5 | 296 | 56 | 326 | 45 | 358 | 54 | 29 |
| 30 | 32 | 13 | 62 | 7 | 90 | 0 | 117 | 53 | 147 | 47 | 180 | 0 | 212 | 13 | 242 | 7 | 270 | 0 | 297 | 53 | 327 | 47 | 360 | 0 | 30 |
| | A | | B | | C | | D | | E | | F | | G | | H | | I | | J | | K | | L | | |

**Table A-07 of oblique ascensions**

Like the Table of right ascensions, Tables of oblique ascensions are also a common part of different sets of astronomical tables.[10] The mathematical basis of the oblique ascensions is explained in Chapter 26. Their tables are different for different geographical latitudes. In manuscripts with the TR we mostly find Tables of oblique ascensions for the latitude 50° or 51°, sometimes for both.[11] In the manuscripts with sets of the TR we also find the tables with the latitudes for Toledo, for the fifth and sixth climate or even all climates from one to six. These climates correspond to latitudes of southern Europe, and therefore it is not surprising that their tables are present in manuscripts with copies of the TR for Salamanca and Bologna. In our overview in Chapter 17 we list examples of these tables regardless of the latitudes for which they are composed.

**Example of Table A-07 taken from Oxford, Bodl., Canon. misc. 499, f. 160v/150v**

| Tabula ascensionum signorum ad orisontem Pragensem, cuius latitudo est 50 graduum, 7 minutorum, dies vero longior 16 hore, minuta 12 | | | | | | | | | | | | | | | | | | | | | | | | |
|---|---|---|---|---|---|---|---|---|---|---|---|---|---|---|---|---|---|---|---|---|---|---|---|---|
| gradus equales | Aries | | | | Thaurus | | | | Gemini | | | | Cancer | | | | Leo | | | | Virgo | | | |
| | ascensiones | | partes horarum | | ascensiones | | partes horarum | | ascensiones | | partes horarum | | ascensiones | | partes horarum | | ascensiones | | partes horarum | | ascensiones | | partes horarum | |
| | G | M | G | M | G | M | G | M | G | M | G | M | G | M | G | M | G | M | G | M | G | M | G | M |
| 1 | 0 | 26 | 15 | 5 | 14 | 15 | 17 | 26 | 32 | 19 | 19 | 25 | 59 | 39 | 20 | 14 | 97 | 23 | 19 | 19 | 139 | 24 | 17 | 17 |
| 2 | 0 | 53 | 15 | 10 | 14 | 46 | 17 | 30 | 33 | 3 | 19 | 28 | 60 | 45 | 20 | 14 | 98 | 45 | 19 | 15 | 140 | 49 | 17 | 12 |
| 3 | 1 | 19 | 15 | 14 | 15 | 17 | 17 | 35 | 33 | 48 | 19 | 31 | 61 | 53 | 20 | 14 | 100 | 7 | 19 | 12 | 142 | 14 | 17 | 7 |
| 4 | 1 | 46 | 15 | 19 | 15 | 48 | 17 | 39 | 34 | 33 | 19 | 34 | 63 | 1 | 20 | 14 | 101 | 30 | 19 | 9 | 143 | 39 | 17 | 3 |
| 5 | 2 | 12 | 15 | 24 | 16 | 20 | 17 | 44 | 35 | 19 | 19 | 37 | 64 | 10 | 20 | 13 | 102 | 53 | 19 | 5 | 145 | 4 | 16 | 58 |
| 6 | 2 | 39 | 15 | 28 | 16 | 51 | 17 | 48 | 36 | 6 | 19 | 40 | 65 | 20 | 20 | 12 | 104 | 16 | 19 | 1 | 146 | 28 | 16 | 53 |
| 7 | 3 | 5 | 15 | 33 | 17 | 23 | 17 | 53 | 36 | 53 | 19 | 43 | 66 | 30 | 20 | 11 | 105 | 40 | 18 | 58 | 147 | 52 | 16 | 49 |
| 8 | 3 | 32 | 15 | 38 | 17 | 55 | 17 | 57 | 37 | 42 | 19 | 45 | 67 | 40 | 20 | 10 | 107 | 4 | 18 | 54 | 149 | 16 | 16 | 44 |
| 9 | 3 | 58 | 15 | 43 | 18 | 28 | 18 | 1 | 38 | 31 | 19 | 48 | 68 | 52 | 20 | 9 | 108 | 27 | 18 | 50 | 150 | 40 | 16 | 39 |

10 See, e.g., Chabás and Saby, *The Tables of 1322 by John of Lignères*, pp. 66–72, or *Almagest* II,8 (see Toomer, *Ptolemy's Almagest*, pp. 100–03).

11 The former value is often described as for 'the middle of the seventh climate' or for Prague, and the latter value as for 'the end of the seventh climate' or for Cracow, despite the fact that the true latitude of Cracow is also 50°.

| Tabula ascensionum signorum ad orisontem Pragensem, cuius latitudo est 50 graduum, 7 minutorum, dies vero longior 16 hore, minuta 12 | | | | | | | | | | | | | | | | | | | | | | | | |
|---|---|---|---|---|---|---|---|---|---|---|---|---|---|---|---|---|---|---|---|---|---|---|---|---|
| gradus equales | Aries | | | | Thaurus | | | | Gemini | | | | Cancer | | | | Leo | | | | Virgo | | | |
| | ascensiones | | partes horarum | | ascensiones | | partes horarum | | ascensiones | | partes horarum | | ascensiones | | partes horarum | | ascensiones | | partes horarum | | ascensiones | | partes horarum | |
| | G | M | G | M | G | M | G | M | G | M | G | M | G | M | G | M | G | M | G | M | G | M | G | M |
| 10 | 4 | 25 | 15 | 48 | 19 | 2 | 18 | 5 | 39 | 21 | 19 | 50 | 70 | 4 | 20 | 8 | 109 | 51 | 18 | 46 | 152 | 4 | 16 | 35 |
| 11 | 4 | 52 | 15 | 52 | 19 | 36 | 18 | 9 | 40 | 11 | 19 | 52 | 71 | 16 | 20 | 7 | 111 | 15 | 18 | 42 | 153 | 28 | 16 | 30 |
| 12 | 5 | 19 | 15 | 57 | 20 | 10 | 18 | 14 | 41 | 2 | 19 | 55 | 72 | 30 | 20 | 6 | 112 | 39 | 18 | 39 | 154 | 53 | 16 | 25 |
| 13 | 5 | 46 | 16 | 2 | 20 | 44 | 18 | 18 | 41 | 53 | 19 | 57 | 73 | 44 | 20 | 4 | 114 | 3 | 18 | 39 | 156 | 17 | 16 | 21 |
| 14 | 6 | 13 | 16 | 7 | 21 | 18 | 18 | 22 | 42 | 45 | 19 | 59 | 74 | 59 | 20 | 2 | 115 | 27 | 18 | 31 | 157 | 41 | 16 | 16 |
| 15 | 6 | 40 | 16 | 11 | 21 | 52 | 18 | 26 | 43 | 39 | 20 | 1 | 76 | 15 | 20 | 1 | 116 | 52 | 18 | 26 | 159 | 6 | 16 | 11 |
| 16 | 7 | 7 | 16 | 16 | 22 | 27 | 18 | 31 | 44 | 33 | 20 | 2 | 77 | 31 | 19 | 59 | 118 | 17 | 18 | 22 | 160 | 29 | 16 | 7 |
| 17 | 7 | 35 | 16 | 21 | 23 | 3 | 18 | 35 | 45 | 28 | 20 | 4 | 78 | 47 | 19 | 57 | 119 | 42 | 18 | 18 | 161 | 53 | 16 | 2 |
| 18 | 8 | 3 | 16 | 25 | 23 | 39 | 18 | 39 | 46 | 24 | 20 | 6 | 80 | 4 | 19 | 55 | 121 | 6 | 18 | 14 | 163 | 17 | 15 | 57 |
| 19 | 8 | 30 | 16 | 30 | 24 | 16 | 18 | 42 | 47 | 20 | 20 | 7 | 81 | 22 | 19 | 52 | 122 | 30 | 18 | 9 | 164 | 40 | 15 | 52 |
| 20 | 8 | 58 | 16 | 35 | 24 | 53 | 18 | 46 | 48 | 18 | 20 | 8 | 82 | 39 | 19 | 50 | 123 | 54 | 18 | 5 | 166 | 4 | 15 | 48 |
| 21 | 9 | 26 | 16 | 39 | 25 | 31 | 18 | 50 | 49 | 16 | 20 | 9 | 83 | 57 | 19 | 48 | 125 | 18 | 18 | 1 | 167 | 27 | 15 | 43 |
| 22 | 9 | 54 | 16 | 44 | 26 | 9 | 18 | 54 | 50 | 14 | 20 | 10 | 85 | 16 | 19 | 45 | 126 | 43 | 17 | 57 | 168 | 51 | 15 | 38 |
| 23 | 10 | 22 | 16 | 49 | 26 | 48 | 18 | 58 | 51 | 14 | 20 | 11 | 86 | 35 | 19 | 43 | 128 | 7 | 17 | 53 | 170 | 15 | 15 | 33 |
| 24 | 10 | 51 | 16 | 53 | 27 | 27 | 19 | 1 | 52 | 14 | 20 | 12 | 87 | 55 | 19 | 40 | 129 | 31 | 17 | 48 | 171 | 39 | 15 | 28 |
| 25 | 11 | 20 | 16 | 58 | 28 | 8 | 19 | 5 | 53 | 15 | 20 | 13 | 89 | 15 | 19 | 37 | 130 | 56 | 17 | 44 | 173 | 2 | 15 | 24 |
| 26 | 11 | 49 | 17 | 3 | 28 | 48 | 19 | 9 | 54 | 17 | 20 | 14 | 90 | 35 | 19 | 34 | 132 | 21 | 17 | 39 | 174 | 26 | 15 | 19 |
| 27 | 12 | 18 | 17 | 7 | 29 | 29 | 19 | 12 | 55 | 20 | 20 | 14 | 91 | 56 | 19 | 31 | 133 | 45 | 17 | 35 | 175 | 49 | 15 | 14 |
| 28 | 12 | 47 | 17 | 12 | 30 | 11 | 19 | 15 | 56 | 23 | 20 | 14 | 93 | 17 | 19 | 28 | 135 | 10 | 17 | 30 | 177 | 13 | 15 | 10 |
| 29 | 13 | 16 | 17 | 17 | 30 | 53 | 19 | 19 | 57 | 27 | 20 | 14 | 94 | 39 | 19 | 25 | 136 | 35 | 17 | 26 | 178 | 36 | 15 | 5 |
| 30 | 13 | 45 | 17 | 21 | 31 | 35 | 19 | 22 | 58 | 33 | 20 | 14 | 96 | 1 | 19 | 22 | 137 | 59 | 17 | 21 | 180 | 0 | 15 | 0 |

A B C D E F G H I J K L

Remaining part of Table A-07 (Oxford, Bodl., Canon. misc. 499, f. 161r/151r)

| Residuum Tabule ascensionum signorum ad orisontem Pragensem, cuius latitudo 50 graduum, 7 minutorum, dies longior 16 horarum, 12 minutorum | | | | | | | | | | | | | | | | | | | | | | | | |
|---|---|---|---|---|---|---|---|---|---|---|---|---|---|---|---|---|---|---|---|---|---|---|---|---|
| gradus equales | Libra | | | | Scorpio | | | | Sagittarius | | | | Capricornus | | | | Aquarius | | | | Pisces | | | |
| | ascensiones | | partes horarum | | ascensiones | | partes horarum | | ascensiones | | partes horarum | | ascensiones | | partes horarum | | ascensiones | | partes horarum | | ascensiones | | partes horarum | |
| | G | M | G | M | G | M | G | M | G | M | G | M | G | M | G | M | G | M | G | M | G | M | G | M |
| 1 | 181 | 24 | 14 | 55 | 223 | 25 | 12 | 34 | 265 | 21 | 10 | 35 | 302 | 33 | 9 | 46 | 329 | 7 | 10 | 41 | 346 | 44 | 12 | 43 |
| 2 | 182 | 47 | 14 | 50 | 224 | 50 | 12 | 30 | 266 | 43 | 10 | 32 | 303 | 37 | 9 | 46 | 329 | 49 | 10 | 45 | 347 | 13 | 12 | 48 |
| 3 | 184 | 11 | 14 | 46 | 226 | 15 | 12 | 25 | 268 | 4 | 10 | 29 | 304 | 40 | 9 | 46 | 330 | 31 | 10 | 48 | 347 | 42 | 12 | 53 |
| 4 | 185 | 34 | 14 | 41 | 227 | 39 | 12 | 21 | 269 | 25 | 10 | 26 | 305 | 43 | 9 | 46 | 331 | 12 | 10 | 51 | 348 | 11 | 12 | 57 |
| 5 | 186 | 58 | 14 | 36 | 229 | 4 | 12 | 16 | 270 | 45 | 10 | 23 | 306 | 45 | 9 | 47 | 331 | 52 | 10 | 55 | 348 | 40 | 13 | 2 |
| 6 | 188 | 21 | 14 | 32 | 230 | 29 | 12 | 12 | 272 | 4 | 10 | 20 | 307 | 46 | 9 | 48 | 332 | 32 | 10 | 59 | 349 | 9 | 13 | 7 |
| 7 | 189 | 45 | 14 | 27 | 231 | 53 | 12 | 7 | 273 | 25 | 10 | 17 | 308 | 46 | 9 | 49 | 333 | 12 | 11 | 2 | 349 | 38 | 13 | 11 |
| 8 | 191 | 9 | 14 | 22 | 233 | 17 | 12 | 3 | 274 | 44 | 10 | 15 | 309 | 46 | 9 | 50 | 333 | 51 | 11 | 6 | 350 | 6 | 13 | 16 |
| 9 | 192 | 33 | 14 | 17 | 234 | 42 | 11 | 59 | 276 | 3 | 10 | 12 | 310 | 44 | 9 | 51 | 334 | 29 | 11 | 10 | 350 | 34 | 13 | 21 |
| 10 | 193 | 56 | 14 | 12 | 236 | 6 | 11 | 55 | 277 | 21 | 10 | 10 | 311 | 42 | 9 | 52 | 335 | 7 | 11 | 14 | 351 | 2 | 13 | 25 |
| 11 | 195 | 20 | 14 | 8 | 237 | 30 | 11 | 51 | 278 | 39 | 10 | 8 | 312 | 40 | 9 | 53 | 335 | 44 | 11 | 18 | 351 | 30 | 13 | 30 |
| 12 | 196 | 43 | 14 | 3 | 238 | 54 | 11 | 46 | 279 | 56 | 10 | 5 | 313 | 36 | 9 | 54 | 336 | 21 | 11 | 21 | 351 | 57 | 13 | 35 |
| 13 | 198 | 7 | 13 | 58 | 240 | 18 | 11 | 42 | 281 | 13 | 10 | 3 | 314 | 32 | 9 | 56 | 336 | 57 | 11 | 25 | 352 | 25 | 13 | 39 |
| 14 | 199 | 31 | 13 | 53 | 241 | 43 | 11 | 38 | 282 | 29 | 10 | 1 | 315 | 27 | 9 | 58 | 337 | 33 | 11 | 29 | 352 | 53 | 13 | 44 |
| 15 | 200 | 55 | 13 | 49 | 243 | 8 | 11 | 34 | 283 | 45 | 9 | 59 | 316 | 21 | 9 | 59 | 338 | 8 | 11 | 34 | 353 | 20 | 13 | 49 |
| 16 | 202 | 19 | 13 | 44 | 244 | 33 | 11 | 29 | 285 | 1 | 9 | 58 | 317 | 15 | 10 | 1 | 338 | 42 | 11 | 38 | 353 | 47 | 13 | 53 |
| 17 | 203 | 43 | 13 | 39 | 245 | 57 | 11 | 25 | 286 | 15 | 9 | 56 | 318 | 7 | 10 | 3 | 339 | 16 | 11 | 42 | 354 | 14 | 13 | 58 |
| 18 | 205 | 7 | 13 | 35 | 247 | 21 | 11 | 21 | 287 | 30 | 9 | 54 | 318 | 58 | 10 | 5 | 339 | 50 | 11 | 46 | 354 | 41 | 14 | 3 |
| 19 | 206 | 32 | 13 | 30 | 248 | 45 | 11 | 18 | 288 | 44 | 9 | 53 | 319 | 49 | 10 | 8 | 340 | 24 | 11 | 51 | 355 | 8 | 14 | 8 |
| 20 | 207 | 56 | 13 | 25 | 250 | 9 | 11 | 14 | 289 | 56 | 9 | 52 | 320 | 39 | 10 | 10 | 340 | 58 | 11 | 55 | 355 | 35 | 14 | 12 |
| 21 | 209 | 20 | 13 | 21 | 251 | 33 | 11 | 10 | 291 | 8 | 9 | 51 | 321 | 29 | 10 | 12 | 341 | 32 | 11 | 59 | 356 | 2 | 14 | 17 |

**Residuum Tabule ascensionum signorum ad orisontem Pragensem, cuius latitudo 50 graduum, 7 minutorum, dies longior 16 horarum, 12 minutorum**

| gradus equales | Libra | | | | Scorpio | | | | Sagittarius | | | | Capricornus | | | | Aquarius | | | | Pisces | | | |
|---|---|---|---|---|---|---|---|---|---|---|---|---|---|---|---|---|---|---|---|---|---|---|---|---|
| | ascensiones | | partes horarum | | ascensiones | | partes horarum | | ascensiones | | partes horarum | | ascensiones | | partes horarum | | ascensiones | | partes horarum | | ascensiones | | partes horarum | |
| | G | M | G | M | G | M | G | M | G | M | G | M | G | M | G | M | G | M | G | M | G | M | G | M |
| 22 | 210 | 44 | 13 | 16 | 252 | 57 | 11 | 6 | 292 | 19 | 9 | 50 | 322 | 18 | 10 | 15 | 342 | 5 | 12 | 3 | 356 | 28 | 14 | 22 |
| 23 | 212 | 8 | 13 | 11 | 254 | 20 | 11 | 2 | 293 | 30 | 9 | 49 | 323 | 7 | 10 | 17 | 342 | 37 | 12 | 7 | 356 | 55 | 14 | 27 |
| 24 | 213 | 32 | 13 | 7 | 255 | 44 | 10 | 59 | 294 | 40 | 9 | 48 | 323 | 54 | 10 | 20 | 343 | 9 | 12 | 12 | 357 | 21 | 14 | 32 |
| 25 | 214 | 56 | 13 | 2 | 257 | 7 | 10 | 55 | 295 | 50 | 9 | 47 | 324 | 41 | 10 | 23 | 343 | 40 | 12 | 16 | 357 | 48 | 14 | 36 |
| 26 | 216 | 21 | 12 | 57 | 258 | 30 | 10 | 51 | 296 | 59 | 9 | 46 | 325 | 27 | 10 | 26 | 344 | 12 | 12 | 21 | 358 | 14 | 14 | 41 |
| 27 | 217 | 46 | 12 | 53 | 259 | 53 | 10 | 48 | 298 | 7 | 9 | 46 | 326 | 12 | 10 | 29 | 344 | 43 | 12 | 25 | 358 | 41 | 14 | 46 |
| 28 | 219 | 11 | 12 | 48 | 261 | 15 | 10 | 45 | 299 | 15 | 9 | 46 | 326 | 57 | 10 | 32 | 345 | 14 | 12 | 30 | 359 | 7 | 14 | 50 |
| 29 | 220 | 36 | 12 | 43 | 262 | 37 | 10 | 41 | 300 | 21 | 9 | 46 | 327 | 41 | 10 | 35 | 345 | 45 | 12 | 34 | 359 | 34 | 14 | 55 |
| 30 | 222 | 1 | 12 | 39 | 263 | 59 | 10 | 38 | 301 | 27 | 9 | 46 | 328 | 25 | 10 | 38 | 346 | 15 | 12 | 39 | 360 | 0 | 15 | 0 |

M N O P Q R S T U V W X

# Overview of the Tables in the *Tabule resolute*

In this overview we list which folios in the selected manuscripts and printed editions contain individual tables from the set of *Tabule resolute* (TR-01 – TR-20) and most frequent additional tables (A-01 – A-07). The variants for different meridians are not distinguished, but they can be recognised from the letters of the cities given in the second line.

| | 74 | X B 3 | 499 | 6361a | 6361b | VIII G 24 | IV F 19 |
|---|---|---|---|---|---|---|---|
| | W 1424 | P 1428 | P 1428 | P 1428 | P 1428 | P 1428 | P 1428 |
| TR-01 | 251r-252r | 79r-80v | 72r-73r | 47r-48v | 135r-136v | 2r-3v | 47r-49r |
| TR-02 | 252v-253v | 81r-82v | 73v-74v | 49r-50v | 137r-138v | 4r-5v | 49v-50v |
| TR-03 | 251r-252r | 83r-84v | 75r-76v | 51r-51v | 139r-140v | 6r-7v | 47r-49r |
| TR-04 | 252v-253v | 83r-84v | 75r-76v | 51r-51v | 139r-140v | 6r-7v | 49v-50v |
| TR-05 | 254r-255r | 85r-86v | 77r-78r | 53r-54v | 141r-142v | 8r-9v | 51r-52r |
| TR-06 | 255v-256r | 87r-88r | 78v-79v | 55r-56r | 143r-144r | 10r-11r | 52v-53r |
| TR-07 | 256v | 92v-93r | 83v-84r | 59v-60r | 159v-160r | 15v-16r | 54v |
| TR-07a | | 93r-94v | 84v-85v | 60r-61v | 160r-161v | 16r-17v | 54v-55r |
| TR-07b | | | | | | | 76v |
| TR-07c | | | 85v | | | | 77r |
| TR-07d | | | 86r | | | | |
| TR-08 | 257r | 95r | 86r | 62r | 162r | 18r | 55v |
| TR-09 | 257v | 88v-89r | 80r-80v | 56v-57r | 144v-145r | 11v-12r | 53v |
| TR-10 | 257v | 89v-90r | 81r | 57r-57v | 145r-145v | 12v-13r | 53v |
| TR-11 | 258r | 90v-91r | 81v-82r | 57v-58r | 145v-146r | 13v-14r | 54r |
| TR-12 | 258r | 91v-92r | 82v-83r | 58v-59r | 146v 159r | 14v-15r | 54r |
| TR-13 | 258v | 95v-96r | 86v-87r | 62v-63r | 162v-163r | 18v-19r | 56r-56v |
| TR-14 | 259r | 117r-117v | 164v-165r | | | 59v-60r | |
| TR-15 | 259v-260v | 96v-99r | 87v-90r | 63v-66r | 163v-166r | 19v-22r | 57r-60v |
| TR-16 | 261r-262r | 99v-102r | 90v-93r | 66v-69r | 166v-169r | 22v-25r | 61r-63v |
| TR-17 | 262v-263v | 102v-105r | 93v-96r | 69v-72r | 169v-171v<br>147r | 25v-28r | 64r-66v |
| TR-18 | 264r-265r | 105v-108r | 96v-99r | 72v-75r | 147v-150r | 28v-31r | 67r-69v |
| TR-19 | 265v-266v | 108v-111r | 99v-102r | 75v-78r | 150v-153r | 31v-34r | 70r-72v |
| TR-20 | 267r-268r | 111v-114r | 102v-105r | 78v-81r | 153v-156r | 34v-37r | 73r-75v |
| A-01 | | 114v | 159v | | | 37v | 151v |
| A-02 | | 115r | 105v-106r | 81v | 156v | | |
| A-03 | | | 106v-118r | | | 38v-50r | |
| A-04 | | | 41v-47r | 83r-89r | 158r-158v | 50v-57r | |
| A-05 | | 144v | 134r | | | | 143v |
| A-06 | | | 53v-54r | | | | 144v-145r |
| A-07 | | | 160v-161r | | | | 145v-146v |

| 553 | 5240 | 1381 | 617 | 560 | 1852 | 5227 | 5245 |
|---|---|---|---|---|---|---|---|
| **P 1428** | **P 1428** | **P 1388** | **PC 1428** | **PC 1428** | **PC 1428** | **PC 1428** | **CP 1428** |
| 68r-69v | 1r-2v | 113r-114v | 13r-14r | 13-16 9-12 | 7-10 11-14 | 41r-42v<br>37r-38v | 62r-63v |
| 70r-71v | 3r-4v | 115r-116v | 14v-15r | 17-20 | 15-18 | 43r-44v | 64r-65v |
| 72r-73v | 5r-6v | 113r-114v | 15v-16r | 21-24 25-27 | 19-26 | 45r-46v | 66r-67v |
| 72r-73v | 5r-6v | | 15v-16r | 21-24 25-27 | 19-26 | 45r-46v | 66r-67v |
| 74r-75v | 7r-8v | 117r-118v | 16v-17r | 31-34 27-29 | 27-33 | 47r-48v | 68r-69v |
| 76r-77r | 9r-10r | 119r-120v | 17v-18r | 35-37 | 34-36 | 49r-50r | 71r-72r |
| 81v-82r | 14v-15r | | | 50 53 | 58-60 | 58-60 | 76v-77r |
| 82r-83v | 15r-16v | | | 53-56 | 61-64 | 55v-56v | 77v-78v |
| | | | 18r | 51 | 60 | 57r | 126v |
| | | | | 51 | 60 | 57r | 128v |
| | | | | | | | |
| 84r | 17r | | | 61 | 65 | 57v | 79r |
| 77v-78r | 10v-11r | | | 38-39 45-46 | 50-51 46-47 | 50v-51r<br>39r-39v | 72v-73r |
| 78v-79r | 11v-12r | | | 40-41 47-48 | 52-53 48-49 | 51v-52r<br>40r-40v | 73v-74r |
| 79v-80r | 12v-13r | | | 42-43 | 54-55 | 52v-53r | 74v-75r |
| 80v-81r | 13v-14r | | | 44 49 | 56-57 | 53v-54r | 75v-76r |
| 84v-85r | 17v-18r | 100r | 18v-19r | 62-63 | 66-67 | 58r-58v | 79v-80r |
| 108v-109r<br>104r | 50v-51r | 102v | 1r | | 128-129 | 91r-91v | 110v-111r |
| 85v-88r | 18v-21r | 100v-101r | 19r-22r | 64-69 | 68-73 | 59r-62r | 80v-83r |
| 88v-91r | 21v-24r | 72r-73r | 22v-25r | 70-75 | 74-79 | 62v-65v | 83v-86v |
| 91v-94r | 24v-27r | 73v-74v | 25v-28r | 76-81 | 80-85 | 66r-68v | 86r-89r |
| 94v-97r | 27v-30r | 103v-104v | 28v-31r | 82-87 | 86-91 | 69r-71v | 89v-92r |
| 97v-100r | 30v-33r | 75r-76r | 31v-34r | 88-93 | 92-97 | 72r-74v | 92v-95r |
| 100v-103r | 33v-36r | 102v | 34v-37r | 94-99 | 98-103 | 75r-77v | 95v-98r |
| | 48v | 105r | | | 149 | | 125v |
| 103v | | | | 100 | | 78r-78v | 124v-125r |
| | 37v-48r | | | | 104-127 | 79r-90v | 98v-110r |
| | 52v-59r | 124r-131r | 93v-96r | | 132-149 | 93r-99v | 112v-119r |
| 108r | 51v-52r | | 2v | | 130-131 | 92r-92v | 111v-112r |
| 149v-150v | 59v-60r | | 6r | | 150-151 | 102r-102v | 121v-122r |
| 152r-153r | 60v-64r | | | | 152-155 | 103r-104v | 122v-124r |

| | 5275 | 1481 | 600 | 605 | 611 | 614 | 616 |
|---|---|---|---|---|---|---|---|
| | CP 1428 | CP 1428 | C 1428 | C 1428 | C 1428 | C 1428 | C 1428 |
| TR-01 | 88r-89v | 1r-2v | 86r | 16r-17v | 1r-2v | 1r-2v | 1 |
| TR-02 | 90r-91v | 3r-4v | 86v | 18r-19v | 3r-4v | 3r-4v | 2 |
| TR-03 | 92r-93v | 5r-6v | 87r | 20r-21v | 5r-6v | 5r-6v | 3 |
| TR-04 | 92r-93v | 5r-6v | 87r | 20r-21v | 5r-6v | 5r-6v | 3 |
| TR-05 | 94r-95v | 7r-8v | 87v-88r | 23r-25v | 7r-9r | 7r-9r | 4 |
| TR-06 | 96r-97r | 9r-10r | 88v | 26r-27r | 9v-10v | 9v-10v | 5-6 |
| TR-07 | 101v-102r | 16v-17r | 89v | 29v | 14v-15r | 14v-15r | 8 |
| TR-07a | 102r-103v | 17r-18v | 90r | 30r-31r | 15v-17r | 15v-17r | 9 |
| TR-07b | 102v | 17v | 90r 90v | 31v | 16v | | |
| TR-07c | 102v-103r | 17v-18r | 90v | | 16v | | |
| TR-07d | | | | | 17r | | |
| TR-08 | 104r | 19r | | 32r | 17v | 17v | 10 |
| TR-09 | 97v-98r | 13r-13v<br>10v-11r | 89r | 27v | 11r-11v | 11r-11v | 7 |
| TR-10 | 98v-99r | 14r-14v<br>11v-12r | 89r | 28r | 11v-12r | 11v-12r | 7 |
| TR-11 | 99v-100r | 12v 15r | 89r | 28v | 12v-13r | 12v-13r | 7 |
| TR-12 | 100v-101r | 15v-16r | 89v | 29r | 13v-14r | 13v-14r | 8 |
| TR-13 | 104v-105r | 19v-20r | 91r | 32v-33r | 18r-18v | 18r-18v | 11 |
| TR-14 | 135v-136r | 50v-51r | 105v | 65v-66r | 49r-49v | 58r-58v | |
| TR-15 | 105v-108r | 20v-23r | 91v-92v | 33v-37r | 19r-21v | 19r-21v | 12-19 |
| TR-16 | 108v-111r | 23v-26r | 93r-94r | 37v-40r | 22r-24v | 22r-24v | 20-23 |
| TR-17 | 111v-114r | 26v-29r | 94v-95v | 40v-43r | 25r-27v | 25r-27v | 24-29 |
| TR-18 | 114v-117r | 29v-32r | 96r-97r | 43v-47r | 28r-30v | 28rv 38v | 30-35 |
| TR-19 | 117v-120r | 32v-35r | 97v-98v | 47v-50r | 31r-33v | 39r-42v | 36-41 |
| TR-20 | 120v-123r | 35v-38r | 99r-100r | 50v-53r | 34r-36v | 43r-45v | 42-47 |
| A-01 | 150v | | | | | | |
| A-02 | 149v-150r | | 107r | | 62r | | |
| A-03 | 123v-135r | 38v-50r | 102v-105r | 53v-65r | 37r-48v | 46r-57v | |
| A-04 | 137v-144r | 52v-59r | 100v-102r | 66v-72r | 51r-57v | 29r-35v | 48-51 |
| A-05 | 136v-137r | 51v-52r | 105v | | 50r-50v | 59r-59v | |
| A-06 | 146v-147r | 61v-62r | 106r | | 58r-58v | 60r-60v | |
| A-07 | 147v-149r | 62v-64r | 106r-106v | | 59r-60r | 61r-62r | |

| 1846 | 1858 | 1864 | 1865 | 1916 | 1917 | 1920 | 1966 |
|---|---|---|---|---|---|---|---|
| C 1428 | C 1428 | C 1428 | C 1428 | C 1428 | C 1428 | C 1428 | C 1428 |
| 1-7 | 121-124 | 71-74 | 72r-73v | 57-60 | 1r-2v | 25-28 | 13-16 |
| 2-8 | 125-128 | 75-78 | 78r-79r | 61-64 | 3r-4v | 29-32 | 17-20 |
| 9-12 | 129-133 | 79-82 | 80r-81v | 65-68 | 5r-6v | 33-36 | 21-24 |
| 9-12 | 129-133 | | | | | | |
| 13-16 | 134-136 | 83-86 | 82r-83v | 69-72 | 7r-8v | 37-40 | 25-30 |
| 17-19 | 137-139 | 87-89 | 84r-85r | 73-75 | 9r-10v | 41-43 | 31-33 |
| | 147-148 | 89-99 | 84r-85r | | 13r-13v | 52-56 | 41-42 |
| | 148-149 | 100-102 | 90v-91v | 86-89 | 14r-14v | | 43-46 |
| | 152 | | | 90-91 | 15r | | 47 |
| 19 | 152 | | | | | | |
| | | | | | | | |
| 27 | 153 | 103 | 92r | 92 | 15v | 59 | 49 |
| 20 | 140-141 | 90-93 | 74r-75v | 76-77 | 11r-11v | 44-47 | 34-37 |
| 21 | 142-143 | | | 78-79 | 11r-11v | | |
| 22 | 144-145 | 94-95 | 87v-88r | 80-81 | 12r | 48-49 | 38-39 |
| 22 | 146 | 96-97 | 89r | 82-83 | 12v | 50-51 | 40 |
| 28-29 | 154-155 | 104-105 | 92v-93r | 93-94 | 16r-16v | 60-61 | 50-51 |
| 90-91 | | | | | | | |
| 30-35 | 156-161 | 106-111 | 90v-96r | 95-100 | 17r-19v | 62-67 | 52-57 |
| 36-41 | 162-167 | 112-117 | 96v-99r | 101-104 | 20r-22v | 68-73 | 58-63 |
| 42-47 | 168-173 | 118-123 | 99v-102r | 105-110 | 23r-25v | 74-79 | 64-69 |
| 48-53 | 174-179 | 124-129 | 102v-105r | 111-116 | 26r-28v | 80-86 | 70-75 |
| 54-59 | 180-185 | 130-135 | 105v-108r | 117-122 | 29r-31v | 87-92 | 76-81 |
| 60-65 | 186-191 | 136-141 | 108v-111r | 123-128 | 32r-34v | 93-98 | 82-87 |
| 120 | 208 | | | | | | |
| 118-119 | 206-207 | | | | | | |
| 66-89 | | 142-165 | 112r-123v | 129-15371-94 | 35r-46v | | 88-111 |
| 94-106 | | | | | | | |
| 92-93 | | 168-169 | 125r-125v | 155-156 | 48r-48v | 129-130 | 114-115 |
| 108-109 | | 184-185 | 133v-134r | 171-172 | 58r-58v | 145-146 | 130-131 |
| 110-113 | | 186-189 | 134v-135r | 173-176 | 59r-59v | 147-150 | 132-135 |

| | 2252 | 2478 | 2480 | 2650 | 2664 | 214 | 51 |
|---|---|---|---|---|---|---|---|
| | C 1428 | C 1428 | C 1428 | C 1428 | C 1428 | C 1428 | C 1408 |
| TR-01 | 47-50 | 1r-2v | 1-4 | 75-78 | 1r-1v | 18r | 124r |
| TR-02 | 51-54 | 3r-4v | 5-8 | 79-82 | 1v | 18v | 124r-124v |
| TR-03 | 55-58 | 5r-6v | 9-12 | 83-86 | 2r | 19r | 124v |
| TR-04 | 55-58 | 5r-6v | | | | 19r | 124v |
| TR-05 | 59-62 | 7r-8v | 13-18 | 87-88 | 3v-4r | 19v | 125r |
| TR-06 | 63-65 | 10r-11r | 19-21 | 95-97 | 4r-5r | 20v | 125v |
| TR-07 | 147-148 | 74-75 | 29-30 | 89-90 | 6v | 21v | 125v |
| TR-07a | 76-79 | 16v-17v | 31-34 | 108-111 | 6v | 22r | |
| TR-07b | 80 | 18r | 35-36 | 93-94 | 7r-7v | 22r-22v | |
| TR-07c | | | | | | 22v | |
| TR-07d | | | | | | | |
| TR-08 | 82 | 19r | 37 | 114 | 7v | 22v | |
| TR-09 | 66-67 | 11v-12r | 22-25 | | 5v-6r | 21r | |
| TR-10 | 68-69 | 12v-13r | | 98-101 | | 21r | |
| TR-11 | 70-71 | 13v-14r | 26-27 | 102-103 | | 21r | |
| TR-12 | 72-73 | 14v | 28 | 103-105 | | 21v | |
| TR-13 | 83-84 | 19v-20r | 37-39 | 115-116 | | 23r | 126r |
| TR-14 | 121-122 | 50v-51r | | | | 37v | |
| TR-15 | 85-90 | 20v-23r | 40-45 | 117-122 | | 23v-24v | 126v-127v |
| TR-16 | 91-96 | 23v-26r | 46-51 | 123-128 | 9v-11v | 25r-26r | 128r-129r |
| TR-17 | 97-102 | 26v-29r | 52-57 | 129-134 | | 26v-27r | 129v-130v |
| TR-18 | 103-108 | 29v-32r | 58-63 | 135-140 | | 28r-29v | 131r-132r |
| TR-19 | 109-114 | 32v-35r | 64-69 | 141-148 | | 29v-30v | 132v-133v |
| TR-20 | 115-120 | 35v-38r | 70-75 | 149-154 | | 31r-32r | 134r-135r |
| A-01 | | | | | | 39v | |
| A-02 | | | | | | 39r | |
| A-03 | | 38v-50r | 76-99 | 170-193 | | 34v-37r | |
| A-04 | 126-139 | 50v-59r | | | | 32v-34r | |
| A-05 | 123-124 | 51v-52r | 102-103 | 196-197 | | 37v | |
| A-06 | 140-141 | 59v-60r | 118-119 | 198-199 | | 38r | |
| A-07 | 142-145 | 60v-62r | 120-123 | 204-207 | | 38r-38v | |

| 1375 | 1473 | 2115 | 27 | C278b | Lacher | Schöner | Virdung |
|---|---|---|---|---|---|---|---|
| **C 1428** | **L 1408** | **S 1348** | **S 1348** | **B 1428** | **L 1388** | **N 1388** | **H 1510** |
| 21r | 122r-122v | 19r-20v | 33r-38v | 1r-9v | F3r-F4v | G3v-G4v | L4v-M4v |
| 21v | 122r-122v | 21r-22v | 33r-38v | 1r-9v | G1r-G2r | H1r-H2r | L4v-M4v |
| 22r | 123r-123v | 23r-24v | 33r-38v | 1r-9v | G2v-G3v | H2v-H3v | L4v-M4v |
| 22r | 123r-123v | 23r-24v | 33r-38v | 1r-9v | G2v-G3v | H2v-H3v | L4v-M4v |
| 22v | 124r-124v | 25r-26v | 33r-38v | 1r-9v | G4r-H1r | H4r-I1r | L4v-M4v |
| 23r | 125r-125v | 27r-28v | 33r-38v | 1r-9v | H1v-H2v | I1v-I2v | L4v-M4v |
| 24v | 125v | 33v-34r | 39r | 10r | H3r-H3v | I3r | N3r |
| 24v | 125v | 34v-36v | 39v-40r | 10v-11r | H3v-I1v | I3v-K1v | N3r |
| | | 37r | 40v | | I1v | K1v | N3r |
| | | 37r | 40v | | I1v | K1v | N3r |
| | | | | | | | |
| 25r | | 37v | 43r | | K2r | N1r | P3v |
| 23v | 126r-126v | 29v-30r | 40v | 11v | I2v | K2v | O2r |
| 23v | | 30v-31r | | | I2r | K2r | |
| | | 31v-32r | 42r | 12r | I3r | K3r | O2r |
| | 126v | 32v-33r | 41r 42v | 12r | I3v | K3v | O2v |
| 25v | | 38r-38v | 43v-44r | 13r-13v | K2v-K3r | N1v-N2r | N1r |
| 40r | | 84v-85r | 88r | 42r | I4r | K4r | |
| 26r-27r | | 39r-41v | 44v-47r | 14r-16v | K3v-L2r | N2v-O1r | N1r |
| 27v-28v | | 42r-44v | 47v-50r | 17r-19v | O2v-O5r | R1v-R4r | N1r-N2r |
| 29r-30r | | 45r-47v | 50v-53r | 20r-22v | N3v-O2r | Q2v-R1r | N1r-N2r |
| 30v-31v | | 48r-50v | 53v-56r | 23r-25v | M4v-N3r | P3v-Q2r | N1r-N2v |
| 32r-33r | | 51r-53v | 56v-59r | 26r-28v | M1v-M4r | O4v-P3r | N1v-N2v |
| 33v-34v | | 54r-56v | 59v-62r | 29r-31v | L2v-M1r | O1v-O4r | N1v-N3r |
| | | | 78v | | | | O2v |
| | | | 77r-77v | | | | |
| 37r-39v | | 70v-82r | 66v-76v | | | | |
| 35r-36v | | 59v-66r | 62v-66r | 58v-64r | E1v-F3r | F1v-G3r | |
| 40r | | 82v | 79r | 43r | I4v | K4v | P1v |
| | | 101r-101v | | 42v | | | |
| | | 102r-103r | 103v-104r | 46v-47r | | | |

**Explanatory notes to the columns, first line:**

| | |
|---|---|
| 74 | Toruń, BU, 74/III (olim: Königsberg, 1735) |
| X B 3 | Prague, NK, X B 3 |
| 499 | Oxford, Bodl., Canon. misc. 499 |
| 6361a | Mikulov, RM, Mik 6361 |
| 6361b | Mikulov, RM, Mik 6361 |
| VIII G 24 | Prague, NK, VIII G 24 |
| IV F 19 | Wrocław, BU, IV F 19 |
| 553 | Cracow, BJ, 553 |
| 5240 | Vienna, ÖNB, Cod. 5240 |
| 1381 | Vatican, BAV, Pal. lat. 1381 |
| 617 | Cracow, BJ, 617 |
| 560 | Cracow, BJ, 560 |
| 1852 | Cracow, BJ, 1852 |
| 5227 | Vienna, ÖNB, Cod. 5227 |
| 5245 | Vienna, ÖNB, Cod. 5245 |
| 5275 | Vienna, ÖNB, Cod. 5275 |
| 1481 | Leipzig, UB, 1481 |
| 600 | Cracow, BJ, 600 |
| 605 | Cracow, BJ, 605 |
| 611 | Cracow, BJ, 611 |
| 614 | Cracow, BJ, 614 |
| 616 | Cracow, BJ, 616 |
| 1846 | Cracow, BJ, 1846 |
| 1858 | Cracow, BJ, 1858 |
| 1864 | Cracow, BJ, 1864 |
| 1865 | Cracow, BJ, 1865 |
| 1916 | Cracow, BJ, 1916 |
| 1917 | Cracow, BJ, 1917 |
| 1920 | Cracow, BJ, 1920 |
| 1966 | Cracow, BJ, 1966 |
| 2252 | Cracow, BJ, 2252 |
| 2478 | Cracow, BJ, 2478 |
| 2480 | Cracow, BJ, 2480 |
| 2650 | Cracow, BJ, 2650 |
| 2664 | Cracow, BJ, 2664 |
| 214 | Munich, BSB, Clm 214 |
| 51 | Munich, BSB, Clm 51 |
| 1375 | Vatican, BAV, Pal. lat. 1375 |
| 1473 | Leipzig, UB, 1473 |
| 2115 | Lisbon, ANTT, ML, 2115 |
| 27 | Oxford, Bodl., Canon. misc. 27 |
| C278b | Milan, BA, C 278 b Inf. |
| Lacher | Lacher, Ambrosius, *Tabule resolute de motibus planetarum aliorumque supercelestium mobilium* (Frankfurt an der Oder: Johannes Hanau, 1511) |

| | |
|---|---|
| Schöner | Schöner, Johannes, *Tabulae astromonicae, quas vulgo… resolutas vocant* (Nuremberg: Johannes Petreius, 1536) |
| Virdung | Virdung, Johannes, *Tabulae resolutae de supputandis siderum motibus* (Nuremberg: Johannes Petreius, 1542) |

**Second line:**
The letter in the second line of columns indicates the city to whose meridian the tables are adapted: W = Wrocław, P = Prague, C = Cracow, L = Leipzig, S = Salamanca, B = Bologna, N = Nuremberg, H = Heidelberg. Some of the manuscripts contain tables for two cities (PC = Prague, Cracow). The number following the letter of the city gives the year of the first cyclical radix.

**Explanatory notes to the lines:**

| | |
|---|---|
| TR-01 | Mean motions in collected years |
| TR-02 | Mean motions in expanded years |
| TR-03 | Mean motions in months in non-leap years |
| TR-04 | Mean motions in months in leap years |
| TR-05 | Mean motions in days |
| TR-06 | Mean motions in hours and their fractions |
| TR-07 | True planetary apogees in collected years |
| TR-07a | Apogees of planets in expanded years |
| TR-07b | Apogees of planets in months |
| TR-07c | Apogees of planets in days |
| TR-07d | Apogees of planets in hours |
| TR-08 | Equation of access and recess |
| TR-09 | Mean conjunctions of the Sun and Moon in collected years |
| TR-10 | Mean oppositions of the Sun and Moon in collected years |
| TR-11 | Mean conjunctions and oppositions of the Sun and Moon in expanded years |
| TR-12 | Mean conjunctions and oppositions of the Sun and Moon in months |
| TR-13 | Equation of the Sun |
| TR-14 | True motion of the Sun and Moon in one hour |
| TR-15 | Equations of the Moon |
| TR-16 | Equations of Saturn |
| TR-17 | Equations of Jupiter |
| TR-18 | Equations of Mars |
| TR-19 | Equations of Venus |
| TR-20 | Equations of Mercury |
| A-01 | Stations and retrogradation |
| A-02 | Mean motions of planets in selected numbers of days |
| A-03 | Interpolation to intermediate days for making an almanac |
| A-04 | Sexagesimal multiplication |
| A-05 | Equation of time |
| A-06 | Right ascensions |
| A-07 | Oblique ascensions |

# III

# Commentary

In the Part III we give a commentary on both the *Tabule resolute* and the canons for their use. The purpose of the TR was to facilitate computations of planetary motions according to the model derived from Ptolemy's *Almagest*. This model was generally accepted throughout the Latin Middle Ages with only a few minor changes, mostly consisting in adjustment of free parameters of the model, such as the radii of epicycles. These changes were intended to improve the agreement of the model with the observed motions of the celestial bodies. Unlike the use of the tables, their construction is not described in the canons or in other contemporary sources. Therefore, in an attempt to recognise and understand the procedures by which the authors constructed the tables, we must draw on a comparison of Ptolemy's explanation of his model, the explanation of the use of the tables in the canons, and on the numerical values found in the tables themselves.

The individual copies of the tables suffer from various scribal errors. Some of them can be discerned by comparing different copies with each other, but others are common to many witnesses due to the tradition of copying. However, some scribal errors can be identified and eliminated as outliers using the least squares method. This method is also useful for verifying what function and with what values of free parameters a particular table expresses. In order to apply the least squares fitting to the tabulated values, algebraic expressions for the expected functions are needed. We thus derive them in a compact form suitable for programming in the subsequent chapters, which deal with individual problems and the corresponding tables. It is obvious that Ptolemy and his medieval followers performed the relevant calculations in successive steps. Some of these steps required an interpolation in tables of sines or tangents (shadows), which caused further errors in addition to the errors of rounding (or truncation) of intermediate results. Because we lack any information about the sampling and precision of the used tables of trigonometric functions, we compute here the errors and show the residuals of the analysed tables with respect to values computed according to the presented formulae.

It is worth noting that computations by means of the tables did not follow precisely the Ptolemaic model, but used different approximations which should provide roughly equivalent results. As we know today, the Ptolemy's model itself was only a crude approximation to the actual motion of the celestial bodies as it would later be described with more precision by Kepler's laws and Newtonian dynamics. Yet, with an appropriate adjustment of free parameters of the underlying Ptolemaic model, the method provided quite satisfactory agreement with the then available observations. To show the capabilities and limitations of the Ptolemaic model we also present a comparison with some results of the later development of celestial mechanics. It is interesting to follow how our predecessors struggled with the treatment of observational errors and the modelling and theoretical explanations of laws of nature. Because such problems on a more sophisticated level are even more important for contemporary science, it is instructive to see that an agreement of a theory with experiment does not necessarily confirm its correctness and disagreement does not necessarily imply its incorrectness. It is, however, important to realise that it was the effort to find this agreement, with its failures and successes, that paved the way for the development of the science.

# Mean Motions

In agreement with the conception of Aristotelian physics, the perpetual planetary motions were assumed in Middle Ages to be uniform and circular. The non-uniformity observed for instance in the retrograde motion of planets but also in changes of angular velocity, which are due to the conservation of angular momentum in an eccentric orbit and are described by Kepler's second law for elliptical motion, was thus explained by a superposition of several uniform circular motions around different centres. Accepting this presumption, the calculation of planetary positions consisted of two main steps. *(i)* It was necessary first to compute the so called mean motions, i.e., phases of all the circular motions. This step requires a straightforward computation of proportions using simple multiplication, but, regarding the large number of sexagesimal digits necessary for the required precision, it was also a laborious task. *(ii)* In the second step, the dominant mean motion was 'equated', it means corrected for the influence of the other circular motions. Its computation is a task of trigonometry dependent on the concrete geometry of the problem in question. In this chapter we shall analyse the tables assembled for solving the first step.

The purpose of tables of mean motions, represented in the above edition by Tables TRW-01 – TRW-06, is to find phases

$$\varphi=\varphi_0+\omega\,(t\text{-}t_0) \tag{18.1}$$

of the uniform circular motions of planets in a chosen time $t$. The reference time $t_0$, the epoch of the tables, is here usually the same for all planets and their motions, while the initial values $\varphi_0$ at $t = t_0$, the so-called 'radices' (roots), and the angular velocities $\omega$ are parameters different for each motion. The primary mean motion of each planet is its mean ecliptical longitude $\lambda$, for which the desired precision (since Ptolemy) was at least a minute of arc, which was a limiting precision of pre-telescopic observations. The tables were supposed to be sufficiently precise for a time interval of about 1000 years. This means that the desired precision of $\omega$ would be of the order of $10^{-8}$ or $60^{-5}$ degrees per day. Ptolemy gives mean motions of all planets with a precision of six sexagesimal parts of a degree,[1] the Parisian Alfonsine Tables from 5 to 7 (see Tab. 1).

The first concern in applying the tables of mean motions is to convert the time $t$ of interest from the calendric dates, hours and their fractions into a linear scale enabling the computation of the time difference from the epoch $t_0$ of the table. To achieve the desired precision of the product with the daily rate $\omega$, this time difference $t - t_0$ should be expressed also with a precision of at least 5 sexagesimal digits. This makes the direct multiplication

1 Ptolemy, *Almagest* III,2; IV,4, and IX,4 (see Toomer, *Ptolemy's Almagest*, pp. 142–43; 182–87, and 427–41).

**Table 1. Daily rates and radices at the Incarnation of mean motions according to the PAT (in degrees). The radix of apogee corresponds to the apogee of the Sun.**

| | rate | radix | |
|---|---|---|---|
| A | 0;0,0,4,20,41,17,12 | 1,11;25,23 | apogees and fixed stars |
| B | 0;0,0,30,24,49 | 5,59;12,34 | access and recess |
| C | 0;59,8,19,37,19,13,56 | 4,38;21,0,30,28 | longitude of the Sun, Venus, and Mercury |
| D | 13;10,35,1,15,11,4,35 | 2,2;46,50,16,39 | longitude of the Moon |
| E | 13;3,53,57,30,21,4,13 | 3,19;0,14,31,16 | anomaly of the Moon |
| F | 13;13,45,39,22,25,53,45 | 3,34;28,42,38,28 | argument of latitude of the Moon |
| G | 0;3,10,38,7,14,49,10 | 1,31;55,52,41 | head of dragon |
| H | 0;2,0,35,17,40,21 | 1,14;5,20,12 | longitude of Saturn |
| I | 0;4,59,15,27,7,23,50 | 3,0;37,20,43 | longitude of Jupiter |
| J | 0;31,26,38,40,5 | 0,41;25,29,43 | longitude of Mars |
| K | 0;36,59,27,23,59,31 | 2,9;22,2,36 | anomaly of Venus |
| L | 3;6,24,7,42,40,52 | 0,45;23,58,0 | anomaly of Mercury |

somewhat laborious. Ptolemy thus provided in his *Almagest* separated tables of increments $\Delta\varphi = \omega\Delta t$ of mean motions for time steps $\Delta t$ in hours, days, months, years from 1 to 18, and for multiples of 18 years. The user of these tables can thus decompose his time $t$ into these time steps, to find in the tables the corresponding increments of $\varphi$ and to add them together with the radix $\varphi_0$.

The standard PAT chose an alternative approach and a corresponding layout of the tables of mean motions. The question what are the 'standard PAT' does not have a simple answer because many manuscript copies of the PAT include also different tables taken from other sets of tables, e.g. from the Toledan Tables or tables by John of Lignères. The diversity of the PAT has been recently thoroughly studied by R. L. Kremer on the basis of more than 180 manuscript copies.[2] For our present purpose of comparison with the *Tabule resolute*, the first part of the PAT dealing with planetary motions is crucial. This part was included into the edition by E. Poulle according to *edicio princeps* and its form will be taken here as the standard.[3] It provides tables facilitating a conversion of dates in different calendric systems into a number of days in a sexagesimal notation and similarly also the hours and their fractions into sexagesimal fractions of a day. It is then sufficient to present only one table of sixty multiples of the daily rate $\omega$ for each mean motion, in which users can find the increments of $\varphi$ for each sexagesimal digit of the decomposed time, shift them according to their order and sum them together with the radix. This consistent sexagesimalization is mathematically elegant and quite versatile for well-trained users. An analogy of such a procedure is used in the present-day astronomy, where the so-called Julian dates, i.e. the sequence number of days and their fractions, facilitate a treatment of changes of various quantities with time and a search for their periodicities.

2 See Kremer, 'The Parisian Alfonsine Tables in Manuscript, 1351–1525'.

3 See Poulle, *Les Tables Alphonsines*, according to Ratdolt 1483.

The computations of mean motions in the PAT, however, still require a large number of numerical operations and the sexagesimal notation of time was too abstract for less experienced users. Medieval practitioners were mostly interested in computations of the planetary motions for the preparation of almanacs or horoscopes in a narrow interval of years and hence they did not need to repeat every time the calculation of increments of the mean motions from the beginning of era until their epoch. The *Tabule resolute* thus restored the Ptolemaic practice and adapted it to the Julian calendar. Because the Julian calendar has a period of four years in which the leap years are repeated, the TR employed a cycle of 20 years (instead of the Ptolemy's 18 years) for which the increments of mean motions were calculated. In these so called collected years the increments of mean motions were directly added to the radix at the beginning of the Christian era, so that a set of cyclical radices was created. We can see an example in Table TRW-01 (see also Fig. 1). The usage of cyclical radices of mean motions is the most striking characteristic feature of the *Tabule resolute*. However, the TR are not unique in this respect. Cyclical radices can be found already in al-Battānī's tables both with a step of 30 years for the Arabic calendar and 20 years for Julian calendar.[4] They are given with a step of 20 years for the years 1340–1600 in tables of William Rede[5] or for the years 1460–2600 in the encyclopaedia by Paulerinus.[6] A step of 60 years is used in the tables by Hermann of Saxony for the years 1300–3040[7] and several other combinations of ranges and steps of cyclical radices appear in other sets of tables. The increments of mean motions found in tables for the so-called expanded years (i.e., single years from 1 to 20 – see the Table TRW-02), months (presented in two versions for the non-leap and leap years – TRW-03 and TRW-04), days (TRW-05) and hours (and their sexagesimal parts – TRW-06) could then be added to the closest previous cyclical radix. It was also sufficient to present, for these increments, a smaller number of sexagesimal digits, usually to tertias (the third sexagesimal parts of a degree), so that compared to the PAT, there were to be added fewer and shorter numbers. The resolution of date and time into the collected and expanded years, the months, days and hours was also more intuitive than the sexagesimalization in the PAT. Undoubtedly for these advantages, the *Tabule resolute* became popular across Central Europe in the fifteenth century. In the preface to his canons *Mirabilis in altis Dominus*, Petrus Cruciferus explains that his desire to facilitate the use of the Alfonsine tables led him to compile a modified set of tables. John of Głogów justifies in the introduction to his canons *Pro (introduccione) generali* that these tables deserve to be named *resolute*.

Closely related to the computations of mean motions is another problem, namely, the adaptation of the tables to a local meridian.[8] It was common in the Middle Ages to use a local time and hence the computations of positions of planets, e.g. for casting horoscopes or predicting eclipses, were also needed in the local time. The tables of planetary motions were, however, composed for a one chosen meridian, very often (even in the Parisian Alfonsine Tables) for the meridian of Toledo. This is related to the fact that the original

4 See Hadravová and Hadrava, 'Cyclical radices in the *Tabule resolute*', note 5.

5 See Chabás, *Computational Astronomy in the Middle Ages*, pp. 207–08 or, e.g., MS Oxford, Bodl., Ashmole 191, f. 62r–v.

6 Pavel Žídek – see Chapter 2.1.4. The unfinished table is in MS Cracow, BJ, 257, f. 143r.

7 See, e.g., MS Metz, BM, 287, ff. 53r–58v.

8 See Hadravová and Hadrava, 'Cyclical radices in the *Tabule resolute*'.

Castilian Alfonsine Tables were computed for Toledo and were heir to the Toledan Tables, also computed for Toledo. So Toledo was seen as the natural reference of all such computations. To apply the tables for another city, it was necessary to take into account the difference of local time between the local and the reference meridian. The geographical coordinates of various cities were known, despite often with great errors or ambiguities. It was thus possible to calculate the needed differences in geographical longitudes and local times. An easy way to make this correction would be to convert the desired local time to the time at the reference meridian. Most canons, however, give a more complicated advice to compute the increments of all mean motions corresponding to the time shift and to subtract or add them as additional terms to the above-described increments in years, months etc. Alternatively, if the computations for the same meridian are to be performed repeatedly, these corrections can be added to the radices and this new set of radices for the local meridian thus completely adapts the tables to the needed location.

Given the intention of the *Tabule resolute* to simplify the computation of planetary positions as much as possible, it is thus not surprising that they were composed for a particular city to avoid the additional correction for the local time. This motivation and procedure is well described by Petrus Cruciferus in the introduction to his canons *Mirabilis in altis Dominus*. The distance of Wrocław from Toledo, which he presented in the canons, agrees with the values found in the tables in MS Toruń, BU, 74/III, which are also explicitly inscribed to correspond to the meridian of Wrocław. Similarly, other copies of the TR usually indicate for which meridian they have been composed (see, e.g., Table TRP-01 for Prague and TRC-01 for Cracow). In the cases where such an inscription is missing, we can find the intended destination from the numerical values of the radices. The composition of the tables for a particular city simplified their routine usage for this meridian but, on the other hand, limited their applicability if the user wanted to compute planetary motions for other places or if the tables were transported, for instance, when their owners engaged in the common *peregrinacio academica*. We can thus often find marginal notes in the tables which give corrections of the radices to various cities. An example of such glosses can be seen already in Table TRW-01, where the corrections between the Wrocław and Prague are given. These notes may thus help us to trace the links between the centres of education and spreading of the tables.

## Table of Mean Motions in Collected Years (Table TRW-01)

The cyclical radices are given in the table entitled Table of mean motions of planets in Christian collected years at the meridian of Wrocław, which is split into three grids written on upper halves of ff. 251r–252r in MS Toruń, BU, 74/III. This table presents cyclical radices for the years 1424–1704 with the step of 20 years, while all other preserved sets of manuscript copies of the TR use cycles shifted for 4 years beginning for 1428. The cyclical radix for the year 1424 is mentioned in the example in Rule 2 of the canons *Mirabilis in altis Dominus*, which is one of several clues suggesting, that this uniquely preserved set of tables corresponds to the canons by Petrus Cruciferus.

The cyclical radices are given in zodiacal signs and degrees with their sexagesimal parts to tertias. There are four columns on each page with headings identifying the corresponding

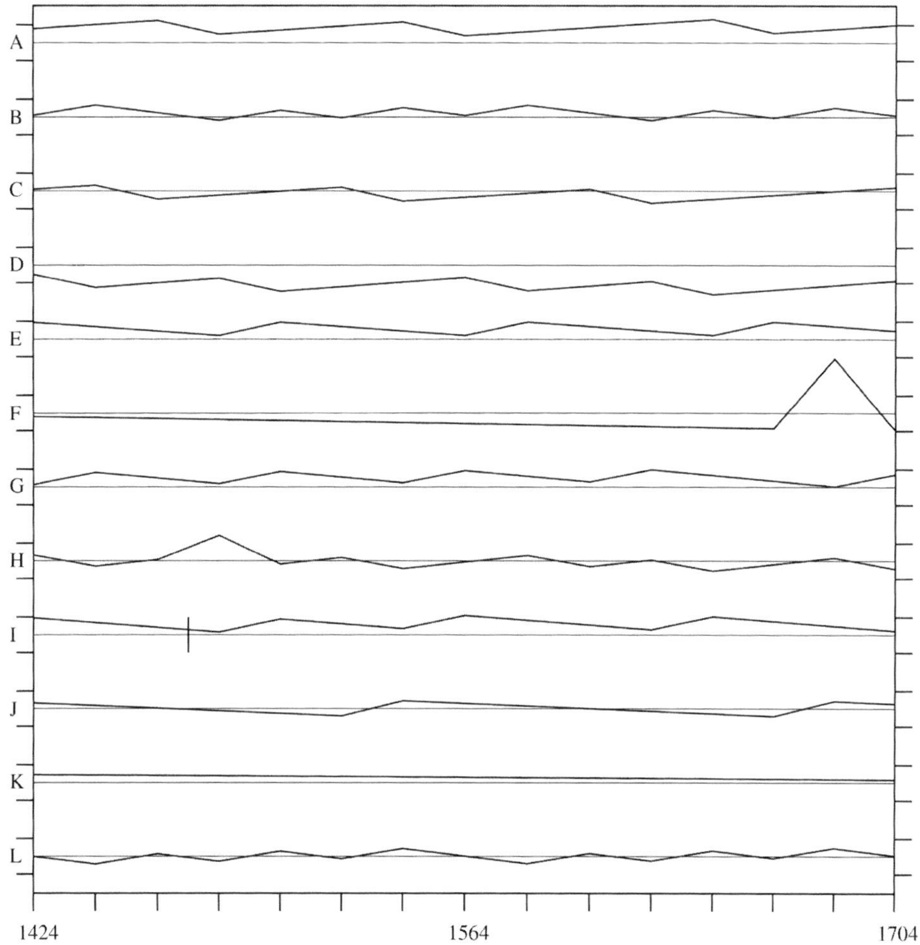


**Figure 30.** Errors in the Table TRW-01 of mean motions in collected years.

mean motions – see Tab. 1. We can verify by recalculation that the mean motions were computed according the daily rates and radices to the Incarnation presented in the PAT (as they can be found, e.g., in Ratdolt's edition – see Tab. 1). Only in the case of mean motion of apogees and fixed stars in the column A, the radix at the Incarnation is chosen to be zero. This corresponds to the mean motion of the 'common' apogee, because the different planets have different radices. The cyclical radices in collected years have mostly errors up to one tertia. However, the anomaly of Venus has a systematic shift of +20 tertias, and the anomaly of Mercury is shifted by −4 seconds. The mean motion of Jupiter is shifted for +10 minutes starting from the cyclical radix for the year 1484. This indicates that the values were computed by subsequent adding the constant increment of mean motion $\Delta\varphi$ in the time step $\Delta t$ of twenty years (naturally with a precision to more sexagesimal digits than to the tabulated tertias). A few errors which have not progressed to subsequent values are obviously scribal and may suggest that MS Toruń, BU, 74/III was copied from another nowadays unknown source. The errors are shown in Fig. 30. The lines of individual motions are denoted by letters corresponding to the columns in the Table TRW-01. The marks on the ordinate above and below the letters denote the value of error ± 1 tertia, or between 21 and 19 tertias in the case of the anomaly of Venus (letter K), and between −239 and −241 tertias in anomaly of Mercury (L). The vertical line, which crosses the line of the mean motion of Jupiter (I) between the collected years 1464 and 1484, separates the region with errors ±1 tertia on the left and between 36001 and 35999 tertias on the right. The sawtooth shape of most of the lines confirms that the mean motions were calculated with a precision higher than tertias and were rounded only afterwards. Systematic shifts of some of them (e.g. the mean motion of the Moon – D) indicate, that the corresponding radices at the Incarnation do not precisely agree with the values at the PAT. The excess of the line F (argument of the latitude of Moon) at the year 1684 is an example of an outlier most probably due to an error in copying the tables.

Recalculation also proves that the radices were adapted to a meridian that is 1 hour and 28 minutes east of Toledo, i.e., for the value for Wrocław named in the introduction of the canons *Mirabilis in altis Dominus*. This value of the time shift is found only in MS Toruń, BU, 74/III, which also confirms the correspondence of this set of the *Tabule resolute* to these canons. In the bottom halves of the folios are copied mean motions in collected years, with written notes with corrections to be added to the cyclical radices to convert them from the meridian of Wrocław to the meridian of Prague. This agrees with the Prague origin of this manuscript and its later ownership by Prague scholar Wenceslas Koranda the Younger. The fact that the oldest TR and the canons *Mirabilis* were composed by Petrus Cruciferus for Wrocław and then spread first to Prague supports the hypothesis by Jerzy Dobrzycki that Petrus Cruciferus was a member of the Order of the Crusaders with the Red Star which had its headquarters in Prague and a commandery in Wrocław, and he was not thus identical with Petrus Rein of Zittau.

## Table of Mean Motions in Expanded Years (Table TRW-02)

The tables of mean motions in expanded years give increments of the mean motions from the beginning of the twenty-year long cycle until the end of the given year. It means that they are increased in the subsequent lines for $\Delta\varphi$ corresponding to the step $\Delta t$ of 365 days, with the exception of every fourth leap year in which, in agreement with the Julian calendar, it is increased for motions in 366 days. In MS Toruń, BU, 74/III we can find these tables in the upper halves of ff. 252v–253v.[9] Unlike other copies of the TR, these tables are not denoted as leap (*bisextiles*) years. There is one significant outlier for mean motion of anomaly of Mercury in year 12, where the scribe originally wrote '34' minutes (instead of the correct 40 minutes) and it was then changed by scrambling to '20' minutes.

9 See Chapter 12, Table TRW-02.

## Table of Mean Motions in Months (Tables TRW-03 and TRW-04)

The tables of mean motions in months give the increments of mean motions from the beginning of a year to the end of corresponding month. They are prepared in two versions, one for non-leap and the other for leap years. The former can be found in MS Toruń, BU, 74/III, in bottom halves of ff. 251r–252r, the latter in bottom halves of ff. 252v–253v.[10]

## Table of Mean Motions in Days of Month (Table TRW-05)

The increments of mean motions from 1 to 31 days are given in the table split into three grids on ff. 254r–255r in MS Toruń, BU, 74/III.[11] We have identified only one scribal error in this table (namely 54 instead of 52 tertias in the column C for day 21).

## Table of Mean Motions in Hours and their Fractions (Table TRW-06)

The table of mean motions in hours gives increments of the mean motions for one to thirty hours. The extension of the table beyond twenty-four hours enables determination of the increments also in minutes, seconds, and other sexagesimal fractions of hours. For a case of more than thirty (i.e. up to fifty nine) of these units, they must be split into thirty plus the remaining number of units and the corresponding increments are to be added. The small triangular matrices at bottom of the columns facilitate the proper shift of sexagesimal orders of the motion for corresponding fractions of the hour, as explained and illustrated by an example in Rule 1 of the canons *Mirabilis*. Thanks to the fact that even the fastest mean motion of Moon (as well as of its anomalies) is only slightly more than thirteen degrees per day and less than seventeen degrees in thirty hours, no one of the mean motions reaches a non-zero value of the zodiacal signs and hence a column corresponding to signs can be skipped. All twelve motions can thus be written on two pages, namely ff. 255v and 256r in MS Toruń, BU, 74/III.[12] The tabulated values have only a small number of errors in rounding the tertias. There is, however, a scribal error of 4 minutes for the mean motion of Moon at twelve hours, which is corrected in an interlinear note, testifying to a practical use of the tables.

10 See Chapter 12, Tables TRW-03 and TRW-04, respectively.
11 See Chapter 12, Table TRW-05.
12 See Chapter 12, Table TRW-06.

19

# Apogees of Planets

The equations of planetary motions were computed as functions of the anomalies, which were measured from the apogees of the orbits. The knowledge of ecliptical longitudes of apogees of all planets was thus needed to equate the true positions of the planets. It was supposed in the Ptolemaic model that the positions of apogees of planets (with an exception of the Moon) are fixed with respect to the eighth sphere of stars[1] and they share its precession. This is also why the first mean motion in columns A of the above treated Tables TRW-01 – TRW-06 deals with the apogees and fixed stars. However, the main deviation of the Alfonsine model from the Ptolemaic one consists in an assumption that precession is not uniform but is at least partly subjected to a periodic motion of the so-called 'access and recess of the eighth sphere', the phase of which is given by the second subtables (in columns B) of the mean motions.

Ptolemy modelled precession as a rearward motion (with respect to the equinoxes) of the sphere of stars around the poles of ecliptic with a constant angular velocity of 1° in 100 years.[2] This picture is qualitatively quite good, but quantitatively it significantly underestimates the rate of the precession which in reality is 1° in about 72 years, so that the equinoxes arrive at the same point of sky after approximately 26 000 years.[3] The disagreement of the observed position of equinoxes with the extrapolation from Ptolemy's theory became obvious to astronomers within a few centuries, but instead of improving the value of the precession angular velocity which is a free parameter of the theory, it was speculated that the rate of the precession had changed in the course of the centuries. Because later measurements of the inclination of ecliptic to the equator also showed smaller values than 23;51,20° given by Ptolemy,[4] a completely different theory of precession (later also named as theory of trepidation) was introduced, in which both the inclination of ecliptic and the ecliptical longitudes of stars and planetary apogees are subject to a periodic motion of access and recess. Several tricky variants of geometrical models of trepidation were elaborated by Islamic astronomers.[5]

1 This assumption, supported by an analysis of historical observations (Ptolemy, *Almagest* IX,7, see Toomer, *Ptolemy's Almagest*, pp. 449–53), appears to be consistent with the solution of Newtonian two-body problem in which the line of apsides is fixed in the inertial reference frame. It is also an acceptable first approximation to the real motion of planets since the actual shifts of the perihelia of planets due to their mutual perturbations are slower than the precession of equinoxes.

2 Ptolemy, *Almagest* VII,3 (see Toomer, *Ptolemy's Almagest*, p. 329).

3 See, e.g., Petr Hadrava, 'Medieval Catalogues of Stars', in *The Stars in the Classical and Medieval Tradition*, ed. by Alena Hadravová, Petr Hadrava, and Kristen Lippincott (Prague: Scriptorium, 2019), pp. 115–33.

4 Ptolemy, *Almagest* I,15 (see Toomer, *Ptolemy's Almagest*, p. 72).

5 See Julio Samsó, *On Both Sides of the Strait of Gibraltar: Studies in the History of Medieval Astronomy in the Iberian Peninsula and the Maghrib* (Leiden and Boston: Brill, 2020), pp. 579–654.

One of these models described in the Latin treatise *De motu octave spere* became influential in medieval Latin Europe.[6] This treatise is a translation of an unpreserved Arabic text ascribed to Thābit ibn Qurra.[7] The basic idea of the model, often mentioned also in other medieval sources, is that the head of Aries and head of Libra, which are two opposite points of a movable ecliptic firmly connected to the eighth sphere of stars, circumscribe two small circles around the mean vernal points, i.e., the intersections of a fixed ecliptic with the equator. The orientation of the movable ecliptic around these points is not unambiguously described in the treatise, but it seems as if the solstitial points of the movable ecliptic were supposed to slide on the fixed ecliptic. This geometry determines changes of both the inclination of the movable ecliptic and the ecliptical longitudes[8] of the heads of Aries and Libra, the planetary apogees, as well as the longitudes of all stars.[9]

In addition to this description of the pseudo-Thābitan model of trepidation, *De motu octave spere* also contains a recipe for calculating the change of ecliptical longitudes of the head of Aries using three included tables. The first table gives the mean motion of the argument of the trepidation. The desired ecliptical longitude of the head of Aries is tabulated in the second as a function of the argument and the declination of the head of Aries is given as a function of the argument in the third table. The ecliptical longitude can then be found using a table of declinations of the degrees of ecliptic. There was no table for a change of the inclination of ecliptic. The calculation of trepidation using the third table of declinations of the head of Aries obviously assumed the inclination of the ecliptic to be fixed. By analysing the values of longitude in the second table we can find that they also were not calculated according to the above described geometrical model but according to its simplified version in which the heads of Aries and Libra circumscribe the small circles while the inclination of the ecliptic is fixed. These tables of trepidation were included also in the Toledan Tables.[10]

In agreement with the Parisian Alfonsine Tables, the *Tabule resolute* employ a model of trepidation which combines the original Ptolemaic precession around the pole of ecliptic and the pseudo-Thābitan trepidation in its simplified version with a constant inclination of ecliptic analogous to the tables in *De motu octave spere* and the Toledan Tables. The ecliptical longitude $\lambda$ of the head of Aries is thus

$$\lambda = \lambda_P + \lambda_T, \tag{19.1}$$

where the Ptolemaic part of the precession

6 The edition of the Latin text was prepared by Carmody, *The Astronomical Works of Thābit ben Qurra*, pp. 84–113. English translation can be found in Otto Neugebauer, 'Thābit ben Qurra On the Solar Year and On the Motion of the Eighth Sphere', *Proceedings of the American Philosophical Society*, 106 (1962), 290–99.

7 Thābit ibn Qurra was active in Baghdad c. 830–901. The attribution of the model and treatise to him is, however, doubted now because the text seems to be of a later, probably al-Andalusian origin. The model is thus denoted as Pseudo-Thābitan.

8 That is the angular distances from the intersection of the movable ecliptic with the equator.

9 See Petr Hadrava, 'Mathematical Treatment of Alfonsine Trepidation and its Sources', in *Alfonsine Astronomy: Computational Practices*, ed. by Matthieu Husson, José Chabás, and Richard L. Kremer (Turnhout: Brepols, forthcoming).

10 See Pedersen, *The Toledan Tables*, p. 1554 and 1559.

$$\lambda_{\mathrm{P}} = \lambda_{\mathrm{P}_{\mathrm{o}}} + \omega_{\mathrm{P}}(t - t_{\mathrm{o}}) \quad (19.2)$$

has a rate $\omega_{\mathrm{P}} = 360° / 49000$ years $= 1° / 136.11$ years. The trepidation, i.e., the second part of the precession, is given by the relation

$$\lambda_{\mathrm{T}} = \arcsin(\sin\lambda_{\max}\sin\theta)\,, \quad (19.3)$$

where the amplitude of the access and recess

$$\lambda_{\max} = \arcsin\frac{\sin\rho}{\sin\varepsilon_{\mathrm{o}}}\,, \quad (19.4)$$

$\rho$ being the radius of the small circle on which the head of Aries orbits the mean solstitial point and $\varepsilon_{\mathrm{o}}$ is the fixed inclination of the ecliptic. In the Alfonsine tables, both the PAT and the TR, $\lambda_{\max}$ was taken 9°, unlike the Toledan Tables where it was 10;45°. The phase of the trepidation

$$\theta = \theta_{\mathrm{o}} + \omega_{\mathrm{T}}(t - t_{\mathrm{o}}) \quad (19.5)$$

has a rate $\omega_{\mathrm{T}} = 360° / 7000$ years $= 1° / 19.44$ years. The values of $\lambda_{\mathrm{P}}$ and $\theta$ can be computed using the first two columns A and B of the Tables TRW-01 – TRW-06 of mean motions and correspond to the values of radices $\lambda_{\mathrm{P}_{\mathrm{o}}}$, $\theta_{\mathrm{o}}$ and angular velocities $\omega_{\mathrm{P/T}}$ given in the PAT, as they are summarized here in Tab. 1. The value of $\lambda_{\mathrm{T}}$ is to be found in Table TRW-08 as a function of the mean motion $\theta$ given by Eq. (19.3).

## Table of True Planetary Apogees in Collected Years (Table TRW-07)

The computation of true apogees of planets is explained by Petrus Cruciferus in Rule 2 of his canons *Mirabilis*. In agreement with his intention to simplify computations for users of his tables, he starts first with an approximate method based on the Table of true apogees in collected years,[11] which gives the ecliptical longitudes of the apogees between the years 1424 and 1704 with a step of 20 years. This table (f. 256v in MS Toruń, BU, 74/III) thus formally resembles the Table of mean motions in collected years. However, as we have shown above, the motion of apogees was thought to be subjected to the 'trepidation' which is a non-uniform motion. The computation and usage of this table thus differs from the Tables of mean motions. Because of the non-linearity of the trepidation, it is not possible to find the planetary apogees at a given time as a simple sum of increments tabulated in collected and expanded years, months, days, and hours. On the other hand, a linear interpolation can provide a satisfactory approximation, because this motion is slow and its non-linearity (i.e. a change of its rate) is moderate. In this context the canons *Mirabilis* offer an example for the year 1430, describing in detail the method of interpolation (see Chapter 27).

11 See Chapter 12, Table TRW-07.

The apogees of planets differ for additive constants only, given by the radices tabulated for the beginning of Christian epoch in the last lines of the upper and bottom parts of the table on f. 256v. The first column A of Table TRW-07 gives the resulting true motion of the 'common' apogee, that is, for a zero radix at the Incarnation. Next columns with the planetary apogees are then simply shifted for the corresponding radices. All these shifts are calculated precisely up to tertias, with the only exception of the year 1464 where 24' instead of 25' is written in column A of the common apogee, obviously due to a scribal error. It is interesting to note that MS Toruń, BU, 74/III gives the final sexagesimal digit (i.e. the tertia) of the radix of the apogee of the Sun and Venus equal to 4 as it is for all other planets, while this digit is zero in the early editions (and Poulle's edition)[12] as well as in many manuscripts. This value of 4 tertias is really taken into account in the computation of the solar apogee in collected years.

If we recalculate the apogees precisely according to the Alfonsine model of the motion of access and recess, we find that the tabulated values have errors up to 152 tertias (in the year 1624). However, the medieval astronomers had at their disposal tables of sines with a limited sampling and precision only, in which they had to interpolate. It is thus most likely that for computation of the apogees in collected years they interpolated directly in the subsequent Table TRW-08 of equations of access and recess. However, this table borrowed from the Parisian Alfonsine Tables gives the equation only up to seconds and moreover it has its own errors. If we replace the precise computation of the equation by interpolation in this table, then the discrepancies of the tabulated apogees decrease mostly to one tertia, with the exception of 10 tertias again for the year 1624. This signature in errors reveals the procedure how the tables were computed and that a precision to tertias was, in principle, unattainable.

To facilitate an interpolation of the position of apogees to dates between the collected years as described in Rule 2 of the canons *Mirabilis*, the last two columns D and E in both upper and lower parts of the Table TRW-07 give differences between the current and the subsequent value (i.e. an increment in twenty years) and its twentieth part. These differences are the same for all planets because their apogees are only mutually shifted as mentioned above. The meaning of the last column E, which is somewhat confusingly inscribed as *differencia differencie*, is confirmed by a marginal note written at the upper right corner of the table: *Differencia differencie est motus in anno, sicut differencia est motus in 20 annis, verum motum subtrahendo unum ab alio, et exibit differencia. Differenciam reduc ad quarta, multiplicando per 60 usque in quarta, post divide per 20 annos et resultabit differencia diffrencie 35 secunda, 0 tercia, 24 quarta, que est motus unius anni.*[13] – 'The difference of difference is the motion in one year, because the difference is the motion in 20 years got by subtracting of one true motion from the other and yields the difference. This difference is reduced to quartas by multiplying by 60 up to quartas, then dividing by 20 years and the difference of difference results in 35 seconds, 0 tertias, 24 quartas which is the motion in one year.'[14]

A marginal note at the bottom of f. 256v states: *1 addatur, cum praxis sit ad meridianum Pragensem in omnibus augibus.* – '1 is to be added in all apogees when working with the

12 See Poulle, *Les Tables Alphonsines*, p. 124 (Ratdolt 1483, f. C8r).
13 Toruń, BU, 74/III, f. 256v.
14 The numerical example is done for the first difference $11^m\,40^s\,8^t / 20 = 3 \times 11^s\,40^t\,8^q = 35^s\,0^t\,24^q$ in Table TRW-07.

meridian of Prague.' This value of one tertia is, however, significantly overestimated. Taking the greatest annual increase of the trepidation, which is 35 seconds of a degree in the year 1424, the increase in 14 minutes of an hour, which is the time difference between the meridians of Wrocław and Prague, should result in a correction of about 3.4 quartas of a degree.

## Tables of True Planetary Apogees in Expanded Years, Months, Days, and Hours (Tables TRP-07a and TRC-07b, c, d)

In later versions of the *Tabule resolute*, the Table of true apogees in collected years was recalculated for years starting from 1428. Because of the negligible influence of the meridian, this table differs in various sets of the TR only due to scribal errors and we thus give here only one example from MS Prague, NK, X B 3, ff. 92v–93r.[15] The column A of the common apogee is skipped in these tables, as are the columns D and E. These two columns designed for the interpolation between the collected years are replaced by a subsequent Table of apogees in expanded years (e.g. in MS Prague, NK, X B 3, ff. 93r–94v).[16] This table gives increments of the true apogees from the previous collected year to the beginning of each of the following twenty years, so that it could be used similarly as the Tables of mean motions in expanded years. However, because of the non-linearity of the true motion of the apogees, this Table of expanded years must be different for each collected year. By recalculation we can find that this table has not been constructed by an interpolation in Table TRP-07, but by a quite precise computation of trepidation to each of the years 1428–1808. The trepidation component of the true motion of planetary apogees for each year in Table TRP-07a was calculated using linear interpolation in the Table TRW-08 of equation of access and recess. The residual errors of the overall calculation of the true motion of apogees by means of Table TRW-08 are shown in the bottom panel of Fig. 31; the upper panel shows the residual errors with respect to the precise calculation according to Eq. (19.3). The scales of both panels are labelled in tertias, so that it is obvious that computation using Table TRW-08 agrees with the tabulated data with about two orders better precision. The outliers of 4 tertia in the first twenty expanded years are due to a scribal error in the first line (to the collected year 1428) of the apogee of the Sun and Venus in Table TRP-07, which does not occur in the apogees of other planets and in other copies of the TR.

The sets of the TR for Cracow then extended this analogy with the Tables of mean motions by adding also Tables of true apogees in months, days and even in hours. In Rule 5 of the canons *Girum recensendo* by Andrzej Grzymała we can read: *Verum locum augium cuiuslibet planete per thabulas invenire. Intra Thabulam annorum collectorum et expansorum, cuius titulus est Tabula augium planetarum et cetera iuxta modum iam dictum de mediis motibus. In annis tamen expansis habendo respectum super annos collectos, qui in sumitatibus predictarum tabularum scribuntur, et patebit propositum. In Tabula autem mensium et dierum precise intratur, sicut in mediis motibus planetarum. Unde et Tabule augium precise corriguntur, sicut de mediis motibus, licet non sic continuantur in annis expansis.* – 'To find the true position of the apogee of any planet through the tables. Enter the table of collected and expanded years, the title of which is the Table of the apogees of the planets, and the

15 See Chapter 13, Table TRP-07.
16 See Chapter 13, Table TRP-07a.

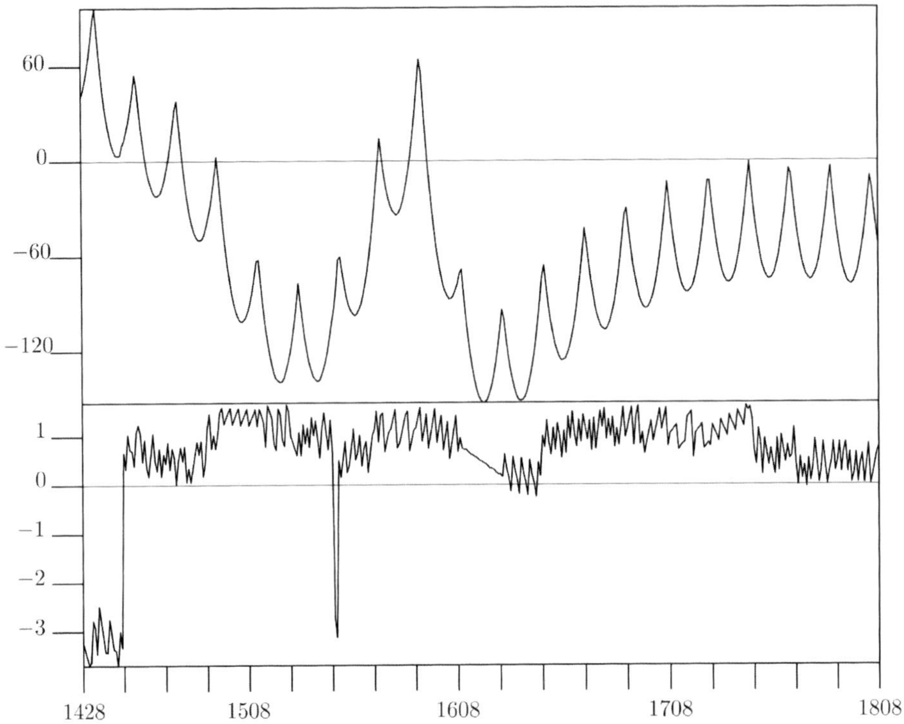


**Figure 31.** Residual errors (in tertias of a degree) in Tables TRP-07 and TRP-07a of the true motions of apogees.

rest make according to the manner already noted for the mean motions. In expanded years, however, take into account the collected years which are written in the heading of the above tables, and the purpose will be clear. But the Tables of the months and days are to be entered precisely as in the mean motions of the planets. This is also why the Tables of the apogees are corrected precisely as those of the mean motions, although they are not continued in this way in extended years.'[17] Unlike Petrus Cruciferus in the canons *Mirabilis*, Andrzej Grzymała does not deal in his *Girum recensendo* with the procedure of interpolation, because he takes supposedly avoided its need by usage of the Tables of increments in the expanded years, months and days.

The copy of canons *Nemo integre sapit* in MS Vatican, BAV, Pal. lat. 1381 does not contain any passage on the calculation of apogees, while the copy in MS Cracow, BJ, 614 includes on f. 69rb a text nearly identical with the above quoted Rule 5 of the canons *Girum recensendo*. The canons *Quoniam Tabularum Alfonsi* give the same explanation in only slightly different words.[18]

John of Głogów in Rule 3 of his canons *Pro (introduccione) generali* describes the computation of true apogees as a straightforward analogy of computation of mean motions

17 Cracow, BJ, 573, ff. 229r–v. Similarly, in Cracow, BJ, 2252, pp. 24–25, or Prague, NK, VIII G 24, f. 61v.

18 See Rule 5 in edition by Porres de Mateo and Chabás, 'Los cánones', pp. 67–68, according to Oxford, Bodl., Canon. misc. 27, ff. 124rb–va, or in Cambridge, Trinity College, O.3.13, ff. 76r–v, and Lisbon, ANTT, ML, 2115, ff. 11v–12r.

without mentioning the variety of the tables in expanded years and he gives an example for the year 1486.[19] He mentions that for a high precision, leap day could be taken into account, but he justifies that the motion of apogees in hours can be neglected due to its smallness.

Ambrosius Lacher in his 1511 print of the *Tabule resolute* and after him in practically the same words also Johannes Schöner in his editions of 1536, 1551, 1587, and 1588 present the calculation of apogees by means of Tables of true apogees in collected and expanded years, in months, and in days in more extensive explanation, but without any substantial change: *De planetarum augibus. Propositio II. Si augem alicuius planetae ad tempus quodcunque scire desideras, praescribe primo illud distinctae per suas differentias, ut assolet. Et intra cum annis Christi completis Tabulam radicum augium planetarum annorum collectorum, quaerendo in linea numeri annum tuum finitum, quem si praecise inveneris, recipe de tabula planetae tui motum iuxta positum et habes intentum. Si vero annos Christi transactos praecise non inveneris, tunc fiat introitus cum minore viciniori, circa quem accipe motum extra scribendo. Non aliter cum residuis annis, quos expansos vocamus, et Tabula an||norum expansorum te facere oportet. Tabulam, dico, hanc, in cuius fronte numerus annorum collectorum primi introitus fuerit repertus. Motum ergo ibidem scriptum priori adiicias et invenisti radicem augis anni propositi. Notandum tamen erit, quod Tabulae augium in annis collectis sunt tales, quod cuilibet planetae seorsum una ordinata est, praeterquam Soli et Veneri, quibus simul una est propter eundem utriusque augis locum. In expansis vero semper una duntaxat planetis omnibus ad 20 inserviet annos, illa scilicet, quae in fronte annos tuos collectos gestare cernitur. Eodem modo una et eadem Tabula mensium alteraque dierum cunctis planetis suffecturae censentur, cum nullam temporis post dies ipsos differentiam in praesentiarum ob augium motus nimiam tarditatem dignum duximus esse curandam. Deinde cum mensibus et diebus transactis succedentibus completis annis assignati temporis redeat operatio canonis praecedentis. Illo secluso, quod si contigerit te practicare per annum bisextilem, et dies propositus sequatur Februarium, intrandum erit cum die sequente. Hac etiam utere cautela, si placet, nec inde subcrescit error sensibilis.* – 'On the planetary apogees. Proposition II. If you desire to know the apogee of any planet for any time, first prescribe it by distinguishing it by its differences, as usual. And with the completed years of Christ, enter the Table of radices of planetary apogees in collected years, looking in the line of numbers for your finished year, which if you find precisely, take from the table the motion of your planet placed next to it and you have the desired. If, however, you do not find exactly the passed years of Christ, then make an entrance with a neighbouring smaller year around which you take the motion by writing it aside. Not otherwise you must do it with the remaining years, which we call expanded, and with the Table of expanded years. I mean that table in the headings of which is found the number of collected years for the first entry. Then add the motion written in the same place to the previous one, and you have found the radix of the apogees of the proposed year. It must be noted, however, that the Tables of the apogees in collected years are such, that each planet is treated separately, except for the Sun and Venus, which are together in one, because they have the same position of the apogee. But in expanded years, only one value will always serve all planets at 20 years, namely that, which is seen in the heading to carry your accumulated years. In the same way, one and the same Table of months and another of days are considered to

19 See the edition and translation of the canons *Pro (introduccione) generali* in Chapter 8.

support all the planets. Since there was no difference in presence of any time after the days themselves due to the excessive slowness of the motion of apogees, we thought that it was not worth to take care of it. Then, with the successive months and days that have passed since the completed years of the assigned time, let the operation of the preceding canon return. With the exception of that, if it happens that you practice during a leap year and the proposed day follows February, you will enter with the following day. Use this also cautiously, if you like, so that the perceptible error does not increase.'[20]

The corresponding Tables of true apogees in months and days can be found in several versions in many manuscripts, and sometimes there are also Tables of true apogees in hours.[21] While Tables TRW-07, TRP-07, and TRP-07a are given up to tertia of a degree like the mean motions, Tables TRC-07b, c, d are given to quarta or even quinta, obviously with an intention to facilitate the precision of the data to their overall value and not to increase the error of their sum. However, the main problem of application of Tables TRC-07b, c, d is that the rate of motion of the true motion of the apogees is changing and hence, similar to Table TRP-07a in the expanded years, there should be a different table in months for each year, a different table in days for each month of each year, etc.

Comparing the value 0;0,34,40,37° of the increment of the apogee at December according to version A in Table TRC-07b with the annual increments according to Table TRP-07a, we can see that it corresponds to the values around the year 1448. Similarly the value 0;0,34,14,57° in version B fits with the motion of apogees around the year 1468 and in this sence it thus really deserves to be denoted as *Tabula augium in mensibus nova*. The motions 0;0,2,56,42° and 0;0,2,54,31° in 31 days in versions A and B, respectively, according to Table TRC-07c practically agree with the motions in January according to Table TRC-07b. This indicates that versions A and B in both these tables are mutually consistent. The version C in Table TRC-07c gives the motion 0;0,0,5,25,29° in one day, which agrees with the motion in 24 hours in Table TRC-07d. It thus confirms mutual consistency of these versions in both tables. Extrapolating the daily motion, we find that it corresponds to an anual motion of about 33 seconds of a degree and thus to a rate of trepidation supposed at the beginning of the sixteenth century. This version C is thus even newest, but none of the currently known versions would be appropriate for the later dates, for which tables TRP-07 and TRP-07a were designed.

The tables of true apogees in collected years in Lacher's edition do not give the 'common' apogees but directly those of planets for the years 1388–2008. The apogees in collected and expanded years are given to tertia and in months and days to quarta. The rate of precession in the table in months (0;0,34,14,56° for December) in the Lacher's print corresponds to that around the year 1468 like the above mentioned version B in the Cracovian manuscripts. In Schöner's edition all values are rounded to seconds.

Johannes Virdung explains in his Rule 17 *De augibus planetarum computandis*[22] the calculation of apogees using his Table 12[23] of apogees of the Sun and Venus (*Tabula augis Solis et Veneris*).[24] Similarly to Virdung's tables of mean motions which give always on one

20 Schöner, 1536, f. B1v–B2r.
21 See Chapter 14, Tables TRC-07b, TRC-07c, and TRC-07d.
22 Virdung, 1542, f. G1r–v (25r–v).
23 Virdung's tables are numbered in accordance with the contents at the beginning of his edition.
24 F. N3r.

page the motions for 1 to 20 years continued with a step of 20 years up to 100 and then with a step of 100 up to 1000, as well as the motions in months in leap and non-leap years, in days and hours and also the cyclical radices with a step of 20 years, the Table of apogees of the Sun and Venus gives on the same page the motion of apogees in collected and expanded years, in months and days, all values rounded to seconds. While the cyclical radices of mean motions are given for the years 1510–90, the true apogees in collected years run from 1508 to 1628 and are obviously accepted from Table TRP-07 in older sets of the *Tabule resolute*. The same is true for the apogees in expanded years which are taken from the Table TRP-07a, and similarly also the apogees in months go up to the value 34 seconds in December, like Table TRC-07b, despite that it is greater than all values after the year 1508. To find the apogees of other planets, Virdung includes in his Rule 17 a table of differences between their apogees and the apogees of the Sun and Venus. He also gives an example for the date 1525, December 29.

## Table of Equation of Access and Recess (Table TRW-08)

In the second part of Rule 2 of canons *Mirabilis*, Petrus Cruciferus presents, as a more precise alternative to the above described calculation of true planetary apogees by means of Table TRW-07, its detailed calculation using the Table of access and recess.[25] This procedure is equivalent to that described in Rule 16 of canons *Tempus est mensura motus*.[26] Table TRW-08 of equation of access and recess gives the trepidation according to Eq. (19.3). This table is taken from the Parisian Alfonsine Tables,[27] but there are two scribal errors in MS Toruń, BU, 74/III, f. 256v: the PAT values for $\theta = 8°$ is $\lambda_T = 1;14,48°$ and for $\theta = 62°$ is $\lambda_T = 7;56,19°$. But there are also many errors in the PAT.[28]

Like the canons *Mirabilis*, the canons *Girum recensendo* describe the approximate method by means of the tables of apogees in collected and expanded years, months and days, followed by an explanation of the direct calculation: *Aliter verum locum augium invenire poteris querendo motum augium et stellarum fixarum ad annum, quem volueris, iuxta doctrinam de mediis motibus traditam. Deinde motum accessus et recessus octavi circuli, cum quo intra Tabulam equacionum motus accessus et recessus argumenti capitis Arietis sive octave spere secundum modum, qui paulo post dicetur in canone sequenti, recipiendo equacionem ibidem, quam adde motui augium et stellarum fixarum, si motus accessus et recessus, cum quo recepisti equacionem, fuerit minor sex signis, vel subtrahe, si fuerit maior, ut liquet in eadem Tabula equacionum. Et quod post addicionem vel subtraccionem provenerit, adde radici cuiuslibet planete, que ponuntur in tabula sequenti, et habe verum locum omnium augium omnium planetarum.* – 'In another way, you will be able to find the true position of the apogees by searching for the motion of the apogees and the fixed stars for the year you want, according to the traditional teaching about the mean motions. Then the motion of the access and recess of the eighth sphere, with which enter into the Table of equations of the motion of the access and recess of the argument of the head of Aries or the eighth

25 See Chapter 12, Table TRW-08.
26 See Poulle, *Les Tables Alphonsines*, pp. 62–65 (Ratdolt 1483, ff. A6r–v).
27 See Poulle, *Les Tables Alphonsines*, p. 131, Table 12 (Ratdolt 1483, f. D3v).
28 See Hadrava, 'Mathematical Treatment of Alfonsine Trepidation and its Sources'.

sphere according to the method, that will be discussed a little later in the following canon, receiving the equation there which is added to the motion of the apogees and the fixed stars if the motion of the access and recess with which you received the equation was less than six signs, or subtracted if it was greater, as it is clear in the same Table of equations. And what comes out after the addition or subtraction, add to the radices of each planet, which are placed in the following table and will have the true position of all the apogees of all the planets.'[29] A separate table follows with the radices of apogees for all planets at the Incarnation, the values in which are identical with values given at the bottom of the Tables TRW-07 or TRP-07 of apogees in collected years. It is also identical with that in Rule 17 of the canons *Pro (introduccione) generali*.[30]

The canons *Nemo integre sapit* in MS Cracow, BJ, 614 continue also this second part of the rule on the true apogees with practically identical words and on f. 69vb is the table of radices of the apogees at the Incarnation. The canons *Quoniam Tabularum Alfonsi* give the same explanation on the direct computation using Table TRW-08 of the equation of access and recess.[31]

In canons *Astrorum* only the method of calculation by means of the table of access and recess is presented. It is illustrated by an example for the beginning of the year 1482, which means the completed (that is, to the end of the) year 1481: *In exemplo. – Volo invenire augem communem ad principium anni currentis 1482. Inveni quidem per primum canonem motum augium et stellarum 0 signa, 10 gradus, 52 minuta, 50 secunda, 55 tercia, motum vero accessus et recessus octavi circuli 2, 15, 22, 30, 25, intrando autem Tabulas equacionum octave spere respectu duorum signorum et graduum 15 occurit mihi equacio 8 gradus, 41 minuta, 28 secunda, titulus vero circa ipsam est 'a' in fronte tabule, similiter et signum. Demum et differencia secundum processum graduum eundo 2 minuta, 22 secunda, cum titulo eciam 'a'. Attende, quatenus titulus semper accipias, scilicet ubi signum fuerit inventum tam equacionis, quam differencie. Ducam ergo predictam differenciam in omnes fracciones, que sunt in motu octavi circuli ultra gradus, et proveniunt mihi 53 secunda, 16 tercia, et cum differencia titulum habeat 'a', adiecta ipsa hoc productum addam equacioni dudum servate provenietque equacio equata 8 gradus, 42 minuta, 21 secunda, 16 tercia, hoc est equacio correspondens gradibus et ceteris fraccionibus in motu octavi circuli positis. Hanc iam addendo motui augium et stellarum erit aux communis reperta ista: 0 signa, 19 gradus, 35 minuta, 12 secunda, 11 tercia. Si itaque habere voles augem specialem cuiuslibet planete ad eundem annum, hanc augem communem augi cuiuslibet in tabella subscripta posite adde et factum est.* – 'For example. – I want to find the common apogee at the beginning of the current year 1482. By the first canon I actually found the motion of the apogees and the stars to be 0 signs, 10 degrees, 52 minutes, 50 seconds, 55 tertias, and the motion of the access and recess of the eighth sphere to be 2, 15, 22, 30, 25; and entering the Table of equations of the eighth sphere, I encountered beside two signs and 15 degrees the equation of 8 degrees, 41 minutes, 28 seconds, and the label around it is 'a' in the front of the table, and likewise the sign. And finally, the difference according

29 MS Cracow, BJ, 573, f. 229v. Similarly, in MS Cracow, BJ, 2252, p. 25, or MS Prague, NK, VIII G 24, f. 61v.

30 See Chapter 8.

31 See the Rule 5 in edition by Porres de Mateo and Chabás, 'Los cánones', p. 68, according to MS Oxford, Bodl., Canon. misc. 27, ff. 124rb–va, or in Cambridge, Trinity College, O.3.13, ff. 76r–v, and Lisbon, ANTT, ML, 2115, ff. 11v–12r.

to the increase of degrees, going 2 minutes, 22 seconds with the label 'a'. Pay attention, always when you take the label, that for both the equation and for the difference it is where the sign was found.[32] I will then multiply the aforesaid difference by all the fractions that are in the motion of the eighth circle beyond the degrees, and I will get 53 seconds, 16 tertias[33] and the difference has the title 'a'. After adding this product to the equation, keep it for a while and it will result in an equation equal to 8 degrees, 42 minutes, 21 seconds, 16 tertias; this is the equation corresponding to the degrees and the other fractions placed in the motion of the eighth circle. Now by adding this to the motion of the apogees and the stars, the common apogee found will be: 0 signs, 19 degrees, 35 minutes, 12 seconds, 11 tertias. If, therefore, you wish to have the special apogee of any planet for the same year, add this common apogee to that what is for each of them given in the table written below.'[34] By recalculation we can verify that the mean motions of both the apogees and of the access and recess correspond to 13 expanded years added to the cyclical radix at the collected year 1468 and the equation found in the table of access and recess for the latter is added to the former one.

John of Głogów describes the direct computation of the true apogees in his Rule 17 of canons *Pro (introduccione) generali*, independent of the method using The tables of true apogees in collected and expanded years, the months, and days.[35] Similarly, both Lacher and Schöner describe in Rule 8 on true motions of their canons how to calculate the apogees of planets using the Table of equation of access and recess without any reference or explanation of the procedure given in Rule 2 quoted above. They demonstrate the procedure for the example in the year 1502, April 4th, 17 hours, 34 minutes and 10 seconds, used already in the first Rule on mean motions. Lacher refers his examples to the meridian of Leipzig, Schöner to Nuremberg. The values in the Lacher's example do not agree precisely with those found in his tables.

Johannes Virdung pays great attention also to the calculation of the motion of apogees using the Table of equation of access and recess (included as his Table 31) in his Rule 33 *De motuum augium et fixorum siderum, tum octavi eciam orbis inquirendo*.[36] In the subsequent rule he calculates the precession from the epoch of Alfonso's star catalogue to the year 1526 and in Rule 35 he gives Alfonso's catalogue recalculated to that year. In his Table 28 Virdung gives radices of the mean motion of apogees and fixed stars for each year from 1525 to 1604 and the motion for each day of a year. Similarly in Table 29 are given the radices and daily motions of the access and recess and in Table 30 are annual increments of these motions. Finally in Table 32 are listed true motions of the eighth sphere from the years 1520 to 1700 with a step of 20 years.

32 Owing to the symmetry of the equations, a half of signs is written at the top and half at the bottom of the table, and the labels indicating the addition or subtraction have to be taken at the same side of the table.

33 0;2,22 × 0;22,30,25 = 0;0,53,15,59,10. For the meaning of the verb *ducere* compare the notes to the Rule 37 of canons *Pro (introduccione) generali*.

34 Munich, BSB, Clm 51, f. 53va–vb. Similarly, in MS Cracow, BJ, 600, f. 142r. The same table of radices of planetary apogees at the Incarnation mentioned above follows.

35 See Chapter 8.

36 F. K3r (39r) – K4r (40r).

# Mean Conjunctions and Oppositions of the Sun and Moon

Computation of mean syzygies, that is, the mean conjunctions and oppositions of the luminaries – the Sun and Moon, is the basis for finding the times of true syzygies (the new or full Moon, respectively) and hence also the circumstances of eclipses of the Sun and Moon, or at least the possibility of their occurence. The mean elongation between the Sun and Moon is the difference between their mean longitudes, $\lambda_{SM} = \lambda_M - \lambda_S$, and hence its computation is an equally simple linear task as in the case of the other mean motions,

$$\lambda_{SM} = \lambda_{SMo} + \mu_{SM}(t - t_o)\,, \tag{20.1}$$

where the radix $\lambda_{SMo} = \lambda_{Mo} - \lambda_{So}$ is the difference between the radices of the Moon and the Sun, the velocity of the mean elongation $\mu_{SM} = \frac{d}{dt}\lambda_{SM} = \mu_M - \mu_S$ is the difference between mean velocities of the luminaries and $t$ is the time elapsed from the epoch $t_o$ of the tables. To find the syzygies we must solve this equation with respect to $t$ for $\lambda_{SM} = 360°\,l$ or $\lambda_{SM} = 360°\,l + 180°$ for the conjunction or opposition, respectively. The integer number $l$ is here the sequence number of lunation from the period of time in which we are interested. A straightforward solution

$$t = t_o + \frac{\lambda_{SM} - \lambda_{SMo}}{\mu_{SM}} \tag{20.2}$$

could be obtained using the division by $\mu_{SM}$, which, however, was a computationally laborious task because of the high number of sexagesimal digits needed to be taken into account.

The Parisian Alfonsine Tables use the values of the mean daily motions in longitude of the Moon $\mu_M = 13;10,35,1,15,11,4,35°$ and of the Sun $\mu_S = 0;59,8,19,37,19,13,56°$, and hence for their elongation $\mu_{SM} = 12;11,26,41,37,51,50,39°$. The table of mean motion with this velocity in the same form as the other mean motions in the standard PAT, that is, a table of its sixty multiples in the sexagesimal notation,[1] facilitates the division needed in Eq. (20.2), as explained in the Rule 13 of canons *Tempus est mensura motus* by John of Saxony.[2] It is done in two steps. As the sequence number $l$ of the lunation from the initial epoch $t_o$ is not *a priori* known, there is first chosen an estimate $t'$ of the time of syzygy close to the period of interest, and the elongation $\lambda'_{SM}$ at this time is calculated according to Eq. (20.1). This chosen time and the corresponding elongation are then used as an auxiliary epoch and a radix, which replace $t_o$ and $\lambda_{SMo}$ in Eq. (20.2), where $360° = 6,0°$ or $180° = 3,0°$

1 See Poulle, *Les Tables Alphonsines*, p. 138 (Ratdolt 1483, f. D7r).

2 See Poulle, *Les Tables Alphonsines*, pp. 54–59, and 195–97 (Ratdolt 1483, f. A5r–v).

is substituted for $\lambda_{SM}$, if we want to find the first conjunction or opposition, which follows after the time $t'$. The arithmetics of the division then consists in finding and subtracting the closest smaller multiple of $\mu_{SM}$, whose number of line gives the first sexagesimal digit of the quotient, and the remainder shifted for one order enables finding recursively subsequent digits. To find the time of the next mean syzygy of the same kind, one has to add to the already found time the duration of the lunation $t_l = \frac{360^\circ}{\mu_{SM}}$ in which the relative motion of the Sun and Moon circumscribes 360°, or its half or quarter in order to find the opposite mean syzygy or the quarters of the Moon. With the same precision of eight sexagesimal digits, the lunation $t_l$ should be 29;31,50,7,37,27,8,25 days. John of Saxony gives the half of the lunation equal to 14;45,55,3,48 days,[3] which is truncated (not rounded) to the fourth sexagesimal part of a day.

Such a procedure, however, would not be consistent with the intention and layout of the *Tabule resolute*. In order to simplify the computations and to make them better understandable to less specialized users, the *Tabule resolute* keep the tradition which dates back to Ptolemy's *Almagest*. In Book VI of the *Almagest* devoted to syzygies of the Sun and Moon we find four tables for computations of the mean syzygies.[4] The first gives the times (that is, the number of day and its two sexagesimal parts) of the conjunction in the Egyptian month Thoth in every 25$^{th}$ year from 1 to 1101 of the Nabonassar (Nebuchadnezzar) era. Together with the time are also listed the anomalies of the Sun and Moon and the argument of the latitude of Moon, all in degrees, minutes and seconds. In the second table, the same quantities are given for oppositions of the luminaries. The next table gives the increments of these quantities between the Thoth syzygies[5] in 25 subsequent years, and finally the fourth table lists these increments in all months during one year. Adding the appropriate values from the last two tables to the cyclical radices from the first or second table, we can thus easily calculate the required data for any mean syzygy in the entire range covered by the first two tables.

This arrangement of the tables of syzygies, also based on the Egyptian calendar, was used by al-Battānī.[6] The only significant change is that instead of the anomaly of the Sun there is tabulated the mean motion of the Sun at the time of the syzygies, that is, the ecliptical longitude, where the coincidence of the mean Moon with the mean Sun occurs in the case of conjunction, or opposite to which the mean Moon is in the case of opposition. The same layout of tables of syzygies, reworked to the Arabic calendar for periods of 30 years for the collected years, was also used in the Toledan Tables.[7] The times of syzygies are given here in days, hours and minutes of hours, the angular values in signs, degrees and minutes.

The same approach has been adapted in the *Tabule resolute* for the Julian calendar. The epochs at the ends of every twentieth year, for which the cyclical radices are listed in the Table of mean motions in collected years, are chosen as $t_0$ in Eq. (20.2) and the times $t$ of the

3 See Poulle, *Les Tables Alphonsines*, p. 58 (Ratdolt 1483, f. A5v).
4 See Toomer, *Ptolemy's Almagest*, pp. 278–80.
5 This means the syzygies in the month of Thoth, which was the first month of the year in the Egyptian calendar. The Egyptian calendar was also used by Ptolemy in his *Almagest*.
6 See MS Paris, BnF, lat. 8322, ff. 40v–42r, or the edition by Nallino, *Al-Battānī sive Albatenii Opus astronomicum II*, pp. 29–32.
7 See Pedersen, *The Toledan Tables*, p. 1332 and 1335. An example can be found, e.g., in MS Prague, NK, XIV A 18, ff. 32r–33v.

first following mean syzygies are computed in the Table of mean conjunctions of the Sun and Moon in collected years[8] and in the Table of mean oppositions of the Sun and Moon in collected years[9] (or their variants for other meridians) in the *Tabule resolute*. These times are thus always in January of the first year after a leap year for which the cyclical radices of mean motions are tabulated. They are given in days, hours and their sexagesimal parts up to quartas. Together with the syzygy times are listed the mean motions of the Sun at those instants. Next, there are also given the anomaly of the Moon, which is needed to find the correction for the true syzygies, and the argument of latitude of Moon, required to verify the possibility of an eclipse. All these angular quantities are given in zodiacal signs, degrees and their parts up to tertias. Because the *Tabule resolute* are adjusted for a particular city, the times of the syzygies are given in the local times, which are obtained by addition of the time difference from the meridian of Toledo to the times calculated using the PAT compiled for Toledo. Because these times are quite sensitive to the assumed geographical longitudes, we reprint the Tables of syzygies in collected years for different sets of the *Tabule resolute*. The other three quantities are independent of the meridian, because they are related to the same instant for all observers and they are not dependent on the position of their zenith. The next Table of mean conjunctions and oppositions of the Sun and Moon in expanded years,[10] which is common for both syzygies and also for all meridians, gives the increments of the same quantities between the first syzygies in January for the twenty subsequent years. Finally, the Table of mean conjunctions and oppositions of the Sun and Moon in months[11] lists the increments of these quantities between the first and all subsequent syzygies in a single year. Unlike the tables of mean motions, the tables of mean syzygies in months are presented for non-leap years only. The canons to the *Tabule resolute* advise users how to proceed with these tables also in leap years – see Rule 3 of canons *Mirabilis* in Chapter 2, or Rules 28–30 of the canons *Pro (introduccione) generali* in Chapter 8.

## Tables of Mean Conjunctions and Oppositions of the Sun and Moon in Collected Years (Tables TRW-09 and TRW-10)

The tables of mean conjunctions or oppositions of the Sun and Moon in Christian collected years for the meridian of Wrocław can be found in MS Toruń, BU, 74/III at the upper and lower half of f. 257v, respectively.[12] They give the times of the first mean syzygies which occur after the times for which the cyclical radices are tabulated in the Tables of mean motions in collected years and the mean motion of the Sun and Moon, mean anomaly of the Moon and mean argument of latitude of the Moon at the instant of the mean syzygy. The mean motions in the collected years are given in the *Tabule resolute* always to the end of a leap year, hence the first subsequent syzygies are always in the January of the following year, that is in 1425, 1445 etc. in the case of MS Toruń, BU, 74/III. The time is thus tabulated

8 See Chapter 12, Table TRW-09.
9 See Chapter 12, Table TRW-10.
10 See Chapter 12, Table TRW-11.
11 See Chapter 12, Table TRW-12.
12 See Chapter 12, Tables TRW-09 and TRW-10.

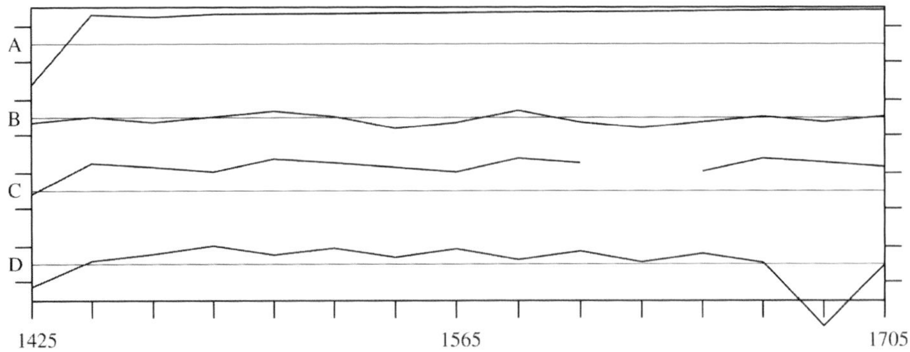


**Figure 32.** Errors in Table TRW-09 of mean conjunctions of the luminaries in collected years.

by the year, the day number of January and hours and their fractions up to quartas. The time is for Wrocław, it means that a constant 1 hour and 28 minutes is added to the time computed according to the Parisian Alfonsine Tables for the meridian of Toledo. The mean motions in the subsequent columns are computed for the times of the syzygies and they are thus independent of the choice of the meridian.

The times of the mean conjunctions of the Sun and Moon have errors of the order of tertias of an hour, despite being tabulated with a precision of quartas of an hour – see the line A in Fig. 32, for which the marks on the ordinate denote errors ±1 tertia of an hour. The angular values B, C, and D have errors of the order of tertias of a degree, with the exception of the anomaly of the Moon (line C) in the year 1625, where is an outlier of 10 seconds of a degree (the marks in Fig. 32 denote errors ±1 tertia of a degree for these quantities). The errors of the oppositions of the Sun and Moon are very similar, because the times of the oppositions and the corresponding motions are simply shifted for a half of the lunation with respect to the conjunctions.

## Table of Mean Conjunctions and Oppositions of the Sun and Moon in Expanded Years (Table TRW-11)

This table of mean syzygies in expanded years lists the increments of time between the first syzygy in a collected year given in the previous table and the first syzygy in any of the subsequent 20 years. Before the sequence numbers of the expanded years in the second column, the letter 'B' in the untitled first column indicates which of these years is leap (*bisextilis*) – because the first expanded year follows after a leap year, the next leap year is the third one and then every fourth year. The intervals of time between the subsequent syzygies in this and the following table in MS Toruń, BU, 74/III are computed from the rounded value $29^{d}$ $12;44,3,2,59^{h}$ of the lunation, which is more than 8 quintas above the correct value. Consequently, the duration of twelve lunations, which is the value of increment of time in the first expanded year, is already overestimated for nearly 2 quartas, and this discrepancy increases in subsequent years to about 36 quartas in the twentieth year. Apart from this systematic error, there are also several scribal errors, the most important of which is '12' instead of '20' days in the year 9. The Julian year has about 12.36 synodic months. This is why approximately every third year has 13 instead of 12 lunations and has thus a larger increment of time with respect to the previous one than the other two have. These years are denoted by E (for *embolismalis*) in the fourth column between columns A and B.

## Table of Mean Conjunctions and Oppositions of the Sun and Moon in Months (Table TRW-12)

The table of the syzygies in the months in the lower half of Toruń, BU, 74/III, f. 258r gives the shifts of the time and mean motions of the subsequent syzygies in the course of a year with respect to the first syzygy of the year. The times are thus the first twelve integer multiples of the synodic month. As shown above, the synodic month (the lunation) should be 29;31,50,7,37,27,8,25 days or $29^{d}$ 12;44,3,2,58,51,22,$8^{h}$ according to the PAT. However, this value is rounded in the *Tabule resolute* to $29^{d}$ 12;44,3,2,$59^{h}$, i.e. to the quartas of the hour, which corresponds to 29;31,50,7,37,27,30 days. The values for the next eleven months are computed as integer multiples of this rounded value of the mean lunation (to which also Rule 3 of the canons *Mirabilis*, or Rule 28 of canons *Pro /introduccione/ generali* refer). This causes already the fourth value, i.e. for May, to be incorrectly rounded and the last two values have an error of two quartas.

# Equation of the Sun

The simplest model of non-uniform motion is for a celestial body $S$ on an eccentre, i.e. on a circle $ASP$ with radius $R = CS = CA = CP$, the centre $C$ of which is displaced for $OC = eR$ from the position of an observer $O$ on the Earth ($e$ being the numerical eccentricity; see Fig. 33). If $M = \angle ACS$ is the (mean) anomaly, i.e. the angular distance of the body from the apogee $A$ (the point of the eccentre most distant from the Earth) as seen from the centre $C$ of the eccentre, then the true anomaly $v = \angle AOS$, i.e. the angular distance of the body $S$ from the apogee A as seen by the observer $O$, is given by formula

$$v = \arctan \frac{\sin M}{e + \cos M}, \tag{21.1}$$

which can be derived from the right triangles $CSL$ and $OSL$. The mean anomaly $M$ is supposed to be a linear function of time and the true anomaly $v = M - \varepsilon$ is then 'equated', i.e. corrected for a nonlinear 'equation' $\varepsilon = \angle OSC$. From the law of sines in the triangle $OSC$ or from the right triangle $SKO$ we can find that $\sin \varepsilon = OC/OS \sin M = OK/OS$.

If we calculate $OS$ from the triangle $OLS$, we find

$$\varepsilon = \arcsin \frac{e \sin M}{\sqrt{1 + 2e \cos M + e^2}} \tag{21.2}$$

for the equation, or equivalently from the triangles $SKO$ and $CKO$

$$\varepsilon = \arctan \frac{e \sin M}{1 + e \cos M}. \tag{21.3}$$

It reaches maximum value

$$\varepsilon_{max} = \arcsin e \tag{21.4}$$

at $M = \arccos(-e)$, where $v = 90°$ and the line of sight $OS$ is thus perpendicular to $ES$ and tangential to the epicycle.

Using an expansion of Eq. (21.1) into Taylor series in the eccentricity $e$, we can find[1] an approximation for small $e$

$$v = M - e \sin M + e^2 \sin M \cos M + o(e^3). \tag{21.5}$$

1 Both Ptolemaic and Keplerian motions are equally precisely uniform for zero eccentricity. The expansion into a power series in the eccentricity thus enables to compare how much these motions differ for a small eccentricity. See Alena Hadravová and Petr Hadrava, 'Cosmological Theories at the Time of Tycho Brahe', in *Mysterium cosmographicum 1596–1996. Acta historiae rerum naturalium necnon technicarum, New series*, 2 (1998), p. 232.

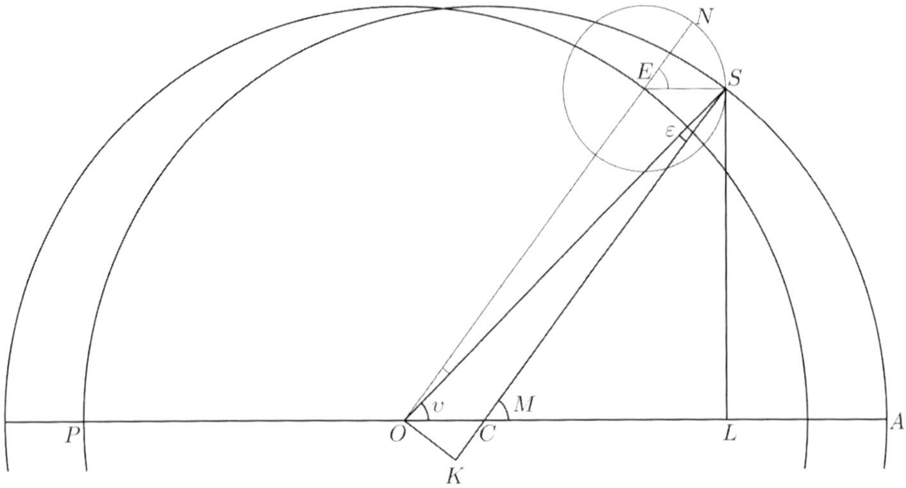


**Figure 33.** Motion on eccentre.

If we make a similar expansion for the Keplerian elliptical motion, which is a precise solution of the Newtonian two-body problem, we find

$$v = M + 2\,e_K \sin M + \tfrac{5}{2}\,e_K{}^2 \sin M \cos M + o(e^3)\,. \tag{21.6}$$

By a choice

$$e = -2\,e_K \tag{21.7}$$

we can thus get agreement of the true anomaly of the motion on eccentre with the Keplerian motion in the linear term of expansion in $e$,[2] and the quadratic term is overestimated for 60%.

A similar Taylor expansion of the distance $r = OS = \sqrt{1 + 2e \cos M + e^2}$ of the body $S$ from $O$ in the framework of the eccentric model gives

$$r \;=\; R\left(1 + e \cos M + \tfrac{1}{2}\,e^2 \sin^2 M\right) + o(e^3)\,, \tag{21.8}$$

while for the Keplerian motion

$$r \;=\; R\,(1 - e_K \cos M + e_K{}^2 \sin^2 M) + o(e^3)\,. \tag{21.9}$$

The doubled Keplerian eccentricity thus gives double the variation of distance in both the linear and quadratic terms. However, the distance was not directly measurable and it played a role only in combination of the Earth's motion with motions of the planets, for which thus a more precise model with equant was used; see Eqs (23.1) and (23.2).

2 The minus sign at the doubled Keplerian eccentricity reflects the habit to measure the anomaly of Keplerian motion from the pericentre $P$ instead of from the apocentre $A$ used in Ptolemaic astronomy.

The same position of the body $S$ can also be explained by an equivalent epicyclic model,[3] i.e. a uniform motion in the opposite direction ($\angle NES = \angle AOE$) on a small circle with radius $ES = OC$ around the centre $E$ orbiting around the observer $O$ with the mean anomaly $M$ (see the thin lines in Fig. 33). A choice of different ratios of the diameters $ES$ of the epicycle and $OE$ of the orbit of its centre $E$ around $O$ and different angular velocities of these motions can relatively well model the mutual motions of the Earth and planets orbiting around the Sun. The combinations of uniform circular motions are thus a relatively flexible mathematical tool for fitting the observed motions of celestial bodies. It can be compared to the Fourier series used in our present dynamical astronomy to calculate the perturbations of planetary orbits due to their mutual interactions. This explains why the Ptolemaic model was successful and quite sufficient for medieval astronomy.

## Table of Equation of the Sun (Table TRW-13)

The table of equation of the Sun in the *Tabule resolute* in MS Toruń, BU, 74/III, f. 258v[4] is identical with the corresponding table in the PAT, up to a few scribal errors,[5] and naturally the usage of the zodiacal signs. The later versions of the TR differ by adding columns with differences of the equations between the subsequent lines.[6] This innovation of layout of tables of equations (of the Sun as well as of the other planets) is also reflected in the corresponding canons. Rule 4 of the canons *Mirabilis in altis Dominus* describes that in order to interpolate between the integer degrees of the argument it is necessary to enter the table also with the subsequent degree, to subtract the found equations, and to calculate and to add or subtract the proportional part of the difference. But Rule 6 of canons *Girum recensendo* refers already to the column of differences: *Modum generalem in Tabulas equacionum <intrandi> proponere. Unde in Tabulas equacionum intratur cum signis communibus, quorum 6 sunt posita in superiori margine, 0, 1, 2, 3, 4, 5 eorumque gradus ponuntur in prima linea descendendo versus sinistram, et alia sex posita sunt in margine inferiori, videlicet 6, 7, 8, 9, 10, 11 retrorsum, eorumque gradus ponuntur in ultima linea ascendendo versus dextram. Si ergo signum, cum <quo> intratur, sit in superiori margine, tunc gradus signi accipiuntur versus sinistram descendendo et ex directo eiusdem gradus accipitur equacio, que scribitur extra. Deinde accipitur differencia, que immediate sequitur in linea descendendo, et consideratur litera scripta super differenciam in superiori margine, que litera scribitur extra circa differenciam. Et eandem literam, si fuerit 'a', significat, quod pars proporcionalis <sit> addenda equacioni accepte. Si autem fuerit 's', significat, quod <pars> proporcionalis sit subtrahenda ab equacione accepta. Si autem signum, cum quo intratur, sit in inferiori margine, tunc gradus signi accipitur versus dextram ascendendo et ex directo eiusdem gradus et hoc versus sinistram accipitur equacio sive minuta proporcionalia sive diversitas diametri, que scribitur extra. Deinde accipitur differencia, que immediate sequitur in linea ascendendo, et consideratur litera scripta sub differencia in*

3 The equivalence of both models has been proved by Ptolemy in *Almagest* III,3.

4 See Chapter 12, Table TRW-13.

5 See the Poulle, *Les Tables Alphonsines*, pp. 145–47, Table 26. According to Ratdolt's edition of 1483 (ff. E2v–E3v) should be $\varepsilon(6°) = 0;12,53°$, $\varepsilon(60°) = 1;50,44°$, and $\varepsilon(126°) = 1;47,40°$.

6 See, e.g., MS Prague, NK, X B 3, ff. 95v–96r.

*inferiori margine, que litera scripta extra circa differenciam est. Et hoc si fuerit 'a', significat, quod pars proporcionalis sit addenda equacioni accepte, ut dictum est. Que equacio addatur vel subtrahatur a medio motu planete, secundum quod litere circa equacioni posite ostendunt. Si autem in Tabulis equacionum ponitur paragraphus rubeus vel niger, ibi in differencia tam in superiori quam in inferiori margine ponuntur due litere, videlicet 'a', 's' vel 's', 'a' sic, quod una ponitur super aliam. Quarum superior litera in utroque margine respicit equaciones, que sunt supra*[7] *paragrapho, et inferior similiter in utroque margine respicit equaciones, que sunt sub paragrapho.* – 'To propose a general way how to enter the tables of equations. Enter the tables of equations with the common signs, of which six are placed in the upper margin, 0, 1, 2, 3, 4, 5, and their degrees are placed descending in the first column towards the left, and the other six are placed in the lower margin, namely 6, 7, 8, 9, 10, 11 backward, and their degrees are placed ascending in the last column towards the right. If, therefore, the sign, with which it is entered, is in the upper margin, then the degrees of the sign are taken at left going down, and directly from the same degree is taken the equation, which is written aside. Then the difference is taken, which immediately follows in the column descending, and the letter written in the upper margin above the difference is considered, which letter is written beside at the difference. And that letter, if it is 'a', means that the proportional part must be added to the equation. And if it is 's', it means that the proportional part must be subtracted from the accepted equation. But if the sign, with which it is entered, is in the lower margin, then the degree of the sign is taken at the right and directly from the same degree towards the left is taken either the equation or proportional minutes or the variations of the diameter, which is written outside. Then the difference is taken, which immediately follows in the column ascending, and the letter written in the lower margin under the difference is considered, which letter is written beside at the difference. And if this is 'a', it means that a proportional part must be added to the accepted equation, as has been said. That equation should be added to or subtracted from the mean motion of the planet, according to which letter is placed at the equation. If, however, a red or black paragraph mark is placed in the Tables of equations, there are two letters placed in difference both in the upper and in the lower margin, that is, namely 'a', 's' or 's', 'a' so that one is placed above the other. The upper letter which in each margin refers to the equations that are above the paragraph mark, and the lower letter similarly in each margin refers to the equations that are below the paragraph mark.'[8]

This table of the equations corresponds to motion on a simple eccentre, for which the equation $\varepsilon$ is given by formula (21.2). By least-squares fit to the values in the Ratdolt's edition (which is only insignificantly better than the fit for MS, Toruń, BU, 74/III, f. 258v) we find the best agreement for the value of eccentricity $e \simeq 0.0378 \simeq 2;16,5^{p}$, which according to Eq. (21.4) also agrees (within an error of 0.00001) with the maximum value of the equation $\varepsilon_{max} = \arcsin e = 2;10°$. This differs from the value $e = 2;29,30^{p}$ given in Ptolemy's *Almagest* III,4 and the consequent value $\varepsilon_{max} = 2;23°$ found in his table in *Almagest* III,6.[9] While the mean motions are tabulated in the *Tabule resolute* to tertias of a degree, the values of the equation of the Sun are given in the Alfonsine tables – both the PAT and the

---

7 *supra* Prague, NK, VIII G 24, f. 62r : *sub* Cracow, BJ, 573, f. 229v *(per err.)*
8 Cracow, BJ, 573, f. 229v.
9 See Toomer, *Ptolemy's Almagest*, pp. 155 and 167.

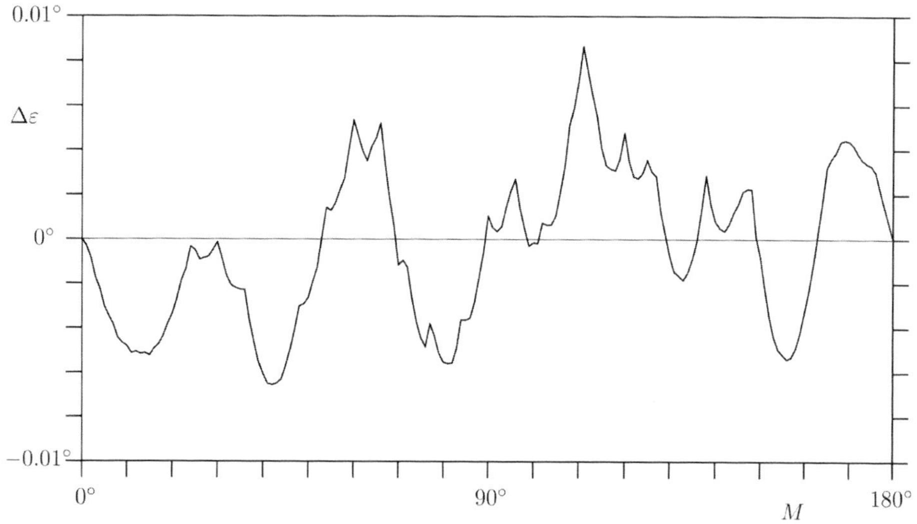


**Figure 34.** Residuals of a least-squares fit of Ratdolt's table of equation of the Sun.

TR – to seconds of degree only (or even to minutes for the other bodies). However, the recalculation reveals that the root mean square error is nearly 13″ and the maximum error reaches over 31″ ≃ 0.0086° – see Fig. 34.[10] For a comparison, the same fit of the Peurbach's table of equations of the Sun found in manuscripts Cracow, BJ, 617 and Munich, BSB, Clm 51[11] gives $e \simeq 2;16,6^{P}$, r.m.s. error 1.6″, and only one outlier with an error of 19″, while the errors at other points are less than 3″.

10 See also Figure 13 in Kremer, 'The Parisian Alfonsine Tables in Manuscript, 1351–1525'.
11 See Chapter 11.

# Equations of the Moon

As shown above, the eccentric model or the equivalent model of motion on an epicycle can relatively well fit the variations of angular velocity of Keplerian motion. However, due to perturbations by the Sun, the lunar orbit deflects appreciably from the Keplerian ellipse. Significant corrections of the model are thus needed to fit the observed motion of the Moon.

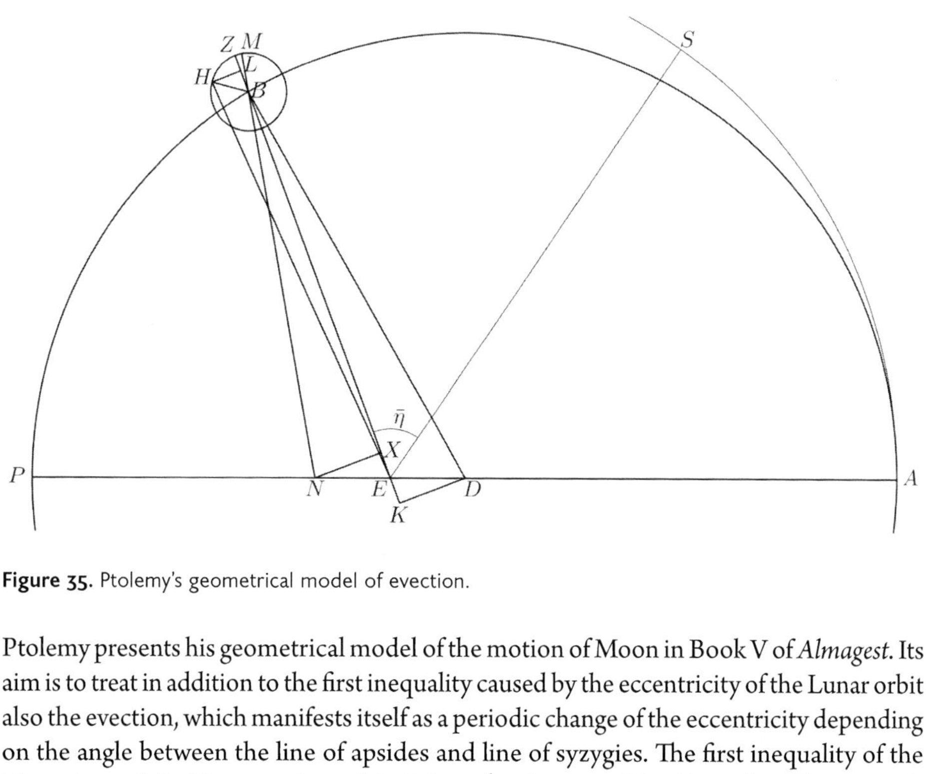


**Figure 35.** Ptolemy's geometrical model of evection.

Ptolemy presents his geometrical model of the motion of Moon in Book V of *Almagest*. Its aim is to treat in addition to the first inequality caused by the eccentricity of the Lunar orbit also the evection, which manifests itself as a periodic change of the eccentricity depending on the angle between the line of apsides and line of syzygies. The first inequality of the Moon is modelled by a motion of the Moon (at the point $H$ in Fig. 35) on the epicycle $MZH$ with a radius $r \equiv BH$ around the mean position $B$ of the Moon. The value of the numeric eccentricity of the Moon is given by the ratio $r / d$, where $d \equiv EB$ is the distance of the centre $B$ of the epicycle from the Earth $E$. The variability of this eccentricity which corresponds to the evection is then explained by motion of the centre $B$ on an eccentre $ABP$ with a radius $R \equiv DB$ and centre $D$, which is at a constant distance $e \equiv ED$ from the Earth. The centre $D$ and hence also the apogee $A$ of the eccentre move so that the mean anomaly of the Moon $\angle AEB = 2\bar{\eta}$, where $\bar{\eta} \equiv \angle SEB = \lambda_M - \lambda_S$ is the mean elongation of the Moon from the Sun (which is in the direction of the point $S$ from the Earth). We thus find from the law of cosines[1] in the triangle $DEB$ that

$$d(\bar{\eta}) = \sqrt{R^2 - e^2 \sin^2 2\bar{\eta}} + e \cos 2\bar{\eta} . \tag{22.1}$$

The true anomaly of the first inequality is the angle $a \equiv \angle ZBH$ between the true apogee $Z$ of the epicycle and the Moon $H$. From a comparison with observations Ptolemy concluded that this $a$ is not a linear function of time, but it has a phase shift (*prosneusis*) with respect to a uniformly increasing mean anomaly $\bar{a} = \angle MBH$ measured from a mean apogee $M$, which is a projection on the epicycle of the centre $B$ from a point $N$ twice as distant from the centre $D$ of the eccentre than the Earth $E$. To get the true anomaly $a$, the mean anomaly $\bar{a}$ has to be corrected for the so-called equation of centre (*equacio centri*), i.e. the angle $\theta = \angle ZBM = \angle NBE$ which is found from the right triangle $BNX$, e.g., in the form

$$\theta(\bar{\eta}) = \arctan \frac{e \sin 2\bar{\eta}}{\sqrt{R^2 - e^2 \sin^2 2\bar{\eta}} + 2e \cos 2\bar{\eta}} . \tag{22.2}$$

The final equation $\varepsilon \equiv \angle BEH$ for which the mean longitude of the Moon has to be corrected, can be similarly calculated from the right triangle $EHL$ in the form

$$\varepsilon(\bar{\eta}, a) = \arctan \frac{r \sin a}{d(\bar{\eta}) + r \cos a} , \tag{22.3}$$

where the 'true anomaly' (*argumentum verum*)

$$a = \bar{a} + \theta(\bar{\eta}) . \tag{22.4}$$

The computations according this geometrical model are both lengthy and tedious. Ptolemy thus presented in *Almagest* V,8 tables to facilitate this task. However, as is obvious from the formula (22.3), the equation $\varepsilon$ is a function of two independent variables $\bar{\eta}$ and $\bar{a}$, which is not convenient for a representation in single-argument tables. It was thus approximated by formula

$$\varepsilon(\bar{\eta}, a) \simeq \varepsilon_0(a) + \Delta(a)\, \mu(\bar{\eta}) , \tag{22.5}$$

where the equation of anomaly (*equacio argumenti*)

$$\varepsilon_0(a) = \arctan \frac{r \sin a}{R + e + r \cos a} \tag{22.6}$$

is the equation $\varepsilon\,(0, a)$ at the apogee of the eccentre, in which, according to Eq. (22.1), $d = R + e$ is the largest and hence the amplitude of $\varepsilon$ the smallest.[2] The so called 'variation of diameter' (*diversitas diametri*)

1 Ptolemy solved this problem using the right triangles $BDK$ and $EDK$ where $EK = e \cos 2\bar{\eta}$ and $DK = e \sin 2\bar{\eta}$.
2 This equation is similar to Eq. (21.3), if the unit radius is replaced by the distance $R + e$ of the lunar apogee and the radius $r$ of the epicycle is substituted instead of the eccentricity $e$ of the solar orbit.

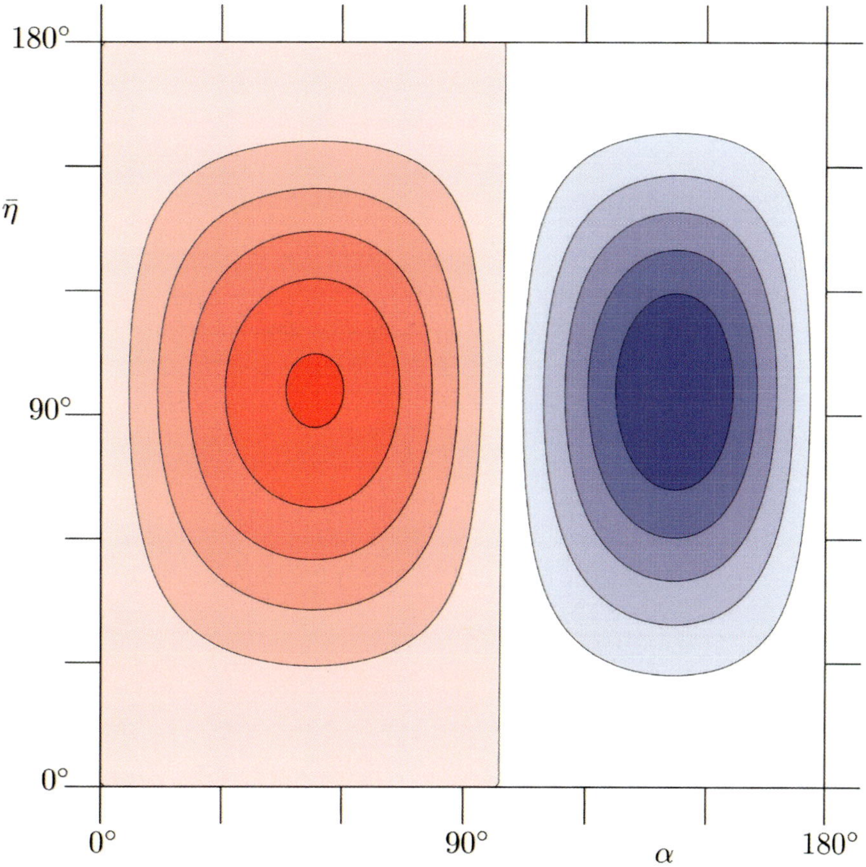


**Figure 36.** Residual errors $\varepsilon - \varepsilon_0 - \Delta\,.\mu$ of Ptolemy's approximation (22.5) to the model (22.3) of the equation of the Moon.

$$\Delta(a) = \varepsilon(90°,a) - \varepsilon_0(a) \tag{22.7}$$

is the difference of the equation $\varepsilon$ between its maximum at the perigee and its minimum at the apogee of the eccentre. The 'proportional minutes' (*minuta proporcionalia*)

$$\mu(\bar{\eta}) = \frac{\varepsilon_{\max}(\bar{\eta}) - \varepsilon_{\max}(0)}{\varepsilon_{\max}(90°) - \varepsilon_{\max}(0)}, \tag{22.8}$$

where

$$\varepsilon_{\max}(\bar{\eta}) = \arcsin\frac{r}{d(\bar{\eta})} \tag{22.9}$$

is the maximum value of the equation for a given elongation $\bar{\eta}$, which occures at $a = \arccos(-r/d)$ when the line *EH* is tangent to the epicycle, so that the triangle *EBH* has a right angle at *H*. It can be verified by computation[3] that the approximation (22.5) has a maximum negative error of about one arcminute with respect to Eq. (22.3) at $a \simeq 142°$, $\bar{\eta} \simeq 96°$ and maximum positive error $\simeq 0.85''$ at $a \simeq 53°$, $\bar{\eta} \simeq 96°$ – see Fig. 36, where the contours of the residual error are drawn with a step of 10″. The approximation (22.5) was thus satisfactory given the precision of naked-eye observations.

## Table of the True Motion of the Sun and Moon in One Hour (Table TRW-14)

The table of the true motion of the Sun and Moon in one hour,[4] which can be found in MS Toruń, BU, 74/III on f. 259r, gives angular velocities of the luminaries needed for a correction of true syzygies as described in Rule 7 of the canons *Mirabilis*. For the Sun, the time derivative of Eq. (21.1) gives the instantaneous angular velocity of a motion on the eccentre

$$\dot{\upsilon} = \frac{1 + e \cos M}{1 + 2e \cos M + e^2} \dot{M} \tag{22.10}$$

and the same result can also be found judging the projection of composition of the velocities of the centre *E* of the epicycle and of the point *S* on its circumference in Fig. 33. It is obvious that to compute the velocities (true hourly motions) the ancient and medieval astronomers used finite differences, as it is described in Ptolemy's *Almagest* VI,4,[5] instead of the derivatives.

The equation of the Moon, which is given by formula (22.3), and hence also its time derivative are more complicated. However, as stated in Rule 7 of the canons *Mirabilis*, at the syzygies 'there is no centre of the Moon', which means $\bar{\eta} = 0$ and hence Eq. (22.3) simplifies to Eq. (22.6), which has the same form as Eq. (21.2) valid for the Sun. This is why the second sentence of Rule 7 states that there is no difference in equating the Moon and the Sun at mean syzygies.

**Table 2. Minimum and maximum hourly motions of luminaries at syzygies in different sets of tables.**

| source | Sun | | Moon | |
|---|---|---|---|---|
| Toruń, BU,74/III | 0;2,23° | 0;2,33° | 0;30,18° | 0;36,4° |
| Tables of 1322 | 0;2,23° | 0;2,32° | 0;30,18° | 0;36,4° |
| Toledan Tables | 0;2,22° | 0;2,34° | 0;30,12° | 0;35,40° |
| Prague, NK, X B 3 | 0;2,23° | 0;2,34° | 0;29,37° | 0;36,53° |
| JoG, FV | 0;2,23° | 0;2,34° | 0;29,38° | 0;36,53° |
| JoG, TM | 0;2,22,30° | 0;2,33,46° | 0;29,37,11° | 0;36,53,21° |
| JoG, TB | 0;2,22° | 0;2,34° | 0;29,37° | 0;36,53° |
| Prague, NK, XIII C 17 | 0;2,22,30° | 0;2,33,40° | 0;29,37,13° | 0;36,58,54° |
| Ratdolt 1483 | 0;2,23° | 0;2,34° | 0;30,21° | 0;36,25° |

3 See Hadrava and Hadravová, 'Tabulka ekvací Měsíce'.
4 See Chapter 12, Table TRW-14.
5 See Toomer, *Ptolemy's Almagest*, p. 282.

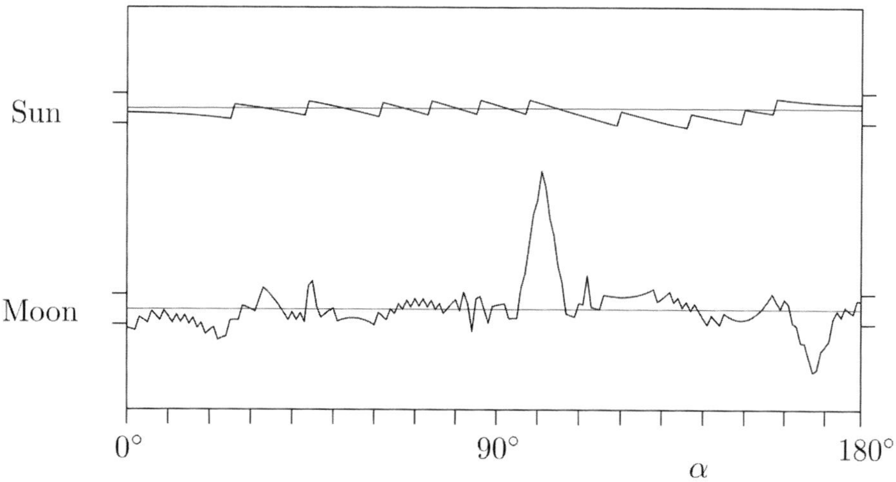


**Figure 37.** Residual errors of hourly motions of luminaries at syzygies.

Tables of hourly motions of the luminaries similar to Table TRW-14 can be found in many manuscripts, some of which also contain the *Tabule resolute*, but the tables are either included in other sets of tables or they were copied as independent works. It is interesting to note, that tables of the motion of luminaries in an hour are also present in some manuscripts with sets of the Parisian Alfonsine Tables, despite the fact that the hour is not consistent with the otherwise systematically used sexagesimal fractions of a day. The numerical values in these tables split into several different versions. We give some examples in Tab. 2, where they are characterised by the ranges of values of the Sun and Moon hourly motion. The same or within the variant readings very similar values which appear in MS Toruń, BU, 74/III, f. 259r,[6] are found in the Tables of 1322 by John of Lignères,[7] where they are joined into a single table with equations of the Sun and Moon. The same range as in Table TRW-14 but with slight differences in some intermediate values are found already in al-Battānī's tables,[8] which give the velocities of luminaria with a step of 6°. Toledan Tables use slightly different values borrowed from al-Khwārizmī.[9] Substantially different values, especially for the Moon, are given in the version represented in Tab. 2 by MS Prague, NK, X B 3, f. 117r–v, which is the most frequent in the manuscripts with the *Tabule resolute*.[10] Very

6 See also Chapter 17, Overview of the tables, line TR-14.

7 See Chabás and Saby, *The tables of 1322 by John of Lignères*, pp. 108-13. The table can be found, e.g., in MS Cracow, BJ, 551 or Prague, NK, X A 23, ff. 30v–31v.

8 See Nallino, *Al-Battānī sive Albatenii Opus astronomicum II*, p. 88, and Pedersen, *The Toledan Tables*, p. 1412. The same table is also in MS Oxford, Bodl., Canon. misc. 27, f. 88r.

9 See Pedersen, *The Toledan Tables*, p. 1416.

10 This table is also contained, e.g., in MSS Prague, NK, VIII G 24, ff. 59v–60r; Oxford, Bodl., Canon. misc. 499, ff. 51v and 164v–165r (two copies); Cracow, BJ, 553, ff. 108v–109r; BJ, 600, f. 105v; BJ, 605, ff. 65v–66r; BJ, 610, ff. 80r–v and 137v–138r; BJ, 611, ff. 49r–v; BJ, 614, ff. 58r–58v; BJ, 617, f. 1r; BJ, 1846, pp. 90–91; BJ, 1852, pp. 128–129; BJ, 2252, pp. 121–122; BJ, 2478, ff. 50v–51r; Leipzig. UB, 1481, ff. 50v–51r; Lisbon, ANTT, ML, 2115, ff. 84v–85r; Munich, BSB, Clm 214, f. 37v; Vatican, BAV, Pal. lat. 1375, f. 40r; Vienna, ÖNB, Cod. 5227, ff. 91r–v; Cod. 5240, ff. 50v–51r; Cod. 5275, ff. 135v–136r; and the editions by Lacher (1511), f. I4r and Schöner (1536), f. K4r.

similar values are also used by John of Gmunden (JoG) in the first version (FV) of his tables.[11] In his *Tabule maiores* (TM), he gives the hourly motion with a precision to tertias,[12] while in the *Tabule breviores* (TB), he returns to the precision in seconds with slightly different values and steps of 6°.[13] A table of hourly motions with a step of 6° and precision to tertias was computed already by John of Genoa,[14] represented in Tab. 2 by MS Prague, NK, XIII C 17.[15] A table of hourly motions is also included in the Ratdolt's edition of the Parisian Alfonsine Tables,[16] and it is followed with tables of the motion of the Moon in a minute of a day and another one with a step of 3 degrees for the motion of the Sun in a minute of a day. The table of hourly motions in the prints of the Parisian Alfonsine Tables by Johannes Lucilius Santritter and Petrus Liechtenstein has the same range of values as does the MS Toruń, BU, 74/III, but a slightly different progress of the values.[17]

Using a least-square fit, we can find the best coincidence of the hourly motions in MS Toruń, BU, 74/III for the eccentricities of the Sun and Moon 1;54$^p$ and 5;11$^p$, respectively. The residuals are drawn in Fig. 37, where the marks on the ordinate denote an error of one second of degree. We can see that for the Sun the errors only rarely exceed this resolution of the tabulated values (the largest error is −1.3″ for $a = 137°$, r.m.s. error is 0.5″). For the Moon, however, the r.m.s. error is 1.7″ which is mainly due to the outlier around $a = 102°$, where the error reaches 9.2″. Another error of −4.1″ is reached at $a = 168°$. The linear progress in the neighbouring points reveals, that the values in Table TRW-14 were obtained by a linear interpolation in a table with a step of six degrees as mentioned above, where the values in these points were erroneous. Compared to the best fit, we find that Table TRW-14 gives the hourly motions of the Moon 33′27″ instead of 33′17″ at $a = 102°$, and 35′54″ instead of 35′58″ at $a = 168°$. These errors of Table TRW-14 also appear in the Tables of 1322 and in the *Tabule magne* by John of Lignères.[18] On the contrary, the al-Battānī's tables, and also the table in MS Oxford, Bodl., Canon. misc. 27 as well as both editions by Santritter and Liechtenstein, have correct values at these points.

11 Beatriz Porres de Mateo in her thesis 'Les tables astronomiques de Jean de Gmunden' distinguishes three versions of tables by John of Gmunden: the first version, the *Tabule maiores*, and the *Tabule breviores*. These versions differ significantly in the tables of hourly motion (see pp. 531–37). An example of the table in FV is, e.g., in MS Vatican, BAV, Pal. lat. 1411, ff. 48r–49r.

12 See Porres de Mateo, 'Les tables astronomiques de Jean de Gmunden', pp. 329–33. The third version of the table of hourly motions in MS Cracow, BJ, 610, f. 272v, equally as the table in MS Prague, NK, X B 3, f. 119v, has the same region of values (given in tertias), but with a step of 6°.

13 See, e.g., MS Prague, NK, XIV F 10, f. 115r.

14 See Miolo, 'Retracing the Tradition of John of Genoa's Opus astronomicum through Extant Manuscripts', 351–58; José Chabás and Bernard R. Goldstein, 'Computing Solar and Lunar Velocities and Radii: The Table of John of Genoa and his Predecessors', in *Alfonsine Astronomy: Computational Practices*, ed. by Matthieu Husson, José Chabás, and Richard L. Kremer (Turnhout: Brepols, forthcoming).

15 See, e.g., MS Prague, NK, XIII C 17, f. 169r. The same region for the Moon, but a bit different region 0;2,22,33° – 0;2,33,46° for the Sun is in MS Prague, KMK, M.CLXII, f. 47r.

16 Ratdolt 1483, ff. G6r–G7r; this and the following tables are not contained in the edition by Poulle, *Les Tables Alphonsines*.

17 Santritter 1492, ff. H5v–H6r, Liechtenstein 1518/21, ff. N5v–N6r. Both these prints have the same typographic error at the first three values of the Sun motion, where is printed 0;2,33° instead of 0;2,23°.

18 Examples of these tables can be found in manuscripts Bernkastel-Kues, BNH, Cus 212, ff. 83v–84r, or Vatican, BAV, Pal. lat. 1412, ff. 110r–111r. These copies have also some scribal errors and they differ from TRW-14 for one or two seconds also in some other points.

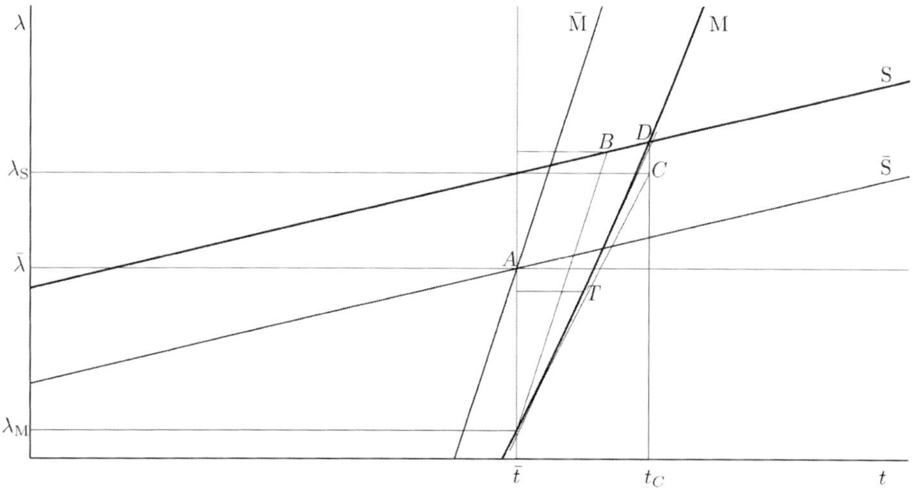


**Figure 38.** Computation of true syzygies.

## Computation of True Syzygies

Rule 7 of canons *Mirabilis in altis Dominus* or Rule 31 of canons *Pro (introduccione) generali*, as well as Rule 22 of canons *Tempus est mensura motus*,[19] describe how to find the true syzygies. While the mean syzygies can be calculated directly using the radices and rates of mean motions of the Sun and Moon, search for the true syzygies requires in fact a solution of non-linear equations for the time or position of the conjunction or opposition of the Sun and Moon because of non-linearity of their true motions. The solution is thus iterative, using the mean syzygies as the first approximation. When the time $\bar{t}$ and ecliptical longitude $\bar{\lambda}$ of the mean syzygy are found, true longitudes $\lambda_S$ of the Sun and $\lambda_M$ of the Moon at this time $\bar{t}$ are calculated – see Fig. 38, where the thick lines denoted by *S* and *M* show the true motion of the Sun and Moon, medium lines $\overline{S}$ and $\overline{M}$ are trajectories of their mean motions, and their intersection *A* is the event of the mean syzygy. To pass over the elongation $\Delta\lambda = \lambda_S - \lambda_M$ between the true longitudes of the luminaries at the time of mean syzygy requires a time interval, which can be estimated first again from the mean velocities of both bodies. The mean angular velocity of the Moon is approximately 13 times greater than that of the Sun and hence their relative velocity is 12 times higher than the velocity of the Sun. With these velocities the luminaries would meet at event *B*, distant by $\Delta\lambda$ / 12 from $\lambda_S$ and by 13 $\Delta\lambda$ / 12 from $\lambda_M$. The velocities, however, are not constant. They are given in Table TRW-14 as functions of the anomalies of each body. We can see from this table that especially the angular velocity of the Moon varies significantly. That is why the

19 See Poulle, *Les Tables Alphonsines*, p. 80 (Ratdolt 1483, ff. A8r–B1v). Various methods of computation of true syzygies are described, e.g., in Richard L. Kremer, 'Cracking the *Tabule permanentes* of John of Murs and Firmin of Beauval with Exploratory Data Analysis', in *Editing and Analysing Numerical Tables: Towards a Digital Information System for the History of Astral Sciences*, ed. by Matthieu Husson, Clemency Montelle and Benno van Dalen (Turnhout: Brepols, 2021), pp. 363–422 (pp. 371–77); Kremer, 'John of Murs, Wenzel Faber and the Computation of True Syzygy'.

rules recommend finding the velocity of the Moon at the middle of the estimated interval between the mean and true syzygy, where the tangent fits the trajectory of the Moon with a higher order of acuracy (see point *T* at Fig. 38). On the other hand, the smaller and less variable velocity of the Sun is taken simply at $\lambda_S$. Using the difference of these velocities and the elongation of the luminaries, the position of point *C* is then calculated, which gives the next estimate of time $t_C$ of the true syzygy. There could also be calculated in a linear approximation longitude of the event *D* of this new estimate of the true syzygy. However, because of the nonlinearities, there is no guarantee that true positions of both luminaries will coincide at the time $t_C$. If a high degree of accuracy is needed (e.g. for calculation of eclipses), the canon *Mirabilis in altis Dominus* recommends repeating one more iteration, starting with the true positions of the luminaries at the time $t_C$.

The canons *Mirabilis in altis Dominus* suggest in the first paragraph of Rule 7 one more 'correction' to the velocity of the Moon, namely to add or subtract from it as many seconds per hour as there are degrees (minus one) in the true distance between the luminaries at the mean syzygy. This suggestion is not included in later canons to the TR, but a quite similar text can found in Rule 22 of the canons *Tempus est mensura motus* to the PAT[20] or in Rule 31 of canons *Priores astrologi*,[21] and already in al-Battānī's canons to his tables.[22] A simple special table *Tabula equacionis diversi motus Lune in una hora*, which gives the correction of the velocity of the Moon as a function of its elongation from the Sun, is found in al-Battānī's tables,[23] the Toledan Tables[24] and the Tables of 1322 by John of Lignères.[25]

The meaning of this correction is questionable. C. A. Nallino repeated a Schiaparelli's explanation that the aim of this correction was to take into account the evection, which is exactly zero only in syzygies, but not in the whole time between the mean and true syzygies. However, as is obvious from Fig. 35 or from Eqs. (22.3) or (22.5) and their derivatives with respect to $\bar{\eta}$, the increase of $d(\bar{\eta})$ and hence also the evection is not linear but rather quadratic around the syzygies, so that its influence on the velocity of the Moon cannot be taken as a mean between the syzygies and quarters. This was obvious already to Ptolemy and is why the proportional minute $\mu(\bar{\eta})$ according to Eq. (22.8) was introduced into the tables of equations of the Moon.

The assumed proportionality of the suggested correction to the distance of luminaries at the mean syzygy indicates that this could be an attempt for a second-order correction for the change of the lunar velocity in the time between the mean and true syzygy. Really, within half of one lunation, that is approximately 330.5 hours (which is 27 days and 13 hours divided by two), during which the lunar anomaly increases from 0 to 180°, the speed of the Moon increases from 0;30,18° to 0;36,4° per hour, i.e., by a total of 0;5,46° or 346″ per hour (according to Table TRW-14). The one degree of distance between the Sun and Moon is

20 See Poulle, *Les Tables Alphonsines*, p. 82 (Ratdolt 1483, ff. A8v–B1r).
21 See Saby, 'Les canons de Jean de Lignères', pp. 221–22.
22 See Nallino, *Al-Battānī sive Albatenii Opus astronomicum I*, p. 94, line 9–11, and p. 150, appendix H.
23 See Nallino, *Al-Battānī sive Albatenii Opus astronomicum II*, p. 88.
24 See Pedersen, *The Toledan Tables*, pp. 1413–14; Toomer, 'A survey of the Toledan Tables', *Osiris* 15 (1968), 85.
25 See Chabás and Saby, *The tables of 1322 by John of Lignères*, pp. 106–07. The only difference is that al-Battānī does not subtract 1, so that the numbers 1 to 7, which cover the whole possible range of the luminaries elongation at the mean coniunction, are identical in the columns of degrees of the elongation and seconds per hour of the lunar velocity.

travelled by the Moon in about 2.04 hours, and in that time its velocity increases in the mean by 2.14″ per hour, so that in middle of that time, the velocity at the mean syzygy should increase by 1.07″ per hour, in agreement with the texts of the canons. However, this acceleration is not uniform; it reaches a $\pi/2$ times greater value when the anomaly is around 90° and reaches zero at the apsides of the lunar orbit. It is positive in the first half of the epicyclic motion (from the apogee to the perigee) and negative in the second half (from the perigee to the apogee), not around the perigee and apogee (where the velocity has the minimum and maximum while its time derivative is zero) as claimed in the canons. In fact, the abrupt change of sign of a non-zero value of the correction between the upper and lower part of the epicycle, as is described in the canons[26] cannot be consistent with any mathematical reasoning. It thus seems likely that, if this correction of the lunar velocity had a mathematical origin, its description was distorted in the preserved testimonies. In any case, the value of the order of seconds per hour has a negligible influence on the resulting true syzygies (especially in comparison with numerical errors in the calculation), so that the omission of this point from the later canons is surely justified.

## Table of Equations of the Moon (Table TRW-15)

The table of equations of the Moon in the *Tabule resolute* in MS Toruń, BU, 74/III, ff. 259v–260v,[27] is, in its layout and most of the values, practically identical with that table in the PAT.[28] In the first column C after the degrees of anomaly (which are, however, counted in the zodiacal signs), it gives the equation of centre according to Eq. (22.2), then the proportional minutes according to Eq. (22.8) in column M, followed by the variation of diameter according to Eq. (22.7) in column D, and finally the equation of anomaly according to Eq. (22.6) in column A. As in the Table of equation of the Sun, the later versions of the *Tabule resolute* add also to the Table of equations of the Moon columns with differences between values of all quantities in subsequent lines.[29] The same order of columns is found also in the Toledan Tables, where, however, there is one more column for the latitude of the Moon.[30] The first three columns (C, M, and D in Table TRW-15) have in Toledan Tables similar values as in the *Tabule resolute* and the PAT, but the equation of anomaly (the column A) differs significantly. The Toledan Tables agree in their values with al-Battānī's tables.[31] Going back in the tradition of the tables, we can find that in *Almagest* V,8, the Table of lunar anomaly gives first the column C with the equation of centre, then the column A with the equation of anomaly (and values followed by al-Battānī and the Toledan Tables), the column D with the variation of diameter (increment in epicyclic equation), then the column M with the proportional minutes (the 'sixtieths', which are given with a higher precision also to seconds – a practice abandoned by the followers),

26 Note that in agreement with the al-Battānī's appendix H, the manuscripts **O**, **L**, and **R** with the canons *Mirabilis in altis Dominus* specify this boundary to be at 95° from the apogee.
27 See Chapter 12, Table TRW-15.
28 See Poulle, *Les Tables Alphonsines*, pp. 148–53, Table 27 (Ratdolt 1483, ff. E4r–E6v).
29 See, e.g., MS Prague, NK, X B 3, ff. 96v–99r.
30 See Pedersen, *The Toledan Tables*, pp. 1253–58, or, e.g., MS Prague, NK, XIV A 18, ff. 4r–7v.
31 See Nallino, *Al-Battānī sive Albatenii Opus astronomicum II*, pp. 78–83.

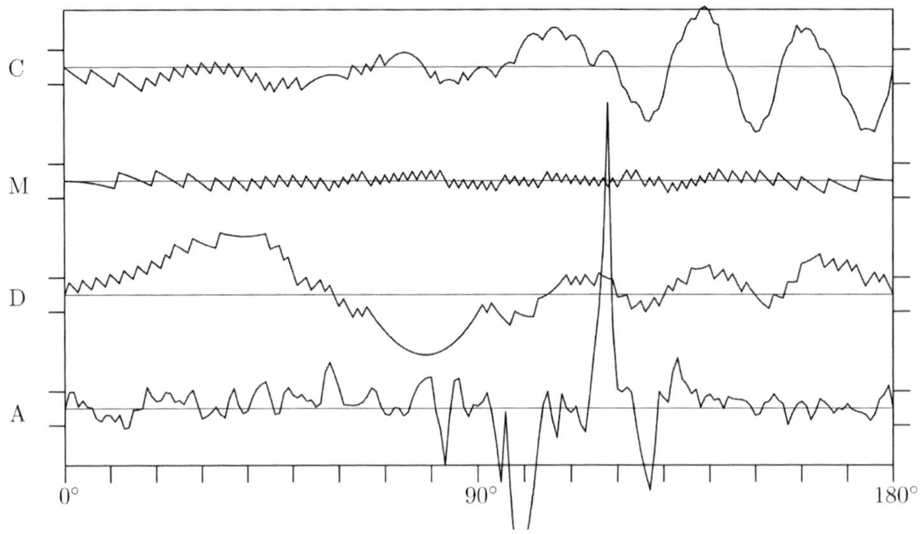


**Figure 39.** Residuals of least-squares fit of Table TRW-15 of equations of the Moon.

and finally the latitude of the Moon. The anomaly is given in the *Almagest* with a step of 6° up to 90°, and then with a step of 3° up to 180°.[32]

By a least-square fit to the equation of centre $\theta$ in the column C of the Table TRW-15, we can find the ratio $e/R \simeq 0.20788 \pm 0.00004$ in Eq. (22.2). This is in a relatively good agreement with the value 0.20765, which follows from the Ptolemy's *Almagest* V,4.[33] The residuals of the fit have r.m.s. error about 1.5′, but in the second quadrant they reach nearly 4′, as can be seen in Fig. 39, where the marks on the ordinate at the letter C display an error of ±1′.

According to Eq. (22.8), the proportional minutes in column M of the Table TRW-15 are dependent on both parameters – the eccentricity $e$ and the radius $r$ of the epicycle. However, the dependence on $r$ is quite weak.[34] Regarding the poor precision of the proportional minutes (given by a single sexagesimal digit), it is not surprising that a least square fit cannot robustly find the parameters. It gives the ratio $e/R \simeq 0.230 \pm 0.015$, while $r$ converges to zero. In any case, the r.m.s. errors of $\mu$ are $0.3^{p}$ maximum error is $0.75^{p}$, i.e. below one minute for both these values of the parameters (see the line M in Fig. 39, where $\pm 1^{p}$ is marked on the ordinate) as well as for their proper Ptolemaic values.[35]

The variation of diameter in column D also depends on both parameters $e$ and $r$. Its least-square fit leads to values $e/R \simeq 0.228 \pm 0.003$, $r/R \simeq 0.096 \pm 0.002$ (the parameters

32 See Toomer, *Ptolemy's Almagest*, p. 238.

33 See Toomer, *Ptolemy's Almagest*, p. 226, where $\frac{e}{R} = \frac{ED}{DA} = \frac{10;19^{p}}{49;41^{p}}$ in his Fig. 5.3 as well as in our Fig. 35.

34 If a Taylor expansion of arcsin in Eq. (22.9) is substituted into Eq. (22.8), $r$ can be cancelled in the linear terms and it will remain only in the cubic terms.

35 Let us note that the least-square fit to the values of proportional minutes in the Almagest yields $e/R \simeq 0.181 \pm 0.021$, $r/R \simeq 0.41 \pm 0.11$ (the parameters being anticorrelated), residual r.m.s. error $0.13^{p}$, maximum error is $0.31^{p}$.

being anticorrelated), residual r.m.s. error 1.8′, maximum error is 3.6′; see line D in Fig. 39, where the marks denote an error of ±1′.

According to Eq. (22.6), the equation of anomaly in column A also depends on both $e$ and $r$, but only through the ratio $e' \equiv r/(R + e)$. It is thus a one-parametric function equivalent to Eq. (21.3) and (21.2) if we substitute there $e'$ for the $e$. In the least-square fit of column A from Table TRW-15 we exclude two obvious outliers, namely the values at 95° = $3^s$ 5° and 118° = $3^s$ 28°. Instead of 4;55,0°, the former should read 4;56,0° as it is, e.g., in the PAT or later in the *Tabule resolute*.[36] A note in the margin of MS Toruń, BU, 74/III, f. 260r at this value indicates that it is a maximum value of the equation of anomaly; yet it is smaller than the surrounding values. An attempt to correct this scribal error is done by an interlinear gloss '55' above the zero in the column of seconds, which would make the value equal to the previous one for 94°. The latter value should read 4;31,34° according to the PAT instead of 4;31,44°. In MS Prague, NK, X B 3, the opposite error of 10″ is found, i.e., 4;31,24°, but, e.g., MS Prague, NK, VIII G 24, f. 21r, MS Mikulov, RMM, Mik 6331, ff. 65r and 165r, and others contain the correct value 34″. The results of the least-square fit give $e' = 0.085993 \pm 0.000001$, the residual r.m.s. 2.0″. In addition to the above mentioned outliers due to scribal errors, there is a maximum error exceeding 10″ at 99° – see line A in Fig. 39, drawn for 4;55,55° at 95° and the mark on the ordinate now denoting an error of 1″. Let us note that the same least-square fit to the equation of anomaly in Toledan Tables[37] results in $e' = 0.087493 \pm 0.000002$ (mean residual 3.4″ and maximum 13.4″), which is in a good agreement with the value of 0.0875, which follows from the parameters of lunar orbit according to the *Almagest*.[38]

## Double-argument Tables of the Equations of the Moon

As mentioned in Chapter 2.1.2, some manuscripts of Prague origin contain extended Tables of equations of the Moon attributed to Nicolaus Mülhus. These tables give the equation $\varepsilon(\bar{\eta}, a)$ calculated according Eq. (22.5) as a function of two arguments $a$ and $\mu$. Subtables of $\varepsilon$ and their differences rounded to degrees and minutes for the true argument $a$ running from 1° to 180° with a step of 1° are given separately for each integer value of the proportional minutes $\mu$ from 0 to 60. Additional small tables give the equation of centre $\theta(\bar{\eta})$ computed according to Eq. (22.2). The preserved copies slightly differ in their layout: common (zodiacal) signs are used in MS Prague, NK, adlig. 14 B 10, ff. 1r–31v and Mikulov, RM, Mik 6361, ff. 99r–129v, while physical signs are used in the other copies (Vatican, BAV, Pal. lat. 1374, ff. 87r–102r, Prague, NK, X B 3, ff. 42v–72v, and Cracow, BJ, 610, ff. 47v–77v, but the numbers of zodiacal signs are additionally written also in the latter). Two subtables consisting of three columns of sixty rows are written on one page in MS Vatican, BAV, Pal. lat. 1374, while the other manuscripts always present only one subtable per page with six columns of thirty rows.

A short canon on use of the table is written in MS Prague, NK, adlig. 14 B 10, f. 32r:

36 See Poulle, *Les Tables Alphonsines*, p. 151 (Ratdolt 1483, f. E5v), or MS Prague, NK, X B 3, f. 98r.

37 See Pedersen, *The Toledan Tables*, pp. 1253–58.

38 See Toomer, *Ptolemy's Almagest*, p. 226, where $e' = \frac{r}{R+e} = \frac{G\Theta}{EA} = \frac{5;15^P}{60^P}$ in his Fig. 5.3 or $BH/EA$ in our Fig. 35.

*Ad inveniendum breviter et leviter motum Lune verum per has tabulas, habeas primo medium motum Lune et centrum et argumentum medium, que invenies per Tabulas mediorum motuum Alfoncii aut alias. Hiis habitis intra primo cum centro medio in tabulam parvam subscriptam et quere equacionem centri cum duplici introitu, si oporteat, quam equacionem centri adde ad argumentum medium, si centrum est minus 6 signis, vel subtrahe, si est maius, et proveniet argumentum verum, cum quo sic examinato intra in magnam tabulam parve subscriptam et quere equacionem argumenti Lune similiter cum duplici introitu, si oporteat. Quam equacionem argumenti adde medio motui Lune, si argumentum equatum fuerit maius sex signis, vel minue a medio motu Lune, si fuerit minus sex signis, et quod provenerit post addicionem vel subtraccionem, erit verus motus Lune.*

*Tempus vere coniunccionis et opposicionis Solis et Lune invenire. Scias tempus medie coniunccionis, si coniunccionem queris, aut opposicionis, si opposicionem. Scias eciam argumentum Solis et Lune ad idem tempus et intra primo cum argumento Solis ad tabulam suprascriptam et accipe equacionem Solis et scribe super eam 'm', si argumentum Solis fuerit minus 6 signis, vel 'a', si fuerit plus 6 signis. Accipe eciam ibi diversitatem equacionis Solis atque minuta proporcionalia et scribe super ea 'Solis', quibus servatis. Postea cum argumento Lune intra eandem tabulam et accipe equacionem Lune et scribe super eam 'a', si argumentum fuerit minus 6 signis, vel 'm', si maius. Accipe eciam diversitatem Lune et minuta proporcionalia et scribe super ea 'Lune'. Hiis habitis accipe partem proporcionalem de diversitate equacionis Solis secundum proporcionem minutorum proporcionalium Lune ad 60, quam partem proporcionalem minue ab equacione Solis prius servata, et habebis eam equatam. Accipe similiter partem proporcionalem de diversitate equacionis Lune secundum proporcionem minutorum proporcionalium Solis ad 60, quam adde super equacionem Lune prius servatam, et habebis eam equatam. Postea vide, si super ambas equaciones, scilicet Solis at Lune, scriptum sit 'a', et tunc adde eas simul iunctas tempori medie coniunccionis, aut subtrahe, si super utramque scriptum fuerit 'm'. Si autem super unam scriptum fuerit 'a' et super alteram 'm', tunc minue minorem de maiore et residuum adde cum tempore medie coniunccionis vel opposicionis secundum quod super maiorem scriptum fuerit 'a'; vel si enim fuerit scriptum 'a', tunc adde, si autem 'm', tunc subtrahe, et habebis veram coniunccionem vel opposicionem Solis et Lune et hoc diebus non equatis.*

– 'To find quickly and easily the true motion of the Moon using these tables, you must first have the mean motion of the Moon, the centre, and the mean anomaly, which you will find using the Tables of the mean motions of Alfonso or others. Having that, enter first with the mean centre in the small attached table and search for the equation of the centre with a double entry, if necessary. Then add the equation of the centre to the mean anomaly, if the centre is less than six signs, or subtract it, if it is greater, and the true anomaly will result. With this, having thus been examined, enter into the large table written with the small one, and search for the equation of the anomaly of the Moon, with a double entry if necessary. Add this equation of the anomaly to the mean motion of the Moon, if the equated anomaly is greater than six signs, or subtract it from the mean motion of the Moon, if it is less than six signs, and what results after the addition or subtraction will be the true motion of the Moon.

To find the time of the true conjunction and opposition of the Sun and Moon. Find out the time of the mean conjunction if you search for the conjunction, or of opposition if for opposition. Find the anomaly of the Sun and the Moon at that time, and enter first with the anomaly of the Sun into the above table; find the equation of the Sun and write 'm' above it if the anomaly of the Sun is less than six signs, or 'a', if it is more than six signs. Take there also the difference of the equation of the Sun and the proportional minutes, write above it the 'Sun' and keep it. Then enter with the anomaly of the Moon into the same table and take the equation of the Moon and write above it 'a' if the anomaly is less than six signs, or 'm' if it is greater. Also find out the difference of the Moon and the proportional minutes and write the 'Moon' above it. Having these, take the proportional part of the difference of the equation of the Sun according to the proportion of the proportional minutes of the Moon to 60, and subtract the proportional part from the equation of the Sun previously kept, and you will have it equated. Similarly take the proportional part of the difference of the equation of the Moon according to the proportion of the proportional minutes of the Sun to 60, and add to the equation of the Moon previously kept, and you will have it equated. Then see if 'a' is written above both equations, that is, of the Sun and the Moon, and then add them together with the time of the mean conjunction, or subtract it if 'm' is written above both. If, however, 'a' is written above one and 'm' above the other, then subtract the smaller from the greater and add the remainder to the time of the mean conjunction or opposition according to whether 'a' is written above the greater, or if 'a' is written, then add, and if 'm', then subtract, and you will have a true conjunction or opposition of the Sun and Moon, and this on days that are not equated.'

A different double-argument tabular representation of the equations of the Moon was chosen by Nicolaus Wodka. In his tables in MS Milan, BA, C 278 b Inf., ff. 33v–39r, we find six double-page tables for individual common signs. Each one gives in thirty lines first the equations of anomaly $\varepsilon_0(a)$, rounded to minutes and computed according to Eq. (22.6) for individual degrees of the true anomaly $a$ and their differences. In thirty columns are then given the products of the variation of diameter $\Delta(a)$, computed according Eq. (22.7), and the proportional minutes $\mu(\bar{\eta})$, computed according to Eq. (22.8), for the mean elongation $\bar{\eta}$ of the Moon from the Sun corresponding to the day of lunation inscribed to the column heading. These products are then to be added to $\varepsilon_0$ to get the resulting equation $\varepsilon(\bar{\eta}, a)$ of the Moon according to Eq. (22.5).

# 23

# Equations of Planets

The retrogradation of planets is an obvious high non-uniformity of the planetary motions. It is due to the annual parallax, i.e., the observation of a planet from the moving Earth, whose motion combines with the intrinsic orbital motion of the planets around the Sun. The Ptolemaic geocentric model of planetary motion thus needs to combine at least two circular motions with periods corresponding to the heliocentric orbits of the planet and Earth. Moreover, the amplitude of the parallax depends on the radial distance of the planet from the centre, which varies in the course of its orbit on Keplerian ellipse. It thus influences the observed ecliptical longitudes of the planets. The variation of radial distance from the centre in Keplerian ellipse can be better fit by the model with the so-called equant than by the simple eccentre or epicycle. This model, which still presumes circular motion of the planet, will be described below, but to justify its advantage with respect to the model of simple eccentre, let us present in advance the Taylor expansion in the eccentricity of its true anomaly

$$v = M - 2e \sin M + 2e^2 \sin M \cos M + o(e^3) \tag{23.1}$$

and the radial distance

$$r = R\left(1 + e \cos M + \tfrac{3}{2} e^2 \sin^2 M\right) + o(e^3)\,. \tag{23.2}$$

For

$$e = -\,e_{\mathrm{K}} \tag{23.3}$$

the model with equant thus agrees with the Keplerian model (21.6) and (21.9) in linear terms of both $v$ and $r$, while the quadratic terms in $e$ are 20% lower for $v$ and 50% higher for $r$.

In the equant model, which Ptolemy proposed for all planets with the exception of Mercury (*Almagest* IX), the planet moves with the point $Z$ on the circumference of epicycle $UVXTZ$ with radius $p$ and centre $S$, which moves on the deferent $ASP$ (see Fig. 40). The mean centre, i.e. the mean anomaly $M = \angle ACS$ of the centre $S$, increases linearly in time with respect to the point of equant $C$, which is placed at distance $DC = OD$, symmetrically on the other side of the centre $D$ of the deferent $ASP$, from the observer $O$ on Earth. The numerical eccentricity is $e = OD/R$, where the radius $R = DA = DS = DP$ of the deferent can be taken as a unit size ($R = 1$, see Fig. 40). As follows from the right triangle $SDL$, the distance $q \equiv CS$ of the centre $S$ of epicycle from the equant is thus

$$q = \sqrt{1 - e^2 \sin^2 M} - e \cos M \tag{23.4}$$

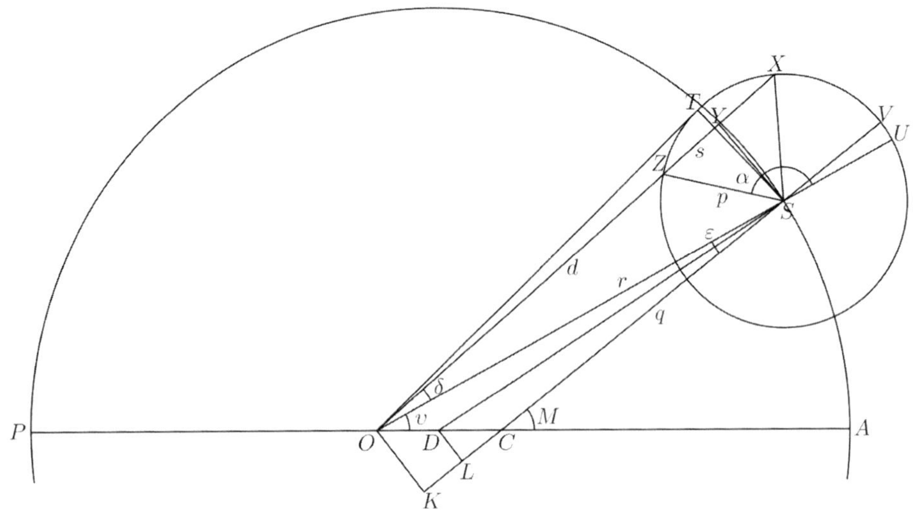


**Figure 40.** Motion with equant.

and according to the right triangle $SOK$, the distance $r \equiv OS$ of the epicycle centre from the Earth $O$

$$r = \sqrt{q^2 + 4e^2 + 4eq\cos M} \tag{23.5}$$

varies between $1 + e$ at the apogee $A$ of the centre for $M = 0$ and $1 - e$ at the perigee $P$ for $M = 180°$. The equation $\varepsilon = \angle CSO$ of the centre $S$ of the epicycle can be easily found from the same triangle as

$$\varepsilon = \arcsin\frac{2e\sin M}{r} = \arctan\frac{2e\sin M}{q + 2e\cos M} \tag{23.6}$$

and the true centre $v \equiv \angle AOS$ is

$$v = M - \varepsilon = \arctan\frac{q\sin M}{2e + q\cos M}\,. \tag{23.7}$$

The equation reaches its maximum

$$\varepsilon_{\max} = \arcsin\frac{2e}{1 + e^2} \tag{23.8}$$

at $M = \arccos\left(-e/\sqrt{1 + e^2}\right)$ when the line $DS$ is perpendicular to $AP$.

The planet at point $Z$ moves uniformly on the epicycle. It means that the mean anomaly of the epicycle $\bar{a} \equiv \angle VSZ$ is a linear function of time. It is counted from the apocentre $V$ of the epicycle with respect to the equant $C$ because of its uniform rotation relative to the static system in which the apogee $A$ is fixed. The true anomaly of the epicycle $a \equiv \angle USZ$ counted from the apogee $U$ of the epicycle (i.e., its point most distant from the Earth $O$) thus increases non-uniformly,

$$a = \bar{a} + \varepsilon \ , \tag{23.9}$$

because $\angle USV = \angle OSC = \varepsilon$.[1] The second anomaly $\delta \equiv \angle SOZ$ of the planet is thus a complicated function of the two mean motions $M$ and $\bar{a}$,

$$\delta = \arctan \frac{p \sin a}{r(M) + p \cos a} \ . \tag{23.10}$$

Similarly to Eq. (22.5) for the lunar evection, Ptolemy introduced an approximation

$$\delta(M,a) = \delta_o(a) + \Lambda_{\pm}(a)\,\mu(v) \tag{23.11}$$

to facilitate the calculation of the second anomaly of planets using tables. The equation of anomaly

$$\delta_o(\alpha) = \arctan \frac{p \sin a}{1 + p \cos a} \tag{23.12}$$

is here the second anomaly $\delta$ at a unit distance $r = 1$ of the epicycle.[2] The variation of diameter (*diversitas diametri*)

$$\Lambda_{\pm} = \left| \arctan \frac{p \sin a}{1 \pm e + p \cos a} - \delta_o(a) \right| \tag{23.13}$$

is the difference of the second anomaly in the unit distance and in the apogee $A$ of the centre if $r > 1$, that is in the farther distance (*longitudo longior*) when the upper sign is +, or in the perigee $P$ if $r < 1$, that is, in the nearer distance (*longitudo propior*) when the lower sign is –. The proportional part is

$$\mu(v) = \frac{\delta_{max}(r(v)) - \delta_{max}(1)}{\delta_{max}(1 \pm \mathrm{e}) - \delta_{max}(1)} \ , \tag{23.14}$$

where

$$\delta_{\max}(r) = \arcsin \frac{p}{r} \tag{23.15}$$

is the maximum second anomaly at the given distance $r$ of the epicycle centre, which occurs when the line of sight to the planet is tangential to its epicycle (i.e., at the vertex $T$ of the right triangle $OST$ in Fig. 40), and in analogy with Eq. (23.4),

$$r(v) = \sqrt{1 - e^2 \sin^2 v} + e \cos v \ . \tag{23.16}$$

Unlike the Parisian Alfonsine Tables, the tables of planetary equations in the *Tabule resolute* also contain a column F for the first station, i.e., the true anomaly $a = \angle USZ$ of the epicycle of planet at such a point $Z$ in which the projection of its motion on

1 For $M > 180°$, $\varepsilon$ should be taken as negative, i.e., $v$ should be increased over $M$ and $\bar{a}$ decreased for absolute value of $\varepsilon$.

2 It occurs when $v = \arccos(e/2)$ and $M = \arccos(-1.5e/\sqrt{1+2e^2})$.

the epicycle (around its apogee, see Fig. 40) compensates the overall motion of the epicycle on the deferent. For higher $a$, the decrease of $\delta$ is faster than the increase of $v$ and the resulting ecliptical longitude $\lambda$ of the planet decreases, i.e., the planet retrogrades, until a second station, where the two motions compensate each other again and the direct motion restarts. The second stations were (incorrectly) assumed to be placed symmetrically with respect to the apogee $U$ of the epicycle, so that in this approximation it was sufficient to give the first stations only. Such a symmetry $a_2 = 360° - a_1$ is explicitly declared, e.g., in the last paragraph of Rule 6 of canons *Mirabilis in altis Dominus*. A detailed mathematical treatment of the retrogradation will be given in Chapter 24, where the subtables with their values will also be analysed. The addition of the first stations to the tables of the planetary equations in the *Tabule resolute* restores the tradition of Toledan Tables[3] which also give the first station together with the planetary equations.

## Table of Equations of Saturn (Table TRW-16)

The Table of equations of Saturn in MS Toruń, BU, 74/III, ff. 261r–262r reproduced here in Chapter 12, Table TRW-16, gives in the first column A after the degrees of anomaly the equation of centre $\varepsilon$ computed according Eq. (23.6). A least-squares fit of the only free parameter – the eccentricity $e$ of the deferent – results in value $e = 0.0570 = 3;25^p$. The residuals of the fit of all quantities are displayed in Fig. 41, in which the marks on the ordinate denote an error $\pm 1'$ at the lines A, C, D, and E, and $\pm 1^p$ at the line B. The fit of the

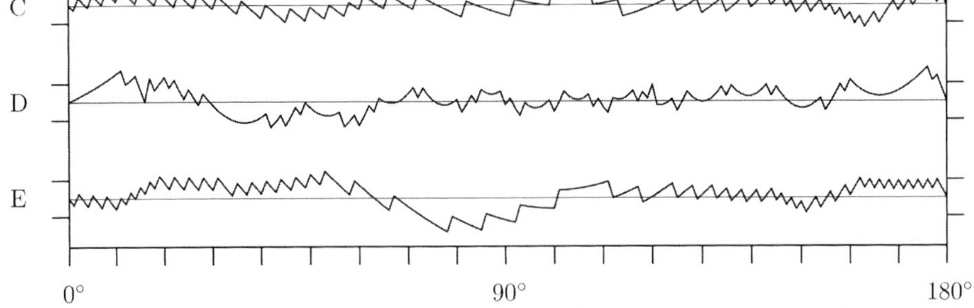


**Figure 41.** Residuals of least-squares fit of Table TRW-16 of equations of Saturn.

3 For its roots in Arabic works, see Bernard R. Goldstein and José Chabás, 'Joseph Ibn Waqār and the Treatment of Retrograde Motion in the Middle Ages', *Archive for History of Exact Sciences*, 77 (2023), 175–99.

equation of centre displayed in the line A has the r.m.s. error 1.0′ and maximum error 2.3′. The fit of proportional minutes $\mu$ in column B of Table TRW-16 according to Eq. (23.14) is not much sensitive to its parameters ($e$ and $p$) due to the low precision of the minutes given in one sexagesimal digit only. It results in the eccentricity $e = 0.1108$ and radius of the epicycle $p = 0.6664$. The residuals shown in the line B of Fig. 41 have r.m.s. error $0.7^p$. and maximum error $1.6^p$. In this column B of Table TRW-16, there is a mark (for the sign = 2) denoting from which value ($2^s + 28° = 88°$) of the true anomaly $v$ the shorter distance $\Lambda_-$ should be taken instead of the farther distance $\Lambda_+$ as the function of $a$ in Eq. (23.11). The next column C gives the farther distance $\Lambda_+$ according to Eq. (23.13) for the upper sign +. Its least-squares fit gives $e = 0.1098$, $p = 0.1156$, residual r.m.s. error 0.5′ and maximum error 1.3′. The equation of anomaly $\delta_0$ in column D is given by Eq. (23.12). Its least-squares fit results in the radius of the epicycle $p = 0.1084 = 6;30^p$, r.m.s. error 0.7′ and maximum error 1.9′. Finally, the shorter distance $\Lambda_-$ in column E leads to $e = 0.0834$, $p = 0.0905$, r.m.s. error 0.7′ and maximum error 1.9′.

## Table of Equations of Jupiter (Table TRW-17)

The Table TRW-17 presents equations of Jupiter in the same arrangement as the table for Saturn and the other planets. A least-square fit of equation of centre $\varepsilon$ in column A gives the eccentricity of the deferent $e = 0.0520 = 3;7^p$. The residuals displayed in line A of Fig. 42 have r.m.s. error 0.7′ and maximum error 1.8′. The fit of proportional minutes $\mu$ in column B results in (again unreliable) values of the eccentricity $e = 0.0305$ and epicycle radius $p = 0.5508$. The residuals have r.m.s. error $0.6^p$ and maximum error $1.6^p$. The fit to the farther distance $\Lambda_+$ in column C results in $e = 0.1984$ and $p = 0.2067$, r.m.s. error of the residuals 0.4’ and maximum error 1.1′. The fit of the equation of anomaly $\delta_0$ in column D yields the radius of the epicycle $p = 0.1917 = 11;30^p$, r.m.s. error 0.6′, and maximum error

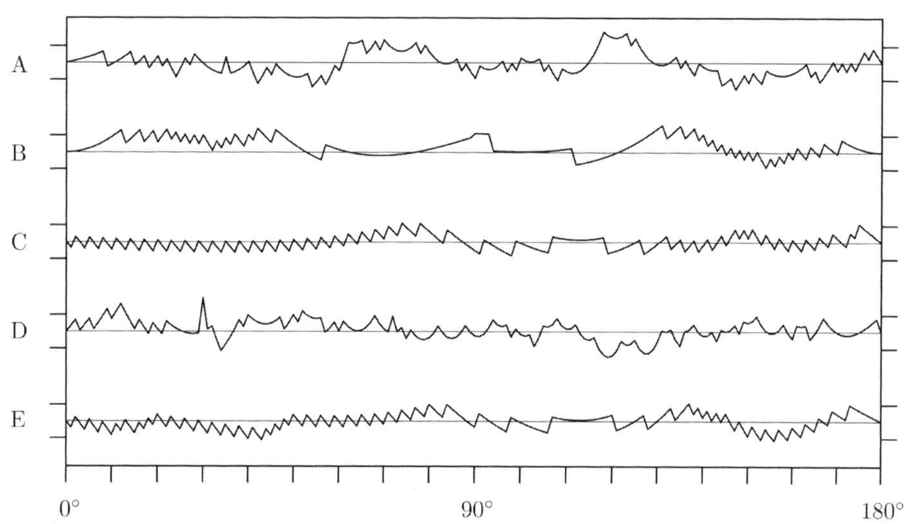


**Figure 42.** Residuals of least-squares fit of Table TRW-17 of equations of Jupiter.

2.0′. The shorter distance $\Lambda_-$ in column E leads to values $e = 0.1977$ and $p = 0.2069$, r.m.s. error 0.5′ and maximum error 1.1′.

## Table of Equations of Mars (Table TRW-18)

A least-square fit of equation of centre $\varepsilon$ in column A of Table TRW-18 of equations of Mars results in the eccentricity of deferent $e = 0.0999 = 6;0^p$. The r.m.s. error of the residuals is 1.2′ and the maximum error reaches up to 4.0′[4] – see Fig. 43. The fit of the proportional minutes $\mu$ in column B gives the eccentricity $e = 0.0955$ and epicycle $p = 0.5281$, r.m.s. error $0.5^p$ and maximum error $1.2^p$. The fit of the farther distance $\Lambda_+$ in column C leads to values of the parameters $e = 0.5980$ and $p = 0.6577$, r.m.s. error of the residuals 1.0′, and maximum error 2.6′. The least-square fit of the equation of anomaly $\delta_0$ in column D results in the radius of the epicycle $p = 0.6583 = 39;30^p$, r.m.s. error 1.1′, and maximum error 3.7′. From the fit of the shorter distance $\Lambda_-$ in column E we get $e = 0.6570$ and $p = 0.7301$, r.m.s. error 1.5′ and maximum error 8.0′.

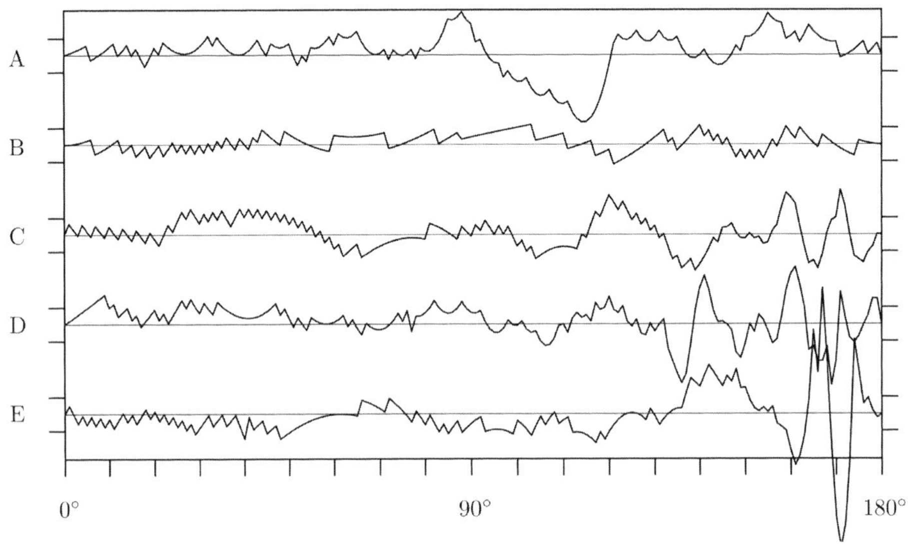


**Figure 43.** Residuals of least-squares fit of Table TRW-18 of equations of Mars.

## Table of Equations of Venus (Table TRW-19)

From a least-squares fit of equation of centre $\varepsilon$ in column A of Table TRW-19 of equations of Venus we can find the eccentricity of deferent $e = 0.0189 = 1;8^p$. The r.m.s. error of the residuals is 1.0′ and maximum error 2.2′ – see Fig. 44. The fit of proportional minutes $\mu$ in column B gives unreliable values of the parameters $e = 0.0497$ and $p = 2.439 \times 10^{-7}$.

4 In agreement with the PAT, there is equation 10°45′ instead of 10°49′ at the anomaly 114°; see Poulle, *Les Tables Alphonsines*, p. 169 (Ratdolt 1483, f. F6v).

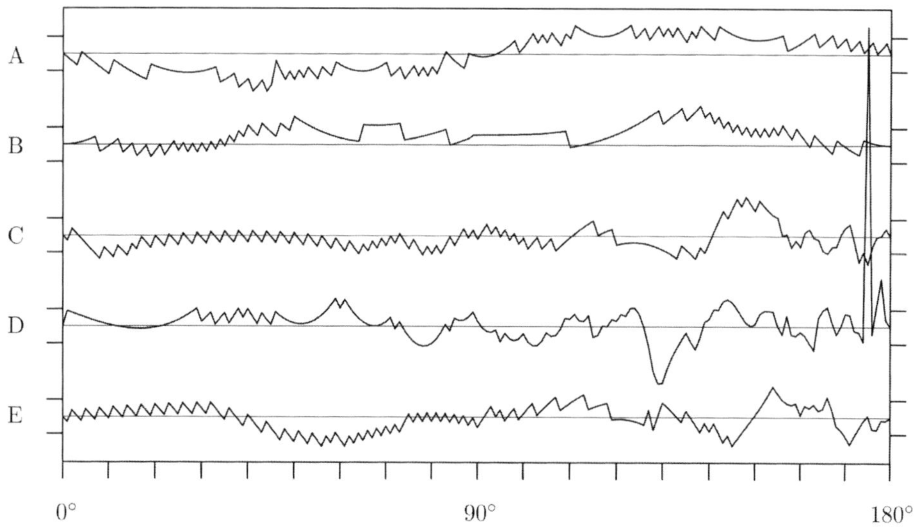


**Figure 44.** Residuals of least-squares fit of Table TRW-19 of equations of Venus.

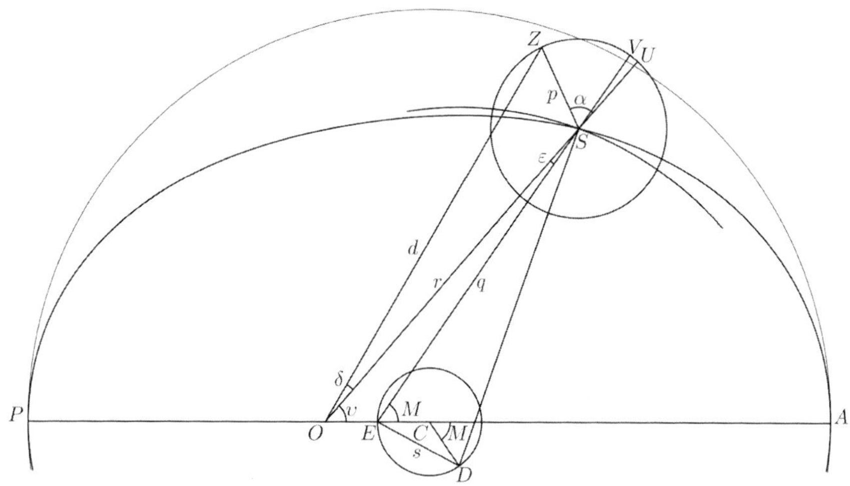


**Figure 45.** Model of motion of Mercury.

The residuals have a large r.m.s. error 0.8$^p$ and maximum error 2.3$^p$. The fit of the farther distance $\Lambda_+$ in column C leads just opposite to very large values $e = 0.7055$ and $p = 0.7200$, r.m.s. error 0.7′ and maximum error 2.3′. The least-square fit of the equation of anomaly $\delta_0$ in column D results in the radius of the epicycle $p = 0.7193 = 43;10^p$. The residuals have r.m.s. error 0.8′, and maximum error 17.6′ at the value of the anomaly $5^s\,25° = 175°$. The value 12;46° given in the MS Toruń, BU,74/III on f. 266v is a scribal error; MS Prague, NK, X B 3, f. 111r gives value 12;27°, while the value 12;26° is given in the PAT.[5] This data point

5 See Table 31 in Poulle, *Les Tables Alphonsines*, p. 177 (Ratdolt 1483, f. G2v).

has been omitted from the fit as an outlier – see line D in Fig. 44. The fit of the shorter distance $\Lambda_-$ in column E gives values of the parameters $e = 0.7176$ and $p = 0.7330$, r.m.s. error 0.7′ and maximum error 1.8′.

## Table of Equations of Mercury (Table TRW-20)

Ptolemy had chosen a more complicated model to explain the motion of Mercury. The equant $E$ is displaced at a distance of only one $e$ from the observer $O$ on the Earth, while the centre $D$ of the deferent moves on a circle of radius $e$ around a point $C$ at a distance of $2e$ from $O$ – see Fig. 45. The point $D$ moves around $C$ in the opposite direction with the same constant angular velocity of mean motion $M$ as the centre $S$ of the epicycle $UVZ$ around the equant $E$. It means that in the isosceles triangle $ECD$, the base angle $\angle CED = M/2$ and the base

$$s \equiv ED = 2e\cos\frac{M}{2}. \tag{23.17}$$

Because $\angle SED = 3M/2$ and the radius $DS$ of the equant is set equal to 1, the distance of the epicycle centre $S$ from the equant point $E$

$$q \equiv ES = s\cos\frac{3M}{2} + \sqrt{1 - s^2\sin^2\frac{3M}{2}} =$$

$$= e\,(\cos M + \cos 2M) + \sqrt{1 - e^2\,(\sin M + \sin 2M)^2}\ . \tag{23.18}$$

Due to the combination of uniform motion of the deferent centre $D$ and uniform increase of $M$ around the equant point $E$, the centre $S$ of the epicycle describes a non-circular curve $ASP$. The distance of $S$ from the observer $O$ is

$$r(v) \equiv OS = \sqrt{e^2 + q^2 + 2eq\cos M}\ . \tag{23.19}$$

At the apogee $A$, it is equal to $r_A = 1 + 3e$, while at the opposite point $P$ it has another local maximum $r_P = 1 - e$. The minimum distance $r$ of $S$ from $O$ is reached around $M = \pm 120°$ and it was thus accepted that Mercury has two perigees at these directions. We can find from the above given equations that $r = \sqrt{1 - 3e + 3e^2} \simeq 1 - \frac{3}{2}e + \frac{3}{8}e^2$ at $M = 120°$, however the derivative $\frac{dr}{dM} = -\frac{e^2\sqrt{3}}{4r}$, so that it is not a precise minimum. The error is, however, negligible; by a numerical treatment we can find the minimum at 120.472° for the Ptolemaic value $e = 0.05$ of the eccentricity of Mercury.[6]

The equation $\varepsilon \equiv \angle OSE$ of the centre $S$ is given by

$$\varepsilon = \arcsin\frac{e\sin M}{r} \tag{23.20}$$

and the true anomaly $v \equiv \angle AOS$ of $S$ by

6 It is in a good agreement with the value 120;28,20° found by Olaf Pedersen, *A Survey of the Almagest with Annotation and New Commentary by Alexander Jones* (New York: Springer, 2011), p. 323, Eq. (10.39).

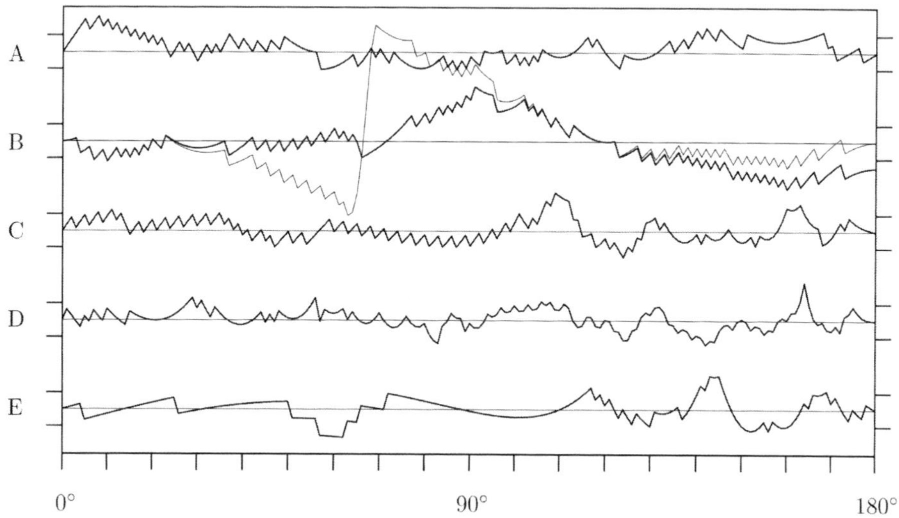


**Figure 46.** Residuals of least-squares fit of Table TRW-20 of equations of Mercury.

$$v = \arctan\frac{q \sin M}{e + q \cos M} \,. \qquad (23.21)$$

The second anomaly $\delta \equiv \angle SOZ$ of Mercury is again given by Eq. (23.10).

In the case of Mercury, the least-squares fit of equation of the centre $\varepsilon$ must be thus computed according to Eq. (23.20), where $r$ is given by Eq. (23.19). For the values of equations of centre of Mercury given in the column A of the Table TRW-20 we find the eccentricity of the deferent $e = 0.0501 = 3;0^p$. The r.m.s. error of the residua is 0.7′, the maximum error 2.1′. A least-squares fit of proportional minutes in column B of the Table TRW-20 leads to unrealistic values of the parameters $e$ and $p$ (they converge to zero and a negative value, respectively), while improving the r.m.s. of residuals only to $2.4^p$ from $2.6^p$ which corresponds to the values $e = 0.05 = 3^p$ and $p = 0.375 = 22;30^p$ given in Ptolemy's *Almagest*.[7] The residuals have a jump from $-4.1^p$ to $+5.8^p$ in vicinity of the anomaly 65°, where the shorter lengths should be used instead of the farther length – see the thin line B in Fig. 46. An appreciably better fit (with r.m.s. residual error $1.4^p$– see the thick line B in Fig. 46) can be achieved if we keep the values of $e$ and $p$ in agreement with the *Almagest* and we converge instead the value of the mean radius at which the boundary between the farther and shorter distance is set. This procedure results in the value of mean radius equal to 1.0093 times the radius $R$ of the deferent, which really corresponds to the boundary between the values of anomaly 64° and 65° as indicated in the PAT and the TR. The fit of the farther distance $\Lambda_+$ in column C yields again unreliable values of the parameters $e = 0.3246$ and $p = 0.3735$, but relatively acceptable r.m.s. error of the residuals 0.7′, and maximum error 2.3′. The least-squares fit of the equation of anomaly $\delta_0$ in column D, which is for Mercury also given by Eq. (23.12), results in the radius of the epicycle $p = 0.3749 =$

7 See Toomer, *Ptolemy's Almagest*, p. 546. The same fit to $\mu$ tabulated in column 8 of *Almagest* XI,11 on p. 553 of the Toomer's edition leads to $e = 0.0527$, $p = 0.2184$ and residual r.m.s. error $0.2^p$ and maximal error $0.7^p$ only.

22;30$^p$, r.m.s. error 0.6′ and maximum error 2.3′ at the anomaly 164°. The MS Toruń, BU, 74/III gives at this point the equation 9;13°, while according to the PAT or MS Prague, NK, X B 3, it should be 9;12°.[8] Finally, the fit of the shorter distance $\Lambda_-$ in column E gives $e = 0.3759$ and $p = 0.4057$, r.m.s. error 0.7′ and maximum error 2.0′.

Ptolemy devoted a great effort in his *Almagest* to derive values of the orbital parameters from the observations. He summarized his results in a form of a table in *Almagest* XI,10.[9] In Tab. 3, we compare his values with the values obtained here from the least-square fits of the tables of planetary equations of centre and equations of the anomaly in the *Tabule resolute*. We can see a good agreement, which indicates, that the planetary equations in the PAT and consequently also in the TR, which accepted the values from the PAT, were computed with the Ptolemaic values of the parameters. There is, however, an appreciable difference in the eccentricity of the deferent for the planets Jupiter and Venus. This was noted and discussed already by E. Poulle and O. Gingerich.[10] They pointed out that the eccentricity of Venus is one half of the eccentricity of the Sun, which is naturally needed to give similar motions on the eccentre and on the deferent, as it follows from the comparison of Eqs (21.7) and (23.3). It can also be seen from the equal maximum values of the equations of the Sun and Venus, which are 2;23° in the *Almagest*[11] and 2;10° in the TR,[12] both in agreement with Eqs (21.4) and (23.8) for the Sun and Venus, respectively. Similarly for Jupiter we find in agreement with Eq. (23.8) the maximum equation of centre 5;16° in the *Almagest*[13] and 5;57° in the TR.[14] Contrary to the values of equations $\varepsilon$ and $\delta_0$, the tabulated values of the proportional minutes $\mu$ and the values of farther and shorter distances $\Lambda_\pm$ are not reliable indicators of the values of parameters, for which these tables were computed.

**Table 3. Parameters of planetary orbits in the *Almagest* and the *Tabule resolute*.**

| | deferent eccentricity $e$ | | epicycle radius $p$ | |
|---|---|---|---|---|
| **Planet** | **Almagest** | **fit of $\varepsilon$** | **Almagest** | **fit of $\delta_0$** |
| Saturn | 3;25 | 3;25 | 6;30 | 6;30 |
| Jupiter | 2;45 | 3;7 | 11;30 | 11;30 |
| Mars | 6;0 | 6;0 | 39;30 | 39;30 |
| Venus | 1;15 | 1;8 | 43;10 | 43;10 |
| Mercury | 3;0 | 3;0 | 22;30 | 22;30 |

8 See Poulle, *Les Tables Alphonsines*, p. 183 (Ratdolt 1483, f. G5v), or MS Prague, NK, X B 3, f. 114r.

9 See Toomer, *Ptolemy's Almagest*, p. 546. Ptolemy gives distances of apogees, which are $R(1+e)$ for the planets Saturn, Jupiter, Mars, and Venus, and $R(1+3e)$ for Mercury, in units of the radius $R = 60^p$. of the deferent (e.g. 66$^p$. for Mars or 69$^p$ for Mercury).

10 Emmanuel Poulle and Owen Gingerich, 'Les positions des planètes au Moyen Âge: Application du calcul électronique aux tables alphonsines', in *Comptes rendus des séances de l'Académie des Inscriptions et Belles-Lettres* 111, 1967, pp. 531–48.

11 See Toomer, *Ptolemy's Almagest*, pp. 167 and 552.

12 See Tables TRW-13 and TRW-19.

13 See Toomer, *Ptolemy's Almagest*, p. 550.

14 See Table TRW-17.

## Latitudes of planets

Tables of latitudes of planets are not contained in the original *Tabule resolute*; nevertheless some canons to the TR include rules for calculation of the latitudes[15] using the corresponding tables contained, e.g., in the Tables of 1322,[16] Toledan Tables,[17] and al-Battānī's tables.[18] These tables follow the tradition established by Ptolemy in his *Almagest* XIII,5.[19] The editions by Ambrosius Lacher, Johannes Schöner, and Johannes Virdung include both the Tables of planetary latitudes (with a step of one degree) and rules for their use.

We shall not deal here with the Ptolemaic theory of planetary latitudes in its completeness.[20] However, it should be noted, that some medieval authors brought various errors into this theory, as can be seen, e.g., in Rule 39 of the canon *Pro (introduccione) generali* on the latitude of Venus by John of Głogów.

The latitude $\beta$ of any planet can be expressed as a function of the true anomaly $a$ and true centre $v$. Ptolemy approximated its dependence on $v$ as a linear superposition of three components $\beta_i(a)$,

$$\beta(a,v) \;=\; \beta_1(a)\cos(v+v_1) + \beta_2(a)\cos(v+v_2) + \beta_3(a)\cos^2(v+v_3)\;, \qquad (23.22)$$

with constant phase shifts $v_i$ (this expression is in fact equivalent to a Fourier expansion up to the second-order term). For the inner planets, $\beta_1$ is called *declinacio* (declination, inclination), $\beta_2$ *refleccio* (reflection, slant, obliquity), and $\beta_3$ *deviacio* (deviation). The cosine functions in this expression are called proportional minutes and they are expressed in sexagesimal minutes and seconds. Ptolemy showed from observations, that the directions $\lambda_A$ to the apogee and $\lambda_\Omega$ to the ascending node of the inner planets are nearly parallel for Venus and nearly antiparallel in the case of Mercury. This is why $v_1 = 90°$ for Venus and $v_1 = 270°$ for Mercury, and $v_2 = v_3 = 0$ for both planets.[21] It means that the proportional minutes are sine of $v$ at the declination and cosine at the reflection. These phase shifts are kept in canons by al-Battānī[22] as well as in canons to the Toledan Tables.[23]

The canons *Priores astrologi* by John of Lignères describe the calculation of the latitude of Venus as follows:

15 See the Chapter 8, Rules 38–40 by John of Głogów.
16 See Chabás and Saby, *The Tables of 1322 by John of Lignères*, pp. 76–79.
17 See Pedersen, *The Toledan Tables*, pp. 1325–26.
18 See Nallino, *Al-Battānī sive Albatenii Opus astronomicum II*, pp. 140–41.
19 See Toomer, *Ptolemy's Almagest*, pp. 632–34.
20 We can refer in this respect to Otto Neugebauer, *A History of Ancient Mathematical Astronomy I–III* (Berlin, Heidelberg, New York: Springer, 1975), pp. 206–26, and Noel M. Swerdlow, 'Ptolemy's Theories of the Latitude of the Planets in the Almagest, Handy Tables, and Planetary Hypotheses', in *Wrong for the Right Reasons*, ed. by J. Z. Buchwald and A. Franklin (Dordrecht: Springer, 2005), pp. 41–71.
21 See *Almagest* XIII,6 (see Toomer, *Ptolemy's Almagest*, pp. 635–36).
22 See Nallino, *Al-Battānī sive Albatenii Opus astronomicum I*, pp. 115–16. Carlo A. Nallino mentions in a footnote that Plato Tiburtinus (Plato of Tivoli, twelfth cent.) gives in his translation 60° instead of 90°. In the first edition of this translation (*Rudimenta astronomica Alfragrani, item Albategnius… de motu stellarum etc.* /Nuremberg: Johannes Petreius, 1537/, f. t. 1v), we read: *post hoc equato centro Veneris 60, Mercurii 270 gradus superadde.*
23 See Pedersen, *The Toledan Tables*, p. 515.

'How to find the latitude of Venus using tables. First find the equated centre and the equated anomaly and write each one down separately. Then, with the anomaly, go into the Table of the latitude of Venus and write down what you find in the Table of declination and reflection separately. Then add 60 degrees to the centre of Venus, and if that exceeds 360 degrees after the addition, discard the 360 degrees and go to the same place in the table with the remainder (or what is left after subtracting the 360). And how many proportional minutes you find there, take those and keep them. Then take the proportional part of the previously preserved declination according to the proportion of the proportional minutes found earlier to 60 using the ninth rule, and this will be the first latitude found that comes out of the declination of the epicycle, and preserve that.

But if the centre with which you have worked and the true anomaly, i.e., the true centre, together with that which has been added, are in the same part of the circle, that is, if both (these values) were either in the upper half or both in the lower half, it would be an equated southern latitude; if one of them was in one half and the other value in the other, it would be a northern latitude; and so you will know the part of it and write down the name of that part.

It is generally said to be the upper half of both the centre and the anomaly when the centre or anomaly is 0 signs and 0 degrees to 3 full signs, or from 9 full signs to 12. The lower is from three to nine signs. Then with the simple centre of Venus, which you found earlier, namely before you added 60, go into the table and write how many proportional minutes you find, separately in two places. Then take the proportion from the reflection previously preserved according to the ratio of these minutes to 60, using the ninth rule, and this will be the reflection found; write it outside the table. And then if the simple centre of Venus, over which you have taken the proportional minutes, be in the upper half and if simultaneously the true anomaly of the planet be less than 180 degrees, it will be an ascertained northern reflection; but if the true anomaly be more than 180, it will be a southern reflection. If the true centre be in the lower half, and the true anomaly of the planet be less than 180 degrees, it will be a southern reflection. If the true anomaly was greater than 180, it would be a northern reflection. And that would be the second latitude found, which would come from the reflection of the epicycle. Keep it and write the name of the part above it.'[24]

The choice of $v_1 = 60°$ suggests, that John of Lignères could have been influenced by Plato Tiburtinus's translation of al-Battānī's canons in writing his canons. John of Głogów repeated the same mistake in his Rule 39. However, there is another error in this rule, namely in the acceptance of $v_2 = -60°$. This error can already be found in the canons *Astrorum* by Albertus of Brudzewo. The canons *Astrorum* were apparently a template for John of Głogów, as already mentioned in Chapter 10, and this coincidence brings a further confirmation.

The amplitude and $\beta_3$ of the deviation was taken constant by Ptolemy and his followers, namely 0;10° for Venus and −0;45° for Mercury. The medieval authors, however, replaced the square of the proportional minutes in Eq. (23.22) by their absolute value, keeping still the orientation ever northern for Venus and ever southern for Mercury.

24 Rule 22 of canons *Priores astrologi* was translated from the edition by Saby, 'Les canons de Jean de Lignères', pp. 207–09.

# 24
# Retrogradation

The computation of planetary stations in medieval astronomy – including the *Tabule resolute* – is based on Ptolemy's *Almagest*, book XII. It also contains the Table of planetary stations in *Almagest* XII,8[1] which gives both the first and second stations of all planets with a step of 6 degrees of the anomaly. The second stations were computed on the assumption of their symmetry with respect to the first ones. We shall treat here the retrogradation in the framework of the Ptolemy's model of planetary motions in a full analytical form, which was not accessible in his contemporary mathematics, and then we shall show the approximations, which Ptolemy and his followers used to tackle the problem.[2]

Generally, to find the stations means to find a position of the planet where the time derivative of the equated longitude is zero,

$$0 = \frac{d\lambda}{dt} = \frac{d}{dt}(v + \delta) = \frac{d}{dt}(M - \varepsilon + \delta) \ . \qquad (24.1)$$

In a case of zero eccentricity, in which both angular velocities $\dot{v} \equiv \frac{d\alpha}{dt} \equiv \frac{dv}{dt}$ and $\dot{\alpha} \equiv \frac{d\alpha}{dt}$ of the motion on the deferent and the epicycle are constant, it is easy to find the stations analytically according to the Apollonius theorem[3]

$$\frac{s}{d} = \frac{\dot{v}}{\dot{\alpha}} \equiv \xi, \qquad (24.2)$$

which is valid because the velocity $\dot{v}d$ of $Z$ due to its motion with $S$ on the deferent must be equal to the tangential projection $\dot{\alpha}p\sin\angle YSZ = \dot{\alpha}s$ of the orbital velocity on the epicycle; here $d \equiv OZ$ is the distance of the planet $Z$ from the Earth $O$ and $s \equiv ZY = ZX/2$ (see Fig. 40). The distance $d$ can then be found by substituting for $s = \xi d$ from Eq. (24.2) into the relation

$$d^2(1 + 2\xi) = d(d + 2s) = (r - p)(r + p) = OT^2, \qquad (24.3)$$

which follows from $r^2 - (d + s)^2 = p^2 - s^2 = YS^2$ in the right triangles $OSY$ and $ZSY$. The true anomaly $\alpha$ can be calculated from the law of cosines

$$d^2 = p^2 + r^2 + 2pr\cos\alpha \qquad (24.4)$$

1 See Toomer, *Ptolemy's Almagest*, p. 588.

2 For a similar approach see Donn MacMinn, 'An analysis of Ptolemy's treatment of retrograde motion', *Journal for the History of Astronomy* 29 (1998), 257–70.

3 *Almagest* XII,1 (see Toomer, *Ptolemy's Almagest*, pp. 555–62).

in the triangle $OSZ$. It has a solution if $\xi$ is a sufficiently small parameter,

$$\xi \leq \frac{p}{r-p}, \tag{24.5}$$

that is, if the angular velocity $\dot{a}$ of the epicyclic motion is sufficiently fast. This condition is satisfied by all planets because the orbital velocity of a planet closer to the Sun, which is in the Ptolemy's model described by the epicycle, is always higher than the velocity of the more distant planet.

For a general eccentric motion with equant, however, there are constant time derivatives only of the mean longitude $M$ and the mean anomaly $\bar{a}$, while $v$ and $a$ are their complicated functions given by Eqs. (23.7) and (23.9). In the case of the upper planets and Venus, they are given by Eqs (23.7) and (23.10). We can then find the derivatives of Eqs (23.4), (23.5), and (23.6)

$$\frac{dq}{dM} = \frac{eq \sin M}{q + e \cos M}, \tag{24.6}$$

$$\frac{dr}{dM} = -\frac{e\,q^2 \sin M}{r\sqrt{1 - e^2 \sin^2 M}}, \tag{24.7}$$

$$\frac{d\varepsilon}{dM} = \frac{2e(e + q\cos M)(q + 2e\cos M)}{r^2(q + e\cos M)}. \tag{24.8}$$

Similarly, regarding Eq. (23.9), the derivative of Eq. (23.10) reads

$$\frac{d\delta}{dM} = \frac{p}{d^2}\left[(p + r\cos a)\left(\frac{\dot{\bar{a}}}{\dot{M}} + \frac{d\varepsilon}{dM}\right) - \sin a\frac{dr}{dM}\right], \tag{24.9}$$

where $d$ is given by Eq. (24.4). For a chosen time $t$ and its corresponding mean motion $M$ we can evaluate the derivatives (24.7) and (24.8) and substituting Eqs (24.8) and (24.9) into the condition (24.1), we arrive at a simple algebraic equation

$$pr\left(2 + \frac{\dot{\bar{a}}}{\dot{M}} - \frac{d\varepsilon}{dM}\right)\cos a - p\frac{dr}{dM}\sin a = r^2\frac{d\varepsilon}{dM} - (p^2 + r^2) - p^2\frac{\dot{\bar{a}}}{\dot{M}} \tag{24.10}$$

for the true anomaly $a$. It can be seen from here that the second term on the left-hand side with $\sin a$ violates the symmetry of the first and second station with respect to the line $OS$. This violation is well understandable, taking into account that in the first semicircle of the deferent depicted in Fig. 40, the apparent size of the epicycle as viewed from $O$ increases in the course of its anticlockwise motion from the apogee $A$ to the perigee $P$. This increase results in a positive additional angular velocity of the ecliptical longitude around the first station, which supports $\dot{M}$ in its balance with $\dot{\bar{a}}$, and a negative velocity which weakens it around the second station. Both stations are thus shifted to larger values of $a$. Using the values of eccentricity and radius of the epicycle according to the *Almagest*, this effect in the case of Mars exceeds the value of 4° in the difference $\Delta = a_1 + a_2 - 360°$ between the true anomaly $a_1$ of the first station and the reflected second station $360° - a_2$ (see Fig. 47).[4] In the case of Venus, the difference $\Delta$ only slightly exceeds 0.9°, for Jupiter 0.4°, and for Saturn 0.2°.

4 The approximate value of the stations is drawn according to MS Toruń, BU, 74/III, ff. 264r–265r (see the Table TRW-18, column F), where is due to a scribal error given 5s 9;6° for the anomaly 45° instead of 5s 9;9° found, e.g., in MS Prague, NK, X B 3, f. 106r.

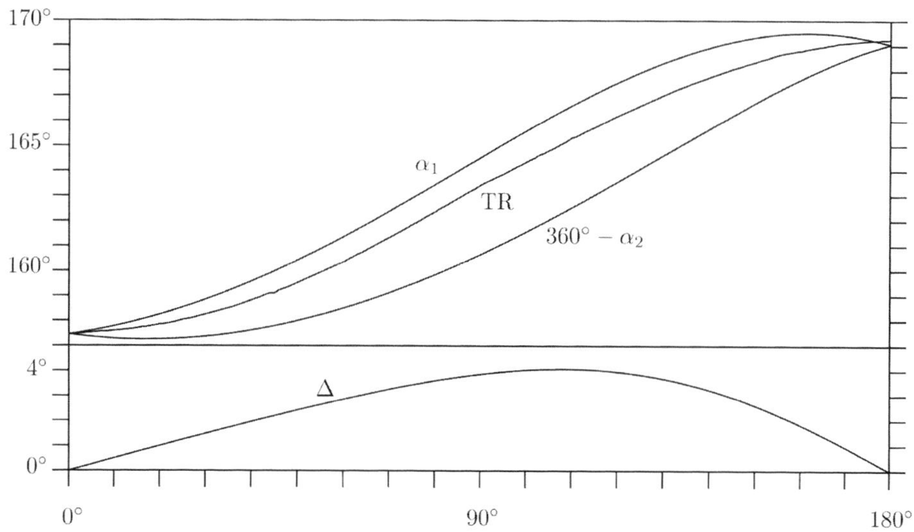


**Figure 47.** The precise values of true anomaly of the first and second station of Mars (the upper and lower curve in the upper panel), their difference (in the lower panel), and the approximate value given in the TR (middle curve in the upper panel).

It is obvious that Ptolemy and his medieval followers could not proceed according to the above outlined precise treatment of the model presented in the *Almagest*. Ptolemy, however, utilized one approximate solution attributed to a 'mean' distance of the epicycle and two precise solutions for the extremal distances $r$. Let us investigate first the cases of $S$ being at the apogee $A$ of the deferent or its perigee $P$, where

$$\cos M = \pm 1 \Rightarrow q = 1 \mp e, r = 1 \pm e, \varepsilon = 0, \frac{dq}{dM} = \frac{dr}{dM} = 0, \frac{d\varepsilon}{dM} = \pm\frac{2e}{1 \pm e}. \quad (24.11)$$

Eq. (24.10) has then the form

$$(1 + 2\xi)pr \cos a = -\xi r^2 - (1 + \xi)p^2, \quad (24.12)$$

where

$$\xi = \frac{q\dot{M}}{r(\dot{\bar{\alpha}} + \dot{\varepsilon})}. \quad (24.13)$$

These are equivalent to Eqs (24.4), (24.3), and (24.2), taking into account that at the apogee and perigee $\dot{\upsilon} = q\dot{M}/r$ and $\dot{\alpha} = \dot{\bar{\alpha}} + \dot{\varepsilon}$. The problem of retrogradation in the apsides is thus equivalent to that with zero eccentricity, which simplifies its precise solution.

Ptolemy's third solution is only approximate. It is based on the assumption that in a mean distance, i.e. $r = 1$, the angular velocity $\dot{\upsilon} \simeq \dot{M}$ and $\dot{\alpha} \simeq \dot{\bar{\alpha}}$. From Eq. (24.8) we can see that these equalities are satisfied precisely when $e + q\cos M = 0$, where $\dot{\varepsilon} = 0$. It is satisfied when $DS$ is perpendicular to $PA$; it is obvious also from symmetry that at this point, where $r = q = \sqrt{1 + e^2}$, $\upsilon$ reaches its maximum and the mean motion $M$ increases in the same rate as $v$. This approximate solution also neglects the above explained effect of $\dot{r}$ and $\alpha$ is calculated from Eq. (24.4) as in the case of zero eccentricity. In a general value of the mean motion $M$, Ptolemy interpolates the anomaly of the first station between these extremal and mean values of $r$ in the same way as above using the proportional minutes given by Eq. (23.14).

The sixth rule, however, claims that the table of the first stations is to be entered with the true centre, i.e. with $v$. It means, that the distance $r$ of the centre $S$ from the observer $O$ is to be calculated directly from the relation

$$r = \sqrt{1 - e^2 \sin^2 v} + e \cos v \tag{24.14}$$

instead of from Eqs (23.4) and (23.5). There is no difference in the apogee and perigee, where $v = M$, and only a small difference at the mean distance. Yet the tabulated values have smaller residuals when entered with the true centre.

Due to the more complicated model of the motion of Mercury, the explicit form of condition (24.1) for the stations of Mercury also differs from that of the other planets. The derivatives of Eqs (23.18) and (23.19) read

$$\frac{dq}{dM} = -e(\sin M + 2 \sin 2M) - \frac{e^2(\sin M + \sin 2M)(\cos M + 2 \cos 2M)}{\sqrt{1 - e^2 (\sin M + \sin 2M)^2}} \tag{24.15}$$

and

$$\frac{dr}{dM} = \tfrac{1}{r}\left((q + e \cos M)\frac{dq}{dM} - eq \sin M\right). \tag{24.16}$$

The derivative of Eqs (23.20) and (23.21) are then

$$\frac{d\varepsilon}{dM} = \frac{e\left(r \cos M - \sin M \frac{dr}{dM}\right)}{r(q + e \cos M)} \tag{24.17}$$

and

$$\frac{dv}{dM} = \frac{q^2 + 2eq \cos M + e \sin M \frac{dq}{dM}}{e^2 + q^2 + 2eq \cos M}. \tag{24.18}$$

We can find from them the condition (24.1) of station in the form

$$q(q + e \cos M) + e \sin M \frac{dq}{dM} = 0. \tag{24.19}$$

The precise solution of the first and the second station of Mercury (using again the Ptolemaic values of $e$ and $p$) is shown in Fig. 48 together with the values given in the TR as they can be found in Table TRW-20, column F. The precise values have a more complicated behaviour than in the other planets. The difference $\Delta$ between the two stations exceeds 3°, that is, similar to Mars and more than in the other three planets.

It is interesting to note that Table A-01 of stations and retrogradations, which is one of the tables sometimes joining the *Tabule resolute*, gives the values of the first and second stations just for the apogees and perigees of the centre of epicycle. Nevertheless, its full title *Tabula retrogradacionis, direccionis, stacionis planetarum et intratur cum argumento vero*

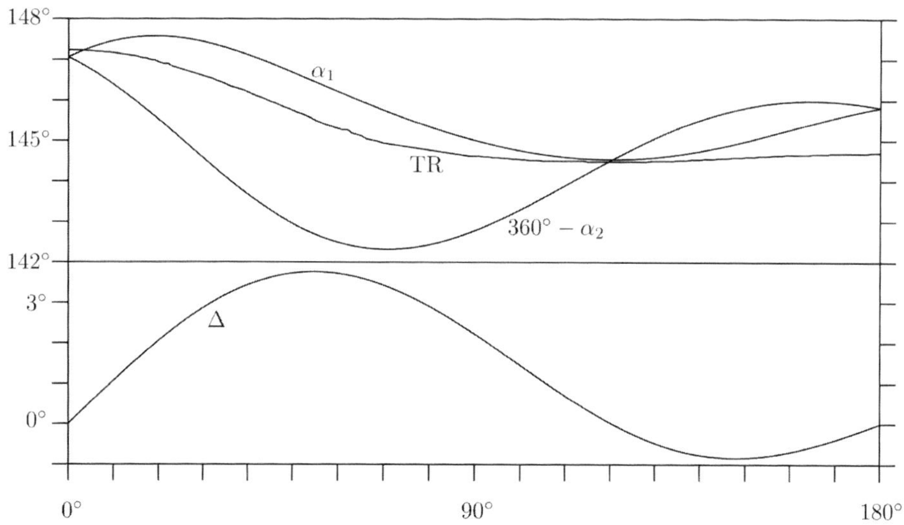


**Figure 48.** The precise values of true anomaly of the first and second station of Mercury ($\alpha_1$ and $\alpha_2$ in the upper panel), their difference (in the lower panel), and the approximate value given in the TR (in the upper panel).

*eorum* also specifies that the stations are given for the true and not the mean anomaly, despite the fact that they are equivalent at the apogee and periee of the centre. These tables, however, give the true anomalies of the epicycle. They also give a somewhat confusing description of the stations: *inicium prime stacionis* for the first station at the apogee, *inicium retrogradacionis* at the perigee, *finis retrogradacionis* for the second station at the perigee, and *finis stacionis secunde* for the second station at the apogee. The tabulated values are – up to the scribal errors – identical with the upmost and bottom values of stations in the tables of equations in the *Tabule resolute*, so these tables are superfluous in a codex that includes the TR.

25

# Equation of Time and Right Ascension

The equation of time is the difference between the apparent solar time shown by sundials and mean solar time which is measured by mechanical clocks and which is also a basis for astronomical calculations of motion of celestial bodies. This difference is due to a combination of two effects: the eccentricity of the annual orbit of the Sun (see Chapter 21) and the projection of the true Sun from the ecliptic onto the equator. Both effects cause an advance or delay of the true Sun with an amplitude about 10 minutes of an hour, the former calculated in Table TRW-13 with a period of one year, the latter calculated in the Table A-06 with a period of half a year. Their superposition results in a curve with two maxima and two minima during the year and an amplitude about ± 15 minutes. It is presented here in the Table A-05. While today we name this correction of the apparent solar time simply the equation of time, medieval authors referred it as an 'equation of days with (their) nights' to distinguish it clearly from the much larger seasonal changes of the length of day.

The effect of eccentricity of the solar orbit can be calculated according to Eq. (21.2) or Eq. (21.3) and its result is presented here in Table TRW-13. We can then find the ecliptic longitude $\lambda = \Upsilon S$ of the Sun $S$ from the vernal point $\Upsilon$

$$\lambda = \lambda_A + \upsilon = \lambda_A + M - \varepsilon \,, \tag{25.1}$$

where $\lambda_A$ is the ecliptical longitude of the apogee of the solar orbit calculated as described in Chapter 19 – see Fig. 49.[1]

To find the effect of the ecliptic longitude $\lambda = \Upsilon S$ and its irregularities on the apparent solar time, it is necessary to calculate its projection $a = \Upsilon T$ from the ecliptic $\Upsilon SQ$ on the equator $\Upsilon TR$ from the pole $P$ along the declination circle $PST$. This problem of spherical trigonometry can be solved using the Menelaus theorem

$$1 = \frac{\sin \Upsilon S \sin QP \sin RT}{\sin \Upsilon T \sin QS \sin RP} \tag{25.2}$$

for the declination circle $PST$ and the spherical triangle $\Upsilon QR$, where $PQR$ is the solstitial colur, so that $RQ = \varepsilon$ is the obliquity of the ecliptic and $\Upsilon R = \Upsilon Q = RP = 90°$. We arrive from this relation to the formula

$$\tan a = \cos \varepsilon \tan \lambda \,, \tag{25.3}$$

1 Alternatively, the true anomaly $\upsilon$ can be calculated according to Eq. (21.1).

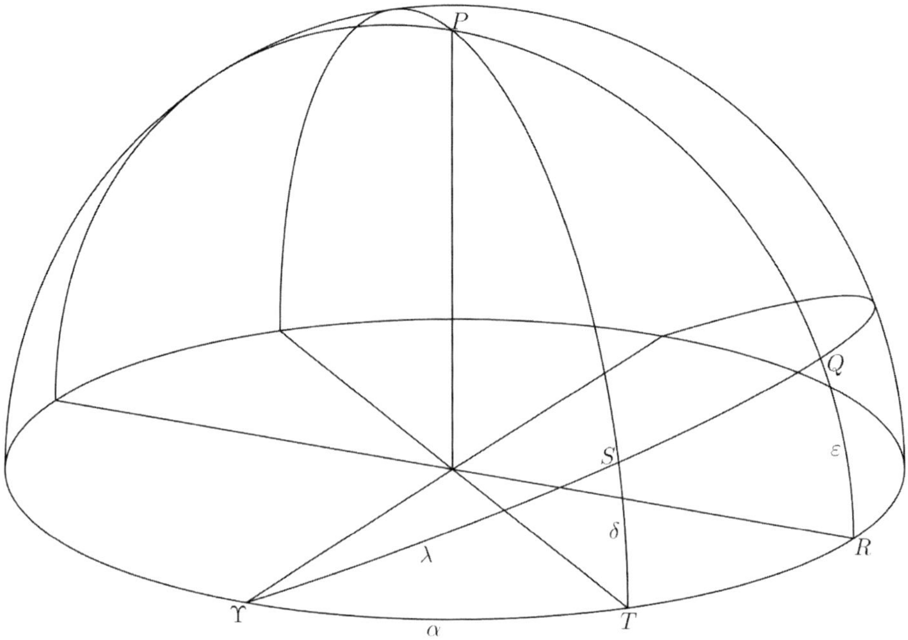


**Figure 49.** Projection of ecliptic on the equator.

which could also be derived from the laws of sines and cosines in the right triangle $\Upsilon TS$. The value $a$ of this projection determines when the point $S$ culminates and passes through the local meridian. For an observer at the geographic equator, where the celestial equator is upright, i.e. perpendicular to the horizon, and the celestial pole is on the horizon, each point of sky rises six hour before and sets six hours after its culmination. This is why the quantity $a$ is named right ascension.

The Table of right ascensions, presented in Chapter 16 as Table A-06, gives values $a' = a(\lambda) + 90°$, where $a(\lambda)$ is calculated according to Eq. (25.3) for a fixed value of the obliquity $\varepsilon$ of the ecliptic to the equator.[2] This obliquity is its only free parameter which can be determined from the tabulated values. A least square fit to Table A-06 of right ascensions in MS Cracow, BJ, 617, f. 6r, gives the obliquity of the ecliptic $\varepsilon \simeq$ 23;34,45°. The r.m.s. residual noise is 0;0,20° and the maximum error is 1.4′. Several other values have an error of 1′ due to an incorrect rounding.

The values of equation of time are then the difference $\Delta a = a(\lambda(M)) - M$ given either in degrees or in units of time, usually minutes of an hour. One degree corresponds to $24 \times 60 / 360 = 4$ minutes. While today we give the equation of time as a deviation of the true Sun from mean Sun with both positive and negative values and a semiamplitude about 15 minutes, our predecessors calculated the values with a zero point at the minimum (or maximum) and values from 0 up to about 30 minutes of an hour.

2 The right ascensions are given in *Almagest* II,8 without the shift of 90° (see Toomer, *Ptolemy's Almagest*, p. 100).

Tables of the equation of time have a long and rich tradition. They differ in their layout as well as in the tabulated values. In some sources they are given in the units of angle, in others in units of time, and they are also calculated for different free parameters corresponding to different epochs. An explanation of the equation of time can be found in Ptolemy's *Almagest* III. The calculation of the anomaly of the Sun given by Eq. (21.1) is presented in *Almagest* III,5[3] and the resulting values are given in a table in *Almagest* III,6.[4] In *Almagest* III,7, Ptolemy determines the position of the apogee of Sun $\lambda_A = 65;30°$ and the position of the mean Sun at the reference epoch of the *Almagest*.[5] The calculation of $\lambda(t)$ of the Sun in any time $t$ is summarized in *Almagest* III,8[6] and finally in *Almagest* III,9 Ptolemy discusses the consequences of the eccentricity of the solar orbit and obliquity of the ecliptic for the length of day and points out that despite the fact that the influence on the length of a single day is negligible, its accumulation in the course of the year may be significant.[7] A table of equation of time itself is not contained in *Almagest*, but in Ptolemy's Handy Tables, where it is incorporated into the table of right ascensions.[8]

Similarly, al-Battānī's tables and the Toledan Tables include the equation of time in the table of right ascensions.[9] These two sources give nearly identical values of the equation of time. A separate table of equation of time is included in the Tables of 1322 by John of Lignères.[10] This table is also often attached to other Alfonsine tables including the *Tabule resolute*. For instance, Rosińska lists 28 copies of this table in her census of astronomical manuscripts in the Jagellonian Library in Cracow.[11] We present here in Table A-05 a copy of this table from MS Prague, NK, X B 3, f. 144v. While the *editio princeps* of Alfonsine tables includes the equation of time again in the tradition of Toledan Tables in the table of right ascensions,[12] the next editions published by Johannes Lucilius Santritter and Petrus Liechtenstein contain two versions of separate tables of equation of time.[13] The first one in both prints is inscribed as 'old' (*tabula vetus*) and it gives the equation of time in degrees and minutes (usually attributed to Peter St Omer), while the second one is inscribed 'modern' (*tabula moderna*) and it gives the values in minutes and seconds of an hour. The latter table is entitled *Tabula equationis dierum cum noctibus: moderna verificata ad annos Christi 1456* and first appeared in Bianchini's *Tabula primi mobilis*. A simple table of equation of time with values rounded to minutes of an hour is also contained in the incunable of Tables of conjunctions of the Sun and Moon by Wenceslas Faber of Budweis.[14] Formally similar

3 See Toomer, *Ptolemy's Almagest*, pp. 157–66.
4 See Toomer, *Ptolemy's Almagest*, p. 166–69.
5 See Toomer, *Ptolemy's Almagest*, p. 167.
6 See Toomer, *Ptolemy's Almagest*, p. 169.
7 See Toomer, *Ptolemy's Almagest*, pp. 169–72.
8 See Tihon, *Πτολεμαίου Πρόχειροι κανόνες*, pp. 97–100, and Mercier, *Πτολεμαίου Πρόχειροι κανόνες*, pp. 10–13 and 89–96.
9 See Nallino, *Al-Battānī sive Albatenii Opus astronomicum II*, pp. 61–64, and Pedersen, *The Toledan Tables*, pp. 972–75.
10 See Chabás and Saby, *The Tables of 1322 by John of Lignères*, pp. 72–75.
11 Rosińska, *Scientific Writings*, p. 519.
12 Ratdolt 1483, ff. K1r–K2r.
13 Santritter 1492, ff. E4r–v, Liechtenstein 1518/21, f. E3r–v.
14 *Opusculum tabularum utile verarum Solis et Lunae coniunctionum per magistrum Wenceslaum Fabri de Budweiss, baccalarium medicine, compositum* (Leipzig: Martin Landsberg, 1494–95 and 1499), f. B1r. See Hadrava and Hadravová, 'Nová akvizice Národní knihovny', 198–205.

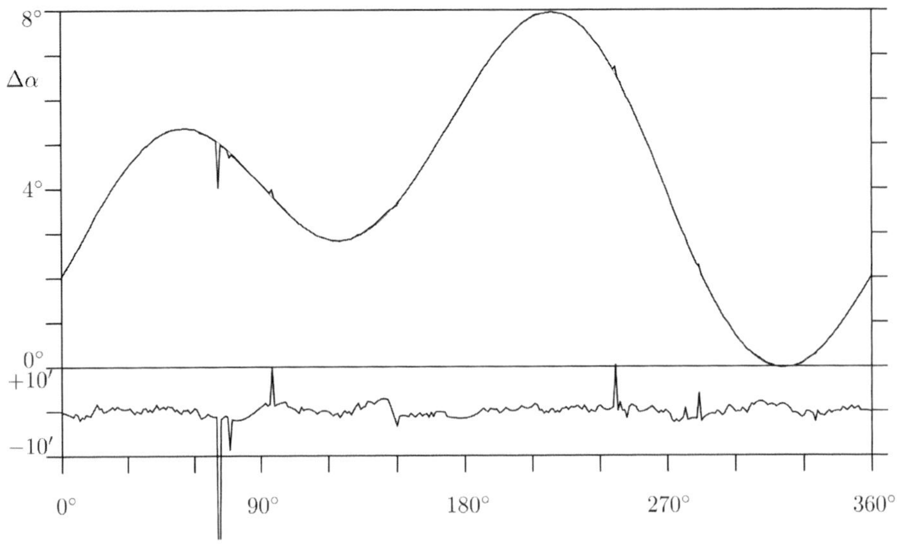


**Figure 50.** Equation of time in Table A-05.

to the 'modern' table from the Santritter's and Liechtenstein's editions but with updated values is also the table of equation of time published in the *Tabule eclypsium* by Georg Peurbach.[15] This table is entitled *Tabula equationis dierum novissime constituta, presupponens augem Solis in principio Cancri et declinationem Almeonis* (The newly established Table of the equation of days, assuming the apogee of the Sun at the beginning of Cancer and the obliquity according to Almeon).

A least square fit to the tables of equation of time can reveal the parameters for which they were calculated. The values of the parameters resulting from this fitting of several examples of such tables are given in Tab. 4. These parameters are the eccentricity $e$ of the solar orbit, ecliptical longitude $\lambda_A$ of its apogee, and the obliquity $\varepsilon$ of the ecliptic. Regarding the traditional shift of the zero value to the minimum, it is useful to add this shift $\Delta_0$ as an additional free parameter, and we can also test a possible shift $\lambda_0$ in the ecliptical longitude. The tables which are given in units of time are converted first into the angular units. To yield information on which units are used in which table, we give in Tab. 4 also the maximum value $\Delta_{max}$ of the equation of time found in the table in the original units used in the source. In the case of the table from manuscript Prague, NK, X B 3, which is presented here in Table A-05, we dropped from the fit the obvious outliers (scribal errors) at the angles 71, 76, 95, 247, and 284 degrees. They are, however, involved in the calculation of r.m.s. of the residuals $\sigma$ presented in the corresponding column denoted as TR in Tab. 4. Without these points $\sigma$ would be 0.0176°. The values tabulated in Table A-05 are displayed in the upper panel of Fig. 50 by the thick line, the best fit by the thin line, and the residuals of the fit are shown in the bottom panel. The negative value of $\Delta_0$

15 *Tabulae eclypsium magistri Georgii Peurbachii, Tabula primi mobilis Joannis de Monteregio.* Ed. Georg Collimitius Tannstetter (Vienna: Johannes Winterburger, 1514), f. d5r.

for the Handy Tables in the column inscribed HT is due to the fact that these tables use inverse orientation of the equation of time. The value of $\lambda_A$ greater for half a degree than the value 65;30° derived in the *Almagest* may be due to a later updating of the Handy Tables by Theon of Alexandria. A more significant increase of this value till the epoch of al-Battānī can be seen in the column TT of Toledan Tables in which the al-Battānī's table was adopted and later also into Ratdolt's edition. A farther increase to over 87° is common to the *Tabula vetus* (in the column denoted TV) and the TR, which means also the Tables of 1322 by John of Lignères. Similar values also are found for the table of Wenceslas Faber. The *Tabula moderna* in the column TM gives a greater longitude of the apogee than the Peuerbach's table in the column P, which very precisely corresponds to the declared value 90°. The smaller value of $\varepsilon$ is also close to the declared obliquity according to Almeon, i.e., al-Ma'mūn, to whom the value 23;33° or 23;33,30° is ascribed.[16]

**Table 4. Parameters of the tables of equation of time.**

| | TR | HT | TT | TV | TM | P |
|---|---|---|---|---|---|---|
| $\Delta_0$ | 4.017° | −3.428° | 4.099° | 4.014° | 4.082° | 4.099° |
| | 4;1,1° | −3;25,41° | 4;5,56° | 4;0,50° | 4;4,55° | 4;5,56° |
| $\lambda_0$ | −0.1650° | 0.0178° | −0.1486° | −0.2429° | 0.0817° | 0.0290° |
| | −0;9,54° | 0;1,4° | −0;8,55° | −0;14,34° | 0;4,54° | 0;1,44° |
| $e$ | 0.03485 | 0.04164 | 0.03441 | 0.03488 | 0.03785 | 0.03779 |
| | 2;5,28$^p$ | 2;29,54$^p$ | 2;3,53$^p$ | 2;5,34$^p$ | 2;16,16$^p$ | 2;16,3$^p$ |
| $\varepsilon$ | 23.656° | 23.712° | 23.612° | 23.640° | 23.597° | 23.570° |
| | 23;10,56° | 23;42,43° | 23;36,43° | 23;38,24° | 23;35,49° | 23;34,12° |
| $\lambda_A$ | 87.340° | 66.088° | 82.890° | 87.628° | 90.635°f | 90.008° |
| | 87;20,24° | 66;5,17° | 82;53,24° | 87;37,41° | 90;38,6° | 90;0,27° |
| $\sigma$ | 0.0593° | 0.0132° | 0.0268° | 0.0235° | 0.0252° | 0.0085° |
| $\Delta_{max}$ | 7;57° | 0;33,23h | 7;54° | 7;57° | 0;32,52h | 0;32,48h |

16 See Pedersen, *The Toledan Tables*, p. 961.

# Oblique Ascensions and Visibility of Planets

While the celestial equator and all parallels of declination cross the horizon at right angles at the geographic equator and hence any celestial body sets and falls relatively fast below the horizon, in other climates, that is, at different geographic latitudes, the parallels are inclined with respect to the horizon. Consequently, there are seasonal differences between the length of day and night increasing with the geographic latitude, as well as prolonging the dusk and dawn. This effect is of great importance for a daily life and timekeeping. The inclination of ecliptic to the horizon and its diurnal variations are also dependent on the geographic latitude of the observer. This had to be taken into account in calculating astrological houses or the visibility of planets. Dedicated tables of oblique ascensions for different climates, represented here by Table A-07, were compiled for this purpose. Their computation is a bit more tricky problem of spherical astronomy, which we shall summarize here before analysing Table A-07 and an example from the table of visibility of planets mentioned in Chapter 9.1.

Let the obliquity of the ecliptic *♈SDNA* to the equator *♈TCWMBE* be $\varepsilon = \angle M♈N$ – see Fig. 51. Similar to the relation (25.2), the declination $\delta = TS$ of a general point $S$ on the ecliptic with an ecliptical longitude $\lambda = ♈S$ can be found from the Menelaus theorem

$$1 = \frac{\sin S♈ \sin QR \sin PT}{\sin ST \sin Q♈ \sin PR} \qquad (26.1)$$

applied now to the triangle $SPQ$ and the equator circle $♈TR$ in Fig. 49 (where $QR = \varepsilon$ and $PT = Q♈ = PR = 90°$). It results in the law of sines

$$\sin \delta = \sin \varepsilon \sin \lambda \qquad (26.2)$$

in the right triangle $♈TS$.

From the law of cosines

$$\sin h = \sin \varphi \sin \delta + \cos \varphi \cos \delta \cos \tau \qquad (26.3)$$

in the nautical triangle $ZPS$ (formed by the point $S$, the zenith $Z$, and the pole $P$, where $\varphi = MZ = 90° - ZP$ is the geographic latitude of the site of observer in Fig. 51), we can solve the hour angle $\tau = \angle MPT$ and with the right ascension $a = ♈T$ calculated according to Eq. (25.3), we can find the corresponding sidereal time

$$\Theta = \tau + a \qquad (26.4)$$

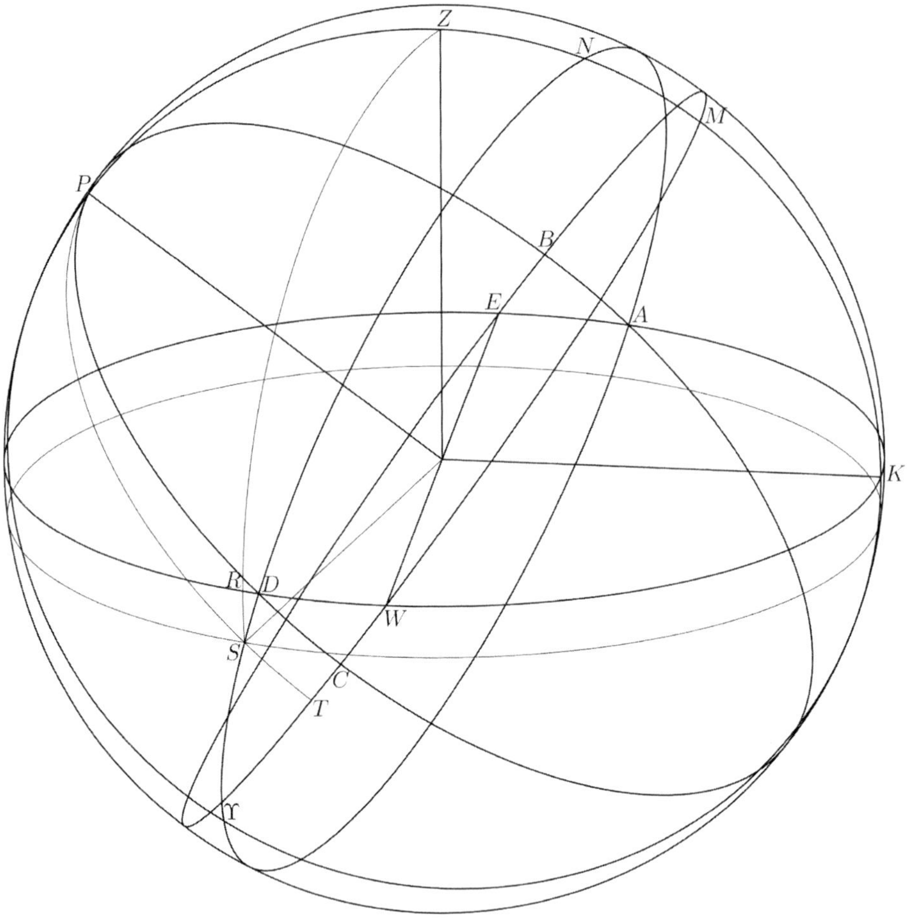


**Figure 51.** Visibility of planets.

at which the point $S$ reaches an altitude $h = RS$ above or below (as it is the case of $h < 0$ in Fig. 51) the horizon $RDWKAE$.

The oblique ascension $\vartheta = \vartheta(\lambda)$ given in Table A-07 is the right ascension $\vartheta = \Upsilon E$ of the (eastern) point $E$ of the equator, which rises simultaneously with the point $A$ (the ascendant) of the ecliptic with the longitude $\lambda = \Upsilon A$. The right ascension $a = \Upsilon B$ of the ascendant $A$ is given by Eq. (26.2). The difference $\Delta = a - \vartheta = BE$ of right ascensions of $A$ and $E$ can be found from the right triangle $ABE$ (which is similar to the triangle $DCW$) as

$$\sin\Delta = \tan\varphi \tan\delta . \qquad (26.5)$$

Together with the oblique ascensions in Table A-07 are also tabulated the so-called parts of hours (*partes horarum*)

$$\psi = 15^\circ + \frac{\Delta(\lambda)}{6} , \qquad (26.6)$$

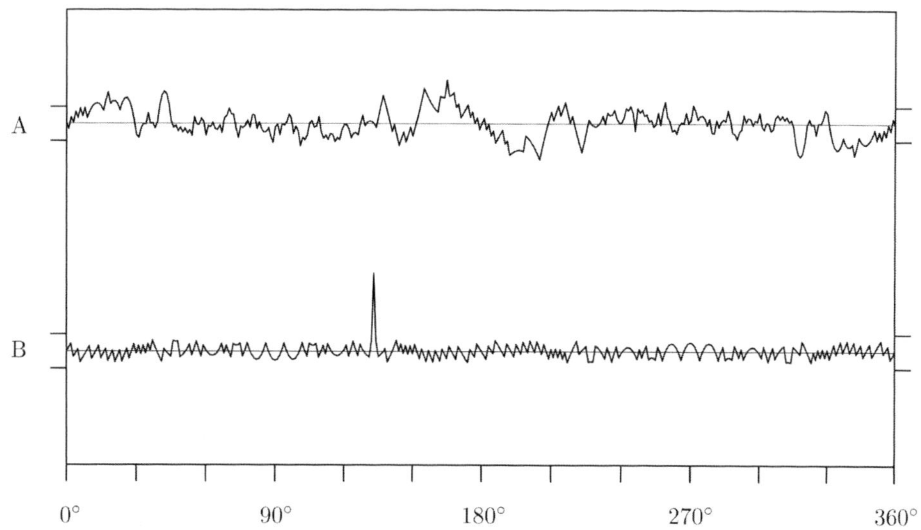


**Figure 52.** Residuals of least-squares fit of Table A-07 of oblique ascensions.

which give the number of degrees of the horizon (that is, the part of horizon), which rise during one unequal hour when the Sun is at the ecliptical longitude $\lambda$. Its multiple $4^{\mathrm{h}}\psi/5°$ shows the length of the corresponding day in equal hours.

A least-square fit of the oblique ascension in Table A-07 gives the obliquity of the ecliptic $\varepsilon \simeq 23;35,36°$ (close to Battani's value of 23;35°) and the geographic latitude $\varphi \simeq 50;3,55°$, which is in a good agreement with the declared latitude 50;7°. The maximum error is 2.6′ and the r.m.s. residual error 0.8′ – see the line A in Fig. 52. The least-square fit to the parts of hours in the same table gives values $\varepsilon \simeq 23;43,7°$ (unattested in historical sources) and $\varphi \simeq 49;54,9°$. The r.m.s. residual error is only 0.3′ (see the line B in Fig. 52), while the maximum error is 4.5′ at the anomaly 133°, where the copyist obviously wrote by an error the same value as above (see line 13, column J in the Table A-07). We thus skipped this point from the fit.

The first visibility of a planet, dealt with by Ambrosius Lacher,[1] corresponds to the situation when the planet sets at the point $D$ and the Sun is at the same time in the point $S$ at a sufficient depth $h_{\odot}$ below the horizon. The Table of visibility gives the distance in longitude $\Delta\lambda = \lambda_D - \lambda_S$ of the planet from the Sun, when the planet is at the beginning of different zodiacal signs (and zero ecliptical latitude), that is at a given longitude $\lambda_D$. Inserting its declination $\delta_D$ from Eq. (26.1) into Eq. (26.3) and taking into account that $h_D = 0$, we find the hour angle

$$\tau_D = \arccos(-\tan\varphi \tan\delta_D) . \tag{26.7}$$

1 See Chapter 9.1.

From Eqs (26.2) and (26.4) we can then compute also $a_D$ and $\Theta$. To find $\Delta\lambda$ as a function of $\lambda_D$, we must solve the set equations (26.1) – (26.4) with respect to $\lambda_S$ (and $\tau$, $a$, and $\delta$ for chosen $h$ and fixed $\varepsilon$ and $\varphi$). An explicit algebraic solution is complicated; however, the longitude $\lambda_S$ can be found numerically in an itterative process, estimating first, e.g., $\lambda_S \simeq \lambda_D + h_\odot$, computing $\delta_S$ from Eq. (26.1), $a_S$ from Eq. (26.2), $\tau_S$ from Eq. (26.4), $h_S$ from Eq. (26.3), and correcting $\lambda_S \to \lambda_S + (h_\odot - h_S)$.

This algorithm is invariant with respect to the change $\lambda_D \to 180° - \lambda_D$. This symmetry is satisfied in *Almagest* XIII,10[2] by the visibilities of Saturn and Jupiter, that is, they are precisely the same, e.g., at Taurus and Pisces. As an example, we give in Tab. 5 the values from the *Almagest* of the first visibility of Jupiter at evening setting. Its least-squares fit with $\varepsilon$ fixed to the Ptolemy's value 23;51,20° gives the most probable value of geographic latitude $\varphi \simeq$ 21;12° and the altitude of the Sun below the horizon $h_\odot \simeq -10;38°$. The r.m.s. error of the fit is 17′ and the maximum error is 30′ at Libra – see the recomputation in Tab. 5. The symmetry is satisfied in Lacher's edition by the visibility of Mars. A least-square fit with all three parameters free leads to residuals comparable with those for Jupiter in the *Almagest*, but the obliquity of the ecliptic is nearly 27°. When we fix $\varepsilon$ = 23;33,30°, which was used by John of Lignères and the Toledan Tables, then the latitude $\varphi \simeq$ 35;38° and $h_\odot \simeq -14;13°$. The r.m.s. error of the fit is 22′ and the maximum error 34′ at Sagittarius – see Tab. 5.

**Table 5. Examples of the first visibility $\Delta$ (in degrees) from the *Almagest* and Lacher's edition of the TR, and their recomputations.**

| sign | $\lambda_D$ | $\Delta$ of Jupiter | | $\Delta$ of Mars | |
|---|---|---|---|---|---|
| | | *Almagest* | fit | Lacher | fit |
| Ari | 0 | 10;19 | 10;39 | 14;12 | 14;33 |
| Tau | 30 | 10;29 | 10;43 | 15; 8 | 14;47 |
| Gem | 60 | 11;10 | 11; 6 | 16; 7 | 15;52 |
| Cnc | 90 | 12;46 | 12;18 | 18;15 | 18;33 |
| Leo | 120 | 14;31 | 14;18 | 22;14 | 22;48 |
| Vir | 150 | 16;12 | 16;16 | 27;11 | 26;56 |
| Lib | 180 | 16;34 | 17; 4 | 29; 0 | 28;39 |
| Sco | 210 | 16;12 | 16;16 | 27;11 | 26;56 |
| Sgr | 240 | 14;31 | 14;18 | 22;14 | 22;48 |
| Cap | 270 | 12;46 | 12;18 | 18;15 | 18;33 |
| Aqr | 300 | 11;10 | 11; 6 | 16; 7 | 15;52 |
| Psc | 330 | 10;29 | 10;43 | 15; 8 | 14;47 |

2 See Toomer, *Ptolemy's Almagest*, p. 647.

# 27 Interpolation

Rules for the use of astronomical tables often require interpolation between the tabulated values. They mostly recommend linear interpolation between the two neighbouring values. The canons *Mirabilis in altis Dominus* denote the finding of these values as 'double entry' into the tables.

The procedure of the linear interpolation is described in various rules in a different level of detail. For instance, in the tenth rule on finding the position of apogee of any planet in canons *Quia ad inveniendum loca planetarum* by John of Lignères we can find the following commentary: 'It is necessary to realise here that there are two methods to the proportional part of some number in its ratio to another number, either by division or multiplication. Using division, if some number is a third or a quarter of the other number, take the third or the quarter of that number. Using multiplication, multiply the second number by the third and divide by the first and the result will be the required unknown number that you were seeking. Then the first number is that which is known and also its part is known; the second number is the known part. And the third number is the one that is known, but its part is unknown.'[1]

The method by means of multiplication corresponds to the nowadays common method of linear interpolation: if we want to find a value $y$ of a function $y(x)$ at value $x$ of the independent variable and we find in the table two values $y_1 = y(x_1)$ and $y_2 = y(x_2)$, where $x_1 < x < x_2$, then we approximate

$$y \simeq y_1 + \delta y \,, \tag{27.1}$$

where using the rule of three we calculate the 'proportional part'

$$\delta y = \delta x \frac{\Delta y}{\Delta x} \,. \tag{27.2}$$

Here the 'first number' $\Delta x = x_2 - x_1$, the 'second number' $\Delta x = x - x_1$, and the 'third number' $\Delta y = y_2 - y_1$.

The method by division may seem not to be applicable generally, but only in the rare cases when $\delta x / \Delta x$ is a simple ratio like 1:3 or 1:4. However, the example and explanation given by Petrus Cruciferus in the second rule (interestingly devoted also to the planetary apogees) of his canons *Mirabilis in altis Dominus* shows that this method can also be universal and precise, despite seeming clumsy from the present-day point of view. The straightforward proportionality (27.2) is replaced by its equivalent

1 See Hadravová and Hadrava, 'John of Lignères, Quia ad inveniendum', p. 289.

$$\delta y = \frac{\Delta y}{r} - \frac{\Delta y}{\Delta x} \frac{\Delta x - r\delta x}{r}, \qquad (27.3)$$

where $r = \left[\frac{\Delta x}{\delta x}\right]$ is the integer part of the ratio of the tabular step $\Delta x$ to the needed increment $\delta x$ of the independent variable (i.e. inverse to the ratio mentioned by John of Lignères). The first term on the right-hand side of Eq. (27.3) gives the result if this ratio is precisely integer. If it is not, the second term gives a correction for the reminder $\Delta x - r\,\delta x$. The expression (27.3) thus offers a precise linear interpolation that is equivalent to the simpler formula (27.2). Their precision was, however, degraded in practice by rounding of intermediate results of the division.

In his example of interpolation between the years 1424 and 1444 to the year 1430 Petrus Cruciferus has $\Delta x = 20$ (years), $\delta x = 6$ (years) and $\Delta y = 42008'''$. The proportional part, i.e. the increment of the apogee, should thus be $\delta y = 12602.4'''$. However, with the integer part of the ratio $r = [20/6] = 3$ Petrus Cruciferus truncates the first term $\Delta y/r = 14002;40'''$ in the right-hand side of (27.3) to 14002‴. Similarly, instead of the annual increment $\Delta y/\Delta x = 2100.4'''$ he uses the rounded value $35'' = 2100'''$ tabulated in the column D of Table TRW-07 in Chapter 12, entitled *differencia differencie*. The second term is thus equal to 1400‴ and the resulting proportional part is then also rounded to integer tertias, namely to 12602‴. The error of neglected 24 quartas in this complicated method of interpolation by division is small compared to the error of the same example calculated by Petrus Cruciferus using the simple method by multiplication. He uses here again the rounded value 2100‴ of the annual motion, which multiplied by the whole increment $\delta x$ of 6 years gives the resulting proportional part of 12600‴ with a six-times higher error of 2.4‴.

# Editorial Note

The editorial principles stated here are applied to all Latin texts published in this book. Most of the texts are retained in their medieval orthography which was used in Central Europe. That is, we transcribe the occasional *-ct-* (e.g. *coniunctio, oppositio*) as *-cc-* (*coniunccio, opposicio*), the very rare *ae* (e.g. *planetae, tabulae*) as *e* (*planete, tabule*) etc. The only exception being those early modern prints from the humanist period, which again leaned towards a more classical mode of their writing.

The spelling of other Latin words is left unchanged even if it sometimes varies throughout the text. This concerns, for example, the inconsistent use of *i* and *y* (e.g. *diameter* × *dyameter*), double versus single letters (*agrego* × *aggrego*), and other phenomena (*nihil* × *nichil, era* × *hera, Toletum* × *Tolletum* × *Tholetum*). In the critical apparatus we do not note the following and similar orthographic variants and deviations: *mihi* × *michi, nihil* × *nichil, titulum* × *tytulum, congnoscere* × *cognoscere, oporteat* × *opporteat, repperi* × *reperi, relucet* × *rellucet, admiramur* × *ammiramur, Wratislavia* × *Wratisslavia, bisextilis* × *bissextilis* etc.

Common paleographic abbreviations are transcribed in full, as is customary. The spelling is chosen so as to match the full forms in the surrounding text (*numquam* × *nunquam*). The question mark *(?)* denotes unreadable places of the manuscripts.

< > Letters in pointed brackets are, for various reasons, missing from the manuscript (e.g. in initials) and have been added, based on our conjecture, to fill in the lacunae (*<C>anon; Incarnac<i>onis; me<n>sium; quociente<m>*).

<2> In pointed brackets are also chapter numbers, added by us for clarity of the text.

{ } Composite brackets are used to reject redundant letters or words, e.g. *cum {cum}; ingredi{t}or.*

[…] = lacuna; [ ] = place left out for the initial letter

The critical apparatus is deliberately given in the form from which the binary correlations were computed.

# Abbreviations

| | |
|---|---|
| *add.* | addidit (added) |
| *al. m.* | alia manus (other hand) |
| *cap.* | caput, capitulum (chapter) |
| *corr.* | correxit (corrected) |
| *corr. ex* | correxit ex (corrected from) |
| *corr. in* | correxit in (corrected to) |
| *del.* | delevit (deleted) |
| *expl.* | explicit (the end of a text) |
| *f., ff.* | folio, folia |
| *gl.* | glossa (gloss) |
| *in mg.* | in margine (on the margin of the folio) |
| *in mg. dext.* | in margine dextro (on the right margin) |
| *in mg. infer.* | in margine inferiore (at the bottom margin) |
| *in mg. sin.* | in margine sinistro (on the left margin) |
| *in mg. super.* | in margine superiore (at the upper margin) |
| *inc.* | incipit (the beginning of a text) |
| *infrascr.* | infrascripsit, infrascriptum (napsal dole, napsáno dole) |
| *ms.; mss.* | codex manu scriptus (manuscript); codices manu scripti (manuscripts) |
| *om.* | omisit (omitted) |
| *p., pp.* | pagina, paginae (page, pages) |
| *per err.* | per errorem (erroneously) |
| *sic* | sic (thus) |
| *suprascr.* | suprascripsit, suprascriptum (written above, wrote above) |

| | |
|---|---|
| **S** | signum (sign) |
| **G** | gradus (degree) |
| **M** | minutum (minute), angular or time unit |
| **S** or **2** | secunda (second), angular or time unit |
| **T** or **3** | tercia (tertia), angular or time unit |
| **4** | quarta (quarta) |
| **5** | quinta (quinta) |
| **6** | sexta (sexta) |
| **D** | dies (day) |
| **H** | hora (hour) |

# Bibliography

## List of Manuscripts


Augsburg, Universitätsbibliothek (UB), Cod. II 1 Quarto 74: 107, 178
Berlin, Staatliche Museen (StM), Kupferstichkabinett, 78 D 12 (olim: Hamilton 556): 28
Berlin, Staatsbibliothek (SB), Preußischer Kulturbesitz (PK), 4° 175: 86
Bernkastel-Kues, Bibliothek des St. Nikolaus-Hospitals (BNH), Cusanus (Cus) 212: 512
Cambridge, Trinity College, O.3.13: 37, 116, 117, 488, 492
Cracow, Biblioteka Jagiellońska (BJ), 257: 44, 477
Cracow, Biblioteka Jagiellońska (BJ), 459: 280
Cracow, Biblioteka Jagiellońska (BJ), 550: 57, 244
Cracow, Biblioteka Jagiellońska (BJ), 551: 57, 511
Cracow, Biblioteka Jagiellońska (BJ), 553: 271, 315, 316, 407, 463, 468, 511
Cracow, Biblioteka Jagiellońska (BJ), 555: 172
Cracow, Biblioteka Jagiellońska (BJ), 557: 172
Cracow, Biblioteka Jagiellońska (BJ), 560: 57, 315, 407, 417, 463, 468
Cracow, Biblioteka Jagiellońska (BJ), 570: 86, 88, 89, 95, 97, 172
Cracow, Biblioteka Jagiellońska (BJ), 573: 19, 37, 85, 87, 90, 93, 96, 115, 116, 119, 157, 488, 492, 504
Cracow, Biblioteka Jagiellońska (BJ), 600: 37, 107, 112, 178, 182, 225, 229, 235, 236, 247, 261–63, 265, 266, 271, 275, 280, 309, 316, 415, 417, 418, 438, 439, 464, 468, 493, 511
Cracow, Biblioteka Jagiellońska (BJ), 602: 280
Cracow, Biblioteka Jagiellońska (BJ), 605: 108, 316, 464, 468, 511
Cracow, Biblioteka Jagiellońska (BJ), 609: 26, 27
Cracow, Biblioteka Jagiellońska (BJ), 610: 23, 32–36, 38, 49–51, 59, 163, 227, 511, 512, 517
Cracow, Biblioteka Jagiellońska (BJ), 611: 259, 316, 409–14, 416–19, 438, 464, 468, 511
Cracow, Biblioteka Jagiellońska (BJ), 614: 99, 101, 103, 316, 464, 468, 488, 492, 511
Cracow, Biblioteka Jagiellońska (BJ), 616: 316, 464, 468
Cracow, Biblioteka Jagiellońska (BJ), 617: 261, 316, 317, 320, 454, 455, 463, 468, 505, 511, 540
Cracow, Biblioteka Jagiellońska (BJ), 1846: 108, 179, 318, 465, 468, 511
Cracow, Biblioteka Jagiellońska (BJ), 1848: 99, 101, 106
Cracow, Biblioteka Jagiellońska (BJ), 1852: 57, 88, 318, 401, 417, 463, 468, 511
Cracow, Biblioteka Jagiellońska (BJ), 1858: 109, 318, 465, 468
Cracow, Biblioteka Jagiellońska (BJ), 1864: 88, 109, 318, 465, 468
Cracow, Biblioteka Jagiellońska (BJ), 1865: 89, 318, 465, 468
Cracow, Biblioteka Jagiellońska (BJ), 1915: 57
Cracow, Biblioteka Jagiellońska (BJ), 1916: 89, 97, 179, 318, 465, 468
Cracow, Biblioteka Jagiellońska (BJ), 1917: 179, 318, 465, 468
Cracow, Biblioteka Jagiellońska (BJ), 1920: 89, 109, 179, 318, 465, 468

Cracow, Biblioteka Jagiellońska (BJ), 1927: 57
Cracow, Biblioteka Jagiellońska (BJ), 1966: 109, 318, 465, 468
Cracow, Biblioteka Jagiellońska (BJ), 2252: 37, 85, 87, 90, 94, 116, 228, 318, 466, 468, 488, 492, 511
Cracow, Biblioteka Jagiellońska (BJ), 2478: 511
Cracow, Biblioteka Jagiellońska (BJ), 2480: 110, 318, 466, 468
Cracow, Biblioteka Jagiellońska (BJ), 2650: 180, 319, 466, 468
Cracow, Biblioteka Jagiellońska (BJ), 2664: 319, 466, 468
Cracow, Biblioteka Jagiellońska (BJ), 2855: 280
Heidelberg, Universitätsbibliothek (UB), 2503: 110, 319
Jena, Thüringer Universitäts- und Landesbibliothek, Ms. El. f. 77: 110, 111
Kalocsa, Fószékesegyházi Könyvtár (FK), 326: 57
Leipzig, Universitätsbibliothek (UB), 1473: 156, 309, 319, 421–25, 467, 468
Leipzig, Universitätsbibliothek (UB), 1476: 23, 36, 59
Leipzig, Universitätsbibliothek (UB), 1481: 319, 464, 468, 511
Leipzig, Universitätsbibliothek (UB), 1482: 97, 99, 101, 104, 106, 322
Leipzig, Universitätsbibliothek (UB), 1491: 319
Lisbon, Arquivo Nacional da Torre do Tombo (ANTT), Manuscritos da Livraria (ML), 2115: 37, 116, 117, 309, 319, 425–31, 433, 467, 468, 488, 492, 511
London, British Library (BL), Add 22317: 19, 34, 121, 124, 125, 157, 163, 309, 319, 320, 433–36
Mainz, Wissenschaftliche Stadtbibliothek, HS I 536: 281
Metz, Bibliothèque municipale (BM), 287: 477
Mikulov, Regionální muzeum v Mikulově (RM), Mik 6361: 23, 34, 45, 51–56, 58, 59, 163, 319, 320, 401, 462, 468, 517
Milan, Biblioteca Ambrosiana (BA), C 278 b Inf.: 34, 123, 125, 163, 167–70, 172, 175, 309, 320, 433, 437, 467, 468, 519
Munich, Bayerische StaatsBibliothek (BSB), Clm 51: 110, 112, 309, 320, 466, 468, 493, 505
Munich, Bayerische StaatsBibliothek (BSB), Clm 214: 320, 438, 466, 468, 511
Munich, Bayerische StaatsBibliothek (BSB), Clm 25011: 320
Oxford, Bodleian Library (Bodl.), Ashmole 191: 477
Oxford, Bodleian Library (Bodl.), Canon. misc. 27: 37, 115, 117, 309, 320, 429, 432, 433, 439, 443, 467, 468, 488, 492, 511, 512
Oxford, Bodleian Library (Bodl.), Canon. misc. 499: 23, 36, 37, 39–41, 44–46, 54, 59, 321, 393, 401, 417, 443, 447, 448, 456–59, 462, 468, 511
Padua, Biblioteca universitaria (BU), MS 643: 111, 180
Paris, Bibliothèque nationale de France (BnF), lat. 7427: 115, 123, 159–61
Paris, Bibliothèque nationale de France (BnF), lat. 8322: 498
Prague, Knihovna Královské kanonie premonstrátů na Strahově (Strahov), DA II 13: 30
Prague, Knihovna Metropolitní kapituly (KMK), M.CLXII (1524): 443, 512
Prague, Národní knihovna (NK), I G 6: 433
Prague, Národní knihovna (NK), III C 2: 57, 88
Prague, Národní knihovna (NK), VIII E 9: 43
Prague, Národní knihovna (NK), VIII E 27: 42
Prague, Národní knihovna (NK), VIII G 24 (Cim K 120): 37, 50, 85, 90, 94, 116, 309, 321, 393, 401, 449–51, 462, 468, 488, 492, 504, 511, 517
Prague, Národní knihovna (NK), X A 23: 244, 511

Prague, Národní knihovna (NK), X B 3 (Cim C 20): 23, 34–36, 45, 48–50, 59, 163, 254, 260, 271, 321, 393–406, 441, 443–45, 452, 468, 487, 503, 510–12, 515, 517, 530, 534, 541
Prague, Národní knihovna (NK), XIII C 17: 510, 512
Prague, Národní knihovna (NK), XIII F 17: 42
Prague, Národní knihovna (NK), XIII F 25: 57
Prague, Národní knihovna (NK), XIV A 18: 496, 515
Prague, Národní knihovna (NK), XIV F 10: 512
Prague, Národní knihovna (NK), adlig. 14 B 10: 34, 163, 517
Toruń, Biblioteka Uniwersytecka (BU), 74/III (olim: Königsberg, 1735): 18, 19, 23–25, 28–30, 32, 33, 44, 45, 48, 59, 76, 97, 115, 233, 242, 244, 245, 256, 260, 309, 313, 321, 323–91, 407, 443, 468, 478–81, 485, 486, 497–99, 503, 504, 510, 512, 515, 517, 524, 527, 530, 534
Trento, Fondazione Biblioteca San Bernardino, MS 313: 433
Vatican, Biblioteca Apostolica Vaticana (BAV), Pal. lat. 1374: 34, 163, 244, 517
Vatican, Biblioteca Apostolica Vaticana (BAV), Pal. lat. 1375: 321, 467, 468, 511
Vatican, Biblioteca Apostolica Vaticana (BAV), Pal. lat. 1381: 99, 101–04, 106, 309, 321, 463, 468, 488
Vatican, Biblioteca Apostolica Vaticana (BAV), Pal. lat. 1409: 99, 104, 105
Vatican, Biblioteca Apostolica Vaticana (BAV), Pal. lat. 1411: 512
Vatican, Biblioteca Apostolica Vaticana (BAV), Pal. lat. 1412: 425, 512
Vatican, Biblioteca Apostolica Vaticana (BAV), Pal. lat. 1439: 19, 37, 178, 180, 182, 227, 235, 275, 280, 292
Vatican, Biblioteca Apostolica Vaticana (BAV), Vat. lat. 3099: 123
Vatican, Biblioteca Apostolica Vaticana (BAV), Vat. lat. 4572: 123, 159
Vienna, Österreichische National Bibliothek (ÖNB), Cod. 2352: 16, 28, 244
Vienna, Österreichische National Bibliothek (ÖNB), Cod. 5145: 57
Vienna, Österreichische National Bibliothek (ÖNB), Cod. 5227: 90, 111, 321, 463, 468, 511
Vienna, Österreichische National Bibliothek (ÖNB), Cod. 5240: 26, 91, 96, 97, 322, 463, 468, 511
Vienna, Österreichische National Bibliothek (ÖNB), Cod. 5245: 19, 91, 322, 415, 463, 468
Vienna, Österreichische National Bibliothek (ÖNB), Cod. 5275: 92, 96, 97, 322, 464, 468, 511
Wrocław, Biblioteka Universytecka (BU), IV F 19: 37, 181, 182, 225, 275, 280, 322, 393, 407, 462, 468

## Early Modern Printed Sources

*Astrolabii, quo primi mobilis motus deprehenduntur, canones* (Venice: Paganinus de Paganinis, *c.* 1497–98 or 1494). – ISTC ia01171000.

*Astrolabii, quo primi mobilis motus deprehenduntur, canones* (Venice: Petrus Liechtenstein, 1512). – GW 2759.

*Astrolabii, quo primi mobilis motus deprehenduntur, canones… diligentia Petri Cattena* (Padua: Iacobus Fabrianus, 1549).

Bianchini, Giovanni (Blanchinus, Johannes), *Tabulae celestium motuum earumque canones*, ed. by Augustinus Moravus, additions by Johannes Basilius Augustonus (Venice: Simon Bevilaqua, 1495). – ISTC ib00697000. GW 4410.

Lacher, Ambrosius, *Tabule resolute de motibus planetarum aliorumque supercelestium mobilium* (Frankfurt an der Oder: Johannes Hanau, 1511). – VD16 L 12.

*Opusculum tabularum utile verarum Solis et Lunae coniunctionum per magistrum Wenceslaum Fabri de Budweiss, baccalarium medicine, compositum* (Leipzig: Martin Landsberg, 1494–95 and 1499). – ISTC if00009000; if00009500.

*Pomponii Gaurici Neapolitani De sculptura* (Nuremberg: Johannes Petreius, 1542). – VD16 G 561.

*Roberti Anglici, viri astrologia prestantissimi, De astrolabio canones incipiunt* (Perugia: Petrus Petri de Colonia, Fridericus Ebert et Johannes Conradi, 1477–1479). – ISTC ir00203000.

*Rudimenta astronomica Alfrag{r}ani, item Albategnius… de motu stellarum etc.* (Nuremberg: Johannes Petreius, 1537). – VD16 A 1202.

Schöner, Johannes, *Tabulae astromonicae, quas vulgo… resolutas vocant* (Nuremberg: Johannes Petreius, 1536). – VD16 S 3505; VD16 M 6553.

—, 'Tabulae resolutae astronomicae', in *Opera mathematica Ioannis Schoneri… in unum volumen congesta* (Nuremberg: Johannes Montanus and Ulricus Neuber, 1551). – VD16 S 3465.

—, *Tabulae resolutae astronomicae* (Wittenberg: Mattheus Welack, 1587). – Holding: Prague, NK, 14 J 000184. – VD16 3506. VD16 M 6554.

—, *Tabulae resolutae astronomicae* (Wittenberg: Mattheus Welack, 1588). – Holding: Prague, NK, 14 J 000201/adl.1. – VD16 S3507. VD16 M 6555.

*Tabulae astronomicae Alfonsi Regis. Canones in Tabulas Alphonsi.* With additions by Augustinus Moravus Olomucensis. Ed. Johannes Lucilius Santritter (Venice: Johannes Hamman, 1492). – ISTC ia00535000.

*Tabulae astronomicae Alphonsinae. Johannes de Saxonia: Canones in Tabulas Alphonsi* (Venice: Erhard Ratdolt, 1483). – ISTC ia00534000.

*Tabulae astronomicae divi Alfonsi, regis Romanorum et Castelle* (Venice: Petrus Liechtenstein, 1518/21).

*Tabulae eclypsium magistri Georgii Peurbachii, Tabula primi mobilis Joannis de Monteregio.* Ed. Georg Collimitius Tannstetter (Vienna: Johannes Winterburger, 1514). – VD16 P 5208.

Virdung, Johannes, *Tabulae resolutae de supputandis siderum motibus* (Nuremberg: Johannes Petreius, 1542). – Holding: Prague, NK, 14 F 000007/adl. 1. – VD16 V 1320.

Wenceslas Faber of Budweis, *Commentary on Iohannes Sacrobosco's Spera mundi* (Leipzig: Martin Landsberg, 1495). – ISTC ij00416000.

## Literature

Ackermann, Silke, *Sternstunden am Kaiserhof. Michael Scotus und sein Buch von den Bildern und Ziechen des Himmels* (Frankfurt am Main: Peter Lang, 2009).

Alfonso X el Sabio, *Tablas de las constelaciones – Alfonsine Tables – Alfonsinische Tafeln*, ed. by Ana Domínguez Rodríguez et al. (Valencia: Patrimonio, 2006).

Almási, Gábor, 'Rethinking Sixteenth-Century Lutheran Astronomy', *Intellectual History Review*, 24 (2014), 5–20.

Avelar de Carvalho, Helena, and Luis Ribeiro, *On the Heavenly Spheres. A Treatise on Traditional Astrology* (Tempe: American Federation of Astrologers, 2010).

Bartoš, František Michálek, *Soupis rukopisů Národního musea v Praze. – Catalogus codicum manu scriptorum Musaei nationalis Pragensis, I–II* (Prague: Melantrich, 1926–27).

Birkenmajer, Aleksander, 'Andrzej Grzymała z Poznania, astronom i lekarz z XV wieku', *Kwartalnik Historii Nauki i Techniki*, 3 (1958), 409–21.

—, 'Andrzej Grzymała de Poznań, astronome et medecin du XVe siecle', in *Aleksander Birkenmajer, Études d'histoire des sciences et de la philosophie du Moyen Age* (Wrocław: Ossolineum, 1972), pp. 515–26.

Birkenmajer, Ludwik A., *Mikołaj Wodka z Kwidzyna zwany Abstemius. Lekarz i astronom Polski XV-go stulecia* (Toruń: Towarzystwo naukowe, 1926).

Blume, Dieter, Mechthild Haffner, and Wolfgang Metzger, *Sternbilder des Mittelalters und der Renaissance. Der gemalte Himmel zwischen Wissenschaft und Phantasie, Band II: 1200–1500*, Teilband II.I: *Text und Katalog der Handschriften*. Unter Mitarbeit von Katharina Glanz (Berlin: De Gruyter, 2016).

Boudet, Jean-Patrice, and Jean-Charles Coulon, 'La version arabe (Ghāyāt al-hakīm) et la version latine du Picatrix. Points communs et divergences', *Cahiers de recherches médiévales et humanistes / Journal of Medieval and Humanistic Studies*, 33 (2017), 67–101.

Brodský, Pavel, *Iluminované rukopisy českého původu v polských sbírkách. – The Illuminated Manuscripts of Czech Origin in the Polish Collections* (Prague: Archiv Akademie věd ČR, 2004).

Buben, Milan M., *Rytířský řád křižovníků s červenou hvězdou* (Mělník: LINPA, 1996).

—, *Encyklopedie řádů, kongregací a řeholních společností katolické církve v českých zemích. I. díl: Řády rytířské a křižovníci* (Prague: Libri, 2002).

Burnett, Charles, 'Teaching the Science of the Stars in Prague University in the Early Fifteenth Century: Master Johannes Borotin', *Aither*, 2 (2014), 9–50.

—, Keiji Yamamoto, and Michio Yano, *Al-Qabīṣī (Alcabitius): The Introduction to Astrology*, Editions of the Arabic and Latin texts and an English translation (London, Turin: The Warburg Institute – Nino Aragno Editore, 2004).

Carmody, Francis J., *The Astronomical Works of Thābit ben Qurra* (Berkeley, Los Angeles: University of California Press, 1960).

*Catalogue of Additions to the Manuscripts in the British Museum in the year 1854–1860. Additional Mss. 19,720–24,026* (London: The Trustees, 1875).

*Catalogus codicum manuscriptorum medii aevi Latinorum, qui in Bibliotheca Jagellonica (BJ) Cracoviae asservantur*, Vol. I–XI (Wrocław: Ossolineum, 1980–2016).

Chabás, José, 'Astronomy in Salamanca in the Mid-Fifteenth Century: The Tabulae resolutae', *Journal for the History of Astronomy*, 29 (1998), 167–75.

—, 'The Diffusion of the Alfonsine Tables: The Case of the Tabulae resolutae', *Perspectives on Science*, 10 (2002), 168–78.

—, 'Characteristics and Typologies of Medieval Astronomical Tables', *Journal for the History of Astronomy*, 43 (2012), 269–86.

—, 'An Early Witness of Alfonsine Astronomy: The London Tables for 1336', *Journal for the History of Astronomy*, 48 (2017), 324–28.

—, *Computational Astronomy in the Middle Ages: Sets of Astronomical Tables in Latin* (Madrid: Consejo Superior de Investigaciones Científicas, 2019).

—, 'Computational practices in Alfonsine Astronomy: Hermann of Saxony's *practica motuum planetarum*', in *Alfonsine Astronomy: Computational Practices*, ed. by Matthieu Husson, José Chabás, and Richard L. Kremer (Turnhout: Brepols, forthcoming).

—, and Bernard R. Goldstein, 'Nicholaus de Heybech and his Table for Finding True Syzygy', *Historia Mathematica*, 19 (1992), 265–89.

—, and —, 'Some Astronomical Tables of Abraham Zacut Preserved in Segovia', *Physis*, 35 (1998), 1–10.

—, and —, *The Alfonsine Tables of Toledo* (Dordrecht: Kluwer, 2003).

—, and —, 'Early Alfonsine Astronomy in Paris: The Tables of John Vimond (1320)', *Suhayl*, 4 (2004), 207–94.

—, and —, *The Astronomical Tables of Giovanni Bianchini* (Leiden: Brill, 2009).

—, and —, *A Survey of European Astronomical Tables in the Late Middle Ages* (Leiden, Boston: Brill, 2012).

—, and —, *Essays on Medieval Computational Astronomy* (Leiden: Brill, 2015).

—, and —, 'The Moon in the Oxford Tables of 1348', *Journal for the History of Astronomy*, 47 (2016), 159–67.

—, and —, 'New Evidence on Abraham Zacut's Astronomical Tables', *Archive for History of Exact Sciences*, 72 (2018), 21–62.

—, and —, 'The Medieval Moon in a Matrix: Double Argument Tables for Lunar Motion', *Archive for History of Exact Sciences*, 73 (2019), 335–59.

—, and —, 'Computing Solar and Lunar Velocities and Radii: The Table of John of Genoa and his Predecessors', in *Alfonsine Astronomy: Computational Practices*, ed. by Matthieu Husson, José Chabás, and Richard L. Kremer (Turnhout: Brepols, forthcoming).

—, and —, 'Nicholaus de Heybech of Erfurt, Table-maker and Copyist', *Aestimatio* n.s., 4 (2023), 1–21.

—, and Marie-Madeleine Saby, *The Tables of 1322 by John of Lignères. An Edition with Commentary* (Turnhout: Brepols, 2022).

Cikhart, Roman, 'Páni z Borotína', *Jihočeský sborník historický*, 11 (1938), 91–96.

Čornejová, Ivana et al., *Dějiny univerzity Karlovy I (1347/48–1622)* (Prague: Karolinum, 1995).

Coxe, Henry O., *Catalogi codicum manuscriptorum Bibliothecae Bodleianae pars tertia codices Grae cos et Latinos Canonicianos complectens* (Oxonii: e typographo academico, 1854).

Curtze, Maximilian, 'Urkunden zur Geschichte der Trigonometrie im christlichen Mittelalter', in *Bibliotheca mathematica*, ser. 3, 1 (1900), 321–416.

Czyżak, Marta, Monika Jakubek-Raczkowska, and Arkadiusz Wagner, *Katalog rękopisów średniowiecznych Biblioteki Uniwersyteckiej w Toruniu* (Toruń: Wydawnictwo Naukowe Uniwersytetu Mikałaja Kopernika, 2016).

Dalen van, Benno, *Ancient and Mediaeval Astronomical Tables: Mathematical Structure and Parameter Values* (Utrecht: Universiteit Utrecht, 1993).

Dobrzycki, Jerzy, *Astronomia przedkopernikowska* (Toruń: Wydawnictwo Naukowe, 1971).

—, 'The Tabulae resolutae', in *De Astronomia Alphonsi Regis*, ed. by Mercè Comes, Roser Puig, and Julio Samsó (Barcelona: Universitat de Barcelona, 1987), pp. 71–77. – Reprinted in: 'Selected Papers on Medieval and Renaissance Astronomy', ed. by Jarosław Włodarczyk, and Richard L. Kremer (Warsaw: Institute for the History of Science, 2010), pp. 129–35.

Dokoupil, Vladislav, *Soupis rukopisů mikulovské dietrichsteinské knihovny. Catalogus codicum manu scriptorum bibliothecae dietrichsteinianae Nicolspurgensis* (Prague: SPN, 1958).

Doležalová, Lucie, and Michal Dragoun, *Kříž z Telče (1434–1504). Písař, sběratel a autor* (Prague: Scriptorium, 2020).

Durand, Dana Bennett, *The Vienna-Klosterneuburg Map Corpus of the Fifteenth Century. A Study in the Transition from Medieval to Modern Science* (Leiden: Brill, 1952).

Fernández Fernández, Laura, 'Alfonsine Astronomy, Salamanca and the Oxford Canon. misc. 27', in preparation.

Gabriel, A. L., *A Summary Catalogue of Microfilms of One Thousand Scientific Manuscripts in The Ambrosiana Library* (Milan: Notre Dame, 1968).

Gellner, Gustav, *Jan Černý a jiní lékaři čeští do konce doby jagellovské* (Prague: Královská česká společnost nauk, 1934).

*Gesamtkatalog der Wiegendrucke* (GW), Berlin, Staatsbibliothek. See <https://www.gesamtkatalogderwiegendrucke.de/>.

Goddu, André, 'Johannes de Glogovia', Routledge Resources Online, Medieval Studies, 31 Oct. 2022. DOI: 10.4324/9780415791182-RMEO16-1.

Goldstein, Bernard R., and José Chabás, 'Continuity and Change: The Planetary Equation Tables of John Vimond and Nicolaus Copernicus', *Physis, International Journal for the History of Science* 58 (2023), 61–83.

—, and —, 'Joseph Ibn Waqār and the Treatment of Retrograde Motion in the Middle Ages', *Archive for History of Exact Sciences*, 77 (2023), 175–99.

Graesse, J. G. Theodor, and Friedrich Benedict, *Orbis Latinus oder Verzeichnis der wichtigsten lateinischen Orts- und Ländernamen* (Berlin: Richard Carl Schmidt, 1909).

Graf, Klaus, 'Matthias von Kemnat', in *Neue Deutsche Biographie*, 16 (1990), pp. 410–11.

Grant, Edward, 'On Saving the Phenomena and the Reality or Unreality of Epicycles and Eccentrics', in *A Source Book in Medieval Science*, ed. by Edward Grant (Cambridge: Harvard University Press, 1974), pp. 516–29.

—, *Planets, Stars and Orbs* (Cambridge: Harvard University Press, 1994).

Grimm, Heinrich, 'Lacher, Ambrosius', in *Neue Deutsche Biographie*, 13 (1982), pp. 367–68.

Günther, Siegmund, 'Virdung, Johann', in *Allgemeine Deutsche Biographie*, 40 (1896), pp. 9–10.

Hadrava, Petr, 'Medieval Catalogues of Stars', in *The Stars in the Classical and Medieval Tradition*, ed. by Alena Hadravová, Petr Hadrava, and Kristen Lippincott (Prague: Scriptorium, 2019), pp. 115–33.

—, 'Mathematical Treatment of Alfonsine Trepidation and its Sources', in *Alfonsine Astronomy: Computational Practices*, ed. by Matthieu Husson, José Chabás, and Richard L. Kremer (Turnhout: Brepols, forthcoming).

—, and Alena Hadravová, 'Tabulka ekvací Měsíce z latinské encyklopedie Pavla Žídka a její srovnání s Almagestem, Toledskými a Alfonsinskými tabulkami', *Dějiny věd a techniky*, 41 (2008), 65–84.

—, and —, 'Cristannus de Prachaticz's Treatises on the Astrolabe', in *Certissima signa. A Venice Conference on Greek and Latin Astronomical Texts*, ed. by Filippomaria Pontani (Venezia: Edizioni Ca'Foscari, 2017), pp. 295–312.

—, and —, 'Nová akvizice Národní knihovny a její význam pro dějiny astronomie', *Dějiny věd a techniky*, 50 (2017), 192–208.

Hadravová, Alena, *Kniha dvacatera umění mistra Pavla Žídka. Část přírodovědná* (Praha: Academia, 2008).

—, *Sphaera octava – Mýty a věda o hvězdách III. Středověká pojednání o souhvězdích.* Chapter on iconography of planets by Lenka Panušková (Prague: Artefactum – Academia, 2013).

—, 'Meteorologie v Knize dvacatera umění mistra Pavla Žídka a ve vybraných pramenech zvláště české provenience', *Studie o rukopisech*, 44 (2014), 439–77.

—, and Petr Hadrava, 'Cosmological Theories at the Time of Tycho Brahe', in *Mysterium cosmographicum 1596–1996. Acta historiae rerum naturalium necnon technicarum, New series*, 2 (1998), pp. 225–233.

—, and —, *Křišťan z Prachatic. Stavba a Užití astrolábu* (Prague: Filosofia, 2001).

—, and —, 'Tycho Brahe and Iohannes Šindel', in *Tycho Brahe and Prague: Crossroads of European Science*, ed. by Johan Robert Christianson, Alena Hadravová, Petr Hadrava, and Martin

Šolc, Acta Historica Astronomiae, vol. 16 (Frankfurt am Main: Harri Deutsch Verlag, 2002), pp. 237–47.

—, and —, 'John of Saxony', in *Medieval Science, Technology and Medicine: An Encyclopedia*, ed. by Thomas F. Glick, Steven J. Livesey, and Faith Wallis (New York, London: Routledge, 2005), p. 292.

—, and —, 'Astronomy in Paulerinus's Fifteenth-Century Encyclopaedia Liber viginti arcium', *Journal for the History of Astronomy*, 38 (2007), 305–24.

—, and —, *Sphaera octava – Mýty a věda o hvězdách IV. Katalogy hvězd a přemyslovský nebeský glóbus* (Prague: Artefactum – Academia, 2013).

—, and —, *Sféra Iohanna de Sacrobosco. Středověká učebnice základů astronomie* (Prague: Akropolis, 2019).

—, and —, 'The Eclipse Instrument by John Šindel', in *Studying the Arts in Late Medieval Bohemia: Production, Reception and Transmission of Knowledge*, ed. by Ota Pavlíček, Studia Artistarum, Études sur la faculté des arts dans les universités médiévales, 48 (Turnhout: Brepols, 2021), pp. 305–40.

—, and —, 'John of Lignères, Quia ad inveniendum loca planetarum. An Edition and Translation', in *Alfonsine Astronomy. The Written Record*, ed. by Richard L. Kremer, Matthieu Husson, and José Chabás (Turnhout: Brepols, 2022), pp. 257–302.

—, and —, 'Cyclical Radices in the Tabulae resolutae and Their Adaptation to the Meridian of Another City', in *Alfonsine Astronomy: Computational Practices*, ed. by Matthieu Husson, José Chabás, and Richard L. Kremer (Turnhout: Brepols, forthcoming).

—, and —, 'Spreading of Treatises on the Astrolabe by Cristannus of Prachatice', in *Alfonsine Astronomy: Expanding the Scene*, ed. by José Chabás, Matthieu Husson, and Richard L. Kremer (Turnhout: Brepols, forthcoming).

—, —, and Jole R. Shackelford (trans.), *Tycho Brahe, Instruments of the Renewed Astronomy* (Prague: KLP, 1996).

Halm, Karl Felix, *Catalogus codicum Latinorum Bibliothecae regiae Monacensis. Vol. 2.4: Clm 21406–27268* (Munich: sumptibus Bibliothecae regiae, 1881).

—, *Catalogus codicum Latinorum Bibliothecae regiae Monacensis. Vol. 1.1: Clm 1–2329* (Munich: sumptibus Bibliothecae regiae, 1892).

Hamel, Jürgen, 'Brudzewski, Albertus de', in *The Biographical Encyclopedia of Astronomers*, ed. by Thomas Hockey (New York: Springer, 2007), pp. 177–78.

—, 'Virdung, Johann', in *The Biographical Encyclopedia of Astronomers*, ed. by Thomas Hockey (New York: Springer, 2007), p. 1180.

Hartfelder, Karl, 'Georg Tannstetter', in Allgemeine Deutsche Biographie 37 (1894), pp. 388–89.

Hausenblasová, Jaroslava, 'Ferdinand Hoffman z Grünbühelu a Strechau – podnikatel, politik a mecenáš rudolfínské doby', in *Mezi kulturou a uměním*, ed. by Ivana Ebelová, Jiří Pešek, Tomáš Sekyrka, and Vít Vlnas (Prague: NLN, 2013), pp. 86–95.

Hilg, Hardo, *Lateinische mittelalterliche Handschriften in Octavo der Universitätsbibliothek Augsburg. Die Signaturengruppe Cod. I.2.8° und Cod. II.1.8°* (Wiesbaden: Harrassowitz, 2014).

Hirsch, August, 'Curio, Jakob', in *Allgemeine Deutsche Biographie* 4 (1876), pp. 646–47.

Holá, Mlada, Martin Holý et al., *Profesoři pražské utrakvistické univerzity v pozdním středověku a raném novověku (1457/1458–1622)* (Prague: Academia – Historický ústav, 2022).

Horský, Zdeněk, *Historické vědecké přístroje v mikulovských sbírkách. Katalog vědeckých přístrojů z 16. až 19. století ve sbírkách Regionálního muzea v Mikulově – Historical Scientific Instruments in the Mikulov Collections. Catalogue of Scientific Instruments from the 16th to 19th Centuries in the Collections of the Regional Museum in Mikulov*, ed. by Stanislava Vrbková et al. (Mikulov: Regionální muzeum, 2011).

—, and Dana Martínková, 'Jan Šindel: Úvod k tabulkám pohybu planet', in *Výbor z české literatury husitské doby II*, ed. by Bohuslav Havránek, Josef Hrabák, Jiří Daňhelka, et al. (Prague: Nakladatelství ČSAV, 1964), pp. 591–97, 608–11.

Husson, Matthieu, 'Ways to Read a Table: Reading and Interpolation Techniques in Canons of Early Fourteenth-Century Double-Argument Tables', *Journal for the History of Astronomy*, 43 (2012), 299–319.

—, 'La Tabula tabularum de Jean de Murs et les modèles de l´arithmétique médiévale', *Cahiers de recherches médiévales et humanistes*, 27 (2014), 97–122.

—, 'Work Cohesion as a Test of Manuscipt Transmission: The Case of John of Lignères' Tabule magne', in *Alfonsine Astronomy. The Written Record*, ed. by Richard L. Kremer, Matthieu Husson, and José Chabás (Turnhout: Brepols, 2022), pp. 317–42.

—, Clemency Montelle, and Benno van Dalen (eds.), *Editing and Analysing Astronomical Tables: Towards a Digital Information System for the History of Astral Sciences* (Turnhout: Brepols, 2021).

*Inventario ceruti dei manoscritti della Biblioteca Ambrosiana*, Vol. 1 (Trezzano: Editrice Etimar, 1973). *The Illustrated Incunabula Short-Title Catalogue* (ISTC), London, British Library. See <https://data.cerl.org/istc/_search>.

James, Montague Rhodes, *The Western Manuscripts in the Library of Trinity College, Cambridge. A Descriptive Catalogue. I–IV* (Cambridge: Cambridge University Press, 1900–1904).

Jensen, Derek, 'Schöner, Johannes', in *The Biographical Encyclopedia of Astronomers*, ed. by Thomas Hockey (New York: Springer, 2007), pp. 1027–28.

Juste, David, *Catalogus codicum astrologorum Latinorum. I. Les manuscrits astrologiques latins conservés à la Bayerische Staatsbibliothek de Munich* (Paris: CNRS Éditions, 2011), pp. 5–236.

Kästner, Ingrid, 'Paulus Paulirinus (Pavel Židek, 1413–1471) und die Medizin im Liber viginti arcium: ein Projektvorschlag', in *Padua als Europäisches Wissenschaftszentrum von der Renaissance bis zur Aufklärung*, ed. by Dietrich von Engelhardt, and Gian Franco Frigo (Aachen: Shaker Verlag, 2017), pp. 153–66.

*Kniha památní na sedmisetleté založení českých křižovníků s červenou hvězdou: 1233–1933* (Prague: Čs. akciová tiskárna, 1933).

Knoll, Paul W., 'The Arts Faculty at the University of Cracow at the End of the Fifteenth Century', in *The Copernican Achievement*, ed. by Robert S. Westman (Berkeley: University of California Press, 1975), pp. 137–56.

Kremer, Richard L., 'Abbreviating the Alfonsine Tables in Cracow: The Tabulae aureae of Petrus Gaszowiec (1448)', *Journal for the History of Astronomy*, 38 (2007), 283–304.

—, 'John of Murs, Wenzel Faber and the Computation of True Syzygy in the Fourteenth and Fifteenth Centuries'. *Acta Historica Leopoldina*, 54 (2008), 147–60.

—, 'Cracking the *Tabule permanentes* of John of Murs and Firmin of Beauval with Exploratory Data Analysis', in *Editing and Analysing Numerical Tables: Towards a Digital Information System for the History of Astral Sciences*, ed. by Matthieu Husson, Clemency Montelle and Benno van Dalen (Turnhout: Brepols, 2021), pp. 363–422.

—, 'The Parisian Alfonsine Tables in Manuscript, 1351–1525', in *Alfonsine Astronomy: Expanding the Scene*, ed. by José Chabás, Matthieu Husson, and Richard L. Kremer (Turnhout: Brepols, forthcoming).

—, and Jerzy Dobrzycki, 'Alfonsine Meridians: Tradition versus Experience in Astronomical Practice c. 1500', *Journal for the History of Astronomy*, 29 (1998), 187–99.

—, Matthieu Husson, and José Chabás (ed.), *Alfonsine Astronomy. The Written Record* (Turnhout: Brepols, 2022).

Kunitzsch, Paul, *Claudius Ptolemäus. Der Sternkatalog des Almagest. Die arabisch-mittelalterliche Tradition. II. Die lateinische Übersetzungen Gerhards von Cremona* (Wiesbaden: Otto Harrasowitz, 1990).

*Liber decanorum facultatis philosophicae Universitatis Pragensis ab anno Christi 1367 usque ad annum 1585*. Facsimile, ed. by Karel Beránek et al. (Pragae: Universitas Carolina, 1983). – Older edition: Monumenta historica Universitatis Carolo-Ferdinandeae Pragensis /MHUP/, tom. I, pars 2 (Prague: Ioannes Nepomuk Gerzabek, 1832).

Lorch, Richard, 'The Treatise on the Astrolabe by Rudolf of Bruges', in *Between Demonstration and Imagination*, ed. by Lodi Nauta, and Arjo Venderjagt (Leiden, Boston, Köln: Brill, 1999), pp. 55–100.

MacMinn, Donn, 'An Analysis of Ptolemy's Treatment of Retrograde Motion', *Journal for the History of Astronomy*, 29 (1998), 257–70.

Malpangotto, Michela, *Theoricae novae planetarum Georgii Peurbachii dans l'histoire de l'astronomie* (Paris: CNRS Éditions, 2020).

Marek, Jindřich, 'Schriften des utraquistischen Konsistoriums aus den Jahren 1471–1489 in der Nationalbibliothek der Tschechischen Republik', in *Sacri canones servandi sunt*, ed. by Pavel Krafl (Prague: Historický ústav, 2008), pp. 545–56.

—, *Václav Koranda mladší. Utrakvistický administrátor a literát* (Prague: NLN, 2017).

Markowski, Mieczysław, *Astronomica et astrologica Cracoviensia ante annum 1550* (Florence: Leo S. Olschki, 1990).

McCarthy, Daniel P., and John G. Byrne, 'Al-Khwārizmī's Tables and a Western Table with the Hindu Norm of R = 150', *Archive for History of Exact Sciences*, 57 (2003), 243–66.

Mercier, Raymond, *Πτολεμαίου Πρόχειροι κανόνες. Ptolemy's Handy Tables, Volume Ib. Tables A1–A2* (Louvain-la-Neuve: Université Catholique de Louvain, Institut Orientaliste, 2011).

Miolo, Laure, 'Retracing the Tradition of John of Genoa's Opus astronomicum through Extant Manuscripts', in *Alfonsine Astronomy. The Written Record*, ed. by Richard L. Kremer, Matthieu Husson, and José Chabás (Turnhout: Brepols, 2022), pp. 343–80.

Nallino, Carlo Alfonso, *Al-Battānī sive Albatenii Opus astronomicum I–III* (Milan: Ulrichus Hoeplius, 1899–1907).

Neugebauer, Otto, 'Thābit ben Qurra On the Solar Year and On the Motion of the Eighth Sphere', *Proceedings of the American Philosophical Society*, 106 (1962), 264–99.

—, *A History of Ancient Mathematical Astronomy I–III* (Berlin, Heidelberg, New York: Springer, 1975).

North, John D., *Richard of Wallingford: An Edition of his Writings, with Introductions, English Translation and Commentary* (Oxford, Clarendon Press: 1976).

Nothaft, C. Philipp E., *Scandalous Error: Calendar Reform and Calendrical Astronomy in Medieval Europe* (Oxford: Oxford University Press, 2018).

—, 'Ephemerides in High Medieval Europe: The Textual Evidence', *Journal for the History of Astronomy*, 51 (2021), 33–52.

Patera, Adolf, and Antonín Podlaha, *Soupis rukopisů knihovny metropolitní kapitoly Pražské I–II* (Prague: Nákladem České akademie věd a umění, 1910 /vol. I/, 1922 /vol. II/).

Pedersen, Fritz S., *The Toledan Tables. A Review of the Manuscripts and the Textual Versions, with an Edition* (Copenhagen: C. A. Reitzels Forlag, 2002).

Pedersen, Olaf, *A Survey of the Almagest with Annotation and New Commentary by Alexander Jones* (New York: Springer, 2011).

Petr, Stanislav, 'Rukopisné fondy zámeckých a hradních knihoven na Moravě a ve Slezsku', in *Problematika historických a vzácných knižních fondů Čech, Moravy a Slezska* (Brno: Sdružení knihoven ČR, 1995), 17.

—, 'Soupis rukopisů zámecké knihovny v Mikulově', in *Sborník Národního muzea v Praze*, řada C, 1994–95, sv. 39–40 (Prague: Národní muzeum, 1997), pp. 29–77.

Porres de Mateo, Beatriz, 'Les Tables astronomiques de Jean de Gmunden: Édition et étude comparative', unpublished PhD thesis, Paris, École Pratique des Hautes Études, 2003.

—, 'Šíření středověkých astronomických tabulek ve střední Evropě v 15. století', in *Astronomie ve středověké vzdělanosti*, trans. and ed. by Alena Hadravová, and Petr Hadrava (Prague: ÚSD a AsÚ, 2003), pp. 39–51.

—, and José Chabás, 'Los cánones de las Tabulae resolutae para Salamanca: origen y transmisión', *Cronos*, 1 (1998), 51–83.

—, and —, 'John of Murs's Tabulae permanentes for Finding True Syzygies', *Journal for the History of Astronomy*, 32 (2001), 63–72.

Poulle, Emmanuel, 'John of Lignères', in *The Dictionary of Scientific Biography* (New York: Charles Scribner's Sons, 1973), VII, pp. 122–28.

—, 'Jean de Murs et les tables alphonsines', *Archives d'Histoire doctrinale et litteraire du Moyen Age*, 47 (1980), 241–71.

—, *Les sources astronomiques (textes, tables, instruments)* (Turnhout: Brepols, 1981).

—, *Les tables Alphonsines avec les canons de Jean de Saxe* (Paris: Centre national de la recherche scientifique, 1984).

—, and Owen Gingerich, 'Les positions des planètes au Moyen Âge: Application du calcul électronique aux tables alphonsines', in *Comptes rendus des séances de l'Académie des Inscriptions et Belles-Lettres* 111 (1967), 531–48.

Ramírez-Weaver, Eric, 'Bohemian King Wenceslas IV's Copy of the Alfonsine Tables and their Place within his Astronomical and Astrological Corpus', in *Alfonsine Astronomy. The Written Record*, ed. by Richard L. Kremer, Matthieu Husson, and José Chabás (Turnhout: Brepols, 2022), pp. 199–240.

Říhová, Milada, 'Na okraj latinské středověké lékařské literatury', in *Septuaginta Paulo Spunar oblata (70 + 2)*, ed. by Jiří K. Kroupa (Prague: KLP, 2000), pp. 221–34.

Rosińska, Grażyna, *Instrumenty astronomiczne na Uniwersytecie Krakowskim w XV wieku*. Studia Copernicana 11 (Wrocław: Ossolineum, 1974).

—, *Scientific Writings and Astronomical Tables in Cracow. A Census of Manuscript Sources (XIVth–XVIth Centuries)* (Wrocław: Ossolineum, 1984).

Saby, Marie-Madeleine, 'Les canons de Jean de Lignères sur les tables astronomiques de 1321. Édition critique, traduction et étude', unpublished thesis, Paris, École Nationale des Chartes, 1987. – A summary appeared as: 'Les canons de Jean de Lignères sur les tables astronomiques de 1321' (Paris, École Nationale des Chartes 1987: Positions des theses), pp. 183–90.

Samsó, Julio, *On Both Sides of the Strait of Gibraltar: Studies in the History of Medieval Astronomy in the Iberian Peninsula and the Maghrib* (Leiden: Brill, 2020).

Schipke, Renate, *Die lateinischen Handschriften in Quarto der Staatsbibliothek zu Berlin – Preußischer Kulturbesitz, Teil 1: Ms. lat. quart. 146–406* (Wiesbaden: Harrassowitz, 2007).

Schmeidler, Felix, 'Schöner, Johannes', in *Neue Deutsche Biographie*, 23 (2007), pp. 405–06.

Schuba, Ludwig, *Die Quadriviums-Handschriften der Codices Palatini Latini in der Vatikanischen Bibliothek* (Wiesbaden: Dr Ludwig Reichert Verlag, 1992).

Šmahel, František, 'Knihovní katalogy koleje Národa českého a koleje Rečkovy', *Acta Universitatis Carolinae – Historia Universitatis Carolinae Pragensis (AUC – HUCP)* 2/1 (1961), 59–85.

—, *Husitská revoluce I–IV* (Prague: Univerzita Karlova, 1995–1996).

—, *Alma mater Pragensis. Studie k počátkům Univerzity Karlovy* (Prague: Karolinum, 2016).

Spunar, Pavel, 'Několik doplňků k rekonstrukci knihovnyVáclava Korandy ml.', *Listy filologické*, 91 (1968), 147–50.

—, *Repertorium auctorum Bohemorum provectum idearum post universitatem Pragensem conditam illustrans I*. Studia Copernicana 25 (Wrocław: Ossolineum, 1985).

—, *Repertorium auctorum Bohemorum provectum idearum post universitatem Pragensem conditam illustrans II*. Studia Copernicana 35 (Warsaw, Prague: Akademie věd ČR, Polska Akademia nauk, 1995).

Sroka, Stanisław A., 'Mikołaj z Tuchowa. Zapomniany astronom i lekarz z XV wieku oraz jego europejskie peregrynacje', *Roczniki historyczne*, 86 (2020), 177–85.

Stahlman, W. D., 'The Astronomical Tables of Codex Vaticanus Graecus 1291', unpublished PhD dissertation, Brown University, 1960.

Studničková, Milada, 'Drehknoten und Drachen: Die Orden Wenzels IV. und Sigismunds von Luxemburg und die Polysemantik ihrer Zeichen', in *Kunst als Herrschaftsinstrument. Böhmen und das Heilige Römische Reich unter den Luxemburgern im europäischen Kontext*, ed. by Jiří Fajt, and Andrea Langer (Berlin: Deutscher Kunstverlag, 2009), pp. 377–87.

Svátek, Josef, 'Organizace řeholních institucí v českých zemích a péče o jejich archivy', *Sborník archivních prací*, 20 (1970), 503–624.

Swerdlow, Noel M., 'The Empirical Foundations of Ptolemy's Planetary Theory', *Journal for the History of Astronomy*, 35 (2004), 249–70.

—, 'Ptolemy's Theories of the Latitude of the Planets in the Almagest, Handy Tables, and Planetary Hypotheses', in *Wrong for the Right Reasons*, ed. by J. Z. Buchwald and A. Franklin (Dordrecht: Springer, 2005), pp. 41–71.

*Tabulae codicum manu scriptorum praeter Grae cos et orientales in Bibliotheca Palatina Vindobonensi asservatorum I–VIII* (Vienna: Gerold, 1864–1893).

*The Cracow Circle of Nicholas Copernicus* (Cracow: The Jagiellonian University Press, 1973).

Thoren, Victor E., and Edward Grant, 'Extracts from the Alfonsine Tables and Rules for their Use (John of Saxony)', in *A Source Book in Medieval Science*, ed. by Edward Grant (Cambridge: Harvard University Press, 1974), pp. 465–87.

Thorndike, Lynn, 'Johann Virdung of Hassfurt Again', *Isis* 25 (1936), 363–71.

—, 'Some Little Known Astronomical and Mathematical Manuscripts', *Osiris* 8 (1948), 41–72.

—, 'Notes on Some Astronomical, Astrological and Mathematical Manuscripts of the Bibliothèque nationale, Paris', *Journal of the Warburg and Courtauld Institutes*, 20 (1957), 112–72.

—, and Pearl Kibre, *A Catalogue of Incipits of Medieval Scientific Writings in Latin* (Cambridge: The Mediaeval Academy of America, 1963).

—, and George Sarton, 'Question No. 10 – Tacuinum', *Isis* 10 (1928), 489–90; 'Answer No. 10 – Tacuinum, taqwīm', *Isis*, 10 (1928), 490–93.

Tihon, Anne, *Πτολεμαίου Πρόχειροι κανόνες. Les Tables faciles de Ptolémée, Volume Ia. Tables AI–A2* (Louvain-la-Neuve: Université Catholique de Louvain, Institut Orientaliste, 2011).

Toomer, Gerald James, 'A Survey of the Toledan Tables', *Osiris* 15 (1968), 5–174.

— (trans. and annot.), *Ptolemy's Almagest* (London: Duckworth, 1984).

Truhlář, Josef, 'Paběrky z rukopisů Klementinských XLVII. Šindelovy astronomické tabulky', Věstník *České akademie*, 9 (1900), 473–74.

—, *Catalogus codicum manu scriptorum Latinorum, qui in c. r. bibliotheca publica atque Universitatis Pragensis asservantur I–II* (Pragae: Česká Akademie věd a umění, 1905–1906).

Unterkircher, Franz, *Die datierten Handschriften der Österreichischen Nationalbibliothek von 1401 bis 1450. 1. Teil: Text. 2. Teil: Tafeln* (Wien: Verlag d. Österreichischen Akademie der Wissenschaften, 1971).

—, *Die datierten Handschriften der Österreichischen Nationalbibliothek von 1451 bis 1500. 1. Teil: Text. 2. Teil: Tafeln* (Wien, Verlag d. Österreichischen Akademie der Wissenschaften, 1974).

—, *Die datierten Handschriften der Österreichischen Nationalbibliothek von 1501 bis 1600. 1. Teil: Text. 2. Teil: Tafeln* (Wien: Verlag d. Österreichischen Akademie der Wissenschaften, 1976).

Urbánková, Emma, 'Zbytky knihovny mistra Václava Korandy mladšího v universitní knihovně v Praze', in *Ročenka universitní knihovny v Praze 1956* (Prague: Univerzitní knihovna, 1956), pp. 135–61.

—, 'Zbytky knihovny snad M. Jana Šindela v Universitní knihovně', in *Ročenka universitní knihovny (RUK) v Praze 1960–1961* (Prague: Univerzitní knihovna, 1962), pp. 87–97.

VD16: *Verzeichnis der im deutschen Sprachbereich erschienenen Drucke des 16. Jahrhunderts*. See <https://opacplus.bib-bvb.de/TouchPoint_touchpoint/start.do?SearchProfile = Altbestand &SearchType = 2>.

Wisłocki, Władysław, *Catalogus codicum manuscriptorum Bibliothecae Universitatis Jagellonicae Cracoviensis I–II* (Kraków: Universitas Jagellonica, 1877–1881).

Wolf, Marek et al., *Astronomická příručka* (Prague: Academia, 1992).

Zinner, Ernst, *Verzeichnis der astronomischen Handschriften des deutschen Kulturgebietes* (München: C. H. Beck, 1925).

—, *Leben und Wirken des Johannes Müller von Königsberg, genannt Regiomontanus*. 2nd rev. ed. (Osnabrück: O. Zeller, 1968).

# Index nominum

# Glossary of chosen Latin Terms

'a' = abbreviation for *adde* (add)
abacus = abacus, mechanical device facilitating counting
accessus et recessus octave spere (octavi circuli) = access and recess of the eighth sphere (of the eight circle), i.e. supposed periodical component of precession
addere = to add
addicio = addition
aldidode = alidade, deliberate ruler of an astronomical instrument
almanach = almanac; see also tacuinum
altitudo = altitude, height above the horizon
  altitudo poli = altitude of the pole
annus = year
  anni (Christi) collecti = collected years (of Christ) in the Julian calendar
  anni expansi = expanded years
  anni residui = remaining years
  annus bissextilis = leap year
  annus communis = common year, non-leap year
  annus completus = completed year
  annus currens = current year
  annus embolismalis = intercalary year
  annus incompletus = incompleted year
  annus perfectus = completed year
Aquarius = zodiacal sign of Aquarius (Water-bearer)
arcus = arc; angle
  arcus diei (diurnus) = arc of a day
  arcus noctis = arc of a night
argumentum = argument, anomaly
  argumentum capitis Arietis = argument of the head of Aries
  argumentum equatum = equated anomaly
  argumentum horarum (horarium) = anomaly of hours
  argumentum latitudinis Lune (planetarum) = argument of the latitude of the Moon (of the planets)
  argumentum Lune = anomaly of the Moon
  argumentum medium planete = mean anomaly of a planet
  argumentum Solis = anomaly of the Sun
  arumentum verum = true anomaly
Aries = zodiacal sign of Aries (Ram)
ascendens = ascendant, i.e. the intersection of the ecliptic with eastern horizon

aspectus planete = aspect of the planet, i.e. conjunction, opposition, square (quartile), trine, sextile etc.
augmentacio = addition
augmentum = increment, increase, growth
aux = apogee
 aux (vera) planete = (true) apogee of a planet
 aux communis = common apogee, i.e. common precessional shift of all planets
 aux specialis = special apogee, i.e. apogee of individual planet

bensahar (genzahar, zamzar, zanzar, zenzahar, zenzar) = see genzahar
bissextilis (annus) = leap (year)
buth = velocity of the true motion of the Sun, of the Moon, or of a planet (from Arabic)

Cancer = zodiacal sign of Cancer (Crab)
canon = some authors use it in the meaning of individual rule
 canones = canons, set of rules, i.e. set of instructions how to work with astronomical tables
caput draconis = head of dragon, i.e. ascending node of the lunar path
caput tabule = heading (header) of a table
cardo (cardines) celi = cardinal point(s), i.e. ascendant, imum celi, descendant, medium celi; it means beginnings of the first, fourth, seventh, and tenth astrological house
cauda draconis = tail of dragon, i.e. descending node of the lunar path
centrum medium planete = mean centre of a planet, i.e. mean centre of an epicycle of a planet
cifra = digit zero
circulus = circle
 circulus obliquus = oblique circle
 circulus rectus = upright circle
clima = climate, i.e. one of the seven geographical zones
congressus = conjunction
coniunccio = conjunction
 coniunccio media = mean conjunction
 coniunccio vera = true conjunction
corda = chord (trigonometric function)
cursus = course, motion
 cursus medius = mean motion
cuspis = cusp, i.e. beginning of an astrological house

de decem in decem = at steps of ten
declinacio = declination 1. angular distance from the equator; 2. declination, inclination, i.e. the first component of the ecliptical latitude of a planet
defalcacio = subtraction
denominacio = denomination, i.e. naming of a sexagesimal order (minute, second, tertia etc.)
descendens = descendant, i.e. the intersection of the ecliptic with western horizon
deviacio = deviation, i.e. the third component of the ecliptical latitude of a planet
dies = day
 dies adaucti = augmented days, i.e. interval of several days in a table
 dies intermedii = intermediate days

dies lunares (dies Lune) = days of a lunation
dies (non) equati = (un)equated days
dies solares (dies Solis) = days of a calendric month
dies superfluus = superfluous, intercalary day
differencia = increment (difference) in a table
differencia differencie = second difference (between subsequent increments)
diminucio = subtraction
direccio planete = direction of a planet
diversitas diametri = variation of a diameter
domus = astrological house
ducere = to double
duplare, duplicare = to double

ecentricus = eccentric circle
eclipsis (Solis, Lune) = eclipse (of the Sun, of the Moon)
effectus planete = influence of a planet
effregion (meridionale, septentrionale) = auxiliary variable for the latitude of the upper planets (southern or northern)
elevacio = height
elevacio poli = height of the pole
elongacio = elongation
epicyclus = epicycle
equacio = equation, i.e. the difference between the mean and the true motion
equacio accessus et recessus octave spere = equation of the access and recess of the eighth sphere
equacio centri = equation of the centre of epicycle
equacio dierum cum noctibus = equation of time, i.e. the difference between the true and mean solar time
equacio equata = equated equation, i.e. corrected equation
equacio primo (secundo) examinata = first (second) examined equation
equacio perfecta = finished equation
equacio Solis (Lune, planete) = equation of the Sun (of the Moon, of the planet)
equare = to equate; to compute, spec. to compute the true position of the planet; to correct
equator = equator
era = era, i.e. the time moment

feria = a day of a week (one of the seven days of the week), serial number of the day of the week; see dies
figura celi = horoscope
fraccio = fraction, part, portion

Gemini = zodiacal sign of Gemini (Twins)
genus = order in sexagesimal notation, i.e. minutes, seconds, tertias etc.
genzahar (bensahar, zamzar, zanzar, zenzahar, zenzar) = ascending node (of a planet); in the case of the Moon, it is the head of dragon
gradus = degree

hora = hour
hora equalis = equal hour, i.e. twenty-fourth part of a natural day
hora equinoccialis = equal hour
hora horologii (horalogii) = horological hour, i.e. equal hour counted from sunset
hora inequalis = unequal hour; i.e. the twelfth part of the day or night from sunrise to sunset
hora meridiei = noon hour, i.e. equal hour counted from noon and midnight
hora planetarum = planetary hour, unequal hour, seasonal hour
horologium = clock counting equal hours from sunset
horologium medium = clock counting equal hours from noon

imum celi = the intersection of the ecliptic with the meridian below the horizon
influencie planetarum = (astrological) influences of the planets
introitus = entry
introitus duplex = double entry (i.e. that one table is entered repeatedly for the second time)
introitus Solis in Arietem etc. = entry of the Sun into Aries etc.
iungere = to add

latitudo = latitude
latitudo regionis = geographical latitude of a given place
Leo = zodiacal sign of Leo (Lion)
Libra = zodiacal sign of Libra (Balance)
linea = line; column (in the table)
linea communis = column with general values (general variables), i.e. independent variables (in today's terminology)
linea latitudinalis = transverse line
linea longitudinalis = longitudinal line
linea numeri = column with units
locus = position, place
locus certus (coniunccionis, Solis) = true position (of the conjunction, Sun) etc.
locus verus (augis, Lune, Solis) = true position (of apogee, Moon, Sun) etc.
longitudo = longitude
longitudo longior = farther distance, apogee
longitudo propior (propinquior, brevis) = nearer distance, perigee
luminaria = luminaries (i.e. Sun and Moon)
Luna = Moon
lunacio (media) = (mean) lunation, lunacy

'm' = abbreviation for *minue* (subtract), see also 's'
mediacio = dividing in half
medium celi = midheaven, the intersection of the ecliptic with the meridian above the horizon
medium noctis = midnight
mensis = month
menses bissextiles = months in leap year
menses communes = months in non-leap year
mensis imperfectus = an unfinished month; a month that has not yet passed entirely

meridianus = meridian
  meridianus civitatis Pragensis (Wratislaviensis) = meridian of the city of Prague (Wrocław) etc.
meridies = noon
meridionalis = southern
minutum = minute
  minuta proporcionalia = proportional minutes
motus = motion
  motus accessus et recessus octave spere = motion of the access and recess of the eighth sphere
  motus certus = true motion
  motus (medius) augis = (mean) motion of the apogee
  motus medius Lune (Solis, planete, stellarum fixarum) = mean motion of the Moon (of the Sun, of the planet, of the fixed stars)
  motus verus (capitis draconis, planete) = true motion (of the head of dragon, of the planet)
multiplicacio = multiplication
  multiplicare = to multiplicate

nadir = nadir
nota = day of the week
numerator = term
numerus = number
  numerus quociens = quotient
  numerus servativus = significant number; valid non-zero digit (opposite of zero)

occidentalis = western
octava spera = the eighth sphere, i.e. the sphere of the fixed stars
opposicio = opposition
  opposicio media = mean opposition
  opposicio vera = true opposition
ordo signorum = order of the signs
orientalis = eastern
orizon = horizon
  orizon obliquus = oblique horizon
  orizon rectus = right horizon

pars = part
  pars proporcionalis = proportional part
passio (planete) = property of a planet
pes tabule = foot of the table
Pisces = zodiacal sign of Pisces (Fishes)
planeta = planet (often including luminaries, i.e. the Sun and Moon)
  planeta inferior = lower, inner planet, i.e. Mercury and Venus
  planeta superior = upper, outer planet, i.e. Mars, Jupiter, and Saturn
prima = prime, the first order in sexagesimal system (esp. one day, i.e. 24 hours)
progressus = progress, i.e. direct motion of a planet
proporcio = proportion

quarta = quarta, the fourth order in sexagesimal system; it is either sixty tertias (two hundred and sixteen thousand primes, i.e. it is a multiple), or it is one sixtieth of a tertia (i.e. the fourth division when dealing with fractions, where a tertia is one sixtieth of a second of a day or degree)
quinta = quinta, the fifth order in sexagesimal system

radix (medii motus) = radix, root (of the mean motion)
radicem reducere = to convert (adapt, recalculate) the radix (to a new meridian)
reducere = to adapt, convert, recalculate
refleccio = reflection, slant, obliquity, i.e. second component of the ecliptical latitude of a planet
regressus = regress, i.e. retrograde motion of a planet
residuum = remainder
resolvere = to resolve, decompose, convert
retrogradacio planete = retrogradation of a planet
revolucio annorum mundi = revolution of the years, i.e. difference between the length of the tropical year and of a calendric year
riga = row (in a table)

's' = abbreviation for *subtrahe* (subtract), see also 'm'
Sagittarius = zodiacal sign of Sagittarius (Archer)
Scorpio = zodiacal sign of Scorpius (Scorpion)
secunda = second, the second order in sexagesimal system; it is either sixty days (sixty primes, i.e. it is a multiple) or it is one sixtieth of a minute (it is a second division when dealing with fractions, where a minute is one sixtieth of a prime of day or degree)
septentrionalis = northern
signum = zodiacal sign
signum cardinale = cardinal sign, i.e. one of the four zodiacal signs (Aries, Cancer, Libra, and Capricorn)
signum commune = common sign, i.e. usual zodiacal sign of thirty degrees
signum phisicum = physical sign, i.e. sign of sixty degrees
sinus (rectus, versus) = sine (right sine, versed sine /versine/)
spera = sphere
spera nona = ninth sphere
spera octava = eighth sphere
spera recta = right sphere, i.e. celestial sphere as seen from the equator
stacio planete (prima, secunda) = (the first, second) station of a planet
stella fixa = fixed star
subtraccio = subtraction
subtrahere = to subtract
successio signorum = order of the signs
superacio, superlacio = *superacio, superlacio*, i.e. excess of the velocity of the Moon with respect to the Sun

tabula = table, spec. astronomical table
tabula loricata = loricated table, i.e. prolonged (extended) table
tac(h)uinum = table, i.e. almanac (from Arabic)

Taurus = zodiacal sign of Taurus (Bull)
tempus = time
  tempus perfectum = elapsed, completed time
tercia = tertia, the third order in sexagesimal system; it is either sixty seconds (three thousand six hundred primes, i.e. it is a multiple) or it is one sixtieth of a second (i.e. third division, if fractions are involved, where a second is one sixtieth of a minute of the day or degree)
titulus = title, caption (of the table)

umbra = shadow
  umbra recta = direct shadow, cotangent
  umbra versa = reversed shadow, tangent

variare = to vary
Virgo = zodiacal sign of Virgo (Virgin)

zanzar (bensahar, genzahar, zamzar, zenzahar, zenzar) = see genzahar